Biomedical Photoacoustics

Wenfeng Xia
Editor

Biomedical Photoacoustics

Technology and Applications

Springer

Editor
Wenfeng Xia
School of Biomedical Engineering &
Imaging Sciences
King's College London
London, UK

ISBN 978-3-031-61410-1 ISBN 978-3-031-61411-8 (eBook)
https://doi.org/10.1007/978-3-031-61411-8

This Springer imprint is published by the registered company Springer Nature Switzerland AG
The registered company address is: Gewerbestrasse 11, 6330 Cham, Switzerland

Preface

Biomedical Photoacoustics: Technology and Applications invites you into the dynamic realm of photoacoustic imaging and sensing, an exciting field where optics, acoustics, and biomedical sciences converge. At its core, photoacoustic imaging, also known as optoacoustic imaging, harnesses the interaction between light and tissue to produce high-resolution, anatomical, molecular, and functional images with remarkable depth penetration. This hybrid modality, utilising the generation and detection of ultrasound waves induced by pulsed or modulated light, offers a powerful synthesis of optical and ultrasound techniques.

The exponential growth of photoacoustic imaging over the past two decades underscores its transformative potential in biomedical research and clinical practice. Continuous advancements in hardware, software, and contrast agents have expanded its capabilities and applications. Chapters 1, 2, and 3 delve into innovations in imaging systems, including microscopy, tomography, and endoscopy, enabling high-resolution imaging across various scales and unprecedented access to inaccessible body areas. Furthermore, Chaps. 4 and 5 review recent advancements in model-based reconstructions and deep learning algorithms, enhancing image quality and speed, and facilitating real-time and quantitative imaging in clinical settings.

The integration of novel contrast agents, discussed in Chap. 6, has enabled deep-tissue molecular imaging and targeted therapy monitoring, particularly in oncology and cardiovascular medicine. Technical validation methodologies, as outlined in Chap. 7, have ensured the reliability and reproducibility of photoacoustic imaging systems, vital for their clinical translation.

Chapters 8, 9, 10, and 11, comprising the Emerging Techniques in Photoacoustics section, illuminate novel approaches such as optical wavefront shaping and optical ultrasound sensors and devices, showcasing the innovative frontier of photoacoustic imaging. Additionally, Chaps. 12 and 13 emphasise the compatibility and complementarity of photoacoustic imaging with established clinical modalities and emerging techniques, positioning it as a promising tool for integration into routine clinical practice. Chapters 14, 15, 16, 17, 18, 19, 20, 21, and 22 encompass a wide range of preclinical and clinical applications, from cancer research and bacterial cell identification to tissue characterisation, brain imaging, and surgical guidance.

Together, these chapters underscore the versatility and promising prospects of photoacoustic imaging in biomedical research and clinical practice.

Born out of a collective aspiration to bridge the gap between theory and practice, this book aims to meet the growing demand for a comprehensive resource encapsulating the latest technologies, methodologies, and applications in photoacoustic imaging. As the editor, I am inspired by the rapid pace of innovation and the expanding horizons in this field, recognising the necessity of consolidating this wealth of knowledge into a single volume that can serve as a guiding beacon for researchers, practitioners, and students alike.

Tailored for a diverse audience encompassing postgraduate students and researchers across physics, engineering, biomedical sciences, and clinical disciplines, this book serves as a vital link between theoretical understanding and practical application in photoacoustic imaging. Whether you are embarking on your journey into this dynamic field or seeking to deepen your existing knowledge and expertise, this volume stands as a reservoir of information and inspiration, poised to propel you forward in your pursuits.

I extend my heartfelt gratitude to the contributors for their generous sharing of expertise and perspectives. I also express my appreciation to the team at Springer Nature, especially senior editor Merry Stuber and production editor Vinodhini Srinivasan, for their dedication and support in bringing this project to fruition.

It is my sincere hope that *Biomedical Photoacoustics: Technology and Applications* will serve as a catalyst for further exploration, collaboration, and innovation in photoacoustic imaging, ultimately leading to advancements that enhance healthcare outcomes and improve the quality of life for patients worldwide.

London, UK Wenfeng Xia

Contents

Part I Principles and Imaging Systems

1 **Photoacoustic Microscopy** 3
Qiangzhou Rong, Lucas Humayun, and Junjie Yao

2 **Principles and Applications of Photoacoustic Computed Tomography** 75
Vijitha Periyasamy, Katherine Gisi, and Manojit Pramanik

3 **Photoacoustic Endoscopy** 109
Hao Yang, Sean Aleman, and Huabei Jiang

Part II Image Reconstruction and Deep Learning

4 **Model-Based Reconstructions for Quantitative Imaging in Photoacoustic Tomography** 133
Andreas Hauptmann and Tanja Tarvainen

5 **Deep Learning-Based Methods for Photoacoustic Imaging Reconstruction: Concepts, Promises, Pitfalls, and Futures** 155
Amir Gholampour, Kalloor Joseph Francis, Min Wu, Nastaran Mohammadian Rad, Richard G. P. Lopata, and Navchetan Awasthi

Part III Contrast Agents and Tissue Phantoms

6 **Photoacoustic Imaging and Applications with Reversibly Switchable Contrast Agents** 181
Donghyeon Oh, Chulhong Kim, and Byullee Park

7 **Technical Validation of Photoacoustic Imaging Systems Using Phantoms** 213
Lina Hacker and James Joseph

Part IV Emerging Techniques in Photoacoustics

8 Optical Wavefront Shaping in Biomedical Photoacoustics 231
Tianrui Zhao and Wenfeng Xia

9 Optical Ultrasound Imaging Device Development and Characterisation: A Literature Review 253
Semyon Bodian, Sacha Noimark, Adrien E. Desjardins, and Ivan P. Parkin

10 Optical Ultrasound Sensors for Biomedical Photoacoustics 299
Yuecheng Shen and Jun Ma

11 Protoacoustics and Its Applications in Radiation Oncology 319
Kaitlyn Kim and Shawn (Liangzhong) Xiang

Part V Photoacoustic Imaging Combined with Other Modalities

12 Hybrid Photoacoustic and Laser-Induced Ultrasound Computed Tomography 333
David Thompson, Michael Jaeger, Damien Gasteau, and Srirang Manohar

13 Multimodal Optoacoustic Imaging 353
Zhenyue Chen, Irmak Gezginer, Quanyu Zhou, and Daniel Razansky

Part VI Pre-clinical and Clinical Applications

14 Multifaceted Potential of Photoacoustic Imaging for Preclinical Cancer Research 377
Deeksha Sankepalle, Allison Sweeney, and Srivalleesha Mallidi

15 Ghost-Particle-Enhanced Photoacoustic Flow Cytometry for Identification of Bacterial Cells in Body Fluids 405
Robert H. Edgar and John A. Viator

16 Photoacoustic Spectrum Analysis for Soft Tissues 413
Yingna Chen, Shiying Wu, Mengjiao Zhang, Jiayan Li, Menglu Qian, and Qian Cheng

17 Photoacoustic Techniques for Bone Characterization 433
Ting Feng, Weiya Xie, Wenyi Xu, Ya Gao, Teng Liu, Dean Ta, Menglu Qian, and Qian Cheng

18 Miniaturized Brain Imaging Apparatus Employing Light, Sound, and Magnetic Fields 477
Shuai Na, Jinyan Zhang, and Bangxu Fan

19 Ultraviolet Photoacoustic Microscopy for Histopathology 499
Bingxin Huang, Ivy H. M. Wong, Yan Zhang, and Terence T. W. Wong

20 Recent Advances in Photoacoustic Imaging of Breast Cancer 533
Huijuan Zhang, Emily Zheng, and Jun Xia

21 Photoacoustic Imaging of Interventional Devices for Guiding Minimally Invasive Medical Procedures 547
Mengjie Shi, Simeon J. West, Tom Vercauteren, Sacha Noimark, Adrien E. Desjardins, and Wenfeng Xia

22 The Evolution of LED-based Photoacoustic Imaging: From Labs to Clinics 573
Mithun Kuniyil Ajith Singh, Naoto Sato, Fumiyuki Ichihashi, Wenfeng Xia, and Yoshiyuki Sankai

Index 609

Contributors

Sean Aleman Department of Medical Engineering, University of South Florida, Tampa, FL, USA

Navchetan Awasthi Department of Biomedical Engineering, Eindhoven University of Technology, Eindhoven, The Netherlands
Informatics Institute, University of Amsterdam, Amsterdam, The Netherlands
Department of Biomedical Engineering and Physics, Amsterdam UMC, Amsterdam, The Netherlands

Semyon Bodian Department of Chemistry, University College London, London, UK
Department of Medical Physics and Biomedical Engineering, University College London, London, UK

Yingna Chen Institute of Acoustics, School of Physics Science and Engineering, Tongji University, Shanghai, China
College of Science and Technology, Ningbo University, Zhejiang, China

Zhenyue Chen Institute for Biomedical Engineering and Institute of Pharmacology and Toxicology, Faculty of Medicine, University of Zurich, Zurich, Switzerland
Institute for Biomedical Engineering, Department of Information Technology and Electrical Engineering, ETH Zurich, Zurich, Switzerland

Qian Cheng Institute of Acoustics, School of Physics Science and Engineering, Tongji University, Shanghai, China

Adrien E. Desjardins Department of Medical Physics and Biomedical Engineering, University College London, London, UK
Wellcome/EPSRC Centre for Interventional and Surgical Sciences, University College London, London, UK

Robert H. Edgar Department of Bioengineering, University of Pittsburgh, Pittsburgh, PA, USA

Bangxu Fan College of Future Technology, Peking University, Beijing, China

Ting Feng Institute of Acoustics, School of Physics Science and Engineering, Tongji University, Shanghai, China
Academy for Engineering and Technology, Fudan University, Shanghai, China

Kalloor Joseph Francis Faculty of Science and Technology, University of Twente, Enschede, The Netherlands

Ya Gao Institute of Acoustics, School of Physics Science and Engineering, Tongji University, Shanghai, China

Damien Gasteau Multi-Modality Medical Imaging group (M3I), Technical Medical Centre, University of Twente, Enschede, The Netherlands

Irmak Gezginer Institute for Biomedical Engineering and Institute of Pharmacology and Toxicology, Faculty of Medicine, University of Zurich, Zurich, Switzerland
Institute for Biomedical Engineering, Department of Information Technology and Electrical Engineering, ETH Zurich, Zurich, Switzerland

Amir Gholampour Department of Biomedical Engineering, Eindhoven University of Technology, Eindhoven, The Netherlands

Katherine Gisi Department of Electrical and Computer Engineering, Iowa State University, Ames, IA, USA

Lina Hacker Department of Oncology, University of Oxford, Oxford, UK

Andreas Hauptmann Research Unit of Mathematical Sciences, University of Oulu, Oulu, Finland
Department of Computer Science, University College London, London, UK

Bingxin Huang Translational and Advanced Bioimaging Laboratory, Department of Chemical and Biological Engineering, Hong Kong University of Science and Technology, Clear Water Bay, Hong Kong, China

Lucas Humayun Department of Biomedical Engineering, Duke University, Durham, NC, USA

Fumiyuki Ichihashi Research and Development Division, CYBERDYNE INC, Tsukuba, Japan

Michael Jaeger Optoacoustic Imaging Group, University of Bern, Bern, Switzerland

Huabei Jiang Department of Medical Engineering, University of South Florida, Tampa, FL, USA

James Joseph School of Science and Engineering, University of Dundee, Dundee, UK

Chulhong Kim Pohang University of Science and Technology, Department of Electrical Engineering, Convergence IT Engineering, Medical Science and Engineering, Mechanical Engineering, School of Interdisciplinary Bioscience and

Bioengineering, Graduate School of Artificial Intelligence, Medical Device Innovation Center, Pohang, Republic of Korea

Kaitlyn Kim Department of Biomedical Engineering, University of California, Irvine, CA, USA

Jiayan Li Institute of Acoustics, School of Physics Science and Engineering, Tongji University, Shanghai, China

Teng Liu School of Electronic and Optical Engineering, Nanjing University of Science and Technology, Nanjing, China

Richard G. P. Lopata Department of Biomedical Engineering, Eindhoven University of Technology, Eindhoven, The Netherlands

Jun Ma Guangdong Provincial Key Laboratory of Optical Fiber Sensing and Communications, Institute of Photonics Technology, Jinan University, Guangzhou, China

Srivalleesha Mallidi Department of Biomedical Engineering, Tufts University, Medford, MA, USA
Wellman Center for Photomedicine, Massachusetts General Hospital, Harvard Medical School, Boston, MA, USA

Srirang Manohar Multi-Modality Medical Imaging group (M3I), Technical Medical Centre, University of Twente, Enschede, The Netherlands

Shuai Na College of Future Technology, Peking University, Beijing, China

Sacha Noimark Department of Chemistry, University College London, London, UK
Department of Medical Physics and Biomedical Engineering, University College London, London, UK
Wellcome EPSRC Institute of Surgical Science, University College London, London, UK

Donghyeon Oh Pohang University of Science and Technology, Department of Electrical Engineering, Convergence IT Engineering, Medical Device Innovation Center, Pohang, Republic of Korea

Byullee Park Sungkyunkwan University, Department of Biophysics, Institute of Quantum Biophysics, Suwon, Republic of Korea

Ivan P. Parkin Department of Chemistry, University College London, London, UK

Vijitha Periyasamy Department of Electrical and Computer Engineering, Iowa State University, Ames, IA, USA

Manojit Pramanik Department of Electrical and Computer Engineering, Iowa State University, Ames, IA, USA

Menglu Qian Institute of Acoustics, School of Physics Science and Engineering, Tongji University, Shanghai, China

Nastaran Mohammadian Rad Department of Biomedical Engineering, Eindhoven University of Technology, Eindhoven, The Netherlands
Department of Precision Medicine, Maastricht University, Maastricht, The Netherlands

Daniel Razansky Institute for Biomedical Engineering and Institute of Pharmacology and Toxicology, Faculty of Medicine, University of Zurich, Zurich, Switzerland
Institute for Biomedical Engineering, Department of Information Technology and Electrical Engineering, ETH Zurich, Zurich, Switzerland
Zurich Neuroscience Center (ZNZ), Zurich, Switzerland

Qiangzhou Rong Department of Biomedical Engineering, Duke University, Durham, NC, USA

Yoshiyuki Sankai Research and Development Division, CYBERDYNE INC, Tsukuba, Japan

Deeksha Sankepalle Department of Biomedical Engineering, Tufts University, Medford, MA, USA

Naoto Sato Research and Development Division, CYBERDYNE INC, Tsukuba, Japan

Yuecheng Shen School of Electronics and Information Technology, Sun Yat-sen University, Guangzhou, China

Mengjie Shi School of Biomedical Engineering & Imaging Sciences, King's College London, London, UK

Mithun Kuniyil Ajith Singh Research and Business Development Division, CYBERDYNE INC, Rotterdam, the Netherlands

Allison Sweeney Department of Biomedical Engineering, Tufts University, Medford, MA, USA

Dean Ta Department of Electronic Engineering, Fudan University, Shanghai, China

Tanja Tarvainen Department of Technical Physics, University of Eastern Finland, Kuopio, Finland
Department of Computer Science, University College London, London, UK

David Thompson Multi-Modality Medical Imaging group (M3I), Technical Medical Centre, University of Twente, Enschede, The Netherlands

Tom Vercauteren School of Biomedical Engineering & Imaging Sciences, King's College London, London, UK

John A. Viator Department of Bioengineering, University of Pittsburgh, Pittsburgh, PA, USA
Department of Engineering, Duquesne University, Pittsburgh, PA, USA

Simeon J. West Department of Anaesthesia, University College Hospital, London, UK

Ivy H. M. Wong Translational and Advanced Bioimaging Laboratory, Department of Chemical and Biological Engineering, Hong Kong University of Science and Technology, Clear Water Bay, Hong Kong, China

Terence T. W. Wong Translational and Advanced Bioimaging Laboratory, Department of Chemical and Biological Engineering, Hong Kong University of Science and Technology, Clear Water Bay, Hong Kong, China

Min Wu Department of Biomedical Engineering, Eindhoven University of Technology, Eindhoven, The Netherlands

Shiying Wu Institute of Acoustics, School of Physics Science and Engineering, Tongji University, Shanghai, China

Jun Xia Department of Biomedical Engineering, University at Buffalo, The State University of New York, Buffalo, NY, USA

Wenfeng Xia School of Biomedical Engineering & Imaging Sciences, King's College London, London, UK

Shawn (Liangzhong) Xiang Department of Biomedical Engineering, University of California, Irvine, CA, USA
Beckman Laser Institute & Medical Clinic, University of California, Irvine, CA, USA
Department of Radiological Sciences, University of California, Irvine, CA, USA

Weiya Xie Institute of Acoustics, School of Physics Science and Engineering, Tongji University, Shanghai, China

Wenyi Xu Institute of Acoustics, School of Physics Science and Engineering, Tongji University, Shanghai, China

Hao Yang Department of Medical Engineering, University of South Florida, Tampa, FL, USA

Junjie Yao Department of Biomedical Engineering, Duke University, Durham, NC, USA

Huijuan Zhang Department of Biomedical Engineering, University at Buffalo, The State University of New York, Buffalo, NY, USA

Jinyan Zhang College of Future Technology, Peking University, Beijing, China

Mengjiao Zhang Institute of Acoustics, School of Physics Science and Engineering, Tongji University, Shanghai, China

Yan Zhang Translational and Advanced Bioimaging Laboratory, Department of Chemical and Biological Engineering, Hong Kong University of Science and Technology, Clear Water Bay, Hong Kong, China
School of Applied and Engineering Physics, Cornell University, Ithaca, NY, USA

Tianrui Zhao School of Biomedical Engineering & Imaging Sciences, King's College London, London, UK

Emily Zheng Department of Biomedical Engineering, University at Buffalo, The State University of New York, Buffalo, NY, USA

Quanyu Zhou Institute for Biomedical Engineering and Institute of Pharmacology and Toxicology, Faculty of Medicine, University of Zurich, Zurich, Switzerland
Institute for Biomedical Engineering, Department of Information Technology and Electrical Engineering, ETH Zurich, Zurich, Switzerland

Part I
Principles and Imaging Systems

Chapter 1
Photoacoustic Microscopy

Qiangzhou Rong, Lucas Humayun, and Junjie Yao

Abstract Biomedical imaging allows us to explore the complex biology of living organisms and better understand the clinical progression of diseases through the visualization of dynamic, functional, and molecular events. Among the modern biomedical imaging technologies, photoacoustic microscopy (PAM) provides unique optical absorption contrast and high-spatial resolution at depths beyond the penetration limit of traditional optical microscopy. Over the last decades, PAM has become an increasingly popular anatomical, functional, and molecular information. In this book, we introduce the basic principles and typical system designs of PAM, including optical-resolution PAM and acoustic-resolution PAM. We also discuss the common characteristics of PAM, including spatial resolutions, penetration depth, PA signal detectors, and the scanning approach. Finally, we present the major biomedical applications of PAM, including functional measurement, anatomical imaging from cellular to organismal level, label-free functional imaging using endogenous biomolecules, molecular imaging using exogenous contrast agents, and preclinical and clinical applications.

Keywords Photoacoustic imaging · Molecular imaging · Targeted contrast agent · Preclinical and clinical applications

1.1 Introduction

Biomedical imaging first rose to popularity when medical professionals began using X-ray to diagnose and detect fractures in foreign bodies [1–7]. Thanks to its unique ability to visualize biological structures and function in vivo, biomedical imaging has become a staple in patient care [8–11]. Today, some of the most common biomedical imaging modalities include digital X-ray radiography [12],

Q. Rong · L. Humayun · J. Yao (✉)
Department of Biomedical Engineering, Duke University, Durham, NC, USA
e-mail: junjie.yao@duke.edu

W. Xia (ed.), *Biomedical Photoacoustics*,
https://doi.org/10.1007/978-3-031-61411-8_1

X-ray computed tomography (CT) [13], nuclear imaging (e.g., positron emission tomography (PET) and single photon emission computed tomography (SPECT)) [14], ultrasound imaging [15], optical imaging [16] and magnetic resonance imaging (MRI) [17]. Optimization of these modalities as well as the development of new imaging technologies have even shown potential to capture visual data of microscopic objects while maintaining high resolution [18–21]. As biomedical imaging continues to expand our understanding of tissue composition, morphology, function, and dynamic biological processes, it has become a necessity for pushing the boundaries of scientific inquiry.

Optical imaging is a unique branch of the biomedical imaging family and has taken great strides over the past two decades [22–24]. Thanks to their characteristic high resolution, high sensitivity, specificity, and practicality, optical imaging techniques are currently able to offer information ranging from a large 3D bulk to finite molecular-grade information (e.g., cancer biomarkers, cell metabolic state, and atherosclerotic lesions). These features align well with the increased demand for personalized health care envisaged for diagnosing patient subpopulations with increased precision [25]. Indeed, optical diagnostic imaging is already proving to revolutionize clinical care [26].

One of the main challenges facing optical imaging is strong light scattering in biological tissue as this decreases spatial resolution and restricts imaging depth [27–31]. Fortunately, several developing optical imaging modalities, including, confocal microscopy [32], two-photon microscopy [33, 34], and optical coherence tomography (OCT) [35–37] offer higher-resolution images. Through the detection of singly backscattered photons for OCT processing and ballistic photons these technologies appear to alleviate the challenge of maintaining high image quality. Unfortunately, by increasing resolution, these optical imaging techniques sacrifice depth, evidenced by their typical limitation to less than 1 mm of penetration. Beyond this limit, existing modalities are unable to produce meaningful visualizations. While researchers have made a tremendous effort to overcome the hurdle of imaging depth [38, 39] through the development of technologies such as diffuse optical tomography (DOT), optical imaging alone is largely unable to achieve depth-dependent spatial resolution.

Photoacoustic imaging (PAI), or optoacoustic imaging (OAI), is a hybrid imaging modality that merges optical illumination and ultrasound detection [40–42]. In the last two decades, PAI has attracted attention for its biomedical pre-clinical and clinical applications. PAI first relies on a pulsed laser with pico-second or nano-second cycles to illuminate the biological sample. The sample then absorbs the laser-derived optical energy and converts it into heat, making local temperature rise. The subsequent thermoelastic expansion results in the release of US waves. US signals can be collected by a range of sensors including piezoelectric transducers and optical interferometers [43, 44].

In contrast to other optical imaging techniques, PAI relies on the detection of US waves, which undergoes less scattering and attenuation in tissue. PAI can therefore image deeper targets with better resolution than comparable optical imaging modalities [45, 46]. In order to deliver clinically relevant information,

PAI apparatuses are often designed to target biomaterial with the intrinsic optical absorption of chromophores, such as hemoglobin, melanin, lipids, and water [47, 48]. Thanks to the adaptability of the pulsed laser component of PAI, each of these targets can be utilized to capture physiological data simply by adjusting the laser's emitted range of wavelengths [49, 50]. In addition, the label-free nature of endogenous contrast PAI makes it appropriate for patients' compliance in long-term longitudinal monitoring.

The PA effect was initially discovered by Alexander Bell in 1880 [51], who observed a correlation between acoustic energy emission frequency and light modulation frequency when light energy was rapidly interrupted while illuminating different solid substances. This discovery electrified the field of light-induced sonorous effects, attracting many notable scientists to develop theories and conduct further experimental studies [52–54]. This early research was instrumental in constructing the foundations of modern PAI. The subsequent development of related optical and acoustic technologies, such as effective and affordable laser sources [55, 56], miniaturized US detection devices [57, 58], novel computational platforms and advanced inversion models [59–61], and refined PAI, making it a valuable tool for modern biological research of clinical diagnostic techniques.

Today, the field of PAI can be well divided into two categories: photoacoustic computed tomography (PACT) [62] and photoacoustic microscopy (PAM) [63]. PACT can provide deep penetration up to several centimeters with high spatial resolution (hundreds of micrometers) by using wide-field optical illumination and parallel acoustic detection [64, 65]. In contrast, PAM can provide high spatial resolution (several micrometers), but it sacrifices penetration depth (one to several millimeters) by using a co-focused optical illumination and acoustic detection method [66]. When used together, PACT and PAM systems offer extremely valuable visualization data with many biomedical applications [66, 67]. While PACT is an equally exciting and valuable field of PAI, we will primarily discuss the PAM system in this chapter.

PAM is implemented by a confocal microscopy method of optical excitation and acoustic detection. The confocal configuration can offer large detection sensitivity and high SNR. According to the size of the focused laser spot on the target, the PAM system can be divided into two groups: optical-resolution PAM (OR-PAM) [68] and acoustic-resolution PAM (AR-PAM) [69]. In OR-PAM, the illuminating light is tightly focused as a source so that the spot size of focused light significantly determines the system's lateral resolution by $0.51 \times \lambda_0 = \mathrm{NA}_0$ [70, 71], where λ_0 is the wavelength of light and NA_0 is the numerical aperture (NA) of the optical objective. For example, when using an objective with NA of 0.1 and light wavelength of 532 nm, the OR-PAM can offer a resolution of 3 μm. However, due to the limited light penetration, the imaging depth of OR-PAM is approximately 1 mm. In AR-PAM, the illuminated light is weakly focused or collimated; thus, the optical beam spot is larger than the focal spot of acoustic detection. In contrast to OR-PAM, the lateral resolution of AR-PAM relies heavily on the spot size of acoustic detection [72], which can be calculated by the equation of $0.71 \times \lambda_a = \mathrm{NA}_a$ where λ_a is the wavelength of the acoustic wave and NA_a is the NA of the focused ultrasonic transducer. In comparison to OR-PAM, AR-PAM lateral resolution is more than ten times worse. In addition, the bandwidth of the ultrasonic detector determines the axial resolutions of OR- and AR-PAM that can be calculated by the equation of $0.88 \times c/(\Delta f)$, where c is the US speed, and Δf is the response frequency bandwidth of the transducer. For in vivo studies, it is necessary that the optical fluence determined by the optical spot size is lower than the maximum permissible exposure (MPE) regulated by American National Standard for Safe Use of Lasers [73]. This limitation allows much larger pulse energy in AR-PAM and therefore achieves larger penetration depth. It is because PA signals suffer from tissue attenuation that the acoustic penetration depth of AR-PAM system is inversely proportional to the central frequency of the transducer [74]. The large field of view (FOV) of PAM system is usually realized by scanning methods, including moving the sample or the light/sound [75, 76]. The conventional scanning method requires advancing through stepper motor stages with considerable speed. In order for high-quality imaging, the step size of the motor should be at most half of the lateral resolution. However, imaging a large FOV takes time. There have been other developed methods to improve the imaging speed, including the use of commercial Galvanometer mirrors [77], micro-electromechanical system (MEMS) scanning mirrors [76], and multiple-surface polygon mirrors [78].

Despite some of the challenges it currently faces, PAM presents itself as a versatile tool that can accurately detect endogenous and exogenous [48, 79] targets at their absorbing wavelengths better than fluorescence-based methods [80, 81], such as wide-field, confocal, and multi-photon microscopy. Its additional utility in capturing anatomical, functional, molecular, flow dynamic, and metabolic contrasts in vivo makes it an important tool for future scientific study and clinical diagnoses.

1.2 Basics of PAI

1.2.1 General Behavior of Light in Biological Tissue

Most biological tissues exhibit strong optical scattering. Thus, tissue samples used in optical imaging are commonly referred to as scattering or turbid media [82, 83]. While optical scattering is generally quite extensive in biological tissue, optical absorption is relatively weak with absorption in the range of 400–1350 nm. The mean free path between photon scattering is on the order of 0.1 nm, and the mean absorption length is from 10 nm to 100 nm. Figure 1.1 in Ref. [40, 82] displays the schematic of photon propagation in biological tissue. Here the light source employs a temporally Dirac delta pulse with a shape of a pencil beam or an infinitely narrow collimated beam. The optical properties of tissue significantly determine the light propagation in tissue as described above. For example, when using the parameters of absorption coefficient $\mu_a = 1.4\,\text{cm}^{-1}$, scattering coefficient $\mu_a = 350\,\text{cm}^{-1}$, scattering anisotropy $g = 0.8$, and refractive index $n = 1.37$, the mean free path

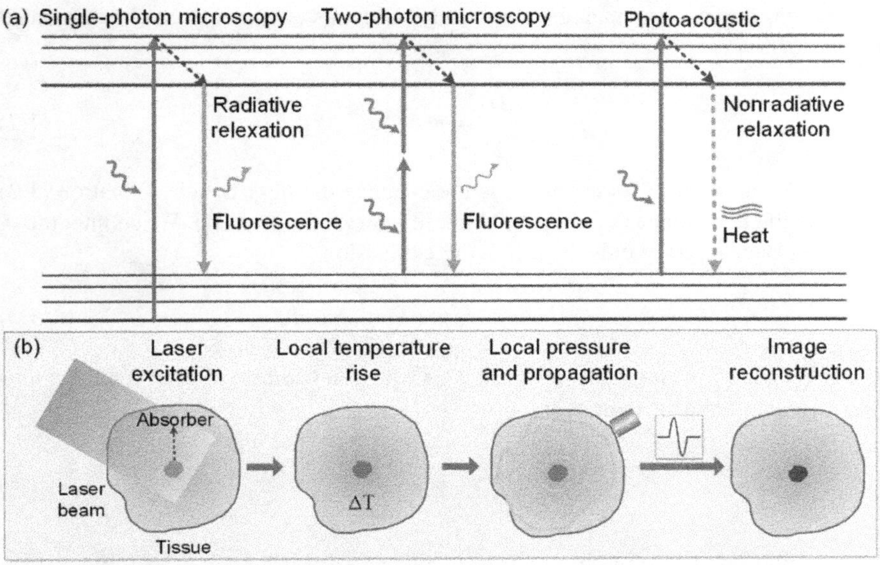

Fig. 1.1 Photon imaging principle. (**a**) Jablonski diagram, photon energy transfer in one-photon fluorescence microscopy, two-photon fluorescence microscopy, and photoacoustic generation. (**b**) basic principle of PAI [40]

is equal to 28 μm with respect to 0.6 ps propagation time. For most optical imaging, the diffusion-like behavior of light is always a key challenge. There have been various methods using various system configurations to contend with the diffusion influence.

1.2.2 Photons Absorption in Tissue

A given photon's absorption is normally characterized by the absorption coefficient μ_a or the probability of photon being absorbed in a medium per unit path length (e.g., 0.1 cm to 1 cm in biological tissue) [84–87]. The mean absorption length can be calculated by taking the reciprocal of μ_a. The absorption cross-sectional area (σ_a) can characterize the absorbing capability of a single absorber. σ_a is related to its geometric cross-sectional area σ_g by the absorption efficiency expression $Q_a : \sigma_a = Q_a \sigma_g$. The density of absorbers within a given volume is represented by N_a; therefore, the absorption coefficient can be calculated by summing the total cross-sectional area. Assuming the absorbers are independent of one another, the μ_a per unit volume can be given by:

$$\mu_a = N_a \sigma_a \tag{1.1}$$

As light propagates in an absorbing-only tissue, its attenuation can be calculated by

$$\frac{dI}{I} = -\mu_a dx \tag{1.2}$$

where I is the light intensity, and x is the distance the light travels. Equation (1.2) displays the percentages of absorbed light in interval $(x,\ x + dx)$. When integrated, invoking Beer's Law yields the following expression

$$I(x) = I_0 \exp(-\mu_a x) \tag{1.3}$$

where I_0 is the light intensity at $x = 0$. This equation is effective even for a tortuous path. Transmittance is defined by

$$T(x) = \frac{I(x)}{I_0} \tag{1.4}$$

Equation (1.4) can be used for the probability of survival after propagation over x.

Within the context of biological tissue, optical absorption is primarily caused by hemoglobin, melanin, and water. Figure 1.2a displays the absorption spectrum of some primary absorbing biological tissue components. In the figure, the absorption coefficients are plotted as a function of wavelengths. This provides an important reference for the laser source option to improve the PA signal. Because of their dramatically different optical absorbing properties, the reported detection sensitivity (NEC) for exogenous contrast agents varies from millimolar to picomolar. Roughly, the reported NEC is on the level of millimolar for microbubbles, micromolar for organic dyes, picomolar for nanoparticles, and nanomolar for fluorescent proteins [85]. Figure 1.2b summarizes the reported PAM detection sensitivity (NEC) of representative endogenous and exogenous contrast agents.

As mentioned above, the oxy- and deoxyhemoglobin present different absorption wavelengths (e.g., 532 nm and 558 nm) but have a few intersections, termed isosbestic points. At these points, the two types of hemoglobin's absorption coefficients are therefore only relative to the total concentration. The concentrations of two kinds of hemoglobin can be obtained as follows:

$$\mu_a(\lambda_1) = \ln(10)\varepsilon_{\mathrm{ox}}(\lambda_1)C_{\mathrm{ox}} + \ln(10)\varepsilon_{\mathrm{de}}(\lambda_1)C_{\mathrm{de}} \tag{1.5}$$

$$\mu_a(\lambda_2) = \ln(10)\varepsilon_{\mathrm{ox}}(\lambda_2)C_{\mathrm{ox}} + \ln(10)\varepsilon_{\mathrm{de}}(\lambda_2)C_{\mathrm{de}} \tag{1.6}$$

where λ_1 and λ_2 are the two corresponding wavelengths, $\varepsilon_{\mathrm{ox}}$ and $\varepsilon_{\mathrm{de}}$ are the known molar extinction coefficient of oxy- and deoxyhemoglobin, C_{ox} and C_{de} are the molar concentrations of oxy- and deoxyhemoglobin, respectively, in the tissue. According to the Eqs. (1.4) and (1.6), the oxygen saturation (sO_2) and the total

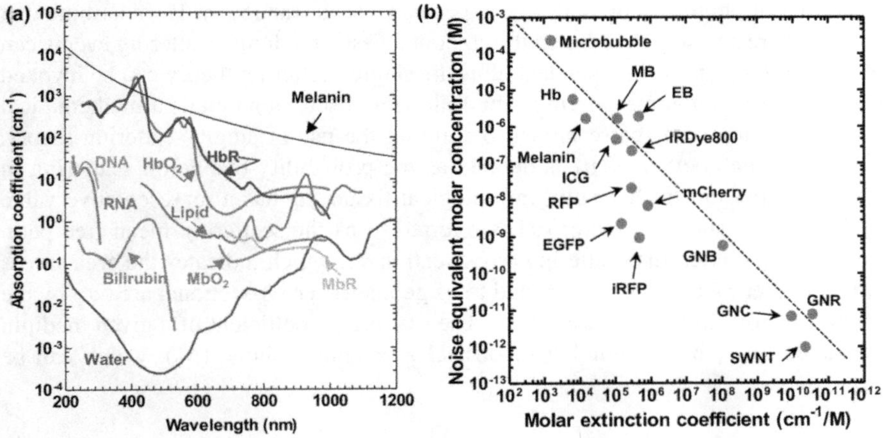

Fig. 1.2 (**a**) Absorption spectra of major endogenous contrast agents in biological tissue at normal concentrations. (**b**) reported noise equivalent molar concentrations (NEC) of major endogenous and exogenous contrast agents, versus their molar extinction coefficients. Due to the lack of complete information, the incident fluence is not corrected here. HbO_2 oxy-hemoglobin, HbR deoxy-hemoglobin, MbO_2 oxymyoglobin, MbR reduced myoglobin, EB evens blue, EGFP enhanced green fluorescent protein, GNB gold nanobeacon, GNC gold nanocage, GNR gold nanorod, Hb hemoglobin, ICG indocyanine green, IRDye800 near-infrared Dye800, iRFP near-infrared red fluorescent protein, MB methylene blue, mCherry monomeric cherry protein; Melanin; RFP red fluorescent protein, SWNT single walled nanotube. (Reprinted with permission from [85] for (a) and (b))

concentration (C_{Hb}) of hemoglobin can be calculated by the C_{ox} and C_{de}, which can be expressed as [88–90].

$$SO_2 = \frac{C_{\text{ox}}}{C_{\text{ox}} + C_{\text{de}}} \tag{1.7}$$

$$C_{\text{Hb}} = C_{\text{ox}} + C_{\text{de}} \tag{1.8}$$

This theoretical analysis can be used for understanding the pulse oximetry and functional imaging. For example, C_{Hb} rising can analyze angiogenesis, whereas sO_2 decreasing can predict tumor hypermetabolism.

1.2.3 Photon Scattering in Tissue

Light scattering effect can be modeled by the Mie theory and the Rayleigh theory if the light wavelength is much larger than the scattering spherical particle [82, 91–95]. As thousands of obstacles are randomly assorted throughout a medium, the scattering of photons should also be expected to occur randomly. If we assume that obstacles are sparsely distributed throughout a tissue medium, scattering events can also be assumed to be independent, thus the single-scattering theory can be invoked to characterize photon scattering. The following discussion details a mathematical implementation of a sparse model, permitting the use of single-scattering theory. The scattering coefficient μ_s is defined as the probability of photon scattering in a medium per unit path length. In biological tissue, μ_s has a representative value of 100 cm^{-1}, and μ_s's reciprocal is referred to as the scattering mean free path. For a single scatter, the scattering cross-section σ_s, which indicates the area across which scattering can occur, is related to its geometric cross-sectional area σ_g by the scattering efficiency $Q_s : \sigma_s = Q_s\sigma_g$. The scattering coefficient of a given medium is related to σ_s by the number of obstacles per unit volume (N_s), which can be expressed as

$$\mu_s = N_s\sigma_s \tag{1.9}$$

After a photon propagates over path length x, the probability of no scattering (or ballistic transmittance T) can be expressed by Beer's law as

$$T(x) = \exp(-\mu_s x) \tag{1.10}$$

There are two factors that strongly influence photon scattering: biological structure size concordance with the optical wavelength and refractive index discordance with the surrounding medium [96–99]. Interestingly, each component of biological tissue presents a different refractive index: 1.35–1.36 for extracellular fluid, 1.36–1.375 for cytoplasm, and 1.38–1.41 for nuclei and mitochondria. The difference in refractive indices ultimately contributes to the highest degree of scattering.

The total interaction-induced extinction coefficient μ_t is the sum of the absorption coefficient and scattering coefficient, which can be expressed as

$$\mu_t = \mu_a + \mu_s \tag{1.11}$$

Finally, the mean free path between interaction events can be computed by taking the reciprocal of μ_t (Fig. 1.3).

1.2.4 Light Transport Through Tissue

Light transport in tissue can be characterized by two kinds of models: an essentially discrete model of individual photon interactions that can be analyzed by a Monte Carlo method and a continuous model based on a differential equation approximation that can be analyzed by the diffusion equation [100–102]. These two methods can be derived by an integro-differential equation expressed as

Fig. 1.3 Biological structures of various sizes for photon scattering [99]

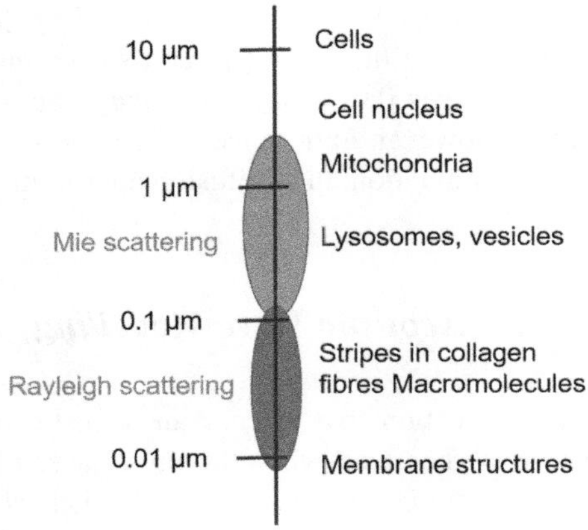

$$\left\{\hat{\text{s}}\cdot\nabla+\mu_{\text{a}}+\mu_s+\frac{1}{\nu}\frac{\partial}{\partial t}\right\}\phi\left(r,\hat{\text{s}}',t\right)=\frac{1}{\nu}q\left(\hat{\text{s}},\hat{r},t\right)+\mu_s\int f\left(\hat{\text{s}}',\hat{\text{s}}\right)\phi\left(r,\hat{\text{s}}',t\right)d^2\hat{\text{s}} \tag{1.12}$$

where μ_a, μ_s are absorption and scattering coefficients, respectively, $\hat{r}$,$\hat{s}$,$\hat{s}'$ are direction vectors, ν is the speed of light in the medium, $\phi\left(r,\hat{s}',t\right)$ is the photon density in direction $\hat{s}'$,q is the source term, and $f\left(\hat{s}',\hat{s}\right)$ is the probability of scattering from direction $\hat{s}'$, into direction $\hat{s}$.

Photon diffusion approximation in a second-order partial differential equation can characterize the time behavior of photon fluence rate distribution in a medium with low absorption but high scattering [103, 104]. Compared with diffusion equation in physics, the photon diffusion equation has an absorption term, which can be expressed as

$$\left\{\nabla\cdot\kappa\nabla-\mu_{\text{a}}c-\frac{\partial}{\partial t}\right\}\Phi\left(r,t\right)=-q_0\left(r,t\right) \tag{1.13}$$

where Φ is the photon fluence rate, q_0 is the source term, and $\kappa(r)$ is the diffusion coefficient. The mathematical form of $\kappa(r)$ can be written as

$$\kappa=\frac{\nu}{3\left(\mu_a+(1-g)\,\mu_s\right)} \tag{1.14}$$

where g is the average cosine of the scattering angle distribution and $(1-g)\mu_s$ can also be represented by the reduced scattering coefficient μ_{s}'.

The photon fluence rate can be calculated by

$$\Phi\left(r,t\right)=\int\phi\left(r,\hat{s}',t\right)d^2\hat{s} \tag{1.15}$$

The Eq. (1.13) has been identified by the theory and experiment demonstrations in the case of the near-infrared (NIR) transillumination of tissue with the representative ranges $0.01 < \mu_a < 0.1\,mm^{-1}$ and $1.0 < \mu_s' < 10.0\ mm^{-1}$. It should be noted, however, in instances where the scattering effect does not dominate the light transillumination, this diffusion approximation may not be exact.

1.2.5 Acoustic Wave Modelling

In this section, we discuss an acoustic model of a fluidic region. The acoustic wave is generated when the pulse laser illuminates the sample and cause thermal expansion. The generation mechanism is based on thermoelastic effect, and the effects of viscosity and thermal conductivity could be ignored [105–107]. The acoustic pressure equation can be expressed as

$$\left\{\nabla^2 - \frac{1}{c^2}\frac{\partial}{\partial t^2}\right\} p = -\frac{\beta}{C_p}\frac{\partial H}{\partial t} \tag{1.16}$$

where c is the sound speed, β is the volume thermal expansivity, C_p is the constant pressure specific heat capacity, p is the acoustic pressure and H is the heat energy per unit volume and per unit time deposited in the fluid. p and H normally rely on the position $r = (x, y, z)$ and time t.

1.2.6 PA Effect on Tissue

In accordance with the photoacoustic mechanism mentioned above, the photoacoustic signal in biomedical tissue is highly dependent on the absorption of the pulsed laser. When absorbed light energy is transferred to thermal energy local temperature is raised, causing tissue expansion followed by sharp contraction. This process produces acoustic pressure. Thus, the generation of photoacoustic signals can be characterized by a thermodynamic energy relation and state equation [82, 107], which can be written as

$$\rho C_p \frac{\partial \tau}{\partial t} = H \tag{1.17}$$

$$p = \frac{1}{\kappa_T}\left(\frac{\delta}{\rho} + \beta\tau\right) \tag{1.18}$$

where κ_T is the isothermal compressibility, ρ is density, and τ is change in temperature. Because the illuminated light pulse is extremely short (ns) in tissue, the pressure rise associated with local temperature increase can be assumed to occur instantaneously without expansion (i.e., adiabatic heating). This instantaneous heating brought about by pulsed laser illumination can be modeled as [105],

$$H(r,t) = H(r)\delta_D(t) \tag{1.19}$$

where $H(r)$ is the function that could describe the heat distribution per unit volume fluid (also called the absorbed energy map) and $\delta_D(t)$ is the Dirac delta. In this case, the optical energy will be absorbed before the fluid density has time to change [105]. For pressure generation to be regarded as instantaneous, the duration t_p of the pulsed laser should satisfy the condition below [106].

$$t_p \ll \frac{1}{\mu_a c} \tag{1.20}$$

Optical energy is therefore absorbed before fluid density changes. Considering the optical absorption coefficient of the medium, the spatial part of the heating function can be expressed as

$$H(r) = \mu_a(r)\Phi(r, \mu_a) \tag{1.21}$$

By the analysis above, the acoustic pressure after initial pressure distribution $p_0(r)$ is therefore proportional to the absorbed energy map.

$$p_0(r) = \left(\frac{\beta c^2}{c_p}\right) H(r) = \Gamma \mu_a(r)\Phi(r, \mu_a) \tag{1.22}$$

where Γ is the Gruneisen coefficient, a dimensionless constant that represents the efficiency of the conversion of heat to pressure.

According to Eq. (1.22), the initial pressure can be recast as an initial value problem, then propagate away as acoustic waves. Consequently, the photoacoustic wave can be performed by the equation below [107].

$$\left\{\nabla^2 - \frac{1}{c^2}\frac{\partial}{\partial t^2}\right\} p = 0,\ p|_{t=0} = p_0(r),\ \left.\frac{\partial p}{\partial t}\right|_{t=0} = 0 \tag{1.23}$$

As the linear acoustic wave propagation is modeled in the soft biological tissue, the propagation medium can be assumed to be isotropic. Therefore, the net flow does not generate and shear waves can be neglected. Three first-order equations, shown below, can be used to bridge acoustic pressure p, acoustic particle velocity u, and acoustic density δ, which correspond to the momentum conservation, mass conservation, and equation of state.

$$\frac{\partial u}{\partial t} = -\frac{1}{\rho}\nabla p,\ \ \frac{\partial \delta}{\partial t} = -\rho \nabla \cdot u,\ \ p = c^2\delta \tag{1.24}$$

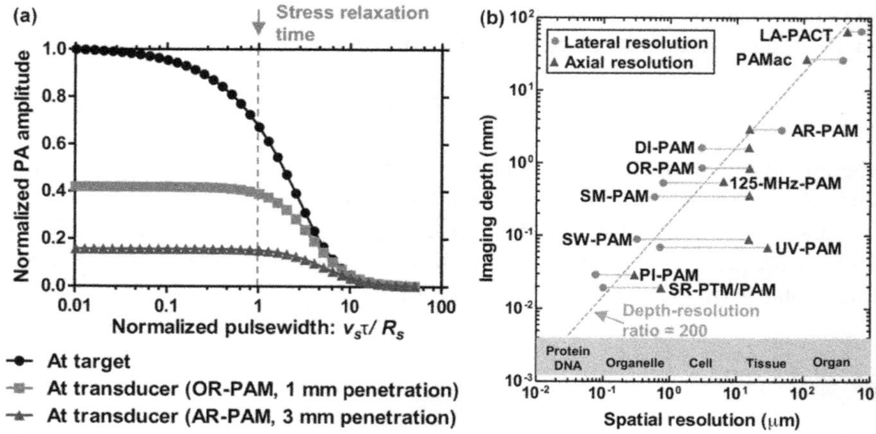

Fig. 1.4 (**a**) Simulated maximum PA pressure amplitude at the target and at the ultrasonic transducer surface as a function of the excitation pulse width in OR-PAM and AR-PAM. The excitation pulse width is normalized by the stress relaxation time, while the pulse energy is a constant. The PA amplitude of all curves is normalized by the maximum PA amplitude at the target (black curve). R_S is the axial resolution. The target is 6 mm from the transducer. In OR-PAM, the target is 1 mm under the tissue surface, while in AR-PAM the target is 3 mm under the tissue surface. Imaging depth versus spatial resolution in multi-scale photoacoustic imaging. The red circles denote lateral resolution, and the blue squares denote axial resolution. (**b**) imaging depth versus spatial resolution in multi-scale PACT. The red circles denote lateral resolution, and the blue squares denote axial resolution. DI-PAM double-illumination PAM, LA-PACT linear-array PA computed tomography, 125-MHz-PAM PAM with a 125 MHz ultrasonic transducer, PAMac deep PA macroscopy, PI-PAM photo-imprint PAM, SR-PTM/PAM super-resolution photothermal, SM-PAM submicron PAM, SW-PAM subwavelength PAM, UV-PAM ultraviolet PAM. (Reprinted with permission from [85])

This equation implies that the sound speed and ambient density rely on the position, and the pressure is related to two parameters: time and position.

The excitation pulse width is important in optimizing PA signal generation. In PAM, the stress confinement on the axial direction requires a pulse width to be less than the acoustic transit time across the resolution voxel (typically on the scale of nanoseconds). The excitation pulse width can affect the axial and lateral resolutions of AR-PAM, and the axial resolution of OR-PAM. In addition, the pulse width is related to the generated PA signal amplitude at the target. As shown in Fig. 1.4a, when the stress confinement is not satisfied, i.e., the excitation pulse is long compared with the stress confinement, the resultant PA pressure at the target is roughly inversely proportional to the square of the pulse width [85, 101]. In this sense, a shorter pulse is more efficient in generating PA signals than a longer pulse. However, the increased PA signal mostly falls into the high-frequency region. As we know, after the low-pass filtering by tissue, high-frequency acoustic components will not be captured by the ultrasonic transducer. As can be seen from Fig. 1.4a, once the stress confinement is satisfied, the excitation pulse can be approximated as a delta function, and the resultant PA pressure at the transducer surface relies only on the excitation pulse energy and is not sensitive to the pulse width anymore. Therefore, it is not necessary to further reduce the pulse width once the targeted signal bandwidth is matched. The pulse width needs to be optimized according to the desired spatial resolution or imaging depth. Furthermore, because excitation intensity is pulse width related, pulse width also affects PA signal generation through the absorption saturation effect [101].

Moreover, the biological systems were usually studied on both macroscopic and microscopic scales. By adjusting the excitation and detection configurations, all of the key imaging parameters of PAM, including spatial resolution, imaging depth, and detection sensitivity, can be scaled over a wide range with the same optical absorption contrast. Such high scalability is critical for comprehensive study of biological phenomena over different length scales and for the translation of laboratory discovery to clinical practice. Comprehensive details about PAM characterizations and biomedical applications can be found in previous review articles [40, 62, 85], and a summary of the scalable imaging performance of representative PAM systems is shown in Fig. 1.4b. To achieve the ultimate detection sensitivity, PAM needs to optimize its optical illumination and acoustic detection, based on the desired imaging specifications.

1.3 PAM Implementations

As discussed in the introduction, PAM systems are usually divided into two categories: AR-PAM and OR-PAM, as shown in Fig. 1.5. Both systems present different lateral resolutions and penetration depths. In this section, we describe several representative system configurations and discuss their characteristics, which may help to understand the two available PAM systems in more detail. While we present acoustic-resolution and optic-resolution systems as distinct PAM systems here, it should be noted that in some cases, OR- and AR-PAM are interchangeable, and merging the two systems together is possible to achieve a dual-illumination system [108, 109]. In PAM, the PA signals are formed at the co-focal point of light and acoustic wave. Aligning two beams coaxially is important to maximize the sensitivity of PAM, which has been realized through the design of many technologies. Various types of PAM systems are summarized in Fig. 1.6.

Off-axis alignment is an easy way to implement an acoustic-optic combiner. However, this method system has limited detection sensitivity and FOV, so it is difficult to implement in AR-PAM systems. Additionally, the misalignment of light and acoustic waves further limits the axial resolution of imaging. Due to these features, the off-axial PAM system is mainly used for the imaging of flat plane samples or very small FOV [110, 111]. Dark-field confocal PAM is another popular system in which a conical lens is used to form an annular-shaped light beam that can bypass the acoustic detector [112, 113]. The light illuminates the sample at a tilted angle, and the transducer collects the generated acoustic wave above the sample.

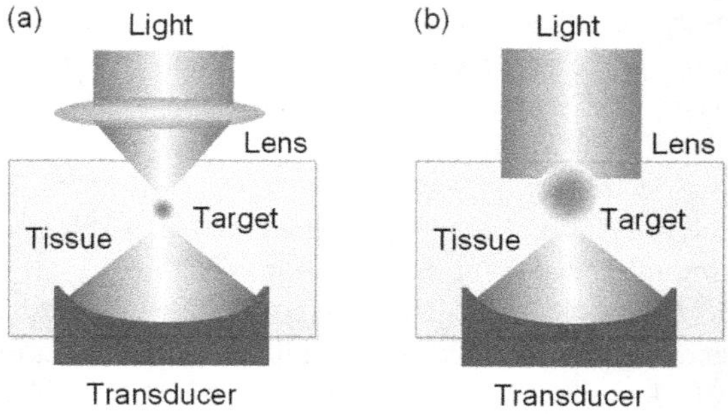

Fig. 1.5 Typical setups of (**a**) OR-PAM and (**b**) AR-PAM

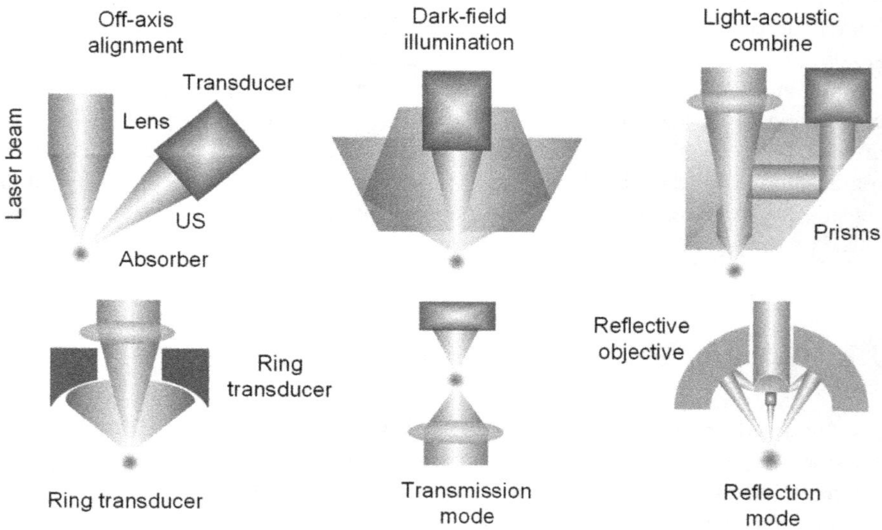

Fig. 1.6 Features of various PAM implementation types. NA numerical aperture, AR acoustic resolution, and OR optical resolution [109]

This coaxial configuration can result in a high SNR. However, because the light cannot be focused tightly, it is only appropriate for use in AR-mode imaging. Like AR-PAM, AR-mode imaging is usually used to visualize larger and deeper targets. The second-generation combiner typically utilizes a glass plane (transparent to light and reflective of acoustic waves) to combine light and acoustic waves together, making them coaxial [114, 115]. This method can focus tightly to get a high lateral resolution less than 5 μm, and presents a considerable imaging depth of 1.2 mm. One of the primary drawbacks of this method is that some acoustic energy will likely be lost thanks to the acoustic impedance mismatch of the combiner. In spite of its drawbacks, this method has been a popular technology used in OR-PAM, AR-PAM, and even a switchable system in in vivo microscopy. In the fourth method, a ring transducer is used, which has a hole for light to pass through [116]. The PA signal is then detected on the other side of light source. This method aligns light and acoustic waves, achieving a lateral resolution of less than 5 μm. However, the presence of a central hole may decrease the sensitivity of the transducer and restrict the optical NA. Its transmission mode in the transmission mode can maximize the optical NA up to 1.0 [117, 118], making it useful in cell imaging. Similar to other transmission modes, this method cannot be used in in vivo imaging. The reflective objective lens-based PAM can be an alternative to solve the problem of the transmission mode [119, 120]. As this method employs the use of a small transducer equipped below an optical lens with very high NA, the reflection mode can improve the lateral resolution up to subwavelength. By restricting the focal length, however, the work distance is also limited.

In AR-PAM mode, the lateral resolution is related to the acoustic properties, which can be given as

$$\mathrm{LR}_{\text{AR-PAM}} = 0.71\frac{\lambda_a}{\mathrm{NA}_a} = 0.71\frac{\nu}{\mathrm{NA}_a \times f_c} \tag{1.25}$$

Where λ_a is the center wavelength of acoustic wave, NA_a is the acoustic NA, ν is the speed of sound, and f_c is the center frequency of the PA signals. Choosing a US transducer with high center frequency and large NA improves the lateral resolution of AR-PAM. However, the lateral resolution outside of the focal region can be deteriorated, and the strong attenuation to high-frequency ultrasound of tissue will limit the imaging depth. The tradeoff between imaging depth and lateral resolution should be carefully considered when building an AR-PAM system.

The lateral resolution of OR-PAM is determined by the optical focal spot size, which is written as.

$$\mathrm{LR}_{\text{OR-PAM}} = 0.51\frac{\lambda_o}{\mathrm{NA}_o} \tag{1.26}$$

where λ_o is the optical wavelength, and NA_o is the optical NA. In comparison, when using a tightly focused light, the OR-PAM has a higher lateral resolution than that of AR-PAM. However, the mean optical path length significantly limits its imaging depth.

Because the axial resolution is only determined by the acoustic bandwidth, both AR-PAM and OR-PAM present similar axial resolution expressed as

$$\mathrm{AR}_{\text{OR-/AR-PAM}} = 0.88\frac{v}{\Delta f_c} \tag{1.27}$$

where Δf_c is the bandwidth of the PA signal. The PA signal has a very wide bandwidth ranging from 1 MHz to 100 MHz (larger than the bandwidth of a commercial piezoelectric transducer). The axial resolution of PAM is thus determined by the transducer used in the system.

With the development of optical technology and ultrasound detection technology, PAM technology has also experienced rapid development. In fact, many PAM systems have shown to improve spatial resolution, penetration depth, and compatibility.

For example, a double excitation process-based photo-imprint PAM (PI-PAM) system was configured and performed [120–124]. In experiment processing, this system can produce PA waves and modulate heterogeneous photobleaching, which normally affects the PA signal amplitude. The photobleaching effect provides a smaller point-spread function (PSF) than the conventional OR-PAM in image processing. Consequently, the lateral resolution is improved to 120 nm. A similar design is used for a switchable PAM modality by using a transient bacterial phytochrome called BphP1 [125]. BphP1 exists in two states (ON and OFF) and provides a subdiffractional lateral resolution of 141 nm and a maximum axial resolution of 410 nm. A Grueneisen relaxation PAM (GR-PAM) system can also be designed to improve spatial resolution based on the nonlinear PA signal generation [126]. In this system, the GR effect ensures that the PA signal is nonlinearly proportional to the optical fluence. The GR signal therefore declines as the axial distance between a given absorber and the point of optical focus increases. In practice, this method is used to visualize hemoglobin with improved axial resolution from 45 μm to 2.3 μm [126]. GR-based PAM systems have also demonstrated great utility in other biological imaging contexts [127, 128]. A dual-view OR-PAM has also been developed to achieve high spatial resolution [129]. In this system, the laser beam is split into two beams that are concentrated on a spot perpendicular to one another. The resulting double-view images are then processed by a double-view Richardson-Lucy deconvolution algorithm. As a result, the lateral resolution can be improved from the original 3.0 μm to 1.8 μm in the x-z plane [129].

1.4 PAM Systems

1.4.1 Light Excitation

Laser sources with duration of nanoseconds are usually used to excite tissue samples [82, 124, 130]. They have been developed based on different methods (e.g., using different pump sources, gain medium, and cavities). Several typical laser configurations are summarized in Fig. 1.7. In this section we detail the pulsed lasers that meet PAM criteria and are widely used in practice.

The Nd:YAG (neodymium-doped yttrium aluminum garnet; Nd:Y3Al5O12) laser is a popular solid-state laser that provides high-intensity output. Nd:YAG lasers utilize Nd:YAG crystal as a medium across which an optical pump source, using flash lambs or laser diodes, emits light with a typical wavelength of 1064 nm [131]. This laser operates in a pulsed fashion, utilizing a Q-switching mode with an optical switch modulating the emitted light waves interacting with medium in the cavity [132]. The pulsed laser is ultimately able to achieve high power output with small pulse duration. By the efficiently doubled frequency technique, Nd:YAG lasers can generate light at the wavelengths of 532 nm, 355 nm, 266 nm and 213 nm.

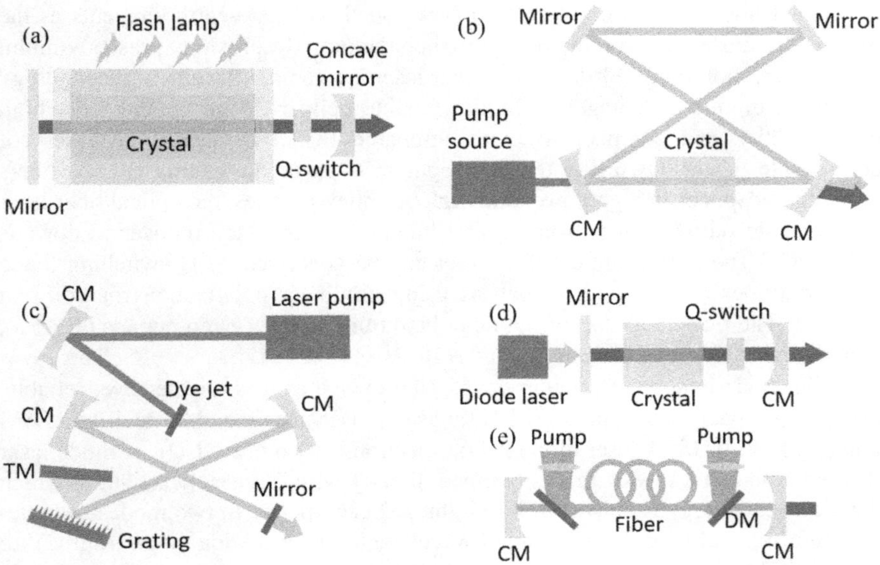

Fig. 1.7 Schematic illustrations of laser sources used for PAM. The typical configurations of (**a**) flash lamp-pumped solid-state laser, (**b**) diode-pumped solid-state laser, (**c**) dye laser, (**d**) an optical parametric oscillator, (**e**) optical fiber laser. TM tuning mirror, CM concave mirror, DM dichroic mirror [130]

Dye laser is another laser commonly seen in PAM systems. Dye lasers use an organic dye, typically assuming liquid state, as the lasing medium [133, 134]. Compared to solid and gas state lasing media, a dye can be used to generate a wide wavelength range from 50 nm to 100 nm or more [135]. Flexibility in bandwidth makes dye lasers an attractive option for PAM as tunable, pulsed laser. However, solid state dye has also been integrated into the design of solid state dye lasers (SSDL), which characteristically utilizes dye-doped organic matrices as gain medium [136]. In addition to the wide bandwidth, dye lasers can also provide large energies or high average power. Flash lamp-pumped dye lasers, for example, can produce hundreds of Joules per pulse.

Ti: sapphire lasers typically use a crystal of sapphire (Al_2O_3) that is doped with Ti^{3+} ions as a lasing medium [137, 138] and operate most efficiently at wavelengths near 800 nm. Various Ti: sapphire lasers have been developed, allowing Ti: sapphire lasers to emit light in the range of 650 to 1100 nm. Pulsed Ti: sapphire laser can be created by passive mode-locking techniques through the use of Kerr lens mode locking (KLM) and semiconductor saturable absorber mirrors (SESAM) [138]. Pulsed Ti: lasers emit ultrashort pulse outputs ranging from several hundreds of nanoseconds to several femtoseconds [138, 139]. Mode-locked Ti: sapphire laser provides relatively high output power on the order of 0.3–4 W and pulse repetition rate up to several tens of MHz to multi-GHz.

Optical fiber lasers utilize optical fibers doped with rare-earth elements as the gain medium, such as erbium, ytterbium, neodymium, dysprosium, praseodymium, thulium, and holmium [139]. Optical fiber laser is a rapidly developing technology due to its unique advantages including flexible delivery, high optical gain, high beam quality, and compact structure compared to solid-state or gas lasers of comparable power. Owing to the development of large mode area (LMA) fibers as well as advances in high power and high brightness diodes, the optical fiber laser can generate output power from several hundred watts to ten thousand kilowatts [140, 141]. The pulsed optical fiber laser can be generated by Q-switching mode and mode-locking techniques, such as using nonlinear polarization rotation and SESAM. The pulse duration of light can be on the order of nanoseconds (10^{-6} s), picoseconds (10^{-12} s), or femtoseconds (10^{-15} s) [142–145].

Laser diode is another popular developing device as it is inexpensive, reliable, and ultracompact and can provide thousands of pulses in a second for real-time PAM imaging [146–150]. A laser diode (LD), injection laser diode (ILD), or diode laser is a semiconductor device that is pumped directly with electrical current, in which the electrical energy is converted into light. LD can operate in two modes—single- and multi-optical modes. The emitted wavelength covers a wide range from UV to mid-NIR [146, 151]. A pulsed laser can be produced by using a pulsed laser diode driver that controls the time units in the hundreds of picoseconds, nanoseconds, and microseconds pulse width range, and high frequencies in the range of single shot [152, 153].

Continuous-wave (CW) lasers may also be modified for use in PA imaging as it is low-cost, compact, durable, and able to access a wide range of possible wavelengths [154–158]. In fact, previous studies have shown that the linear frequency-modulated laser output of a CW laser can be loaded on carrier waves of varying pulse widths [155]. Additionally, the axial resolution of PA images can be increased by using a matching filter and shortening laser pulse duration. High-power CW laser diodes can also be employed as excitation sources while maintaining the unique advantages of CW lasers mentioned above. The use of a CW laser with a narrow-band ultrasound transducer and a narrow-band or lock-in amplifier can offer high sensitivity and strong noise rejection [156, 157]. A lock-in detection-based PAM method has already been designed to measure the surface of solids, and the use of single- and dual-frequency detection methods is still being researched to improve PA imaging quality [158].

The development of these key technologies would potentially provide an opportunity for a variety of multispectral PA investigations, such as real-time sO_2 mapping, tissue features observation, and molecular contrast extraction. However, there are several challenges that still remain to be overcome for the advanced PAM: (1) high pulse energy (from several nJ to hundreds of μJ) with high PRR, (2) fast and tunable output wavelength in the region of UV to mid-infrared, and (3) fast enough scanning and image processing time. In the near future, by overcoming these challenges, the application areas of PAI can be significantly expanded to include not only preclinical applications but also clinical studies that require fast imaging speed.

1.4.2 PA Signal Detection

Compared to full ultrasound imaging, PA-based imaging technology faces a host of new challenges. Ultrasound typically acquires 50% of the central frequency which is a relatively narrow range of frequencies [159]. In contrast, the frequencies resulting from pulsed laser tissue excitation in PAI span a much broader range of frequencies (sub-MHz to hundreds of MHz) [102–112, 160]. Another factor constraining PAI is the fact that PA signals in vivo are three orders of magnitude weaker than those used in medical ultrasound. In practice, PA signals are unlikely to increase in strength as the maximum light fluence legally permissible for tissue illumination constrains the strength of the laser pulse. Thus, in order to be applied in a clinical setting PAI must forgo using greater laser power to avoid burning a patient's tissue. Despite these challenges, PAI's inherent ability to provide a broader range of frequencies allows the implementation of wide-range detectors, which can improve image quality and resolution as well as minimize image artifacts. Ultrasound structure design is therefore another key factor influencing PA imaging. These key performance parameters of ultrasound detectors for PA imaging will be discussed in greater detail in the following section.

1.4.2.1 Bandwidth and Central Frequency

In PAI, PAM, and PACT, both central frequency and bandwidth are critical to determine the resolution and size range of target structures [82]. Thus, the imaging of a micron scale target requires that the ultrasound detector is able to detect central frequencies and bandwidths on the order of hundreds of MHz. For example, a target with the size of 10 μm and 300 μm requires an ultrasound detector with frequency responses from several MHz to 150 MHz to collect the associated PA signals. Therefore, the parameters of the ultrasound detector are highly dependent on the intended application. The degree of attenuation on the order of 0.5 dB per MHz per cm in biological tissue, for example, relies on ultrasound wave frequency. Additionally, imaging targets at depth requires narrowing the ultrasound transducer's bandwidth, which will likely result in the loss of superficial structure information. For example, for the attenuation of 25 dB, a 5 MHz ultrasound will propagate through a depth of 10 cm, and a 50 MHz ultrasound only propagates through 1 cm. Thus, while higher imaging resolution can be realized through the use of higher ultrasound frequencies, this comes at the cost of penetration depth.

1.4.2.2 Sensitivity

Sensitivity of ultrasound transducer is the minimum detectable signal pressure amplitude. It is determined in relation to the noise level of the detector expressed in pressure units and is referred to as noise-equivalent pressure (NEP). The pressure magnitude of acoustic pulse that is reflected on a tissue interface is in the range of kPa. In PA imaging, the maximum laser fluence permitted by ANSI safety standards is 100 mJ/cm^2 in the near-infrared (NIR) region [73]. The generated PA signal is weak on the surface of tissue, especially the outgoing PA signal from deep tissue. Therefore, the sensitivity of detector should be in the Pa or sub-Pa range.

1.4.2.3 Size

Detector size is a critical parameter that affects detection sensitivity and spatial resolution. Minimizing detector size can improve the imaging resolution and the compactness of the system, but this will decrease the sensitivity with the reducing active transducer area. In addition, considering the piezoelectric element, preamplifiers and electrical isolation measurement, there is a strict limit on minimum size of the piezoelectric transducer. In recent years, optical detectors have gained attention in PA signal detection because they can be miniaturized to micrometer scale [161–163]. For example, the optical fiber detectors can minimize the sensing area down to 0.13 × 0.27 mm [164, 165], while maintaining high sensitivity (100 Pa NEP) and wide bandwidth (77 MHz). Minimized ultrasound detectors might be extremely beneficial in minimally invasive procedures such as medical endoscopy, inspection of small hollow structures, and visualization of the lumen in nondestructive testing.

1.4.2.4 Detector Aperture

Detector aperture is another critical parameter affecting PAI and is related to the acceptable range of angles for collecting PA signals [166, 167]. It is typically defined as the angle formed between (a) the axis perpendicular to the detector area and (b) the axis on which the signal from an ultrasound source is detected with 6 dB attenuation with respect to the signal detected at angle zero. For example, in PACT, a large aperture and well-characterized angle-dependent frequency response are essential for accurate image reconstruction and quantitative measurements [168, 169]. Piezoelectric elements enhance the transducer's sensitivity to ultrasound detection, and therefore its acceptance angles are typical ±20°. In contrast, the optical detectors have inherently larger acceptance angles, allowing for a diverse array of PA imaging modalities to be developed.

As mentioned above, to conduct PAM, an integrative imaging setup consisting of both optical and acoustic components is required. There are two kinds of US detectors: the piezoelectric transducer (PZT) and the optic-based US sensor. Selection of US detector is largely dependent on the parameters for PAI that have been briefly discussed above. In the following section, we summarize these PA signal detectors.

1.4.2.5 Electric-Based Ultrasound Transducer

Ultrasound transducers or ultrasound sensors are commercially produced with piezoelectric elements that can convert electrical signals into ultrasound and convert ultrasound into electrical signals [170–172]. In ultrasonography, the transducers act as both the transmitters and receivers. In PAM, however, the transducers only function as recipients of PA signals. The following discussion will reveal the acoustic detection mechanism of piezoelectric-based transducers and their applications in PAM.

Direct transduction from mechanical to electrical domains and vice versa is achieved through the use of piezoelectric materials. The reversible and linear piezoelectric effect manifests as the production of a charge (voltage) upon application of stress (direct effect) and/or strain upon application of an electric field (converse effect). Exploiting this effect, transducers can be configured either as actuators by using the converse piezoelectric effect or as sensors by using the direct piezoelectric effect in response to mechanical inputs, like acoustic waves.

The frequency and amplitude of piezoelectric transducers in response to acoustic emission are highly dependent on physical dimensions and component materials. Favorable scaling of piezoelectricity with miniaturized size has caused a growing interest in piezoelectric films for microsystem applications. Most piezoelectric and ferroelectric materials are formed by nitrides and oxides of metals and semiconductors [173–175]. In addition, piezoelectric properties are related to the stoichiometry and the morphology of these materials and therefore need appropriate seeding as well as controlled nucleation, growth, and crystallization processes. Novel, low-cost piezoelectric integration methods such as sputtering of aluminum nitride films and chemical solution deposition of ferroelectric lead zirconate titanate (PZT) films with patterning using plasma processes are paving the way for the use of piezoelectric microsensors in PA systems.

Micro-machined ultrasonic transducers (MUTs) are mainly fabricated by silicon micro-machining (MEMs) methods based on the diaphragm principle [176]. The diaphragm vibration can be detected or excited electronically by controlling the capacitance between the diaphragm and a closely spaced backing plate (CMUT), or the thin layer of piezo-electric material on the diaphragm (PMUT). Ultrasonic transducers can also be designed using a non-piezoelectric mechanism. The mechanical change in size is typically caused by a magnetic field or ambient acoustic wave. For example, a capacitor ("condenser") microphone with a thin diaphragm can detect ultrasonic waves by interrogating the changes in the electric field between the diaphragm while a closely spaced backing plate converts sound signals into electric currents [177].

For PA signal collection, the US transducers can completely block the light delivery to the target when they are fabricated by optically-opaque material layers. Because glass has a larger acoustic impedance than air and water [178], glass-based optical components (for light focusing and delivery) can reflect or scatter PA signals. To address this issue, many methods have been developed to deliver optical and acoustic beams along different paths to avoid mutual signal obstruction. One successful implementation of this method utilized an acoustic-optic prism combiner formed by a right-angle prism and a rhomboid prism pressed tightly to a thin layer of silicone oil to achieve coaxial alignment [115]. Focused light was then passed through the acoustic-optic prism combiner and refocused by a correction lens. PA waves then propagated through the rhomboid prism and reflected off of the silicone oil layer into an ultrasound detector attached to the prism. This method faces some drawbacks, however, including slow acquisition speed, limited field of view (FOV), and significant loss of acoustic signal. Another, more optimal method sandwiches aluminum foil between the right angle and rhomboid so that light is reflected rather than the acoustic waves [179]. This permits two-dimensional galvo mirror scanning of the dual-axis optic, which can effectively improve the image acquisition speed and generate a wide FOV. However, this method makes the greater PAM system complex and bulky, making beam alignment and propagation path stabilization more difficult. Moreover, acoustic loss in this system is still significant because acoustic waves must travel through the large coupling medium and the prism combiner before being detected by a transducer. The transmission-mode setup is an alternative approach to address the problems facing the reflection-mode. In this method, the optical illumination and acoustic detection are placed on opposite sides of the target. As mentioned above, this mode is not appropriate for in vivo applications as it has limited light penetration depth. Thus, the reflection-mode remains the most popular method for PAM. Recently, focused hollow or ring ultrasound transducers have been designed to allow light and acoustic waves to share the same path during reflection-mode. These transducers can simplify and compact the reflection-mode PAM system, making co-alignment and co-registration of optical excitation and subsequent acoustic detection easier. In fact, implementations of reflection-mode PAM with ring transducers have already seen great success [180, 181]. This setup has also shown potential for use in configuring handheld, wearable, and even endoscopic systems [182, 183]. Hollow structure significantly decreases the detection area of the transducer, however, limiting the focusing capability of the transducer thereby reducing the greater detectable FOV.

Laser-scanning PAM (LS-PAM) is a branch of PAM, which can realize high-speed in vivo imaging of biological tissues without scanning the ultrasonic transducer or the imaging subject [109]. In LS-PAM systems, a laser beam is scanned by an X-Y galvanometer, and a stationary unfocused ultrasonic transducer is used for PA signal detection [184]. However, as discussed above, the unfocused transducer will decrease the SNR of the imaging system. To address this issue, light and ultrasound beams are scanned together. One method to accomplish dual scanning is the use of an ultrasonic phased array, which consists of multiple small transducer elements [185–187]. These rectangular elements can be constructed on a straight line in one dimension as the detection signal of each element in the array is independent. Detected signals can then be delayed and summed as expected, earning its namesake: delay-sum beamforming technique. The laser beam can be steered to and focused at different spatial positions, resulting in SNR improvement. This component has been successfully used in mechanical scanning PAM systems, where multifocal optical illumination is demonstrated to improve the imaging speed.

The transducers mentioned above are common piezoelectric types that are opaque and block optical paths. Another alternative method is to use flat or planar piezoelectric transducers made of optically transparent sensing and electrode materials [188]. These transducers allow light to pass through transparent structure, allowing it to share the same transmission path as the resulting ultrasound waves. This study demonstrates the feasibility of a transparent piezoelectric transducer consisting of two indium tin oxide (ITO) electrodes sputtered on both sides of a polyvinylidene fluoride (PVDF) film [189, 190]. From these results, we can see that a sample can be illuminated almost homogeneously without significant shadowing along the optical axis. Flat, transparent piezoelectric transducers can be used to retain the original pressure without alterations caused by acoustic diffraction in very close vicinity to the sample. However, the unfocused design of these transducers significantly decreases acoustic detection sensitivity and creates poor detection conditions for PA signals.

In this section, we briefly discussed several kinds of piezoelectric transducers with varying unique features. In most instances, these transducers are not well modified for use in PA detection, especially more modern PAM systems. As piezoelectric transducers present relatively limited bandwidths and acceptance angles, they are unable to detect many useful PA signals. As a result, many novel optical methods are being considered as promising alternatives to ultrasound detection in PA systems to address the limitations of piezoelectric and capacitive technologies. Merging optic-based acoustic detection, both acoustic generation and acoustic detection can be realized by alternative optical components [161, 191–194]. In fact, some studies have confirmed that the use of full-optic methods may offer higher sensitivity and broader detection bandwidths with smaller form factors than piezoelectric transducers, PMUTs, or CMUTs.

1.4.2.6 Optical Ultrasound Detection

Common optical methods for detecting ultrasound signals include refractometry and interferometry. Refractometry is realized by interrogating light reflection intensity invoking the photoelastic principle [191, 192]. In PA systems, ultrasound waves induced by a pulsed laser can cause mechanical stress. This ultrasound-induced pressure causes the refractive index (RI) to change proportionally with the illuminated light fluence and target absorption. An interrogating or probe beam can then be used to measure RI changes in response to propagating ultrasound waves. Synthesizing information about probe beam intensity, deflection angle or phase, and change with RI, conclusions can be made about the ultrasound signals being interrogated. Another popular ultrasound detection method is optical interferometric which detects the optical interference pattern caused by ultrasound waves. Ultrasound waves can alter the interference condition by interacting directly with an optical beam, by vibrating a reflector or by altering the resonance frequency of a resonator [193]. Optical perturbations can also be triggered by changes in the mean free path, which is determined by the interferometric configuration employed.

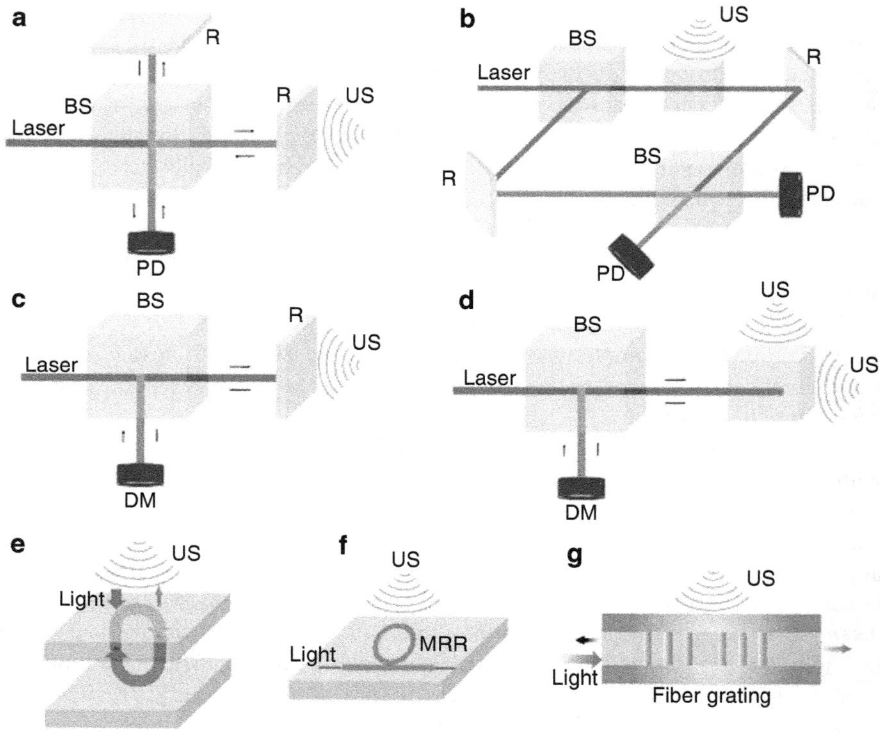

Fig. 1.8 Optical detection of the acoustic signals. (**a–d**) phase detection-based sensing mechanisms: (**a**) Michelson interferometer, (**b**) Mach–Zehnder interferometer, (**c**) Doppler-based sensing, and (**d**) resonator-based sensing. (**e–g**) Common resonator geometries demonstrated for ultrasound sensing: (**e**) planar Fabry-Pérot, (**f**) micro-ring, and (**g**) π-phase shifted fiber Bragg grating. BS beam splitter, D detector, DM demodulator, LA laser, R reflector, US ultrasound [161]

The ultrasound information can finally be synthesized through analysis of variations in the optical signal. With this information above we can now discuss several types of optical detection methods used in PAM, as shown in Fig. 1.8.

1.4.2.6.1 Non-contact PA Detector

In order to address the limitations of traditional contact PA detection imaging modalities, a novel technique known as photoacoustic remote sensing microscopy (PARS) was developed. PARS is based on the mechanism of RI-induced light reflection intensity change [194–196]. In a PARS system, pulsed laser induced ultrasound waves can cause mechanical stress (up to one hundred-order MPa while maintaining ANSI exposure limits). This large local pressure results in change in refractive index (RI) proportional to illuminated light fluence and target absorption. An interrogating or probe beam can then be used to measure RI changes in response to propagating ultrasound waves. PARS exhibits exceptional sensitivity and resolution and has thus seen use in phantom and in vivo imaging [197]. Moreover, PARS systems are not limited by traditional PA detection design constraints such as interferometric detectors or an external resonator, making PARS a more compact method for PA signal detection.

1.4.2.6.2 Fabry-Perot (FP) Etalons

The FP etalon detection method is a representative surface-coupled optical resonance technology for ultrasound detection [164, 193]. This method has seen widespread use in PA imaging and provides prominent results. FP-based devices can be used in an AR system using a non-focal excitation beam or an OR system using a focal excitation beam. In this system, detection relies on the acoustic coupling of ultrasound waves to an FP etalon structure. By coupling ultrasound waves to an FP etalon structure, ultrasound signal can be focused as multiple beam interference funnels the signal into a narrow band resonance spectrum. When acoustically coupling such a device to the sample, ultrasound propagating into the etalon can cause subtle changes to the physical and optical properties of the internal layer. This interaction results in subtle changes in the path length and the phase of the signal, which is then amplified by the high-fitness structure, shifting the resonant wavelength, finally perturbing the intensity of back-reflected and transmitted light. Unlike a piezoelectric device, this optical approach allows for decoupling of the effective aperture size, element size, and any desired raster scanning step size. FP etalons usually involve high-fitness and low-fitness structures. The former presents a stronger resonance effect and higher sensitivity to the ultrasound perturbation. Diverse methods have been developed to form high-fitness embodiment, such as a single transparent layer between two highly reflective mirrors. The back-reflected intensity of FP etalon can be demonstrated by either the spectral-side filtering technology or the use of a more refined approach such as the transfer-matrix method.

1.4.2.6.3 Micro-Ring Resonators (MRR)

MRR is another kind of typical optical resonator, involving ring-shaped waveguides [198–200]. Using nanotechnology, diverse MRR platforms have been widely developed for many physical or chemical parameter measurements. In PAI, MRR can be acoustically coupled to the sample. The promotion of ring cavity makes MRR present extremely high Q factors on the order of 10^4 to 10^7. Additionally, optical resonances in MRR are highly sensitive to cavity deformation resulting from ultrasound pressure. As a result, the transmitted optical intensity at a given wavelength is modulated sensitively. Previous studies have characterized the beneficial properties of MRR, such as noise equivalent pressures of 29 Pa reported from an early device featuring a 60 μm diameter ring [201]. Using focused excitation, PAM systems can achieve a lateral resolution of 5 μm and an axial resolution of 8 μm. After that, another optically transparent ring resonator was formed by using a fused quartz substrate with a total thickness of only 250 μm. It has an improved sensitivity with a noise-equivalent pressure of 6.8 Pa over a 140 MHz bandwidth [202]. It is because this type of optical detector has an extremely high bandwidth that the axial resolution of the PAM system can be improved to 5.3 μm. The MRR detector is commonly placed on the sample in the water tank, and the excitation beam passes through and focuses on the sample. Owing to its high sensitivity, MRR is also able to detect the PA signal by overcoming significant impedance mismatching resulting from air coupling and the high attenuation of ultrasound in air. In contrast to the sensing mechanisms that rely on ultrasound to modulate the refractive index of the resonator, MRR detectors rely on mechanical deformation to relay ultrasound signal information.

1.4.2.6.4 Fiber Bragg Grating (FBG) Sensors

FBG is very popular in acoustic wave detection owing to its narrow spectral band (<0.2 nm) [203]. π-phase-shifted FBGs, called π-FBGs have a notch arising from a π-phase discontinuity in the center of the grating, which narrows the spectrum bandwidth to several pico-meters. Previous studies have demonstrated the utility of π-FBGs-based sensing systems for endoscopic and PAM frameworks [204, 205]. This system can offer an acoustic sensitivity of 100 Pa and bandwidths up to 77 MHz. π-FBGs have a small sensing region much smaller than the given physical length of the grating, and thus, they can feature 1D cavities with the size of 10 × 270 μm. They therefore can operate as the line detectors for PACT, providing lower resolution and allowing less straightforward image reconstruction than point-like detectors. In fact, miniaturized π-FBGs were recently created in an optical waveguide, generating greater RI contrast than optical fibers [206, 207].

1.4.2.6.5 Other Interferometers

Michelson interferometers (MIs) and Mach-Zehnder interferometers (MZIs) are two of the earliest generation of interferometers and have also seen use in PAI [208–210]. They both are based on the two-beam interference principle, where a light beam is split into two optical paths: one being the reference path and the other being the sensing path perceiving ultrasound signals. The divided beams unite at the output of the interferometer thereby generating an interference pattern. Ultrasound will induce the sensing path perturbation (RI and length), making the interference phase difference change, causing the interference pattern change. Modulation in interference pattern is eventually translated into light intensity change, which is proportional to the ultrasound intensity. MIs typically use the beam from the sensing arm or a reflector at the sensing arm to detect ultrasound signals instead of the sensing arm itself; whereas, MZIs typically use the sensing arm to interact with the ultrasound beam. Therefore, an acoustic coupling medium must be used when ultrasound interacts with the sensing arms of MIs or MZIs. MIs and MZIs are typically constructed using fiber-based or free-beam configurations. Previous studies have demonstrated that MZI-based ultrasound detectors have sensitivity down 180 Pa [211], which is considerable for PAI. However, MZIs narrow bandwidth of 5 MHz is insufficient for PAI, especially for PAM systems. Another MZI-based approach utilizes a free-beam configuration to achieve a sensitivity of 100 Pa and a bandwidth of 17 MHz [212]. Free-beam MZI is highly sensitive to ambient vibrations and electromagnetic noise, however, so further development is needed to bolster this approach.

Doppler interferometers (DI) are another type of sensor that detects ultrasound signals by interrogating Doppler shifts in light frequency [213]. In DIs, similar to the MI configuration, the ultrasound signal interacts with a reflector. The reflected probe beam does not interfere with a reference beam, however; instead, Doppler shifts in the probe beam induced by pressure-induced oscillations of the reflector are recorded by a wavelength demodulator. The DI method thus provides advantages over the MI method, especially when using reflectors with a rough surface for greater interferometric contrast, making it a strong candidate for noncontact optoacoustic imaging.

1.4.3 Imaging Formation

1.4.3.1 Mechanical Motor Scanning System

In order to perfect PAM, many seek to find ways to increase imaging speed and quality. As a result, many diverse optical scanning mirror-based units are being developed to produce high-speed, large FOV imaging systems [65, 214–216]. Because light and ultrasound waves need to be reflected and scanned simultaneously, traditional optical mirrors cannot be directly used for PAM systems. Thus, 3D mechanical motors have been constructed to move the target at a frequency of a few Hz to several thousand Hz, and a large FOV 3D scanning has been realized by stably controlling the motors. In one study, a microtomy-assisted PAM (mPAM) system demonstrated the implementation of the mechanical motor method to scan mouse brains and other organs [217]. As shown in Fig. 1.9, this system generated micrometer-resolution 3D images of paraffin- or agarose-embedded organs with high fidelity, achieved by label-free visualization of DNA/RNA, hemoglobin, and lipids. mPAM's imaging speed is limited, however. As scanning square millimeter section takes several minutes, mPAM may not see use in future functional imaging applications.

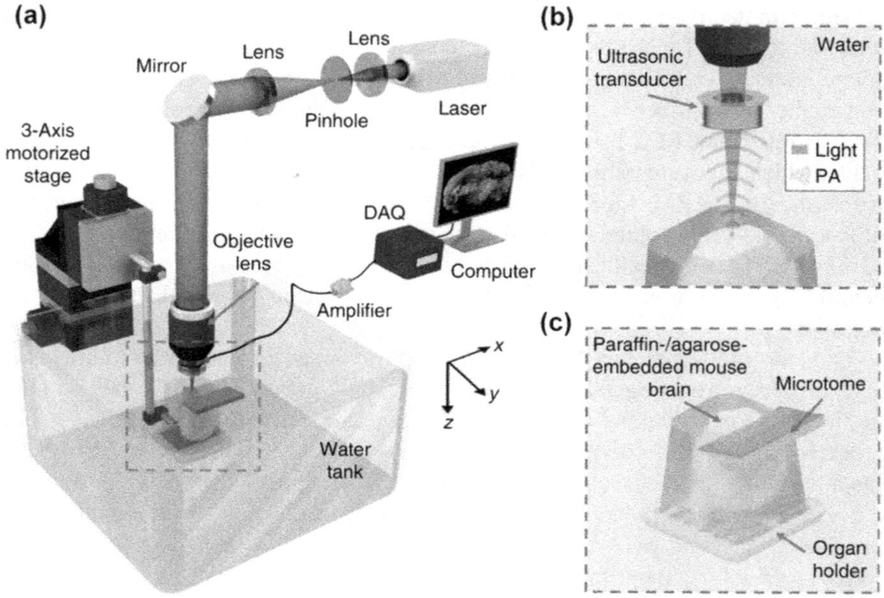

Fig. 1.9 Schematic of the mPAM system for whole-organ imaging and sectioning. (**a**) schematic of system setup. (**b**) close-up of the blue dashed region in (**a**) during imaging, and (**c**) sample preparation. (Reprinted with permission from [217])

1.4.3.2 Micro-Electro-Mechanical Systems (MEMS)

To address the problem of imaging speed, micro-electro-mechanical systems (MEMS) have been developed for use as high-speed scanning systems in air and water environments [218, 219]. MEMS scanners are smaller and more stable, improving the greater PAM system's design and imaging quality. Prior to MEMS scanners, light beam scanning was restricted to the region of acoustic focus, which has limited FOV. By using an opto-acoustic combiner and a water-immersible MEMS scanner, as shown in Fig. 1.10a, light beams and PA waves are coaxially aligned by a combiner and simultaneously steered by the MEMS scanner [220]. Thus, through the use of MEMS, PAM achieved high-speed, high-resolution imaging of vascular morphology, blood oxygenation, blood flow dynamics, and oxygen metabolism in a mouse brain [220]. More specifically, PAM has achieved a 1D time-resolved imaging rate of 500 kHz for morphological imaging and 100 kHz for blood oxygenation imaging. As the volumetric in MEMS PAM is scanned by a linear motor, however, it is difficult to miniaturize the system. A potential solution to this constraint is to use a 2-axis MEMS scanner with a size of 15 mm × 15 mm × 15 mm, providing a wide scanning range of 9 mm × 4 mm and an imaging speed of 50 Hz and 30 Hz in fast and slow scanning, respectively,

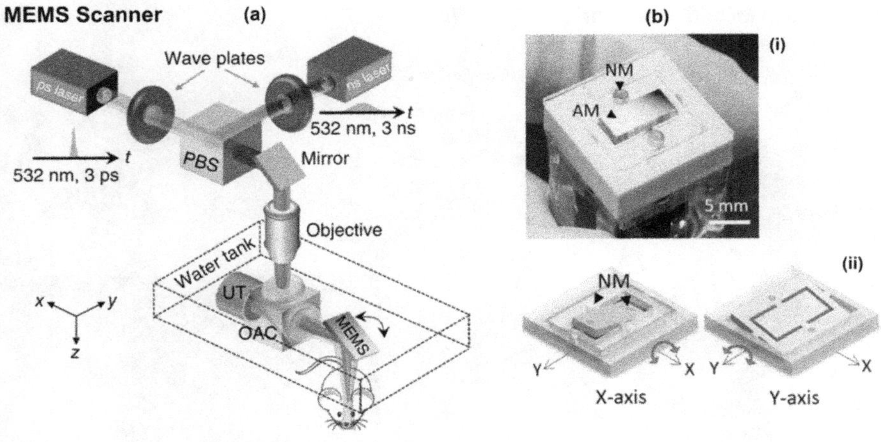

Fig. 1.10 MEMS-based PAM system. (**a**) schematic of the PAM system. (**b**) (**i**) photograph, (**ii**) scanning mirror schematic, and (**iii**) system setup of the 2-axis MEMS scanner OR-PAM. OAC optical-acoustic combiner, PBS polarizing beam splitter, UT ultrasonic transducer. (Reprinted with permission from [220] for (a) and [221] for (b))

as shown in Fig. 1.10b [221]. Due to its unique capabilities, this compact 2D-axis MEMS scanner has become quite popular for use in PAM and has been extended to hand-held PAM systems.

1.4.3.3 Hand-Held MEMS

Hand-held OR-PAM systems attract wide interest and are developed to image various anatomical sites [222–227]. By compacting MEMS, small, mobile OR-PAM systems have seen successful development. For example, an 80 mm × 115 mm × 150 mm OR-PAM system with a 1-axis MEMS scanner and a linear stage was developed with an imaging range of 2.5 mm × 2.0 mm and a scanning speed of 220 Hz along the lateral axis and 1 Hz in along vertical axis, as shown in Fig. 1.11a [224]. To further miniaturize the OR-PAM system, a 2-axis MEMS scanner can be employed instead of linear stage, as shown in Fig. 1.11b [225]. All acoustical, optical, and mechanical components are integrated into a single probe with a diameter of 17 mm and a weight of 162 g. The measured B-scan and volumetric imaging rates were respectively 35 Hz and 0.05 Hz. The maximum FOV was 2.8 mm × 2 mm. It has been demonstrated to image phantom and in vivo of various samples acquired with the probe, including carbon fibers, electrospun microfibers, and the ear, iris, and brain of a living mouse.

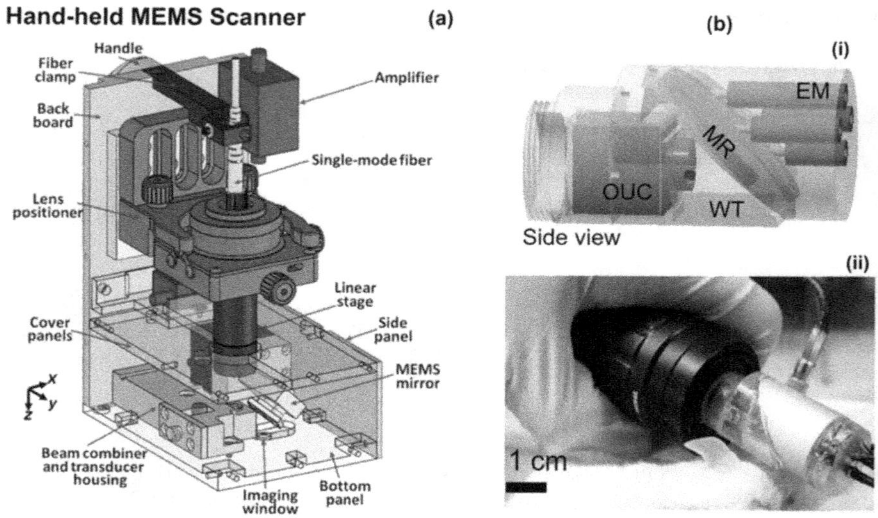

Fig. 1.11 Hand-held PAM system. (**a**) schematic of the PAM system using a 1-axis MEMS scanner. (**b**) scanner schematic (top) and photograph (bottom) of handheld PAM with a 2-axis MEMS scanner. EM electromagnetics, MR mirror, OUC opto-acoustic combiner, and WT water tank. (Reprinted with permission from [224] for (a) and [225] for (b))

1.4.3.4 Galvanometer Scanner

A galvanometer scanner provides comparable stability and imaging speed to PAM systems and is thus a good alternative to the MEMS scanner [228–237]. One such PAM system demonstrated significant enhancement of temporal resolution using a stable and commercial galvanometer scanner with a custom-made scanning mirror (L-PAM-GS), as shown in Fig. 1.12a [231]. The system presents a high SNR and fast.

B-mode rate (500 Hz). Using the L-PAM-GS system, functional hemodynamics, namely blood flow rate in the microvasculature, were successfully monitored and quantified in vivo. In order to improve imaging speed and the SNR, another galvanometer scanner-based PAM was developed using a cylindrically focused ultrasound transducer [232]. Using this system, an imaging speed of 35 s with an FOV of approximately 7 mm in diameter was achieved. To increase FOV, one research group moved the axis of rotation of the ultrasound transducer to one end of the cylindrical focus achieving an FOV of 40 mm in diameter [233]. To further improve the performance of commercial galvanometers, a non-conducting liquid, can be employed as an acoustic impedance matching medium instead of water, as shown in Fig. 1.12b [236, 237]. In this method, two beams are coaxially aligned via an optoacoustic beam combiner, which captures images with a high SNR of 44 dB and a wide FOV of 4 mm $\times$ 8 mm within 2 s. This system has been demonstrated to image the microvasculature in a mouse ear in vivo.

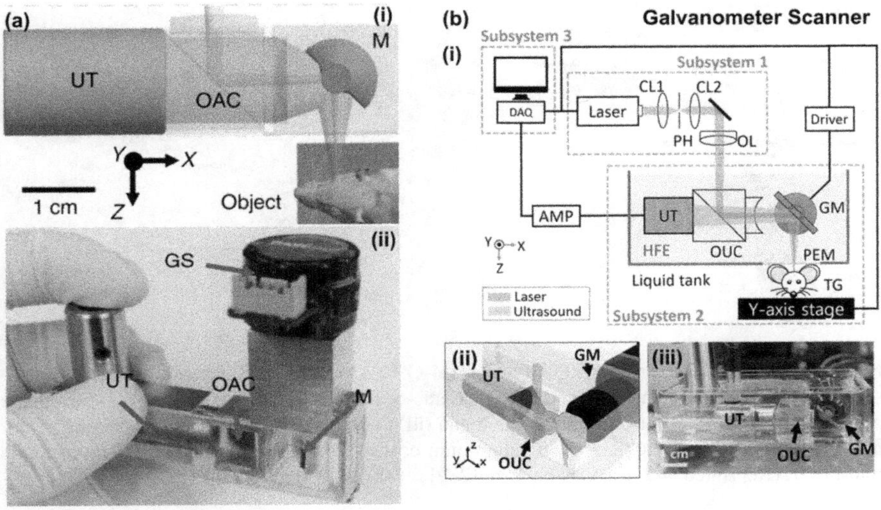

Fig. 1.12 Galvanometer scanner-based PAM system. (**a**) Localization PAM using a galvanometer scanner: (**i**) close-up view of the scanning part outlined by the red dashed box and (**ii**) photograph of the scanning part. (**b**) (**i**) Schematic of the galvanometer-based OR-PAM in non-conducting liquid: (**ii**) 3D design and (**iii**) photograph of the GM-OR-PAM in NCL system. BS beam splitter, OL objective lens, PD photodiode, GS galvanometer scanner, UT ultrasound transducer, MS motor stage, OAC optical-acoustic combiner, M mirror, FWHM full width at half maximum, GM galvanometer, OUC opto-ultrasound combiner, UT ultrasound transducer, CL condenser lens, PH pinhole, PEM polyethylene membrane, TG target. (Reprinted with permission from [231] for (a) and [237] for (b))

1.4.3.5 Polygon-Mirror Scanner

Polygon mirror scanners have also been utilized in the development of OR-PAM systems, demonstrating imaging speeds up to 900 Hz B-scan [238]. In these systems, a waterproof motor is used to rotate a hexagonal mirror at a constant speed. Each facet of the hexagonal mirror has dimensions of several millimeters, which can be adjusted to match the target of interest. The mirror's surface is coated with aluminum to reflect light beams and acoustic waves simultaneously. The polygon scanner system does not require repeated acceleration and deceleration. Thus, this system's scanning speed is only dependent on the motor speed, allowing for faster and more reliable scanning than most MEMS PAM systems. As the DC motor can operate in water, it is capable of steering coaxially aligned beams, improving the overall imaging SNR. For example, a functional OR-PAM was developed using a polygon-scanning mirror to steer both the laser and the ultrasonic beams simultaneously, ensuring a high sensitivity and fast scanning, as shown in Fig. 1.13a [239]. This polygon scanner offers a B-scan rate of 477.5 Hz over a 12-mm range. Meanwhile, using a 1 MHz dual-wavelength (532 nm and 558 nm) pulsed SRS laser as a source, in vivo sO_2 imaging at a 1 MHz A-line rate over

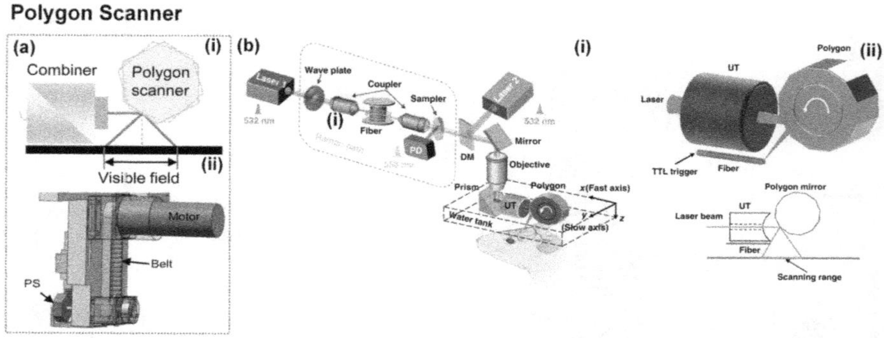

Fig. 1.13 Hexagon-mirror based OR-PAM (HM-OR-PAM). (**a**) (**i**) Schematic of the polygon-scanning OR-PAM B-scan range and (**ii**) schematic of the belt-driving system. (**b**) (**i**) Schematic of the HM-OR-PAM system. (**ii**) 3D drawing and (**iii**) photograph of the hexagon scanning mirror driven by a high-speed DC-motor. AC aluminum coating, HM hexagon mirror, UST ultrasonic transducer. (Reprinted with permission from [239] for (a) and [116] for (b))

a 12 × 12 mm^2 scanning area was realized with a volumetric imaging speed of ~1 Hz within 12 × 5 mm^2. In polygon systems, the light beam is often blocked by the transducer, meaning only 60% of A-lines can be used. This inefficiency can be avoided by decreasing the facet size of the mirror or increasing the number of polygon faces. An improved polygon-based PAM system was developed, which presents a maximum time-resolved A-line frame rate of 2 MHz, a B-scan frame rate of 2 kHz over 11 × 1.5 mm^3, and a C-scan frame rate of 2 Hz over an 11 × 7.5 mm^2 FOV, as shown in Fig. 1.13b [116]. Meanwhile, A lab-made start of scan detection system uses a multi-mode fiber to detect the starting position of the scanning laser beam by each facet. This system has been demonstrated to capture the mouse brain's fast morphological and functional changes over the entire cortex, and the vasoconstriction and hypoxia simultaneously in the spreading depolarization wave.

1.4.3.6 Micro-Lens Array

Single ultrasound transducer OR-PAM systems have limited imaging speed since scanning must be conducted point by point [240–245]. One study presents a solution to this problem by using a linear ultrasound array transducer with 48 elements [240], which can collect PA signals from a large area in a timely manner, as shown in Fig. 1.14a. In this method, a micro-lens array is enabled to optically stimulate 20 distinct locations simultaneously. After scanning the target, a volumetric image was obtained with a size of 1000 × 500 × 200 voxels and a 10 μm lateral resolution. Moreover, a high-throughput multifocal ultraviolet (UV)-PAM was developed for label-free histology-like images with high resolution, as shown in Fig. 1.14b

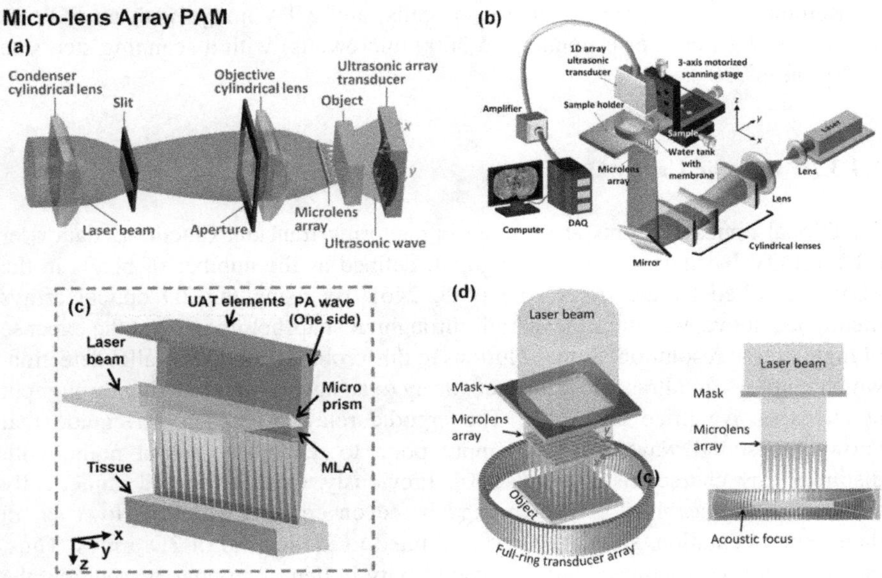

Fig. 1.14 2D micro array-based multifocal OR-PAM. (**a**) Schematics of system setup. (**b**) Multifocal ultraviolet laser PAM. (**c**) 3D view of the acoustic-optics combiner in reflection-mode multifocal OR-PAM system. (**d**) 3D view of the multifocal OR-PAM system, and cross-sectional view, especially highlighting the acoustic focus. UAT ultrasonic array transducer, MLA microlens array. (Reprinted with permission from [240] for (a), [241] for (b), [242] for (c) and [243] for (d))

[241]. This system contained a microlens array and a one-dimensional (1-D) array ultrasonic transducer.

It presents a 1.6 ± 0.2 μm lateral resolution and high scanning speed, 40 times faster than the previously developed point-by-point scanning UV-PAM. However, this transmission-mode PAM system is difficult to use in in vivo imaging. A reflection-mode multifocal OR-PAM system based on a microlens array that provides multiple foci as well as an ultrasonic array transducer that receives the excited PA waves from all foci simultaneously is presented, as shown in Fig. 1.14c [242]. This reflection-mode multifocal OR-PAM is capable of imaging microvessels in vivo, and it can image a 6 × 5 × 2.5 mm^3 volume at 16 μm lateral resolution in ~2.5 min, which was limited by the signal multiplexing ratio and laser pulse repetition rate. Another multi-focus PA computed microscopy system that uses a 2D microlens array (transmitting 1800 optical foci) and a ring array transducer was developed, as shown in Fig. 1.14d [243]. In this method, PA signal was digitized by a 512-element ring array transducer and a 48-channel digitizer, and an FOV of 10 mm × 10 mm with lateral resolution of 29.4 μm was achieved in 36 s. A single-cell metabolic PAM system (SCM-PAM) was also developed by combining a high-density microwell array with wide-field fast-scanning functional OR-PAM [244, 245]. This system provides label-free, high-throughput, single-cell measurements for a large population of cells, and a PA image with an FOV of 7.2 mm × 7.2 mm (corresponding to 3600 microwells) with a scanning step size of 20 μm in 720 s.

1.4.3.7 Ergodic Cavity Detection

Traditional optical systems are limited in capturing transient functional data over a large FOV by their imaging throughput, defined as the number of pixels in the FOV multiplied by the frame rate [245, 246]. By using the 2D optical arrays mentioned above, we can achieve high-throughput snapshot imaging at the expense of high section resolution. Some solutions to this problem include parallel detection, which employs an ultrasonic transducer array to achieve a high imaging throughput at an expensive price tag [247]. The ergodic relay (ER) is a waveguide that allows the sound wave from any input point to reach any output point, with distinct delay characteristics [248–250]. Previously, Cox and Beard numerically simulated the feasibility of PA imaging based on ergodicity [248]. However, in their experimentation, simulations were limited to 2D instead of 3D arrays. Thus, the ergodicity was achieved with a closed cavity in lieu of an open waveguide, the object was located inside the cavity rather than outside the waveguide, and the image reconstruction was based on acoustic modeling instead of experimental calibration. For instance, an ER-based, low-cost, high-throughput PA imaging method was developed, demonstrating the ability to observe the changes in oxygen saturation in a mouse brain and demonstrated high-speed recognition of vascular patterns for biometric authentication, as shown in Fig. 1.15a [251]. In this method, the ER—as an encoder—effectively transforms each dimension. As spatial information was encoded with randomized temporal signatures through ergodicity, only a single-element ultrasonic transducer with one laser pulse was required to capture a wide-field image. Moreover, an acoustic ER was used in a 2D multifocal PAM system, as shown in Fig. 1.15b [244]. As mentioned above, the acoustic ER is ideally a low-loss acoustic propagation medium that allows input sound waves to be sufficiently scrambled inside the medium, resulting in distinct delay characteristics for each input position. Combining the powerful capability of an ER with a single-element ultrasonic transducer to detect multiple PA signals in parallel with a simple system setup, this system is a low-cost alternative to transducer array. Combining the microlens array with the ER can also improve the acoustically defined image resolution of PAM to the optically defined image resolution and shorten the scanning time.

1.5 Contrast in PAM

PA contrast agents improve PA imaging by changing both acoustic and optical features of local tissues, significantly enhancing the ability to discern pertinent

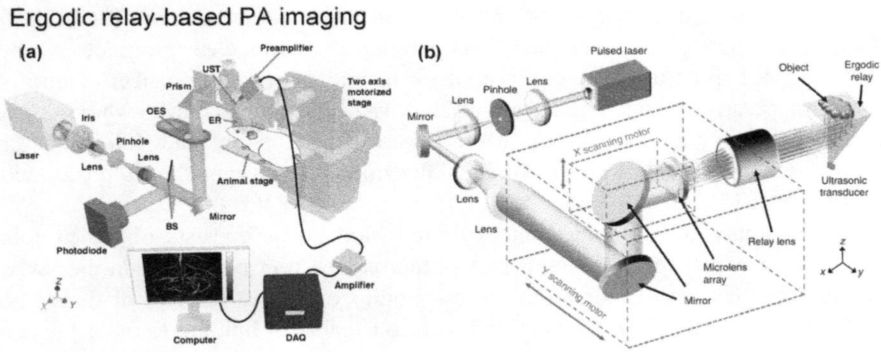

Fig. 1.15 ER relay-based PA imaging system. (**a**) Schematic of the system. (**b**) System schematic of multifocal OR-PAM. The multi-focal plane of the microlens array is imaged onto the imaging face of the ER prism to illuminate the imaging object. A single-element ultrasonic transducer is attached to a corner of the ER to detect the encoded PA signals. BS beam splitter, DAQ data-acquisition unit, ER ergodic relay, OEW optical-element wheel, and UST ultrasonic transducer. (Reprinted with permission from [251] for (a) and [244] for (b))

details from PAI visualizations. In biomedical studies, the advantages of endogenous contrast agents are undeniable. They do not induce perturbations to original tissue micro-environments, are non-toxic, and do not require costly and time-consuming regulatory approval. However, many diseases and physiological processes do not contain intrinsic PA contrasts, such as tumor cell receptor expression and protease activity [252, 253]. Thus, various exogenous PA contrast agents have also been developed to enhance the specificity and sensitivity of PAM: small-molecule dyes, gold, polymer, and carbon nanostructures to liposome encapsulations [252–257]. PA contrast agents are of great importance to advancing the molecular imaging applications of PAM, especially for detecting early-stage cancers and tracking immunotherapy mechanisms against diseases [258–261]. In this section, endogenous and exogenous PA contrast agents will be discussed in detail.

1.5.1 Endogenous Contrast

Endogenous contrast exploits the natural absorption properties of common biomolecules. As each biomolecule has a distinct absorption wavelength, the optical source of a PAM system can be manipulated to analyze specific biomolecules, providing contrast. In this section, we will discuss some of the common biomolecules used for endogenous contrast in PAM.

In the wavelength bands of ultraviolet (UV) from 180 nm to 400 nm and visible from 400 nm to 700 nm, the primary absorbers in PA imaging consist of DNA/RNA [262, 263], cytochrome c [264], myoglobin [265], hemoglobin [266] and melanin [267]. By exciting DNA and RNA using UV pulse laser, PAM has achieved noninvasive imaging of cell nuclei in vivo without staining [262, 268]. As cancer cells typically have abnormal cell density and nuclear morphology, the PAM technology is thus a potential tool for histology-free early cancer diagnosis and intraoperative cancer cell detection [269]. PAM has even achieved whole-organ imaging by exciting DNA/RNA in the tissue using 266 nm and 420 nm laser wavelength pulses, followed by PA signal detection using a ring-shaped ultrasound transducer [217].

In mammals, 95% of the content of red blood cells consists of the protein hemoglobin, which plays a major part in the oxygen transport chain in the body. Hemoglobin carries oxygen from the respiratory organs to the rest of the body, where it is released [270]. Hemoglobin is also capable of binding to other ligands (e.g., carbon dioxide (CO_2)) and transports them towards the respiratory organs. Hemoglobin has been extensively imaged by PAM systems for a number of hemodynamic parameters, such as total hemoglobin concentration, oxygen saturation, blood flow speed, and the metabolic rate of oxygen [200, 271, 272]. Compared to other chromophores naturally occurring in tissue, hemoglobin presents relatively high absorption in the visible spectrum. Strong PA signals can therefore be generated for use in various PAM applications such as neurological imaging, angiogenesis, blood oxygen saturation measurements, microvasculature damage induced by diabetes, neovascularization, hypoxia, and hypermetabolism.

Melanin is a ubiquitous natural pigment that can be found in the skin, hair, and eyes. Its optical absorption enables sensitive characterization of primary melanoma and metastatic melanoma cells [273]. Its absorption is strong in the ultraviolet range and decreases gradually as wavelengths approach near-infrared [274, 275]. Images of subcutaneous melanomas in nude mice have been obtained using PA imaging at multiple wavelengths by imaging both the surrounding tissue (764 nm) and the melanoma itself (584 nm) [274–276]. The overall attenuation of the natural pigment makes melanin an ideal contrast agent for early melanoma detection [276–280]. In fact, PAM has been applied for in vivo detection of circulating melanoma cells in the bloodstream and chronic monitoring of melanoma growth, with penetration depth up to 4 mm and an SNR of 30 dB [278, 280].

In the near-infrared spectrum region from 700 nm to 1400 nm, lipid, water, and glucose are the major absorbers for PA imaging [84, 86, 262]. Abnormal concentrations of these biomolecules can indicate signs of tissue damage. Lipids, for example, are a common constituent in atherosclerotic plaques, the location and area of which are closely related to the progression of atherosclerosis [281]. PAM can therefore exploit the absorption properties of lipids and water to generate high-resolution, relatively deep visualizations of early signs of clotting and stroke [282]. Blood glucose level is a standard monitoring tool for diabetes diagnosis and treatment evaluation. Previous PAM studies have demonstrated success in tracking blood glucose concentration in vitro and in humans with good correlations to clinical measurements [283, 284]. As lipid, water, and glucose possess such low-frequency optical absorption properties, only OPO or Ti: Sapphire lasers can provide required wavelength and excitation energy. These lasers typically have low repetition rate, reducing overall image quality. With further development of laser technology,

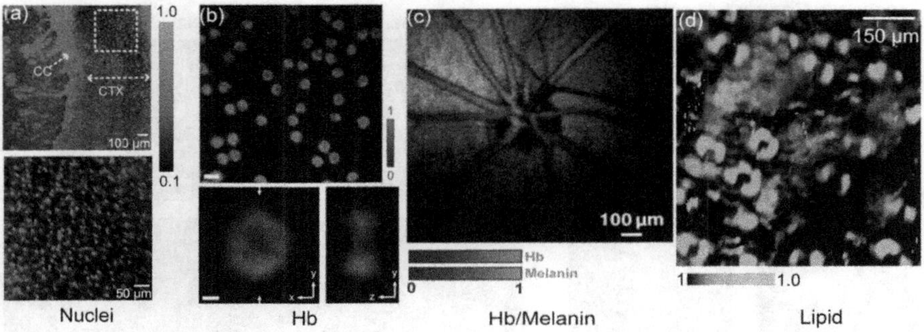

Fig. 1.16 Representative multi-endogenous contrast PAM images. (**a**) UV-PAM of cell nuclei (zoomed-in yellow dash region) in mouse brain slices at 266 nm. (**b**) SW-PAM of red blood cells (RBC), the signals of RBCs come from hemoglobin (Hb). (**c**) Pseudo-colored PA ophthalmoscopy images of the retinal vessels and the retinal pigment epithelium. (**d**) PAM of sebaceous glands at 1210 nm. CC corpus callosum, CTX cerebral cortex. (Reprinted with permission from [262] for (a) and (d), [200] for (b), and [277] for (c))

however, PAM is expected to engage more with biomarkers in the near-infrared absorption range.

As described above, endogenous chromophores have been successfully employed for in vivo PA imaging of the vasculature, melanomas, and brain perfusions. With the help of endogenous contrast agents, PA imaging has the capability to visualize on a molecular scale. By tuning the absorption of contrast agents to the region where the endogenous chromophores exhibit minimal absorption, greater detail can be obtained from PA images. However, intrinsic chromophores access only a limited range of biological processes. Thus, further PA contrast agents must be developed to access targets that are not characterized by endogenous contrast agents (Fig. 1.16).

1.5.2 Exogenous Contrast

In this section, we summarize the results of various studies using exogenous contrast agents in PAM systems.

One study demonstrates the use of X-gal, a molecular probe absorbing light wavelengths ranging from 605 nm to 655 nm, to simultaneously image a subcutaneous lacZ-expressing tumor and its surrounding microvasculature networks in vivo using AR-PAM technology, as shown in Fig. 1.17a [258, 259]. After injecting the lacZ-expressing tumor cells into the mouse, the malignant tumor and surrounding tissues were clearly imaged at 635 nm and 584 nm. With the help of X-gal, the spatial distributions of both the tumor and surrounding microvasculature were observed by the fused MAP image, demonstrating the utility of exogenous contrast in monitoring the expression of reporter gene systems.

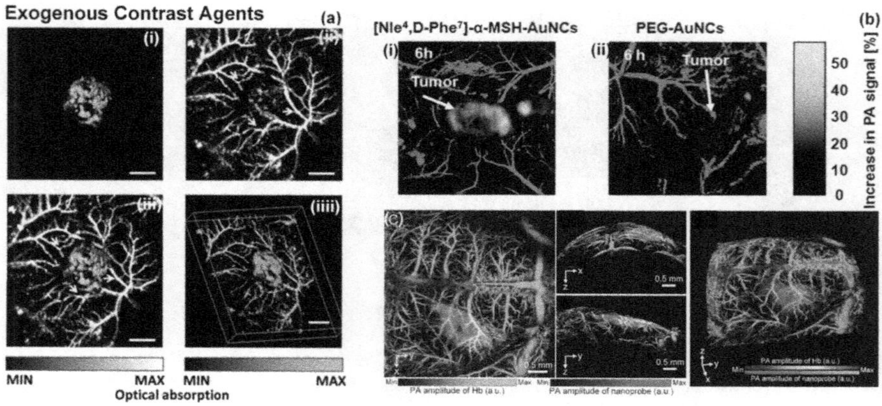

Fig. 1.17 Exogenous contrast agents-based PA imaging. (**a**) In vivo AR-PAM images showing the lacZ-expressing tumor and the surrounding microvascular networks. (**b**) In vivo AR-PAM images of the melanomas targeted by [Nle4, D-Phe7]-a-MSH-AuNCs and PEG-AuNCs, respectively. (**c**) The merged PA images of microvasculature and nanoparticles distribution in a complete mouse skull state, and 3D PA images. (Reprinted with permission from [258] for (a), [254] for (b) and [287])

Previous studies have also quantified contrast enhancement of active targeting (bioconjugating AuNCs with [Nle4, D-Phe7]-a-melanocyte-stimulating hormone ([Nle4, D-Phe7]-a-MSH-AuNCs)) and passive targeting (bioconjugating AuNCs with poly (ethylene glycol) (PEG-AuNCs)) of AuNCs to B16 melanomas [254, 260]. Figure 1.15b displays the MAP images of B16 melanomas targeted by [Nle4, D-Phe7]-a-MSH-AuNCs and PEG-AuNCs. The results obtained by dual-wavelength AR-PAM (570 nm for microvasculature and 778 nm for AuNCs) indicate that [Nle4, D-Phe7]-a-MSHAuNCs experienced three times more contrast enhancement than PEG-AuNCs. With further material development, other nanostructures present themselves as effective PA contrast agents, such as the palladium nanosheets [285], semiconducting polymer nanoparticles [257], silver nanoplates [253], and carbon nanotubes [286].

Nanoparticles are a kind of a reported organic dye with strong NIR absorption, multi-color capability, high spectral stability, and has no potential heavy toxicity [261, 287]. Den-RGD/Cy7 is a kind of nanoparticle, which has been verified to present good photothermal performance, as shown in Fig. 1.17c. The particles were injected into the mouse brain via intravenous injection. The PAM system was then used to provide the morphology and quantify the distribution of microvasculature and nanoparticles in 4 T1 tumor model. The nanoparticles are good candidates to guide the implementation of photothermal therapy.

1.6 Functional PAM

As the clinical course of an afflicted individual is seldom stagnant, many seek to develop functional imaging modalities to capture the dynamic nature of these individuals' disorders. Functional PAM has seen recent success in capturing dynamic information, and we will discuss two predominant areas of study in functional PAM (blood oxygenation and temperature change) in this section.

1.6.1 Blood Oxygenation and Blood Flow

Oxygen saturation reflects the oxygen in the blood that is bound to hemoglobin [288]. Imaging hemoglobin oxygen saturation sO_2 is important for understanding hemodynamics in response to sensory stimulations [289]. It is also invaluable in various medical settings, such as evaluating the effects of chemotherapy and radiotherapy on tumors, monitoring the healing of wounds, and studying gene expression [290]. Since HbR and HbO_2 are primary optical absorbers in biological tissues (isosbestic wavelengths at 498 nm, 568 nm, and 794 nm), PAM is well suited for imaging blood vessels within an appropriate spectral region [288, 289]. PAM measures sO_2 in the same way near-infrared spectroscopy (NIRS) does, where HbR and HbO_2 are treated as the dominant absorbing compounds at each wavelength (λ_i). Thus, the blood absorption coefficient $\mu_a(\lambda_i)$ (in cm^{-1}) can be expressed as [288, 289]

$$\mu_a(\lambda_i) = \varepsilon_{\mathrm{HbR}}(\lambda_i)\,[\mathrm{HbR}] + \varepsilon_{\mathrm{HbO_2}}(\lambda_i)\,[\mathrm{HbO_2}] \tag{1.28}$$

where $\varepsilon_{\mathrm{HbR}}(\lambda_i)$ and $\varepsilon_{\mathrm{HbO_2}}(\lambda_i)$ are known molar extinction coefficients (in cm^{-1} M^{-1}) of HbR and HbO_2 at wavelength λ_i and [HbR] and [HbO_2] are the concentrations of the two types of hemoglobin, respectively. Since the amplitude of the acquired localized PA signal $\phi(\lambda_i, x, y, z)$ is proportional to the local optical energy deposition, we can replace $\mu_a(\lambda_i)$ with $\phi(\lambda_i, x, y, z)$ to calculate the [HbR] and [HbO_2] in relative values. The sO_2 image can then be calculated using

$$\mathrm{sO}_{2(x,y,z)} = \frac{[\mathrm{HbO_2}]_{(x,y,z)}}{[\mathrm{HbO_2}]_{(x,y,z)} + [\mathrm{HbR}]_{(x,y,z)}} \tag{1.29}$$

where two wavelengths are selected to determine the sO_2 map. Previous studies using fast functional PAM achieved a 2D frame rate of 400 Hz over an ~3 mm scanning range and a 3D volumetric rate of 1 Hz over a 3 × 2 mm^2 FOV. Using this system, multiple functional parameters were characterized, such as vascular morphology, blood oxygenation, blood flow, and oxygen metabolism in both resting and activated tissue states [65]. In addition, another Raman-laser–based dual-wavelength high-speed functional PAM (fPAM) was also developed, and it was capable of capturing capillary-level hemodynamics with a C-scan frame rate of 2 Hz [116]. This method allowed monitoring the local initiation of the spreading depolarization (SD) waves displayed by sO_2 changes after performing a permanent stroke procedure, as well as tracking the propagation pattern and vasoconstriction over the entire event across the whole cortex, as shown in Fig. 1.18a. Therefore, it can potentially be used by the stroke research community to discern stroke core and penumbra, as well as expansion of the penumbra in real time, and assess neuroprotective therapies in future studies. Compared to the linear PAM, a nonlinear OR-PAM technique demonstrated the ability to compensate for the absorption saturation in sO_2 imaging of a mouse ear [291]. In this method, a nonlinear model was used to compute sO_2 levels from PA measurements, as shown in Fig. 1.18b. Interestingly, the nonlinear sO_2 results were closer to these values than the linear results. Thus, utilizing the nonlinear sO_2 imaging method can reduce the systemic error caused by absorption saturation. Based on three modalities of PAM, high resolution, deep-penetrating, and high-speed PAM, an interesting research focus on noninvasively measuring RBC perfusion of the whole body and specific tissues during rest and exercise [292]. In this work, Taboada et al. have captured dynamic changes in RBC circulation in real-time with precision and high resolution, and to measure oxygenation by measuring saturation of hemoglobin (sO_2). They found that the glassfrogs "hide" RBCs in their liver during sleep, allowing them to achieve high levels of transparency on their ventral side. As the glassfrogs wake and become active, the amount of RBCs in circulation increases considerably. Figure 1.18c displays the frog's sO_2 under awake and sleep. In another work, to study the performance of iGECI in vivo, a hybrid PA and fluorescence microscopy system capable of simultaneously imaging brain oxygenation and neural activity was developed, as shown in Fig. 1.18d [293]. As seen from the examples given above, it is clear that PAM has proven effective for quantifying blood oxygenation with high sensitivity, using only hemoglobin as an endogenous probe. Doppler PAM has also shown the ability to measure the blood flow speed, based on the frequency shift in the emitted PA signal of moving RBCs. As many chronic diseases are accompanied by abnormal blood flow, including diabetes [294], hypertension [295], and dementia [296]. Doppler PAM is also of great interest for implementation in a clinical setting [297].

1.6.2 Temperature Measurement

In the fields of thermotherapy and cryotherapy, it is necessary to monitor the temperature distribution in biological tissues to ensure safe distribution of heat energy in healthy tissue and efficient destruction of tumor cells [298]. To meet this need, real-time temperature monitoring with high spatial resolution and high temperature sensitivity is needed. While existing methods such as thermocouple and thermistor can measure temperature directly, they are highly invasive and therefore unfavorable in the eyes of medical professionals. Thus, several noninvasive temperature

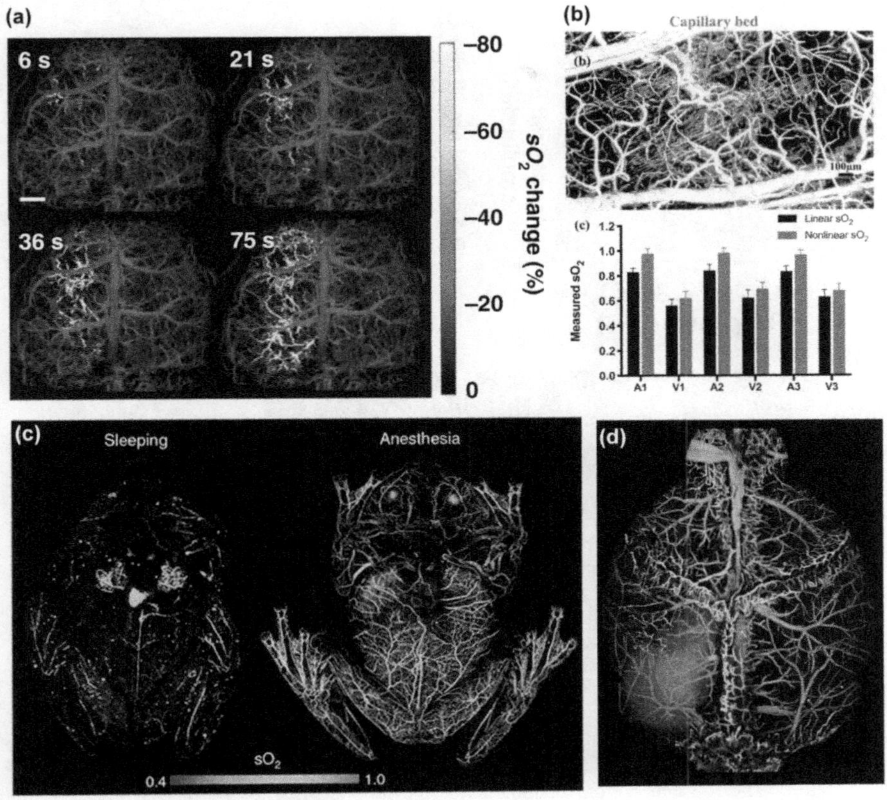

Fig. 1.18 Blood oxygenation PAM research. (**a**) The change in sO_2 during the stroke-induced SD waves propagation. (**b**) In vivo imaging of the total hemoglobin concentration in the mouse ear. (**c**) High-resolution PAM showing RBC perfusion within the vasculature of the same frog while asleep and under anesthesia. (**d**) Cortex vasculature and blood oxygenation resolved by PAM at a single-vessel level. (Reprinted with permission from [116] for (a), [291] for (b), [292] for (c) and [293] for (d))

monitoring methods have been developed for real-time, high-resolution temperature measurements, such as infrared thermography, ultrasound, and MRI [299–304]. Temperature measurement utilizing the PA effect has also shown potential as a non-invasive, high sensitivity method for real-time thermal monitoring [305, 306]. To achieve thermal monitoring PAI exploits the relationship between temperature and the tissue's Grüneisen parameter. Because these parameters are proportionally related, PA thermometry can measure relative changes in tissue temperature [307, 308]. To determine absolute temperature in tissue, ex vivo calibrations of tissues with known optical properties must be performed by measuring relative changes in PA signals at multiple reference temperatures [309]. To visualize temperature change at greater tissue depth, another temperature measurement method was

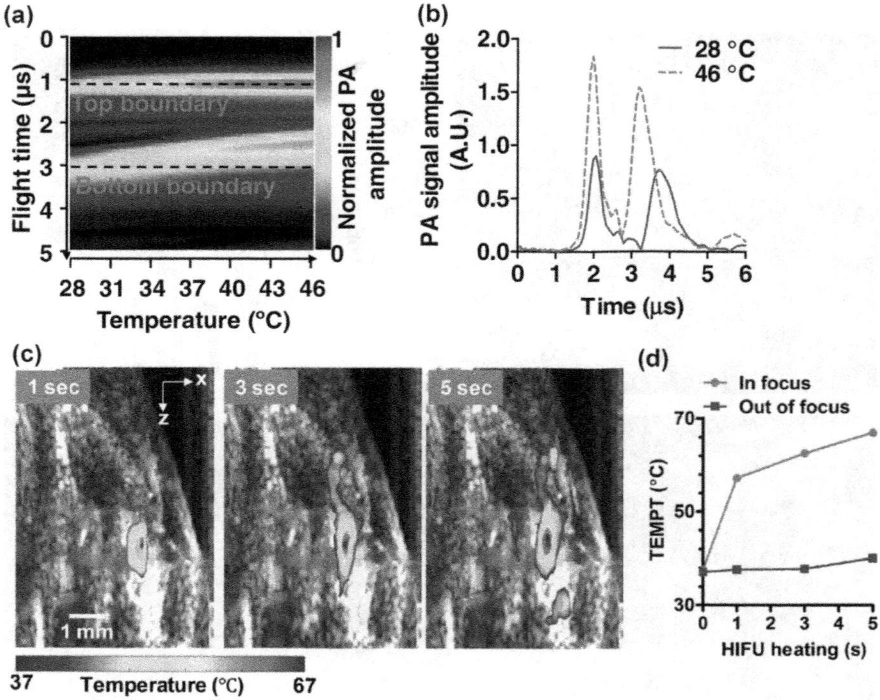

Fig. 1.19 Systematic calibrations for absolute temperature measurement. (**a**) Assembly of the time-resolved A-line signals of a 3 mm diameter blood-filled tube in a clear medium with changing temperature. The top and bottom boundaries of the tube are marked by dashed lines. As the temperature increased, the PA signal amplitude increased, and the acoustic flight time between the two boundaries decreased. The A-lines were aligned at the top boundary for better visualization of the change in acoustic flight time. (**b**) Representative A-line signals at 28 °C and 46 °C, showing the changes in PA signal amplitude and the acoustic flight time. (**c**) TEMPT temperatures (shown in color) overlaid on the corresponding ultrasound images (shown in gray). (**d**) TEMPT temperatures in the HIFU focus [red box in (a)] and in a representative surrounding tissue area [blue box in (a)] as a function of the heating time. (Reprinted with permission from [310] (a) and (b), and [311] for (c) and (d))

developed using the dual-temperature dependences of the Grüneisen parameter and the speed of sound in tissue, as shown in Fig. 1.19a and b [310]. A thermal-energy-memory-based PA thermometry (TEMPT) was also developed, which exploited the temperature dependence of lingering thermal energy on the corresponding PA signals acquired within thermal confinement, as shown in Fig. 1.19c and d [311]. Self-normalized ratiometric measurement cancels out temperature-irrelevant quantities and estimates the Grüneisen parameter so that temperature can be evaluated using the Grüneisen parameter, mass density, and specific heat capacity.

1.7 Preclinical and Clinical Applications

While PAM has yet to transition into a clinical setting, PAM-based systems have been and are being widely developed for use in areas such as neuroscience, cancer biology for microangiography, functional imaging, histology, and dermatology [46, 82, 109].

1.7.1 Microangiography in Cancer Biology and Neuroscience

One of the hallmarks of cancer biology and neurobiology is the generation of neovascularization to supply nutrients to growing cancer cells and damaged neural tissue. PAM can be used as a tool to monitor this angiogenesis. For example, it has been used for monitoring the response of tumor vascular network to single exposures to varying doses of ionizing radiation by employing a murine colon carcinoma CT26 model on eighteen 6-week-old female Balb/c mice, as shown in Fig. 1.20a [312]. In the PAM, vascular density is clearly visualized as a percentage of tumor area occupied by vessels. Moreover, PAM has also shown success in visualizing the angiogenesis of a growing tumor of the human oral vessel network and a mouse LS174 T tumor [313, 314]. The PAM method has also shown utility in monitoring blood vessel networks in human tongue and lips [313] and brain hemodynamic activities [314–316]. Compared to conventional imaging methods, such as functional magnetic resonance imaging (fMRI), two-photon microscopy, and wide-field optical microscopy, OR-PAM methods provide greater spatial resolution, imaging depth, and allow the use of external contrast agents. With these advantages, PAM is poised to provide greater detail about the subject, especially functional parameters such as sO_2, cerebral blood flow (CBF), flow speed, oxygen extraction fraction (OEF), and cerebral metabolic rate of oxygen ($CMRO_2$) [315, 316]. As shown in Fig. 1.20b, merging the scanning method mentioned above, fast functional OR-PAM presented excellent capability of imaging the whole mouse brain, while also providing information about the CBF, OEF, and $CMRO_2$ of each cerebral vessel [315].

1.7.2 Breast Cancer PAM

PAM has also shown utility in expediting tumor resection procedures in breast cancer. Breast-conserving surgery (BCS) minimizes the risk of recurrence and cosmetic side effects because intraoperative frozen section analysis (FSA) allows immediate analysis of the resected tissue allowing removal of residual diseased tissue [317]. Despite its utility, it is difficult to freeze and interpret adipose-rich

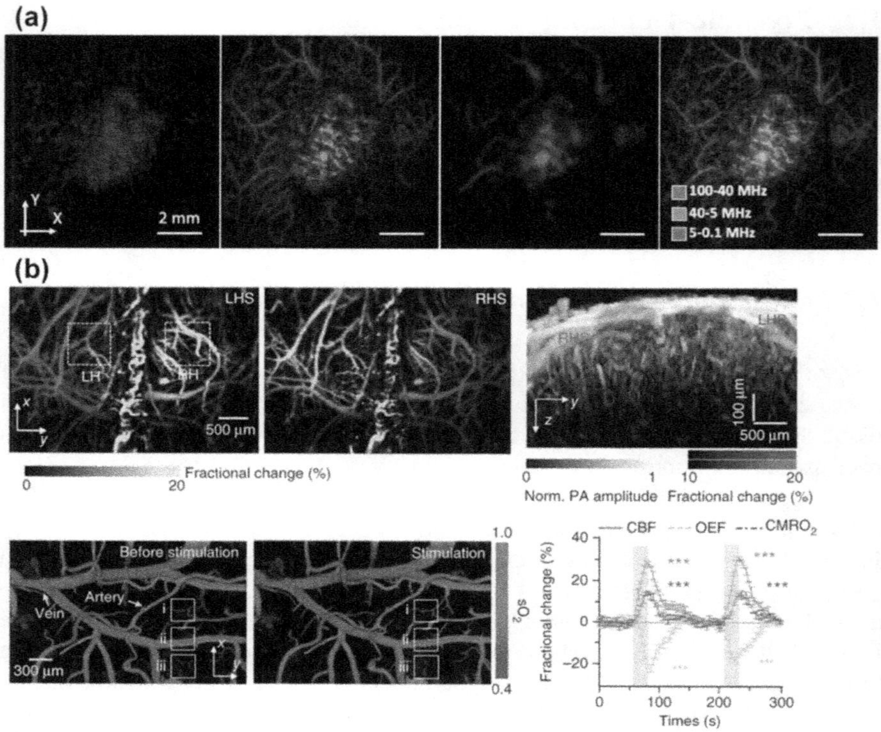

Fig. 1.20 Cancer biology and neuroscience PAM. (**a**) PAM of CT26 tumor reconstructed in three frequency ranges and merged: 100–40 MHz range (small vessels of <40 μm in diameter); 40–5 MHz range (vessels of middle sizes of 40–300 μm in diameter); 5–0.1 MHz range (large structures of >300 μm in diameter). (**b**) PAM of brain responses to electrical stimulations of the hindlimbs of mice ($n = 6$). (Reprinted with permission from [312] for (a) and [315] for (b))

breast tissue using FSA, especially in BCS [318]. PAM might ease this process as it is able to image cell nuclei using an ultra-violet (UV) wavelength excitation that is strongly absorbed by DNA and RNA in cell nuclei. In fact, a label-free UV-PAM imaging of formalin-fixed breast tissue was conducted, in which the morphology of individual cell nuclei was imaged with the high resolution of 330 nm and differentiated the tumor from normal tissue (Fig. 1.21) [319]. Ductal carcinoma in situ (DCIS) and invasive ductal carcinoma (IDC) were also observed in the tumor region. However, this conventional PAM system is quite slow (100 mins to image 5 mm × 5 mm FOV), but further development using the scanning methods mentioned above might solve this problem.

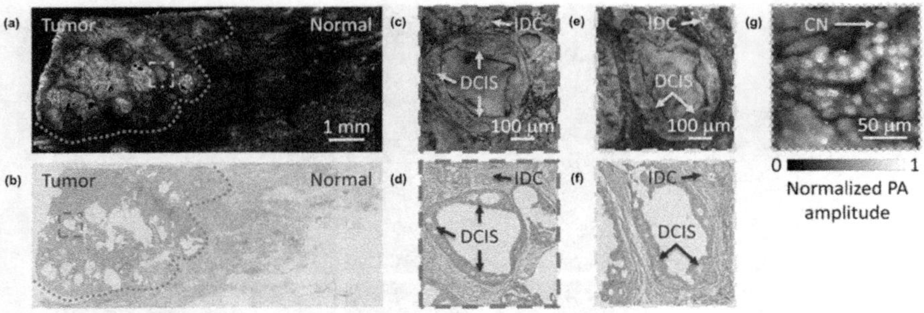

Fig. 1.21 PAM of breast tumor from the first patient. (**a**) UV-PAM of the fixed, unprocessed breast tumor. (**b**) H&E-stained histologic image of the same area shown in (**a**) acquired after processing, sectioning, and staining the excised breast tissue, (**c**) and (**d**) zoomed-in UV-PAM and H&E-stained images of the red dashed regions in (**a**) and (**b**), respectively. (**e**) and (**f**) zoomed-in UV-PAM and H&E images of the yellow-dashed regions in (**a**) and (**b**), respectively. (Reprinted with permission from [319])

1.7.3 Retinal Neovascularization PAM

Retinal neovascularization (RNV) is one of the major factors contributing to vision loss and blindness, and it results in many retinal diseases, including proliferative diabetic retinopathy, retinopathy of prematurity, sickle cell retinopathy, and retinal vein occlusions [320–322]. Because the eye has a high optical transparency, PAM can be considered a potential tool for retinal imaging. Previous studies acquired retinal PAM images comparing the growth of RNV before and after vascular endothelial growth factor (VEGF) injection with a scan range of 10.5 mm × 10.5 mm and corresponding resolutions of 20.5 μm and 41 μm [323]. In Fig. 1.22, gathered PAM images clearly show retinal vasculature in the targeted area. Due to PAM's high resolution and sensitivity to hemoglobin, it can be used to quantify RNV by extracting information about retinal vessels from different layers. Joint high-resolution multimodal PAM and OCT are being further developed to improve the efficiency for visualizing newly developed RNV and to monitor dynamic changes of retinal vein occlusions (RVO).

1.7.4 Dermatology PAM

The methods of PAM can also be applied to dermatology. Dermoscopy and human vision are conventional methods for diagnosing skin diseases, such as skin tumors, warts, or fungal infections [324]. These inspection methods have limited imaging depth and low optical contrast. In contrast, PAM systems can visualize chromophores (e.g., melanin) and hemoglobin, making them an attractive option for new dermoscopy systems [54, 325]. Previous studies have used PA to image

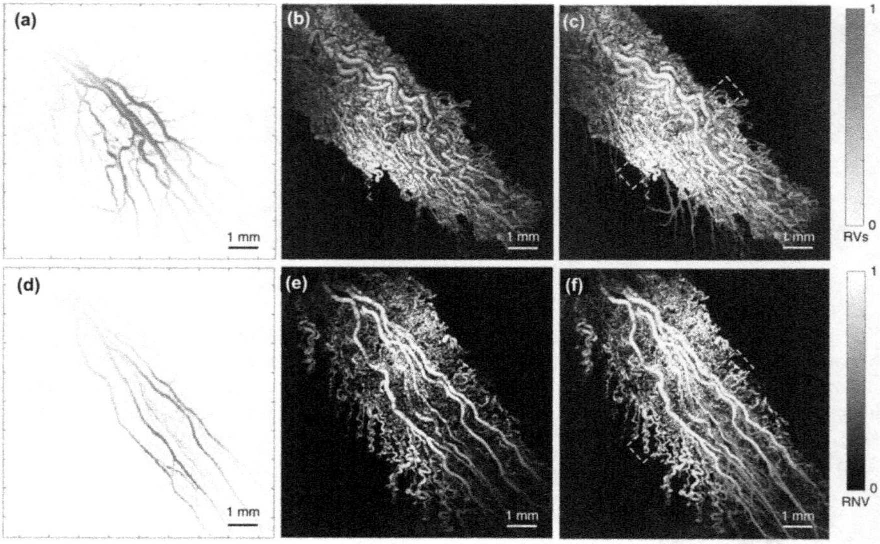

Fig. 1.22 Comparison of PAM images between the normal retinal vessels and VEGF-induced RNV. (**a**) Normal retinal vessels in albino rabbit. (**b**) RNV induced by VEGF in albino rabbit. (**c**) composite pseudo color image of albino rabbit showing the retinal vessels in the same area before and after VEGF injection. (**d**) normal retinal vessels in pigmented rabbit. (**e**) RNV induced by VEGF injection in a pigmented rabbit. (**f**) composite pseudo color image of pigmented rabbit showing the retinal vessels in the same area before and after VEGF injection. (Reprinted with permission from [323])

dermatological microvasculature at a depth of approximately 4 mm, distinguishing the papillary dermis, superficial vascular plexus, and subcutaneous tissue from one another [325]. As many skin abnormalities are associated with vascular irregularities, monitoring vascular change with PAM may be useful in describing clinical course. In fact, an ultra-broadband raster scan PA mesoscropy (UB-RSOM) was developed to characterize active dermal vascularization commonly seen in psoriatic skin [326]. Similarly, the hyperemic response after locally heating the human forearm was also observed using the UB-RSOM system [327]. In addition to its utility as a clinical diagnostic tool, PAM has also been identified as a useful tool in the study of the biology and physiology of the skin [328, 329]. A raster-scan PA mesoscopy, for instance, was used for skin imaging at 100 MHz (RSOM100) and 50 MHz (RSOM50) central frequency. This work demonstrates the quantitative account of the relation between the excitation energy, detection bandwidth, and imaging depth achieved with PA dermoscopy (Fig. 1.23a) [330]. Notably, the RSOM50 image reveals vessels of the reticular dermis and deep horizontal plexus not visible on the RSOM100 image (green and white arrows). On the other hand, the higher resolution is achieved by RSOM100, compared to RSOM50. These images reveal fine details of the skin surface and dermal papillae, with single capillary loops clearly resolved with RSOM100 but not with RSOM50. A similar PA mesoscopy

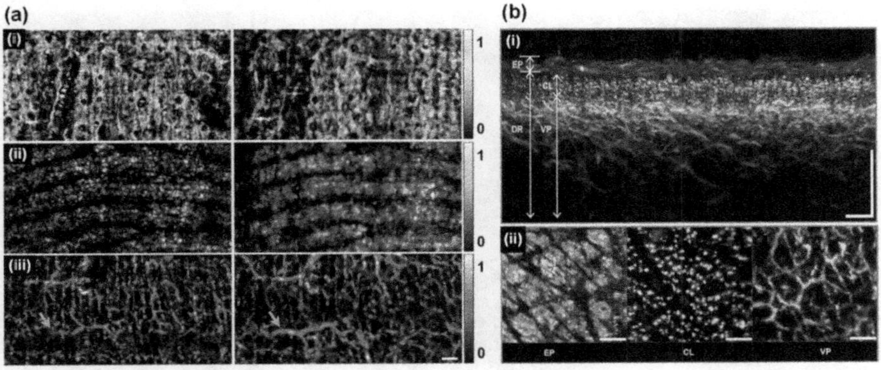

Fig. 1.23 PAM of skin. (**a**) Comparison of in vivo contrasts with RSOM100 and RSOM50. The left column shows images acquired with RSOM100, and the right column depicts results from RSOM50. (**i**) Lateral imaging of the stratum corneum depicting a slice located 0–210 μm below the skin surface. (**ii**) Lateral imaging of the epidermal-dermal junction 210–350 μm below the surface. (d) Lateral imaging of the dermis below the capillary loops at a depth of 350–1200 μm. Scale bars: 500 μm. (**b**) Raster-scan PA mesoscopy (RSOM): illustration of the system and imaging of healthy skin. (**i**) RSOM images of healthy skin in transverse cross section. Smaller structures (e.g., small vessels) emit ultrasound signals rich in high-frequency components (depicted in green); larger structures (e.g. larger vessels) emit signals rich in low-frequency components (depicted in red). Scale bar, 500 μm. (**ii**) RSOM images of skin layers in enface sections. CL capillary loops, DR dermis, EP epidermis, VP vascular plexus. Scale bar, 500 μm. (Reprinted with permission from [330] for (a) and [331] for (b))

technology is also able to image the vascular tree and melanin distribution with virtually no background signal, providing excellent contrast without the need for any exogenous label, as shown in Fig. 1.23b [331]. Segmenting images to isolate the vascular tree is often straightforward since melanin-rich structures usually do not overlap morphologically with vascular structures.

1.7.5 Single Cell PAM

The visualization of cellular processes in vivo enables the investigation of important biological questions such as understanding the function of genes and gene products, and how they regulate cellular processes [332–338]. As cell populations are heterogeneous, acquiring spatial and temporal information is fundamental to understanding cellular mechanisms. Monitoring the cellular processes therefore requires non-invasive methods that can visualize a given cell in action. Thanks to the advances in laser technology, ultrasound detection, and signal processing techniques discussed above, PAM systems have refined imaging resolution to less than 300 nm [333]. Sub-cellular structures can thus be visualized by targeting either endogenous

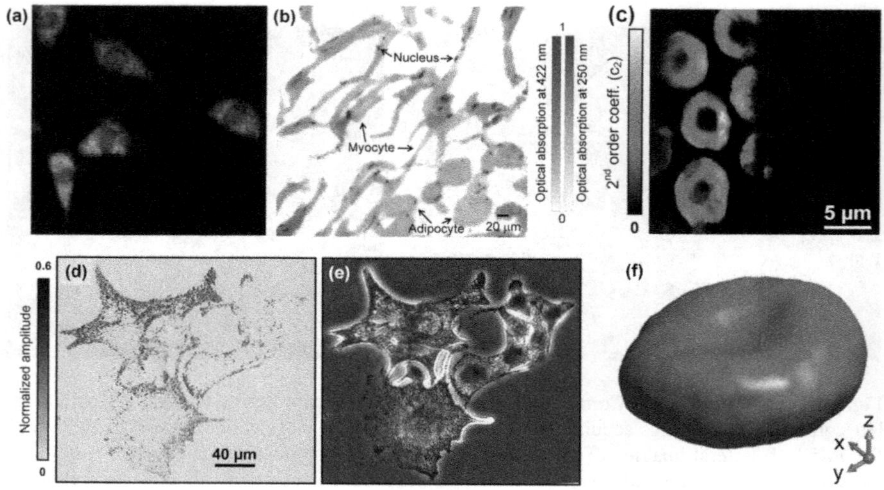

Fig. 1.24 Cell PAM. (**a**) Superimposed label-free PAM of fixed but unstained fibroblasts at 422 nm (green) and 250 nm (blue). (**b**) Dual-wavelength label-free PAM image of a mouse ear section, where absorption at 422 nm is shown in pink, and absorption at 250 nm is shown in blue. (**c**) Second-order PA nanoscopy of red blood cells. (**d**) and (**e**) Imaging melanoma cells before and after PA nanoscopy. (**f**) 3D reconstruction of the RBC with a micro-ring resonator. (Reprinted with permission from [337] for (a) and (b), [338] for (c)-(e), and [200] for (f))

chromophores such as hemoglobin and melanin, or exogenous contrast agents such as nanoparticles and dyes. Advances in PAM technology have also made it possible for novel quantitative analysis to offer insight into functional information, allowing us to infer structural and biomechanical properties on a unicellular scale [334]. For example, a label-free PAM has been used to image fixed fibroblasts on slides at 422 nm and 250 nm to reveal cytoplasms and nuclei, as shown in Fig. 1.24a and b [337]. As well it was also applied to label-free histology of mouse ear sections by showing cytoplasms and nuclei of various cells. With a larger NA objective lens for light focus, the lateral resolution was improved up to 88 nm [338]. This PA nanoscopy technique is able to observe cross-sectional images of red blood cells at 1 μm depth and melanoma cells with ejecting background from out-of-focus absorbers, as shown in Fig. 1.24c–e. Moreover, another study used an optical detector (microring resonator) to achieve broad ultrasound bandwidth of 200 MHz. It breaks the acoustically determined axial resolution limitation in traditional PAM systems and is also an attractive technique in single-cell imaging (e.g., 3D blood cell shown in Fig. 1.24f).

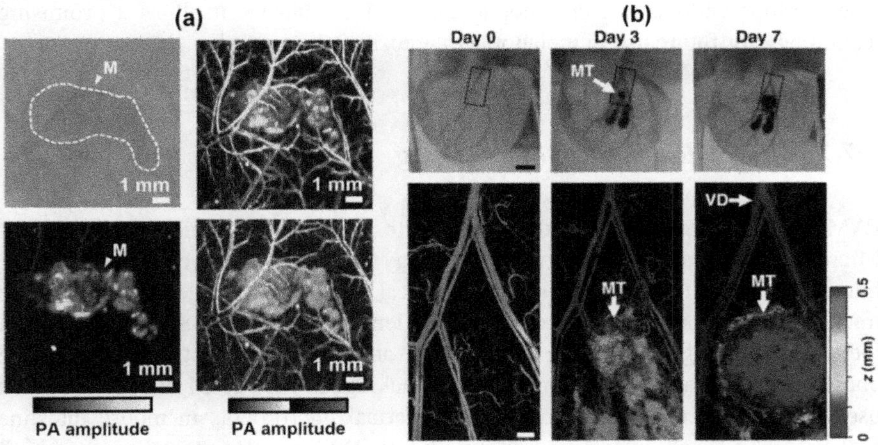

Fig. 1.25 Melanoma PAM. (**a**) In vivo PA images of a subcutaneously inoculated B16-melanoma in an immunocompromised nude mouse (on the back). (**b**) mPAM images of the tumor region at 584 nm. z is coded by colors: blue (superficial) to red (deep). Scale bar: 125 μm. (Reprinted with permission from [342] for (a) and [343] for (b))

1.7.6 Melanoma PAM

Melanoma is the most volatile type of skin cancer and is highly lethal. In modern clinical settings, cutaneous melanomas are imaged using positron emission tomography (PET) [339], optical imaging modalities [340], and high-frequency ultrasound [341]. As discussed above, PAM can function as a sensitive blood detector with high contrast and specificity, granting the ability to image the volumetric morphology of subcutaneous microvasculature associated with tumor growth and metastasis, as shown in Fig. 1.25a [342]. Using multiple wavelengths, various pigments can be distinguished from one another using their different absorption spectra [40, 342]. Because both melanin and hemoglobin have comparably strong absorptions at visible wavelength band, the morphological relationship between a melanoma and the surrounding blood vessels can be imaged by PAM in the x-y view. However, it is difficult for visible light to penetrate melanin-rich tumors, complicating tumor thickness measurement. In contrast, near-infrared light can infiltrate the tumor because of decreased optical absorption of this wavelength of light in melanin and blood. Using these wavelengths, the 3D morphology of both the melanoma and surrounding vasculature can be visualized. A label-free metabolic PAM system (mPAM) for early-stage melanoma imaging measuring metabolic rate of oxygen (MRO_2) was also developed, showcasing the flexibility of PAM systems in targeting various absorption wavelength spectra [343]. Other groups have also demonstrated melanoma PAM imaging using high frequency (UHF) wide-bandwidth transducers and a nonlinear incident optical illumination method [344–346]. Due to its unique ability to synthesize absorption information from various endogenous chromophores, PAM establishes itself as a promising technology for future use in melanoma diagnosis and monitoring.

1.7.7 Micro-Rocket Robot Tracking

PAM is also being considered for use with other novel clinical technologies. Micro/nanorobots are an exciting technology expected to travel throughout the human body through blood vessels for medical treatment or surgery [347]. Precise tracking of microrobots is an ongoing challenge, however [348]. Because PAM methods have high resolution and optical absorption contrast, they are being considered as a candidate for microrobot tracking. In fact, PAM has already been used to track microcapsules containing spherical microrobots in mouse intestine [349]. Tracking a singular microrobot with OR-PAM has also seen success in both clear media and blood. In this study, the micro-rocket was clearly observed in PA images because the average PA amplitude of the microrocket was 33% higher than that of background blood. Expected transport of the ~50 μm micro-rocket was further mimicked by a 250 μm-diameter rubber tube filled with bovine blood, and even under these dynamic conditions, PAM recorded the microrobot's structure, depth, and position, as shown in Fig. 1.26 [349].

1.8 Conclusion

Since the discovery of the PA effect in 1880, PAI technology has made great strides, solidifying itself as a well-recognized biomedical imaging modality. Through the development of suitable laser sources, sensitive and broadband ultrasound detection technology, as well as fast data acquisition and processing capacities, PAI systems have become more efficient and more practical imaging devices.

Today, PAM represents one of the most popular divisions of PA inquiry and has proven to be a promising field of study in PAI. In PAM, PA signals are formed at the co-focus point of light and acoustic waves. To align two beams coaxially, sensitivity must be maximized, which has been realized through the use of various methods, including off-axis alignment, dark-field illumination, second-generation systems, ring transducers, transmission modes, and reflection modes. Novel PAM systems utilizing pulsed excitations and wideband ultrasonic transducers have great potential for detecting minor changes in tissue structure, which is proving to help in detecting early-stage cancer and weak physiological processes, such as neural transmission.

Purely optical PAM bypasses the limitations of standard PAM systems, setting the peak performance standards for all-optical PAI. All-optical sound detectors such as a phase detection-based interferometer, Doppler-based sensing, and resonator-based sensing yield a higher SNR per unit area than piezoelectric detectors and

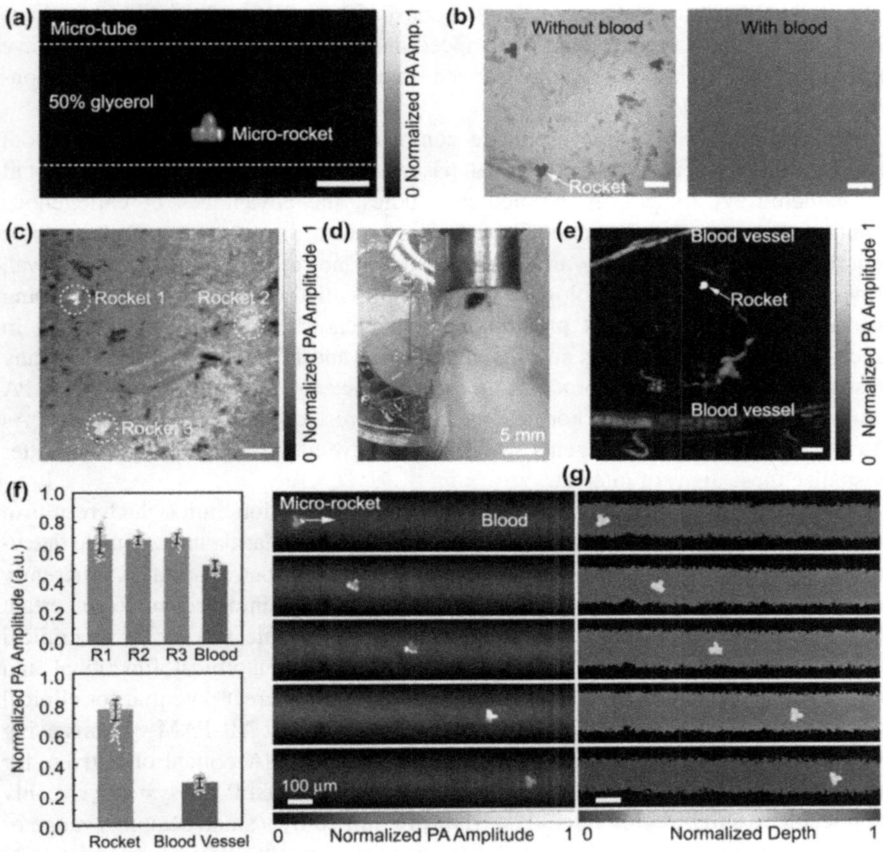

Fig. 1.26 PAI of a single micro-rocket in blood, and penetration of living mouse ear. (**a**) Clear high-resolution PA imaging of a single micro-rocket in the transparent solution. (**b**) Optical microscopy of the micro-rockets without blood and within blood. (**c**) Corresponding visible PA images of these micro-rockets in thick blood. (**d**) Mouse ear covering a micro-rocket, (**e**) PAM of the micro-rocket through the living mouse ear. (**f**) Contrast of the normalized PA amplitude between the micro-rockets and the blood or the living blood vessel. (**g**) High-resolution PA tracking of a single micro-rocket in the blood vessel model. (Reprinted with permission from [349])

access wide detection bandwidths that may be more appropriate for PA applications. Additionally, optical sensing of sound is less sensitive to electromagnetic noise, making it appropriate for a broader spectrum of environments. Despite the clear advantages of all-optical PAM, piezoelectric transducers are generally more economical than optical technologies, especially for large-scale manufacturing; therefore, piezoelectric transducers remain the standard for PAM systems.

Diverse optical scanning mirror-based units are developed to improve imaging speed and increase the size of the FOV in PAM systems. Traditional optical mirrors typically cannot be used in PAM systems because the light and generated US waves need to be reflected and scanned simultaneously. Thus, 3D mechanical motors have been constructed to increase imaging speed and FOV. Several other methods have been developed to improve scanning speed such as MEMS, galvanometer, polygon-mirror, and micro-lens array.

PA contrast agents improve image contrast and resolution by changing both acoustic and optical features of local tissues, significantly enhancing the detail of gathered PA images. In biomedical studies, the advantages of endogenous contrast agents are undeniable. They do not perturb the original sample tissue's microenvironment, and they are nontoxic and do not require regulatory approval. Endogenous absorbers in biological tissue are classified by their primary absorbing wavelengths with the most predominant endogenous contrast agents falling in accessible absorption ranges such as hemoglobin and DNA/RNA. However, many diseases and physiological processes are not characterized by sufficient intrinsic PA contrast activity such as tumor cell receptor expression and protease activity. As a result, many exogenous agents such as dyes have also been developed to better visualize these areas of interest.

PAM is uniquely capable of bypassing the optical diffusion limit (which results in achieving scalable spatial resolution based on optical excitation and acoustic detection), achieving 100% sensitivity to optical absorption contrast (that is inherently capable of functional and molecular imaging), and remaining sensitive to certain key functional parameters, making PAM extremely viable for use in preclinical applications. With promising data as a powerful tool for anatomical, functional, and molecular imaging for preclinical studies, PAM also has great potential for clinical applications. However, PAM is not without its faults. Most AR-PAM systems using weakly focused light and OR-PAM systems using low-NA optical objectives, for example, have limited view. Moreover, most of the reported PAM systems are only applicable to small-region imaging as they are limited by a small scanning range or slow scanning speed. Finally, many of the published results may be unrealistically optimized as miniaturization studies are based on table-top PAM systems, and many animal model studies are done with the assistance of anesthesia to eliminate any motion artifacts. Despite these challenges, PAM presents its utility as a biomedical imaging modality and will likely experience many more technical advances in near future.

References

1. Rajan S. Gurjar, Vadim Backman, Lev T. Perelman, Irene Georgakoudl, Kamran Badizadegan, Irving Itzkan, Ramachandra R. Dasarl, and Michael S. Feld, Imaging human epithelial properties with polarized lightscattering spectroscopy, Nature Medicine, Vol. 7, No. 11, 2001.
2. James G. Fujimoto, Costas Pitris, Stephen A. Boppart, and Mark E. Brezinski, Optical Coherence Tomography: An Emerging Technology for Biomedical Imaging and Optical Biopsy, Neoplasia Vol. 2, Nos. 1–2, 2000, pp. 9–25.
3. Benjamin J. Vakoc, Dai Fukumura, Rakesh K. Jain and Brett E. Bouma, Cancer imaging by optical coherence tomography: preclinical progress and clinical potential, Nature Reviews, Cancer, Vol. 12, 2012, pp. 363–368.

4. Martin Dierolf, Andreas Menzel, Pierre Thibault, Philipp Schneider, Cameron M. Kewish, Roger Wepf, Oliver Bunk, and Franz Pfeiffer, Ptychographic X-ray computed tomography at the nanoscale, Nature, Vol. 467, 2010, pp. 436–439.
5. Qiang Huang, Malkiel A. Cohen, Fernando C. Alsina, Garth Devlin, Aliesha Garrett, Jennifer McKey, Patrick Havlik, Nikolai Rakhilin, Ergang Wang, Kun Xiang, Parker Mathews, Lihua Wang, Cheryl Bock, Victor Ruthig, Yi Wang, Marcos Negrete, Chi Wut Wong, Preetish K. L. Murthy, Shupei Zhang, Andrea R. Daniel, David G. Kirsch, Yubin Kang, Blanche Capel, Aravind Asokan, Debra L. Silver, Rudolf Jaenisch, Xiling Shen, Intravital imaging of mouse embryos, Science, Vol. 368, No. 6487, 2020, pp. 181–186.
6. Claudia Errico, Juliette Pierre, Sophie Pezet, Yann Desailly, Zsolt Lenkei, Olivier Couture, and Mickael Tanter, Ultrafast ultrasound localization microscopy for deep super-resolution vascular imaging, Nature, Vol. 527, 2015, pp. 499–502.
7. Jung Ho Yu, Seung-Hae Kwon, Zdenek Petrášek, Ok Kyu Park, Samuel Woojoo Jun, Kwangsoo Shin, Moonkee Choi, Yong Il Park, Kyeongsoon Park, Hyon Bin Na, Nohyun Lee, Dong Won Lee, Jeong Hyun Kim, Petra Schwille, and Taeghwan Hyeon, High-resolution three-photon biomedical imaging using doped ZnS nanocrystals, Nature Materials, Vol 12, 2013, pp. 359–366.
8. Fanghao Hu, Lixue Shi, and Wei Min, Biological imaging of chemical bonds by stimulated Raman scattering microscopy, Nature Methods, Vol. 16, 2019, pp. 830–842.
9. X. Wu, M. Delbianco, K. Anggara, T. Michnowicz, A. Pardo-Vargas, P. Bharate, S. Sen1, M. Pristl, S. Rauschenbach, U. Schlickum, S. Abb, P. H. Seeberger, and K. Kern, Imaging single glycans, Nature, Vol. 582, 2020, pp. 375–378.
10. Vipul Gujrati, Jaya Prakash, Jaber Malekzadeh-Najafabadi, Andre Stiel, Uwe Klemm, Gabriele Mettenleiter, Michaela Aichler, Axel Walch, and Vasilis Ntziachristos, Bioengineered bacterial vesicles as biological nanoheaters for optoacoustic imaging, Nature Communication, Vol. 10, 1114, 2019.
11. Yicong Wu, and Hari Shroff, Faster, sharper, and deeper: structured illumination microscopy for biological imaging, Nature Methods, Vol. 15, 2018, pp. 1011–1019.
12. Bhosale, Y.H., Patnaik, K.S. Bio-medical imaging (X-ray, CT, ultrasound, ECG), genome sequences applications of deep neural network and machine learning in diagnosis, detection, classification, and segmentation of COVID-19: a Meta-analysis & systematic review, Multimedia Tools and Applications (2023).
13. Anton du Plessis, and Chris Broeckhoven, Looking deep into nature: A review of micro-computed tomography in biomimicry, Acta Biomaterialia, Vol. 85, 2019, pp. 27–40.
14. Pak Wing Calvin Cheng, Wing Chung Chang, Gladys G. Lo, Kit Wa Sherry Chan, Ho Ming Edwin Lee, Lai Ming Christy Hui, Yi Nam Suen, Yim Lung Eric Leung, Kai Ming Paul Au Yeung, Sirong Chen, Ka Fung Henry Mak, Pak Chung Sham, Barbara Santangelo, Mattia Veronese, Chi-Lai Ho, Yu Hai Eric Chen, and Oliver D. Howes, The role of dopamine dysregulation and evidence for the transdiagnostic nature of elevated dopamine synthesis in psychosis: a positron emission tomography (PET) study comparing schizophrenia, delusional disorder, and other psychotic disorders, Neuropsychopharmacology, 0:1–7, 2020.
15. Gajendra S. Shekhawat, Steven M. Dudek, and Vinayak P. Dravid, Development of ultrasound bioprobe for biological imaging, Science Advances, Vol. 2017, No. 3, 2017.
16. Yicong Wu, and Hari Shroff, Faster, sharper, and deeper: structured illumination microscopy for biological imaging, Nature Methods, Vol. 15, 2018, pp. 1011–1019.
17. Zijian Zhou, Hongzhang Deng, Weijing Yang, Zhantong Wang, Lisen Lin, Jeeva Munasinghe, Orit Jacobson, Yijing Liu, Longguang Tang, Qianqian Ni, Fei Kang, Yuan Liu, Gang Niu, Ruiliang Bai, Chunqi Qian, Jibin Song, and Xiaoyuan Chen, Early stratification of radiotherapy response by activatable inflammation magnetic resonance imaging, Nature Communications, Vol. 11, No. 3032, 2020.
18. Peiwu Qin, Mahmut Parlak, Cem Kuscu, Jigar Bandaria, Mustafa Mir, Karol Szlachta, Ritambhara Singh, Xavier Darzacq, Ahmet Yildiz, and Mazhar Adli, Live cell imaging of low- and non-repetitive chromosome loci using CRISPR-Cas9, Nature Communications, Vol. 8, No. 14725, 2017.

19. Norihide Yoneda, Osamu Matsui, Satoshi Kobayashi, Azusa Kitao, Kazuto Kozaka, Dai Inoue, Kotaro Yoshida, Tetsuya Minami, Wataru Koda, Toshifumi Gabata, Current status of imaging biomarkers predicting the biological nature of hepatocellular carcinoma, Japanese Journal of Radiology, Vol. 37, 2019, pp. 191–208.
20. Catarina Costa Moura, Konstantinos N. Bourdakos, Rahul S. Tare, Richard O. C. Oreffo and Sumeet Mahajan, Live-imaging of bioengineered cartilage tissue using multimodal non-linear molecular Imaging, Scientific Reports, Vol. 9, 5561, 2019.
21. Yves F. Dufrêne, Toshio Ando, Ricardo Garcia, David Alsteens, David Martinez-Martin, Andreas Engel, Christoph Gerber and Daniel J. Müller, Imaging modes of atomic force microscopy for application in molecular and cell biology, Nature Nanotechnology, Vol. 12, 2017, pp. 295–307.
22. Fanghao Hu, Chen Zeng, Rong Long, Yupeng Miao, Lu Wei, Qizhi Xu, and Wei Min, Supermultiplexed optical imaging and barcoding with engineered polyynes, Nature Methods, Vol.15 No. 3, 2018, pp. 194–200.
23. Siân Culley, David Albrecht, Caron Jacobs, Pedro Matos Pereira, Christophe Leterrier, Jason Mercer & Ricardo Henriques, Quantitative mapping and minimization of super-resolution optical imaging artifacts, Nature Methods, Vol.15, No.4, 2018, pp. 263–266.
24. Seokchan Yoon, Moonseok Kim, Mooseok Jang, Youngwoon Choi, Wonjun Choi, Sungsam Kang, and Wonshik Choi, Deep optical imaging within complex scattering media, Nature Reviews Physics, Vol. 2, 2020, pp. 141–158.
25. Sean J. Miller, Thomas Philips, Namho Kim, Raha Dastgheyb, Zhuoxun Chen,Yi-Chun Hsieh, J. Gavin Daigle, Malika Datta, Jeannie Chew, Svetlana Vidensky, Jacqueline T. Pham, Ethan G. Hughes, Michael B. Robinson, Rita Sattler, Raju Tomer, Jung Soo Suk, Dwight E. Bergles, Norman Haughey, Mikhail Pletnikov, Justin Hanes, and Jeffrey D. Rothstein , Molecularly defined cortical astroglia subpopulation modulates neurons via secretion of Norrin, Nature Neuroscience, Vol. 22, 2019, pp. 741–752.
26. Fanghao Hu, Chen Zeng, Rong Long, Yupeng Miao, Lu Wei, Qizhi Xu, and Wei Min, Supermultiplexed optical imaging and barcoding with engineered polyynes, Nature Methods, Vol. 15, No. 3, 2018, pp. 194–199.
27. I. Sokolova, M. E. Dokukina, V. Kalaparthia, M. Miljkovica, A. Wanga, J. D. Seigned, P. Grivase, and E. Demidenko, Noninvasive diagnostic imaging using machinelearning analysis of nanoresolution images of cell surfaces: detection of bladder cancer, PNAS, Vol. 115, No. 51, 2018, pp. 12920–12925.
28. Hamid Pahlevaninezhad, Mohammadreza Khorasaninejad, Yao-Wei Huang, Zhujun Shi, Lida P. Hariri, David C. Adams, Vivien Ding, Alexander Zhu, Cheng-Wei Qiu, Federico Capasso, and Melissa J. Suter, Nano-optic endoscope for high-resolution optical coherence tomography in vivo, Nature Photonics, Vol. 12, 2018, pp. 540–547.
29. Joseph P. Angelo, Sez-Jade Chen, Marien Ochoa, Ulas Sunar, Sylvain Gioux, Xavier Intes, Review of structured light in diffuse optical imaging, Journal of Biomedical Optics, Vol. 24, No. 7, 071602, 2019.
30. Jianfeng Wang, Yang Xu, Stephen A. Boppart, Review of optical coherence tomography in oncology, Journal of Biomedical Optics, Vol. 22, No. 12, 121711, 2017.
31. Valery V. Tuchin, Light interaction with biological tissues: overview, Proc. SPIE 1884, Static and Dynamic Light Scattering in Medicine and Biology, (21 July 1993).
32. Pawel R. Laskowski, Moritz Pfreundschuh, Mirko Stauffer, Zöhre Ucurum,Dimitrios Fotiadis, and Daniel J. Müller, High-resolution imaging and multiparametric characterization of native membranes by combining confocal microscopy and an atomic force microscopy-based toolbox, ACS Nano, Vol. 11, 2017, pp. 8292–8301.
33. Devin R. Beaulieu, Ian G. Davison, Kıvılcım Kılıç, Thomas G. Bifano, and Jerome Mertz, Simultaneous multiplane imaging with reverberation two-photon microscopy, Nature Methods, Vol. 17, 2020, pp. 283–286.
34. Eirini Papagiakoumou , Emiliano Ronzitti, and Valentina Emiliani, Scanless two-photon excitation with temporal focusing, Nature Methods, Vol 17, 2020, pp. 571–581.

35. Roosje M. Ruis, Arne Leinse, Ronald Dekker, Rene G. Heideman, Ton G. van Leeuwen, and Dirk J. Faber, Decreasing the size of a spectral domain optical coherence tomography system with cascaded arrayed waveguide gratings in a photonic integrated circuit, IEEE Journal of Selected Topics in Quantum Electronics, Vol. 25, No. 1, 2019, pp. 1–9.
36. Amaury Badon, Dayan Li, Geoffroy Lerosey, A. Claude Boccara, Mathias Fink, and Alexandre Aubry, Smart optical coherence tomography for ultra-deep imaging through highly scattering media, Science Advances, Vol. 2, e1600370, 2016.
37. Zhao Wang, Benjamin Potsaid, Long Chen, Chris Doerr, Hsiang-Chieh Lee, Torben Nielson, Vijaysekhar Jayaraman, Alex E. Cable, Eric Swanson, and James G. Fujimoto, Cubic meter volume optical coherence tomography, Optica, Vol. 3, No. 12, 2016, pp. 1496–1503.
38. Zhixing Chen, Lu Wei, Xinxin Zhu, and Wei Min, Extending the fundamental imaging-depth limit of multi-photon microscopy by imaging with photo-activatable fluorophores, Optics Express, Vol. 20, No. 17, 2012, pp. 18525–18536.
39. T. Gefen, A. Rotem, and A. Retzker, Overcoming resolution limits with quantum sensing, Nature Communications, Vol. 10, 4992, 2019.
40. Junjie Yao, and Lihong V. Wang, Photoacoustic molecular imaging: principles and practice, molecular imaging, 2021, pp. 233–244.
41. Eric M. Strohm, Michael J. Moore, Michael C. Kolios, High resolution ultrasound and photoacoustic imaging of single cells, Photoacoustics, Vol. 4, 2016, pp. 36–42.
42. Lihong V. Wang, and Song Hu, Photoacoustic tomography: in vivo imaging from organelles to organs, Science, Vol. 335, 2012, pp. 1458–1462.
43. Fu B, Cheng Y, Shang C, Li J, Wang G, Zhang C, Sun J, Ma J, Ji X, He B, Optical ultrasound sensors for photoacoustic imaging: a narrative review, Quantitative Imaging in Medicine and Surgery, Vol. 12, No. 2, 2022, pp. 1608–1631.
44. Malcolm C Finlay, Charles A Mosse, Richard J Colchester, Sacha Noimark, Edward Z Zhang, Sebastien Ourselin, Paul C Beard, Richard J Schilling, Ivan P Parkin, Ioannis Papakonstantinou, and Adrien E Desjardins, Through-needle all-optical ultrasound imaging in vivo: a preclinical swine study, Light: Science & Applications, Vol. 6, e17103, 2017.
45. Ben Cox, Jan G. Laufer, Simon R. Arridge, Paul C. Beard, Quantitative spectroscopic photoacoustic imaging: a review, Journal of Biomedical Optics, Vol. 17, No. 6, 2012.
46. Seung Hee Han, Review of Photoacoustic Imaging for Imaging-Guided Spinal Surgery, Neurospine, Vol. 15, No. 4, 2018, pp. 306–322.
47. Manoj Kumar Dasa, Christos Markos, Michael Maria, Christian R. Petersen, Peter M. Moselund, and Ole Bang, High-pulse energy supercontinuum laser for high-resolution spectroscopic photoacoustic imaging of lipids in the 1650-1850 nm region, Biomedical Optics Express, Vol. 9, No. 4, 2018, pp. 1762–1770.
48. Judith Weber, Paul C Beard, and Sarah E Bohndiek, Contrast agents for molecular photoacoustic imaging, Nature Methods, Vol. 13, No. 8, 2016, pp. 639–649.
49. Shriram Sethuraman, James H. Amirian, Silvio H. Litovsky, Richard W. Smalling, and Stanislav Y. Emelianov, Spectroscopic intravascular photoacoustic imaging to differentiate atherosclerotic plaques, Optics Express, Vol. 16, No. 5, 2008, pp. 3362–3367.
50. Janggun Jo, Chang H. Lee, Raoul Kopelman, and Xueding Wang, In vivo quantitative imaging of tumor pH by nanosonophore assisted multispectral photoacoustic imaging, Nature Communications, Vol. 8, 2017, pp. 471.
51. Alexander Graham Bell, The Photophone, Science, 1880, Vol. 1, No. 11, pp. 130–134.
52. D. F. Nelson, K. W. Wecht, and D. A. Kleinman, Photophone performance, J. Acoust. Soc. Am., Vol. 60, No. 1, 1976, pp. 251–255.
53. Minghua Xu, and Lihong V. Wang, Photoacoustic imaging in biomedicine, Rev. Sci. Instrum. Vol. 77, 041101, 2006.
54. Lihong V. Wang, Multiscale photoacoustic microscopy and computed tomography, Nature Photonics, Vol. 3, 2009, pp. 503–508.
55. Thomas J. Allen and Paul C. Beard, Pulsed near-infrared laser diode excitation system for biomedical photoacoustic imaging, Optics Letters, Vol. 31, No. 23, 2006.

56. Ali Hariri, Afreen Fatima, Nafiseh Mohammadian, Sadreddin Mahmoodkalayeh, Mohammad Ali Ansari, Nicholas Bely, Mohammad R. N. Avanaki, Development of low-cost photoacoustic imaging systems using very low-energy pulsed laser diodes, Journal of Biomedical Optics, Vol. 22, No. 7, 2017, 075001.
57. Sri-Rajasekhar Kothapalli, Te-Jen Ma, Srikant Vaithilingam, Omer Oralkan, Butrus T. Khuri-Yakub, and Sanjiv Sam Gambhir, Deep Tissue Photoacoustic Imaging Using a Miniaturized 2-D Capacitive Micromachined Ultrasonic Transducer Array, IEEE Transactions on Biomedical Engineering, Vol. 59, No. 5, 2012, pp. 1199–1204.
58. Warren Jackson and Nabil M. Amer, Piezoelectric photoacoustic detection: Theory and experiment, Journal of Applied Physics, Vol. 51, No. 6, 1980, pp. 3343–3353.
59. Kenji Mitsuhashi, Kun Wang, Mark A. Anastasio, Investigation of the far-field approximation for modeling a transducer's spatial impulse response in photoacoustic computed tomography, Photoacoustics, Vol. 2, 2014, pp. 21–32.
60. Minghua Xu and Lihong V. Wang, Analytic explanation of spatial resolution related to bandwidth and detector aperture size in thermoacoustic or photoacoustic reconstruction, Physical Review, Vol. 67, 2003.
61. Bradley E. Treeby, B. T. Cox, k-Wave: MATLAB toolbox for the simulation and reconstruction of photoacoustic wave fields, Journal of Biomedical Optics, Vol. 152, 2010.
62. Lihong V Wang, and Junjie Yao, A practical guide to photoacoustic tomography in the life sciences, Nature Methods, Vol.13, No.8, 2016, pp. 627–635.
63. Yang Li, Terence T. W. Wong, Junhui Shi, Hsun-Chia Hsu, and Lihong V. Wang, Multifocal photoacoustic microscopy using a single-element ultrasonic transducer through an ergodic relay, Light: Science & Applications, Vol. 9, 135, 2020.
64. Pengfei Zhang, Lei Li, Li Lin, Junhui Shi, and Lihong V. Wang, In vivo superresolution photoacoustic computed tomography by localization of single dyed droplets, Light: Science & Applications, Vol. 8, 36, 2019.
65. Wonseok Choi and Chulhong Kim, Toward in vivo translation of superresolution localization photoacoustic computed tomography using liquid-state dyed droplets, Light: Science & Applications, Vol. 8, 57, 2019.
66. Lihong V. Wang, Multiscale photoacoustic microscopy and computed tomography, Nature Photonics, Vol. 3, 2009, pp. 503–507.
67. Junjie Yao, Lidai Wang, Joon-Mo Yang, Konstantin I Maslov, Terence T W Wong, Lei Li, Chih-Hsien Huang, Jun Zou, and Lihong V Wang, high-speed label-free functional photoacoustic microscopy of mouse brain in action, Nature Methods, Vol.12, No.5, 2015, pp. 407–409.
68. Hao F Zhang, Konstantin Maslov, George Stoica, and Lihong V Wang, Functional photoacoustic microscopy for high-resolution and noninvasive in vivo imaging, Nature Biotechnology, Vol. 24, No. 7, 2006, pp. 848–851.
69. Donald B. Conkey, Antonio M. Caravaca-Aguirre, Jake D. Dove, Hengyi Ju, Todd W. Murray, and Rafael Piestun, Super-resolution photoacoustic imaging through a scattering wall, Nature Communications, Vol. 6, No. 7902, 2015.
70. Song Hu, Konstantin Maslov, and Lihong V. Wang, Second-generation optical-resolution photoacoustic microscopy with improved sensitivity and speed, Optics Letters, Vol. 36, No. 7, 2011, pp. 1134–1136.
71. Geng Ku, Konstantin Maslov, Li Li, and Lihong V. Wang, Photoacoustic microscopy with 2-m transverse resolution, Journal of Biomedical Optics, Vol. 152, 2010, 021302.
72. Joanna Brunker, and Paul Beard, Acoustic resolution photoacoustic Doppler velocimetry in bloodmimicking fluids, Scientific Reports, Vol. 6, 20902, 2016.
73. Laser Institute of America, American National Standard for Safe Use of Lasers Outdoors, American National Standards Institute, Inc., Approved, 2015.
74. Junjie Yao, and Lihong V. Wang, Photoacoustic microscopy, Laser Photonics Reviews, Vol. 7, No. 5, 2013, pp. 758–778.

75. Jin Woo Baik, Jin Young Kim, Seonghee Cho, Seongwook Choi, Jongbeom Kim, and Chulhong Kim, Super Wide-field photoacoustic microscopy of animals and humans in vivo, IEEE Transactions on Medical Imaging, Vol. 39, No. 4, 2020, pp. 975–983.
76. Junjie Yao, Chih-Hsien Huang, Lidai Wang, Joon-Mo Yang, Liang Gao, Konstantin I. Maslov, Jun Zou, Lihong V. Wang, Wide-field fast-scanning photoacoustic microscopy based on a water-immersible MEMS scanning mirror, Journal of Biomedical Optics, Vol. 17, No. 8, 2012.
77. Zhixing Xie, Shuliang Jiao, Hao F. Zhang, and Carmen A. Puliafito, Laser-scanning optical-resolution photoacoustic microscopy, Optics Letters, Vol. 34, No. 12, 2009, pp. 1771–1773.
78. Bangxin Lan, Wei Liu, Ya-Chao Wang, Junhui Shi, Yang Li, Song Xu, Huaxin Sheng, Qifa Zhou, Jun Zou, Ulrike Hoffmann, Wei Yang, and Junjie Yao, High-speed widefield photoacoustic microscopy of small-animal hemodynamics, Biomedical Optics Express, Vol. 9, No. 10, 2018, pp. 4689–4701.
79. Dipanjan Pan, Benjamin Kim, Lihong V. Wang, and Gregory M Lanza, A brief account of nanoparticle contrast agents for photoacoustic imaging, Wiley Interdisciplinary Reviews: Nanomedicine and Nanobiotechnology, Vol. 5, 2013, pp. 517–543.
80. Jason R. Swedlow, and Melpomeni Platani, Live cell imaging using wide-field microscopy and deconvolution, Cell Structure and Function, Vol. 27, 2002, pp. 335–341.
81. Murat Yildirim, Hiroki Sugihara, Peter T.C. So, and Mriganka Sur, Functional imaging of visual cortical layers and subplate in awake mice with optimized three photon microscopy, Nature Communications, Vol. 10, No. 177, 2019.
82. Lihong V. Wang, and Hsin-i Wu, Biomedical optics principles and imaging, WILEY-INTERSCIENCE, A John Wiley & Sons, Inc., Publication, 2007.
83. Wai-Fung Cheong, Scott A. Prahl, and Ashley J. Welch, A Review of the Optical Properties of Biological Tissues, IEEE Journal of Quantum Electronics, Vol. 26, No. 12, 1990, pp. 2166–2185.
84. G. Hong, A. L. Antaris, and H. Dai, Near-infrared fluorophores for biomedical imaging, Nature biomedical engineering, Vol. 1, No. 1, 2017, 0010.
85. Junjie Yao, Lihong V. Wang, Sensitivity of photoacoustic microscopy, Photoacoustics, Vol. 2, Issue 2, 2014, pp. 87–101. https://doi.org/10.1016/j.pacs.2014.04.002.
86. S. Jacques and S. Prahl, Oregon medical laser center, September 2018, https://omlc.org /spectra/index.html.
87. Luís Manuel Couto Oliveira, Valery Victorovich Tuchin, The optical clearing method-a new tool for clinical practice and biomedical engineering, 2019, https://doi.org/10.1007/978-3-030-33055-2.
88. G. Richard Kelman, Digital computer subroutine for the conversion of oxygen tension into saturation, Journal of applied physiology, Vol. 21, No. 4, 1966, pp. 1375-6.
89. Yu Wang, Song Hu, Konstantin Maslov, Yu Zhang, Younan Xia, and Lihong V. Wang, In vivo integrated photoacoustic and confocal microscopy of hemoglobin oxygen saturation and oxygen partial pressure, Optics Letters, Vol. 36, No. 7, 2011, pp. 1029–1031.
90. Dirk J. Faber, Maurice C. G. Aalders, Egbert G. Mik, Brett A. Hooper, Martin J. C. van Gemert, and Ton G. van Leeuwen, Oxygen saturation-dependent absorption and scattering, Physical Review Letters, Vol. 93, No. 2, 2004.
91. R. L. Longini, and R. Zdrojkowski, A Note on the Theory of Backscattering of Light by Living Tissue, IEEE Transactions on Bio-Medical Engineering, Vol. BME-15, No. 1, 1968.
92. Michael S. Patterson, Brain C. Wilson, Douglas R. Wyman, The propagation of optical radiation in tissue II: optical properties of tissues and resulting fluence distributions, Lasers in Medical Science, Vol. 6, 1991, pp. 379–390.
93. Valery Tuchin, Light Scattering methods and instruments for medical diagnosis, SPIE Optical Engineering Press, 2000. https://doi.org/10.1117/3.1003040.
94. MD Stern, In vivo evaluation of microcirculation by coherent light scattering, Nature, Vol. 254, 1975, pp. 56–58.
95. Tobias Meinert, Olaf Tietz, Klaus J. Palme, and Alexander Rohrbach, Separation of ballistic and diffusive fluorescence photons in confocal light-sheet microscopy of arabidopsis roots, Scientific Reports, Vol. 6, 30378, 2016.

96. Yifeng Zhou, Kenny K. H. Chan, Tom Lai, and Shuo Tang, Characterizing refractive index and thickness of biological tissues using combined multiphoton microscopy and optical coherence tomography, Biomedical Optics Express, Vol. 4, No. 1, 2013, pp. 38–50.
97. Joris J. J. Dirckx, Liesbeth C. Kuypers, Willem F. Decraemer, Refractive index of tissue measured with confocal microscopy, Vol. 104, 2005.
98. P. Y. Liu, L. K. Chin, W. Ser, H. F. Chen, C.-M. Hsieh, C.-H. Lee, K.-B. Sung, T. C. Ayi, P. H. Yap, B. Liedberg, K. Wang, T. Bourouina, and Y. Leprince-Wang, Cell refractive index for cell biology and disease diagnosis: past, present and future, Lab Chip, Vol. 16, 2016, pp. 634–644.
99. Rudolf Steiner, Laser-tissue interactions, laser and IPL technology in dermatology and aesthetic medicine, 2011.
100. A. Edward Profio, Light transport in tissue, Applied Optics, Vol. 28, No. 12, 1989, pp. 2216–2222.
101. Calasso IG, Craig W, Diebold GJ. Photoacoustic point source. Phys Rev Lett, 2001; 86(16):3550–3.
102. Tanja Tarvainen, Computational methods for light transport in optical tomography, Department of Physics University of Kuopio, 29th September 2006.
103. Adam Liebert, Heidrun Wabnitz, Dirk Grosenick, Michael Möller, Rainer Macdonald, and Herbert Rinneberg, Evaluation of optical properties of highly scattering media by moments of distributions of times of flight of photons, Applied Optics, Vol. 42, No. 28, 2003, pp. 5785–5792.
104. V. Venugopalan, J. S. You, and B. J. Tromberg, Radiative transport in the diffusion approximation: An extension for highly absorbing media and small source-detector separations, Physical Review E, Vol. 58, No. 2, 1998, pp. 2395–2407.
105. B. T. Cox, and P. C. Beard, Fast calculation of pulsed photoacoustic fields in fluids using k-space methods, Journal of the Acoustical Society of America, Vol. 117, No. 6, 2005, pp. 3616–3627.
106. B. T. Cox, J. Laufer, K. Köstli and P. Beard, Experimental validation of photoacoustic k-space propagation models, Proc. SPIE 5320, pp 238–248, 2004.
107. B. T. Cox, S. Kara, S. R. Arridge, and P. C. Beard, k-space propagation models for acoustically heterogeneous media: Application to biomedical photoacoustics, Journal of the Acoustical Society of America, Vol. 121, No. 6, 2007, pp. 3453–3464.
108. Wei Liu, Daria M. Shcherbakova, Neel Kurupassery, Yang Li, Qifa Zhou, Vladislav V. Verkhusha, and JunjieYao, Quad-mode functional and molecular photoacoustic microscopy, Scientific Reports, Vol. 8, 11123, 2018.
109. Seungwan Jeon, Jongbeom Kim, Donghyun Lee, Jin Woo Baik, Chulhong Kim, Review on practical photoacoustic microscopy, Photoacoustics, Vol. 15, 100141, 2019.
110. R.L. Shelton, B.E. Applegate, Off-axis photoacoustic microscopy, IEEE Transactions on Biomedical Engineering, Vol. 57, No. 8, 2010, pp. 1835–1838.
111. S. Jiao, Z. Xie, H.F. Zhang, C.A. Puliafito, Simultaneous multimodal imaging with integrated photoacoustic microscopy and optical coherence tomography, Optics Letters, Vol. 34, No. 19, 2009, pp. 2961–2963.
112. Liang Song, Konstantin Maslov, Rachel Bitton, K. Kirk Shung, and Lihong V. Wang, Fast 3D dark-field reflection-mode photoacoustic microscopy in vivo with a 30-MHz ultrasound linear array, Journal of biomedical optics, Vol. 13, No. 5, 2008, 054028.
113. K. Maslov, G. Stoica, L.V. Wang, In vivo dark-field reflection-mode photoacoustic microscopy, Optics Letters, Vol. 30, No. 6, 2005, pp. 625–627.
114. K. Maslov, H.F. Zhang, S. Hu, L.V. Wang, Optical-resolution photoacoustic microscopy for in vivo imaging of single capillaries, Optics Letters, Vol. 33, No. 9, 2008, pp. 929–931.
115. S. Hu, K. Maslov, L.V. Wang, Second-generation optical-resolution photoacoustic microscopy with improved sensitivity and speed, Optics Letters, Vol. 36, No. 7, 2011, pp. 1134–1136.
116. Xiaoyi Zhu, Qiang Huang, Anthony DiSpirito, Tri Vu, Qiangzhou Rong, Xiaorui Peng, Huaxin Sheng, Xiling Shen, Qifa Zhou, Laiming Jiang, Ulrike Hoffmann , and Junjie Yao,

Real-time whole-brain imaging of hemodynamics and oxygenation at micro-vessel resolution with ultrafast wide-field photoacoustic microscopy, Light: Science & Applications, Vol. 11, 138, 2022.
117. Chi Zhang, Konstantin Maslov, and Lihong V. Wang, Subwavelength-resolution label-free photoacoustic microscopy of optical absorption in vivo, Optics Letters, Vol. 35, 2010, pp. 3195–3197.
118. Rui Cao, Scott D. Nelson, Samuel Davis, Yu Liang, Yilin Luo, Yide Zhang, Brooke Crawford, and Lihong V. Wang, Label-free intraoperative histology of bone tissue via deep-learning-assisted ultraviolet photoacoustic microscopy, Nature Biomedical Engineering, Vol. 7, 2023, pp. 124–134.
119. Hui Wang, Xiaoquan Yang, Yanyan Liu, Bowen Jiang, and Qingming Luo, Reflection-mode optical-resolution photoacoustic microscopy based on a reflective objective, Optics Express, Vol. 21, 2013, pp. 24210–24218.
120. Rui Cao, Joseph P. Kilroy, Bo Ning, Tianxiong Wang, John A. Hossack, Song Hu, Multispectral photoacoustic microscopy based on an optical–acoustic objective, Photoacoustics, Vol. 3, No. 2, 2015, pp. 55–59.
121. Lee H, Choi W, Kim C, Park B, Kim J. Review on ultrasound-guided photoacoustic imaging for complementary analyses of biological systems in vivo, Experimental Biology and Medicine. Vol. 248, No. 9, 2023, pp. 762–774.
122. Rui Cao, Jingjing Zhao, Lei Li, Lin Du, Yide Zhang, Yilin Luo, Laiming Jiang, Samuel Davis, Qifa Zhou, Adam de la Zerda, and Lihong V. Wang, Optical-resolution photoacoustic microscopy with a needle-shaped beam, Nature Photonics, Vol. 17, 2023, pp. 89–95.
123. Junjie Yao, Lidai Wang, Chiye Li, Chi Zhang, and Lihong V. Wang, Photo-imprint photoacoustic microscopy for three-dimensional label-free sub-diffraction imaging, Physical Review Letters, Vol. 112, No. 1, 014302, 2014.
124. Junhui Shi, Yuqi Tang, Junjie Yao, Advances in super-resolution photoacoustic imaging, Quant Imaging Med Surg, Vol. 8, No. 8, 2018, pp. 724–732.
125. Junjie Yao, Andrii A. Kaberniuk, Lei Li, Daria M. Shcherbakova, Ruiying Zhang, Lidai Wang, Guo Li, Vladislav V. Verkhusha, and Lihong V. Wang, Multi-scale photoacoustic tomography using reversibly switchable bacterial phytochrome as a near-infrared photochromic probe, Nat Methods., Vol. 13, No. 1, 2016, pp. 67–73.
126. Lidai Wang, Chi Zhang, and Lihong V. Wang, Grueneisen Relaxation Photoacoustic Microscopy, Physical Review Letters, Vol. 113, 174301, 2014.
127. Xiaowei Liu, Terence T. W. Wong, Junhui Shi, Jun Ma, Qing Yang, and Lihong V. Wang, Label-free cell nuclear imaging by Grüneisen relaxation photoacoustic microscopy, Optics Letters, Vol. 43, No. 4, 2018, pp. 947–950.
128. Junhui Shi, Lidai Wang, Cedric Noordam, Lihong V. Wang, Bessel-beam Grueneisen relaxation photoacoustic microscopy with extended depth of field, Journal of Biomedical Optics, Vol. 20, No. 11, 116002, 2015.
129. De Cai, Terence T. W. Wong, Liren Zhu, Junhui Shi, Sung-Liang Chen, and Lihong Wang, Dual-view photoacoustic microscopy for quantitative cell nuclear imaging, Optics Letters, Vol. 43, No. 20, 2018, pp. 4875–4878.
130. Soon-Woo Cho, Sang Min Park, Byullee Park, Do Yeon Kim, Tae Geol Lee, Beop-Min Kim, Chulhong Kim, Jeesu Kim, Sang-Won Lee, Chang-Seok Kim, High-speed photoacoustic microscopy: A review dedicated on light sources, Photoacoustics, Vol. 24, 2021, 100291, https://doi.org/10.1016/j.pacs.2021.100291.
131. Ikesue, A., Aung, Y. Ceramic laser materials. Nature Photon., Vol. 2, 2008, pp. 721–727.
132. ZW Fan, JS Qiu, ZJ Kang, YZ Chen, WQ Ge, XX Tang, High beam quality 5 J, 200 Hz Nd:YAG laser system, Light: Science & Applications, Vol. 6, 2017, e17004.
133. Hammond, P., Hughes, R. Hammond, P., Hughes, R. Search for wide-range tunable dye laser systems, Nature Physical Science, Vol. 231, 1971, Vol. 59–60.
134. Berns, M., Partial cell irradiation with a tunable organic dye laser, Nature, Vol. 240, 1972, pp. 483–485.

135. Alexander J. C. Kuehne, Malte C. Gather, Irwin A. Eydelnant, Seok-Hyun Yun, David A. Weitz, and Aaron R. Wheeler, Lab Chip, Vol. 11, 2011, pp. 3716–3719.
136. A. Costela, I. García-Morenoa, and R. Sastreb, Polymeric solid-state dye lasers: Recent developments, Physical Chemistry Chemical Physics, Vol. 5, 2003, pp. 4745–4763.
137. Herink, G., Jalali, B., Ropers, C., D. R. Solli, Resolving the build-up of femtosecond mode-locking with single-shot spectroscopy at 90 MHz frame rate, Nature Photon, Vol. 10, 2016, pp. 321–326.
138. Ursula Keller, U. Recent developments in compact ultrafast lasers, Nature, Vol. 424, 2003, pp. 831–838.
139. F. Wang, A. G. Rozhin, V. Scardaci, Z. Sun, F. Hennrich, I. H. White, W. I. Milne, and A. C. Ferrari, Wideband-tunable, nanotube mode-locked, fibre laser, Nature Nanotechnology, Vol. 3, 2008, pp. 738–742.
140. Jackson, S. Towards high-power mid-infrared emission from a fibre laser, Nature Photonics, Vol. 6, 2012, pp. 423–431.
141. Hirose, K., Liang, Y., Kurosaka, Y. et al. Watt-class high-power, high-beam-quality photonic-crystal lasers, Nature Photonics, Vol. 8, 2014, pp. 406–411.
142. Michael Kues, Christian Reimer, Benjamin Wetzel, Piotr Roztocki, Brent E. Little, Sai T. Chu, Tobias Hansson, Evgeny A. Viktorov, David J. Moss, and Roberto Morandotti, Passively mode-locked laser with an ultra-narrow spectral width, Nature Photonics, Vol. 11, 2017, pp. 159–162.
143. Xu, C., Wise, F. Recent advances in fibre lasers for nonlinear microscopy, Nature Photonics, Vol. 7, 2013, pp. 875–882.
144. Sergey Kobtsev, Sergey Kukarin, Sergey Smirnov, Sergey Turitsyn, and Anton Latkin, Generation of double-scale femto/pico-second optical lumps in mode-locked fiber lasers, Optics Express, Vol. 17, No. 23, 2009, pp. 20707–20713.
145. David Steinberg, Juan Diego Zapata, Eunezio A. Thoroh de Souza, and Lúcia A. M. Saito, Mechanically exfoliated graphite onto D-shaped optical fiber for femtosecond mode-locked Erbium-doped fiber laser, Journal of Lightwave Technology, Vol. 36, No. 10, 2018, pp. 1868–1874.
146. Harumasa Yoshida, Yoji Yamashita, Masakazu Kuwabara, and Hirofumi Kan, A 342-nm ultraviolet AlGaN multiple-quantum-well laser diode. Nature Photonics, Vol. 2, 2008, pp. 551–554.
147. Sciamanna, M., Shore, K. Physics and applications of laser diode chaos, Nature Photonics, Vol. 9, 2015, pp. 151–162.
148. Yunhao Zhu, Guan Xu, Jie Yuan, Janggun Jo, Girish Gandikota, Hakan Demirci, Toshitaka Agano, Naoto Sato, Yusuke Shigeta, and Xueding Wang, Light Emitting Diodes based Photoacoustic Imaging and Potential Clinical Applications, Scientific Reports, Vol. 8, No. 9885, 2018.
149. Lvming Zeng, Guodong Liu, Diwu Yang, and Xuanrong Ji, Portable optical-resolution photoacoustic microscopy with a pulsed laser diode excitation, Applied Physics Letters, Vol. 102, 053704, 2013.
150. Ekaterina I. Galanzha, Evgeny V. Shashkov, Paul M. Spring, James Y. Suen and Vladimir P. Zharov, In vivo, noninvasive, label-free detection and eradication of circulating metastatic melanoma cells using two-color photoacoustic flow cytometry with a diode laser, Cancer Research, Vol. 69, No. 20, 2009, pp. 7926-7934.
151. Chun-Cai Hou, Hong-Mei Chen, Jin-Chuan Zhang, Ning Zhuo, Yuan-Qing Huang, Richard A Hogg, David TD Childs, Ji-Qiang Ning, Zhan-Guo Wang, Feng-Qi Liu, and Zi-Yang Zhang, Near-infrared and mid-infrared semiconductor broadband light emitters. Light Sci Appl., Vol. 7, 2018, 17170.
152. Shaoqiang Chen, Takashi Ito, Akifumi Asahara, Masahiro Yoshita, Wenjie Liu, Jiangyong Zhang, Baoping Zhang, Tohru Suemoto, and Hidefumi Akiyama, Spectral dynamics of picosecond gain-switched pulses from nitride-based vertical-cavity surface-emitting lasers, Scientific Reports, Vol. 4, 4325, 2014.

153. Takashi Ito, Hidekazu Nakamae, Yuji Hazama, Takahiro Nakamura, Shaoqiang Chen, Masahiro Yoshita, Changsu Kim , Yohei Kobayashi, and Hidefumi Akiyama, Femtosecond pulse generation beyond photon lifetime limit in gain-switched semiconductor lasers, Communications Physics, Vol. 1, 42, 2018.
154. Clemens Grunsteidl, Thomas Berer, Mike Hettich, and Istvan Veres, Determination of thickness and bulk sound velocities of isotropic plates using zero-group-velocity Lamb waves, Applied Physics Letters, Vol. 112, 251905, 2018.
155. T. J. Allen, and P. C. Beard, Pulsed near-infrared laser diode excitation system for biomedical photoacoustic imaging, Optics Letters, Vol. 31, 2006, pp. 3462–3464.
156. T. W. Murray and O. Balogun, High-sensitivity laser-based acoustic microscopy using a modulated excitation source, Applied Physics Letters, Vol. 85, 2004, pp. 2974–2476.
157. Y. Fan, A. Mandelis, G. Spirou, and I. A. Vitkin, Development of a laser photothermoacoustic frequency-swept system for subsurface imaging: theory and experiment, Journal of the Acoustical Society of America, Vol. 116, 2004, pp. 3523–3533.
158. Konstantin Maslov, Lihong V. Wang, Photoacoustic imaging of biological tissue with intensity-modulated continuous-wave laser, Journal of Biomedical Optics, Vol. 132, 2008, 024006.
159. Craig K. Abbey, Nghia Q. Nguyen, and Michael F. Insana, Effects of frequency and bandwidth on diagnostic information transfer in ultrasonic B-mode imaging, IEEE Transactions on Uitrasonics, Ferroelectrics, and Frequency control, Vol. 59, No. 6, 2012, pp. 1115–1126.
160. Stephan Kellnberger, Dominik Soliman, George J. Tserevelakis, Markus Seeger, Hong Yang, Angelos Karlas, Ludwig Prade, Murad Omar, and Vasilis Ntziachristos, Optoacoustic microscopy at multiple discrete frequencies, Light: Science & Applications, Vol. 7, No. 109, 2018.
161. Georg Wissmeyer, Miguel A. Pleitez, Amir Rosenthal, and Vasilis Ntziachristos, Looking at sound: optoacoustics with all-optical ultrasound detection, Light: Science & Applications, Vol. 7, No. 53, 2018.
162. Stefan Preisser, Wolfgang Rohringer, Mengyang Liu, Christian Kollmann, Stefan Zotter, Balthsar Fischer, and Wolfgang Drexler, All-optical highly sensitive akinetic sensor for ultrasound detection and photoacoustic imaging, Biomedical Optics Express, Vol. 7, No. 10, 2016, pp. 4171–4186.
163. Biqin Dong, Cheng Sun, and Hao F. Zhang, Optical detection of ultrasound in photoacoustic imaging, IEEE Transactions on Biomedical Engineering, Vol. 64, No. 1, 2017.
164. James A. Guggenheim, Jing Li, Thomas J. Allen, Richard J. Colchester, Sacha Noimark, Olumide Ogunlade, Ivan P. Parkin, Ioannis Papakonstantinou, Adrien E. Desjardins, Edward Z. Zhang and Paul C. Beard, Ultrasensitive plano-concave optical microresonators for ultrasound sensing, Nature Photonics, Vol. 11, 2017, pp. 714–719.
165. Thomas Berer, Istvan A. Veres, Hubert Grun, Johannes Bauer-Marschallinger, Karoline Felbermayer, and Peter Burgholzer, Characterization of broadband fiber optic line detectors for photoacoustic tomography, Journal of Biophotonics, Vol. 5, No. 7, 2012, pp. 518–527.
166. Chao Tian, Mengliu Pei, Kang Shen, Songde Liu, Zhiming Hu, and Ting Feng, Impact of system factors on the performance of photoacoustic tomography scanners, Physical Review Applied, Vol. 13, 2020.
167. Minghua Xu and Lihong V. Wang, Analytic explanation of spatial resolution related to bandwidth and detector aperture size in thermoacoustic or photoacoustic reconstruction, Physical Review E, Vol. 67, 2003, 056605.
168. Kun Wang, Richard Su, Alexander A Oraevsky, and Mark A Anastasio, Investigation of iterative image reconstruction in three-dimensional optoacoustic tomography, Physics in Medicine & Biology, Vol. 57, 2012, pp. 5399–5423.
169. Chao Huang, Kun Wang, Liming Nie, Lihong V. Wang, and Mark A. Anastasio, Full-wave iterative image reconstruction in photoacoustic tomography with acoustically inhomogeneous media, IEEE Transactions on Medical Imaging, Vol. 32, No. 6, 2013, pp. 1097–1109.
170. Chenhong Lang, Jian Fang, Hao Shao, Xin Ding, and Tong Lin, High-sensitivity acoustic sensors from nanofiber webs, Nature Communications, Vol. 7, 2016.

171. Qiang Zheng, Bojing Shi, Zhou Li, and Zhong Lin Wang, Recent progress on piezoelectric and triboelectric energy harvesters in biomedical systems, Advanced Science, Vol. 4, 2017, 1700029.
172. F. Stuart Foster, Linda K. Ryan and Daniel H. Turnbull, Characterization of lead zirconate titanate ceramics for use in miniature high-frequency (20-80 MHz) transducers, IEEE Transactions on Ultrasonics. Ferroelectrics, and Frequekcy Control, Vol. 3X. No. S, 1991, pp. 446–453.
173. Simon Fichtner, Niklas Wolff, Fabian Lofink, Lorenz Kienle, and Bernhard Wagner, AlScN: A III-V semiconductor based ferroelectric, Journal of Applied Physics, Vol. 125, 114103, 2019.
174. Ram Mohan Verma, Ashwath Rao, and B. R. Singh, Electrical characterization of the metal ferroelectric oxide semiconductor and metal ferroelectric nitride semiconductor gate stacks for ferroelectric field effect transistors, Applied Physics Letters, Vol. 104, 092907, 2014.
175. Chaojie Cui, Fei Xue, Wei-Jin Hu, and Lain-Jong Li, Two-dimensional materials with piezoelectric and ferroelectric functionalities, npj 2D Materials and Applications, Vol. 2, No. 18, 2018.
176. Igal Ladabaum, Xuecheng Jin, Hyongsok T. Soh, Abdullah Atalar, and Butrus T. Khuri-Yakub, Surface micromachined capacitive ultrasonic transducers, IEEE transactions on ultrasonics, ferroelectrics, and frequency control, Vol. 45, No. 3, 1998, pp. 678–690.
177. Huijun Ding, Xiaolan Shu, Yukun Jin, Taojian Fan, and Han Zhang, Recent advances in nanomaterial-enabled acoustic devices for audible sound generation and detection, Nanoscale, Vol. 11, 2019, pp. 5839–5860.
178. Kai Zhang, Chu Ma, Qi He, Shaoting Lin, Yi Chen, Yu Zhang, Nicholas X. Fang,and Xuanhe Zhao, Metagel with broadband tunable acoustic properties over air–water–solid ranges, Advanced Functional Materials, Vol. 29, 1903699, 2019.
179. Jin Young Kim, Changho Lee, Kyungjin Park, Sangyeob Han, and Chulhong Kim, High-speed and high-SNR photoacoustic microscopy based on a galvanometer mirror in non-conducting liquid, Scientific Reports, Vol. 6, 34803, 2016.
180. Hao Li, Biqin Dong, Zhen Zhang, Hao F. Zhang, and Cheng Sun, A transparent broadband ultrasonic detector based on an optical micro-ring resonator for photoacoustic microscopy, Scientific Reports, Vol. 4, 4496, 2014.
181. Jiapu Li, Yuqing Ma, Tao Zhang, K. Kirk Shung, and Benpeng Zhu, Recent Advancements in Ultrasound Transducer: From Material Strategies to Biomedical Applications, BME Frontiers, Vol. 2022, Article ID 976450, 2022.
182. Jianbo Tang, Xianjin Dai, and Huabei Jiang, Wearable scanning photoacoustic brain imaging in behaving rats, Journal of biophotonics, Vol. 9, No. 6, 2016, pp. 571–575.
183. Joon-Mo Yang, Christopher Favazza, Ruimin Chen, Junjie Yao, Xin Cai, Konstantin Maslov, Qifa Zhou, K Kirk Shung, and Lihong V Wang, Simultaneous functional photoacoustic and ultrasonic endoscopy of internal organs in vivo, Nature Medicine, Vol. 18, 2012, pp. 1297–1302.
184. Jaeyul Lee, Sangyeob Han, Daewoon Seong, Junsoo Lee, Sungjo Park, Ruchire Eranga Wijesinghe, Mansik Jeon, and Jeehyun Kim, Fully waterproof two-axis galvanometer scanner for enhanced wide-field optical-resolution photoacoustic microscopy, Optics Letters, Vol. 45, No. 4, 2020.
185. Fan Zheng, Xiangyang Zhang, Chi Tat Chiu, Bill L. Zhou, K. Kirk Shung, Hao F. Zhang, and Shuliang Jiao, Laser-scanning photoacoustic microscopy with ultrasonic phased array transducer, Biomedical Optics Express, Vol. 3, No. 11, 2012, pp. 2694–2699.
186. Maryam Basij, Yan Yan, Suhail S. Alshahrani, Hamid Helmi, Timothy K. Burton, Jay W. Burmeister, Michael M. Dominello, Ira S. Winer, Mohammad Mehrmohammadi, Miniaturized phased-array ultrasound and photoacoustic endoscopic imaging system, Photoacoustics, Vol. 15, 2019, 100139.
187. Roger J. Zemp, Liang Song, Rachel Bitton, K. Kirk Shung, and Lihong V. Wang, Realtime photoacoustic microscopy in vivo with a 30-MHz ultrasound array transducer, Optics Express, Vol. 16, No. 11, 2008, pp. 7915–7928.

188. Jie Deng, Yingxiang Liu, Junkao Liu, Dongmei Xu, and Yun Wang, Development of a planar piezoelectric actuator using bending–bending hybrid transducers, IEEE Transactions on Industrial Electronics, Vol. 66, No. 8, 2019, pp. 6141–6149.
189. C. Fang, H. Hu and J. Zou, A focused optically transparent PVDF transducer for photoacoustic microscopy, IEEE Sensors Journal, Vol. 20, No. 5, 2020, pp. 2313–2319.
190. Graham W. J. Brodie, YongQiang Qiu, Sandy Cochran, Gabriel C Spalding, and Michael P. Macdonald, Optically transparent piezoelectric transducer for ultrasonic particle manipulation, IEEE Transactions on Ultrasonics, Ferroelectrics, and Frequency control, Vol. 61, No. 3, 2014, pp. 389–391.
191. Eric Y. Zhu, Cory Rewcastle, Raanan Gad, Li Qian, Ofer Levi, Refractive-Index-based ultrasound sensing with photonic crystal slabs, Optics Letters, Vol. 44, No. 10, 2019, pp. 2609–2612.
192. Xiaoyi Zhu, Zhiyu Huang, Ziyuan Li, Wenzhao Li, Xi Liu, Zhaolong Chen, Jian Tian, and Changhui Li, Resolution-matched reflection mode photoacoustic microscopy and optical coherence tomography dual modality system, Photoacoustics, Vol. 19, 100188, 2020.
193. Amit P. Jathoul, Jan Laufer, Olumide Ogunlade, Bradley Treeby, Ben Cox, Edward Zhang, Peter Johnson, Arnold R. Pizzey, Brian Philip, Teresa Marafioti, Mark F. Lythgoe, R. Barbara Pedley, Martin A. Pule, and Paul Beard, Deep in vivo photoacoustic imaging of mammalian tissues using a tyrosinase-based genetic reporter, Nature Photonics, Vol. 9, 2015.
194. Parsin Haji REZA, Kevan Bell, Wei Shi, James Shapiro, and Roger J. Zemp, Deep non-contact photoacoustic initial pressure imaging, Optica, Vol. 5, No. 7, 2018, pp. 814–820.
195. Saad Abbasi, Martin Le, Bazil Sonier, Deepak Dinakaran, Gilbert Bigras, Kevan Bell,John R. Mackey, and Parsin Haji Reza, Scientific Reports, Vol. 9, 13392, 2019.
196. Parsin Hajireza, Wei Shi, Kevan L. Bell, Robert J Paproski, and Roger J Zemp, Non-interferometric photoacoustic remote sensing microscopy, Light: Science & Applications, Vol. 6, 2017.
197. Kevan L. Bell, Parsin Hajireza, and Roger J Zemp, Coherence-gated photoacoustic remote sensing microscopy, Optics Express, Vol. 26, No. 18, 2018, pp. 23689–23704.
198. Heming Wei, and Sridhar Krishnaswamy, Polymer micro-ring resonator integrated with a fiber ring laser for ultrasound detection, Optics Letters, Vol. 42, No. 13, 2017, pp. 2655–2658.
199. Tao Ling, Sung-Liang Chen, and L. Jay Guo, Fabrication and characterization of high Q polymer micro-ring resonator and its application as a sensitive ultrasonic detector, Optics Express, Vol. 19, No. 2, 2011, pp. 861–869.
200. Biqin Dong, Hao Li, Zhen Zhang, Kevin Zhang, Siyu Chen, Cheng Sun, and Hao F. Zhang, Isometric multimodal photoacoustic microscopy based on optically transparent micro-ring ultrasonic detection, Optica, Vol. 2, No. 2, 2015, pp. 169–176.
201. S.M. Leinders, W.J. Westerveld, J. Pozo, P.L.M.J. van Neer, B. Snyder, P. O'Brien, H.P. Urbach, N. de Jong, and M.D. Verweij, A sensitive optical micro-machined ultrasound sensor (OMUS) based on a silicon photonic ring resonator on an acoustical membrane, Scientific Reports, Vol. 5, No. 14328, 2015.
202. Li, H., Dong, B., Zhang, Z., Zhang, H. F. & Sun, C. A transparent broadband ultrasonic detector based on an optical micro-ring resonator for photoacoustic microscopy, Scientific Reports, Vol. 4, No. 4496, 2014.
203. Jung-Ryul Lee, and Hiroshi Tsuda, Fiber optic liquid leak detection technique with an ultrasonic actuator and a fiber Bragg grating, Optics Letters, Vol. 30, No. 24, 2005, pp. 3293–3295.
204. Tongqing Liu and Ming Han, Analysis of π-phase-shifted fiber Bragg gratings for ultrasonic detection, IEEE Sensors Journal, Vol. 12, No. 7, 2012, pp. 2368–2372.
205. Amir Rosenthal, Daniel Razansky, and Vasilis Ntziachristos, High-sensitivity compact ultrasonic detector based on a pi-phase-shifted fiber Bragg grating, Optics Letters, Vol. 36, No. 10, 2011, 1833–1835.
206. Amir Rosenthal, Murad Omar, Héctor Estrada, Stephan Kellnberger, Daniel Razansky, and Vasilis Ntziachristos, Embedded ultrasound sensor in a silicon-on-insulator photonic platform, Applied Physics Letters, Vol. 104, 021116, 2014.

207. Tsesses, S., Aronovich, D., Grinberg, A., Hahamovich, E. & Rosenthal, A. Modeling the sensitivity dependence of silicon-photonics-based ultrasound detectors, Optics Letters, Vol. 42, 2017, pp. 5262–5265.
208. Jingshan Jia, Yi Jiang, Liuchao Zhang, Hongchun Gao, and Lan Jiang, Fiber optic dual-ring Michelson interferometer-Based Detection Scheme for the Measurement of Dynamic Signals, Journal of Lightwave Technology, Vol. 37, No. 15, 2019, pp. 3750–3755.
209. Rory W. Speirs and Alexis I. Bishop, Photoacoustic tomography using a Michelson interferometer with quadrature phase detection, Applied Physics Letters, Vol. 103, 053501, 2013.
210. Boling Ouyang, Yanlu Li, Marten Kruidhof, Roland Horsten, Koen W. A. Van Dongen, and Jacob Caro, On-chip silicon Mach–Zehnder interferometer sensor for ultrasound detection, Optics Letters, Vol. 44, No. 8, 2019, pp. 1928–1931.
211. Horacio Lamela, Daniel Gallego, and Alexander Oraevsky, Optoacoustic imaging using fiber-optic interferometric sensors, Optics Letters, Vol. 34, No. 23, 2009, 3695–3697.
212. Paltauf, G., Nuster, R., Haltmeier, M. & Burgholzer, P. Photoacoustic tomography using a Mach-Zehnder interferometer as an acoustic line detector, Applied optics, Vol. 46, 2007, pp. 3352–3358.
213. Fei Gao, Xiaohua Feng, and Yuanjin Zheng, Micro-Doppler photoacoustic effect and sensing by ultrasound radar, IEEE Journal of Selected Toptics in Quantum Electronics, Vol. 22, No. 3, 2016.
214. Zhen Qiu and Wibool Piyawattanametha, MEMS-based medical endomicroscopes, IEEE Journal of Selected Topics in Quantum Electronics, Vol. 21, No. 4, 2015.
215. Xiaoyang Zhang, Sanjeev J. Koppal, Rui Zhang, Liang Zhou, Elizabeth Butler, and Huikai Xie, Wide-angle structured light with a scanning MEMS mirror in liquid, Optics Express, Vol. 24, No. 4, 2016, 3479–3487.
216. Jiamiao Yang, Lei Gong, Xiao Xu, Pengfei Hai, Yuecheng Shen, Yuta Suzuki, and Lihong V. Wang, Motionless volumetric photoacoustic microscopy with spatially invariant resolution, Nature Communications, Vol. 8, No. 780, 2017.
217. Terence T. W. Wong, Ruiying Zhang, Chi Zhang, Hsun-Chia Hsu, Konstantin I. Maslov, Lidai Wang, Junhui Shi, Ruimin Chen, K. Kirk Shung, Qifa Zhou, and Lihong V. Wang, Label-free automated three-dimensional imaging of whole organs by microtomy-assisted photoacoustic microscopy, Nature Communications, Vol. 8, No. 1386, 2017.
218. Seonghee Cho, Jinwoo Baik, Ravi Managuli, and Chulhong Kim, 3D PHOVIS: 3D photoacoustic visualization studio, Photoacoustics, Vol. 18, 2020.
219. Weijian Zong, Runlong Wu, Mingli Li, Yanhui Hu, Yijun Li, Jinghang Li, Hao Rong, Haitao Wu, Yangyang Xu, Yang Lu, Hongbo Jia, Ming Fan, Zhuan Zhou, Yunfeng Zhang, Aimin Wang, Liangyi Chen, and Heping Cheng, Fast high-resolution miniature two-photon microscopy for brain imaging in freely behaving mice, Nature Methods, Vol. 14, 2017, pp. 713–719.
220. Junjie Yao, Lidai Wang, Joon-Mo Yang, Konstantin I Maslov, Terence T W Wong, Lei Li, Chih-Hsien Huang, Jun Zou, and Lihong V Wang, high-speed label-free functional photoacoustic microscopy of mouse brain in action, Nature Methods, Vol.12, No.5, 2015, pp. 407–410.
221. Jin Young Kim, Changho Lee, Kyungjin Park, Geunbae Lim, and Chulhong Kim, Fast optical-resolution photoacoustic microscopy using a 2-axis water-proofing MEMS scanner, Scientific Reports, Vol. 5, No. 7932, 2015.
222. Parsin Hajireza, Wei Shi, and Roger J. Zemp, Real-time handheld optical-resolution photoacoustic microscopy, Optics Express, Vol. 19, No. 21, 2011, pp. 20097–20102.
223. Hao Yang, Lei Xi, Sean Samuelson, Huikai Xie, Lily Yang, and Huabei Jiang, Handheld miniature probe integrating diffuse optical tomography with photoacoustic imaging through a MEMS scanning mirror, Biomedical Optics Express, Vol. 4, No. 3, 2013, pp. 427–432.
224. Li Lin, Pengfei Zhang, Song Xu, Junhui Shi, Lei Li, Junjie Yao, Lidai Wang, Jun Zou, and Lihong V. Wang, Handheld optical-resolution photoacoustic microscopy, Journal of Biomedical Optics, Vol. 22, No. 4, 2017.

225. Kyungjin Park, Jin Young Kim, Changho Lee, Seungwan Jeon, Geunbae Lim, and Chulhong Kim, Handheld photoacoustic microscopy probe, Scientific Reports, Vol. 7, 13359, 2017.
226. Qian Chen, Heng Guo, Tian Jin, Weizhi Qi, Huikai Xie, and Lei Xi, Ultracompact high-resolution photoacoustic microscopy, Optics Letters, Vol. 43, Issue 7, pp. 1615–1618, 2018.
227. Seungwan Jeon, Jongbeom Kim, Donghyun Lee, Jin Woo Baik, Chulhong Kim, Review on practical photoacoustic microscopy, Photoacoustics, Vol. 15, 100141, 2019.
228. Wei Song, Wei Zheng, Ruimin Liu, Riqiang Lin, Hongtao Huang, Xiaojing Gong, Shousheng Yang, Rui Zhang, and Liang Song, Reflection-mode in vivo photoacoustic microscopy with subwavelength lateral resolution, Biomedical Optics Express, Vol. 5, No. 12, 2014, pp. 4235–4241.
229. Yi Yuan, Sihua Yang, and Da Xing, Optical-resolution photoacoustic microscopy based on two-dimensional scanning galvanometer, Applied Physics Letters, Vol. 100, 023702, 2012.
230. Wei Shi, Parsin Hajireza, Peng Shao, Alexander Forbrich, Roger J. Zemp, In vivo near-realtime volumetric opticalresolution photoacoustic microscopy using a high-repetition-rate nanosecond fiber-laser, Optics Express, Vol. 19, No. 18, 2011, pp. 17143–17150.
231. Jongbeom Kim, Jin Young Kim, Seungwan Jeon, Jin Woo BAIK, Seong Hee Cho and Chulhong Kim, Super-resolution localization photoacoustic microscopy using intrinsic red blood cells as contrast absorbers, Light: Science & Applications, Vol. 8, 2019.
232. Wei Qin, Tian Jin, Heng Guo, and Lei Xi, Large-field-of-view optical resolution photoacoustic microscopy, Optics Express, Vol. 26, No. 4, 2018, pp. 4271–4278.
233. T. Jin, H. Guo, H. Jiang, B. Ke, L. Xi, Portable optical resolution photoacoustic microscopy (pORPAM) for human oral imaging, Optics Letters, Vol. 42, No. 21, 2017, pp. 4434–4437.
234. Wei Song, Qiang Xu, Yang Zhang, Yang Zhan, Wei Zheng, and Liang Song, Fully integrated reflection-mode photoacoustic, two-photon, and second harmonic generation microscopy in vivo, Scientific Reports, Vol. 6, 32240, 2016.
235. Lei Li, Chenghung Yeh, Song Hu, Lidai Wang, Brian T. Soetikno, Ruimin Chen, Qifa Zhou, K. Kirk Shung, Konstantin I. Maslov, and Lihong V. Wang, Fully motorized optical-resolution photoacoustic microscopy, Optics Letters, Vol. 39, No. 7, 2014, pp. 2117–2120.
236. Thomas John Allen, Josh Spurrell, Martin O. Berendt, Olumide Ogunlade, Shaif U. Alam, Edward Z. Zhang, David J. Richardson, and Paul C. Beard, Ultrafast laser-scanning optical resolution photoacoustic microscopy at up to 2 million A-lines per second, Journal of Biomedical Optics, Vol. 23, No. 12, 126502, 2018.
237. JinYoung Kim, Changho Lee, Kyungjin Park, Sangyeob Han, and Chulhong Kim, High-speed and high-SNR photoacoustic microscopy based on a galvanometer mirror in nonconducting liquid, Scientific Reports, Vol. 6, 34803, 2016.
238. Bangxin Lan, Wei Liu, Ya-chao Wang, Junhui Shi, Yang Li, Song Xu, Huaxin Sheng, Qifa Zhou, Jun Zou, Ulrike Hoffmann, Wei Yang, and Junjie Yao, High-speed widefield photoacoustic microscopy of small-animal hemodynamics, Biomedical Optics Express, Vol. 9, No. 10, 2018, pp. 4689–4701.
239. Jiangbo Chena, Yachao Zhanga, Linyun He, Yizhi Liang, Lidai Wang, Wide-field polygon-scanning photoacoustic microscopy of oxygen saturationat 1-MHz A-line rate, Photoacoustics, Vol. 20, 100195, 2020.
240. Liang Song, Konstantin Maslov, and Lihong V. Wang, Multifocal optical-resolution photoacoustic microscopy in vivo, Optics Letters, Vol. 36, No. 7, 2011, pp. 1236–1238.
241. Toru Imai, Junhui Shi, Terence T. W. Wong, Lei Li, Liren Zhu, Lihong V. Wang, High-throughput ultraviolet photoacoustic microscopy with multifocal excitation, Journal of Biomedical Optics, Vol. 23, No. 3, 036007, 2018.
242. Guo Li, Konstantin I. Maslov, and Lihong V. Wang, Reflection-mode multifocal optical resolution photoacoustic microscopy, Journal of Biomedical Optics, Vol. 18, No. 3, 2013.
243. Jun Xia, Guo Li, Lidai Wang, Mohammadreza Nasiriavanaki, Konstantin Maslov, John A. Engelbach, Joel R. Garbow, and Lihong V. Wang, Wide-field two-dimensional multifocal optical-resolution photoacoustic-computed microscopy, Optics Letters, Vol. 38, No. 24, 2013, pp. 5236–5239.

244. Yang Li, Terence T. W. Wong, Junhui Shi, Hsun-Chia Hsu, and Lihong V. Wang, Multifocal photoacoustic microscopy using a single-element ultrasonic transducer through an ergodic relay, Light: Science & Applications, Vol. 9, 2020.
245. Pengfei Hai, Toru Imai, Song Xu, Ruiying Zhang, Rebecca L. Aft, Jun Zou, and Lihong V. Wang, High-throughput, label-free, single-cell photoacoustic microscopy of intratumoral metabolic heterogeneity, Nature Biomedical Engineering, Vol. 3, 2019, pp. 381–391.
246. Vakoc, B. J., Fukumura, D., Jain, R. K., and Bouma, B. E., Cancer imaging by optical coherence tomography: preclinical progress and clinical potential, Nature Reviews Cancer, Vol. 12, 2012, pp. 363–368.
247. Gao, L., Liang, J., Li, C., and Wang, L. V, Single-shot compressed ultrafast photography at one hundred billion frames per second, Nature, Vol. 516, 2014, pp. 74–77.
248. Cox, B. & Beard, P. C. Photoacoustic tomography with a single detector in a reverberant cavity, Journal of the Acoustical Society of America, Vol. 125, 2009, pp. 1426–1436.
249. Yuning Guo, Baowen Li, and Xiaobo Yin, Single-shot compressed photoacoustic tomographic imaging with a single detector in a scattering medium, Physical Review Applied, Vol. 13, 044009, 2020.
250. Yang Li, Lei Li, Liren Zhu, Junhui Shi, Konstantin Maslov, Lihong V. Wang, Photoacoustic topography through an ergodic relay for functional imaging and biometric application in vivo, Journal of Biomedical Optics, Vol. 25, No. 7, 2020.
251. Yang Li, Lei Li, Liren Zhu, Konstantin Maslov, Junhui Shi, Peng Hu, En Bo, Junjie Yao, Jinyang Liang, Lidai Wang, and Lihong V. Wang, Snapshot photoacoustic topography through an ergodic relay for high-throughput imaging of optical absorption, Nature Photonics, Vol. 14, 2020, pp. 164–170.
252. Nie LM, Chen M, Sun XL, Rong PF, Zheng NF, Chen XY. Palladium nanosheets as highly stable and effective contrast agents for in vivo photoacoustic molecular imaging, Nanoscale, Vol. 6, 2014, pp. 1271–6.
253. Homan KA, Souza M, Truby R, Luke GP, Green C, Vreeland E, Emelianov S. Silver nanoplate contrast agents for in vivo molecular photoacoustic imaging, ACS Nano, Vol. 6, 2012, pp. 641–50.
254. Kim C, Cho EC, Chen JY, Song KH, Au L, Favazza C, Zhang QA, Cobley CM, Gao F, Xia YN, Wang LHV. In vivo molecular photoacoustic tomography of melanomas targeted by bioconjugated gold nanocages, ACS Nano, Vol. 4, 2010, pp. 4559–64.
255. Li PC, Wang CRC, Shieh DB, Wei CW, Liao CK, Poe C, Jhan S, Ding AA, Wu YN. In vivo photoacoustic molecular imaging with simultaneous multiple selective targeting using antibody-conjugated gold nanorods, Optics Express, Vol 16, 2008, pp. 18605–15.
256. Nie LM, Wang SJ, Wang XY, Rong PF, Bhirde A, Ma Y, Liu G, Huang P, Lu GM, Chen XY. In vivo volumetric photoacoustic molecular angiography and therapeutic monitoring with targeted plasmonic nanostars, Small, Vol. 10, 2014, pp. 1585–93.
257. Pu KY, Shuhendler AJ, Jokerst JV, Mei JG, Gambhir SS, Bao ZN, Rao JH. Semiconducting polymer nanoparticles as photoacoustic molecular imaging probes in living mice. Nature Nanotechnology, Vol. 9, 2014, pp. 233–9.
258. Xin Cai., Li Li, Arie Krumholz, Zijian Guo, Todd N. Erpelding, Chi Zhang, Yu Zhang, Younan Xia, Lihong V. Wang, Multi-scale molecular photoacoustic tomography of gene expression, PLOS ONE, Vol. 7, No. 8, 2012.
259. Li Li, Roger J. Zemp, Gina Lungu, George Stoica, and Lihong V. Wang, Photoacoustic imaging of lacZ gene expression in vivo, Journal of Biomedical Optics, Vol. 122, 2007.
260. Xiaolei Cai, Xiaofeng Liu, Lun-De Liao, Aishwarya Bandla, Ji Min Ling, Yu-Hang Liu ,Nitish Thakor, Guillermo C. Bazan, and Bin Liu, Encapsulated conjugated oligomer nanoparticles for real-time photoacoustic sentinel Lymph node imaging and targeted photothermal therapy, Small, Vol. 12, No. 35, 2016, pp. 4873–4880.
261. Chulhong Kim, Eun Chul Cho, Jingyi Chen, Kwang Hyun Song, Leslie Au, Christopher Favazza, Qiang Zhang, Claire M. Cobley, Feng Gao, Younan Xia, and Lihong V. Wang, In vivo molecular photoacoustic tomography of melanomas targeted by bioconjugated gold nanocages, Vol. 4, No.8, 2010, pp.4559–4564.

262. Qiangzhou Rong, Carlos Taboada, Ángela del Águila, Ilaria Merutka, Nishad Jayasundara, Yushun Zeng, Wei Yang, Qifa Zhou, and Junjie Yao.From Ultraviolet to Near-Infrared: Label-Free Reflection-Mode Hyperspectral Photoacoustic Microscopy for Single-Cell Biochemical Mapping, GEN Biotechnology, 2023, pp. 435–444.
263. Da-Kang Yao, Ruimin Chen, Konstantin Maslov, Qifa Zhou, Lihong V. Wang, Optimal ultraviolet wavelength for in vivo photoacoustic imaging of cell nuclei, Journal of Biomedical Optics, Vol. 17, No. 5, 056004, 2012.
264. Chi Zhang, Yu Shrike Zhang, Da-Kang Yao, Younan Xia, Lihong V. Wang, Label-free photoacoustic microscopy of cytochromes, Journal of Biomedical Optics, Vol. 18, No. 2, 2013.
265. Chi Zhang, Ya-Jian Cheng, Junjie Chen, Samuel Wickline, Lihong V. Wang, Label-free photoacoustic microscopy of myocardial sheet architecture, Journal of Biomedical Optics, Vol. 17, No. 6, 2012.
266. Zhongjiang Chen, Sihua Yang, and Da Xing, In vivo detection of hemoglobin oxygen saturation and carboxyhemoglobin saturation with multiwavelength photoacoustic microscopy, Optics Letters, Vol. 37, No. 16, 2012, pp. 3414–3416.
267. Jacob Staley, Patrick Grogan, Abbas K. Samadi, Huizhong Cui, Mark S. Cohen, and Xinmai Yang, Growth of melanoma brain tumors monitored by photoacoustic microscopy, Journal of Biomedical Optics, Vol. 154, 2010.
268. D. Yao, K. Maslov, K. K. Shung, Q. Zhou, and L. V. Wang, In vivo label-free photoacoustic microscopy of cell nuclei by excitation of DNA and RNA, Optics Letters, Vol. 35, No. 24, 2010, pp. 4139–4141.
269. D. Zink, A. H. Fischer, and J. A. Nickerson, Nuclear Structure in Cancer cells, Nature Reviews Cancer, Vol. 4, No. 9, 2004, pp. 677–687.
270. William A. Eaton, Eric. R. Henry, James Hofrichter and Andrea Mozzarelli, Is cooperative oxygen binding by hemoglobin really understood? Nature Structural Biology, Vol. 6, No. 4, 1999, pp. 351–358.
271. Zhongjiang Chen, Sihua Yang, and Da Xing, In vivo detection of hemoglobin oxygen saturation and carboxyhemoglobin saturation with multiwavelength photoacoustic microscopy, Optics Letters, Vol. 37, No. 6, 2012, pp. 3414–3416.
272. Lei Li, Anton A. Shemetov, Mikhail Baloban, Peng Hu, Liren Zhu, Daria M. Shcherbakova, Ruiying Zhang, Junhui Shi, Junjie Yao, Lihong V. Wang, and Vladislav V. Verkhusha, Small near-infrared photochromic protein for photoacoustic multi-contrast imaging and detection of protein interactions in vivo, Nature Communications, Vol. 9, No. 2734, 2018.
273. Yumiko Saito, Hans-Peter Nothacker, Zhiwei Wang, Steven H. S. Lin, Frances Leslie, and Olivier Civelli, Molecular characterization of the melanin-concentrating-hormone receptor, Nature, Vol. 400, 1999, pp. 265–269.
274. Songzhu An, Gene Cutler, Jack Jiagang Zhao, and Kang Dai, Identification and characterization of a melanin-concentrating hormone receptor, Vol. 98, No. 13, 2001, pp. 7576–7581.
275. I. Alexvitkin, Joshuawoolsey, and Brianc Wilson, Optical and Thermal Characterization of Natureal (Sepia oficinalis) Melanin, Photochemistry and Photobiology, Vol. 59, No. 4, 1994, pp. 455–462.
276. Jacob Staley, Patrick Grogan, Abbas K. Samadi, Huizhong Cui, Mark S. Cohen, and Xinmai Yang, Growth of melanoma brain tumors monitored by photoacoustic microscopy, Journal of Biomedical Optics, Vol. 154, 2010.
277. Yanxiu Li and Yannis Mantas Paulus, Photoacoustic Imaging of the Eye, April 27th, 2019 DOI: https://doi.org/10.5772/intechopen.86050.
278. Ekaterina I. Galanzha, Evgeny V. Shashkov, Thomas Kelly, Jin-Woo Kim, Lily Yang and Vladimir P. Zharov, In vivo magnetic enrichment and multiplex photoacoustic detection of circulating tumour cells, Nature Nanotechnology, Vol. 4, 2009, pp. 855–860.
279. Yu Shrike Zhang, Lihong V. Wang, and Younan Xia, Seeing through the surface: non-invasive characterization of biomaterial-tissue interactions using Photoacoustic Microscopy, Annals of biomedical engineering, Vol. 44, No. 3, 2016, pp. 649–666.

280. Yu Zhang, Xin Cai, Sung-Wook Choi, Chulhong Kim, Lihong V. Wang, and Younan Xia, Chronic label-free volumetric photoacoustic microscopy of melanoma cells in three-dimensional porous scaffolds, Biomaterials, Vol. 31, No. 33, 2010, pp. 8651–8658.
281. Thijs W.H. Pols, Mitsunori Nomura, Taoufiq Harach, Giuseppe Lo Sasso, Maaike H. Oosterveer, Charles Thomas, Giovanni Rizzo, Antimo Gioiello, Luciano Adorini, Roberto Pellicciari, Johan Auwerx, and Kristina Schoonjans, TGR5 Activation Inhibits Atherosclerosis by Reducing Macrophage Inflammation and Lipid Loading, Cell Metabolism Vol. 14, 2011, pp. 747–755.
282. Takashi Buma, Nicole C. Conley, and Sang Won Choi, Multispectral photoacoustic microscopy of lipids using a pulsed supercontinuum laser, Biomedical Optics Express, Vol. 9, No. 1, 2018, pp. 276–288.
283. Jonas Kottmann, Julien M. Rey, Joachim Luginbuhl, Ernst Reichmann, and Markus W. Sigrist, Glucose sensing in human epidermis using mid-infrared photoacoustic detection, Biomedical Optics Express, Vol. 3, No. 4, 2012, pp. 667–680.
284. Manoj Kumar Dasa, Christos Markos, Jakob Janting, and Ole Bang, Multispectral photoacoustic sensing for accurate glucose monitoring using a supercontinuum laser, Journal of the Optical Society of America B, Vol. 36, No. 2, 2019, pp. A61–A65.
285. Liming Nie, Mei Chen, Xiaolian Sun, Pengfei Rong, Nanfeng Zheng, and Xiaoyuan Chen, Palladium nanosheets as highly stable and effective contrast agents for in vivo photoacoustic molecular imaging, Nanoscale, Vol. 6, 2014, pp. 1271–1276.
286. Adamde la Zerda, Zhuang Liu, Sunil Bodapati, Robert Teed, Srikant Vaithilingam, Butrus T. Khuri-Yakub, Xiaoyuan Chen, Hongjie Dai, and Sanjiv Sam Gambhir, Ultrahigh Sensitivity Carbon Nanotube Agents for Photoacoustic Molecular Imaging in Living Mice, Nano Letters, Vol. 10, Nol. 6, 2010, pp. 2168–2172.
287. Zhiyang Wang, Fei Yang, Zhongwen Cheng, Wuyu Zhang, Kedi Xiong, Sihua Yang, Photoacoustic-guided photothermal therapy by mapping of tumor microvasculature and nanoparticle, Nanophotonics, 2021.
288. Yun He, Development of high-speed photoacoustic imaging technology and its applications in biomedical research, Doctor of Philosophy dissertation, Washington University in St. Louis, 2019.
289. Hao F. Zhang, Konstantin Maslov, Mathangi Sivaramakrishnan, Gheorghe Stoica, and Lihong V. Wang, Imaging of hemoglobin oxygen saturation variations in single vessels in vivo using photoacoustic microscopy, Applied Physics Letters, Vol. 90, 053901, 2007.
290. C. Chandrakala and D. L. Fraker, Tumor oxygenation status as a prognostic marker, Cancer Letters, Vol. 211, No. 225, 2005.
291. Chao Liu, Yizhi Liang, and Lidai Wang, Optical-resolution photoacoustic microscopy of oxygen saturation with nonlinear compensation, Biomedical Optics Express, Vol. 10, No. 6, 2019, pp. 3061–3069.
292. Carlos Taboada, Jesse Delia, Maomao Chen, Chenshuo Ma, Xiaorui Peng, Xiaoyi Zhu, Laiming Jiang, Tri Vu, Qifa Zhou, Junjie Yao, Lauren O'Connell, and Sönke Johnsen, Glassfrogs conceal blood in their liver to maintain transparency, Science 378,1315–1320 (2022).DOI:https://doi.org/10.1126/science.abl6620.
293. Anton A. Shemetov, Mikhail V. Monakhov, Qinrong Zhang, Jose Ernesto Canton-Josh, Manish Kumar, Maomao Chen, Mikhail E. Matlashov, Xuan Li, Wei Yang, Liming Nie, Daria M. Shcherbakova, Yevgenia Kozorovitskiy, Junjie Yao, Na Ji and Vladislav V. Verkhusha, A near-infrared genetically encoded calcium indicator for in vivo imaging, Nature Biotechnology, Vol. 39, 2021, pp. 368–377.
294. N. E. Cameron, M. A. Cotter, and P. A. Low, Nerve blood flow in early experimental diabetes in rats: relation to conduction deficits, American Journal of Physiology-Endocrinology And Metabolism, 1991.
295. Juan E. Grunwald, Jody Piltz, Seenu M. Hariprasad, Joan Dupont, Maureen G. Maguire, Optic nerve blood flow in glaucoma: Effect of systemic hypertension, Vol. 127, No. 5, 1999, pp. 516–522.

296. Hachinski VC, Iliff LD, Zilhka E, et al., Cerebral blood flow in dementia, Arch Neurol, Vol. 32, No. 9, 1975, pp. 633–637.
297. Fang, H., K. Maslov, and L.V.J.P.R.L. Wang, Photoacoustic Doppler effect from flowing small light-absorbing particles, Physical Review Letters, Vol. 99, No. 18, 2007.
298. I. V. Larina, K. V. Larin, and R. O. Esenaliev, Real-time optoacoustic monitoring of temperature in tissues, Journal of Physics D: Applied Physics, Vol. 38, No. 15, 2005, pp. 2633–2639.
299. A. J. Welch and M. J. C. Van Gemert, Optical-Thermal Response of Laser-Irradiated Tissue, Plenum, New York 1995.
300. R. Seip and E. S. Ebbini, Noninvasive estimation of tissue temperature response to heating fields using diagnostic ultrasound, IEEE transactions on biomedical engineering, Vol. 428, 1995, pp. 828–839.
301. R. MaassMoreno and C. A. Damianou, Noninvasive temperature estimation in tissue via ultrasound echo-shifts, 1. analytical model, Journal of Acoustical Society America, Vol. 100, No. 4, 1996, pp. 2514–2521.
302. R. Seip, P. VanBaren, C. A. Cain, and E. S. Ebbini, Noninvasive real-time multipoint temperature control for ultrasound phased array treatments, IEEE Transactions on Ultrasonics, Ferroelectrics, and Frequency Control, vol. 43, no. 6, 1996, pp. 1063–1073.
303. S. J. Graham, M. J. Bronskill, and R. M. Henkelman, Time and temperature dependence of MR parameters during thermal coagulation of ex vivo rabbit muscle, Magnetic resonance in medicine, Vol. 392, 1998, pp. 198–203.
304. P. Steiner, R. Botnar, B. Dubno, G. G. Zimmermann, G. S. Gazelle, and J. F. Debatin, Radio-frequency-induced thermoablation: monitoring with T1-weighted and proton-frequency-shift MR imaging in an interventional 0.5-T environment, Radiology, Vol. 206, No. 3, 1998, pp. 803–810.
305. Manojit Pramanik, and Lihong V. Wang, Thermoacoustic and photoacoustic sensing of temperature, Journal of Biomedical Optics, Vol. 145, 054024, 2009.
306. Irina V Larina, Kirill V Larin, and Rinat O Esenaliev, Real-time optoacoustic monitoring of temperature in tissues, Journal of Physics D: Applied Physics, Vol. 38, No. 15, 2005, pp. 2633–2639.
307. Jignesh Shah, Suhyun Park, Salavat Aglyamov, Timothy Larson, Li Ma, Konstantin Sokolov, Keith Johnston, Thomas Milner, Stanislav Y. Emelianov, Photoacoustic imaging and temperature measurement for photothermal cancer therapy, Journal of biomedical optics, Vol. 13, No, 3, 2008, 034024.
308. Elena Petrova, Sergey Ermilov, Richard Su, Vyacheslav Nadvoretskiy, André Conjusteau, and Alexander Oraevsky, Using optoacoustic imaging for measuring the temperature dependence of Grüneisen parameter in optically absorbing solutions, Optics express, Vol. 21, No. 21, 2013, pp. 25077–25090.
309. Liang Gao, Lidai Wang, Chiye Li, Yan Liu, Haixin Ke, Chi Zhang, Lihong V. Wang, Single-cell photoacoustic thermometry, Journal of Biomedical Optics, Vol. 18, No. 2, 2013, 026003.
310. Junjie Yao, Haixin Ke, Stephen Tai, Yong Zhou, and Lihong V. Wang, Absolute photoacoustic thermometry in deep tissue, Optics Letters, Vol. 38, No. 24, 2013, pp. 5228–5231.
311. Yuan Zhou, Mucong Li, Wei Liu, Georgy Sankin, Jianwen Luo, Pei Zhong, and Junjie Yao, Thermal memory based photoacoustic imaging of temperature, Optica, Vol. 6, No. 2, 2019, pp. 198–205.
312. Anna Orlova, Ksenia Pavlova, Aleksey Kurnikov, Anna Maslennikova, Marina Myagcheva, Evgeniy Zakharov, Dmitry Skamnitskiy, Valeria Perekatova, Alexander Khilov, Andrey Kovalchuk, Alexander Moiseev, Ilya Turchin, Daniel Razansky, Pavel Subochev, Noninvasive optoacoustic microangiography reveals dose and size dependency of radiation-induced deep tumor vasculature remodeling, Neoplasia, Vol. 26, 2022, 100778, https://doi.org/10.1016/j.neo.2022.100778.
313. Tian Jin, Heng Guo, Lei Yao, Huikai Xie, Huabei Jiang, and Lei Xi, Portable optical-resolution photoacoustic microscopy for volumetric imaging of multiscale organisms, Journal of Biophotonics, Vol. 11, 2018.

314. T. Jin, H. Guo, H. Jiang, B. Ke, L. Xi, Portable optical resolution photoacoustic microscopy (pORPAM) for human oral imaging, Optics Letters, Vol. 42, No. 21, 2017, pp. 4434–4437.
315. Junjie Yao, Lidai Wang, Joon-Mo Yang, Konstantin I Maslov, Terence T W Wong, Lei Li, Chih-Hsien Huang, Jun Zou, and Lihong V Wang, High-speed label-free functional photoacoustic microscopy of mouse brain in action, Nature Methods, Vol. 12, 2015, pp. 407–410.
316. B. Ning, N. Sun, R. Cao, R. Chen, K.K. Shung, J.A. Hossack, J.-M. Lee, Q. Zhou, S. Hu, Ultrasound-aided multi-parametric photoacoustic microscopy of the mouse brain, Scientific Reports, Vol. 5, 18775, 2015.
317. T. P. Olson, J. Harter, A. Muñoz, D. M. Mahvi, and TM. Breslin, Frozen section analysis for intraoperative margin assessment during breast-conserving surgery results in low rates of re-excision and local recurrence, Annals of Surgical Oncology, Vol. 14, No. 10, 2007, pp. 2953–2960.
318. Juan C Cendán, MD, FACS, Dominique Coco, MD, Edward M Copeland III, MD, FACS, Accuracy of intraoperative frozen-section analysis of breast cancer lumpectomy-bed margins, Journal of the American College of Surgeons, Vol. 201, No. 2, 2005, pp. 195–198.
319. Terence T. W. Wong, Ruiying Zhang, Pengfei Hai, Chi Zhang, Miguel A. Pleitez, Rebecca L. Aft, Deborah V. Novack, and Lihong V. Wang, Fast label-free multilayered histology-like imaging of human breast cancer by photoacoustic microscopy, Science Advances, Vol. 3, 2017, e1602168.
320. Susumu Ishida, Kenji Yamashiro, Tomohiko Usui, Yuichi Kaji, Yuichiro Ogura, Tetsuo Hida, Yoshihito Honda, Yoshihisa Oguchi, and Anthony P Adamis, Leukocytes mediate retinal vascular remodeling during development and vaso-obliteration in disease, Nature Medicine, Vol. 9, 2003, pp. 781–788.
321. Jeffrey De Fauw, Joseph R. Ledsam, Bernardino Romera-Paredes, Stanislav Nikolov, Nenad Tomasev, Sam Blackwell, Harry Askham, Xavier Glorot, Brendan O'Donoghue, Daniel Visentin, George van den Driessche, Balaji Lakshminarayanan, Clemens Meyer, Faith Mackinder, Simon Bouton, Kareem Ayoub, Reena Chopra, Dominic King, Alan Karthikesalingam, Cían O. Hughes, Rosalind Raine, Julian Hughes, Dawn A. Sim, Catherine Egan, Adnan Tufail, Hugh Montgomery, Demis Hassabis, Geraint Rees, Trevor Back, Peng T. Khaw, Mustafa Suleyman, Julien Cornebise, Pearse A. Keane, and Olaf Ronneberger, Clinically applicable deep learning for diagnosis and referral in retinal disease, Clinically applicable deep learning for diagnosis and referral in retinal disease, Nature Medicine, Vol. 24, 2018, pp.1342–1350.
322. Yu Qin, Yixin Yu, Julia Fu, Mingyang Wang, Xinmai Yang, Xueding Wang, Yannis M. Paulus, Photo-mediated ultrasound therapy for the treatment of retinal neovascularization in rabbit eyes, Lasers in Surgery and Medicine, Vol. 54, No. 5, 2022, pp. 747–757.
323. Wei Zhang, Yanxiu Li, Van Phuc Nguyen, Ziyi Huang, Zhipeng Liu, Xueding Wang, and Yannis M. Paulus, High-resolution, in vivo multimodal photoacoustic microscopy, optical coherence tomography, and fluorescence microscopy imaging of rabbit retinal neovascularization, Light: Science & Applications, Vol. 7, No. 103, 2018.
324. Lei Bi, Jinman Kim, Euijoon Ahn, Ashnil Kumar, Michael Fulham, and Dagan Feng, Dermoscopic Image Segmentation via Multistage Fully Convolutional Networks, IEEE Transactions on Biomedical Engineering, Vol. 64, No. 9, 2017, pp. 2065–2074.
325. Yu Wang, Konstantin Maslov, Yu Zhang, Song Hu, Lihmei Yang, Younan Xia, Jian Liu, and Lihong V. Wang, Fiber-laser-based photoacoustic microscopy and melanoma cell detection, Journal of Biomedical Optics, Vol. 16, No. 1, 2011, 011014.
326. Juan Aguirre, Mathias Schwarz, Natalie Garzorz, Murad Omar, Andreas Buehler, Kilian Eyerich, and Vasilis Ntziachristos, Precision assessment of label-free psoriasis biomarkers with ultra-broadband optoacoustic mesoscopy, Nature Biomedical Engineering, Vol. 1, 0068, 2017.
327. A. Berezhnoi, M. Schwarz, A. Buehler, S.V. Ovsepian, J. Aguirre, V. Ntziachristos, Assessing hyperthermia-induced vasodilation in human skin in vivo using optoacoustic mesoscopy, Journal of Biophotonics, 2018, e201700359.

328. J J. W. Baik, J. Y. Kim, S. Cho, S. Choi, J. Kim and C. Kim, Super wide-field photoacoustic microscopy of animals and humans in vivo, IEEE Transactions on Medical Imaging, Vol. 39, No. 4, 2020, pp. 975–984.
329. Joongho Ahn, Jin Woo Baik, Yeonggeon Kim, Karam Choi, Jeongwoo Park, Hyojin Kim, Jin Young Kim, Hyung Ham Kim, Sung Hyun Nam, Chulhong Kim, Fully integrated photoacoustic microscopy and photoplethysmography of human in vivo, Photoacoustics, Vol. 27, 2022, 100374.
330. Mathias Schwarz, Dominik Soliman, Murad Omar, Andreas Buehler, Saak V. Ovsepian, Juan Aguirre, and Vasilis Ntziachristos, Optoacoustic Dermoscopy of the Human Skin: Tuning Excitation Energy for Optimal Detection Bandwidth With Fast and Deep Imaging in vivo, IEEE Transactions on Medical Imaging, vol. 36, no. 6, pp. 1287–1296, June 2017, doi: https://doi.org/10.1109/TMI.2017.2664142.
331. B. Hindelang, J. Aguirre, M. Schwarz, A. Berezhnoi, K. Eyerich, V. Ntziachristos, T. Biedermann, U. Darsow, Non-invasive imaging in dermatology and the unique potential of raster-scan optoacoustic mesoscopy, JEADV 2019, 33, 1051–1061
332. Benedict Anchang, Tom D P Hart, Sean C Bendall, Peng Qiu, Zach Bjornson, Michael Linderman, Garry P Nolan, and Sylvia K Plevritis, Visualization and cellular hierarchy inference of single-cell data using SPADE, Nature Protocols, Vol. 11, 2016, pp. 1264–1279.
333. Yaron M. Sigal, Ruobo Zhou, Xiaowei Zhuang, Visualizing and discovering cellular structures with super-resolution microscopy, Science, Vol. 361, No. 6405, 2018, pp. 880–887.
334. Eric M. Strohm, Michael J. Moore, and Michael C. Kolios, Single cell photoacoustic microscopy: a review, IEEE Journal of Selected Topics in Quantum Electronics, Vol. 22, No. 3, 2016, pp. 137–151.
335. Eric M. Strohm, Elizabeth S.L. Berndl, Michael C. Kolios, High frequency label-free photoacoustic microscopy of single cells, Photoacoustics, Vol. 1, 2013, pp. 49–53.
336. Eno Hysi, Ratan K. Saha, and Michael C. Kolios, Photoacoustic ultrasound spectroscopy for assessing red blood cell aggregation and oxygenation, Journal of Biomedical Optics Vol. 17, No. 12, 2012, 125006.
337. Chi Zhang, Yu Shrike Zhang, Da-Kang Yao, Younan Xia, Lihong V. Wang, Label-free photoacoustic microscopy of cytochromes, Journal of Biomedical Optics, February 2013, Vol. 18(2).
338. Amos Danielli, Konstantin Maslov, Alejandro Garcia-Uribe, Amy M. Winkler, Chiye Li, Lidai Wang, Yun Chen, Gerald W. Dorn, Lihong V. Wang, Label-free photoacoustic nanoscopy, Journal of Biomedical Optics, Vol. 19, No. 8, 086006, 2014.
339. Sanjiv Sam Gambhir, Molecular imaging of cancer with positron emission tomography, Nature Reviews Cancer, Vol. 2, 2002, pp. 683–693.
340. McDonald, D., Choyke, P. Imaging of angiogenesis: from microscope to clinic, Nature Medicine, Vol. 9, 2003, pp. 713–725.
341. L. Serrone, F. M. Solivetti, M. F. Thorel, L. Eibenschutz, P. Donati and C. Catricala, High frequency ultrasound in the preoperative staging of primary melanoma: a statistical analysis, Melanoma Research, Vol. 12, 2002, pp. 287–290.
342. Soon-Woo Cho, Thi Tuong Vy Phan, Van Tu Nguyen, Sang Min Park, Hwidon Lee, Junghwan Oh, Chang-Seok Kim, Efficient label-free in vivo photoacoustic imaging of melanoma cells using a condensed NIR-I spectral window, Photoacoustics, Volume 29, 2023, 100456, ISSN 2213-5979, https://doi.org/10.1016/j.pacs.2023.100456.
343. Junjie Yao, Konstantin I. Maslov, Yu Zhang, Younan Xia, and Lihong V. Wang, Label-free oxygen-metabolic photoacoustic microscopy in vivo, Journal of Biomedical Optics, Vol. 16, No. 7, 076003, 2011.
344. Jung-Taek Oh, Meng-Lin Li, Hao F. Zhang, Konstantin Maslov, Three-dimensional imaging of skin melanoma in vivo by dual-wavelength photoacoustic microscopy, Journal of Biomedical Optics, Vol. 113, 034032, 2006.
345. Chi Zhang, Konstantin Maslov, and Lihong V. Wang, Subwavelength-resolution label-free photoacoustic microscopy of optical absorption in vivo, Optics Letters, Vol. 35, No. 19, 2010, pp. 3195–3197.

346. Yating Wang, Dong Xu, Sihua Yang, and Da Xing, Toward in vivo biopsy of melanoma based on photoacoustic and ultrasound dual imaging with an integrated detector, Biomedical Optics Express, Vol. 7, No. 2, 2016, pp. 279–286.
347. Medina-Sánchez, M. & Schmidt, O. G. Medical microbots need better imaging and control. Nature, Vol. 545, 2017, pp. 406–408.
348. Z Wu, L Li, Y Yang, P Hu, Y Li, SY Yang, LV Wang, W Gao, A microrobotic system guided by photoacoustic computed tomography for targeted navigation in intestines in vivo, Science robotics, Vol. 4, eaax0613, 2019.
349. Dengfeng Li, Chao Liu, Yuanyuan Yang, Lidai Wang, and Yajing Shen, Micro-rocket robot with all-optic actuating and tracking in blood, Light: Science & Applications, Vol. 9, 84, 2020.

Chapter 2
Principles and Applications of Photoacoustic Computed Tomography

Vijitha Periyasamy, Katherine Gisi, and Manojit Pramanik

Abstract In this chapter, we will introduce photoacoustic computed tomography (PACT). A brief history and rationale for doing PACT. In Sect. 2.2, we will describe the different types of PACT systems according to the ultrasound transducer used for imaging. Finally, we will discuss various applications of PACT for pre-clinical small animal imaging and clinical human imaging in Sect. 2.3. In conclusion, we will discuss future developments.

Keywords Photoacoustic computed tomography · Single element transducer · Curved array transducer · Linear array transducer · Planar matrix transducer · Spherical matrix transducer · Pre-clinical imaging

2.1 Introduction

Biomedical imaging is crucial for diagnosis and treatment of diseases [1]. Frequent monitoring of the health parameters is a fast-growing field. From the smallest organelle to the largest organs, the whole body is being imaged for various information. Based on the source of the energy used and the mode of acquisition of the data, the information obtained from the region of interest varies significantly. Electromagnetic waves (such as x-rays, Gamma rays, radio waves, light) and sound waves (ultrasound (US)), are widely used for biomedical imaging. X-rays (used in x-ray projection imaging, and x-ray computed tomography or x-ray CT) and Gamma rays (used in nuclear imaging modalities such as single photon emission computed tomography (SPECT) or positron emission tomography (PET)) are ionizing radiation. Radiofrequency (RF) waves used in magnetic resonance imaging (MRI) are considered safe, however, MRI is one of the most expensive imaging modalities. US imaging is one of the safest modalities, on top of being real-time,

V. Periyasamy · K. Gisi · M. Pramanik (✉)
Department of Electrical and Computer Engineering, Iowa State University, Ames, IA, USA
e-mail: vijitha@iastate.edu; kagisi@iastate.edu; mano@iastate.edu

W. Xia (ed.), *Biomedical Photoacoustics*,
https://doi.org/10.1007/978-3-031-61411-8_2

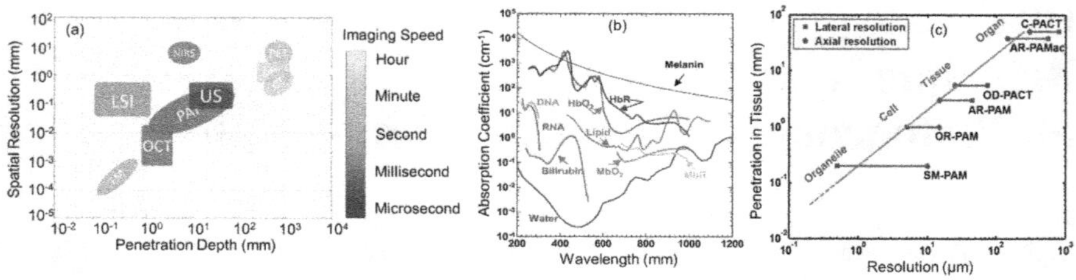

Fig. 2.1 (**a**) Distribution of penetration depth, spatial resolution, and the speed of imaging of commonly used biomedical imaging techniques; Fluorescence microscopy (FM), laser speckle imaging (LSI), optical coherence tomography (OCT), photoacoustic imaging (PAI), ultrasound (US), near infrared spectroscopy (NIRS), positron emission tomography (PET), magnetic resonance imaging (MRI) and computed tomography (CT). Reproduced from [12]. (**b**) Absorption coefficient of endogenous chromophores at wavelength between 200 nm to 1200 nm. HbR—deoxy-hemoglobin, HbO_2—oxy-hemoglobin, MbR- reduced myoglobin MbO_2—oxy-myoglobin, DNA—deoxyribonucleic acid RNA- ribonucleic acid. Reprinted from [13] with permission from Elsevier. (**c**) Resolution vs penetration depth of photoacoustic systems, *SM-PAM* submicron photoacoustic microscopy, *OR-PAM* optical resolution microscopy, *AR-PAM* acoustic resolution microscopy, *OD-PACT* optical detection based photoacoustic computed tomography, *AR-PAMac* acoustic resolution macroscopy, *C-PACT* circular photoacoustic tomography. Reprinted from [14] with permission

portable, and cheap. However, it provides poor soft tissue contrast and image quality can vary because of the operator dependency [2]. All these modalities have their own advantages and disadvantages; therefore, scientists are always looking for new imaging modalities.

Optical imaging can provide high-resolution imaging, but the scattering of light inside biological tissue limits the imaging depth to a few millimeters [3, 4]. This leads to challenges in visualizing tissue beyond the superficial skin layers [5–7]. As shown in Fig. 2.1a, optical imaging techniques such as fluorescence microscopy (FM), and laser spectral imaging (LSI) have the capability of imaging at a resolution of ~1 μm, only up to the depth of 1 mm. Near infrared spectroscopy (NIRS), which is a diffused optical imaging technique, has a deeper imaging depth of up to 10–15 mm with a poorer resolution of 5–10 mm. US imaging, on the other hand, can image up to a few tens of centimeters, but the spatial resolution is compromised to 1 mm. Therefore, by combining both optics and ultrasound through photoacoustics [8–11] one can overcome the limitations of both optical and ultrasound imaging.

Photoacoustic imaging (PAI) works on the basic principle of the photoacoustic (PA) effect, first reported by Alexander Graham Bell in 1880 [15]. During a photophone experiment, he observed that sound waves were produced when a modulated light beam was absorbed by a metal sheet. The early application of the PA effect was in photoacoustic spectroscopy studies primarily focusing on gas analysis with commercial gas spectrophones. The gas molecules were excited with a chopped or periodically interrupted monochromatic light source. The light energy absorbed by the gas molecules would partially or entirely convert into the kinetic motion of the gas particles thereby resulting in pressure fluctuations. Then a very sensitive microphone would record these pressure fluctuations as an audible sound signal. Studies in solids and liquids were conducted in the 1970s majorly focusing on surface heating effects. Manohar et al. have nicely summarized the history and early works of photoacoustics [16]. Only recently (a couple of decades ago), PA phenomena found applications in biomedical imaging. Advancement in high-energy pulsed laser sources, sensitive ultrasound detectors, fast data acquisition electronics, computing power, etc., has expedited this progress.

Photoacoustic imaging can be divided into a few steps; step 1: illumination of tissue with a short-pulsed light source (usually nano seconds lasers are used), step 2: absorption of light by chromophores (intrinsic such as blood, melanin, water, etc., or extrinsic such as contrast agents [17]) results in a local temperature rise due to non-radiative relaxation process, step 3: temperature rise leads to initial pressure rise (via thermoelastic expansion), step 4: the pressure is then propagated as an ultrasound wave (also known as photoacoustic waves or PA waves), step 5: these PA waves are then detected and acquired at the tissue boundary using ultrasound detectors, step 6: these acquired PA signals are used to form images of the internal structure/function of the tissue.

For the efficient generation of PA signals the light excitation should be in thermal and stress confinement [18]. Once the photon is absorbed and the energy is converted to thermal energy in the tissue, it will diffuse out of the heated zone (region where photons are absorbed) into the surrounding tissue. Thermal relaxation time (τ_{th}) is estimated as,

$$\tau_{th} = \frac{{d_c}^2}{\alpha_{th}}, \tag{2.1}$$

where, α_{th} is the thermal diffusivity (m^2/s), and ${d_c}^2$ is the characteristic dimension of the heated region. When an object is illuminated with a laser pulse, it will be considered as in a state of thermal confinement when the laser pulse width (or pulse duration time, t_p) is much shorter than τ_{th}. In such condition ($t_p \ll \tau_{th}$ or $t_p \ll \frac{{d_c}^2}{\alpha_{th}}$) the heat conduction is negligible during the laser pulse, and the heat energy is confined within the heated zone throughout the laser pulse duration (no thermal diffusion happening). This (excitation in thermal confinement) results in the effective generation of the PA signal. Another time scale, known as the stress relaxation time, is defined as,

$$\tau_s = \frac{d_c}{\upsilon_s}, \tag{2.2}$$

where υ_s is the speed of sound (~1500 m/s in soft biological tissue). There will be an increase in local temperature in the heated region. This temperature rise will result in thermoelastic expansion of the tissue resulting in mechanical stress. This stress will also get dissipated into the surrounding tissue (or surrounding medium) with the speed of sound (υ_s). If the laser pulse width (t_p) is much shorter than τ_s, the excitation is in stress confinement. Under stress confinement ($t_p \ll \tau_s$ or $t_p \ll \frac{d_c}{v_s}$) the stress propagation is negligible, and the stress is confined within the heated zone throughout the laser pulse duration. This condition (excitation in stress confinement) is also important for the effective generation of photoacoustic signals.

The d_c is obtained from the area of the heated region (i.e., the area where the light is absorbed, and heat is generated; do not confuse this with the area where light is illuminated). In case the structure of interest that absorbs light (say blood vessels or gold nanoparticles) is identifiable, we can take the size of that object as the absorber dimension. In situations where the absorber is homogeneous (or the structure of the absorber is not identifiable), let us say light falling on ink solution, the decay constant of the optical energy deposition can be considered as the d_c value. In a situation where both are applicable, then d_c takes the value of the smaller quantity, i.e., the smaller among the dimension of the structure of interest and the decay constant of the optical energy deposition.

Once the laser excites and the light is absorbed the heated region will undergo volume expansion. The fractional volume expansion of the heated region can be written as,

$$\frac{dV}{V} = -\kappa p + \beta \Delta T, \tag{2.3}$$

where V is the volume of the heated region, dV is the volume change, κ is isothermal compressibility (Pa^{-1}), p is the change in pressure (Pa), β is the thermal coefficient of volume expansion (K^{-1}), and ΔT is the temperature change (K). If the laser pulse duration is shorter than both the *thermal relaxation time* and *stress relaxation time*, the laser excitation is said to be in thermal and stress confinements ($t_p \ll \tau_{th}\ and t_p \ll \tau_s$). In that case, we can neglect the fractional volume change. Therefore, the initial pressure rise (p_0) can be obtained from Eq. (2.4) as,

$$p_0 = \beta \times \Delta T/\kappa. \tag{2.4}$$

The local temperature rise can be written as:

$$\Delta T = \frac{\eta_{th} A_e}{\rho C_V}, \tag{2.5}$$

where, η_{th} denotes the percentage of light getting converted into heat, A_e denotes specific optical energy absorption (J/m^3), C_V denotes specific heat capacity at constant volume, ρ is the mass density. After substituting Eq. (2.6) into Eq. (2.5) we get,

$$p_0 = \beta \frac{\eta_{th} A_e}{\kappa \rho C_V} = \frac{\beta}{\kappa \rho C_V} \eta_{th} A_e. \tag{2.6}$$

By defining a dimensionless Gruneisen parameter as,

$$\Gamma = \frac{\beta}{\kappa \rho C_V}. \tag{2.7}$$

Equation (2.6) becomes,

$$p_0 = \Gamma \eta_{th} A_e. \tag{2.8}$$

A_e is proportional to the local optical fluence F. Then Eq. (2.8) becomes

$$p_0 = \Gamma \eta_{th} \mu_a F, \tag{2.9}$$

where, μ_a is the optical absorption coefficient. Γ and η_{th} are dependent on type of the tissue, however, they are usually approximated to be constants. Hence, the initial pressure rise p_0 is proportional to μ_a and F. Therefore, if we can measure p_0, and F is known, we can find out μ_a. However, in many situations fluence inside the tissue is unknown, making it difficult to recover the exact values of μ_a.

Another thing to remember, even though we have treated the Gruneisen parameter Γ as constant, in reality, this depends on the isothermal compressibility, the thermal coefficient of volume expansion, the mass density, and the specific heat capacity at the constant volume of the tissue. As a result, Γ changes significantly with temperature and can be written as,

$$\Gamma = \frac{\beta}{\kappa \rho C_V} = \frac{\beta \upsilon_s^2}{C_P} = f(T), \tag{2.10}$$

where υ_s is the velocity of sound, C_P denotes the specific heat capacity at constant pressure, and T is the temperature of the tissue. We used this relation $v_s{}^2 = \frac{1}{\kappa \rho}$ and replacing C_V with C_P. Therefore, Eq. (2.9) can be re-written as,

$$p_0 = f(T) \eta_{th} A_e. \tag{2.11}$$

Note that, here T refers to the base temperature of the object, not the change in temperature due to the laser heating (which is denoted as ΔT). ΔT is in the order of milliKelvin and its effect on the Grueneisen parameter is negligible. The base temperature (T) of the tissue is a slowly varying parameter compared to the transient temperature increase induced by the laser pulse.

As seen earlier on the laser excitation there is the generation of the initial pressure p_0 (Eq. 2.8). This initial pressure rise will then propagate within the tissue at the speed of sound (v_s) as an acoustic wave. The propagation of this acoustic wave in an inviscid medium is described by the general photoacoustic equation in the time domain. At any position $\vec{r}$ and time t the acoustic pressure $p\left(\vec{r}, t\right)$ is governed by the following wave equation [2],

$$\nabla^2 p\left(\vec{r}, t\right) - \frac{1}{v_s{}^2} \frac{\partial^2}{\partial t^2} p\left(\vec{r}, t\right) = -\frac{\beta}{\kappa v_s{}^2} \frac{\partial^2 \Delta T\left(\vec{r}, t\right)}{\partial t^2} \tag{2.12}$$

or

$$\left(\nabla^2 - \frac{1}{v_s{}^2}\frac{\partial^2}{\partial t^2}\right)p\left(\vec{r},t\right) = -\frac{\beta}{\kappa v_s{}^2}\frac{\partial^2 \Delta T\left(\vec{r},t\right)}{\partial t^2} \tag{2.13}$$

The left-hand side of the equation represents the wave propagation, whereas the right-hand side represents the source term. Under the thermal confinement condition (we can ignore the heat conduction, as it is negligible), the heat diffusion equation becomes,

$$\rho C_V \frac{\partial \Delta T\left(\vec{r},t\right)}{\partial t} = H\left(\vec{r},t\right) \tag{2.14}$$

where, $H\left(\vec{r},t\right)$ represents the heating function, defined as the thermal energy deposited per unit volume and per unit time. H is related to the specific optical power deposition A_p by $H = \eta_{th}A_p$, to the specific optical energy deposition (A_e) by $H\left(\vec{r},t\right) = \eta_{th}\frac{\partial A_e(\vec{r},t)}{\partial t}$, and to the optical fluence rate Φ by $H = \eta_{th}\mu_a\Phi$.

Using Eq. (2.13) and Eq. (2.14) we obtain (using this relation $v_s{}^2 = \frac{1}{\kappa\rho}$ and replacing C_V with C_P),

$$\left(\nabla^2 - \frac{1}{v_s{}^2}\frac{\partial^2}{\partial t^2}\right)p\left(\vec{r},t\right) = -\frac{\beta}{C_P}\frac{\partial H\left(\vec{r},t\right)}{\partial t} \tag{2.15}$$

The right-hand side of Eq. (2.15) has the first derivative of the heating function H. Thus, time-invariant heating (e.g., a constant light exposure) does not produce any photoacoustic waves, as the first derivative of a constant is zero. The source term on the right-hand side of the Eq. (2.15) will be zero. Only with time-variant heating function pressure waves (or photoacoustic waves) will be produced and propagated in the medium.

For efficient PA signal generation typically a laser with nano-second pulse width is used. A nano-second laser pulse can be approximated as a delta pulse illumination [$\delta(t)$]. In such a situation the initial pressure rise due to the delta pulse illumination can be written as $p_0\left(\vec{r}\right) = \Gamma H_s\left(\vec{r}\right)$, as for delta pulse excitation, the heating function can be decomposed as $H\left(\vec{r},t\right) = H_s\left(\vec{r}\right)H_t(t)$ and $H_t(t) = \delta(t)$. Therefore, Eq. (2.15) can be reduced to the following photoacoustic wave equation,

$$\left(\nabla^2 - \frac{1}{v_s{}^2}\frac{\partial^2}{\partial t^2}\right)p\left(\vec{r},t\right) = -\frac{1}{v_s{}^2}p_0\left(\vec{r}\right)\frac{\partial\delta(t)}{\partial t}. \tag{2.16}$$

This can be used to find out the pressure $p\left(\vec{r},t\right)$ at any location $\vec{r}$ and time t due to the propagation of the initial pressure $p_0\left(\vec{r}\right)$. For a delta heating function, the solution to the photoacoustic wave equation (Eq. 2.16) is given as,

$$p\left(\vec{r},t\right)=\frac{1}{4\pi {v_s}^2}\frac{\partial}{\partial t}\left[\frac{1}{v_s t}\int d\vec{r}'p_0\left(\vec{r}'\right)\delta\left(t-\frac{\left|\vec{r}-\vec{r}'\right|}{v_s}\right)\right],\qquad(2.17)$$

where, $\vec{r}'$ is the source location. If a delta excitation heats up a sphere of radius R, an initial pressure rise p_0 is generated within the sphere. This pressure will propagate both inward and outward as a spherical wave. The pressure $p\left(\vec{r},t\right)$ is recorded by an ultrasound transducer on the tissue boundary.

There are various photoacoustic systems designed, based on the source and pattern of the light illumination, the distribution and positioning of the ultrasound detectors, and the fabrication of the transducer. The wavelength of light illumination is determined by the absorption spectrum of the chromophore that is being imaged. The absorption spectrum of tissue chromophores is shown in Fig. 2.1b. There are various ways of constructing a photoacoustic imaging system. Mainly, they are categorized into three types known as photoacoustic microscopy (PAM) [19–22], photoacoustic endoscopy (PAE) [23–26], and photoacoustic tomography (PAT)/ photoacoustic computed tomography (PACT) [23, 27–32]. PAM can be further categorized into optical resolution photoacoustic microscopy (ORPAM) and acoustic resolution photoacoustic microscopy (ARPAM). In ORPAM, the laser beam is tightly focused, therefore, the spatial resolution is defined by the focused beam size [33–35]. The resolution of ORPAM is 2–5 μm at an imaging depth of 1 mm (Fig. 2.1c). In ARPAM, a focused ultrasound transducer (UST) is used to acquire the data, and the light illumination is not tightly focused (often donut shape dark field illumination is used, although not necessary). The resolution in ARPAM is defined by the ultrasound focus and it is typically ~30–50 μm at an imaging depth of 2–10 mm [36–38]. Although PAM provides high-resolution imaging, the imaging depth is still limited to a few millimeters to a centimeter. A photoacoustic endoscope can be either in microscopy mode or in computed tomography mode depending on how it is constructed.

In PACT, an expanded homogeneous light illumination and an unfocused UST are used [30, 39–44]. Usually in PACT, there exist various types of reconstruction methods, such as simple back-projection, frequency domain reconstruction, and system matrix-based reconstruction [45–51], which maps the initial pressure rise inside the tissue (p_0) from the acquired boundary PA signals ($p\left(\vec{r},t\right)$). In recent times, aided with deep learning methods the recovery of pressure maps (or absorption maps) has become more accurate, faster, with better resolution, and with fewer artifacts [52–57]. The resolution of PACT is hundreds of microns and the imaging depth reaches a few centimeters allowing whole-body small animal imaging and other human imaging applications. In Sect. 2.2, we discuss the different PACT systems broadly classified based on the transducers used. The different types of transducers used are single-element transducers (SET), curved array transducers, linear array transducers, planar array transducers, and spherical transducers. Special transducers fabricated for the specific desired applications are also discussed here. Applications of PACT for small animal imaging and human imaging are discussed in Sect. 2.3. We conclude with the drawbacks and the scope of improvement in PACT technology.

2.2 Photoacoustic Computed Tomography Systems

2.2.1 Single-Element Transducer (SET)-Based Systems

The first PACT imaging system was developed back in the early 2000s with a single-element transducer [58]. This kind of simple orthogonal configuration (Fig. 2.2a) is the most widely used system in laboratory and small animal imaging studies to date for deep tissue imaging. The illumination source was a tunable dye laser, operated by a short-pulsed Q-switched Nd: YAG laser operating at 532 nm. A concave lens and ground glass were used to expand and homogenize the laser beam such that the incident laser energy is <10 mJ/cm^2 on the rat skin. This illumination led to a

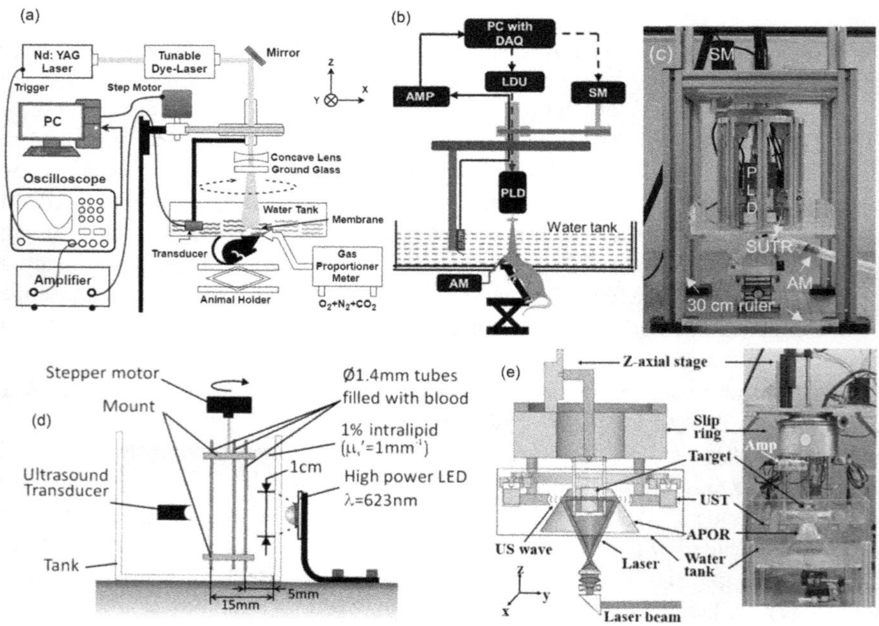

Fig. 2.2 Single element transducer-based PACT systems. (**a**) First developed Nd:YAG based system. Reprinted from [58] with permission. (**b**) Pulsed laser diode (PLD) based system. *AMP* amplifier, *LDU* laser driving unit, *SM* stepper motor, *DAQ* data acquisition card. Reproduced from [63]. (**c**) Multiple SET-based desktop PACT system. *AM* anesthesia machine, *SUTR*, acoustic-reflector-based ultrasound transducer. Adapted with permission from [62] © The Optical Society. (**d**) Light emitting diode (LED) based system. Adapted from [64] with permission. (**e**) Slip-ring-based PACT. *APOR* acoustically penetrable optical reflector, *US* ultrasound. Adapted with permission from [65] © The Optical Society

temperature increase of <20 mK on the skin of the animal which is well within the American National Standard for Safe Use of Lasers (ANSI) [59]. The generated PA waves that propagated through the water medium were acquired by a high-sensitive wide bandwidth ultrasound transducer (88% at −6 dB with a center frequency of 3.5 MHz). The transducer was rotated around the sample using a stepper motor and a PA signal was acquired at different locations around the object (rat brain) in a full circle. This configuration is also known as circular scanning geometry. The acquired signals were amplified by an amplifier (500 PR, Panametrics) before they were recorded and stored in a computer. Later a simple backprojection algorithm was used for the PACT image reconstruction. The functional imaging capability of PACT was demonstrated using the system with whisker stimulation in a small animal model and subsequent imaging of the cortical region of the brain (described in detail in Sect. 2.3.1). This system took ~16 mins for a full scan and the spatial resolution of the PACT images was ~0.2 mm.

The imaging speed and system bulkiness were further optimized in the SET-PACT system by introducing a continuous scan of the transducer (instead of a stop-and-go scan [60]) and the use of an acoustic reflector to position the SET in a vertical direction [61]. The results obtained with the acoustic reflector attached to the SET were very similar to that without using any reflector. The use of a reflector enabled the SET to rotate the scanning system at a smaller radius making the imaging system compact. Further improvements were made to make the system a truly desktop imaging system by using a more compact light delivery system. The bulky Nd: YAG and Optical parametric oscillator (OPO) lasers are not suitable for a compact desktop system. Hence the illumination source was replaced with a pulsed laser diode (PLD) [62, 63]. The PLD can provide a pulse width of ~107 ns at ~816 nm with a pulse energy of ~3.4 mJ per pulse at a repetition rate of 2 kHz. As seen in Fig. 2.2b, the PLD is mounted right at the center of the scanning system over the sample. The rectangular beam produced by the PLD illuminates the sample after passing through a diffuser to homogenize the beam. Even though the PLD energy was much lower than the traditional laser sources, because of this direct illumination (devoid of many optical components like mirrors, lenses, etc. between the light source and the sample surface), the illumination on the sample surface was quite effective. On top, the pulse repetition rate (PRR) for these types of PLDs is higher in the range of a few thousand hertz (compared to tens of hertz in the case of traditional lasers), leading to more signal averaging. As a result, the overall image quality did not suffer significantly on these PLD-based PACT systems compared to the traditional pulse laser-based PACT system.

In circular scanning, the SET rotates around the object in a full 360° rotation and collects PA signals from various positions around the circular path. The scanning and the data collection can be done in two ways [60]. The first method is called stop-and-go, where the SET moves to a desired position, acquires the PA signal, and moves to the next position. The second method is where the SET continuously rotates around the sample and acquires the PA signals, also called continuous scan. In this method, the A-lines are acquired as the transducer is in motion. Sharma et al. explored the quality of images acquired from these two methods for the

Q-switched and PLD illumination PACT system. The results were compared for various numbers of steps and different numbers of A-lines (360, 720, and 1440) collected during the scanning of phantoms and rat brains. Continuous scans had an advantage in terms of scan time and low laser exposure, and they didn't compromise the image quality. This advantage was more prominent when the pulse repetition rate of the laser used was high, especially with PLD where the PRR is a few thousand hertz.

The imaging speed of the PLD-PACT system was further enhanced by using multiple SETs [62, 66] as shown in Fig. 2.2c, the second-generation desktop PLD-PACT system. The 8 unfocused transducers with a center frequency of 5 MHz were attached with an acoustic reflector. This resulted in a rotation of 45° by each SET to make a full 360° data acquisition. The multiple SET system enabled a scan in 0.5 s (2 frames/s cross-sectional imaging speed). The spatial resolution of the system was ~165 μm. The fast-scanning capability is essential for dynamic *in vivo* imaging. This second-generation PACT was portable and ultra-compact with high frame-rate imaging capability. The performance of the system was demonstrated in structural imaging of rat brains and the real-time monitoring of indocyanine green (ICG) wash-out.

Another advancement in PACT is the use of light-emitting diodes (LEDs). Q-switched Nd: YAG lasers are not portable and PLDs are manufactured only at certain wavelengths and not many manufacturers are making them commercially. LEDs are another source of illumination that can be used for low-cost photoacoustic imaging [67]. LEDs are compact and safe to use which is crucial for the clinical applications of PA. Allen et al. demonstrated the feasibility of LED for PACT [64]. The authors used commercially available LEDs of wavelengths 460 nm, 530 nm, 590 nm, and 620 nm to study the signal generated from the blood vessels mimicking phantom as shown in Fig. 2.2d. The SNR of the images obtained from LED-PACT were 11.8, 2.2, 6.8, and 6 for the four wavelengths used. The recently designed full-view LED-based system, which operated at 850 nm was demonstrated for *in vivo* imaging of human fingers [68]. 16 LEDs were distributed on either side of a circular array transducer. The 512-element circular array transducer had an interelement pitch of $0.37 \times 15\ \text{mm}^2$ and the central frequency was 5 MHz. The cylindrical focus of the array was 38 mm. The system was able to resolve blood vessels as small as 0.2 mm in the human finger. The cost of the LED in LED-based PACT was ~2k USD, which is affordable compared to the Nd: YAG laser which costs 100–150k USD.

PACT with multiple SETs have challenges with handling the cables when they are rotating. Deng et al. proposed the use of a slip-ring to assist the rotation of two SETs around the sample [65]. The electric slip-ring transmits the electrical power to the rotators which are widely used in CT, MRI, and other tomographic imaging systems, that require rotation during signal acquisition. The authors used a customized six-internal channel slip-ring with brush coupling. One of the channels delivers electric power to the other channels. The grounded aluminum sheet formed the electromagnetic shielding for the slip-ring. A well-grounded metal plate served as a common junction to decrease the impedance from the transducer. The grounding reduces the electric noise which would otherwise interfere with the

PA signals. Another modification in this set-up was using a conical lens made of acoustically penetrable optical reflectors to reflect the diverging ring-shaped beam. The set-up is shown in Fig. 2.2e. 10 slices were acquired over 2 mm intervals along the abdomen of a ~20 g male nude mouse. The spinal cord, kidneys, liver, and blood vessels were seen in the cross-sectional images. The SNR of the images was 25. With the two transducers, the scanning time was 9 mins, which can be reduced by increasing the number of SETs.

Most of the PACT systems with SETs discussed so far use circular scanning geometry and orthogonal light illumination/acoustic detection. Overall, the improvements we see in the SET-based PACT systems are in the direction of (1) fast imaging using multiple SETs, (2) compact design with PLD/LED light sources, (3) improvement of image quality using deep learning and other image processing techniques, and (4) lowering the cost of the imaging system. Although the SET-based systems are sufficient for PA imaging for preclinical studies, for human imaging or even real-time small animal imaging array-based transducers are preferred. Based on the type of arrays used (e.g., linear, curved array, etc.) the image acquisition, image reconstruction, and light illumination may vary. In the next sections, we will discuss those systems.

2.2.2 *Curved Array Transducer-Based PACT*

Gamelin et al. reported a curved array transducer for photoacoustic imaging which was demonstrated for the imaging of a sacrificed mouse brain [69]. A Q-switched Nd: YAG laser which delivered 8–12 ns pulses at 15 Hz at a tunable wavelength from 700 nm to 950 nm was used for the imaging. The laser beam was expanded using a plano-concave lens and homogenized using a diffuser to produce a 50 mm diameter illumination. The illumination was orthogonal to the imaging plane. Four 128-element array transducers embedded along a 90° arc of a 25 mm radius curvature were designed for a 512-element closed ring image acquisition set-up. The array was fabricated based on a piezo-composite technique for high sensitivity and SNR. The design had a reception bandwidth of 80% and a center frequency of 5 MHz. Each of the elements had an elevation height of 10 mm, azimuthal pitch of 0.308 mm, and a kerf of 0.1 mm. The focus of the array was at 19 mm. The F-number of 1.9 produced an elevation resolution of 600 μm at the focus and 3 mm at the annular ring. Receiver boards were used to acquire and amplify the signal. In-plane resolution of the system was 80 μm, the axial resolution was 150 μm, and the azimuthal resolution was 170 μm, which correlates well with the theoretical value of 160 μm for a transducer of 35 mm effective aperture and center frequency of 5.1 MHz. The capability of the curved array transducer to image blood vasculature was demonstrated on a sacrificed mouse (Fig. 2.3a). The time to acquire a 2D slice was 15 s due to multiplexing in data acquisition. The fine vasculature even at a depth of 2 cm was well defined in the PA images.

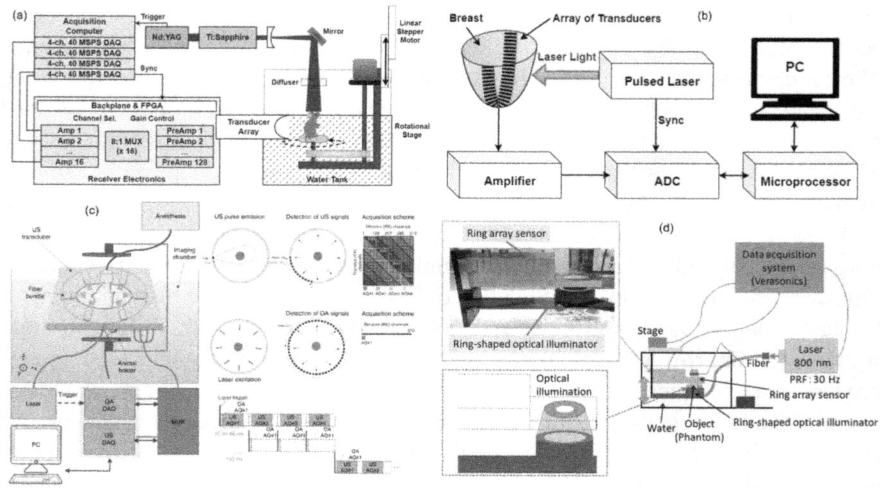

Fig. 2.3 Curved array-based systems. (**a**) Block diagram of the first curved array transducer used for imaging the brain of a sacrificed rat. Reproduced from [69]. (**b**) Laser optoacoustic imaging system that was used to image human breast. Reproduced from [70]. (**c**) The full ring array system which was used for full body imaging of small animals. Reprinted from [71] with permission. (**d**) Ring array system that was designed for human finger imaging. Reprinted from [72] with permission

Ermilov et al. reported a Laser optoacoustic imaging system (LOIS) for breast cancer detection [70]. Q-switched Alexandrite laser was used as the illumination source which was operated at 757 nm at a repetition rate of 10 Hz. The pulse width was 75 ns and energy of 750 mJ/pulse was coupled into an optical fiber bundle for a circular output of 70 mm diameter and fluence of 10 mJ/cm^2. The breast of the subject was suspended in a hemicylindrical cup of radi 70 mm and width of 90 mm. The curved array consisted of 64 piezoelectric elements made of polyvinylidene fluoride. The transducers were 20 × 3 × 0.1 mm. The signal was amplified using a 2-stage amplifier (Fig. 2.3b). The LOIS was used to image human breast which is discussed in detail in Sect. 2.3.2.

Mercep et al. reported a ring array-based PA imaging for small animal imaging [71]. The system—TROPUS (transmission-reflection optoacoustic ultrasound computed tomography) had a hybrid ultrasound and photoacoustic imaging capability. It provided a full view of the cross-sectional slices with information on optical energy, acoustic reflectivity, acoustic attenuation, and speed of sound for mice at a sub-millimeter resolution. The acquisition was through a 512-element cylindrical focused array transducer. The dedicated multiplexer unit was operated in two ways. One was to collect the ultrasound signals generated by laser pulses and the other was to transmit and receive ultrasound signals which enabled both PA and US imaging. The designed transducer covered 360° enabling full view. The full-ring transducer was made of 512 elements. The elements are arranged in 2 sets to form concave sub-arrays of 256 elements. The active angular aperture was 174° and the radius of curvature was 40 mm. The transducer had a 5 MHz center frequency and 60% bandwidth. The interelement pitch was 0.47 mm. 3D volumetric images were obtained by translating the transducer along the vertical axis. A Q-switched Nd: YAG laser operating at 10 Hz repetition rate, producing a ~8 ns laser at 1064 nm, was used for all the experiments. The laser was directed into a fiber bundle that had 620 fibers. The optical bundle was attached to the transducer such that the illumination area was 0.21 × 12.65 mm. The fibers are aligned at an angle of ~24° such that the illumination is maximum at the acoustic detection plane (Fig. 2.3c). Ultrasound images were acquired using synthetic transmit aperture techniques with 128-element active sub-apertures. A statistical weight approach, where a different speed of sound is assigned to water and the sample, was used to reconstruct the acquired images.

Yachao et al. demonstrated simultaneous US-PA imaging [73]. The plane US wave was combined with the PA wave in one shot. A Q-switched Nd: YAG laser operating at 20 Hz producing 5–8 ns pulses was used as an illumination source. The output from the laser was coupled into an optical fiber bundle with an input diameter of 11.7 mm and an output of 35 × 1 mm. The transducer had 128 array elements covering a view of 270°, a radius of 40 mm, an element pitch of 1.47 mm, and a center frequency of 7.5 MHz. The active area was 14 × 1.27 mm and the acoustic numerical aperture was 0.2. The performance of the system was studied using point targets. The full width at half maximum (FWHM) of the point target was 158.5 μm for PA imaging and 148.1 μm for US imaging. The feasibility of the system for *in vivo* imaging was demonstrated in the abdomen scans of mice. Physiological studies such as respiration rate and liver function were monitored. Further, imaging of the human finger joint was done to demonstrate the application of the system for the diagnosis of rheumatoid arthritis.

A full ring array system was proposed by Nishiyama et al. to image human finger vasculature [72]. Firstly, the authors simulated ultrasound wave propagation from the vasculature. The transducer parameters such as element width, center frequency, focal length of the acoustic lens, width, and height of the sensor were studied and fine-tuned with the help of simulations. The k-wave simulations were done using MATLAB [51]. From the simulations, it was evident that the 3 MHz transducer was optimum for the acquisition of signals from blood vessels. The designed system is shown in Fig. 2.3d. The ring transducer was made of 256 elements, with a diameter of 33 mm, 0.4 mm pitch, 0.3 mm width, and 0.5 mm element height. The center frequency was 2.8 MHz with a fractional bandwidth of 85%. A ring-shaped illumination was achieved using optical fiber. A Nd:YAG laser was operated with OPO laser to generate 800 nm pulses, whose pulse width was ~6 ns. The system was able to image objects as small as 0.1 mm in diameter.

Overall, the ring array transducers were able to improve the imaging speed significantly. For a full ring, a single laser pulse was enough to produce a cross-sectional PACT image of the target. Depending on the parallel data acquisition system used, sometimes, multiplexing was needed, hence, more than one laser pulse was necessary to collect all the data. Nevertheless, ring array transducers were significantly better in terms of image acquisition time compared to single-element ultrasound transducers. Some of the things improved over time for the ring array system are the field of view and speed of image acquisition [28]. However, one of the drawbacks of ring array transducers was their geometry. Not all types of objects can be imaged using ring arrays. They are good for circular-shaped objects like brains, breasts, etc. However, for other body parts, a generic linear array provides better suitability. In the next section, we discuss the development of linear array transducer-based PACT.

2.2.3 Linear Array Transducer-Based PACT

Linear array ultrasound probes are quite common in the clinical ultrasound imaging community. Moreover, photoacoustic imaging involves the acquisition of ultrasound signals just like typical pulse-echo ultrasound imaging. Hence, they share a similar signal acquisition setup. Therefore, integration of light illumination sources to an already existing linear array-based ultrasound imaging system may translate to faster adoption of PACT for clinical applications. The flexibility of linear arrays for free-hand scanning enables the scanning of regions that are not accessible with ring array transducers.

Erpelding et al. integrated the light illumination with the commercially available iU22 Philips healthcare US imaging system [74]. The clinical system was modified to communicate with laser trigger, data acquisition and to give access to raw channel RF data, which are essential to do photoacoustic imaging. Post-processing of both US and PA images was done in MATLAB. A delay-and-sum (DAS) algorithm was used for PACT image reconstruction. Images of sentinel lymph nodes of rats were acquired using S5–1 phased array, L8–4, and L15–7 linear array probes. Linear arrays had 128 elements at a 7–15 MHz frequency range. A Q-switched Nd: YAG laser and tunable dye laser operating at 10 Hz (6.5 ns pulse width) were used as the illumination source. The laser was tuned from 606 to 678 nm. This sweep was sufficient to cover the peak absorption of methylene blue (~667 nm). The light incident on the surface of the object was 5 mJ/cm^2 (within ANSI limits). The laser beam was coupled into a fiber bundle which is angled at 45° to the imaging plane. The imaging setup is shown in Fig. 2.4a. The animal was sedated and positioned on the table for sentinel lymph node imaging.

Over the decade, improvisations have been done for the light delivery in hand-held PACT systems [78, 79]. Sivasubramanian et al. performed Monte Carlo simulations of light transport in biological tissues for different angles of illumination ranging from 0° to 60° to optimize the best angle to deliver the light for a particular application of sentinel lymph node (SLN) imaging (Fig. 2.4a). Holders were 3D printed to insert the fiber bundle at different angles to the US probe of E-cube 12 R, Alpinion system. Compared to the iU22 Philips imaging system, which requires modifications to get access to RF data, this study used a commercially available clinical ultrasound imaging platform with access to the RF data (in research mode).

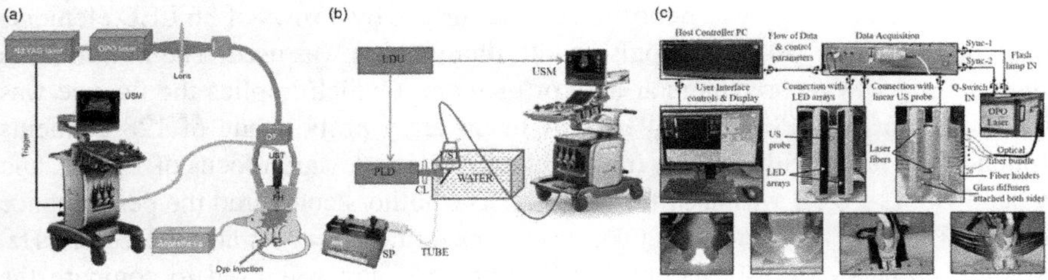

Fig. 2.4 Linear array-based systems (**a**) Photoacoustics integrated with the Philips ultrasound system. Reprinted from [75] with permission. (**b**) Alpinion ultrasound system modified for high-speed PA imaging by integrating it with pulsed laser diode (PLD). Reprinted with permission from [76] © The Optical Society. (**c**) Compact, desktop PACT system, operated with light-emitting diodes from Cyberdyne. Reprinted from [77] with permission

From the simulations, it was evident that an illumination angle beyond 15° was not optimum for the SLN imaging. Hence fiber holders were 3D printed for 0°, 5°, 10° and 15°. The designed fiber bundle was 1 m long and had 1600 fibers. The output end of the fiber bundle was bifurcated with 800 fibers each which were distributed over a rectangular window of 4 × 0.142 cm. The energy deposited on the tissue was 32 mJ per pulse at 675 nm, which implies the energy deposited on the surface was ~6.4 mJ/cm^2 (which is within the ANSI safety limit). 128 element, L3–12 array transducer, operating at 8 MHz was used for the experiments. SLN-mimicking phantoms were made of chicken breast tissue and latex spheres filled with methylene blue. The simulation results correlated well with the experimental results.

Like the ring array transducer, linear array transducers also can image an imaging plane with a single laser pulse. However, most of the high-energy pulse lasers used so far are 10–100 Hz repetition rate lasers, leading to low frames per second imaging. To improve the imaging speed further, one must employ a fast repetition rate light illumination source. The Alpinion PA-US system with L3–12 transducer was integrated with a pulsed diode with a high repetition rate, leading to one of the fastest photoacoustic imaging reported so far (7000 frames per second imaging) [76]. The pulse energy was ~1.4 mJ. The high frame rate imaging system is crucial for monitoring the flow of cells or contrast agents. A flow rate of up to 134.4 m/s (theoretically) can be imaged in real-time (Fig. 2.4b). The limitations of a PLD-PACT system with a hand-held transducer are limited view artifacts, multiple acoustic reflections, and compromise on signal-to-noise ratio due to less energy per pulse produced by PLDs compared to traditional pump lasers.

Agrawal et al. reported a LED-based PA imaging for human vasculature [77]. The commercial dual-mode PA-US system from AcousticX, Cyberdyne Inc., Ibaraki, Japan, housed data acquisition hardware, a host controller, and an interactive graphical user interface (Fig. 2.4c). The ultrasound transducer and the LED arrays are cased in a holder that illuminates the imaging plane at 12° from either side of the probe. For the study of the human vasculature, 805 nm LEDs were used.

200 μJ/pulse of energy was deposited on the surface by 4 rows of 36 LED elements each of 1 × 1 mm. 30–150 ns pulse width illumination was used. The illumination on the surface was a rectangular area of (~ 9 cm^2) which implies the fluence was 0.044 mJ/cm^2 at a PRF of 1–4 kHz. A linear array probe made of 128 elements (pitch of 0.3 mm), with a center frequency of 7 MHz, elevation focus of 15 mm, and a bandwidth of 6 dB was used for imaging. The authors compared the performance of the LED-PACT system with OPO lasers operating at 5–7 ns at a PRF of 10 Hz, and pulse energy of 140 mJ/pulse at 750 nm. 850 nm was used to compare the experimental results. The pulse energy was recorded as 40 mJ/pulse. The laser beam was coupled into a 2 m long optical fiber bundle with an input diameter of 6.5 mm. The illumination pattern on the skin was kept similar to the pattern of LED illumination. With increased averaging of the signal from the LED-PACT system the SNR of the output matched with the output of OPO based system operated at 10 Hz. For human vasculature imaging, the SNR of the images from 10 Hz OPO laser was slightly higher. LED-based PACT system holds an advantage for clinical translation, as the light illumination is not lasers, the safety and regulatory concerns are minimized.

Peng et al. demonstrated the use of a clinical PA-US system for photothermal therapy [80]. The relative change in the PA signal is mapped to the rise in the temperature during the therapy, The images were acquired to study the distribution of the photothermal agents. This information was used to optimize the interval of administration therapy in mice with sub-cutaneous injected tumors. The combined dual-mode PA-US system was combined with a photothermal system. 128-element array transducer (L11-5 V, Verasonics system) was used whose bandwidth, elevation focus, aperture, and pitch were 80%, 18 mm, 38 mm, and 0.3 mm, respectively. An OPO tunable laser which produced 5 ns pulses with a repetition rate of 20 Hz generating a laser energy of 70 mJ at 950 nm was used as the illumination source for PA imaging. Photothermal therapy was done using a continuous wave laser producing energy of 28 mJ/cm^2 and power density of 0.6 W/cm^2 on the tumor. Actual temperature change in the tumor was monitored using a thermocouple probe into the tumor at different locations. The experiments demonstrated a positive correlation between the accumulation of photothermal agents and the effective treatment area.

Even though the linear array-based systems can be easily translated to clinical applications, the data acquired is only two-dimensional. Another limitation of linear array transducers is the limited field of view compared to full-ring array transducers. Doing 3D volumetric imaging is difficult if not impossible. One can use a translation stage or a mechanical robotic arm (with precise position recording) to scan the linear array and do volumetric imaging. Dynamic imaging of a volume is crucial for small animal imaging and clinical translation of PACT. Therefore, for full 3D volumetric photoacoustic imaging matrix transducers can be used. In the next two sub-section, we discuss volumetric PA imaging.

2.2.4 Planar Matrix-Based Systems

Wang et al. designed the matrix array (2D array) of the ultrasound transducer which enabled volumetric photoacoustic imaging [81]. A commercial iU22, Philips healthcare system was modified to accommodate X7–2 planar matrix array. The matrix array was made of 2500 elements and had a bandwidth of 2–7 MHz. 3D registered PA-US images were acquired using a Q-switched Nd: YAG laser. This setup mitigated the need for mechanical scanning when using a linear array probe. Sentinel lymph node imaging was done using this system. Spatial resolution was ~0.67 mm along the azimuthal direction, and ~ 0.9 mm along the axial direction. The light was delivered to the surface using an optic fiber on either side of the probe. A 3D back projection algorithm was used to reconstruct the PACT images.

One of the challenges with the light delivery for both the linear and the matrix array probe is integrating the fiber bundle with the probe. It is sometimes difficult and needs to be done while manufacturing the ultrasound probe itself. Doing the light delivery afterward involves an ad hoc 3D printed fiber bundle holder or separate light delivery holder, which are not very compact, aesthetically appealing, and convenient for handheld scanning. Moreover, side light illumination sometimes makes non-uniform light delivery, especially when the center of the imaging region lacks enough light reaching there, especially for large matrix array probes. Liu et al. designed a planar matrix transducer with an illumination source from the center of the transducer [82]. The transducers are mostly made of piezoelectric material which is opaque. The design of the transducer with a hole in the center is crucial for the uniform distribution of the light across the imaging axis. The lens and collimator were coupled with the transducer to expand the beam. The PA-US hand-held probe weighed only 44 g (Fig. 2.5a). The performance of the center-holed 2D-matrix aperture-based photoacoustic imaging system was demonstrated by tracking a needle in a rat's abdomen. The mean CNR was estimated to be 14.3 ± 3 dB. The handheld probe was used to image the cephalic vein in the human arm, and the CNR of these images was ~5 dB. The lateral resolution of the system at depths 11 mm and 14 mm was 0.70 mm and 0.80 mm, respectively. The axial resolution was 0.72 mm at 11 mm and 0.74 mm at 14 mm.

Vaithilingam et al. described a 2D array probe using a capacitive micromachined ultrasound transducer (CMUT) to acquire volumetric PA images [83]. A 16×16 elements CMUT was integrated for the PA-US imaging. Each element pitch was 250 μm, the membrane diameter was 30 μm, and the cavity thickness was 0.1 μm. The silicon wafer resistivity was more than 1000 Ω-cm. The advantage of CMUT design is that the transducer elements can be customized based on the application. A Q-switched Nd: YAG laser combined with OPO was used as the illumination source. The pulse width was 5 ns and the repetition rate of the laser was 10 Hz. The setup is shown in Fig. 2.5b. The designed transducer was used for volumetric imaging of the fishing line and chicken breast phantom. The lateral resolution for pulse-echo mode was 405 μm, and for PA mode it was 648 μm. The axial resolution for pulse echo

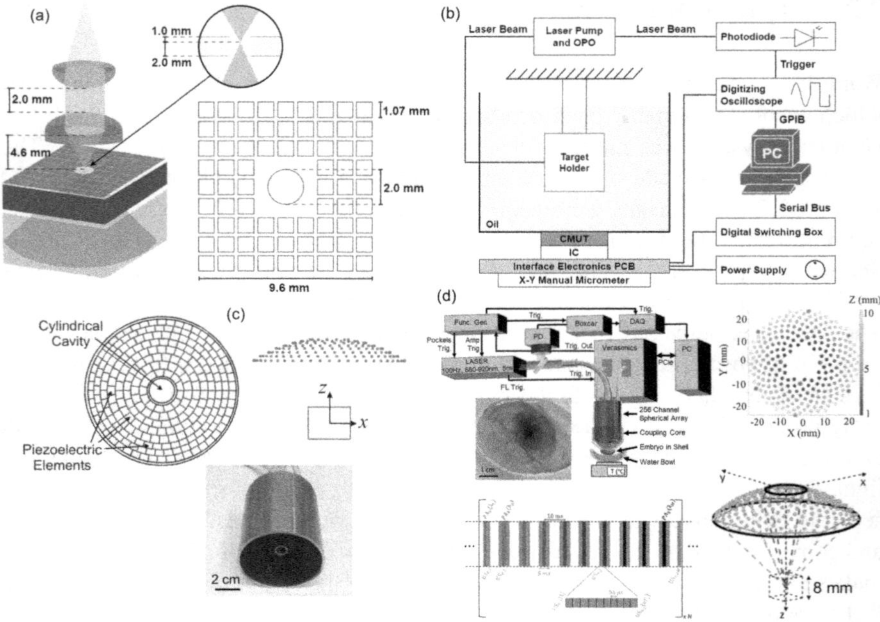

Fig. 2.5 Volumetric PACT systems using planar matrix and spherical distribution of the transducer elements. (**a**) Schematic of the handheld 2D matrix array with the geometric distribution. Reproduced from [82]. (**b**) Experimental set-up for CMUT-based 2D array transducer. Reproduced from [83]. (**c**) 512-element spherical transducer-based PA system where the transducer elements are packed in a matrix. Reprinted from [84, 85] with permission. (**d**) Spherical transducer where the transducer elements were distributed in a sunflower pattern. Reprinted from [86] with permission

mode was 361 μm and for PA imaging it was 300 μm. The SNR of the phantom images was 87 dB.

The developments in the planar matrix-based transducers enable volumetric imaging in a single laser pulse. However, the planar matrix-based arrays have the limitation of angular coverage (leading to limited view artifact) and high cost of fabrication. The limited view problems can be overcome by a hemispherical transducer surface.

2.2.5 Spherical Transducer-Based PACT System

3D photoacoustic imaging using a spherical matrix transducer was reported by Dean-ben et al. a decade ago [84]. The system designed could acquire volumes of PA images at a 10 Hz volume rate. This enabled real-time visualization of a region of interest of $12 \times 12 \times 10$ mm^3. The rate of volume acquisition was limited only by the repetition rate of the laser used. The transducer manufactured by Imasonics had 256 elements distributed on a hemisphere of radius 40 mm and outer diameter of 64 mm (Fig. 2.5c). The field of view of the transducer was 90°. The spherical design of the transducer reduced the limited view artifact of the planar matrix system. The transducer was fabricated using piezo-composite technology. The center frequency and bandwidth of the transducer were 3.9 MHz and 4 MHz, respectively. The spherical transducer was filled with water and a thin plastic film or agar base for the acoustic coupling. The transducer has a central cylindrical cavity for optical fiber of diameter 8 mm. The OPO laser used for human vasculature imaging was operated at 680–900 nm and the pulse width was 10 ns. The illumination area was ~1 cm in diameter at a distance of 3 cm from the center of the transducer. The spatial resolution of the system was 200 μm.

Recently, another 512-element-based spherical array transducer was designed by Godefroy et al. [86, 87]. The designed system could acquire images of blood oxygenation and blood flow dynamics. PA images were used to compute blood oxygenation and ultrasound images were used for the flow dynamics. The performance of the MS-PAFI (multispectral photoacoustic fluctuation imaging) system was demonstrated on chicken embryos. The transducer elements were distributed in a sunflower pattern as shown in Fig. 2.5d. The distribution was sparse and less than the Nyquist spatial sampling rate. Nonetheless, the distribution pattern reduced the grating lobes that were induced due to sub-sampling. The region of interest that could be imaged by this transducer was $8 \times 8 \times 8$ mm^3. OPO laser operated at 100 Hz delivered the illumination to the imaging surface through an optical fiber bundle. The center frequency of the transducer was 8 MHz. The inner diameter of the transducer was 10 mm, and the outer diameter was 50 mm. Each of the 256 elements was 2 mm in diameter and the divergence angle was ~137°. The focus of the transducer was 35 mm. The center frequency of the transducer was 8 MHz, with a − 6 dB relative bandwidth (one-way) of 80%. Laterally the resolution for PA and US imaging modality were 0.27 mm and 0.22 mm, respectively. Axial resolution was 0.35 mm and 0.23 mm for PA and US images, respectively.

The spherical array transducers reduce the problem of limited view. Nonetheless, these transducers are bulky for free-hand human imaging. Another issue that limits the translation of these PA systems into clinics is the coupling medium that needs to be filled between in the hemispherical space. Hence, fabrication technologies are being explored to reduce the size. Compact size enables wearable sensors for continuous monitoring of physiological parameters. Other than the types of array transducers we discussed so far, there are a few other custom-made special-design transducers that are used in PA imaging. In the next section, we will discuss those.

2.2.6 Special Design Transducers

Pan et al. designed a micro-ring sensor array-based PACT system [88]. The authors fabricated 15 high-quality micro-ring array sensors and coupled them with a bus waveguide which was made of chalcogenide material, a composition of $Ge_{25}Sb_{10}S_{65}$. The Young's modulus of the material was 31.9 GPa and the photoelastic coefficient was 0.23. Each sensor had a diameter of 40 μm, height of 850 nm, and width of 400 μm. The bandwidth of the sensors was 175 MHz with an acceptance angle of ±30°. 15 transducers were integrated using a bus waveguide. These transducers were tuned to a slightly offset resonant frequency to generate a digital optical frequency comb. The developed system enables PACT imaging using a single laser pulse and a single photoreceiver. Moving hot spots were generated on black tape by focusing a laser using a galvo mirror set-up to study the PACT capability of the micro-ring array sensor to detect moving objects. Zebrafishes of 7 days to 20 days old were imaged to show the growth of the organs. Even though the resolution of images was acceptable for phantom images, the resolution of the *in vivo* data was not as expected due to the attenuation of high-frequency signals in tissue.

Gao et al. fabricated a patch with a light source and transducers for PACT imaging [89]. Vertical-cavity surface-emitting laser (VCSEL) diodes produced 850 nm which penetrated the tissue up to 2 cm. The power of a single VCESL chip was 40 W. The average power of the patch, considering a 24-chip footprint of 2 × 1.6 cm, was 1.8×10^3 Wm^{-2}. MATLAB-based Monte Carlo simulations and k-wave simulations were done to study the penetration depth. The resolution of images acquired using the patch was 0.4 mm. The center frequency of the fabricated patch was 2.7 MHz. Laser excitation and data collection were managed by the Verasonics Vantage 256 system. The coherence-Factor-weighted-Delay-And-Sum algorithm was used to reconstruct the images. The application of the patch was demonstrated for *in vivo* mapping of hemoglobin in 3D, imaging of veins in the hand, feet, and thigh during venous occlusion test. Even though the SNR of the *in vivo* images is not at par with the traditional PA systems, the advantage of this fabrication is the compactness of the source and the detector.

Interestingly Li et al. proposed an ergodic delay for single-shot acquisition of photoacoustic images, which was called PATER [90]. Zhao and Wang proposed spatiotemporal encoder-based photoacoustic imaging for acquiring PA imaging (PAISE) [91]. These techniques enable imaging of the field-of-view (FOV) without having to mechanically scan the ultrasound transducer around the object or using an array of ultrasound transducers. A FOV of 8×4 mm^2 and 6×6 mm^2 was imaged using the PATER and PAISE system. The FOV is limited by the aperture of the ultrasonic pipe that is used to delay the PA signal.

The fabricated PAI sensors are important for continuous monitoring. Nonetheless, the bulkiness of the data processing set-up still limits the practical use of these configurations. The quality of the images from the fabricated sensors is not good for diagnostic purposes. As a monitoring device, it is feasible to use these to detect the signal changes over time. So far, we have discussed various types of PACT systems based on the transducers they use. In the next section, we will focus on the applications of PACT for pre-clinical and clinical use.

2.3 Applications of PACT

The applications of PACT for preclinical and clinical studies have been discussed in detail by Gu et al. [27]. PACT imaging systems are widely used to study the brain, cardiovascular system, adipose tissue, breast, muscles, stomach, liver, intestine, and kidney of rats, mice, rabbits, and monkeys with or without contrast agents [92–117]. With recent advances in the PACT system, whole-body imaging of small animals is a growing area of interest. PACT has a major role to play in tumor imaging due to its capability of differentiating oxygenated and de-oxygenated blood. Imaging of human extremities, digestive tract, and tumors that are either surficial or accessible through endoscopic probes are the potential uses of PACT systems [118–123]. Here we delve into a few of the interesting applications of PACT for small animal and human imaging.

2.3.1 *Small Animal Imaging*

The first developed SET-PACT was used to image the brain vasculature of the rat with the skull intact [124]. Blood vessels of a minimum thickness of ~30 μm were seen in the cross-sectional images. The acquired images were compared with the histological images to ensure the PA images corresponded well with the anatomy being imaged. The optical contrast between the tissue and the blood vessel was 2.3 and 7.9. The capability of the system for anatomical abnormality imaging was studied on brain lesions. Surficial cortical lesions were induced by piercing the right cortex with a syringe needle. The contrast between the lesion and the surrounding tissue was 1.7–5.2. The capability of PACT for functional imaging was demonstrated using a whisker stimulation experiment. For the change in the activation pattern, the baseline image was subtracted from PA images acquired during the left or right whisker stimulation. Vascular patterns that were seen in the structural images corresponded well with the functional signals, which implies that PACT can detect the change in hemodynamics during an activity.

Lin et al. proposed the volumetric PACT imaging of rat heart to study cardiac hemodynamics [93, 125]. Structural information such as the aorta and coronary artery was visualized using the system. A small region of interest was selected to study the hemodynamics in rats that were healthy, obese, and the ones that had hypertension. The variation in PA signal was observed during ventricular systole (Fig. 2.6). The relative changes in the volume of the left ventricles were also measured. The variations were correlated to the accumulation of fat in the arteries of the diseased rats. Periyasamy et al. reported the real-time monitoring of laser ablation using photoacoustic imaging [126]. Breast cancer cells, 4 T1 were subcutaneously injected in the abdomen of the mice. The developed tumors were ablated using a 100 mJ short-pulsed laser. The photoacoustic signal generated during the ablation was monitored using the spherical array transducer. The PA volume

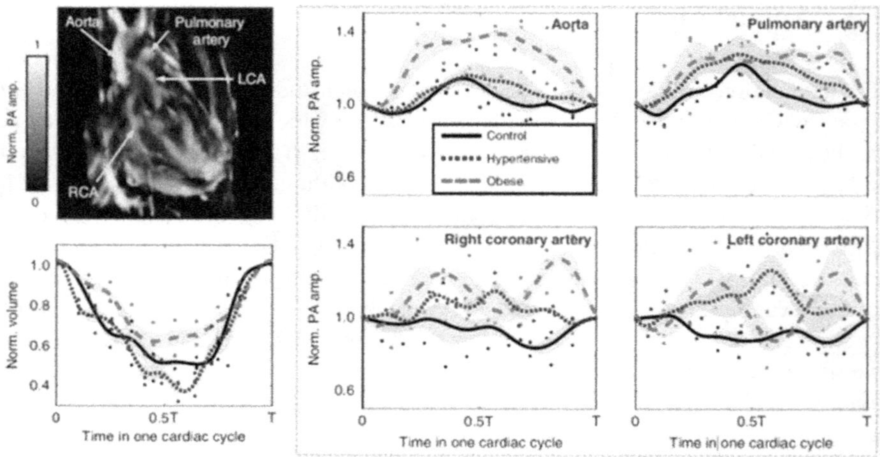

Fig. 2.6 PA signal of cardiac hemodynamics of control, hypertensive, and obese rat. Adapted from [125] with permission

acquired for each pulse enabled real-time visualization of the disintegration of the blood vessels in the tumor. The change in the PA signal corresponded well with the change in the temperature from the thermocouples inserted in the tumor.

Zhang et al. integrated MRI and PACT image acquisition [127]. A commercial multispectral PACT system (MSOT inVision128, iThera Medical, Germany) and a 7-tesla MRI scanner (Pharmascan, Burker, Germany) were used to acquire images of nude mice that were healthy and cancerous. The radial resolution of the system was ~150 μm, while the axial resolution of MRI was ~100 μm but the axial resolution of the PACT was ~800 μm. The authors designed an animal bed that enables dual-modality imaging. The bed consisted of a silicone-based breathing mask and fixing plates on either side of the animal to secure the limbs. This animal bed was mounted into a 3D-printed MRI support or PACT support. Chinese ink was used to mark the anatomical points in the mice to register the images in MRI and PACT. MRI is acquired in the XY plane, and the reconstructed images are sliced in the XZ and YZ planes for coronal and sagittal images. Gaussian fitting of the Chinese ink is computed from the PACT images for axial registration with the MRI data. An automated rigid registration information algorithm was used for transverse registration of the images. The dice similarity coefficient was used to compute the accuracy of co-registration. Spectral unmixing was done to extract the oxygenated and deoxygenated blood characteristics. The registered images from both modalities provided complementary information about the tumor, which is useful for tumor analysis and disease diagnosis.

Tang et al. demonstrated the use of clinically approved microbubbles for improving the PACT signals around the hemoglobin which helps to mitigate the limited-detection-view problem [128]. The proposed technique was called virtual-point source PACT, which enables high-fidelity functional PACT images. A linear (L7–4) array-based PACT system of 10 Hz and 1064 nm was used to image healthy rats. Phospholipid microbubbles filled with perfluorocarbon whose mean diameter was ~4 μm were used to enhance the PACT signals. Singular value thresholding was done to differentiate between the scattered, non-scattered, and white noise. The abdomen of a female Wistar rat was prepared to image the liver during the hypoxia challenge. 100 μl of microbubbles was injected through the tail vein. The accumulating microbubbles enable the visualization of the blood vessels which would otherwise not be visible in the images. The authors also demonstrated that microbubbles act as virtual sources of PA waves using k-wave simulations.

The PLD-based single-element PACT system described in Sect. 2.2.1 was used by Rajendra et al. to study the changes occurring in venous sinuses during intracranial hypotension (IH) [129]. IH was induced in rats by extracting cerebrospinal fluid (CSF) from cisterna magna which provides easy accessibility to sub-arachnoid space. ~45 μl of CSF was withdrawn using a needle with a draw syringe by positioning the rat head at an angle of 45°. 8 images were acquired before the CSF withdrawal and 37 images were acquired after the withdrawal every 4 min. Predominant changes were observed in the cerebral venous sinuses. Since sagittal sinus (SS) is a diagnostic indicator of IH, changes in SS to transverse sinus (in terms of area) and the width of SS was measured in the PA images. The area increased from 1.04 to 1.4 (34%) after the extraction. The width of SS increased from 493 μm to 677 μm in 5 min post-injection and at 149 min it was 705 μm post-injection.

Photoacoustic imaging was done on mice to study ischemic stroke by Li et al. [130]. Ischemic stroke is caused by the blockage of an artery that supplies the blood to the brain. The ischemic stroke model was induced by a set of steps. In the middle cerebral artery occlusion (MCAO) technique, the middle cerebral artery is clogged with a suture. This technique prevents blood flow to the brain from all possible vessels. The common carotid artery (CCA) and external carotid artery are ligated to prevent bleeding. The MCAO was considered complete when the suture reached the access point of the MCA. The tunable laser was operated at 532 nm and 800 nm. The PA signals were acquired by a 512-element full-ring array transducer. ICG was injected for molecular imaging. Images were acquired every 0.2 s. The images were reconstructed using the universal back projection algorithm. The blood vessels on the right hemisphere were not detectable in the images acquired post-surgery as expected. The PA intensity across the left and right brain was 2.15 in the mice that had undergone ligation while it was only 1.01 in the mice that had not undergone the surgery.

The PACT systems are capable of full-body small animal imaging and can be applied for various applications. Although human imaging applications are the final goal, being able to do pre-clinical applications plays a significant role in future clinical translation. In the next section, we will discuss some of the progress in human imaging with the PACT system.

2.3.2 Human Imaging

Pan et al. proposed augmented reality to visualize the blood vessels in a human arm [131]. PACT images were acquired by a hyperbolic-array transducer and a vision-based tracking system was used to project the blood vessel on the arm. The LOIS system described in Sect. 2.2.2 was used for imaging tumors in the breast. 42 patients were selected for the study, out of which 15 of them were scheduled for radical mastectomy. The radiographic images of the samples post-photoacoustic imaging revealed no thermoelastic or laser damage to the tissue. This ensures the safety of the use of PA for clinical applications. In the next phase of this study, subjects who were suspected to have abnormal tissue detected in US or mammography were imaged using PA, to analyze the capability of this imaging modality to detect the abnormality. Post-PA imaging biopsy was performed to study the tissue further. PA was not successful in identifying the tumors in 8 patients. Two of the image sets had insufficient contrast. The rest of the tumors were missed out due to either system malfunction or radiologist error.

Lin et al. studied the application of PACT during neoadjuvant chemotherapy for improving the outcome of breast cancer [95]. The single breath hold (SBH) PACT provided spatial and temporal resolution better than MRI even without the injection of the contrast agent. 1064 nm Nd: YAG laser was expanded using a diffuser such that the laser fluence of the circular beam incident on the breast was 28 mJ/cm^2. The laser was operated at a pulse repetition rate of 20 Hz. A 512-element spherical transducer whose center frequency was 2.25 MHz was used to detect the generated PACT signals. A universal back-projection algorithm was used to reconstruct the images acquired at a delay of 100 μs delay after each laser pulse. 225 μm and 5.6 mm were the in-plane and elevation plane resolution respectively. SBH-PACT was performed on 3 patients for both breasts at 3 different time points. The results demonstrated the feasibility of SBH-PACT for monitoring the effect of chemotherapy after breast cancer imaging.

As with vascular diseases, the capability to image hemoglobin (specifically Hb and HbO_2) provides an avenue for functional imaging of the brain [43, 132, 133]. As a method for functional brain imaging PACT has shown comparative results to conventional fMRI in a developmental study. The configuration was a functional photoacoustic computed tomography (fPACT) system that included 1024 ultrasound transducers in 4 quarter-ring arrays. These arrays were evenly placed in a hemispherical pattern to encircle a portion of a human head. The laser used to target Hb and HbO_2 for PA excitation was a dual-wavelength, 1064 nm/694 nm laser. The scanning was done in a dual-scanning mode. The two modes were termed 'baseline', which provided brain structure through brain angiography, and 'functional' which indicated brain activity. This configuration produced a field of view on the head of 10 cm in diameter, an isotropic spatial resolution of 350 μm, and a temporal resolution of 10 s and 2 s for the baseline and functional modes, respectively. The 3D fPACT system can be configured to image up to ~20 mm from the skin surface. This is deep enough for the entire cortical thickness of the cerebral gyri

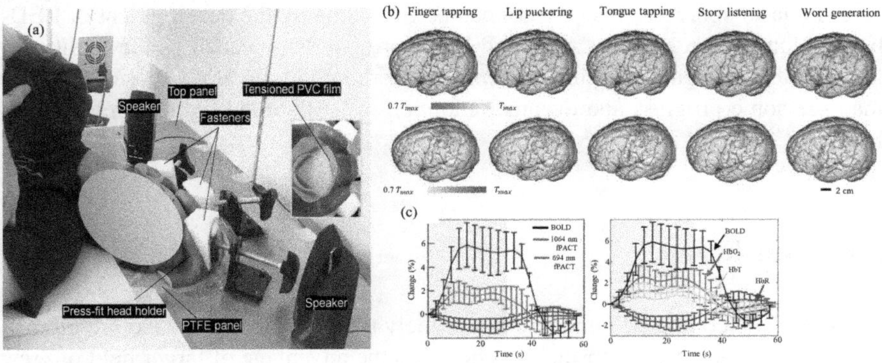

Fig. 2.7 PACT imaging of the human brain. (**a**) Positioning of the subject head for PACT. (**b**) MRI (top row) and PACT images (bottom row) acquired during different activities. (**c**) Temporal activation during the activity in the region of interest. Adapted from [43, 132] with permission

to be imaged. Figure 2.7a shows the positioning of the patient in the PACT setup for brain imaging. Figure 2.7b is the PACT images and MR images obtained during different cognitive tasks (finger tapping, lip puckering, tongue tapping, story listening, and word generation). The fractional changes in the Hb and BOLD signals are shown in Fig. 2.7c. Along with the baseline imaging, standard motor function, and language tests were given to the subjects in both the MRI and the PACT systems. The brain angiography to acquire the baseline image showed strong agreement between PACT and MRI in quantitative measures. The image results from the motor tasks were measured by spatial correspondence between the t-maps and showed moderately strong correspondence in the Dice coefficient, spatial correlation, and center-of-mass error. The language areas were assessed similarly and showed fair to moderate agreement. A patient with a metal implant was successfully imaged using fPACT, which could not be done in the fMRI counterpart without reported levels of discomfort. The current limitations of this technology are depth considerations as the signal agreement was largely demonstrated within a 1 cm depth of the cortical surface. Additionally, technology is continuing to develop to provide the opportunity for transcranial PACT [132, 134].

For a new medical device to be applied for human application, Food and Drug Administration (FDA) approval is necessary. In the year 2022, Seno Medical's Imagio Breast Imaging System [135] received supplementary premarket approval (PMA) from the Center for Devices and Radiological Health (CDHR) of the FDA. Hopefully, with this start other medical device companies working on photoacoustic technology will be encouraged and we will see more and more approval for clinical trials and eventual progress to regular clinical practice. Although we expect breast cancer imaging to be one of the first applications photoacoustic may be used, there are other potential application areas where PA may find its application. Some of the research work described in the earlier sections eventually points towards those. Another safety concern that may arise with the PACT system is the use of traditional class IV laser sources. This concern can be overcome by the development of LED-based illumination sources. Although LEDs are not as powerful as the traditional lasers at the moment, with time and more research going on along those directions may see some progress and eventual adoption of LED-based light sources for PA imaging.

2.4 Conclusion

Photoacoustic computed tomography is widely used for laboratory and pre-clinical studies. Single-element transducer system has the advantage of larger field of view for signal acquisition which plays a major role in the SNR. Nonetheless, the time taken for image acquisition makes it slow, and real-time imaging may not be possible, limiting its use in clinical scenarios. Hence, there are array and matrix-based transducers. An advantage of the linear and curved array transducers is that they are widely used by medical practitioners for ultrasound imaging. Integration of the light source with the transducer remains an area of research. Illumination source for PACT has been developed from bulky expensive Nd: YAG lasers to compact portable LED. With the boom in machine learning and deep learning algorithms, the quality of images from LED-based systems is expected to improve. PACT is widely used in small animals to develop contrast agents and disease models. The application of PACT for clinical imaging is still a developing field. With the recent FDA approval of photoacoustic imaging systems for breast imaging, we expect more and more applications of photoacoustic imaging in the future.

References

1. Acharya, R., R. Wasserman, J. Stevens, and C. Hinojosa, *Biomedical imaging modalities: a tutorial.* Computerized Medical Imaging and Graphics, 1995. **19**(1): p. 3–25.
2. Wang, L.V. and H.-I. Wu, *Biomedical Optics: Principles and Imaging*. 2012, New Jersey: John Wiley & Sons, Inc.
3. Solomon, M., Y. Liu, M.Y. Berezin, and S. Achilefu, *Optical imaging in cancer research: basic principles, tumor detection, and therapeutic monitoring.* Medical Principles and Practice, 2011. **20**(5): p. 397–415.
4. Hillman, E.M., C.B. Amoozegar, T. Wang, A.F. McCaslin, M.B. Bouchard, J. Mansfield, and R.M. Levenson, *In vivo optical imaging and dynamic contrast methods for biomedical research.* Philos Trans A Math Phys Eng Sci, 2011. **369**(1955): p. 4620–43.
5. Arridge, S.R., *Optical tomography in medical imaging.* Inverse Problems, 1999. **15**(2): p. R41-R93.
6. Wang, L.V., *Optical Tomography for Biomedical Applications*, in *IEEE Engineering in Medicine and Biology*. 1998. p. 45–46.
7. Tuchin, V.V., I.L. Maksimova, D.A. Zimnyakov, I.L. Kon, A.H. Mavlyutov, and A.A. Mishin, *Light propagation in tissues with controlled optical properties.* Journal of Biomedical Optics, 1997. **2**(4): p. 401–417.

8. Zeng, Y., X. Da, Y. Wang, B. Yin, and Q. Chen, *Photoacoustic and ultrasonic coimage with a linear transducer array.* Optics Letters, 2004. **29**(15): p. 1760–62.
9. Oraevsky, A., S. Jacques, R. Esenaliev, and F. Tittel. *Laser-based optoacoustic imaging in biological tissues*. in *Proc SPIE*. 1994. SPIE.
10. Kruger, R.A., *Photoacoustic ultrasound.* Medical Physics, 1994. **21**(1): p. 127–131.
11. Gusev, V.E. and A.A. Karabutov, *Laser Optoacoustics*. 1993, New York: American Institute of Physics.
12. Zhu, J., X. He, and Z. Chen, *Perspective: Current challenges and solutions of Doppler optical coherence tomography and angiography for neuroimaging.* APL Photonics, 2018. **3**(12).
13. Yao, J. and L.V. Wang, *Sensitivity of photoacoustic microscopy.* Photoacoustics, 2014. **2**(2): p. 87–101.
14. Hu, S. and L.V. Wang, *Neurovascular photoacoustic tomography.* Front Neuroenergetics, 2010. **2**: p. 10.
15. Bell, A.G., *On the production and reproduction of sound by light.* American Journal of Science, 1880. **20**(118): p. 305–324.
16. Manohar, S. and D. Razansky, *Photoacoustics: a historical review.* Advances in Optics and Photonics, 2016. **8**(4): p. 586–617.
17. Hui, X., M.O.A. Mohammad, and M. Pramanik, *Looking deep inside tissue with photoacoustic molecular probes: a review.* Journal of Biomedical Optics, 2022. **27**(7): p. 070901.
18. Xu, M. and L.V. Wang, *Photoacoustic imaging in biomedicine.* Review of Scientific Instruments, 2006. **77**(4): p. 041101.
19. Cho, S.W., S.M. Park, B. Park, D.Y. Kim, T.G. Lee, B.M. Kim, C. Kim, J. Kim, S.W. Lee, and C.S. Kim, *High-speed photoacoustic microscopy: A review dedicated on light sources.* Photoacoustics, 2021. **24**: p. 100291.
20. Yang, W., W. Wang, L. Jing, and S.L. Chen, *Label-free photoacoustic microscopy: a potential tool for the live imaging of blood disorders in zebrafish.* Biomedical Optics Express, 2021. **12**(6): p. 3643–3657.
21. Zhang, C., H. Zhao, S. Xu, N. Chen, K. Li, X. Jiang, L. Liu, Z. Liu, L. Wang, K.K.Y. Wong, J. Zou, C. Liu, and L. Song, *Multiscale high-speed photoacoustic microscopy based on free-space light transmission and a MEMS scanning mirror.* Optics Letters, 2020. **45**(15).
22. Baik, J.W., J.Y. Kim, S. Cho, S. Choi, J. Kim, and C. Kim, *Super Wide-Field Photoacoustic Microscopy of Animals and Humans In Vivo.* IEEE Transactions on Medical Imaging, 2020. **39**(4): p. 975–984.
23. Guo, H., Y. Li, W. Qi, and L. Xi, *Photoacoustic endoscopy: A progress review.* Journal of Biophotonics, 2020. **13**(12): p. e202000217.
24. Ansari, R., E.Z. Zhang, A.E. Desjardins, and P.C. Beard, *Miniature all-optical flexible forward-viewing photoacoustic endoscopy probe for surgical guidance.* Optics Letters, 2020. **45**(22): p. 6238–6241.
25. He, H., A. Stylogiannis, P. Afshari, T. Wiedemann, K. Steiger, A. Buehler, C. Zakian, and V. Ntziachristos, *Capsule Optoacoustic Endoscopy for Esophageal Imaging.* Journal of Biophotonics, 2019. **12**(10): p. e201800439.
26. Yang, J.-M., C. Favazza, R. Chen, J. Yao, X. Cai, K. Maslov, Q. Zhou, K.K. Shung, and L.V. Wang, *Simultaneous functional photoacoustic and ultrasonic endoscopy of internal organs in vivo.* Nature Medicine, 2012. **18**(8): p. 1297–1302.
27. Gu, Y., Y. Sun, X. Wang, H. Li, J. Qiu, and W. Lu, *Application of photoacoustic computed tomography in biomedical imaging: A literature review.* Bioeng Transl Med, 2023. **8**(2): p. e10419.
28. Yang, J., S. Choi, and C. Kim, *Practical review on photoacoustic computed tomography using curved ultrasound array transducer.* Biomedical Engineering Letters, 2022. **12**(1): p. 19–35.
29. Wang, K., C. Li, R. Chen, and J. Shi, *Recent advances in high-speed photoacoustic microscopy.* Photoacoustics, 2021. **24**: p. 100294.
30. Tian, C., C. Zhang, H. Zhang, D. Xie, and Y. Jin, *Spatial resolution in photoacoustic computed tomography.* Reports on Progress in Physics, 2021. **84**(3).

31. Jeon, S., J. Kim, D. Lee, J.W. Baik, and C. Kim, *Review on practical photoacoustic microscopy.* Photoacoustics, 2019. **15**: p. 100141.
32. Yang, J.M., K. Maslov, H.C. Yang, Q.F. Zhou, K.K. Shung, and L.V. Wang, *Photoacoustic endoscopy.* Optics Letters, 2009. **34**(10): p. 1591–1593.
33. Qi, W., X. Liang, Y. Ji, C. Liu, and L. Xi, *Optical resolution photoacoustic computed microscopy.* Optics Letters, 2021. **46**(2): p. 372–375.
34. Moothanchery, M., A. Sharma, and M. Pramanik, *Switchable Acoustic and Optical Resolution Photoacoustic Microscopy for in vivo small-animal blood vasculature imaging.* Journal of Visualized Experiments, 2017 (124): p. e55810.
35. Moothanchery, M. and M. Pramanik, *Performance Characterization of a Switchable Acoustic and Optical Resolution Photoacoustic Microscopy System.* Sensors, 2017. **17**(2): p. 357.
36. Periyasamy, V., N. Das, A. Sharma, and M. Pramanik, *1064 nm acoustic resolution photoacoustic microscopy.* Journal of Biophotonics, 2019. **12**(5): p. e201800357.
37. Moothanchery, M., K. Dev, G. Balasundaram, R. Bi, and M. Olivo, *Acoustic Resolution Photoacoustic Microscopy based on MEMS scanner.* Journal of Biophotonics, 2019.
38. Park, S., C. Lee, J. Kim, and C. Kim, *Acoustic resolution photoacoustic microscopy.* Biomedical Engineering Letters, 2014. **4**(3): p. 213–222.
39. Yip, L.C., P. Omidi, E. Rascevska, and J.J. Carson, *Approaching closed spherical, full-view detection for photoacoustic tomography.* J Biomed Opt, 2022. **27**(8).
40. Rajendran, P. and M. Pramanik, *High frame rate (~3 Hz) circular photoacoustic tomography using single-element ultrasound transducer aided with deep learning.* Journal of Biomedical Optics, 2022. **27**(6): p. 066005.
41. Rajendran, P. and M. Pramanik. *Multi-transducer photoacoustic tomography imaging without radius calibration using deep learning.* in *Proc SPIE.* 2022.
42. Vu, T., Y. Tang, M. Li, G. Sankin, S. Tang, S. Chen, P. Zhong, and J. Yao, *Photoacoustic computed tomography of mechanical HIFU-induced vascular injury.* Biomedical Optics Express, 2021. **12**(9): p. 5489–5498.
43. Na, S. and L.V. Wang, *Photoacoustic computed tomography for functional human brain imaging [Invited].* Biomedical Optics Express, 2021. **12**(7): p. 4056–4083.
44. Li, L. and L.V. Wang, *Recent Advances in Photoacoustic Tomography.* BME Frontiers, 2021. **2021**: p. 9823268.
45. Poudel, J., S. Na, L.V. Wang, and M.A. Anastasio, *Iterative image reconstruction in transcranial photoacoustic tomography based on the elastic wave equation.* Physics in Medicine & Biology, 2020. **65**(5): p. 055009.
46. Mozaffarzadeh, M., A. Mahloojifar, V. Periyasamy, M. Pramanik, and M. Orooji, *Eigenspace-based minimum variance combined with delay multiply and sum beamforming algorithm: application to linear-array photoacoustic imaging.* IEEE Journal of Selected Topics in Quantum Electronics, 2019. **25**(1): p. 6800608.
47. Prakash, J., D. Sanny, S.K. Kalva, M. Pramanik, and P.K. Yalavarthy, *Fractional Regularization to Improve Photoacoustic Tomographic Image Reconstruction.* IEEE Transactions on Medical Imaging, 2019. **38**(8): p. 1935–1947.
48. Deán-Ben, X. and D. Razansky, *Optoacoustic image formation approaches—a clinical perspective.* Physics in Medicine & Biology, 2019. **64**(18): p. 18TR01.
49. Schoeder, S., I. Olefir, M. Kronbichler, V. Ntziachristos, and W.A. Wall, *Optoacoustic image reconstruction: the full inverse problem with variable bases.* Proceedings of the Royal Society A: Mathematical, Physical and Engineering Sciences, 2018. **474**(2219): p. 20180369.
50. Pramanik, M., *Improving tangential resolution with a modified delay-and-sum reconstruction algorithm in photoacoustic and thermoacoustic tomography.* Journal of the Optical Society of America A, 2014. **31**(3): p. 621–7.
51. Treeby, B.E. and B.T. Cox, *k-Wave: MATLAB toolbox for the simulation and reconstruction of photoacoustic wave fields.* Journal of Biomedical Optics, 2010. **15**(2): p. 021314.
52. Rajendran, P., A. Sharma, and M. Pramanik, *Photoacoustic imaging aided with deep learning: a review.* Biomedical Engineering Letters, 2022. **12**(2): p. 155–173.

53. Awasthi, N., G. Jain, S.K. Kalva, M. Pramanik, and P.K. Yalavarthy, *Deep neural network based sinogram super-resolution and bandwidth enhancement for limited-data photoacoustic tomography.* IEEE Transactions on Ultrasonics, Ferroelectrics, and Frequency Control, 2020. **67**(12): p. 2660–2673.
54. Kim, M.W., G.S. Jeng, I. Pelivanov, and M. O'Donnell, *Deep-learning Image Reconstruction for Real-time Photoacoustic System.* IEEE Transactions on Medical Imaging, 2020. **39**(11): p. 3379–3390.
55. Feng, J., J. Deng, Z. Li, Z. Sun, H. Dou, and K. Jia, *End-to-end Res-Unet based reconstruction algorithm for photoacoustic imaging.* Biomedical Optics Express, 2020. **11**(9): p. 5321–5340.
56. Davoudi, N., X.L. Deán-Ben, and D. Razansky, *Deep learning optoacoustic tomography with sparse data.* Nature Machine Intelligence, 2019. **1**(10): p. 453–460.
57. Gutta, S., V.S. Kadimesetty, S.K. Kalva, M. Pramanik, S. Ganapathy, and P.K. Yalavarthy, *Deep neural network-based bandwidth enhancement of photoacoustic data.* Journal of Biomedical Optics, 2017. **22**(11): p. 116001.
58. Wang, X., X. Xie, G. Ku, L.V. Wang, and G. Stoica, *Noninvasive imaging of hemoglobin concentration and oxygenation in the rat brain using high-resolution photoacoustic tomography.* Journal of Biomedical Optics, 2006. **11**(2): p. 024015.
59. ANSI, *American National Standard for Safe Use of Lasers ANSI Z136.1-2022 (American National Standards Institute, Inc., New York, NY, 2022).* 2022.
60. Sharma, A., S.K. Kalva, and M. Pramanik, *A comparative study of continuous versus stop-and-go scanning in circular scanning photoacoustic tomography.* IEEE Journal of Selected Topics in Quantum Electronics, 2019. **25**(1): p. 7100409.
61. Kalva, S.K. and M. Pramanik, *Use of acoustic reflector to make a compact photoacoustic tomography system.* Journal of Biomedical Optics, 2017. **22**(2): p. 1–7.
62. Kalva, S.K., P.K. Upputuri, and M. Pramanik, *High-speed, low-cost, pulsed-laser-diode-based second-generation desktop photoacoustic tomography system.* Optics Letters, 2019. **44**(1): p. 81–84.
63. Hui, X., P. Rajendran, M.A.I. Zulkifli, T. Ling, and M. Pramanik, *Android mobile-platform-based image reconstruction for photoacoustic tomography.* Journal of Biomedical Optics, 2023. **28**(4): p. 046009.
64. Allen, T.J. and P.C. Beard, *High power visible light emitting diodes as pulsed excitation sources for biomedical photoacoustics.* Biomedical Optics Express, 2016. **7**(4): p. 1260–1270.
65. Deng, Z., W. Li, and C. Li, *Slip-ring-based multi-transducer photoacoustic tomography system.* Optics Letters, 2016. **41**(12): p. 2859–2862.
66. Kalva, S.K., Z.Z. Hui, and M. Pramanik, *Calibrating reconstruction radius in a multi single-element ultrasound-transducer-based photoacoustic computed tomography system.* Journal of the Optical Society of America A, 2018. **35**(5): p. 764–771.
67. Zhu, Y., G. Xu, J. Yuan, J. Jo, G. Gandikota, H. Demirci, T. Agano, N. Sato, Y. Shigeta, and X. Wang, *Light Emitting Diodes based Photoacoustic Imaging and Potential Clinical Applications.* Scientific Reports, 2018. **8**(1): p. 9885.
68. Liu, X., S.K. Kalva, B. Lafci, D. Nozdriukhin, X.L. Dean-Ben, and D. Razansky, *Full-view LED-based optoacoustic tomography.* Photoacoustics, 2023. **31**: p. 100521.
69. Gamelin, J.K., A. Aquirre, A. Maurudis, F. Huang, D. Castillo, L.V. Wang, and Q. Zhu, *Curved array photoacoustic tomographic system for small animal imaging.* Journal of Biomedical Optics, 2008. **13**(2): p. 024007.
70. Ermilov, S.A., T. Khamapirad, A. Conjusteau, M.H. Leonard, R. Lacewell, K. Mehta, T. Miller, and A.A. Oraevsky, *Laser optoacoustic imaging system for detection of breast cancer.* Journal of biomedical optics, 2009. **14**(2): p. 024007–14.
71. Merčep, E., J.L. Herraiz, X.L. Deán-Ben, and D. Razansky, *Transmission–reflection optoacoustic ultrasound (TROPUS) computed tomography of small animals.* Light: Science & Applications, 2019. **8**(1): p. 18.

72. Nishiyama, M., T. Namita, K. Kondo, M. Yamakawa, and T. Shiina, *Ring-array photoacoustic tomography for imaging human finger vasculature.* Journal of Biomedical Optics, 2019. **24**(9): p. 096005.
73. Zhang, Y. and L. Wang, *Video-rate ring-array ultrasound and photoacoustic tomography.* IEEE Transactions on Medical Imaging, 2020. **39**(12): p. 4369–4375.
74. Erpelding, T.N., C. Kim, M. Pramanik, L. Jankovic, K. Maslov, Z. Guo, J.A. Margenthaler, M.D. Pashley, and L.V. Wang, *Sentinel Lymph Nodes in the Rat: Noninvasive Photoacoustic and US imaging with a clinical US system.* Radiology, 2010. **256**(1): p. 102–110.
75. Sivasubramanian, K., V. Periyasamy, and M. Pramanik, *Non-invasive sentinel lymph node mapping and needle guidance using clinical handheld photoacoustic imaging system in small animal.* Journal of Biophotonics, 2018. **11**(1): p. e201700061.
76. Sivasubramanian, K. and M. Pramanik, *High frame rate photoacoustic imaging at 7000 frames per second using clinical ultrasound system.* Biomedical Optics Express, 2016. **7**(2): p. 312–323.
77. Agrawal, S., M. Kuniyil Ajith Singh, K. Johnstonbaugh, C.H. D, R.P. C, and S.R. Kothapalli, *Photoacoustic Imaging of Human Vasculature Using LED versus Laser Illumination: A Comparison Study on Tissue Phantoms and In Vivo Humans.* Sensors (Basel), 2021. **21**(2).
78. Sivasubramanian, K., V. Periyasamy, K.K. Wen, and M. Pramanik, *Optimizing light delivery through fiber bundle in photoacoustic imaging with clinical ultrasound system: Monte Carlo simulation and experimental validation.* Journal of Biomedical Optics, 2017. **22**(4): p. 041008.
79. Sivasubramanian, K., V. Periyasamy, and M. Pramanik, *Hand-held Clinical Photoacoustic Imaging System for Real-time Non-invasive Small Animal Imaging.* Journal of Visualized Experiments, 2017. **128**(128): p. e56649.
80. Peng, K., Y. Wang, L. Li, J. Zhang, H. Chen, and J. Xiao, *In vivo photothermal therapy monitored by multi-position calibrated photoacoustic thermometer.* Photoacoustics, 2023. **31**: p. 100501.
81. Wang, Y., T.N. Erpelding, L. Jankovic, Z. Guo, J.-L. Robert, G. David, and L.V. Wang, *In vivo three-dimensional photoacoustic imaging based on a clinical matrix array ultrasound probe.* Journal of Biomedical Optics, 2012. **17**(6): p. 061208.
82. Liu, S., W. Song, X. Liao, T.T. Kim, and Y. Zheng, *Development of a handheld volumetric photoacoustic imaging system with a central-holed 2D matrix aperture.* IEEE Transactions on Biomedical Engineering, 2020.
83. Vaithilingam, S., T.J. Ma, Y. Furukawa, I.O. Wygant, X. Zhuang, A.D.L. Zerda, O. Oralkan, A. Kamaya, S.S. Gambhir, R.B. Jeffrey, and B.T. Khuri-yakub, *Three-dimensional photoacoustic imaging using a two-dimensional CMUT array.* IEEE Transactions on Ultrasonics, Ferroelectrics, and Frequency Control, 2009. **56**(11): p. 2411–2419.
84. Dean-Ben, X.L., A. Ozbek, and D. Razansky, *Volumetric real-time tracking of peripheral human vasculature with GPU-accelerated three-dimensional optoacoustic tomography.* IEEE Transations in Medical Imaging, 2013. **32**(11): p. 2050–55.
85. Dean-Ben, X.L. and D. Razansky, *Portable spherical array probe for volumetric real-time optoacoustic imaging at centimeter-scale depths.* Optics Express, 2013. **21**(23): p. 28062–71.
86. Godefroy, G., B. Arnal, and E. Bossy, *Full-visibility 3D imaging of oxygenation and blood flow by simultaneous multispectral photoacoustic fluctuation imaging (MS-PAFI) and ultrasound Doppler.* Sci Rep, 2023. **13**(1): p. 2961.
87. Vilov, S., G. Godefroy, B. Arnal, and E. Bossy, *Photoacoustic fluctuation imaging: theory and application to blood flow imaging.* Optica, 2020. **7**(11).
88. Pan, J., Q. Li, Y. Feng, R. Zhong, Z. Fu, S. Yang, W. Sun, B. Zhang, Q. Sui, J. Chen, Y. Shen, and Z. Li, *Parallel interrogation of the chalcogenide-based micro-ring sensor array for photoacoustic tomography.* Nat Commun, 2023. **14**(1): p. 3250.

89. Gao, X., X. Chen, H. Hu, X. Wang, W. Yue, J. Mu, Z. Lou, R. Zhang, K. Shi, X. Chen, M. Lin, B. Qi, S. Zhou, C. Lu, Y. Gu, X. Yang, H. Ding, Y. Zhu, H. Huang, Y. Ma, M. Li, A. Mishra, J. Wang, and S. Xu, *A photoacoustic patch for three-dimensional imaging of hemoglobin and core temperature.* Nat Commun, 2022. **13**(1): p. 7757.
90. Li, Y., L. Li, L. Zhu, K. Maslov, J. Shi, P. Hu, E. Bo, J. Yao, J. Liang, L. Wang, and L.V. Wang, *Snapshot Photoacoustic Topography Through an Ergodic Relay for High-throughput Imaging of Optical Absorption.* Nat Photonics, 2020. **14**(3): p. 164–170.
91. Zhao, Y. and L.V. Wang, *Single-shot photoacoustic imaging with single-element transducer through a spatiotemporal encoder.* J Biomed Opt, 2023. **28**(4): p. 046004.
92. Quiros-Gonzalez, I., M.R. Tomaszewski, M.A. Golinska, E. Brown, L. Ansel-Bollepalli, L. Hacker, D.L. Couturier, R.M. Sainz, and S.E. Bohndiek, *Photoacoustic Tomography Detects Response and Resistance to Bevacizumab in Breast Cancer Mouse Models.* Cancer Res, 2022. **82**(8): p. 1658–1668.
93. Lin, L., P. Hu, X. Tong, S. Na, R. Cao, X. Yuan, D.C. Garrett, J. Shi, K. Maslov, and L.V. Wang, *High-speed three-dimensional photoacoustic computed tomography for preclinical research and clinical translation.* Nature Communications, 2021. **12**(1): p. 882.
94. Chen, R., S. Huang, T. Lin, H. Ma, W. Shan, F. Duan, J. Lv, J. Zhang, L. Ren, and L. Nie, *Photoacoustic molecular imaging-escorted adipose photodynamic–browning synergy for fighting obesity with virus-like complexes.* Nature Nanotechnology, 2021. **16**(4): p. 455–465.
95. Lin, L., X. Tong, P. Hu, M. Invernizzi, L. Lai, and L.V. Wang, *Photoacoustic Computed Tomography of Breast Cancer in Response to Neoadjuvant Chemotherapy.* Adv Sci (Weinh), 2021. **8**(7): p. 2003396.
96. Li, M., T. Vu, G. Sankin, B. Winship, K. Boydston, R. Terry, P. Zhong, and J. Yao, *Internal-Illumination Photoacoustic Tomography Enhanced by a Graded-Scattering Fiber Diffuser.* IEEE Trans Med Imaging, 2021. **40**(1): p. 346–356.
97. Lee, J., H. Cho, S. Park, J. Lee, D. Seong, R.E. Wijesinghe, S. Han, S. Kim, M. Jeon, and J. Kim, *In Vivo Rodent Cervicothoracic Vasculature Imaging Using Photoacoustic Computed Tomography.* Photonics, 2021. **8**(8).
98. Huang, S., A. Blutke, A. Feuchtinger, U. Klemm, R. Zachariah Tom, S.M. Hofmann, A.C. Stiel, and V. Ntziachristos, *Functional multispectral optoacoustic tomography imaging of hepatic steatosis development in mice.* EMBO Mol Med, 2021. **13**(9): p. e13490.
99. Duan, Y., Z. Cheng, T. Qiu, L. Wen, and K. Xiong, *Spherical-matching hyperbolic-array photoacoustic computed tomography.* J Biophotonics, 2021. **14**(6): p. e202100023.
100. Huang, W., R. Chen, Y. Peng, F. Duan, Y. Huang, W. Guo, X. Chen, and L. Nie, *In Vivo Quantitative Photoacoustic Diagnosis of Gastric and Intestinal Dysfunctions with a Broad pH-Responsive Sensor.* ACS Nano, 2019. **13**(8): p. 9561–9570.
101. Zhang, P., L. Li, L. Lin, J. Shi, and L.V. Wang, *In vivo superresolution photoacoustic computed tomography by localization of single dyed droplets.* Light: Science & Applications, 2019. **8**(1): p. 36.
102. Zeng, L., G. Ma, J. Lin, and P. Huang, *Photoacoustic Probes for Molecular Detection: Recent Advances and Perspectives.* Small, 2018. **14**(30): p. e1800782.
103. Reber, J., M. Willershauser, A. Karlas, K. Paul-Yuan, G. Diot, D. Franz, T. Fromme, S.V. Ovsepian, N. Beziere, E. Dubikovskaya, D.C. Karampinos, C. Holzapfel, H. Hauner, M. Klingenspor, and V. Ntziachristos, *Non-invasive Measurement of Brown Fat Metabolism Based on Optoacoustic Imaging of Hemoglobin Gradients.* Cell Metab, 2018. **27**(3): p. 689–701.
104. Li, M., B. Lan, W. Liu, J. Xia, and J. Yao, *Internal-illumination photoacoustic computed tomography.* J Biomed Opt, 2018. **23**(3): p. 1–4.
105. Li, L., L. Zhu, C. Ma, L. Lin, J. Yao, L. Wang, K. Maslov, R. Zhang, W. Chen, J. Shi, and L.V. Wang, *Single-impulse Panoramic Photoacoustic Computed Tomography of Small-animal Whole-body Dynamics at High Spatiotemporal Resolution.* Nature Biomedical Engineering, 2017. **1**(5): p. 0071.
106. Lin, L., J. Yao, L. Li, and L.V. Wang, *In vivo photoacoustic tomography of myoglobin oxygen saturation.* Journal of Biomedical Optics, 2016. **21**(6): p. 61002.

107. Li, J., H. Xiao, S.J. Yoon, C. Liu, D. Matsuura, W. Tai, L. Song, M. O'Donnell, D. Cheng, and X. Gao, *Functional Photoacoustic Imaging of Gastric Acid Secretion Using pH-Responsive Polyaniline Nanoprobes*. Small, 2016. **12**(34): p. 4690–6.
108. Hennen, S.N., W. Xing, Y.B. Shui, Y. Zhou, J. Kalishman, L.B. Andrews-Kaminsky, M.A. Kass, D.C. Beebe, K.I. Maslov, and L.V. Wang, *Photoacoustic tomography imaging and estimation of oxygen saturation of hemoglobin in ocular tissue of rabbits*. Exp Eye Res, 2015. **138**: p. 153–8.
109. Xia, J., W.Y. Chen, K. Maslov, M.A. Anastasio, and L.V. Wang, *Retrospective respiration-gated whole-body photoacoustic computed tomography of mice*. Journal of Biomedical Optics, 2014. **19**(1): p. 016003.
110. Wilson, K.E., S.V. Bachawal, L. Tian, and J.K. Willmann, *Multiparametric spectroscopic photoacoustic imaging of breast cancer development in a transgenic mouse model*. Theranostics, 2014. **4**(11): p. 1062–71.
111. Yao, J.J., J. Xia, K.I. Maslov, M. Nasiriavanaki, V. Tsytsarev, A.V. Demchenko, and L.V. Wang, *Noninvasive photoacoustic computed tomography of mouse brain metabolism in vivo*. Neuroimage, 2013. **64**(1): p. 257–266.
112. Chatni, M.R., J. Xia, R. Sohn, K. Maslov, Z. Guo, Y. Zhang, K. Wang, Y. Xia, M. Anastasio, J. Arbeit, and L.V. Wang, *Tumor glucose metabolism imaged in vivo in small animals with whole-body photoacoustic computed tomography*. Journal of Biomedical Optics, 2012. **17**(7): p. 076012.
113. Xia, J., M.R. Chatni, K. Maslov, Z.J. Guo, K. Wang, M. Anastasio, and L.V. Wang, *Whole-body ring-shaped confocal photoacoustic computed tomography of small animals in vivo*. Journal of Biomedical Optics, 2012. **17**(5): p. 050506.
114. Xi, L., S.R. Grobmyer, G. Zhou, W. Qian, L. Yang, and H. Jiang, *Molecular photoacoustic tomography of breast cancer using receptor targeted magnetic iron oxide nanoparticles as contrast agents*. Journal of Biophotonics, 2012. **7**(6): p. 401–9.
115. Holotta, M., H. Grossauer, C. Kremser, P. Torbica, J. Volkl, G. Degenhart, R. Esterhammer, R. Nuster, G. Paltauf, and W. Jaschke, *Photoacoustic tomography of ex vivo mouse hearts with myocardial infarction*. J Biomed Opt, 2011. **16**(3): p. 036007.
116. Lu, W., Q. Huang, K.B. Geng, X.X. Wen, M. Zhou, D. Guzatov, P. Brecht, R. Su, A. Oraevsky, L.V. Wang, and C. Li, *Photoacoustic imaging of living mouse brain vasculature using hollow gold nanospheres*. Biomaterials, 2010. **31**(9): p. 2617–2626.
117. Yang, X. and L.V. Wang, *Monkey brain cortex imaging by photoacoustic tomography*. Journal of Biomedical Optics, 2008. **13**(4): p. 044009.
118. Chen, Y., C. Xu, Z. Zhang, A. Zhu, X. Xu, J. Pan, Y. Liu, D. Wu, S. Huang, and Q. Cheng, *Prostate cancer identification via photoacoustic spectroscopy and machine learning*. Photoacoustics, 2021. **23**: p. 100280.
119. Goh, Y., G. Balasundaram, M. Moothanchery, A. Attia, X. Li, H.Q. Lim, N. Burton, Y. Qiu, T.C. Putti, C.W. Chan, P. Iau, S.W. Tang, C.W.Q. Ng, F.J. Pool, P. Pillay, W. Chua, E. Sterling, S.T. Quek, and M. Olivo, *Multispectral Optoacoustic Tomography in Assessment of Breast Tumor Margins During Breast-Conserving Surgery: A First-in-human Case Study*. Clinical Breast Cancer, 2018. **18**(6): p. e1247-e1250.
120. Li, L., A.A. Shemetov, M. Baloban, P. Hu, L. Zhu, D.M. Shcherbakova, R. Zhang, J. Shi, J. Yao, L.V. Wang, and V.V. Verkhusha, *Small near-infrared photochromic protein for photoacoustic multi-contrast imaging and detection of protein interactions in vivo*. Nature Communications, 2018. **9**(1): p. 2734.
121. Horiguchi, A., M. Shinchi, A. Nakamura, T. Wada, K. Ito, T. Asano, H. Shinmoto, H. Tsuda, and M. Ishihara, *Pilot Study of Prostate Cancer Angiogenesis Imaging Using a Photoacoustic Imaging System*. Urology, 2017. **108**: p. 212–219.
122. Manohar, S., S.E. Vaartjes, J.C.G.v. Hespen, J.M. Klaase, F.M.v.d. Engh, W. Steenbergen, and T.G.v. Leeuwen, *Initial results of in vivo non-invasive cancer imaging in the human breast using near-infrared photoacoustics*. Optics Express, 2007. **15**(19): p. 12277–12285.

123. Esenaliev, R., A. Karabutov, F. Tittel, B. Fornage, S. Thomsen, C. Stelling, and A. Oraevsky. *Laser optoacoustic imaging for breast cancer diagnostics: limit of detection and comparison with x-ray and ultrasound imaging*. in *Proc SPIE*. 1997. SPIE.
124. Wang, X., Y. Pang, G. Ku, X. Xie, G. Stoica, and L.V. Wang, *Noninvasive laser-induced photoacoustic tomography for structural and functional in vivo imaging of the brain.* Nature Biotechnology, 2003. **21**(7): p. 803–806.
125. Lin, L., X. Tong, S. Cavallero, Y. Zhang, S. Na, R. Cao, T.K. Hsiai, and L.V. Wang, *Non-invasive photoacoustic computed tomography of rat heart anatomy and function.* Light Sci Appl, 2023. **12**(1): p. 12.
126. Periyasamy, V., Ç. Özsoy, M. Reiss, X.L. Deán-Ben, and D. Razansky, *In vivo optoacoustic monitoring of percutaneous laser ablation of tumors in a murine breast cancer model.* Optics Letters, 2020. **45**(7): p. 2006–2009.
127. Zhang, S., Z. Liang, K. Tang, X. Li, X. Zhang, Z. Mo, J. Wu, S. Huang, J. Liu, Z. Zhuang, L. Qi, and W. Chen, *In vivo co-registered hybrid-contrast imaging by successive photoacoustic tomography and magnetic resonance imaging.* Photoacoustics, 2023. **31**: p. 100506.
128. Tang, Y., S. Tang, C. Huang, P. Klippel, C. Ma, N. Caso, S. Chen, Y. Jing, and J. Yao, *High-fidelity deep functional photoacoustic tomography enhanced by virtual point sources.* Photoacoustics, 2023. **29**: p. 100450.
129. Rajendran, P., S. Sahu, R.A. Dienzo, and M. Pramanik, *In vivo detection of venous sinus distension due to intracranial hypotension in small animal using pulsed-laser-diode photoacoustic tomography.* Journal of Biophotonics, 2020. **13**(6): p. e201960162.
130. Li, H., Y. Zhu, N. Luo, and C. Tian, *In vivo monitoring of hemodynamic changes in ischemic stroke using photoacoustic tomography.* J Biophotonics, 2023: p. e202300235.
131. Pan, S., L. Wang, Y. Ma, G. Zhang, R. Liu, T. Zhang, K. Xiong, S. Chen, J. Zhang, W. Li, and S. Yang, *Photoacoustic-enabled automatic vascular navigation: accurate and naked-eye real-time visualization of deep-seated vessels.* Advanced Photonics Nexus, 2023. **2**(04).
132. Na, S., J.J. Russin, L. Lin, X. Yuan, P. Hu, K.B. Jann, L. Yan, K. Maslov, J. Shi, D.J. Wang, C.Y. Liu, and L.V. Wang, *Massively parallel functional photoacoustic computed tomography of the human brain.* Nature Biomedical Engineering, 2021.
133. Zhang, Y., J. Olick-Gibson, A. Khadria, and L.V. Wang, *Photoacoustic vector tomography for deep hemodynamic imaging.* arXiv, 2022. **2209.08706**.
134. Liang, B., S. Wang, F. Shen, Q.H. Liu, Y. Gong, and J. Yao, *Acoustic impact of the human skull on transcranial photoacoustic imaging.* Biomedical Optics Express, 2021. **12**(3): p. 1512–1528.
135. https://senomedical.com/. [10/27/2023].

Chapter 3
Photoacoustic Endoscopy

Hao Yang, Sean Aleman, and Huabei Jiang

Abstract In the rapidly evolving landscape of medical imaging, endoscopy emerged as a revolutionary technique, facilitating non-invasive visualization of internal body structures, and playing a crucial role in the diagnosis, treatment, and monitoring of diverse medical conditions. Photoacoustic endoscopy (PAE) is an advanced endoscopic imaging technique that evolved from photoacoustic imaging and traditional endoscopy. PAE's high-resolution images, which reveal both functional and molecular information, are made possible through the combination of laser light with sound waves. Recent advancements in PAE include miniaturized probes and high-speed imaging capability. These are propelling its clinical translation for early disease detection, accurate diagnosis, and image-guided interventions. PAE's potential applications are vast, and its integration with artificial intelligence and other imaging modalities holds promise for personalized medicine and comprehensive medical imaging in the future. This chapter offers an overview of PAE, encompassing its historical development and recent applications in the medical field. The fundamental principle of PAE is explored, alongside key technological advancements that have propelled its clinical utility. Additionally, the chapter delves into the applications of a typical PAE device in cutting-edge research, showcasing its potential in various medical disciplines. In conclusion, this chapter discusses current challenges in the development of PAE and its potential to reshape the medical landscape through transformative diagnostic and therapeutic possibilities.

Keywords Photoacoustic imaging · Endoscopy · Biomedical imaging · Miniaturized probes · Multimodal imaging · Cancer detection

---

H. Yang · S. Aleman · H. Jiang (✉)
Department of Medical Engineering, University of South Florida, Tampa, FL, USA
e-mail: haoyang@usf.edu; alemans@usf.edu; hjiang1@usf.edu

W. Xia (ed.), *Biomedical Photoacoustics*,
https://doi.org/10.1007/978-3-031-61411-8_3

3.1 Introduction

3.1.1 History

The roots of photoacoustic endoscopy (PAE) can be traced back to the pioneering work in both photoacoustic imaging and traditional endoscopy. Photoacoustic imaging, also known as optoacoustic or thermoacoustic imaging, was first conceptualized by Alexander Graham Bell in the late nineteenth century when he discovered the photoacoustic effect [1]. However, it was not until the twentieth century that significant progress was made in this field. In the 1990s, researchers began exploring the possibilities of using laser-induced ultrasound for biomedical imaging. They found that when a tissue absorbs pulsed laser light, it generates thermal expansion, which leads to the emission of ultrasonic waves. This discovery formed the foundation of modern photoacoustic imaging techniques.

Endoscopy, in parallel, also saw significant advancements [2, 3]. Traditional endoscopy, dating back to the early twentieth century, involved the use of rigid optical fibers to visualize internal body structures. Over time, endoscopic instruments became more flexible and versatile, allowing access to various regions of the body with reduced invasiveness. The convergence of these two fields laid the groundwork for the emergence of photoacoustic endoscopy. In the late 1990s and early 2000s, researchers began exploring the integration of photoacoustic principles into endoscopic devices. The goal was to enhance the capabilities of traditional endoscopy by providing additional functional and molecular information of internal tissues and organs. Early PAE prototypes demonstrated the feasibility of obtaining photoacoustic signals from tissue using miniaturized probes. As technological advancements continued, researchers refined the imaging process, by improving resolution, sensitivity, depth penetration, and scanning speed. These efforts expanded the potential clinical applications of PAE, ranging from early cancer detection to the assessment of vascular diseases and other internal pathologies. In recent years, PAE has seen remarkable progress, with several research groups and medical device companies actively pursuing its development. High-speed imaging, multi-wavelength excitation, and sophisticated data processing techniques have further enhanced PAE's capabilities and its potential impact on medical practice.

Today, PAE stands at the forefront of non-invasive imaging techniques, with ongoing research and clinical trials exploring its applications in various medical specialties. The combination of photoacoustic principles and endoscopic technology continues to unlock new possibilities for personalized and precise medical diagnosis and treatment. As PAE rapidly evolves, it holds promise for transforming the landscape of medical imaging, offering clinicians an invaluable tool for visualizing internal structures and improving patient outcomes.

3.1.2 Applications

PAE has shown great potential in various medical applications, particularly due to its non-invasive imaging capabilities, compact design, and proficiency in delivering both functional and molecular insights. Over time, its versatility has captured the attention of researchers and clinicians alike, leading to its exploration across diverse medical domains and solidifying its clinical value. Among the standout applications of PAE are:

Cancer Detection and Diagnosis Cancer detection and diagnosis have seen remarkable advancements with the integration of innovative imaging technologies. Among these, PAE has emerged as a promising tool for early cancer detection and accurate diagnosis. Its high-resolution imaging capabilities allow clinicians to visualize tumor tissues with greater detail, distinguishing between malignant and benign lesions. Additionally, PAE's ability to detect molecular signatures aids in characterizing tumors, providing valuable insights into their biological properties. The potential for molecular imaging, particularly with targeted contrast agents, offers enhanced specificity. For instance, a pilot study conducted by Horiguchi et al. in 2017 used PAE to visualize prostate cancer in 3 patients, revealing that the intensity of PAE signals might reflect the microvascularity in normal prostatic tissues and index tumors [4]. In 2016, Zhu et al. developed a co-registered PAE and ultrasound (US) prototype system for real-time assessment of ovarian cancer [5].

Gastrointestinal Imaging Gastrointestinal (GI) imaging plays a pivotal role in the diagnosis and management of various GI disorders. In gastroenterology, PAE holds promise for non-invasive imaging of the gastrointestinal tract. Gastrointestinal PAE requires the visualization of internal structures within the esophagus, stomach, and colon. For upper gastrointestinal imaging, a PAE probe with a diameter typically ranging from 2.5 mm to 4.0 mm is commonly used. By providing detailed images of the intestinal lining and mucosal layers, PAE may assist in the early detection and monitoring of gastrointestinal disorders, such as inflammatory bowel diseases and gastrointestinal cancers. In 2015, Wang et al employed an PAE probe, featuring a 3.8 mm diameter, for the purpose of in vivo imaging of the descending colon in a Sprague Dawley rat [6]. Through the integration of PAE and endoscopic ultrasound (EUS) techniques, they subsequently advanced their efforts to create a multimodality system aimed at visualizing the architecture of the rabbit esophagus in vivo [7].

Intravascular Imaging Intravascular imaging has revolutionized our understanding of cardiovascular diseases by providing detailed insights into the structural and functional aspects of blood vessels. Traditional imaging modalities, such as intravascular ultrasound (IVUS) and optical coherence tomography (OCT), offer valuable information, but they each have limitations. Among the emerging techniques, PAE has gained considerable attention due to its ability to combine the advantages of both optical and acoustic imaging modalities. Compared to IVUS and OCT, PAE provides several unique advantages. It offers deeper penetration depths than optical techniques, allowing visualization of structures beyond the vessel wall. PAE also offers superior contrast due to its ability to image endogenous molecules, such as hemoglobin and lipids, which are implicated in various vascular pathologies. In 2015, a confocal intravascular photoacoustic endoscopy (IVPAE) system was developed by Xing et al. Their research encompassed ex vivo experiments involving both normal and atherosclerotic vessels [8]. Subsequently, in 2017, a high-speed hybrid IVPA-ultrasound (US) imaging system was introduced by Soest et al., featuring an outer diameter of 1.3 mm. This system was employed for the ex vivo imaging of human coronary arteries as well as in vivo imaging of swine coronary arteries [9]. In 2019, Yang et al. devised an exceptionally compact IVPA-US probe with a diameter of 0.7 mm, specifically tailored for investigating atherosclerotic plaques within coronary artery branches. They successfully utilized this probe to image the thoracic aorta of a mouse, effectively evaluating the performance of their hybrid imaging approach [10].

Intraoperative Imaging Intraoperative imaging plays a critical role in guiding surgical procedures, enhancing surgical precision, and improving patient outcomes. It has evolved from aiding visualization to becoming an integral part of surgical decision-making. PAE transcends the limitations of traditional intraoperative imaging techniques by offering superior contrast, deeper tissue penetration, and the ability to image both structural and functional information. Additionally, the real-time visualization of blood vessels, oxygenation levels, and tissue composition during surgery has the potential to revolutionize surgical decision-making. Intraoperative PAE may require varying probe sizes depending on the surgical site and the access required. Smaller-diameter probes are advantageous for minimally invasive surgeries, while larger-diameter probes offer enhanced imaging capabilities for open surgeries. In 2010, Xi et al achieved a breakthrough by designing a prototype of an intraoperative PAE probe utilizing a microelectromechanical systems (MEMS) mirror [11]. Building upon this innovation, Yang et al. made notable progress in 2013 by developing a dual-modality probe that combined diffuse optical tomography (DOT) with PAE. To demonstrate its potential for tumor imaging, ex vivo experiments were conducted on tumor-bearing mice that were systematically injected with targeted contrast agents [12]. Although it is promising, many technical challenges, such as probe design, imaging speed, and safety considerations, still need to be addressed to facilitate widespread adoption of PAE in the operating room.

As PAE technology continues to evolve and become more accessible, it is expected that new applications will emerge, further expanding its impact on various medical disciplines. The integration of PAE with other imaging modalities and emerging technologies, such as artificial intelligence, holds the promise of even more sophisticated and comprehensive medical imaging approaches in the future.

3.2 Principle

3.2.1 Categories of PAE

Photoacoustic endoscopy is an innovative imaging technique that combines the capabilities of endoscopy and photoacoustic imaging to visualize tissues and structures within the body at various depths with high resolution and functional information. There are several categories of photoacoustic endoscopy, each with distinct advantages and applications. In this section, we discuss three typical categories.

Photoacoustic Computed Endoscopy (PCE) PCE is an advanced category of photoacoustic endoscopy that aims to provide two- or three-dimensional imaging using computational methods [13–15]. PCE utilizes an endoscope equipped with an array of ultrasound transducers at its distal end. These transducers detect photoacoustic signals from various angles. The detected signals are processed using advanced imaging algorithms, such as filtered back-projection, time-reversal algorithms, or iterative reconstruction techniques, to reconstruct cross-sectional and volumetric images. The spatial resolution in PCE depends on several factors, including the number and arrangement of ultrasound transducers in the endoscope's array, the bandwidth of the transducers, and the reconstruction algorithms used. PCE allows for excellent spatial resolution in the order of tens of micrometers or better, which is sufficient for visualizing small anatomical structures and cellular features. The use of multiple transducers and sophisticated image reconstruction algorithms results in improved imaging depth compared to conventional photoacoustic endoscopy techniques. In general, PCE can achieve depths of several millimeters to centimeters, making it suitable for imaging deep-seated organs and tissues.

Acoustic-Resolution Photoacoustic Endoscopy (AR-PAME) AR-PAME is a category of photoacoustic endoscopy that focuses on achieving high-resolution imaging with enhanced penetration depth [16–19]. In AR-PAME, the endoscope is equipped with a focused ultrasound transducer at its distal end, which plays a crucial role in improving the spatial resolution of the generated images. By combining the advantages of area illumination and a focused transducer, AR-PAME overcomes the limitations of purely optical methods, making it highly applicable for visualizing deeper anatomical structures while maintaining a reasonable spatial resolution. AR-PAME can achieve spatial resolutions in the order of tens of micrometers, enabling the visualization of microvasculature and cellular structures with exceptional detail. Benefiting from the lower attenuation and dispersion of acoustic signals in tissues compared to light, AR-PAME can achieve imaging depths of a few millimeters. AR-PAME strikes a balance between spatial resolution and penetration depth, making it an ideal choice for exploring fine structures, such as microvasculature and cellular details, at significant depths within the body. Its ability to provide high-resolution imaging in deeper tissues holds great potential for advancing medical research, clinical diagnostics, and image-guided interventions.

Optical-Resolution Photoacoustic Endoscopy (OR-PAME) OR-PAME is designed to achieve the highest spatial resolution among the categories of photoacoustic endoscopy [6, 17, 20–22]. Instead of the area illumination in AR-PAEM, OR-PAME uses a highly focused laser beam to increase the spatial resolution of the image. The use of optical focusing techniques, such as optical fibers or miniaturized scanning systems, allows for precise delivery of pulsed laser light into the target tissue, resulting in superior resolution. OR-PAME can achieve spatial resolutions in the order of sub-micrometers to micrometers, enabling visualization of individual red blood cells and detailed cellular structures. Despite its exceptional spatial resolution, OR-PAME has limitations in imaging depth due to the strong scattering of light in tissues. The imaging depth in OR-PAME is generally limited to 1–2 millimeters, making it more suitable for imaging near-surface tissues and small organs.

To summarize, the three photoacoustic endoscopy categories present distinct imaging trade-offs. PCE excels in providing the deepest penetration depth among the three, but it comes at the cost of the lowest spatial resolution. On the other hand, OR-PAME boasts the best spatial resolution, achieving details down to a few micrometers, but its imaging depth is limited to approximately one millimeter. Despite offering moderate spatial resolution and penetration depth, AR-PAME strikes a balance between the other two categories. Furthermore, OR-PAME involves the most complex system design of all three. Therefore, the selection of the appropriate category depends on the specific imaging requirements, with PCE favored for deeper imaging needs, OR-PAME for high-resolution visualization, and AR-PAME for a more balanced approach between spatial resolution and penetration depth.

3.2.2 Scanning Mechanisms

Scanning technology plays a crucial role in photoacoustic endoscopy, as it enables the acquisition of spatial information from different regions within the body. Various scanning techniques allow these systems to visualize tissues in three dimensions, offering valuable insights for medical diagnosis and research. Below are some of the common scanning technologies used in photoacoustic endoscopy:

Mechanical Scanning Mechanical scanning is one of the earliest and most straightforward scanning techniques in photoacoustic endoscopy. It involves physically moving the endoscope's optical or acoustic components to scan the laser beam or ultrasound focus over the area of interest. The movement can be achieved through an external motorized rotator and a linear stage. Mechanical scanning in photoacoustic endoscopy provides high spatial resolution, accurate depth localization, and familiarity as an established technique, minimizing motion artifacts and facilitating precise imaging of tissue structures. However, its limitations include limited imaging speed, possible tissue displacement, invasiveness, limited maneuverability, and probe design complexity, which necessitates careful consideration of its benefits and limitations based on the specific application requirements. Instances of PAE devices employing mechanical scanning can be observed in references [13, 23–27].

Optical Scanning Microelectromechanical systems (MEMS) mirrors have emerged as a captivating technological advancement, revolutionizing optical systems with their precise and dynamic control of light propagation. MEMS mirrors are a testament to the fusion of microfabrication prowess and optical innovation, granting researchers and engineers the ability to manipulate light beams at a microscale level. These mirrors are microscale structures integrated onto silicon substrates using semiconductor fabrication methodologies. Unlike their macroscopic counterparts, MEMS mirrors can be swiftly and accurately manipulated, enabling rapid beam steering and modulation. The operation of MEMS mirrors hinges on their mechanical properties and electrostatic actuation mechanisms. Applying controlled electrical signals induces mechanical movement, causing these structures to tilt or deform, which in turn alters the direction of the reflected light. MEMS mirrors present notable advantages in the context of PAE, including dynamic beam steering for real-time imaging, compact size facilitating endoscope integration, precise control aiding accurate localization, and enhanced imaging speed for reduced motion artifacts. However, their complex design and fabrication, potential cost barriers, reliability concerns related to wear and tear, and integration complexity into existing systems represent challenges that need to be addressed for widespread adoption of MEMS mirror-based PAE, despite their potential to significantly enhance minimally invasive biomedical imaging. Over the past few decades, MEMS mirrors have found an extensive application in the advancement of PAE devices [11, 12, 28, 29].

Acoustic Lens Focusing The acoustic lens focusing method plays a crucial role in advancing photoacoustic imaging through its non-mechanical approach to ultrasound wave manipulation. Using acoustic lenses, this technique precisely focuses and steers beams [30, 31]. This efficient focusing enables the visualization of fine tissue structures and details, vital for accurate diagnostics. The rapid steering and focusing of ultrasound beams allow for high-speed imaging, making this method suitable for real-time intraoperative applications and dynamic tissue monitoring. Unlike mechanical scanning methods, the non-mechanical nature of acoustic lens focusing mitigates the risk of motion artifacts, ensuring stability and consistency in imaging results. However, challenges like calibration complexity, limited imaging depth for specific ranges, system complexity, and costs need to be carefully considered. Despite these challenges, ongoing advancements in material science, lens design, and signal processing techniques hold promise for refining calibration procedures, optimizing performance, and addressing these limitations. Thus, the acoustic lens focusing method underscores its potential as an invaluable component in the evolving landscape of photoacoustic endoscopy, enhancing diagnostic precision and interventional capabilities.

The choice of scanning technology depends on factors such as imaging speed, resolution requirements, system complexity, and the specific application of photoacoustic endoscopy. As research and technology advance, scanning techniques are continuously evolving, driving improvements in image quality, speed, and overall performance in photoacoustic endoscopy.

3.2.3 Configuration

The endoscope's configuration plays a crucial role in determining its imaging capabilities and clinical applications. In this section, we will discuss the side view and forward view configurations of a photoacoustic endoscope.

Side View The side view configuration of a photoacoustic endoscope involves the integration of an imaging probe with a transducer assembly at its side [5, 10, 23, 26–28]. This setup allows the endoscope to capture images in a cross-sectional manner, providing valuable information about tissue structures and abnormalities beneath the surface. In a typical side-view PAE device, the transducers are strategically positioned on the side of the imaging probe and are responsible for detecting the photoacoustic waves generated within the tissue. The received signals are then processed to reconstruct cross-sectional images of the tissue. PAE with a side view configuration is valuable in diagnosing gastrointestinal conditions such as colorectal cancer, gastric tumors, and inflammatory bowel diseases. It allows for the assessment of mucosal and submucosal layers, aiding in early detection and accurate staging. In intravascular applications, the side view photoacoustic endoscope can be used to visualize the interior of blood vessels, detecting vulnerable plaques and guiding interventions like angioplasty or stent placement.

Forward View The forward view configuration of a photoacoustic endoscope involves the integration of imaging and transducer elements at the distal end of the endoscope, providing a view directly ahead of the device [32–36]. This setup enables real-time visualization of tissue structures in front of the endoscope. In the forward view configuration, the imaging and transducer elements are miniaturized and positioned at the tip of the endoscope. These elements work in tandem to illuminate the tissue with laser light and detect the resulting photoacoustic waves. Forward-view photoacoustic endoscopy can be used in urology to visualize the interior of the urinary tract, aiding in the detection and monitoring of bladder tumors and other urological conditions.

In conclusion, both the side view and forward view configurations of a photoacoustic endoscope have their distinct applications, strengths, and limitations. The choice of configuration depends on the specific clinical scenario and the type of tissue or organ being imaged.

3.3 Case Studies of Typical Photoacoustic Endoscopy (PAE) Devices

3.3.1 A Dual-Modality PAE Device Integrating Photoacoustic Microscopy (PAM) and Optical Coherence Tomography (OCT)

In this section, we begin by presenting a miniature 2.3 mm diameter probe that integrates ORPAM with OCT. This configuration is designed to be easily inserted during standard endoscopic examinations.

Figure 3.1a shows the schematic of the integrated probe. In this probe, the illumination beams for both OCT and ORPAM are coupled into one tip of a single-mode fiber (SMF-28e+, Thorlabs) with 0.14 numerical aperture (NA) and 0.9 mm outer diameter. The other tip of the fiber is cut at an 8° angle to minimize back-reflection. Optical UV glue is used to connect the fiber tip with the GRIN lens resulting in a 5 mm working distance. Two micro-prisms (one without coating and the other one coated with aluminum) are glued together and a miniature unfocused ultrasound transducer with 10 MHz central frequency and 2 mm aperture is mounted on the top of the cubic prism pair. The GRIN lens focuses the light beams, which are then reflected by the thin aluminum film onto the tissue surface. The generated ultrasound waves transmit through the cubic prism group and are detected by the transducer. The backscattering photons from the tissue are reflected by the aluminum film and coupled into the same fiber through GRIN lens. Figure 3.1b shows a photograph of the probe, where both the cubic prism pair and GRIN lens are 0.7 mm in diameter. A stainless-steel tube, 1.0 mm in diameter, is used to protect the light path (i.e., the single mode fiber, GRIN lens, and cubic prism pair). The probe is glued to the transducer and protected by a bigger stainless-steel tube (2.3 mm in diameter).

The probe is mounted on a three-dimensional (3D) linear stage (Fig. 3.2) and scanned in two-dimensional (2D) for 3D image formation. The light beam for ORPAM is generated from a pulsed Nd:YAG laser with 532 nm wavelength and a 10 Hz repetition rate. Two neutral density filters are used to attenuate the light

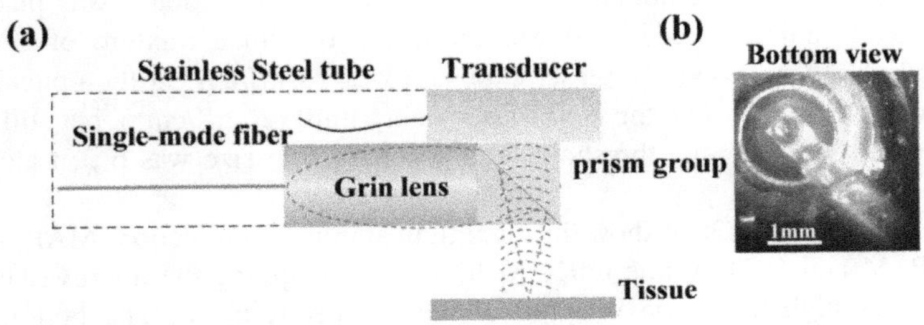

Fig. 3.1 Schematic (**a**) and photograph (**b**) of the hybrid probe

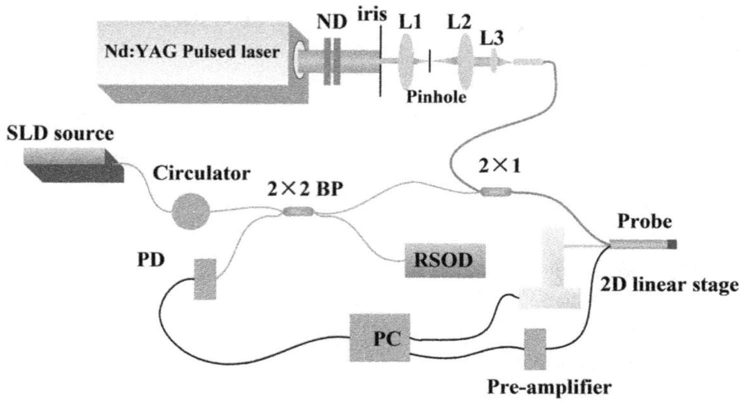

Fig. 3.2 Schematic of the integrated ORPAM and OCT system. ND Neutral Density, L1 lens 1, L2 lens 2, L3 lens 3, BP beam splitter, PD Photodiode

energy and a small iris is utilized to provide homogeneous light distribution. The shaped light beam is focused by a convex lens, which passes through a 50 μm pinhole for spatial filtering and is then coupled into a 2 × 1 beam coupler. The detected acoustic waves are amplified by two wideband amplifiers and digitized by a data acquisition board (NI5152, National Instrument) at a sampling rate of 250MS/s.

A broadband light source (DenseLight, DL-BX9-CS3159A) with a center wavelength of 1310 nm is employed for OCT. The light source has a full-width-at-half-maximum (FWHM) of 75 nm, providing an axial resolution of 10 μm in air. The broadband light is split into the reference arm and the sample arm by a beam splitter and coupled into the same 2 × 1 beam coupler used for ORPAM. The depth scanning from 0 to 1.6 mm at the reference arm is realized by a rapid scanning optical delay line (RSOD) coupled with a galvanometer scanning at 1 kHz. The OCT signal is then detected by a balanced photodetector, whose output is acquired and stored by a DAQ card. The sensitivity of the system is measured to be 74 dB.

To showcase the dual-mode probe's microscopic imaging capabilities, we imaged a mouse's ear. Before starting the experiments, the hair on the ear was gently removed using a human hair-removing lotion. The mouse was placed on a homemade animal holder and was anesthetized with a mixture of ketamine (85 mg/kg) and xylazine. The laser exposure was 25 mJ/cm^2 at the optical focus point which is higher than the ANSI laser safety limit (20 mJ/cm^2), but still below the reported skin damage threshold. The scanning step size was 6 μm along the X-Y plane.

Top rows of Fig. 3.3a, b show the maximum amplitude projection (MAP) images of ORPAM and OCT. While ORPAM focuses on mapping the microvasculature, OCT offers high-resolution imaging of the sebaceous gland. The bottom rows (cross-section of tissue) of Fig. 3.3a, b show the benefits of combining these two

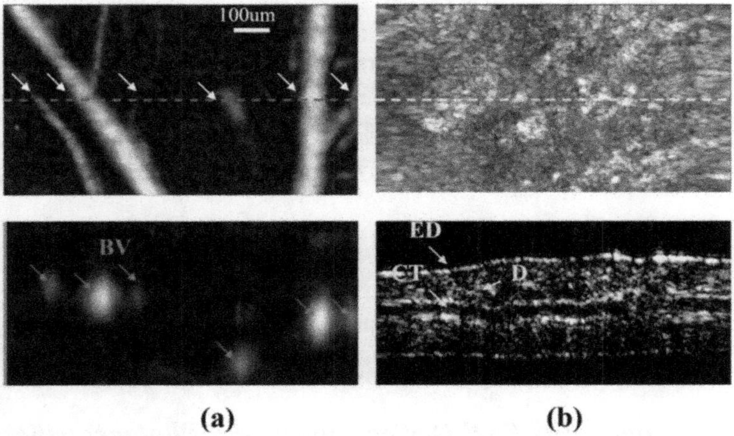

Fig. 3.3 In vivo imaging of mouse ear by the integrated probe. MAP image (top) and cross-section (bottom) of ORPAM (**a**) and OCT (**b**). ED epidermis, CT cartilage, D dermis, BV blood vessel

modalities more clearly. The ear's thickness in the OCT image is from 400 μm to 600 μm. The dermal structure and the sebaceous gland are clearly observed. In the cross-sectional OCT, we can identify epidermis, dermis, and cartilage as indicated with yellow arrows. We have observed that ORPAM is good at locating micro-vessels in the ear with limited surrounding tissue information.

The SNR of the ORPAM is 25 dB, which is lower than that of conventional ORPAM. There are several reasons contributing to the reduced SNR: (1) A 10 MHz ultrasound transducer was used in this probe, while it is known that the strongest generated acoustic signal lies between 30 MHz to 70 MHz; (2) The transducer is flat, resulting in reduced sensitivity compared with a focused transducer commonly used in conventional ORPAM; (3) Throughout the experiments, an 8-bit resolution DAQ card was employed, capable of resolving signals greater than 40 mV. Consequently, certain signals from minor capillaries were not captured. These limitations, however, can be easily overcome using a miniature focused high-frequency transducer (>30 MHz), which will in turn improve the axial resolution of ORPAM and make the whole probe smaller as well. Employing a DAQ card with enhanced resolution will also aid in addressing the previously mentioned issues; and (4) The actual resolution of a GRIN probe tends to be worse than this theoretical limit because the effective NA of GRIN lenses is degraded owing to the spatial aberrations in the lenses and misalignment between optical components. Future enhancements could include a suitable micro-motor for an internal scanning mechanism and a faster laser to reduce scanning duration. Ultimately, the intention is to substitute the existing sluggish time-domain (TD) OCT with rapid frequency domain (FD) OCT. These steps are essential in progressing towards the clinical assessment of the ORPAM/OCT system.

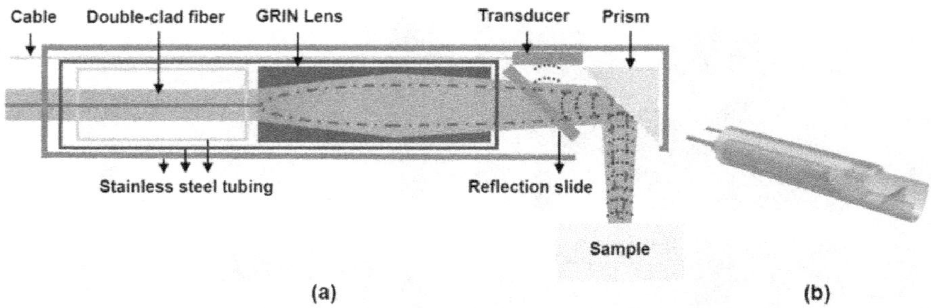

Fig. 3.4 Integrated multimodal probe. (**a**) Schematic of the probe. (**b**) Three-dimensional rendering of the probe

3.3.2 A Trimodality PAE Device Integrating Photoacoustic Imaging (PAI), Optical Coherence Tomography (OCT), and Ultrasound (US)

Within this section, we examine a remarkably compact endoscope, measuring only 1 mm in diameter, which seamlessly combines photoacoustic imaging (PAI), optical coherence tomography (OCT), and ultrasound (US). This unique integration empowers the device to achieve high-resolution imaging capabilities, enabling the visualization of three essential tissue contrasts: PAI, OCT, and US.

The structure of the probe is schematically shown in Fig. 3.4. In this probe, a DCF with a dual cladding structure (DCF13, Thorlabs), in which, single mode light travels in the Ø9 μm core with a numerical aperture (NA) of 0.12 and a spectrum range of 1250–1600 nm, while multimode light propagates in the Ø105 μm inner first cladding (NA = 0.2, 400–2200 nm), is used to deliver light for both PAI (multimode) and OCT (single mode). A custom-designed gradient-index (GRIN) lens with a diameter of 0.25 mm and a working distance of 5 mm (GRINTECH GmbH) is utilized to focus the light beam including single-mode (indicated by a red dotted line) and multimode (indicated by green area) from the tip of the DCF with 8° angle facial cutting for minimizing back reflection. Traveling through a piece of 45°-angle-tilting reflection slide (which is optically transparent but ultrasonically untransparent), the light beam is then reflected into the sample by a 0.5 mm enhanced-aluminum-coated right-angle prism (MPCH-0.5, Tower Optical Corp.). For OCT, the backscattering light from the sample shares the same optical path with light transmission. For PAI, the light-induced photoacoustic signals reflected by the prism and reflection slide are detected by a custom-made unfocused ultrasound transducer with a center frequency of 40 MHz and dimensions of 0.6 mm × 0.5 mm × 0.2 mm (Blatek, Inc., State College, PA, USA) mounted coaxially with respect to the optical path. Although the photoacoustic signal diminishes by approximately 15% after two reflections, it retains sufficient strength for high-contrast PA images. The transducer serves both PAI and US imaging; in the latter, it forms images by detecting its own pulsed ultrasound echoes. Three stainless steel tubes (260, 500, and 1000 μm in diameter) secure all the components.

3.3.2.1 In Vivo Imaging of a Mouse Ear

To demonstrate the capabilities of this integrated multimodal probe, in vivo imaging of a mouse ear (mouse weight = 35 g) was first performed. The laser exposure was about 10 mJ/cm^2 at the optical focus inside the ear tissue, below the American National Standards Institute (ANSI) laser safety limit (20 mJ/cm^2) [37]. An area of 2 mm × 2 mm of the mouse ear was scanned with a scanning step size of 10 μm, which took around 36 min.

The top row of Fig. 3.5 shows the maximum amplitude projection (MAP) images of PAI (Fig. 3.5a), US (Fig. 3.5b), OCT (Fig. 3.5c), and fused multimodal pseudo-color-coded RGB images (Fig. 3.5d) obtained from PAI (R, red), OCT (G, green), and US (B, blue). The middle and bottom rows of Fig. 3.5 show the cross-sectional images of PAI, US, OCT, and the fused multimodality, corresponding to the two dotted lines in the respective MAP images. It is evident that PAI, OCT, and US

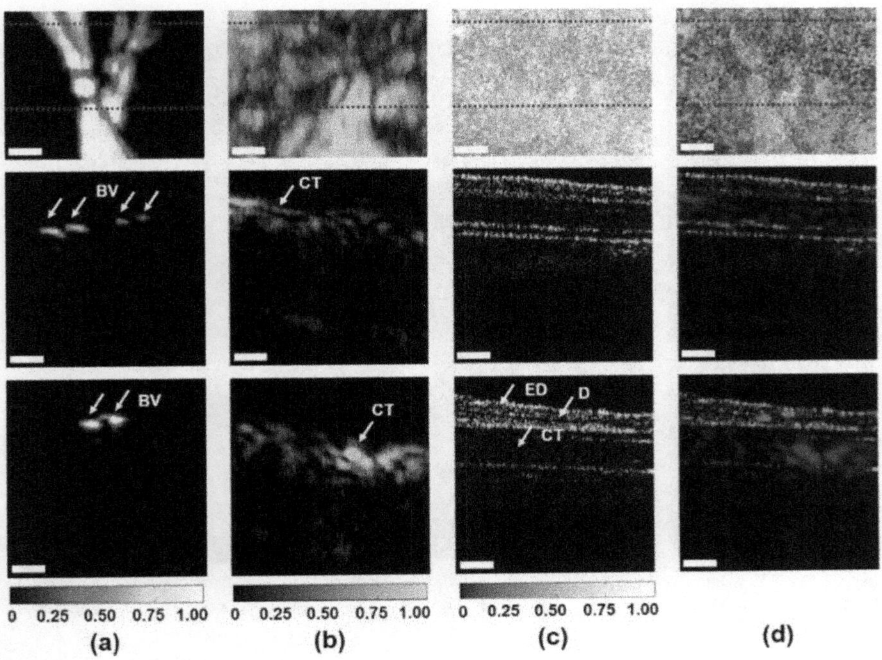

Fig. 3.5 In vivo images of a mouse ear by the integrated multimodal probe. Maximum amplitude projection (MAP) (top row) and cross-sectional images (middle and bottom rows) corresponding to the dotted lines shown in the MAP image of (**a**) PAI, (**b**) US, (**c**) OCT, and (**d**) fused trimodality. BV blood vessel; ED epidermis; D dermis; CT cartilage. The scale bars indicate 0.3 mm in length

succeeded in visualizing the microvasculature, intricate formations encircling blood vessels, and elastic structures, in that order. To closely interpret these images, blood vessels in the mouse ear are mapped by PAI with the highest contrast (Fig. 3.5a), elastic structures such as cartilage, which have high ultrasound reflection coefficients, are clearly visualized by US (Fig. 3.5b), while epidermis, dermis, and cartilage are identified in the OCT image (Fig. 3.5c). By overlapping the PAI, OCT, and US images (Fig. 3.5d), the structures and tissue morphology become more discernible, demonstrating the in vivo imaging capabilities of our integrated multimodal probe.

3.3.2.2 In Vivo Imaging of a Human Hand

Subsequently, in vivo imaging of the back of a human hand was performed to assess the potential clinical capabilities of the integrated multimodal probe. A step size of 10 μm was used for scanning the back of the hand, and it took around 15 seconds to obtain a b-scan (cross-sectional) image. Figure 3.6a–c, respectively, shows the cross-sectional images of the hand from PAI, US, and OCT. Using the same RGB pseudocoding method as the one used for mouse ear imaging, the fused multimodal

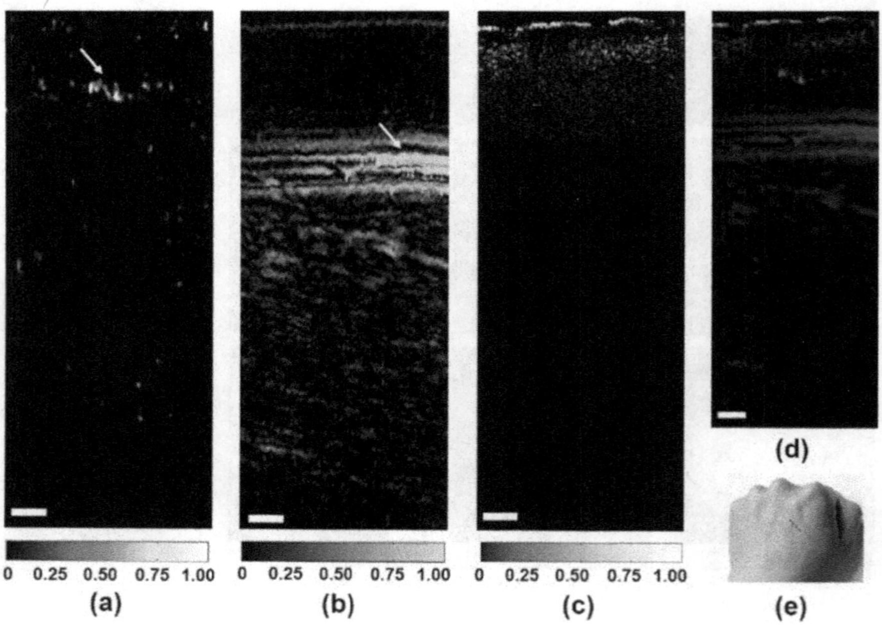

Fig. 3.6 Cross-sectional images of the skin of a human hand. (**a**) PAI image, (**b**) US image, (**c**) OCT image, (**d**) color-coded RGB image and (**e**) photograph of the scanning location. The scale bars indicate 0.3 mm in length

image is given in Fig. 3.6d. Figure 3.6e is the photograph of the scanning location on the surface of the hand indicated by the red dotted line.

From Fig. 3.6, it is observed that blood vessels (as indicated by a white arrow in Fig. 3.6a), superficial fine structures, and elastic structures (such as bone, as indicated by a white arrow in Fig. 3.6b) are identified by PAI, OCT, and US, respectively. Since the typical thickness of a mouse ear is less than 2 mm, the deep penetration ability of the US was not demonstrated from images shown in Fig. 3.6b. Here, in Fig. 3.6b, it is evident that the US can image the structure of tissue up to 5 mm in depth. A thorough analysis of Fig. 3.6b, c reveals that the surface layer of tissue, exhibiting reduced ultrasonic reflection, can be effectively observed through OCT with its superior resolution. The fused image (Fig. 3.6d) vividly demonstrates that this integrated probe excels in mapping the tissue morphology at depths of up to 5 mm with exceptional resolution. Furthermore, the three distinct imaging modalities employed in this system enable effective visualization and differentiation of various tissues.

3.3.2.3 Intravascular Imaging of Human Arteries with Atherosclerotic Plaques

Atherosclerosis, hardening and narrowing of the arteries, is a major cause of deadly diseases such as heart attacks and strokes, primarily due to the rupture of vulnerable atherosclerotic plaques. There is a significant need to develop an effective imaging tool capable of assessing plaque vulnerability. Intravascular ultrasound (IVUS) imaging, which is widely used in clinics, offers high resolution and large penetration depth imaging, but it has limited capacity in characterizing the components of atherosclerotic plaques. Intravascular optical coherence tomography (IV-OCT), an emerging technique, provides very high spatial resolution imaging within a limited penetration (1–2 mm). Intravascular photoacoustic imaging (IVPA), also an emerging technique, has the potential to offer compositional information on vessel walls/plaques with deep penetration depth and high spatial resolution.

To demonstrate the capability of the integrated multimodal probe for intravascular imaging, ex vivo experiments of human arteries with atherosclerotic plaques were conducted. Using rotatory scanning with 1200 samples per circle and a pullback step size of 60 μm, we acquired multimodal images of an artery sample with lipid-rich plaques. The artery sample was fixed using a cylindrical holder made of 2% agar solution and was coupled with the probe by water. For PAI, light with a wavelength of 1210 nm, at which lipid has an absorption peak, was utilized to excite the photoacoustic signal. Figure 3.7a–c, respectively, shows the cross-sectional images of the sample from PAI, US, and OCT. The fused multimodal image using RGB pseudocoding is given in Fig. 3.7d, while the histological image of the sample by H&E staining is shown in Fig. 3.7e.

By closely inspecting these images in comparison with the histology (Fig. 3.7e), regions of plaque rich in lipids are mapped by PAI with high contrast, as indicated by yellow arrows in Fig. 3.7a (the area in the image with high pixel intensity)

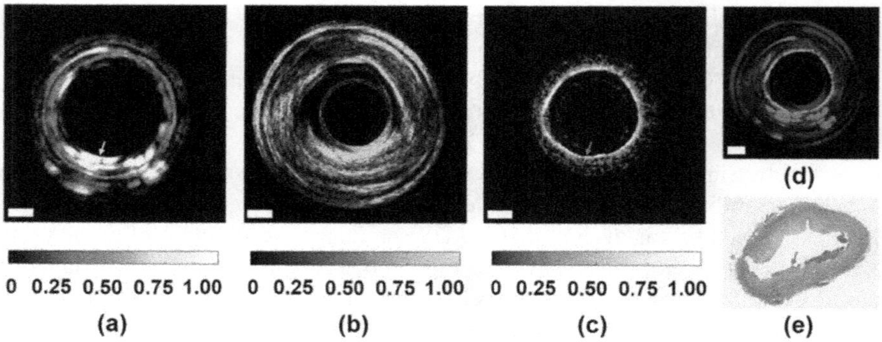

Fig. 3.7 Cross-sectional images of human artery with an atherosclerotic plaque. (**a**) PAI image, (**b**) US image, (**c**) OCT image, (**d**) color-coded RGB image, and (**e**) histology using H&E stain. The scale bars indicate 2 mm in length

and Fig. 3.7e. The OCT distinctly images the vessel wall's superficial structure, resembling a thin fibrous cap, as pointed out by the red arrow in Fig. 3.7c (a thin bright layer contrasted with a dark area) and corroborated in Fig. 3.7e. In contrast, the US captures the comprehensive structure of the entire vessel wall (Fig. 3.7b), thanks to its profound tissue penetration ability. The fused image of PAI, OCT, and US (Fig. 3.7d) undoubtedly presents a more thorough and detailed insight into the structural and compositional aspects of the vessel wall/plaque.

To show the capability of the integrated probe for spectroscopic IVPA, photoacoustic imaging of a human artery sample with atherosclerotic plaques was performed with six wavelengths of 710, 800, 900, 1100, 1150, and 1210 nm (Fig. 3.8). These multispectral PAI images illustrate that multiple chemical components of the plaque can be inferred/obtained for more accurate correlation with plaque vulnerability. For example, elastin and collagen, which have relatively high light absorption coefficients under 1000 nm, would appear as high-intensity pixels in the PAI images (Fig. 3.8a–c), while lipid, which has an absorption peak around 1210 nm, would appear as the brightest areas as shown in Fig. 3.8e, f. Combining this chemical compositional information with the structural images obtained from US (Fig. 3.7b) and OCT (Fig. 3.7c), the properties of the artery wall as well as the plaques can be visualized more completely.

In conclusion, the integrated multimodal probe has proven its capability to simultaneously acquire optical absorption and scattering properties of biological tissue, as well as tissue elastic characteristics in high resolution through both in vivo/ex vivo experiments. It was observed that the imaging speed in this experimental setup was limited by the relatively low repetition frequency of the nanosecond-pulsed Nd:YAG-pumped optical parametric oscillator (OPO) laser (20 Hz) and the relatively slow approach of time-domain OCT. Consequently, this lower imaging speed could lead to image artifacts caused by sample movements during the acquisition process. Nevertheless, the advanced probe showcased here can be easily adapted to any high-speed PAI system and state-of-the-art OCT imaging

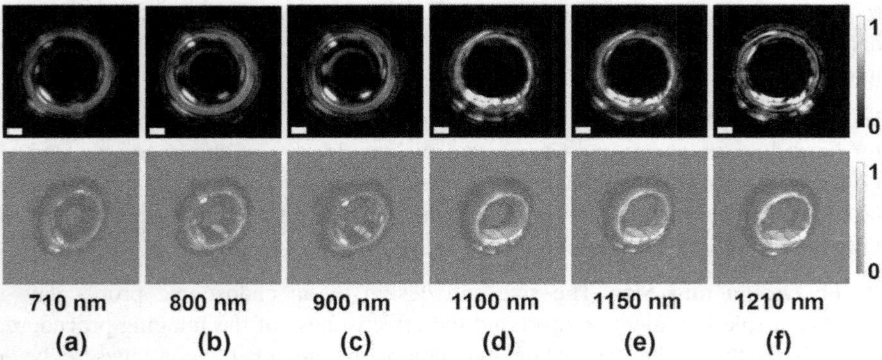

Fig. 3.8 Spectroscopic photoacoustic images of a human artery with atherosclerotic plaques. PA images at 710 nm (**a**), 800 nm (**b**), 900 nm (**c**), 1100 nm (**d**), 1150 nm (**e**), and 1210 nm (**f**). The top and bottom rows, respectively, show cross-sectional and three-dimensional images. The scale bars indicate 2 mm in length

system, such as high-speed swept-source OCT. By utilizing a high repetition rate nanosecond-pulsed laser for PAI and a swept source for OCT, it is expected that imaging speeds could reach as high as hundreds of kHz per scanning point. The highly integrated aspect and miniature size of the probe offer significant potential for impacting various preclinical and clinical applications, including multimodal intravascular coronary imaging, in vivo imaging of internal organs in small animals, ureteroscopy/transurethral imaging, and minimally invasive interventional therapy.

3.4 Current Challenges and Future Prospects

PAE is a promising imaging technique that combines the benefits of endoscopy and photoacoustic imaging, offering the potential for comprehensive and high-resolution visualization of internal tissues and vasculature. Despite its potential, PAE still faces several challenges that need to be addressed for its successful adoption and widespread clinical use. In this section, we will discuss the current challenges and forecast the future developments in PAE.

3.4.1 Technical Challenges

Resolution and Image Quality Achieving high spatial and temporal resolution is crucial for PAE to accurately visualize small structures and dynamic processes within the body. While advancements have been made in improving resolution through reconstruction algorithms and image-enhancement techniques, there is still room for further optimization. Future developments in algorithmic approaches, hardware components, and novel imaging concepts may enable even higher resolution and image quality.

Imaging Speed Real-time imaging is essential for clinical applications, as it allows for immediate diagnosis and treatment decisions. Increasing the imaging speed of PAE systems is an ongoing challenge. Developing faster scanning techniques, using lasers with higher repetition rates, and optimizing data processing methods are areas of focus for future development.

Probe Design and Size The size and design of an endoscopic probe play a significant role in patient comfort and the effectiveness of the imaging procedure. Minimizing the probe size while maintaining image quality continues to be a challenging task. Advances in miniaturization technologies and the incorporation of high-frequency, wide-bandwidth transducers are expected to lead to more compact and efficient probes in the future.

Penetration Depth While PAE offers excellent resolution near the tissue surface, deeper penetration remains challenging. Longer excitation wavelengths, such as near-infrared light, can improve penetration depth. Additionally, wavefront shaping and advanced signal processing techniques may enable deeper imaging capabilities.

Multispectral Functional Imaging Functional PAE, which detects endogenous chromophores sensitive to specific spectra, has great potential for disease characterization. However, the use of multiple wavelengths for functional imaging may decrease the temporal resolution and increase system complexity. Developing more efficient multispectral imaging approaches and optimizing data acquisition and processing will be critical for successful clinical applications.

3.4.2 Clinical Translation Challenges

System Robustness and Cost The transition from research prototypes to clinically viable PAE systems is challenging. Ensuring the robustness, reliability, and stability of technology in real-world clinical settings is essential. Additionally, the cost-effectiveness of PAE systems needs to be improved to make them more accessible to healthcare providers.

Operator Training and Complexity PAE is a complex imaging modality that requires specialized training for operators. Simplifying the operation and user interface of PAE systems will be crucial for wider adoption among medical professionals.

Image Interpretation Interpreting PAE images and distinguishing between normal and abnormal tissue features can be challenging. Developing standardized imaging protocols and providing training and reference databases for image interpretation will improve the clinical utility of PAE.

3.4.3 Forecasting Future Developments

Despite facing challenges, PAE has the potential to significantly reshape the medical landscape. It could usher in a new era, bringing advanced diagnostic and therapeutic possibilities, and holds great promise for the future of medical imaging. Research and development are continually progressing, and these efforts are expected to lead to substantial advancements in the field. Let's consider some potential future developments:

Artificial Intelligence and Image Analysis The application of artificial intelligence (AI) and machine learning algorithms in PAE data analysis holds great potential for automating image interpretation, improving diagnostic accuracy, and streamlining clinical workflows.

Therapeutic Applications PAE's ability to provide real-time imaging and functional information makes it a promising tool for guiding minimally invasive interventions and targeted therapies. Future developments may see PAE being used not only for diagnosis but also for treatment monitoring and guidance.

Multimodal Endoscopy Integrating PAE with traditional endoscopy or other advanced imaging modalities could create powerful multimodal endoscopy systems, allowing for comprehensive imaging and precise disease characterization.

In conclusion, while PAE has already demonstrated its potential for various clinical applications, several technical and clinical challenges need to be addressed for its successful translation and widespread adoption. With ongoing research and development efforts, we can expect significant improvements in resolution, imaging speed, probe design, penetration depth, and image interpretation. Additionally, advancements in AI, multimodal imaging, and therapeutic applications may further enhance the clinical utility of PAE in the future, making it an essential tool for diagnosis, treatment, and monitoring of various diseases and conditions.

References

1. Bell, A.G., *ART. XXXIV.–On the production and reproduction of sound by light.* American Journal of Science (1880–1910), 1880. **20**(118): p. 305.
2. Berci, G. and K.A. Forde, *History of endoscopy.* Surgical endoscopy, 2000. **14**(1): p. 5–15.
3. Kuipers, E. and J. Haringsma, *Diagnostic and therapeutic endoscopy.* Journal of Surgical Oncology, 2005. **92**(3): p. 203–209.
4. Horiguchi, A., et al., *Pilot study of prostate cancer angiogenesis imaging using a photoacoustic imaging system.* Urology, 2017. **108**: p. 212–219.
5. Salehi, H.S., et al., *Coregistered photoacoustic and ultrasound imaging and classification of ovarian cancer: ex vivo and in vivo studies.* Journal of biomedical optics, 2016. **21**(4): p. 046006–046006.
6. Yang, J.-M., et al., *Optical-resolution photoacoustic endomicroscopy in vivo.* Biomedical optics express, 2015. **6**(3): p. 918–932.

7. Yang, J.M., et al., *Three-dimensional photoacoustic endoscopic imaging of the rabbit esophagus.* PLoS One, 2015. **10**(4): p. e0120269.
8. Ji, X., et al., *Intravascular confocal photoacoustic endoscope with dual-element ultrasonic transducer.* Optics express, 2015. **23**(7): p. 9130–9136.
9. Wu, M., et al., *Real-time volumetric lipid imaging in vivo by intravascular photoacoustics at 20 frames per second.* Biomedical optics express, 2017. **8**(2): p. 943–953.
10. Lei, P., et al., *Ultrafine intravascular photoacoustic endoscope with a 0.7 mm diameter probe.* Optics Letters, 2019. **44**(22): p. 5406–5409.
11. Xi, L., et al., *Photoacoustic imaging based on MEMS mirror scanning.* Biomedical optics express, 2010. **1**(5): p. 1278–1283.
12. Yang, H., et al., *Handheld miniature probe integrating diffuse optical tomography with photoacoustic imaging through a MEMS scanning mirror.* Biomedical optics express, 2013. **4**(3): p. 427–432.
13. Yuan, Y., S. Yang, and D. Xing, *Preclinical photoacoustic imaging endoscope based on acousto-optic coaxial system using ring transducer array.* Optics letters, 2010. **35**(13): p. 2266–2268.
14. Zheng, S. and Y. Xiangyang, *Image reconstruction based on compressed sensing for sparse-data endoscopic photoacoustic tomography.* Computers in Biology and Medicine, 2020. **116**: p. 103587.
15. Zheng, S., H. Duoduo, and Y. Yuan, *2-D image reconstruction of photoacoustic endoscopic imaging based on time-reversal.* Computers in Biology and Medicine, 2016. **76**: p. 60–68.
16. Leng, X., et al., *Feasibility of co-registered ultrasound and acoustic-resolution photoacoustic imaging of human colorectal cancer.* Biomedical optics express, 2018. **9**(11): p. 5159–5172.
17. Wang, L.V. and S. Hu, *Photoacoustic tomography: in vivo imaging from organelles to organs.* Science, 2012. **335**(6075): p. 1458–1462.
18. Zhu, Q., et al. *Dual-modality Photoacoustic and Ultrasound Imaging for Assessing Treatment Response in Colorectal Cancer: a pilot study*. in *Clinical and Translational Biophotonics*. 2020. Optica Publishing Group.
19. Park, S., et al., *Acoustic resolution photoacoustic microscopy.* Biomedical Engineering Letters, 2014. **4**: p. 213–222.
20. Guo, Z., et al., *Miniature probe for optomechanical focus-adjustable optical-resolution photoacoustic endoscopy.* arXiv preprint arXiv:1911.09522, 2019.
21. Shao, P., et al., *Integrated micro-endoscopy system for simultaneous fluorescence and optical-resolution photoacoustic imaging.* Journal of Biomedical Optics, 2012. **17**(7): p. 076024–076024.
22. Hajireza, P., et al. *Optical-resolution photoacoustic micro-endoscopy using image-guide fibers and fiber laser technology*. in *Photons Plus Ultrasound: Imaging and Sensing 2011*. 2011. SPIE.
23. Guo, Z., Y. Li, and S.-L. Chen, *Miniature probe for in vivo optical-and acoustic-resolution photoacoustic microscopy.* Optics letters, 2018. **43**(5): p. 1119–1122.
24. Bai, X., et al., *Intravascular optical-resolution photoacoustic tomography with a 1.1 mm diameter catheter.* PloS one, 2014. **9**(3): p. e92463.
25. Xiong, K., et al., *Autofocusing optical-resolution photoacoustic endoscopy.* Optics letters, 2018. **43**(8): p. 1846–1849.
26. Dai, X., et al., *Miniature endoscope for multimodal imaging.* Acs Photonics, 2017. **4**(1): p. 174–180.
27. Dai, X., et al., *Targeted molecular imaging of pancreatic cancer with a miniature endoscope.* Applied Sciences, 2017. **7**(12): p. 1241.
28. Guo, H., et al., *Photoacoustic endomicroscopy based on a MEMS scanning mirror.* Optics Letters, 2017. **42**(22): p. 4615–4618.
29. Chen, S.-L., et al., *Miniaturized all-optical photoacoustic microscopy based on microelectromechanical systems mirror scanning.* Optics letters, 2012. **37**(20): p. 4263–4265.
30. Song, C., L. Xi, and H. Jiang, *Acoustic lens with variable focal length for photoacoustic microscopy.* Journal of Applied Physics, 2013. **114**(19).

31. Xi, L., C. Song, and H. Jiang, *Confocal photoacoustic microscopy using a single multifunctional lens.* Optics Letters, 2014. **39**(11): p. 3328–3331.
32. Ansari, R., et al., *All-optical forward-viewing photoacoustic probe for high-resolution 3D endoscopy.* Light: Science & Applications, 2018. **7**(1): p. 75.
33. Ansari, R., et al., *Miniature all-optical flexible forward-viewing photoacoustic endoscopy probe for surgical guidance.* Optics Letters, 2020. **45**(22): p. 6238–6241.
34. Zhao, T., et al., *Video-rate dual-modal photoacoustic and fluorescence imaging through a multimode fibre towards forward-viewing endomicroscopy.* Photoacoustics, 2022. **25**: p. 100323.
35. Ke, D., et al., *Miniature fiber scanning probe for flexible forward-view photoacoustic endoscopy.* Applied Physics Letters, 2023. **122**(12).
36. Zhao, T., et al., *Wavefront Shaping-Assisted Forward-Viewing Photoacoustic Endomicroscopy Based on a Transparent Ultrasound Sensor.* Applied Sciences, 2022. **12**(24): p. 12619.
37. Institute, A.N.S., *American national standard for safe use of lasers*. 2022: Laser Institute of America.

Part II
Image Reconstruction and Deep Learning

Chapter 4
Model-Based Reconstructions for Quantitative Imaging in Photoacoustic Tomography

Andreas Hauptmann and Tanja Tarvainen

Abstract The reconstruction task in photoacoustic tomography can vary a lot depending on measured targets, geometry, and especially the quantity we want to recover. Specifically, as the signal is generated due to the coupling of light and sound by the photoacoustic effect, we have the possibility to recover acoustic as well as optical tissue parameters. This is referred to as quantitative imaging, i.e., correct recovery of physical parameters and not just a qualitative image.

In this chapter, we aim to give an overview on established reconstruction techniques in photoacoustic tomography. We start with modelling of the optical and acoustic phenomena, necessary for a reliable recovery of quantitative values. Furthermore, we give an overview of approaches for the tomographic reconstruction problem with an emphasis on the recovery of quantitative values, from direct and fast analytic approaches to computationally involved optimisation-based techniques and recent data-driven approaches.

Keywords Quantitative imaging · Tomographic reconstruction · Photoacoustic tomography · Inverse problems

4.1 Introduction

Photoacoustic tomography (PAT) offers the possibility to examine biological tissue on a micrometre scale with high contrast by coupling two physical phenomena: light and sound [49, 91]. The obtained tomographic images carry important structural

---

A. Hauptmann (✉)
Research Unit of Mathematical Sciences, University of Oulu, Oulu, Finland

Department of Computer Science, University College London, London, UK
e-mail: andreas.hauptmann@oulu.fi

T. Tarvainen
Department of Technical Physics, University of Eastern Finland, Kuopio, Finland

Department of Computer Science, University College London, London, UK
e-mail: tanja.tarvainen@uef.fi

W. Xia (ed.), *Biomedical Photoacoustics*,
https://doi.org/10.1007/978-3-031-61411-8_4

and quantitative information, which can be used to derive further biological and clinical markers, such as blood oxygen saturation [8, 17]. For the correctness of such markers, it is important to have a robust and reliable reconstruction procedure available.

While the underlying reconstruction problem is stable under ideal measurement setups, i.e. a full boundary coverage and fine sampling with a large bandwidth, the need for faster and more versatile imaging domains creates the need for more advanced reconstruction methods. In recent years, a multitude of different approaches have been introduced for the reconstruction problem in PAT, tackling specifically the acoustic or optical part, or both combined [17, 64, 92]. In this chapter, we will provide an introductory overview of existing reconstruction approaches, their applicability, and possible limitations. We will put a special emphasis on quantitative reconstruction of correct physical parameters.

To understand the reconstruction task, we first need to define the problem. Mathematically, tomographic reconstructions are understood as an inverse problem, i.e. determining the cause from a set of measurements, under the knowledge of the measurement process, or the involved physics [44]. That is, we aim to recover an image f from noise corrupted measurements g under knowledge of the measurement operator, or forward operator, $A : f \mapsto g$. The measurement equation is then simply given by

$$Af + \varepsilon = g, \tag{4.1}$$

where ε denotes measurement noise, possibly including also mechanical/electronic inaccuracies. The *inverse problem* is now to recover the image f from measurements g under knowledge of the relationship in (4.1). This already tells one important aspect of inverse problems that we need to have a good understanding of the measurement process and the involved physics. For this purpose, we will first review the *forward problem* of photoacoustic tomography in Sect. 4.2, i.e. the physics of the measurement process.

In the following sections, we will first discuss the acoustic reconstruction problem, that is the recovery of a tomographic image representing the initial pressure from the measured acoustic signal. This will provide a comprehensive overview of reconstruction methods and different paradigms available. We will then discuss the specifics of reconstructions for the optical problem and the recovery of optical parameters. We will conclude with a discussion of possible problems that one might encounter in the reconstruction process, such as uncertainties on measurement parameters and the need for computational efficiency. Finally, we give a short outlook on the promise of data-driven techniques to overcome some limitations.

4.2 Forward Modelling

In photoacoustic tomography, a short (ns) pulse of near-infrared light is used to illuminate the region of tissue of interest. As light propagates within the tissue, it is absorbed by light-absorbing molecules (chromophores). This generates localised increase in pressure that propagates through the tissue as an acoustic wave and is detected by ultrasound sensors on the boundary or outside the imaged target. The propagation of the acoustic wave occurs about five orders of magnitude slower than the light absorption, and thus the optical and acoustic parts of the problem can be decoupled and treated separately [49].

The *forward problem* in quantitative PAT is to solve for the photoacoustic time series on the ultrasound sensors when the optical and acoustic properties and the input light are given. Solving this problem consists of computing solutions to the optical and acoustic forward problems.

The process starts with the *optical* forward problem, where we need to compute the absorbed optical energy density when the optical properties of the medium and the input light are given. Let us consider modelling of light propagation and absorption at a single wavelength of light. A widely accepted model for light transport in tissues is given by the radiative transfer equation (RTE) [40]. The RTE is a one-speed approximation of the transport equation, and thus it assumes that the energy of the particles does not change in collisions and that the refractive index is constant in the medium. Let $\Omega \subset \mathbb{R}^d$, $d = 2$ or 3 denote the physical domain with boundary $\partial\Omega$ and let $\hat{s} \in S^{d-1}$ (on the sphere) denote a unit vector in the direction of interest. In quantitative PAT, the time-independent RTE is utilised

$$\hat{s} \cdot \nabla\phi(r, \hat{s}) + (\mu_s + \mu_a)\phi(r, \hat{s}) = \mu_s \int_{S^{d-1}} \Theta(\hat{s} \cdot \hat{s}')\phi(r, \hat{s}')\mathrm{d}\hat{s}', \quad r \in \Omega, \tag{4.2}$$

$$\phi(r, \hat{s}) = \begin{cases} \phi_0(r, \hat{s}), \ r \in \epsilon_j, & \hat{s} \cdot \hat{n} < 0, \\ \quad 0, \quad r \in \partial\Omega \backslash \epsilon_j, & \hat{s} \cdot \hat{n} < 0, \end{cases} \tag{4.3}$$

where $\mu_s(r)$ and $\mu_a(r)$ are the scattering and absorption coefficients of the medium, respectively, $\phi(r, \hat{s})$ is the radiance, $\Theta(\hat{s} \cdot \hat{s}')$ is the scattering phase function, $\phi_0(r, \hat{s})$ is a source at a position $\epsilon_j \subset \partial\Omega$, and $\hat{n}$ is an outward unit normal [4, 82]. The scattering phase function $\Theta(\hat{s} \cdot \hat{s}')$ describes the probability that a photon with an initial direction $\hat{s}'$ will have a direction $\hat{s}$ after a scattering event. In biological tissue, the most commonly used phase function is the Henyey–Greenstein scattering function [37]

$$\Theta(\hat{s} \cdot \hat{s}') = \begin{cases} \frac{1}{2\pi} \frac{1-g^2}{(1+g^2-2g\hat{s}\cdot\hat{s}')}, & d = 2 \\ \frac{1}{4\pi} \frac{1-g^2}{(1+g^2-2g\hat{s}\cdot\hat{s}')^{3/2}}, & d = 3, \end{cases} \tag{4.4}$$

where g is an anisotropy parameter, $-1 < g < 1$.

Due to the computational complexity of the RTE, various approximations are generally utilised in optical imaging. In a highly scattering medium, the RTE is approximated with the diffusion approximation (DA). In the DA, the radiance is approximated by

$$\phi(r, \hat{s}) \approx \frac{1}{|S^{d-1}|}\Phi(r) - \frac{d}{|S^{d-1}|}\hat{s} \cdot (\kappa \nabla \Phi(r)), \tag{4.5}$$

where $\Phi(r)$ is the photon fluence

$$\Phi(r) = \int_{S^{d-1}} \phi(r, \hat{s}) \mathrm{d}\hat{s}. \tag{4.6}$$

Further, $\kappa = \left(d(\mu_a + \mu_s')\right)^{-1}$ is the diffusion coefficient where $\mu_s' = (1 - g_1)\mu_s$ is the reduced scattering coefficient and g_1 is the mean of the cosine of the scattering angle [4, 82] that in the case of the Henyey–Greenstein scattering function is $g_1 = g$. By inserting the approximation (4.5) and similar approximations written for the source term and phase function into equation (4.2) and following the derivation in [4, 40, 82], the DA is obtained

$$-\nabla \cdot \kappa \nabla \Phi(r) + \mu_a \Phi(r) = 0, \qquad r \in \Omega, \tag{4.7}$$

$$\Phi(r) + \frac{1}{2\gamma_d}\kappa \frac{\partial \Phi(r)}{\partial \hat{n}} = \begin{cases} \frac{I_s}{\gamma_d}, & r \in \epsilon_i, \\ 0, & r \in \partial\Omega \setminus \epsilon_i, \end{cases} \tag{4.8}$$

where I_s is a diffuse boundary current and γ_d is a dimension-dependent constant which takes values $\gamma_2 = 1/\pi$ and $\gamma_3 = 1/4$ [82]. The DA is a valid approximation when the radiance is almost a uniform distribution, i.e. in a scattering dominated medium further than a few scattering lengths from the light source [40]. In PAT, however, imaging depth can be small compared to the average scattering length, and thus the DA is not always a valid approximation.

Since light absorption and pressure increase occur significantly faster compared to acoustic wave propagation, they can be modelled instantaneous. The absorbed optical energy density $H(r)$ created by light absorption can be solved from photon fluence as

$$H(r) = \mu_a(r)\Phi(r). \tag{4.9}$$

Further, generated pressure can be approximated as an initial pressure p_0 that is connected to the absorbed optical energy density through the photoacoustic efficiency that can be identified with the Grüneisen parameter G for an absorbing fluid [17]

$$p_0(r) = p(r, t = 0) = GH(r). \tag{4.10}$$

The following *acoustic* forward problem in quantitative PAT is to solve for the photoacoustic time series $p(r, t)$ at the sensors. Time evolution of the photoacoustic wave can be modelled using the equations of linear acoustics [19, 88]. For soft biological tissues, it is generally assumed that the medium is isotropic and quiescent and that shear waves can be neglected. Then, the propagation of initial pressure can be described as an initial value problem in acoustics using an acoustic wave equation

$$(\partial_{tt} - v^2\Delta)p(r,t) = 0, \tag{4.11}$$

$$p(r, t = 0) = p_0(r), \tag{4.12}$$

$$\partial_t p(r, t = 0) = 0. \tag{4.13}$$

The spatial distribution of the speed of sound v depends on the target and is usually not known a priori. Still, it is often assumed to be constant within the medium. The time series $p(r, t)$ on the boundary of the domain now constitutes the measured signal in photoacoustic tomography, from which we will start the reconstruction process.

4.3 Reconstruction of the Photoacoustic Image

The reconstruction task in PAT follows the reverse order of the forward model, that is we are dealing with an inverse problem. Specifically, the forward problem starts with the optical illumination, followed by the generation of the acoustic signal through the photoacoustic effect. The inverse problem consequently starts with the measured pressure wave on the sensor, from which we can recover the initial pressure p_0, only after which we can solve the optical problem. Here, the first acoustic problem is linear, whereas the second optical problem is non-linear.

In the following, we will first discuss the reconstruction task for the acoustic problem in more detail. The measurement is given by the pressure wave $p(r, t)$ on the boundary, i.e. $r \in \partial\Omega$, and a finite time interval $[0, T]$. Usually, the measurement is restricted to only a part of the boundary, and we may only collect sub-sampled data. Additionally, the detector will have an angle-dependent frequency response which needs to be taken into account. For simplicity we assume here that this is encoded in the forward operator A that maps the initial pressure to the measured time series, and we can simply write the measurement equation as

$$Af + \varepsilon = g, \quad \text{where } f(r) = p_0(r), \ r \in \Omega, \tag{4.14}$$
$$g(r,t) = p(r,t), \ r \in \partial\Omega, \ t \in [0, T],$$

where ε denotes additional measurement noise. Below, we give a short overview of different inverse problem solution methodologies applied in PAT. For a more detailed review on reconstruction algorithms, we refer to a recent review [64].

4.3.1 Direct Reconstructions

The acoustic inverse problem is well-posed, when full measurement data exists [96]. That is, the measurement domain surrounds the whole target, and the time series is measured long enough, i.e. for all times where the pressure wave is non-zero. This is clearly easier to achieve in a two-dimensional setting than in 3D, where we often are limited to only a part of the boundary. This leads to what is known as a limited-view setting, see Fig. 4.1 for an illustration.

In case there is sufficiently large part of the boundary covered, we can obtain good reconstructions by *backprojection*-type algorithms. Here, the data g is projected back along a set of spherical shells of radius $t = |r - r_s|/v$ centred on the

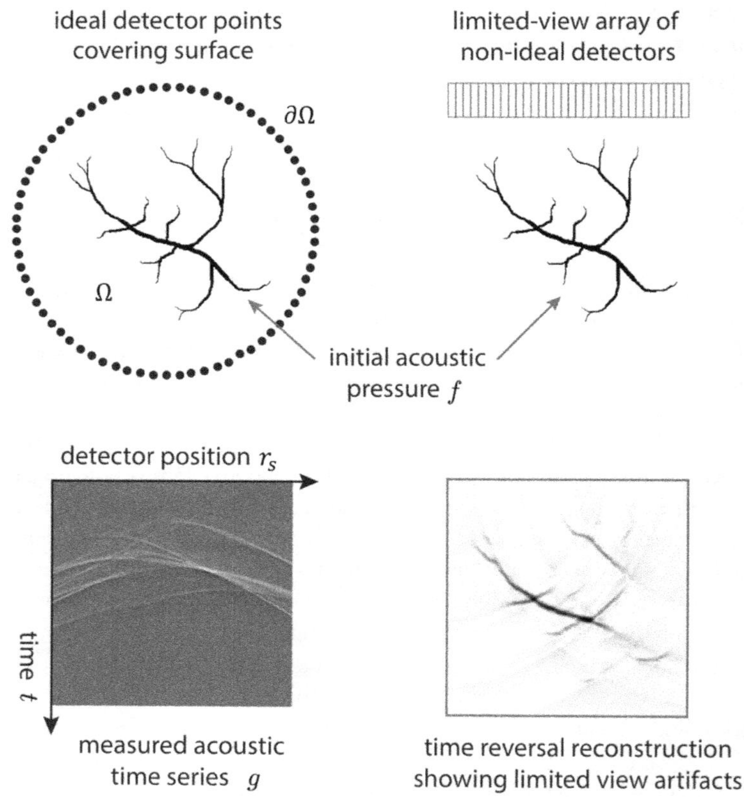

Fig. 4.1 (Top-left) Ideal PAT measurement setup in 2D with point-like omnidirectional detectors covering the surface $\partial\Omega$ surrounding the initial pressure distribution f. (Top-right) A typical measurement setup using a finite-sized linear array of detectors. (Bottom-left) The acoustic time series measured by the linear array. (Bottom-right) A PAT image reconstructed from these time series using the classical time-reversal approach showing arc-like artefacts due to the limited-view detection (figure adapted from [33])

detector points r_s. We can define the general class of backprojection operators $A^{\#}$ that map the measured data $g(r_s, t)$ to the reconstructions $f(r)$ by projecting the data back on these shells by the mapping $t \to |r - r_s|/v$ followed by summation over all detector points r_s. Together with the function $h(r_s, t)$, which represents the measurement data or a transformation that depends on the measurement setup. The general backprojection operator is then given by

$$(A^{\#}g)(x) = \int_{\partial\Omega} [h(r_s, t)]_{t=|r-r_s|/v} \, ds(r_s), \tag{4.15}$$

where ds is an area element on the measurement surface $\partial\Omega$. Clearly, the recovery of quantitative values, in contrast to qualitative reconstructions, depends on the filter function $h(r_s, t)$. For instance, in case of the adjoint operator A^* [5], the backprojection takes the form

$$(A^*g)(x) = \int_{\partial\Omega} \left[\frac{1}{4\pi |r - r_s|} \frac{\partial g}{\partial t}(r_s, t) \right]_{t=|r-r_s|/v} ds(r_s), \tag{4.16}$$

which primarily provides a qualitative reconstruction with slightly over estimated contrast.

In case of the well-known 'universal backprojection' algorithm [94], we obtain exact reconstructions with correct quantitative values for spherical and cylindrical [22, 46] or even planar measurement surfaces that enclose the target.

4.3.1.1 Time Reversal

The wave equation can be stably solved backwards in time, and hence one can obtain a physically intuitive approach to reconstruction by the so-called *time reversal* [39, 95], but one needs to be careful of the boundary conditions. Given the measurement surface $\partial\Omega$ surrounding the measurement domain in which the target is supported $f \subset \Omega$. In the forward model, the photoacoustically generated waves are propagating outwards and are measured on the surface $\partial\Omega$ until after suitably long time T the acoustic field in Ω is zero.

In the time-reversal process, we assume that measured pressure $g(r_s, t)$ is now produced on $\partial\Omega$ in time-reversed order, starting with $g(x_s, T)$. Thus, the acoustic field in Ω created by the *in-going* waves reproduces the out-going wavefield exactly, but backwards in time. Consequently, the acoustic field at $t = 0$ will reproduce the initial acoustic pressure $f(r) = p_0(r)$. Numerically, time-reversal image reconstruction solves the following time-varying boundary value problem for the time-reversed field $p_-(r, t_-)$, from time $t_- = 0$ to T,

$$(\partial_{tt} - v^2\Delta)p_-(r, t_-) = 0, \tag{4.17}$$

$$p_-(r_s, t_-) = g(r_s, T - t_-), \tag{4.18}$$

$$p_- = \partial_t p_-(r, 0) = 0. \tag{4.19}$$

We then obtain at final time T the reconstruction $p_-(r, T) = f(r)$ for $r \in \Omega$. A reconstruction by time reversal is shown in Fig. 4.1 for a limited-view geometry in 2D.

4.3.2 Variational Methods

In limited view or when only sparsely sampled data is available, direct reconstructions often do not perform sufficiently well, qualitatively as well as quantitatively. In such scenarios it is advised to utilise the so-called variational methods [3, 76]. That is, the reconstruction f^* is sought as minimiser of a cost functional that measures data fit together with a regulariser,

$$f^* = \arg\min_f \|Af - g\|_2^2 + \alpha\Psi(f). \tag{4.20}$$

Here, the data-fidelity term $\|Af - g\|_2^2$ naturally ensures that the reconstructed image fits the measured data, and hence f^* also provides correct quantitative values. The regulariser $\Psi(f)$ has a two-fold purpose, first of all it ensures well-posedness of the reconstruction task and stabilises the optimisation, and additionally it allows to model prior knowledge on the target structures by penalising undesired features. Typical regularisers are, for example, Tikhonov regularisation $\Psi(f) = \|f\|_2^2$ that supports smoothness of the reconstruction, the total variation penalty $\Psi(f) = \|\nabla f\|_1$, which enforces reconstructions to be piece-wise constant and hence is highly effective in noise suppression [3, 11, 90], higher-order variants [13], as well as generally regulariser based on the 1-norm promoting sparsity [28]. The parameter $\alpha > 0$ in (4.20) balances both terms and allows one to fine-tune the desired reconstructions: A small parameter allows for emphasis on the data fit (advised under small noise), whereas a large parameter enforces stronger regularity. In other words, we choose how much we trust the measured data.

Reconstructions are then computed by optimisation techniques that minimise (4.20) iteratively. If both data fidelity and regulariser are differentiable, gradient descent methods can be utilised. Whereas the data fidelity using the 2-norm is differentiable, popular choices for the regulariser are based on sparsity and are often based on the non-differentiable 1-norm. Thus, one often uses techniques from convex analysis[1] and optimisation to compute minimiser [10]. One such option are proximal gradient descent schemes. Here, we start with some initialisation $f_0 = A^\# g$, and we then iteratively take the gradient with respect to the differentiable data fidelity, given by $\nabla_f \|Af - g\|_2^2 = 2A^*(Af - g)$, followed by the application

[1] Note that the problem (4.20) [72] is convex if A is linear and the regulariser $\Psi(f)$ is convex.

of a proximal operator that incorporates the regularisation for convex $\Psi(f)$. The iterations are then given by

$$f_{k+1} = \mathrm{prox}_{\Psi,\alpha\lambda_k}\left(f_k - \lambda_k A^*(Af_k - g)\right), \tag{4.21}$$

where the proximal operator projects solutions to the admissible space of the regulariser by solving another simpler optimisation problem

$$\mathrm{prox}_{\Psi,\alpha\lambda_k}(f) = \arg\min_h \left\{\Psi(h) + \frac{1}{2\alpha\lambda_k}\|h - f\|_2^2\right\}. \tag{4.22}$$

The advantage of the proximal operator is two-fold as well, first for some choices of $\Psi(f)$ there is a closed-form solution, such as the 1-norm $\Psi(f) = \|f\|_1$. Second, the problem (4.22) does not involve the forward operator A and hence is a denoising problem for which many efficient algorithms exist, and this is the case for total variation.

An illustration of reconstructions using the variational approach is shown in Fig. 4.2. The backprojection clearly leaves visible limited-view artefacts and loss of contrast in the deeper areas, whereas iterative reconstructions can effectively remove those by iteratively enforcing data consistency. When fully sampled data is available, just enforcing non-negativity as regulariser is sufficient, and for sub-sampled measurements a stronger regulariser is needed, such as total variation enforcing piece-wise constant targets, effectively eliminating the residual noise.

4.3.3 Bayesian Methods

In a Bayesian approach, the inverse problem is solved in the framework of statistical inference [44]. All parameters are considered as random variables, and information about these parameters is expressed by their probability distributions. In the inverse problem, the idea is to obtain information about the parameters of primary interest based on the measurements, the model, and the prior information about the parameters. The solution of the inverse problem is then given by the posterior probability distribution $\pi(f|g)$ that according to Bayes' theorem can be presented as a conditional probability density function of the form

$$\pi(f|g) \propto \pi(g|f)\pi(f), \tag{4.23}$$

where $\pi(g|f)$ is the likelihood density and $\pi(f)$ is the prior density. Considering the observation model with additive noise (4.14), the likelihood probability distribution can be written as

$$\pi(g|f) = \pi_\varepsilon(g - Af), \tag{4.24}$$

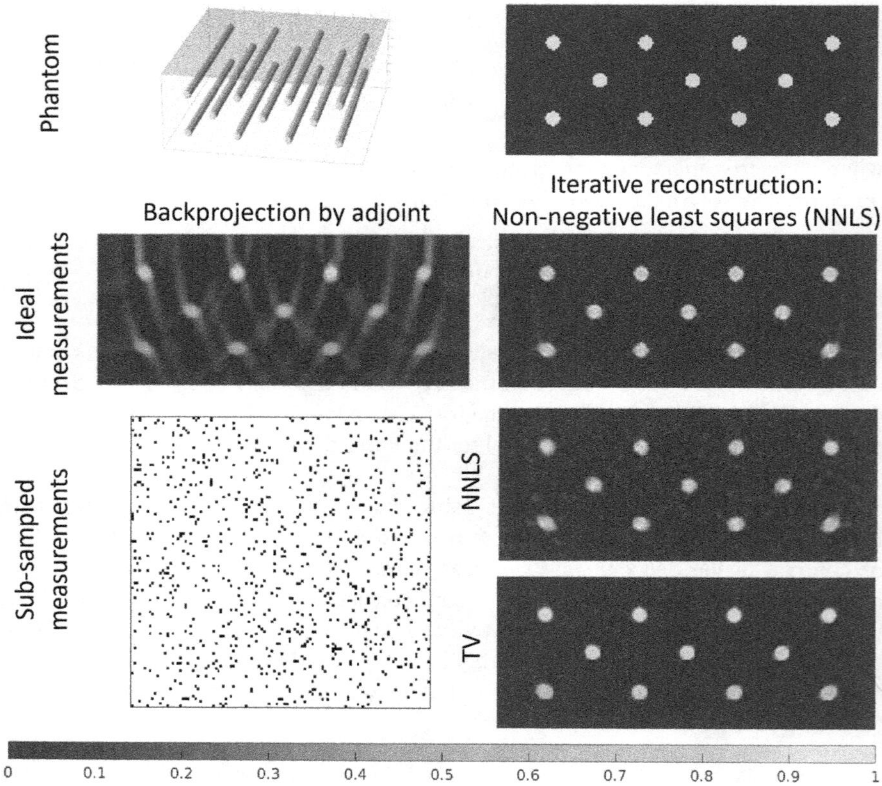

Fig. 4.2 Reconstructions in 3D from fully sampled and sub-sampled limited-view measurements. (Top Row) The phantom in 3D and cross section: The measurement sensor is marked in red on the top. (Middle row) Reconstruction from ideal fully sampled measurements: backprojection by the adjoint on the left and an iterative reconstruction using only non-negativity as regulariser on the right. (Bottom row) Reconstructions from sub-sampled data: The sub-sampling pattern is shown on the left and two reconstructions with non-negativity (NNLS) and total variation (TV) on the right (figure adapted from [35])

where $\pi_\varepsilon(\cdot)$ is the probability density of the noise ε. Let us model noise ε and unknown f as Gaussian distributed, i.e. $\varepsilon \sim \mathcal{N}(\eta_\varepsilon, \Gamma_\varepsilon)$ and $f \sim \mathcal{N}(\eta_f, \Gamma_f)$ where η_ε and Γ_ε are the mean and covariance of the noise and η_f and Γ_f are the mean and covariance of the prior for the unknown f. In this case, the posterior probability density can be written as

$$\pi(f|g) \propto \exp\left\{-\frac{1}{2}\|L_\varepsilon(g - Af - \eta_\varepsilon)\|^2 - \frac{1}{2}\left\|L_f(f - \eta_f)\right\|^2\right\}, \tag{4.25}$$

where $L_\varepsilon^{\mathrm{T}} L_\varepsilon = \Gamma_\varepsilon^{-1}$ and $L_f^{\mathrm{T}} L_f = \Gamma_f^{-1}$ are the square roots such as the Cholesky decompositions of the inverse covariance matrices of the noise and prior, respectively.

In the case of a linear observation model and Gaussian noise and prior, the posterior density is also a Gaussian distribution [44, 86]

$$f|g \sim \mathcal{N}(\eta_{f|g}\Gamma_{f|g}),$$

where the mean $\eta_{f|g}$ and covariance $\Gamma_{f|g}$ can be written in the form

$$\eta_{f|g} = K^{-1}b \text{ and } \Gamma_{f|g} = K^{-1}, \tag{4.26}$$

with

$$K = A^{\mathrm{T}}\Gamma_\varepsilon^{-1}A + \Gamma_f^{-1}, \tag{4.27}$$

$$b = A^{\mathrm{T}}\Gamma_\varepsilon^{-1}(g - \eta_\varepsilon) + \Gamma_f^{-1}\eta_f. \tag{4.28}$$

In case solving the posterior distribution directly using Eqs. (4.26)–(4.28) is computationally too expensive, it can be evaluated using a matrix free approach, for example [85]. The advantage of a Bayesian approach lies in the additional information on mean and variances of the computed solutions, which allows for uncertainty quantification giving estimates on the correctness of solutions.

4.3.4 *Data-Driven Reconstruction Methods*

In recent years, data-driven approaches have gained considerable interest due to their ability to learn data-specific representations. That means, instead of designing hand-crafted analytic priors (in the variational setting) or statistical priors (in the Bayesian setting), one can now utilise large quantities of representative data. In most cases, these are given by ideal reconstructions, but here also lies the primary limitation in photoacoustic tomography, since large quantities of ideal or ground-truth reconstructions are not easy to collect, especially in three dimensions. Additionally, at the current stage there are limited guarantees available on the correctness of the reconstructions apart from empirical evaluations. In the following we give a short summary of the underlying paradigm of learned reconstruction methods in the context of photoacoustic tomography, and for a complete review we refer to [27, 33, 97].

In general terms, the aim of learned image reconstruction is to design a parameterised reconstruction operator R_θ with learnable parameters θ, such that

$$R_\theta(g) \approx f. \tag{4.29}$$

Clearly, there are many ways to define such a learned reconstruction operator, as there are many classical reconstruction approaches available as well. We can roughly group approaches into three classes:

i. The fully learned approach: Here, R_θ is purely given as a neural network $G_\theta : g \mapsto f$ that directly transforms the measurements into reconstructions. See, for examples, [77, 89].
ii. The two-step approach: First a classical reconstruction is obtained by, e.g., a backprojection $A^\# : g \mapsto \widetilde{f}$, which is usually corrupted by noise, undersampling and possible limited-view artefacts. A neural network $G_\theta : \widetilde{f} \mapsto f$ is then trained to remove these artefacts [2].
iii. Learned iterative (model-based): Here, neural networks and the forward/adjoint operator are intertwined in an iterative way [35]. That is, the reconstruction is updated repeatedly motivated by the variational formulation. This can be achieved, for instance, by replacing the proximal operator in (4.21) with a neural network, such that $G_\theta : f_k \mapsto f_{k+1}$.

Generally speaking, while the model information increases from i. to iii., the complexity of the learning task decreases, i.e. the network needs to learn an easier operation. This decrease in complexity does also reduce the quantity of necessary data to train the network but comes with an increase of computational complexity due to the involvement of A. In practice, it is important to choose the most suitable approach for one's needs.

When aiming for correct and reliable quantitative reconstructions with learning-based approaches, there are unfortunately only limited guarantees on correctness. At this stage, the majority of approaches does rely on empirical validation rather than theoretical guarantees. Nevertheless, current theoretical research concentrates on providing reconstruction guarantees. This can be primarily achieved by restricting the networks G_θ but usually also reduces expressivity of the network and hence results in a reduction of performance, see [61] for a recent survey on the subject of provably convergent learned reconstructions.

4.4 The Inverse Problem of Quantitative Photoacoustic Tomography

As the second step of quantitative photoacoustic imaging, optical parameters are estimated from the photoacoustic images obtained as the solution of the acoustic inverse problem. This second inverse problem includes a non-linear forward model describing light propagation in tissues. Most of the approaches for this optical inverse problem have been based on using the DA (4.7)–(4.8) as light transport model. However, the RTE (4.2)–(4.3) [29, 56, 67, 75, 83] and Monte Carlo method for light transport [14, 15, 30, 38, 48, 54] have also been utilised. In addition to modelling light propagation and absorption, photoacoustic efficiency identified by the Grüneisen parameter, Eq. (4.10), needs to be taken into account.

Estimation of more than one optical parameter is a non-unique problem if only one light illumination is used. To overcome this, one approach has been to assume the scattering as known [14, 18, 42, 71]. However, in practical applications scattering is typically not known. It has been shown that the non-uniqueness can be overcome by using multiple optical illuminations [6]. Also combining photoacoustic with diffuse optical tomography (DOT) data has been used to ease the non-uniqueness [51, 63, 70].

4.4.1 Optical Inverse Problem

The aim in quantitative PAT is to estimate the concentrations of chromophores. Generally, the absorption coefficient can be expressed as a linear combination of the chromophore concentrations

$$\mu_a(\lambda) = \sum_{k=1}^{K} \alpha_k(\lambda) C_k, \tag{4.30}$$

where α_k and C_k are the absorption coefficient and concentration of the kth chromophore, and λ is the wavelength of the light. Estimation of the concentrations of chromophores can be achieved either by directly estimating the concentrations from photoacoustic time series obtained using different wavelengths of light [7, 20, 47, 56, 66, 68, 69] or by first recovering the absorption coefficients at different wavelengths and then calculating the concentrations utilising (4.30) for the absorption spectrum [7, 17, 20, 66]. Sometimes, researchers are interested in secondary quantities that can be computed from the optical parameters [12]. One such quantity of high clinical interest is blood oxygen saturation (sO_2) [50], which is related to the concentrations of the two endogenous chromophores, oxy- and deoxy-haemoglobin. Different approaches for solving the optical inverse problem have been proposed. Here, we give a short summary of those approaches. For more information, see, e.g., the recent review [92] and the references therein.

Perhaps the most simple approach for estimating the optical absorption in quantitative PAT has been to assume the photon fluence as known or solvable using some predefined optical parameter values and then solve the absorption utilising the so-called fluence correction methods [100]. The drawback of these approaches is that both absorption and scattering affect photon fluence, and thus approximating these as preassigned parameters creates modelling errors. Alternatively, linearised models leading to simple solvers for estimating absorption [18, 98] or absorption and scattering simultaneously [78] have been utilised.

In most studies on the optical inverse problem, simultaneous estimation of absorption and either scattering or diffusion have been considered. The inverse problem has been approached, for example, using direct methods [6, 7, 56]. Similarly as in the acoustic problem, a commonly utilised approach has been to estimate the optical parameters x by formulating the problem as a regularised minimisation problem

$$x^* = \arg\min_x \|B(x) - y\|_2^2 + \beta\Psi(x), \tag{4.31}$$

where B denotes the optical forward operator, y is the data of the optical inverse problem, β is the regularisation parameter, and Ψ is a regularising penalty functional. Typically, Tikhonov regularisation [75] supporting smoothness of the solution, total variation [24, 32, 83] promoting piece-wise constant parameters, and ℓ_1-norm regularisation [99] promoting sparsity in the solution have been utilised. The minimisation problem (4.31) has been solved using methods of computational optimisation. Alternatively, a Bayesian approach leading to a *maximum a posteriori* (MAP) estimation problem has been taken in [30, 48, 63, 66, 84] enabling inclusion of quantitative prior models and modelling of noise and uncertainties. Furthermore, examples of alternative approaches to the optical inverse problem include [1, 73] where the absorption and the photon fluence were extracted using a sparse signal representation. Further, in [62] boundaries between piece-wise constant optical parameters were estimated, and in [55] a combined reconstruction–classification approach was proposed.

Similarly as for the acoustic problem, data-driven approaches have been utilised recently to solve the optical part, albeit there are fewer approaches proposed. A common approach follows the two-step method ii., where acoustic reconstructions are available. Then, a neural network is trained to recover the absorption coefficient distribution [16]. As mentioned before, there is often interest in computing related quantities such as the blood oxygen saturation sO_2. A few works have proposed to use a neural network similarly as in the previous approach, but instead of recovering the absorption coefficient, the network is trained to directly produce sO_2 maps from multi-wavelength acoustic reconstructions [9, 26], see Fig. 4.3 for an illustration. Currently, there are no equivalent approaches to the model-based iterative paradigm iii. for the optical part, but there are studies for the related

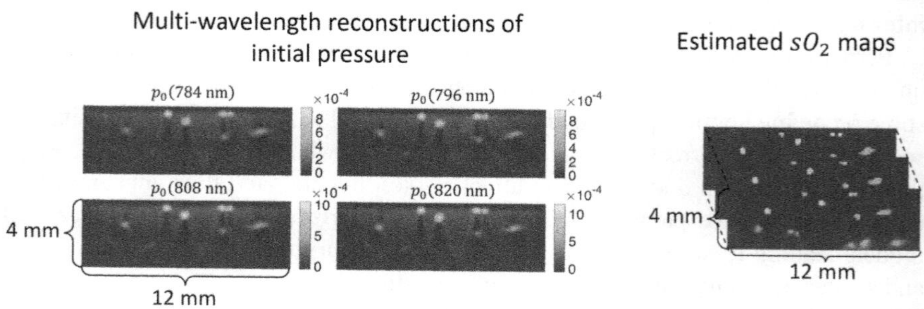

Fig. 4.3 Direct estimation of blood oxygenation saturation maps from multi-wavelength acoustic reconstructions of initial pressure in 3D of simulated human tissue phantoms. (Left) Input slices (through 3D volume) of reconstructed initial pressure at four wavelengths. (Right) Neural network estimation of sO_2 maps in 3D (figure adapted from [9])

problem of DOT. Here, the authors proposed a learned iterative update for the Gauss–Newton algorithm [60].

4.4.2 Single-Stage Approaches

In addition to the two-step approach where acoustic and optical inverse problems are solved one after another, estimation of the optical parameters directly from the photoacoustic measurements has also been studied. This one-step approach was formulated utilising the Born approximation in [79, 81]. Furthermore, one-step approaches using Tikhonov regularisation [29], ℓ_1 sparsity regularisation [23], and total variation regularisation [41] have been considered. Further, a Bayesian approach for the single-stage approach was formulated in [65]. In [21], a one-step approach was used to recover the speed of sound and optical parameters simultaneously.

The authors in [25] investigated the possibility to directly recover optical parameters from the measured acoustic signal using machine learning methods. It was reported that such a naive approach is generally outperformed, especially in terms of generalisation capabilities to unseen data, by methods that use intermediate data or model information.

4.5 Conclusions and Outlook

Much of the theory and analytical approaches on reconstructions for quantitative PAT assume ideal measurement environments. In practice, however, the photoacoustic signal is affected by the measurement system, for example bandwidth and directivity of the acoustic sensors, as well as uncertainties related to the measurement setup. Therefore, a complete modelling of photoacoustic data requires, in addition to modelling optical and acoustic phenomena, modelling of measurement system specifics. Furthermore, regardless of carefully considering all measurement setup specifics, modelling of them always contains uncertainties. Examples of such uncertainties include, e.g., uncertainties in modelling the locations of the ultrasound sensors, variations in the light illumination, uncertainties in the frequency response of the ultrasound sensors, etc. These all can affect the accuracy of the image reconstruction methodology and result into artefacts in images, errors in quantitative estimates, and less reliable confidence intervals. Compensating errors and uncertainties in photoacoustics have been studied, for example, using Bayesian approximation error modelling [31, 74, 84] and utilising learned model correction methods [53].

Additionally, fundamental challenges are given by the physics. One such problem is given by the speed of sound that is often assumed to be constant, and even when assumed constant, the accurate value inside the tissue may still not be known. In principle, a spatially varying speed of sound could be recovered jointly with the acoustic reconstruction, although this problem is inherently unstable and requires additional data or prior knowledge [45, 52, 57, 58]. One possibility here is the use of a separate modality, such as ultrasound tomography [43, 59, 93], to recover a speed of sound map prior to reconstruction. More generally, in the Bayesian setting, errors due to uncertainties in the speed of sound can be compensated for using Bayesian approximation error modelling [87], but the method is limited to small variations in the speed of sound or prior knowledge of the regions with larger variations. Here, data-driven methods can be helpful when the uncertainty on model parameters is incorporated into the training data. In this case, empirical evidence shows that the networks can learn to compensate for this uncertainty in the learned reconstruction.

Finally, the most accurate and stable reconstructions can still be obtained by model-based techniques, but the need to repeatedly evaluate the model equations leads to an immense computational overhead that limits real-time applicability. This is even true for learned methods that do involve physics modelling to some extent, either in a two-step reconstruction or in learned iterative approaches. An area of active research considers how to effectively cut down the computational bottle neck of the involved models by using model reduction techniques (coarser discretisation) or approximate models (making certain simplifying assumptions) [19, 34]. This results in a very similar problem as in model uncertainties, except that these are now introduced as additional complication. Here, the data-driven paradigm is especially promising, as we can effectively simulate training data and compensate the additional introduced approximation errors using neural networks [36, 53, 80].

In conclusion, while the basic mathematical problem of quantitative PAT is solved, there are still many problems that need to be considered to achieve reliable and fast quantitative reconstructions in practice. The primary challenge can be identified as arising uncertainties when imaging *in vivo*, as opposed to ideal simulated studies, as well as computational limitations that one encounters in high-dimensional and possibly even temporal data. Data-driven approaches offer some new opportunities but are certainly not the only solution, as their generalisation capabilities and robustness to changing imaging setups are still not fully understood. This emphasises the need for fundamental research to further our basic understanding, in combination with modern data-driven techniques.

Acknowledgments This work was supported by the European Research Council (ERC) under the European Union's Horizon 2020 research and innovation programme (grant agreement no. 101001417—QUANTOM) and the Academy of Finland (Centre of Excellence in Inverse Modelling and Imaging projects 353093 and 353086, and the Flagship of Advanced Mathematics for Sensing Imaging and Modelling grant 358944, 359186, the Flagship of Photonics Research and Innovation grant 320166, and Academy Research Fellow project 338408). The authors would like to thank the Isaac Newton Institute for Mathematical Sciences, Cambridge, for support and hospitality during the programme Rich and Nonlinear Tomography where work on this paper was undertaken, supported by EPSRC grant no EP/R014604/1.

References

1. G. S. ALBERTI AND H. AMMARI, *Disjoint sparsity for signal separation and applications to hybrid inverse problems in medical imaging*, Applied and Computational Harmonic Analysis, 42 (2017), pp. 319–349.
2. S. ANTHOLZER, M. HALTMEIER, AND J. SCHWAB, *Deep learning for photoacoustic tomography from sparse data*, Inverse problems in science and engineering, 27 (2019), pp. 987–1005.
3. S. ARRIDGE, P. BEARD, M. BETCKE, B. COX, N. HUYNH, F. LUCKA, O. OGUNLADE, AND E. ZHANG, *Accelerated high-resolution photoacoustic tomography via compressed sensing*, Physics in Medicine & Biology, 61 (2016), p. 8908.
4. S. R. ARRIDGE, *Optical tomography in medical imaging*, Inv Probl, 15 (1999), pp. R41–R93.
5. S. R. ARRIDGE, M. M. BETCKE, B. T. COX, F. LUCKA, AND B. E. TREEBY, *On the adjoint operator in photoacoustic tomography*, Inverse Problems, 32 (2016), p. 115012.
6. G. BAL AND K. REN, *Multi-source quantitative photoacoustic tomography in a diffusive regime*, Inv Probl, 27 (2011), p. 075003.
7. G. BAL AND K. REN, *On multi-spectral quantitative photoacoustic tomography in a diffusive regime*, Inv Probl, 28 (2012), p. 025010.
8. P. BEARD, *Biomedical photoacoustic imaging*, Interface Focus, 1 (2011), pp. 602–631.
9. C. BENCH, A. HAUPTMANN, AND B. COX, *Toward accurate quantitative photoacoustic imaging: learning vascular blood oxygen saturation in three dimensions*, Journal of Biomedical Optics, 25 (2020), pp. 085003–085003.
10. M. BENNING AND M. BURGER, *Modern regularization methods for inverse problems*, Acta numerica, 27 (2018), pp. 1–111.
11. M. BERGOUNIOUX, X. BONNEFOND, T. HABERKORN, AND Y. PRIVAT, *An optimal control problem in photoacoustic tomography*, Mathematical Models and Methods in Applied Sciences, 24 (2014), pp. 2525–2548.
12. I. J. BIGIO AND S. FANTINI, *Quantitative biomedical optics: theory, methods, and applications*, Cambridge University Press, 2016.
13. Y. E. BOINK, M. J. LAGERWERF, W. STEENBERGEN, S. A. VAN GILS, S. MANOHAR, AND C. BRUNE, *A framework for directional and higher-order reconstruction in photoacoustic tomography*, Physics in Medicine & Biology, 63 (2018), p. 045018.
14. J. BUCHMANN, B. KAPLAN, S. POWELL, S. PROHASKA, AND J. LAUFER, *Quantitative PA tomography of high resolution 3-D images: experimental validation in a tissue phantom*, Photoacoustics, 17 (2020), p. 100157.
15. J. BUCHMANN, B. A. KAPLAN, S. POWELL, S. PROHASKA, AND J. LAUFER, *Three-dimensional quantitative photoacoustic tomography using an adjoint radiance Monte Carlo model and gradient descent*, Journal of Biomedical Optics, 24 (2019), pp. 066001–066001.
16. T. CHEN, T. LU, S. SONG, S. MIAO, F. GAO, AND J. LI, *A deep learning method based on u-net for quantitative photoacoustic imaging*, in Photons Plus Ultrasound: Imaging and Sensing 2020, vol. 11240, 2020, p. 112403V.
17. B. COX, J. G. LAUFER, S. R. ARRIDGE, AND P. C. BEARD, *Quantitative spectroscopic photoacoustic imaging: a review*, J Biomed Opt, 17 (2012), p. 061202.
18. B. T. COX, S. R. ARRIDGE, K. P. KÖSTLI, AND P. C. BEARD, *Two-dimensional quantitative photoacoustic image reconstruction of absorption distributions in scattering media by use of a simple iterative method*, Applied Optics, 45 (2006), pp. 1866–1875.
19. B. T. COX AND P. C. BEARD, *Fast calculation of pulsed photoacoustic fields in fluids using* k*-space methods*, J Acoust Soc Am, 117 (2005), pp. 3616–3627.
20. B. T. COX, S. R., ARRIDGE, AND P. C. BEARD, *Estimating chromophore distributions from multiwavelength photoacoustic images*, J Opt Soc Am A, 26 (2009), pp. 443–455.
21. T. DING, K. REN, AND S. VALLÉLIAN, *A one-step reconstruction algorithm for quantitative photoacoustic imaging*, Inverse Problems, 31 (2015), p. 095005.

22. D. Finch and R. Sarah K. Patch, *Determining a function from its mean values over a family of spheres*, SIAM Journal on Mathematical Analysis, 35 (2004), pp. 1213–1240.
23. H. Gao, J. Feng, and L. Song, *Limited-view multi-source quantitative photoacoustic tomography*, Inverse Problems, 31 (2015), p. 065004.
24. H. Gao, H. Zhao, and S. Osher, *Bregman methods in quantitative photoacoustic tomography*, CAM Report, 30 (2010), pp. 3043–3054.
25. J. Gröhl, T. Kirchner, T. Adler, and L. Maier-Hein, *Confidence estimation for machine learning-based quantitative photoacoustics*, Journal of Imaging, 4 (2018), p. 147.
26. J. Gröhl, T. Kirchner, T. J. Adler, L. Hacker, N. Holzwarth, A. Hernández-Aguilera, M. A. Herrera, E. Santos, S. E. Bohndiek, and L. Maier-Hein, *Learned spectral decoloring enables photoacoustic oximetry*, Scientific reports, 11 (2021), p. 6565.
27. J. Gröhl, M. Schellenberg, K. Dreher, and L. Maier-Hein, *Deep learning for biomedical photoacoustic imaging: A review*, Photoacoustics, 22 (2021), p. 100241.
28. M. Haltmeier, T. Berer, S. Moon, and P. Burgholzer, *Compressed sensing and sparsity in photoacoustic tomography*, Journal of Optics, 18 (2016), p. 114004.
29. M. Haltmeier, L. Neumann, and S. Rabanser, *Single-stage reconstruction algorithm for quantitative photoacoustic tomography*, Inverse Problems, 31 (2015), p. 065005.
30. N. Hänninen, A. Pulkkinen, S. Arridge, and T. Tarvainen, *Adaptive stochastic Gauss–Newton method with optical Monte Carlo for quantitative photoacoustic tomography*, Journal of Biomedical Optics, 27 (2022), pp. 083013–083013.
31. N. Hänninen, A. Pulkkinen, A. Leino, and T. Tarvainen, *Application of diffusion approximation in quantitative photoacoustic tomography in the presence of low-scattering regions*, Journal of Quantitative Spectroscopy and Radiative Transfer, 250 (2020), p. 107065.
32. A. Hannukainen, N. Hyvonen, H. Majander, and T. Tarvainen, *Efficient inclusion of total variation type priors in quantitative photoacoustic tomography*, SIAM Journal on Imaging Sciences, 9 (2016), pp. 1132–1153.
33. A. Hauptmann and B. Cox, *Deep learning in photoacoustic tomography: current approaches and future directions*, Journal of Biomedical Optics, 25 (2020), pp. 112903–112903.
34. A. Hauptmann, B. Cox, F. Lucka, N. Huynh, M. Betcke, P. Beard, and S. Arridge, *Approximate k-space models and deep learning for fast photoacoustic reconstruction*, in International Workshop on Machine Learning for Medical Image Reconstruction, Springer, 2018, pp. 103–111.
35. A. Hauptmann, F. Lucka, M. Betcke, N. Huynh, J. Adler, B. Cox, P. Beard, S. Ourselin, and S. Arridge, *Model-based learning for accelerated, limited-view 3-d photoacoustic tomography*, IEEE transactions on medical imaging, 37 (2018), pp. 1382–1393.
36. A. Hauptmann and J. Poimala, *Model-corrected learned primal-dual models for fast limited-view photoacoustic tomography*, arXiv preprint arXiv:2304.01963, (2023).
37. L. G. Henyey and J. L. Greenstein, *Diffuse radiation in the galaxy*, AstroPhys. J., 93 (1941), pp. 70–83.
38. R. Hochuli, S. Powell, S. Arridge, and B. Cox, *Quantitative photoacoustic tomography using forward and adjoint Monte Carlo models of radiance*, Journal of Biomedical Optics, 21 (2016), pp. 126004–126004.
39. Y. Hristova, P. Kuchment, and L. Nguyen, *Reconstruction and time reversal in thermoacoustic tomography in acoustically homogeneous and inhomogeneous media*, Inverse problems, 24 (2008), p. 055006.
40. A. Ishimaru, *Wave Propagation and Scattering in Random Media*, vol. 1, Academic Press, New York, 1978.
41. A. Javaherian and S. Holman, *Direct quantitative photoacoustic tomography for realistic acoustic media*, Inverse Problems, 35 (2019), p. 084004.
42. T. Jetzfellner, D. Razansky, A. Rosenthal, R. Schulz, K.-H. Englmeier, and V. Ntziachristos, *Performance of iterative optoacoustic tomography with experimental data*, Applied Physics Letters, 95 (2009).

43. J. JOSE, R. G. WILLEMINK, W. STEENBERGEN, C. H. SLUMP, T. G. VAN LEEUWEN, AND S. MANOHAR, *Speed-of-sound compensated photoacoustic tomography for accurate imaging*, Med. Phys., 39 (2012), pp. 7262–7271.
44. J. KAIPIO AND E. SOMERSALO, *Statistical and Computational Inverse Problems*, Springer, New York, 2005.
45. A. KIRSCH AND O. SCHERZER, *Simultaneous reconstructions of absorption density and wave speed with photoacoustic measurements*, SIAM Journal on Applied Mathematics, 72 (2012), pp. 1508–1523.
46. L. A. KUNYANSKY, *Explicit inversion formulae for the spherical mean radon transform*, Inverse problems, 23 (2007), p. 373.
47. J. LAUFER, B. COX, E. ZHANG, AND P. BEARD, *Quantitative determination of chromophore concentrations form 2D photoacoustic images using a nonlinear model-based inversion scheme*, Appl Opt, 49 (2010), pp. 1219–1233.
48. A. A. LEINO, T. LUNTTILA, M. MOZUMDER, A. PULKKINEN, AND T. TARVAINEN, *Perturbation Monte Carlo method for quantitative photoacoustic tomography*, IEEE Transactions on Medical Imaging, 39 (2020), pp. 2985–2995.
49. C. LI AND L. V. WANG, *Photoacoustic tomography and sensing in biomedicine*, Phys Med Biol, 54 (2009), pp. R59–R97.
50. M. LI, Y. TANG, AND J. YAO, *Photoacoustic tomography of blood oxygenation: A mini review*, Photoacoustics, 10 (2018), pp. 65–73.
51. X. LI AND H. JIANG, *Impact of inhomogeneous optical scattering coefficient distribution on recovery of optical absorption coefficient maps using tomographic photoacoustic data*, Physics in Medicine & Biology, 58 (2013), p. 999.
52. H. LIU AND G. UHLMANN, *Determining both sound speed and internal source in thermo-and photo-acoustic tomography*, Inverse Problems, 31 (2015), p. 105005.
53. S. LUNZ, A. HAUPTMANN, T. TARVAINEN, C.-B. SCHONLIEB, AND S. ARRIDGE, *On learned operator correction in inverse problems*, SIAM Journal on Imaging Sciences, 14 (2021), pp. 92–127.
54. C. M. MACDONALD, S. ARRIDGE, AND S. POWELL, *Efficient inversion strategies for estimating optical properties with Monte Carlo radiative transport models*, Journal of Biomedical Optics, 25 (2020), pp. 085002–085002.
55. E. MALONE, S. POWELL, B. T. COX, AND S. ARRIDGE, *Reconstruction-classification method for quantitative photoacoustic tomography*, Journal of Biomedical Optics, 20 (2015), pp. 126004–126004.
56. A. V. MAMONOV AND K. REN, *Quantitative photoacoustic imaging in radiative transport regime*, Comm Math Sci, 12 (2014), pp. 201–234.
57. T. P. MATTHEWS AND M. A. ANASTASIO, *Joint reconstruction of the initial pressure and speed of sound distributions from combined photoacoustic and ultrasound tomography measurements*, Inverse problems, 33 (2017), p. 124002.
58. T. P. MATTHEWS, J. POUDEL, L. LI, L. V. WANG, AND M. A. ANASTASIO, *Parameterized joint reconstruction of the initial pressure and sound speed distributions for photoacoustic computed tomography*, SIAM journal on imaging sciences, 11 (2018), pp. 1560–1588.
59. E. MERČEP, J. L. HERRAIZ, X. L. DEÁN-BEN, AND D. RAZANSKY, *Transmission–reflection optoacoustic ultrasound (tropus) computed tomography of small animals*, Light: Science & Applications, 8 (2019), p. 18.
60. M. MOZUMDER, A. HAUPTMANN, I. NISSILÄ, S. R. ARRIDGE, AND T. TARVAINEN, *A model-based iterative learning approach for diffuse optical tomography*, IEEE Transactions on Medical Imaging, 41 (2021), pp. 1289–1299.
61. S. MUKHERJEE, A. HAUPTMANN, O. ÖKTEM, M. PEREYRA, AND C.-B. SCHÖNLIEB, *Learned reconstruction methods with convergence guarantees: a survey of concepts and applications*, IEEE Signal Processing Magazine, 40 (2023), pp. 164–182.
62. W. NAETAR AND O. SCHERZER, *Quantitative photoacoustic tomography with piecewise constant material parameters*, SIAM Journal on Imaging Sciences, 7 (2014), pp. 1755–1774.

63. O. Nykänen, A. Pulkkinen, and T. Tarvainen, *Quantitative photoacoustic tomography augmented with surface light measurements*, Biomed Opt Express, 8 (2017), pp. 4380–4395.
64. J. Poudel, Y. Lou, and M. A. Anastasio, *A survey of computational frameworks for solving the acoustic inverse problem in three-dimensional photoacoustic computed tomography*, Physics in Medicine & Biology, 64 (2019), p. 14TR01.
65. A. Pulkkinen, B. T. Cox, S. R. Arridge, H. Goh, J. P. Kaipio, and T. Tarvainen, *Direct estimation of optical parameters from photoacoustic time series in quantitative photoacoustic tomography*, IEEE Transactions on Medical Imaging, 35 (2016), pp. 2497–2508.
66. A. Pulkkinen, B. T. Cox, S. R. Arridge, J. P. Kaipio, and T. Tarvainen, *A Bayesian approach to spectral quantitative photoacoustic tomography*, Inv Probl, 30 (2014), p. 065012.
67. A. Pulkkinen, B. T. Cox, S. R. Arridge, J. P. Kaipio, and T. Tarvainen, *Quantitative photoacoustic tomography using illuminations from a single direction*, Journal of Biomedical Optics, 20 (2015), pp. 036015–036015.
68. D. Razansky, *Multispectral optoacoustic tomography - volumetric color hearing in real time*, IEEE J Sel Topics Quantum Electron, 18 (2012), pp. 1234–1243.
69. D. Razansky, J. Baeten, and V. Ntziachristos, *Sensitivity of molecular target detection by multispectral optoacoustic tomography (MSOT)*, Med Phys, 36 (2009), pp. 939–945.
70. K. Ren, H. Gao, and H. Zhao, *A hybrid reconstruction method for quantitative PAT*, SIAM Journal on Imaging Sciences, 6 (2013), pp. 32–55.
71. J. Ripoll and V. Ntziachristos, *Quantitative point source photoacoustic inversion formulas for scattering and absorbing media*, Physical Review E, 71 (2005), p. 031912.
72. R. T. Rockafellar, *Convex Analysis*, Princeton University Press, Princeton, 1970.
73. A. Rosenthal, D. Razansky, and V. Ntziachristos, *Quantitative optoacoustic signal extraction using sparse signal representation*, IEEE Transactions on Medical Imaging, 28 (2009), pp. 1997–2006.
74. T. Sahlström, A. Pulkkinen, J. Tick, J. Leskinen, and T. Tarvainen, *Modeling of errors due to uncertainties in ultrasound sensor locations in photoacoustic tomography*, IEEE Transactions on Medical Imaging, 39 (2020), pp. 2140–2150.
75. T. Saratoon, T. Tarvainen, B. T. Cox, and S. R. Arridge, *A gradient-based method for quantitative photoacoustic tomography using the radiative transfer equation*, Inv Probl, 29 (2013), p. 075006.
76. O. Scherzer, M. Grasmair, H. Grossauer, M. Haltmeier, and F. Lenzen, *Variational methods in imaging*, vol. 167, Springer, 2009.
77. R. Shang, K. Hoffer-Hawlik, F. Wang, G. Situ, and G. P. Luke, *Two-step training deep learning framework for computational imaging without physics priors*, Optics Express, 29 (2021), pp. 15239–15254.
78. P. Shao, B. Cox, and R. Zemp, *Estimating optical absorption, scattering and Grueneisen distributions with multiple-illumination photoacoustic tomography*, Appl Opt, 50 (2011), pp. 3145–3154.
79. P. Shao, T. Harrison, and R. J. Zemp, *Iterative algorithm for multiple illumination photoacoustic tomography (MIPAT) using ultrasound channel data*, Biomedical Optics Express, 3 (2012), pp. 3240–3249.
80. D. Smyl, T. N. Tallman, J. A. Black, A. Hauptmann, and D. Liu, *Learning and correcting non-Gaussian model errors*, Journal of Computational Physics, 432 (2021), p. 110152.
81. N. Song, C. Deumié, and A. Da Silva, *Considering sources and detectors distributions for quantitative photoacoustic tomography*, Biomedical Optics Express, 5 (2014), pp. 3960–3974.
82. T. Tarvainen, *Computational Methods for Light Transport in Optical Tomography*, PhD thesis, University of Kuopio, Kuopio, Finland, 2006.

83. T. Tarvainen, B. T. Cox, J. Kaipio, and S. R. Arridge, *Reconstructing absorption and scattering distributions in quantitative photoacoustic tomography*, Inverse Problems, 28 (2012), p. 084009.
84. T. Tarvainen, A. Pulkkinen, B. T. Cox, J. P. Kaipio, and S. R. Arridge, *Bayesian image reconstruction in quantitative photoacoustic tomography*, IEEE Transactions on Medical Imaging, 32 (2013), pp. 2287–2298.
85. J. Tick, A. Pulkkinen, F. Lucka, R. Ellwood, B. T. Cox, J. P. Kaipio, S. R. Arridge, and T. Tarvainen, *Three dimensional photoacoustic tomography in bayesian framework*, The Journal of the Acoustical Society of America, 144 (2018), pp. 2061–2071.
86. J. Tick, A. Pulkkinen, and T. Tarvainen, *Image reconstruction with uncertainty quantification in photoacoustic tomography*, The Journal of the Acoustical Society of America, 139 (2016), pp. 1951–1961.
87. J. Tick, A. Pulkkinen, and T. Tarvainen, *Modelling of errors due to speed of sound variations in photoacoustic tomography using a bayesian framework*, Biomedical physics & engineering express, 6 (2019), p. 015003.
88. B. E. Treeby and B. T. Cox, *k-wave: Matlab toolbox for the simulation and reconstruction of photoacoustic wave fields*, Journal of biomedical optics, 15 (2010), p. 021314.
89. D. Waibel, J. Gröhl, F. Isensee, T. Kirchner, K. Maier-Hein, and L. Maier-Hein, *Reconstruction of initial pressure from limited view photoacoustic images using deep learning*, in Photons Plus Ultrasound: Imaging and Sensing 2018, vol. 10494, International Society for Optics and Photonics, 2018, p. 104942S.
90. K. Wang, R. Su, A. A. Oraevsky, and M. A. Anastasio, *Investigation of iterative image reconstruction in three-dimensional optoacoustic tomography*, Physics in Medicine & Biology, 57 (2012), p. 5399.
91. L. V. Wang and J. Yao, *A practical guide to photoacoustic tomography in the life sciences*, Nature Methods, 13 (2016), pp. 627–638.
92. Z. Wang, W. Tao, and H. Zhao, *The optical inverse problem in quantitative photoacoustic tomography: A review*, Photonics, 10 (2023), p. 487.
93. J. Xia, C. Huang, K. Maslov, M. A. Anastasio, and L. V. Wang, *Enhancement of photoacoustic tomography by ultrasonic computed tomography based on optical excitation of elements of a full-ring transducer array*, Optics letters, 38 (2013), pp. 3140–3143.
94. M. Xu and L. V. Wang, *Universal back-projection algorithm for photoacoustic computed tomography*, Physical Review E, 71 (2005), p. 016706.
95. Y. Xu and L. V. Wang, *Time reversal and its application to tomography with diffracting sources*, Physical review letters, 92 (2004), p. 033902.
96. L. W. Y Xu, G. Ambartsoumian, and P. Kuchment, *Reconstructions in limited-view thermoacoustic tomography*, Med Phys, 31 (2004), p. 724.
97. C. Yang, H. Lan, F. Gao, and F. Gao, *Review of deep learning for photoacoustic imaging*, Photoacoustics, 21 (2021), p. 100215.
98. R. J. Zemp, *Quantitative photoacoustic tomography with multiple optical sources*, Appl Opt, 49 (2010), pp. 3566–3572.
99. X. Zhang, W. Zhou, X. Zhang, and H. Gao, *Forward–backward splitting method for quantitative photoacoustic tomography*, Inverse Problems, 30 (2014), p. 125012.
100. X. Zhou, N. Akhlaghi, K. A. Wear, B. S. Garra, T. J. Pfefer, and W. C. Vogt, *Evaluation of fluence correction algorithms in multispectral photoacoustic imaging*, Photoacoustics, 19 (2020), p. 100181.

Chapter 5
Deep Learning-Based Methods for Photoacoustic Imaging Reconstruction: Concepts, Promises, Pitfalls, and Futures

Amir Gholampour, Kalloor Joseph Francis, Min Wu, Nastaran Mohammadian Rad, Richard G. P. Lopata, and Navchetan Awasthi

Abstract The methods currently available for performing deep learning rely a lot on developing models that are reproducible as well as getting the required amount of data for performing deep learning. Here, we have done a review keeping in mind the fact that the data requirement and the model availability and how the computer vision models have translated to the medical imaging scenarios in the current world. This chapter consists of dividing the whole domain into the reconstruction part and then looking forward to how the models from the natural images adapt to the photoacoustic imaging in terms of changing the architecture, working on modifying the architectures as well as looking for the required dataset for performing deep learning on the photoacoustic image reconstruction. We discuss more on the architecture sides of deep learning as well as what are the future directions where deep learning can be utilized for improving photoacoustic imaging.

A. Gholampour · M. Wu · R. G. P. Lopata
Department of Biomedical Engineering, Eindhoven University of Technology, Eindhoven, The Netherlands

K. J. Francis
Faculty of Science and Technology, University of Twente, Enschede, The Netherlands

N. M. Rad
Department of Biomedical Engineering, Eindhoven University of Technology, Eindhoven, The Netherlands

Department of Precision Medicine, Maastricht University, Maastricht, The Netherlands

N. Awasthi (✉)
Department of Biomedical Engineering, Eindhoven University of Technology, Eindhoven, The Netherlands

Informatics Institute, University of Amsterdam, Amsterdam, The Netherlands

Department of Biomedical Engineering and Physics, Amsterdam UMC, Amsterdam, The Netherlands
e-mail: n.awasthi@uva.nl

W. Xia (ed.), *Biomedical Photoacoustics*,
https://doi.org/10.1007/978-3-031-61411-8_5

Keywords Photoacoustic imaging · Deep learning · Image reconstruction · Convolutional neural networks · Inverse problems

5.1 Introduction

In modern clinical practice, ultrasound (US) and optical imaging are commonly used techniques to enable accurate disease diagnosis and to facilitate surgeries. US imaging is fast, cheap, and can well reveal the tissue morphology, however, it has limited contrast to characterize different soft tissues; while optical imaging can be used to image different tissues with high tissue specificity based on the rich optical contrast and at the superior spatial resolution, however, it is typically restricted by light scattering to a limited imaging depth.

To take advantage of both imaging techniques, photoacoustic (PA) imaging has been under intensive development in the past decades. In PA imaging, the light pulse is delivered for tissue illumination and the light can be absorbed in the tissue and converted to heat. Due to the thermoelastic expansion, an ultrasound (US) wave is generated and received by a US transducer to reconstruct PA images. In these images, the amplitude of the PA signal is proportional to the optical absorption of the tissue. By operating at different optical wavelengths, that is, to perform multi-wavelength spectroscopic photoacoustic imaging (sPA), the unique wavelength-dependent behavior of different tissues can be characterized, being a powerful technique to image different tissue compositions [1]. A simple PA imaging process is illustrated in Fig. 5.1. Moreover, PA imaging can be implemented in various styles, such as PA compute tomography (around 10 s to 100 s um resolution) and PA microscopy (can be a few um resolution), presenting the unique feature of multiscale imaging with consistent contrast, which can easily bridge up the gap of the clinical (pre-clinical) translation across different scales [2, 3]. Despite all this great potential, challenges are present for PA techniques to move further toward in vivo clinical applications. One of the major challenges is the relatively low SNR in PA imaging which leads to relatively low contrast and penetration depth.

To improve PA imaging SNR, the development of good imaging reconstruction methods would play an important role. Generally, the reconstructed PA images can be characterized by the convolution of the imaging target, the laser pulse, and the impulse response of the US transducer used for signal receiving. Like in

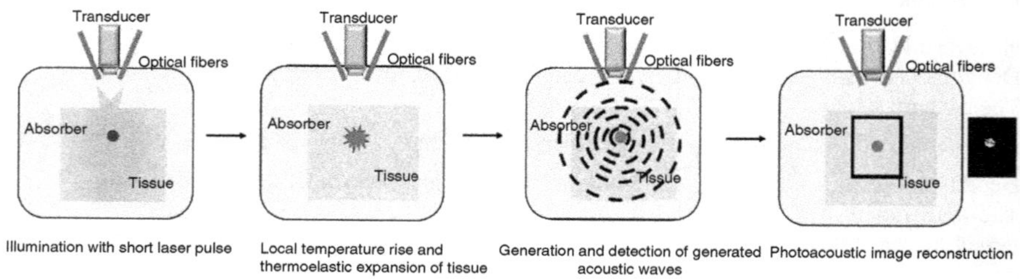

Fig. 5.1 The basic photoacoustic imaging principle

US imaging, conventional delay and sum beamforming have been commonly used for PA imaging reconstruction. However, the laser-induced PA signal usually has a broadband nature, while the practical US transducer can only receive signals at a relatively narrow bandwidth, which leads to signal loss and imaging quality degradation in PA imaging affecting both the resolution and contrast of the imaging system [4]. Moreover, due to the practical system implementation, PA imaging usually suffers from a limited field of view, resulting in extra signal loss. Various modifications based on delay and sum methods [5–10], and other methods such as deconvolution algorithms [11–13] have attempted to improve PA imaging quality. Section 5.2 will briefly address these commonly used reconstruction and beamforming techniques. Despite ongoing efforts, there remains significant scope to further improve PA imaging and accelerate its translation toward in vivo applications, especially with the help of rapidly evolving deep learning (DL) techniques. Many studies have focused on the application of deep learning methods, with a particular emphasis on convolutional neural networks (CNN)-based architectures. While CNNs have demonstrated substantial potential in addressing the limitations of PA imaging, other deep-learning techniques could also offer significant advantages. However, these deep learning methods have not yet been extensively explored in the current body of research, indicating an area ripe for further investigation. Sections 5.3 and 5.4 will elaborate further, providing more detailed information on these promising avenues for enhancing photoacoustic imaging.

5.2 Image Reconstruction in Photoacoustic Imaging

Image reconstruction refers to the process of converting the raw data acquired from the photoacoustic signals into meaningful images. The image reconstruction and beamforming techniques traditionally used for ultrasound imaging can also be adapted for photoacoustic imaging. Among the common techniques for beamforming are Delay-and-Sum (DAS) [14, 15], Time Reversal (TR) [16, 17], Filtered Back-Projection (FBP) [18], Delay-Multiply-and-Sum (DMAS) [6], short-lag spatial coherence (SLSC) [19], and Minimum Variance (MV) [20]. DAS, a fundamental beamforming method, provides rapid results but often at the expense of low resolution and contrast due to increased sidelobes and sensitivity to noise [14]. The DMAS algorithm expands the traditional DAS beamforming technique, aiming to enhance both image contrast and resolution in photoacoustic imaging, albeit at a higher computational cost. SLSC measures the spatial coherence of the received echo signals over short lags, improving image contrast, particularly at shallower imaging depths. Despite its advantages, SLSC has a high computational cost and may not provide as high spatial resolution as some other advanced beamforming techniques [19]. FBP can be implemented efficiently for arc configurations of transducers, yet it relies on homogeneity in the speed of sound and precise measurement geometry which can lead to inaccuracies in complex tissue structures. TR technique can improve the quality of reconstructed images in comparison to DAS, but its time cost is 16 times higher than that of the FBP method [21].

Emerging methodologies using deep learning present a promising avenue to replace or augment these traditional techniques and alleviate their limitations. Deep learning algorithms can learn to recognize complex patterns and make decisions based on them, thus potentially improving the quality of image reconstruction while possibly reducing computational costs after their implementation. They can also model the non-linearity and complexity in tissue better than traditional methods, thereby facilitating more accurate imaging. Despite being computationally intensive, the potential of DL to improve image contrast, resolution, and overall quality, along with the rapid advancements in computational hardware, make it an exciting area of research in the field of photoacoustic imaging.

5.3 Deep Learning

Deep learning is an extension of the neural network (NN) theory that provides the possibility to train an NN architecture with more than two hidden layers using the backpropagation algorithm [22]. It is a representation learning approach with the capability of exploring complex and non-linear patterns in the data [23]. The commonly used deep learning algorithms for PAT image reconstructions include convolution neural networks (CNNs), recurrent neural networks (RNNs), and Generative Adversarial Networks (GANs). In this section, these algorithms are described.

5.3.1 From Classical Feature Extraction to Automatic Feature Learning

Feature extraction is one of the most important steps in the medical data analysis pipeline. Feature extraction affects the performance of the algorithm and reduces its computational complexity. The classical applications of medical image analysis mainly rely on extracting hand-crafted features. Despite the popularity of using hand-crafted features, they suffer from some restricting shortcomings. First, they are based on experts' prior knowledge about the problem, thus the important information in the data may be overlooked due to the lack of domain knowledge [24]. Second, extracting hand-crafted features can be a computationally expensive process [24]. To overcome these limitations, automatic feature learning methods based on deep neural network architectures have recently received considerable attention in the field. Deep learning provides the possibility to learn the mid-level and high-level features from raw data.

5.3.2 Convolutional Neural Networks

CNN is a common deep-learning architecture that is suitable for data with a grid-like topology such as time series and images [25]. CNNs adopt a hierarchical architecture that alternates a convolution layer, a Rectified Linear Unit (ReLU) activation function, and a pooling layer which enables the transformation of large input spaces into lower-dimensional feature spaces. The main advantages of CNNs compared to traditional fully connected neural networks are the possibility of sparse connectivity, parameter sharing, and learning equivariant representations [25]. A CNN architecture starts with a convolutional layer that takes the input data, and convolves it with a set of filters, producing a feature map. The resulting feature map is then fed to an activation function. Generally, a Rectified Linear Unit (ReLU) is used as an activation function to introduce non-linearity to the network and also to prevent the network from the gradient vanishing problem [25]. The resulting outputs are subsequently fed to a pooling layer. A pooling operation downsamples the feature map by computing the average or maximum within a rectangular neighborhood. This downsampling reduces the spatial dimension of the input space, leading to a decrease in the number of parameters. As a result, it reduces the computational costs in both training and prediction processes. Furthermore, the pooling operation provides robustness over shifts and distortions in the input space [25]. A multi-layer CNN architecture comprised of several convolutional, ReLU, and pooling layers stacked sequentially. The reduced feature map of the last CNN layer is flattened as a vector before the classification step. The final output of the last CNN layer is flattened into a vector, forming the learned feature space—a novel representation of the input data. In the classification task, this new representation is typically fed to a fully connected neural network followed by a softmax layer.

5.3.3 Generative Adversarial Networks

Ian Goodfellow and his colleagues introduced Generative Adversarial Networks (GANs) in 2014 [26]. GANs are a type of deep learning-based model that consists of two neural networks: a generator and a discriminator. The generator and discriminator are trained competitively. The generator aims to generate real data while the discriminator's objective is to distinguish between the real data and generated data. During training, the generator improves its performance by learning to generate more realistic data while the discriminator learns to better distinguish real data from the generated data.

5.4 Deep Learning for Photoacoustic Image Reconstruction

Various architectures and approaches were proposed for improved reconstruction of photoacoustic images. The techniques vary from improving the reconstruction

Table 5.1 A timeline of the models developed before U-Net, the year, the type of the network, and the fundamental property of these architectures

Year	Model	Type	Fundamental property
1997 [27]	Long Short-Term Memory (LSTM) Network	Deep Neural Network	Memory cell with input, forget, and output gates
1998 [28]	Convolutional Neural Network (CNN)	Deep Neural Network	Shared weights, local receptive fields
2006 [29]	Autoencoder	Feedforward Neural Network	Nonlinear dimensionality reduction
2008 [30]	Denoising Autoencoder	Feedforward Neural Network	Training for unsupervised learning
2009 [31]	Deep Belief Network (DBN)	Deep Neural Network	Unsupervised layer-wise pre-training
2012 [32]	Stacked Denoising Autoencoder (SDAE)	Feedforward Neural Network	Unsupervised layer-wise learning
2013 [33]	Variational Autoencoder (VAE)	Generative Model	Latent variable model, variational inference
2015 [34]	U-Net	Convolutional Neural Network	Encoder-decoder architecture with skip connections

obtained by removing the artifacts to direct reconstruction of the image from the radio frequency (RF) received data domain data. With the advent of CNNs [52] in 1989, the field of deep learning has evolved a lot due to the development of U-Net architecture in 2015 [34]. In general, for initial experiments, a U-Net or modified U-Net architecture is utilized to improve the photoacoustic (PA) image reconstruction. Later, architectures such as GANs, etc. are utilized for further improvement of the image reconstruction. Table 5.1 gives information about the timeline of various models developed and how they led to the U-Net architecture. As discussed in the following section, U-net and CNN architectures have been dominating the majority of the proposed method. We believe there has been a need to utilize recent more advanced models for further improving the reconstruction. Table 5.2 addresses a handful of deep learning models proposed and developed after U-net architecture (year 2015) that can significantly help enhance the performance of a photoacoustic imaging system.

5.4.1 Deep Learning-Based Direct Image Reconstruction

This section involves getting a reconstruction directly from the raw data using deep learning-based architectures as can be seen from Fig. 5.2. Toward this end, most proposed models are based on U-net and CNN architectures. In [53], this problem was solved by utilizing the modified U-Net-based architecture for PA imaging. In [54], another DL network based on dilated convolutions was proposed to transform RF data directly to reconstruct the photoacoustic image. An end-to-end network based on U-Net was proposed to reconstruct PA images from RF-received data for

Table 5.2 A timeline of the models developed after U-Net, their type, and the fundamental property of these architectures

Year	Model	Type	Fundamental property
2015 [34]	U-Net	Convolutional Neural Network	Encoder-decoder architecture with skip connections
2015 [35]	Dilated Convolutional Neural Network	Deep Neural Network	Dilated convolution, multi-scale context
2015 [36]	Fully Convolutional Network (FCN)	Convolutional Neural Network	Transforms CNN into a fully convolutional network
2015 [37]	Deconvolutional Network (DeconvNet)	Convolutional Neural Network	Upsampling with unpooling and deconvolution
2016 [38]	Residual Network (ResNet)	Deep Neural Network	Skip connections, residual learning
2017 [39]	Deep Convolutional Neural Network (DCNN)	Deep Neural Network	Deep CNN with many layers
2017 [40]	Attention Mechanism	Deep Neural Network	Allows the network to selectively focus on relevant information
2017 [41]	Dual Path Network (DPN)	Deep Neural Network	Two parallel paths of processing for better feature learning
2020 [42]	Generative Adversarial Network (GAN)	Generative Model	Adversarial training, generator, and discriminator networks
2017 [43]	Capsule Network	Deep Neural Network	Hierarchy of capsules for part-whole relationships
2017 [44]	Multi-scale Residual Network (MSRN)	Deep Neural Network	Multi-scale processing, residual connections
2018 [45]	Image Reconstruction Using Variational Networks	Deep Neural Network	Image reconstruction, variational networks
2018 [46]	Deep Image Prior	Deep Neural Network	Image reconstruction, variational networks
2018 [47]	Attention Unet	Deep Neural Network	Image segmentation, attention mechanism
2019 [48]	SinGAN	Deep Neural Network	Image synthesis, multi-scale, and multi-level learning
2019 [49]	Residual Non-Local Attention Networks	Deep Neural Network	Attention mechanism, residual connections
2020 [50]	Diffusion Probabilistic Models for Image Restoration	Deep Neural Network	Image restoration, probabilistic modeling
2021 [51]	NeRF	Deep Neural Network	Neural rendering, scene reconstruction

Fig. 5.2 General diagram of a deep learning-based direct image reconstruction workflow. The received data is directly fed into the model, which subsequently reconstructs the image, enhancing the image features compared to those obtained with conventional image

multi-frequency raw data [55]. With residual networks gaining popularity for image recognition [38] in the year 2017, a residual U-Net architecture was proposed [56] for getting a direct reconstruction from the raw sensor data for getting the final initial pressure distribution and compared with model-based algorithms and was shown to improve the reconstruction by as high as 18% for U-Net++ architecture. A customized model was proposed in [57] for obtaining the photoacoustic image combining the feature projection network with a U-Net architecture utilized as a post-processing scheme and was shown to provide reconstruction rates as high as 15 frames per second.

These models mostly utilize U-Net-based architectures, modified U-Net, and CNN-based methods for improving the RF received data to image domain reconstruction. There has been a lot of improvement in the deep learning models since U-Net was developed in the year 2015 as shown in Table 5.2. There has been a need to utilize these models for further improving the reconstruction using deep learning-based models.

5.4.2 *Improving the Image Domain Data*

A common technique to improve photoacoustic imaging is to employ a deep learning-based model on a reconstructed image [see Fig. 5.3]. For example, in [53], Waibel et al. proposed the use of a modified U-Net architecture to alleviate the limited view problem in the case of a linear probe for either a direct or a post-processed reconstruction. The results of this in silico study show that both cases resulted in improved contrast and resolution, but the post-processing approach achieved a better outcome. In [58], a deep convolutional network based on U-Net was utilized after a Universal Back-Projection (UBP) reconstruction with Dynamic Aperture Length (DAL) correction. This approach showed improvement in experimental vessel PA images in which the sparse detector array, i.e., limited view problem, and finite impulse response degraded the quality. Anas et al. proposed to use of CNN architectures to address low signal-to-noise ratio (SNR) PA images obtained with LED-based systems with linear probes [59, 60]. The results show high noise suppression in the reconstructed experimental data.

Another work used Singular Value Decomposition (SVD) to filter background noise in a PA image where only the larger singular values are preserved, then a CNN was introduced to recover the truncated singular values resulting in a

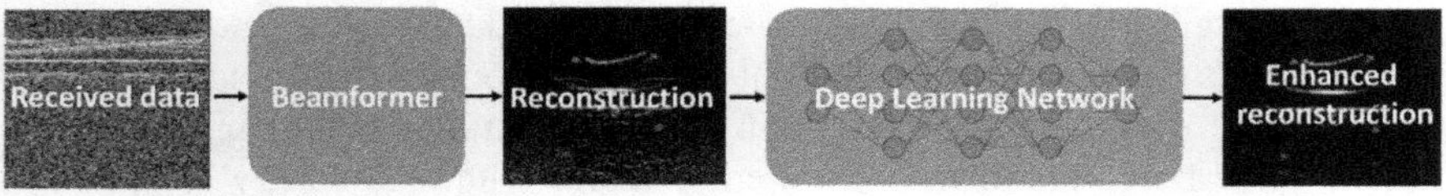

Fig. 5.3 Conceptual diagram for image enhancement using deep learning networks. The received data is first reconstructed using conventional beamformers. Subsequently, the photoacoustic image is refined and improved using a deep learning approach

significant improvement over the original PA image [61]. In [62], authors used the Filter Back-Projection (FBP) based reconstruction as an initial input utilizing a U-Net architecture to improve the PAI system and compared it with state-of-the-art techniques. A modified version of U-Net consisting of feature extraction, artifact reduction, and reconstruction sections was utilized to address the reflection artifacts under different noise conditions [63]. Two methods: (a). U-Net based and (b). Principal component analysis (PCA) followed by a Visual Geometry Group network (VGGnet) was proposed in [64] for improving limited view reconstruction to compensate for missing information. Davoudi et al. utilized a modified U-Net architecture to improve the output of the PCT system with sparse detectors [65]. The proposed approach was tested on both simulation and whole-body mouse imaging data. A fully dense U-Net architecture was proposed in [66] to be applied to the time reversal reconstruction of the PACT imaging system to address sparse data/detector problems. The results were compared to conventional U-Net and showed significant improvements in artifact reduction and image enhancement, especially in resolving smaller structures and vessels.

A deep learning technique based on U-Net was proposed in [67] for the correction of speed of sound (SoS) aberration in PAI. This approach was applied to the in-vivo human arm PA image and successfully corrected the image for aberrations caused by SoS. This work is improved in [68] by utilizing a more complex neural network (based on U-net) to not only correct for SoS aberration but also reduce artifacts such as streaking caused by sparse data. A pixel-wise interpolation was proposed in [69] in which a U-Net architecture is employed for pixel-wise interpolation followed by a CNN to reconstruct a PA image. It was shown that this method is computationally efficient, hence, it can be used in the real-time processing of photoacoustic tomographic images that suffer from limited and sparse view. A joint Generative adversarial network (GAN) and U-net-based approach was developed in [70] for the limited view and limited bandwidth experimental data received by a linear array probe. It was shown that the proposed method performs better in comparison to a standard U-Net architecture that uses a time reversal beamformer as an input. Similar to earlier studies, in [71], a time reversal image is used as an input to a U-net to address the sparse detectors problem in a PACT system, where the results improve over the conventional TR reconstruction. In [72], another CCNN-based model was proposed to address the limited view problem in PACT using a time reversal image as the input for intravascular PA images and was shown to achieve 10% better SSIM when compared to a conventional compressed sensing technique. Manwar et al. proposed the use of a modified U-net to improve the penetration depth in PAI [73]. The method was applied to sheep brain data acquired with low-energy laser pulses (low penetration depth) and the results were comparable to PA images that were obtained with high-energy laser pulses. In [74], the authors proposed a 10-layer CNN architecture to be used for sparse detector artifacts in the case of a three-quarter ring transducer array. The network was trained on simulated UBP PA images and verified on a numerical phantom, and in-vivo brain imaging; it was shown that the method is superior to state-of-the-art compressed sensing algorithms. A multi-level wavelet-convolutional neural network was proposed in [75] for mapping low-fluence images to corresponding high-fluence excitation maps. The method was applied for in vivo mouse data and the results showed great improvements in PSNR, SSIM, and CNR. In [76], a U-net architecture was used to address the limited view and bandwidth problems of a PAI system based on a linear probe. In this work, photographs of the samples were utilized as the ground truth and the results show a great improvement in their morphology. Kim et al. proposed a framework to be used in a modified U-Net architecture for resolving the limited-view problem of linear arrays [77]. This method was assessed using both simulation and in-vivo data showing improvement in terms of contrast and resolution when compared to conventional beamformers and a CS-based approach.

Shahid et al. proposed to use of a Residual U-Net architecture for the PACT system and brain imaging application to improve the results degraded from an under-sampling problem. The results were compared to the conventional compress sensing method and showed great improvement in PSNR values [78]. In [79], a multitask residual dense network is applied on Optical-resolution photoacoustic microscopy system outputs with low laser dosage and successfully estimated the output of the system in case of high laser dosage, e.g., increasing the SNR of deep tissues. a self-supervised network based on 12-layer CNN is proposed in [80] which the total variation of the beamformed data was penalized. The results showed improvement in morphology and accuracy for the proposed deep neural network method when compared to the conventional cross-correlation-based method.

A CNN architecture is proposed in [81] to improve reconstructed PA images from under-sampled data. The proposed method uses a noise map and a sampling mask to estimate the fully sampled image iteratively which outperforms conventional interpolation techniques. In [82], an artifact reduction method based on a modified U-Net architecture with dense connectivity and dilated convolutions is proposed for 3D photoacoustic tomography systems. The results demonstrate an improvement in morphology in the network's output compared to the time reversal reconstructions. Joseph et al. used a GAN-based architecture to address the common limited view and bandwidth problems [83]. The network was trained on simulated data and was tested on both synthetic and experimental data where it could successfully remove the limited view artifacts providing a good estimation of the missing data from limited bandwidth. A fully dense U-net aided with a convolutional long short-term memory block was utilized to improve the output of a ring-PCT system with eight elements [84]. The input to the network was reconstructed images without any positional correction, and the results showed significant improvement in terms of PSNR and general morphology.

A modified U-Net architecture was used to address the challenges in transcranial cerebra vascular imaging [85]. The proposed method extracts and utilizes photoacoustic information from speckle patterns obtained from vascular networks imaged under porous media. Application of this network on both synthetic and chicken embryo data yielded images that strongly resembled the ground truth and that were improved in terms of contrast and resolution when compared to the conventional reconstruction methods. In [86], a full U-Net was used to improve the elevation focuses of linear array probes. The method was tested on 2D slices of both numerical and experimental data and was able to improve the elevational resolution fourfold and successfully recover the targets' true sizes. An innovative approach based on multiple feature selection and CNN procedures was proposed by Guo et al. to address sparse data artifacts in photoacoustic imaging systems [87]. The attention-steered network (AS-Net) uses two feature fusion steps to mix information extracted from the DAS image with a modified U-Net to eliminate the sparse data artifacts. The method was tested on fish and mice experimental data and successfully removed most of the sparse data artifacts. In [88], a modified CNN architecture is proposed to deconvolve AR-PAM images and obtain outputs with improved resolution. The proposed method was evaluated against various targets in experiments and compared to existing approaches including neural networks and non-learning methods.

5.4.3 *Joint Models*

The joint models can perform better as they either utilize the information from various sources or train the model for different outputs which would force the model to better learn the features and thus improve the output [see Fig. 5.4]. Some of these models perform more than one task and hence are suitable if the corresponding tasks are also required for further processing. In [89], a method using CNNs was proposed to fuse the information from two different image reconstructions and improve the reconstruction in a limited data scenario. This utilized back-projection combined with analytical techniques (Lanczos Tikhonov and Total variation regularization) and fused the features to get better reconstruction for limited data scenarios. In [90], an algorithm based on learned primal-dual combined with CNNs was proposed for joint reconstruction and segmentation of photoacoustic images. Since there has to be a pre-processing that needs to be applied even after the reconstruction is obtained, this work performs the joint operation and produces better reconstruction and segmentation output for different settings as well as synthetic and experimental data. Lan et al. proposed to use both raw RF data and DAS-reconstructed images as inputs for encoders in a CNN architecture where a decoder yields an image with fewer artifacts [91]. The results are assessed against a conventional U-net using both simulations and phantom measurements where the proposed method successfully

Fig. 5.4 General framework of deep learning networks for joint modeling. These models typically accommodate multiple inputs and are designed for multi-objective training, resulting in multiple outputs

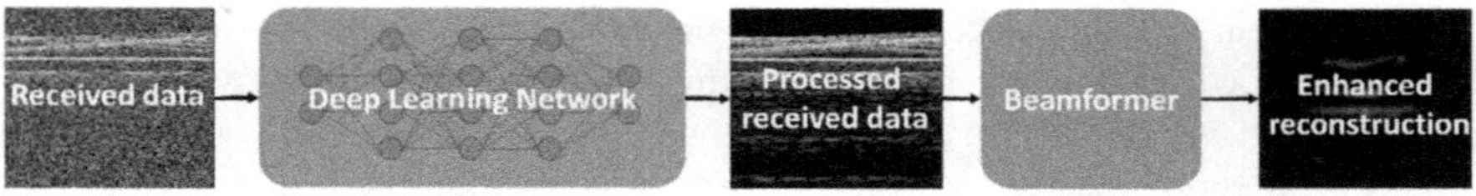

Fig. 5.5 The general framework for RF domain improvement using deep-learning networks. The received data is fed into the deep-learning network, and the photoacoustic image is reconstructed through conventional beamformers

resolves limited view data of a linear array into an artifact-free reconstruction. In [92], limited-view data is fed to two neural networks with multiple losses to further facilitate the fusion of distinct features in the data. The results show significant improvement in the output reconstructed image in terms of SSIM and gCNR when compared to conventional DAS and TV reconstructed images.

5.4.4 Radiofrequency Domain Improvement

Some studies proposed to process radio frequency (RF) received data using deep learning-based techniques prior to image reconstruction as shown in Fig. 5.5. These techniques can be accounted for denoising in the received signal domain as well as bandwidth enhancement in addition to the super-resolution.

In [93], a decoder-encoder network was utilized for bandwidth enhancement of the photoacoustic data in the RF domain. In [94], another work for improving the RF data for super-resolution and denoising was proposed using a 7-layer CNN architecture and it was shown that a k-wave reconstruction was used to give much better results as compared to the limited noisy data. These works deal with the denoising and bandwidth enhancement problems separately and the work in [95] combines these problems and proposes a modified U-Net-based network for denoising, super-resolution as well as bandwidth enhancement. The architecture of the proposed model is shown in Fig. 5.6 which is also a modified version of the U-Net architecture [34].

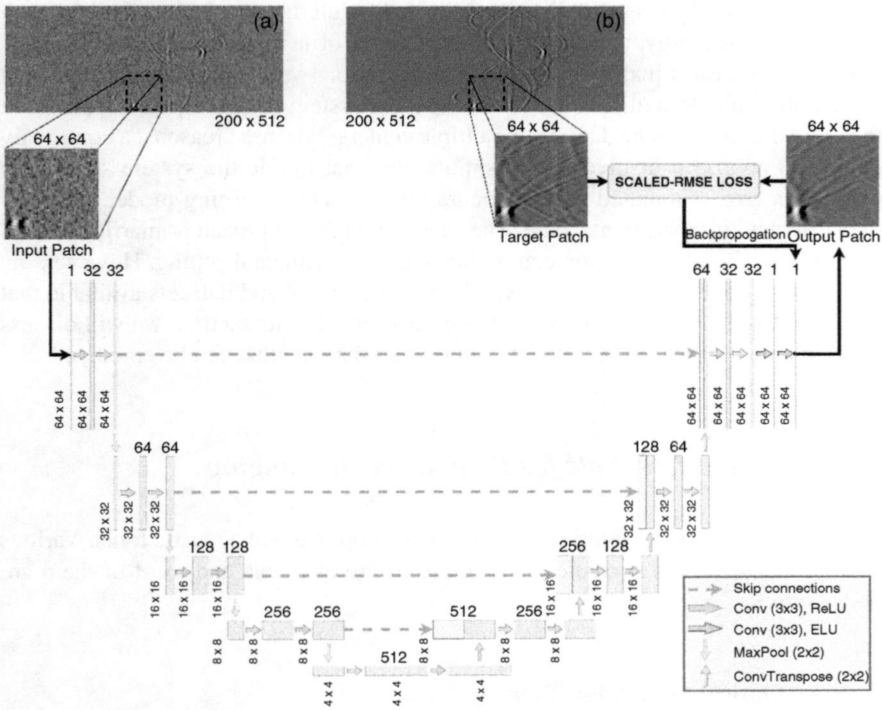

Fig. 5.6 Training configuration for the proposed U-Net-based architecture. Data collected through 100 transducers with 70% bandwidth (BW) is added with variable noise resulting in SNR values of data from 10–70 dB is shown in (**a**). Patches of size 64 × 64 are cropped from (**a**) and input into the U-Net-based model. Output patch of size 64 × 64 is then used to compute the scaled-RMSE loss with the corresponding patch from the data collected through 200 transducers and 100% bandwidth (as shown in (**b**)). This loss is propagated backward to train the proposed model for bandwidth enhancement, super-resolution, and denoising

Dehner et al. proposed using a U-Net architecture to eliminate electrical noise from RF data, thereby improving the contrast of the reconstructed image [96]. This approach was employed for in vivo human breast data resulting in improved spectral contrast and up to 19% higher blood vessel contrast.

5.5 Datasets for Deep-Learning Applications in Photoacoustic

The advances in deep learning-based methods in photoacoustic imaging rely on the availability of large artifact-free datasets. Firstly, deep learning requires data with ground truth information to learn. Obtaining such data using experimental imaging on test objects or in vivo methods is difficult due to a lack of ground truth information. Secondly, measuring a large number of test objects and volunteers is labor intense task. Finally, the system configurations vary significantly, making it even more difficult to utilize data from a specific system for widespread application in the field or to combine data from multiple centers. For these reasons, a commonly employed approach involves using simulations that mimic the system settings to generate a large simulated dataset for training the deep learning model, which is then tested with experimental data. The accuracy of this approach primarily depends on how closely the simulation can replicate the experimental setting. However, the good news is that there are now several simulation tools and datasets available that can be utilized for performing accurate simulations. In this section, we will discuss the available photoacoustic simulation tools and datasets that can be employed.

5.5.1 Simulation Tools for Photoacoustic Imaging

Photoacoustic imaging simulation requires both optical and acoustic tools. Various light propagation and acoustic modeling tools are available and most of them are open-source.

5.5.1.1 Optical Simulation Tools

An optical simulation for photoacoustic imaging should be able to define the light source and model the light-tissue interaction for the defined optical properties of the tissue. An accurate approach for light-tissue interaction is to model the Radiative Transport Equation (RTE). However, due to its computational complexity, approximate solutions or numerical approaches are used. The first-order approximation (P1) to the spherical harmonics expansion of RTE, commonly known as the diffusion approximation, is a widely accepted method. Alternatively, one can use numerical methods such as Monte Carlo simulations to model the light transport.

Software that can be used for diffusion approximation includes NIRFAST [97] and TOAST++ [98]. These were originally developed for diffuse optical tomography using the Finite Element Method (FEM) implementation. However, these toolboxes can be adapted for photoacoustic imaging, which is particularly useful in scenarios requiring optical inversion. Another widely explored software is COMSOL Multi-physics, which can be used to solve diffusion approximation using the FEM method. A wide range of Monte Carlo tools is available for simulating light-tissue interaction. Widely used tools include the following: (i) MCML from Steven Jacques, Marleen Keijzer, Scott Prahl, and Lihong Wang, available at [99], (ii) the GPU version developed by Erik Alerstam, available at [100], (iii) the voxelated version of the MC program (mcxyz) implemented by Boas et al. and Ting et al., available at [101], (iv) the popular MATLAB-based implementation MCMatlab by Marti and Hansen, available at [102], and (v) MCX software, which supports complex geometry along with options for time-varying simulation, pattern illumination, reflectance measurement, and GPU accelerated computation, available at [103]. An overview of the development of Monte Carlo tools can be found in [104].

5.5.1.2 Acoustic Simulation Tools

Once the light propagation is modeled, the next aspect of the photoacoustic simulation is the acoustic part. The open-source toolbox developed by Bradley Treeby and Ben Cox, called k-Wave (k-wave.org), is widely used by the photoacoustic community for acoustic simulations. The toolbox utilizes the pseudospectral method for acoustic modeling. There are specific forward acoustic modeling and reconstruction options available for photoacoustic imaging, which lower the threshold for new groups to start in photoacoustic research. Many deep-learning works have utilized this toolbox for generating synthetic data. This toolbox can be used to generate different simulated data and work with different sensor configurations. An alternative toolbox for acoustic simulation is Field-II, which is primarily designed for ultrasound imaging and requires customization for photoacoustic simulation.

5.5.1.3 Combined Optical and Acoustic Simulation Tools

Combining both optical and acoustic simulations in a photoacoustic pipeline is necessary to generate realistic data for deep learning. Efforts have been made to combine these tools. One such combination is the MCX and k-Wave, available as Simpa [105], which provides combined optical and acoustic modeling for photoacoustic imaging.

Another option is the combination of NIRFAST and k-Wave, as tested by Kothapalli et al. [106]. ValoMC is another Monte Carlo-based simulation package that can be combined with k-Wave for photoacoustic imaging [107]. These are common approaches, but the list provided here is not exhaustive and does not cover all implementations.

5.5.1.4 Photoacoustic Image Reconstruction

Once the forward photoacoustic modeling is completed, the RF data (transducer's received data) can be directly utilized for deep learning applications, or the image can be reconstructed using various reconstruction methods as discussed earlier. The k-Wave toolbox itself serves as a valuable resource for beginners, providing several available methods for photoacoustic image reconstruction. Additionally, the Photoacoustic International Standardization Consortium (IPASC) offers a collection of image reconstruction algorithms [108], making it a valuable reference for researchers in the field.

Table 5.3 Different code bases available as open-source for performing different experiments for deep learning in photoacoustic imaging

Method	Website/Webpage
Source Detection	[109]
Reconstruction	[110]
Segmentation	[111]
Super-resolution and bandwidth-Enhancement	[112]

5.5.1.5 Open Source Programs for Deep Learning in Photoacoustic Imaging

The various code bases available for deep learning applications in photoacoustic imaging, such as source detection, reconstruction, segmentation, and super-resolution are provided in Table 5.3. These code bases can serve as the basis for anyone starting work in deep learning applications in photoacoustic imaging.

5.5.1.6 Data Format

Maintaining a standard format for generated data is crucial for ease of sharing with other researchers in the field. IPASC (International Photoacoustic Standardization Consortium) has made significant efforts to establish a universally accepted standard data format. Currently, an HDF5-based format has been adopted by the community [113]. For the latest updates on the data format, researchers can refer to the IPASC webpage [114].

5.5.2 Dataset Availability for Deep Learning

Most of the photoacoustic imaging applications are based on the visualization of blood vessels. For example, in breast cancer detection, imaging the vasculature in and around the tumor is the intended purpose of photoacoustic imaging. Therefore, accurate vascular imaging is considered important, and most deep-learning methods are proposed with the aim of better visualizing vascular structures in the region of interest. Hence, datasets used for photoacoustic imaging mostly consist of vascular structures. Some datasets also include circular discs, assuming that a cross-section of the blood vessel is being visualized. We can broadly divide available datasets into (i) experimental photoacoustic images and (ii) ground truth digital imaging for photoacoustic simulation.

5.5.2.1 Experimental Datasets

Obtaining an experimental dataset in photoacoustic imaging can be challenging. However, there are a few datasets available that can be used for deep learning purposes. One such dataset is provided by Razansky et al., which offers whole-body mouse images in vivo for both full-view and limited-view cases [65]. This dataset is valuable for training and testing different deep-learning methods in the context of artifact removal from limited-view reconstructions. The group also provides a data repository [115].

Another dataset available for breast imaging is from Manohar et al. at the University of Twente, which can be accessed on the repository [116, 117]. This dataset offers an opportunity to explore various approaches for breast imaging.

Bell et al. from Johns Hopkins University have made available a combination of simulated and experimental data to address reflection artifacts in photoacoustic imaging [118]. This dataset is particularly useful for identifying and mitigating reflection artifacts.

In addition to tomographic imaging datasets, an experimental photoacoustic microscopic imaging dataset is available from Yao et al. at Duke University [119, 120]. This dataset can be utilized for developing and evaluating upsampling methods for photoacoustic microscopes.

5.5.2.2 Ground Truth Digital Datasets

A common approach is to utilize ground truth vascular phantoms and perform photoacoustic simulations to generate sufficiently large datasets (Fig. 5.7). Vascular ground truth images can be obtained from various sources, with segmented retinal images from different diabetic retinopathy screening studies being a significant resource. Some openly available datasets include DRIVE [121], CHASE [122], STARE [123], ROSE [124], and HRF [125]. These vascular images need to be resized to the appropriate range of blood vessels relevant to the specific application. It is also common to subdivide the retinal images and create different rotated versions to augment the dataset's size. Another dataset frequently utilized by the

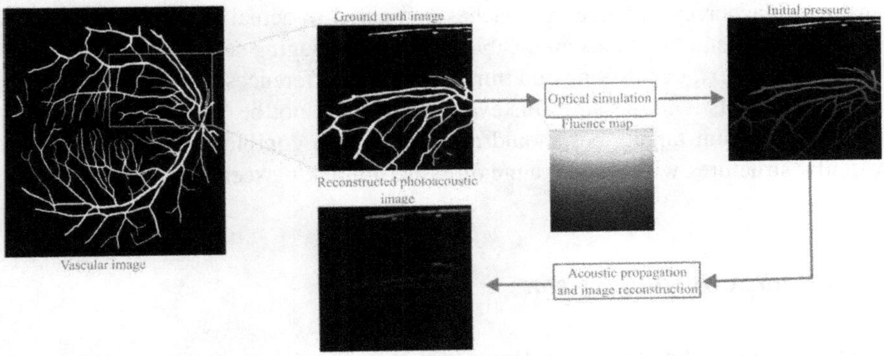

Fig. 5.7 A typical simulation pipeline is used to generate a photoacoustic image dataset. In this specific case, a linear ultrasound transducer array with illumination from slides of the transducer is being considered

photoacoustic community is from Anastasio et al. at the University of Illinois [126]. These segmented MR images of the breast provide a 3D vascular structure. The Neurovascular Network Explorer dataset, consisting of 2-photon measurements of sensory stimulus-induced diameter changes in rat cortical arterioles in vivo, is also used by some groups to simulate vascular structures [127].

Another possibility is to generate ground truth data. For instance, the Virtual Imaging Clinical Trial for Regulatory Evaluation breast image dataset is an example of such an approach [128]. This software aims to create stochastic numerical breast phantoms with anatomically realistic tissue structures and physiologically realistic acoustic properties, which can be used for photoacoustic simulations. A collection of these generated phantoms can be found in this repository [129]. Alternatively, generative deep-learning methods can be employed to generate vascular phantoms, as demonstrated in this work [130, 131].

5.6 Pitfalls and Remedies of Deep Learning for Photoacoustic Image Analysis

5.6.1 Lack of Data and Ground-truth

The performance of a deep neural network model is highly dependent on the number of training samples available. However, meeting this requirement is challenging in current medical photoacoustic (PA) data analysis due to the difficulty in acquiring high-quality medical PA data. Additionally, accurately annotating the collected data poses further challenges, particularly in in-vivo experiments, as there are no gold-standard methods for non-invasively measuring optical absorption, optical scattering, or the speed of sound. This limitation significantly restricts the application of supervised machine learning algorithms for PA data analysis. While simulating and generating photoacoustic data are viable options, it is crucial to ensure the accuracy of these approaches compared to actual experimental data to make deep learning methods applicable to realistic imaging scenarios.

In this context, quantifying and minimizing the differences between simulations and experiments is important. This evaluation should not be limited to simple text objects like point targets but should also encompass complex phantoms, such as vascular structures with a wide range of tissue-mimicking scenarios.

5.6.2 Lack of Comparison

Deep learning techniques have demonstrated impressive performance in various computer vision tasks and have recently found applications in other domains such as medical imaging and remote sensing. In the field of medical imaging, deep learning has been successfully applied to problems in computed tomography, ultrasound, X-ray imaging, diffuse optical tomography, magnetic resonance imaging, and photoacoustic imaging.

When comparing deep learning-based outputs, it is crucial to ensure a fair comparison, considering factors such as speed and accuracy. Unfair comparisons can yield misleading results, where models may appear to perform better but suffer from slow processing speeds or provide inaccurate results.

- Comparisons between different models should consider the variations in datasets used for training and testing. The datasets should be representative and diverse to ensure fair evaluations.
- Tuning the model for individual datasets is crucial when comparing it with a new model. Fine-tuning or retraining may be necessary to optimize performance on specific datasets.
- Fair comparisons require considering the hardware and computational resources used. State-of-the-art algorithms are typically run on CPUs, while deep learning models are often evaluated on GPUs. Comparisons should ensure equal testing conditions for both models.
- Clinical evaluations play a vital role in assessing the effectiveness of deep learning models, particularly when employing techniques such as GANs. Clinicians can identify any adversarial structures or artifacts present in the model's output.
- Explaining how the model's parameters are chosen enhances understanding and transparency. Clear documentation of parameter selection methods helps in reproducibility and allows for informed comparisons.
- The analysis of what the model is learning is valuable in assessing deep learning models. With the development of white box models, understanding the underlying features and representations can provide insights into model behavior.
- Incorporating a clinician's perspective is essential for analyzing the results obtained using neural network architectures. Their expertise can provide valuable insights, interpretations, and clinical relevance to the findings.

5.7 Conclusion

The integration of deep learning techniques with PA imaging provides advancements in imaging technology, data analysis, and clinical applications. Deep learning algorithms have shown their ability to extract meaningful patterns from complex and extensive datasets, enabling enhanced image reconstruction in PA imaging. In this chapter, we provided a comprehensive overview of the recent developments in deep learning for PA image reconstruction. We discussed several key challenges associated with PA data analysis, with a particular focus on the importance of appropriate datasets that can be adapted to the specific requirements of medical imaging scenarios. By tackling some issues such as lack of data through reliance on experimental datasets and establishment of standardized evaluation datasets the field of deep learning in PA image reconstruction can continue to progress and significantly contribute to the advancement of PA imaging as a modality in medical practice.

References

1. B. Cox, J.G. Laufer, S.R. Arridge, P.C. Beard, Journal of biomedical optics 17(6), 061202 (2012)
2. L.V. Wang, S. Hu, science 335(6075), 1458 (2012)
3. L.V. Wang, J. Yao, Nature methods 13(8), 627 (2016)
4. G. Ku, X. Wang, G. Stoica, L.V. Wang, Physics in Medicine & Biology 49(7), 1329 (2004)
5. G. Yang, E. Amidi, Q. Zhu, Biomedical Optics Express 12(4), 2250 (2021)
6. G. Matrone, A.S. Savoia, G. Caliano, G. Magenes, IEEE transactions on medical imaging 34(4), 940 (2014)
7. M. Mozaffarzadeh, A. Hariri, C. Moore, J.V. Jokerst, Photoacoustics 12, 22 (2018)
8. X. Ma, C. Peng, J. Yuan, Q. Cheng, G. Xu, X. Wang, P.L. Carson, IEEE Transactions on Medical Imaging 39(6), 1812 (2019)
9. S. Park, A.B. Karpiouk, S.R. Aglyamov, S.Y. Emelianov, Optics letters 33(12), 1291 (2008)
10. D. Wang, Y. Wang, Y. Zhou, J.F. Lovell, J. Xia, Biomedical optics express 7(5), 1957 (2016)
11. C. Guo, Y. Chen, J. Yuan, Y. Zhu, Q. Cheng, X. Wang, Applied Sciences 8(11), 2113 (2018)
12. J. Prakash, A.S. Raju, C.B. Shaw, M. Pramanik, P.K. Yalavarthy, Biomedical Optics Express 5(5), 1363 (2014)
13. D. Cai, Z. Li, S.L. Chen, Biomedical optics express 7(2), 369 (2016)
14. M. Karaman, P.C. Li, M. O'Donnell, IEEE transactions on ultrasonics, ferroelectrics, and frequency control 42(3), 429 (1995)
15. V. Perrot, M. Polichetti, F. Varray, D. Garcia, Ultrasonics 111, 106309 (2021)
16. D. Finch, S.K. Patch, SIAM journal on mathematical analysis 35(5), 1213 (2004)
17. Y. Hristova, P. Kuchment, L. Nguyen, Inverse problems 24(5), 055006 (2008)
18. M. Xu, L.V. Wang, Physical Review E 71(1), 016706 (2005)
19. M.A. Lediju, G.E. Trahey, B.C. Byram, J.J. Dahl, IEEE transactions on ultrasonics, ferroelectrics, and frequency control 58(7), 1377 (2011)
20. R.G. Lorenz, S.P. Boyd, IEEE transactions on signal processing 53(5), 1684 (2005)
21. S. Zheng, H. Duoduo, Y. Yuan, Computers in Biology and Medicine 76, 60 (2016)
22. D.E. Rumelhart, G.E. Hinton, R.J. Williams, nature 323(6088), 533 (1986)
23. Y. Bengio, in *International Conference on Statistical Language and Speech Processing* (Springer, 2013), pp. 1–37
24. H.P. Martinez, Y. Bengio, G.N. Yannakakis, IEEE Computational intelligence magazine 8(2), 20 (2013)
25. I. Goodfellow, Y. Bengio, A. Courville, *Deep learning* (MIT press, 2016)
26. I.J. Goodfellow, J. Pouget-Abadie, M. Mirza, B. Xu, D. Warde-Farley, S. Ozair, A. Courville, Y. Bengio, arXiv preprint arXiv:1406.2661 (2014)
27. S. Hochreiter, J. Schmidhuber, Neural computation 9(8), 1735 (1997)
28. Y. LeCun, L. Bottou, Y. Bengio, P. Haffner, Proceedings of the IEEE 86(11), 2278 (1998)
29. G.E. Hinton, R.R. Salakhutdinov, science 313(5786), 504 (2006)
30. P. Vincent, H. Larochelle, Y. Bengio, P.A. Manzagol, in *Proceedings of the 25th international conference on Machine learning* (2008), pp. 1096–1103
31. G.E. Hinton, Scholarpedia 4(5), 5947 (2009)
32. P. Vincent, H. Larochelle, I. Lajoie, Y. Bengio, P.A. Manzagol, L. Bottou, Journal of machine learning research 11(12) (2010)
33. D.P. Kingma, M. Welling, arXiv preprint arXiv:1312.6114 (2013)

34. O. Ronneberger, P. Fischer, T. Brox, in *Medical Image Computing and Computer-Assisted Intervention–MICCAI 2015: 18th International Conference, Munich, Germany, October 5-9, 2015, Proceedings, Part III 18* (Springer, 2015), pp. 234–241
35. F. Yu, V. Koltun, arXiv preprint arXiv:1511.07122 615 (2016)
36. J. Long, E. Shelhamer, T. Darrell, in *Proceedings of the IEEE conference on computer vision and pattern recognition* (2015), pp. 3431–3440
37. H. Noh, S. Hong, B. Han, in *Proceedings of the IEEE international conference on computer vision* (2015), pp. 1520–1528
38. K. He, X. Zhang, S. Ren, J. Sun, in *Proceedings of the IEEE conference on computer vision and pattern recognition* (2016), pp. 770–778
39. A. Krizhevsky, I. Sutskever, G.E. Hinton, Communications of the ACM 60(6), 84 (2017)
40. A. Vaswani, N. Shazeer, N. Parmar, J. Uszkoreit, L. Jones, A.N. Gomez, L. Kaiser, I. Polosukhin, Advances in neural information processing systems 30 (2017)
41. Y. Chen, J. Li, H. Xiao, X. Jin, S. Yan, J. Feng, Advances in neural information processing systems 30 (2017)
42. I. Goodfellow, J. Pouget-Abadie, M. Mirza, B. Xu, D. Warde-Farley, S. Ozair, A. Courville, Y. Bengio, Communications of the ACM 63(11), 139 (2020)
43. S. Sabour, N. Frosst, G.E. Hinton, Advances in neural information processing systems 30 (2017)
44. G. Huang, Z. Liu, L. Van Der Maaten, K.Q. Weinberger, in *Proceedings of the IEEE conference on computer vision and pattern recognition* (2017), pp. 4700–4708
45. K. Hammernik, T. Klatzer, E. Kobler, M.P. Recht, D.K. Sodickson, T. Pock, F. Knoll, Magnetic resonance in medicine 79(6), 3055 (2018)
46. D. Ulyanov, A. Vedaldi, V. Lempitsky, in *Proceedings of the IEEE conference on computer vision and pattern recognition* (2018), pp. 9446–9454
47. O. Oktay, J. Schlemper, L.L. Folgoc, M. Lee, M. Heinrich, K. Misawa, K. Mori, S. McDonagh, N.Y. Hammerla, B. Kainz, et al., arXiv preprint arXiv:1804.03999 (2018)
48. T.R. Shaham, T. Dekel, T. Michaeli, in *Proceedings of the IEEE/CVF International Conference on Computer Vision* (2019), pp. 4570–4580
49. Y. Zhang, K. Li, K. Li, B. Zhong, Y. Fu, arXiv preprint arXiv:1903.10082 (2019)
50. J. Ho, A. Jain, P. Abbeel, Advances in Neural Information Processing Systems 33, 6840 (2020)
51. B. Mildenhall, P.P. Srinivasan, M. Tancik, J.T. Barron, R. Ramamoorthi, R. Ng, Communications of the ACM 65(1), 99 (2021)
52. Y. LeCun, B. Boser, J.S. Denker, D. Henderson, R.E. Howard, W. Hubbard, L.D. Jackel, Neural computation 1(4), 541 (1989)
53. D. Waibel, J. Gröhl, F. Isensee, T. Kirchner, K. Maier-Hein, L. Maier-Hein, in *Photons Plus Ultrasound: Imaging and Sensing 2018*, vol. 10494 (International Society for Optics and Photonics, 2018), vol. 10494, p. 104942S
54. E.M.A. Anas, H.K. Zhang, C. Audigier, E.M. Boctor, in *Simulation, Image Processing, and Ultrasound Systems for Assisted Diagnosis and Navigation* (Springer, 2018), pp. 3–11
55. H. Lan, C. Yang, D. Jiang, F. Gao, in *2019 41st Annual International Conference of the IEEE Engineering in Medicine and Biology Society (EMBC)* (IEEE, 2019), pp. 7115–7118
56. J. Feng, J. Deng, Z. Li, Z. Sun, H. Dou, K. Jia, Biomedical optics express 11(9), 5321 (2020)
57. T. Tong, W. Huang, K. Wang, Z. He, L. Yin, X. Yang, S. Zhang, J. Tian, Photoacoustics p. 100190 (2020)
58. J. Schwab, S. Antholzer, R. Nuster, M. Haltmeier, arXiv preprint arXiv:1801.06693 (2018)
59. E.M.A. Anas, H.K. Zhang, J. Kang, E. Boctor, Biomedical optics express 9(8), 3852 (2018)
60. E.M.A. Anas, H.K. Zhang, J. Kang, E.M. Boctor, in *International conference on medical image computing and computer-assisted intervention* (Springer, 2018), pp. 159–167
61. J. Schwab, S. Antholzer, R. Nuster, G. Paltauf, M. Haltmeier, in *Photons Plus Ultrasound: Imaging and Sensing 2019*, vol. 10878 (International Society for Optics and Photonics, 2019), vol. 10878, p. 1087836

62. S. Antholzer, M. Haltmeier, J. Schwab, Inverse problems in science and engineering 27(7), 987 (2019)
63. H. Shan, G. Wang, Y. Yang, Applied Sciences 9(13), 2615 (2019)
64. H. Deng, X. Wang, C. Cai, J. Luo, C. Ma, in *Advanced Optical Imaging Technologies II*, vol. 11186 (International Society for Optics and Photonics, 2019), vol. 11186, p. 111860J
65. N. Davoudi, X.L. Dea'n-Ben, D. Razansky, Nature Machine Intelligence 1(10), 453 (2019)
66. S. Guan, A. Khan, S. Sikdar, P. Chitnis, IEEE Journal of Biomedical and Health Informatics (2019)
67. S. Jeon, C. Kim, in *Photons Plus Ultrasound: Imaging and Sensing 2020*, vol. 11240 (International Society for Optics and Photonics, 2020), vol. 11240, p. 112400J
68. S. Jeon, W. Choi, B. Park, C. Kim, IEEE Transactions on Image Processing 30, 8773 (2021)
69. S. Guan, A.A. Khan, S. Sikdar, P.V. Chitnis, Scientific reports 10(1), 1 (2020)
70. T. Vu, M. Li, H. Humayun, Y. Zhou, J. Yao, Experimental Biology and Medicine 245(7), 597 (2020)
71. P. Farnia, M. Mohammadi, E. Najafzadeh, M. Alimohamadi, B. Makkiabadi, A. Ahmadian, Biomedical Physics & Engineering Express 6(4), 045019 (2020)
72. Z. Sun, X. Yan, Journal of Medical Imaging and Health Informatics 10(11), 2707 (2020)
73. R. Manwar, X. Li, S. Mahmoodkalayeh, E. Asano, D. Zhu, K. Avanaki, Journal of biophotonics 13(10), e202000212 (2020)
74. H. Zhang, L. Hongyu, N. Nyayapathi, D. Wang, A. Le, L. Ying, J. Xia, Computerized Medical Imaging and Graphics 84, 101720 (2020)
75. A. Hariri, K. Alipour, Y. Mantri, J.P. Schulze, J.V. Jokerst, Biomedical optics express 11(6), 3360 (2020)
76. G. Godefroy, B. Arnal, E. Bossy, arXiv preprint arXiv:2006.13096 (2020)
77. M. Kim, G.S. Jeng, I. Pelivanov, M. O'Donnell, IEEE transactions on medical imaging 39(11), 3379 (2020)
78. H. Shahid, A. Khalid, X. Liu, M. Irfan, D. Ta, Frontiers in Neuroscience 15, 598693 (2021)
79. H. Zhao, Z. Ke, F. Yang, K. Li, N. Chen, L. Song, C. Zheng, D. Liang, C. Liu, Advanced Science 8(3), 2003097 (2021)
80. J. Zhang, Q. He, C. Wang, H. Liao, J. Luo, in *2021 IEEE International Ultrasonics Symposium (IUS)* (IEEE, 2021), pp. 1–4
81. T. Vu, A. DiSpirito III, D. Li, Z. Wang, X. Zhu, M. Chen, L. Jiang, D. Zhang, J. Luo, Y.S. Zhang, et al., Photoacoustics 22, 100266 (2021)
82. S. Guan, K.T. Hsu, M. Eyassu, P.V. Chitnis, arXiv preprint arXiv:2104.03130 (2021)
83. F.K. Joseph, A. Arora, P. Kancharla, M.K.A. Singh, W. Steenbergen, S.S. Channappayya, in *Photons Plus Ultrasound: Imaging and Sensing 2021*, vol. 11642 (SPIE, 2021), vol. 11642, pp. 208–213
84. P. Rajendran, M. Pramanik, Optics Letters 46(18), 4510 (2021)
85. Y. Gao, W. Xu, Y. Chen, W. Xie, Q. Cheng, IEEE Transactions on Medical Imaging 41(8), 2191 (2022)
86. H. Zhang, W. Bo, D. Wang, A. DiSpirito, C. Huang, N. Nyayapathi, E. Zheng, T. Vu, Y. Gong, J. Yao, et al., IEEE Transactions on Medical Imaging 41(5), 1279 (2021)
87. M. Guo, H. Lan, C. Yang, J. Liu, F. Gao, IEEE Transactions on Computational Imaging 8, 215 (2022)
88. F. Feng, S. Liang, J. Luo, S.L. Chen, Photoacoustics 26, 100360 (2022)
89. N. Awasthi, K.R. Prabhakar, S.K. Kalva, M. Pramanik, R.V. Babu, P.K. Yalavarthy, Biomedical Optics Express 10(5), 2227 (2019)
90. Y.E. Boink, S. Manohar, C. Brune, IEEE transactions on medical imaging 39(1), 129 (2019)
91. H. Lan, D. Jiang, C. Yang, F. Gao, F. Gao, Photoacoustics 20, 100197 (2020)
92. H. Lan, C. Yang, F. Gao, Photoacoustics 29, 100442 (2023)
93. S. Gutta, V.S. Kadimesetty, S.K. Kalva, M. Pramanik, S. Ganapathy, P.K. Yalavarthy, Journal of Biomedical Optics 22(11), 116001 (2017)
94. N. Awasthi, R. Pardasani, S.K. Kalva, M. Pramanik, P.K. Yalavarthy, arXiv preprint arXiv:2001.06434 (2020)

95. N. Awasthi, G. Jain, S.K. Kalva, M. Pramanik, P.K. Yalavarthy, IEEE Transactions on Ultrasonics, Ferroelectrics, and Frequency Control (2020)
96. C. Dehner, I. Olefir, K.B. Chowdhury, D. Jüstel, V. Ntziachristos, IEEE Transactions on Medical Imaging 41(11), 3182 (2022)
97. H. Dehghani, M.E. Eames, P.K. Yalavarthy, S.C. Davis, S. Srinivasan, C.M. Carpenter, B.W. Pogue, K.D. Paulsen, Communications in numerical methods in engineering 25(6), 711 (2009)
98. M. Schweiger, S. Arridge, Journal of biomedical optics 19(4), 040801 (2014)
99. omlc.org/software/mc/man_mcml.pdf. [Online; accessed 2023]
100. atomic.physics.lu.se/biophotonics/research/monte-carlo-simulations/gpu-monte-carlo [Online; accessed 2023]
101. omlc.org/software/mc/mcxyz. [Online; accessed 2023]
102. gitlab.gbar.dtu.dk/biophotonics/MCmatlab. [Online; accessed 2023]
103. mcx.space. [Online; accessed 2023]
104. S.L. Jacques, Journal of Biomedical Optics 27(8), 083002 (2022)
105. github.com/invimg/simpa. [Online; accessed 2023]
106. github.com/KothapalliLabPSU/US-PA_simulation_codes. [Online; accessed 2023]
107. inverselight.github.io/ValoMC/features.html. [Online; accessed 2023]
108. https://github.com/IPASC/standardised-image-reconstruction. [Online; ac- cessed 2023]
109. https://github.com/derekallman/Photoacoustic-FasterRCNN. [Online; accessed 2023]
110. https://https://github.com/chenyilan/Y-Net. [Online; accessed 2023]
111. https://github.com/szgy66/Vessel-Segmentation-of-photoacoustic-imaging. [Online; accessed 2023]
112. https://github.com/Gaurav1302/DeepSino. [Online; accessed 2023]
113. J. Gröhl, L. Hacker, B.T. Cox, K.K. Dreher, S. Morscher, A. Rakotondrainibe, F. Varray, L.C. Yip, W.C. Vogt, S.E. Bohndiek, et al., Photoacoustics 26, 100339 (2022)
114. https://www.ipasc.science/. [Online; accessed 2023]
115. https://www.research-collection.ethz.ch/handle/20.500.11850/31 (2022). [Online; accessed 2023]
116. https://data.4tu.nl/articles/_/22109687/1 (2023). [Online; accessed 2023]
117. S.M. Schoustra, B. De Santi, T.J. op't Root, C.A. Klazen, M. van der Schaaf, J. Veltman, W. Steenbergen, S. Manohar, Plos one 18(3), e0281434 (2023)
118. D. Allman, A. Reiter, M.A.L. Bell, IEEE transactions on medical imaging 37(6), 1464 (2018)
119. A. DiSpirito, D. Li, T. Vu, M. Chen, D. Zhang, J. Luo, R. Horstmeyer, J. Yao, IEEE transactions on medical imaging 40(2), 562 (2020)
120. https://figshare.com/articles/dataset/Photoacoustic_Microscopy_Dataset/12871805/1 (2020). [Online; accessed 2023]
121. https://www.kaggle.com/datasets/andrewmvd/drive-digital-retinal-images-for-vessel-extraction/code (2020). [Online; accessed 2023]
122. https://blogs.kingston.ac.uk/retinal/chasedb1/ (2016). [Online; accessed 2023]
123. https://cecas.clemson.edu/~ahoover/stare/ (2016). [Online; accessed 2023]
124. https://imed.nimte.ac.cn/dataofrose.html (2021). [Online; accessed 2023]
125. https://www5.cs.fau.de/research/data/fundus-images/ (2013). [Online; accessed 2023]
126. https://anastasio.bioengineering.illinois.edu/downloadable-content/oa-breast-database/ (2013). [Online; accessed 2023]
127. https://sites.bu.edu/nil/data-sharing/ (2016). [Online; accessed 2023]
128. https://github.com/comp-imaging-sci/usct-breast-phantom/tree/0.1.0 (2021). [Online; accessed 2023]
129. https://dataverse.harvard.edu/dataset.xhtml?persistentId= doi:https://doi.org/10.7910/DVN/KBYQQ7 (2021). [Online; accessed 2023]
130. https://zenodo.org/record/7101822#.ZGH7yHZBxaQ (2022). [Online; accessed 2023]
131. M. Schellenberg, J. Gröhl, K.K. Dreher, J.H. Nölke, N. Holzwarth, M.D. Tizabi, A. Seitel, L. Maier-Hein, Photoacoustics 28, 100402 (2022)

Part III
Contrast Agents and Tissue Phantoms

Chapter 6
Photoacoustic Imaging and Applications with Reversibly Switchable Contrast Agents

Donghyeon Oh, Chulhong Kim, and Byullee Park

Abstract Benefiting from both optical and acoustical imaging advantages, photoacoustic imaging (PAI) can noninvasively acquire photochemical features of a live tissue with scalable resolution and imaging depth on demand. Looking over accessible contrast from endogenous chromophores, diverse categories of photoacoustic (PA) contrast agents have been actively excavated to reach a deeper imaging depth. However, conventional PA contrast agents must compete with the endogenous contrast from a background tissue, which might limit the imaging depth to background-equivalent depth and perturb molecular specificity. Here, we introduce novel reversibly switchable PA contrast agents that could effectively mute the background signal and eventually achieve exclusive exogenous contrast. In this chapter, we brief the overall principle and scalable configuration of PAI, summarize conventional PA contrast agents categorized by their compositions, and finally introduce reversibly switchable PA contrast agents. Based on our summary, we further suggest a promising outlook that these novel PA contrasts could lead to new biomedical findings.

Keywords Photoacoustic imaging · Photoacoustic agents · Contrast agent · Preclinical

---

D. Oh
Pohang University of Science and Technology, Department of Convergence IT Engineering, Pohang, Republic of Korea
e-mail: lumin9219@postech.ac.kr

C. Kim
Pohang University of Science and Technology, Department of Electrical Engineering, Convergence IT Engineering, Medical Science and Engineering, Mechanical Engineering, School of Interdisciplinary Bioscience and Bioengineering, Graduate School of Artificial Intelligence, Medical Device Innovation Center, Pohang, Republic of Korea
e-mail: chulhong@postech.edu

B. Park (✉)
Sungkyunkwan University, Department of Biophysics, Institute of Quantum Biophysics, Suwon, Republic of Korea
e-mail: byullee@skku.edu

W. Xia (ed.), *Biomedical Photoacoustics*,
https://doi.org/10.1007/978-3-031-61411-8_6

6.1 Introduction

Photoacoustic imaging (PAI) is a rapidly evolving biomedical imaging technique that combines the merits of both optical and ultrasound imaging modalities [1, 2]. By capitalizing on the photoacoustic (PA) effect, this non-invasive imaging method enables visualization of tissue structures and functional information with high spatial resolution and remarkable depth penetration [3]. The fundamental principle underlying PAI is the PA effect, which involves the absorption of short laser pulses by biological tissues. This absorption leads to conversion of light energy into thermal energy, resulting in transient thermoelastic expansion and subsequent generation of acoustic waves [4]. These acoustic waves are then detected by ultrasonic transducers and utilized for reconstructing high-resolution and deep-tissue PA images (Fig. 6.1) [5].

What sets PAI apart from other conventional medical imaging modalities such as magnetic resonance imaging (MRI), positron emission tomography (PET), optical imaging, and ultrasound imaging is its unique combination of high spatial resolution and deep tissue penetration capabilities [8]. While MRI and PET offer excellent soft tissue contrast, they often lack the spatial resolution required for visualizing intricate anatomical details. Optical imaging, although providing high spatial resolution, is impeded by limited tissue penetration due to light scattering. Although ultrasound imaging offers sufficient tissue penetration, it exhibits inadequate contrast for molecular imaging [9]. Distinctive advantages of PAI lie in its ability to provide both high resolution and substantial tissue penetration, positioning it as a promising modality for a wide range of biomedical applications [10].

PAI can be categorized into optical resolution and acoustic resolution based on system implementation (Fig. 6.2) [11]. These two categories represent distinct approaches to achieving high-resolution imaging with different system configurations. Optical-resolution photoacoustic microscopy (OR-PAM) aims to achieve

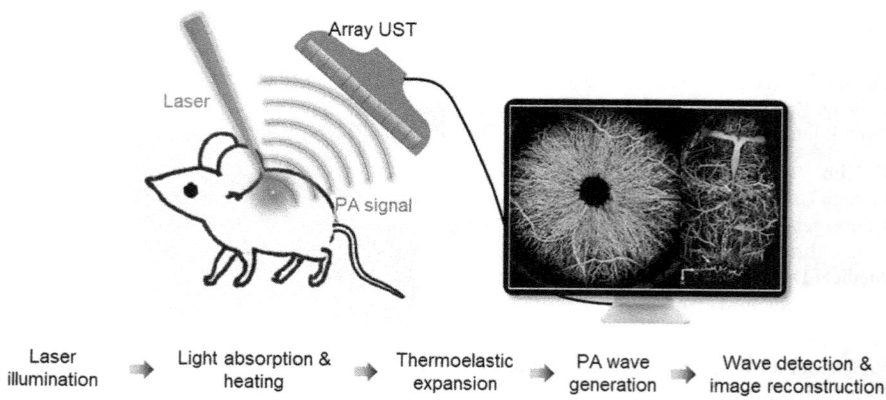

Fig. 6.1 Principle of photoacoustic imaging. (Reproduced with permission from [6, 7])

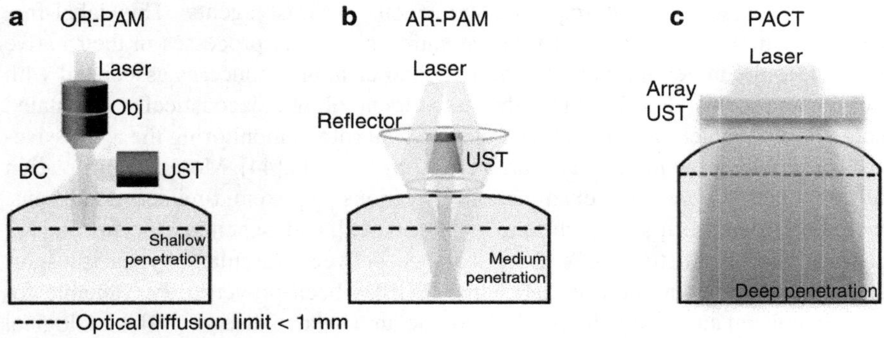

Fig. 6.2 Schematics of optical resolution photoacoustic microscopy (OR-PAM), acoustic resolution photoacoustic microscopy (AR-PAM), and photoacoustic computed tomography (PACT). (**a**) OR-PAM: coaxial alignment of light and ultrasound paths based on an opto-US beam combiner (BC). (**b**) AR-PAM: dark-field illumination method. (**c**) PACT: linear-type array ultrasound transducer (UST) and unfocused light source to allow multiple cm-deep tissue imaging. (Reproduced with permission from [36])

high spatial resolution by tightly focusing the excitation laser beam and using high-frequency ultrasound transducers for signal detection [12]. The focused laser beam illuminates a small area, typically with a numerical aperture (NA) matching the ultrasound transducer's NA (Fig. 6.2a). This configuration allows for localized excitation and detection, resulting in excellent lateral resolution. The use of high-frequency ultrasound transducers can enhance detection sensitivity, enabling visualization of fine anatomical details and microvascular structures [13]. OR-PAM is particularly well suited for imaging superficial tissues such as the skin [14–17], eyes [7, 18, 19], and small biological samples [20–22]. However, due to limited depth of focus and the presence of optical scattering, the imaging depth of OR-PAM is relatively shallow, typically up to 1 mm (i.e., optical diffusion limit). In contrast, acoustic-resolution photoacoustic imaging (AR-PAI) prioritizes deep tissue imaging by sacrificing lateral resolution [23]. AR-PAI can be accomplished through raster scanning using weakly focused optical illumination and a spherically focused ultrasound (US) transducer, known as acoustic-resolution photoacoustic microscopy (AR-PAM) (Fig. 6.2b) [24, 25]. Alternatively, it can be performed by utilizing wide-field optical illumination and parallel acoustic detection with an array type of US transducer, referred to as photoacoustic computed tomography (PACT) (Fig. 6.2c) [26, 27]. While the lateral resolution in AR-PAI is lower than that of OR-PAM, the imaging depth is significantly improved, making it suitable for visualizing structures in deeper tissues and larger organs [28–31]. AR-PAI is commonly employed in preclinical and clinical studies to investigate functional parameters such as oxygen saturation [32, 33] and tissue perfusion [34, 35].

An intriguing attribute of PAI is its inherent label-free nature. By capitalizing on light-absorbing endogenous agents such as hemoglobin [37–39], melanin [40], and lipids [41, 42], PAI enables visualization of tissue structures and physiological processes without requiring exogenous contrast agents. This label-free characteristic is advantageous for investigating biological processes in their native state [43], circumventing potential toxicity and clearance concerns associated with exogenous agents. For instance, the assessment of photoacoustically calculated hemoglobin oxygen saturation (sO_2) plays a vital role in monitoring the aggressiveness of rapidly growing tumors characterized by hypoxia [44]. Moreover, utilization of computed sO_2 enables examination of various hypoxemia-related conditions, including osteoarthritis [45], chronic leg ulcers [46], and ischemia-reperfusion [47] characterized by disturbances in the balance between vascular oxygen transport and consumption. In addition, label-free PAI has been proven to be valuable for visualizing and analyzing blood flow [48], melanin-related disease [49], histological tissues [50, 51], live cells [52], and even mitochondria [12, 53].

However, label-free PAI encounters limitations, particularly in the realms of molecular imaging and specific targeting of biomarkers. In contrast, labeled PAI utilizing exogenous contrast agents provides a significant improvement in image contrast for optically transparent or deep tissues in vivo. Labeled PAI has extensive applications in imaging tumor progression, lymph node metastasis [54–56], gastrointestinal (GI) tract [57, 58], atherosclerotic plaques [59, 60], and so on. Moreover, exogenous agents can be activated by disease microenvironments such as temperature, pH, enzymes, reactive oxygen species (ROS), metal ions, and nitric oxide (NO), thereby enabling specific targeting [50, 61]. To develop optimal exogenous PA agents, the following should be considered: (1) their design should focus on achieving maximum light absorption at specific wavelengths within the near-infrared (NIR)-I or NIR-II window to extend imaging depth and minimize background noise from endogenous chromophores; (2) strategies should be employed to efficiently deliver contrast agents to the target area, including conjugation of PA signal generating compounds with targeting ligands; (3) factors such as size, material composition, and surface encapsulation should be considered for biocompatibility and biodegradability optimization [53, 62].

Although various exogenous contrast agents have been explored, their effectiveness is often limited by abundant endogenous contrast in the surrounding tissue (Fig. 6.3a) [53]. This results in contrast-enhanced PA images to contain both signals from the exogenous contrast agents and prevalent endogenous chromophores. Consequently, even though the original contrast agent has a good PA signal yield in solitary, the imaging depth of PAI is constrained to a background-equivalent depth in an in vivo condition. Such simultaneity degrades a molecular specificity, further raising doubts about PAI as a molecular imaging technique. Although spectral unmixing technique is available, it requires repetitive PA image acquisition under multiple wavelengths, which eventually slows down the imaging process. Additionally, the spectral-unmixed output often contains errors as it often ignores fluence deviation varying by wavelength. A potential solution lies in the use of exogenous contrast agents with reversible discoloring features that can be turned on and off, similar to that of photoactivation and bleaching in photoactivated localization microscopy (Fig. 6.3b) [63, 64]. These switchable contrast agents can exclusively retain their contrast while the differential contrast of intrinsic

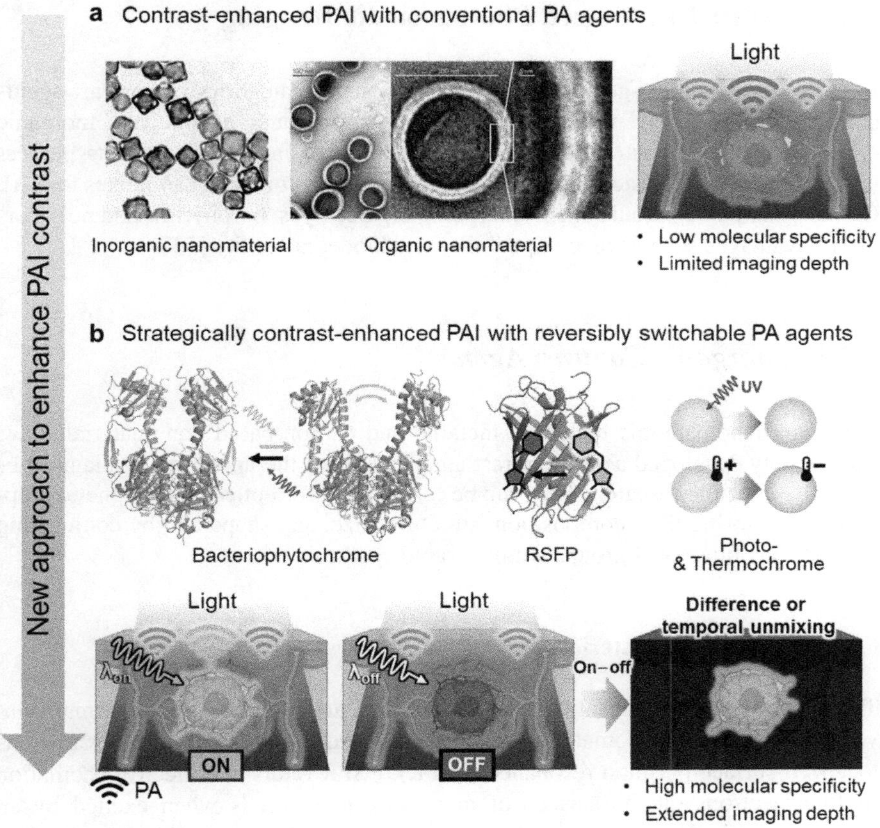

Fig. 6.3 Comparison between conventional PA contrast agent and reversibly switchable PA contrast agent. (**a**) Effective contrast of a conventional exogenous chromophore is decreased by a background PA signal from adjacent tissue. Reproduced with permission from [65, 66]. (**b**) Categories of reversibly switchable PA contrast agents featuring on-off PAI strategy to suppress background tissue signal. Schematic below describes PA image acquisition process with reversibly switchable PA contrast agent. (Reproduced with permission from [67])

chromophores reduces to zero. By muting PA signals from background tissue, this approach can achieve significant enhancement in both contrast-to-background ratio (CBR) and molecular specificity (Fig. 6.3c).

Here, this chapter covers the use of exogenous contrast agents in PAI and their various applications. Section 6.2 provides a brief overview of conventional inorganic and organic agents that have been extensively studied as conventional PA contrast agents. In Sect. 6.3, the main section of this chapter, reversibly switchable PA contrast agents characterized by an “on-off” PA switchability are highlighted.

6.2 Contrast-Enhanced Photoacoustic Imaging

In early studies on exogenous PA agents, two major categories of contrast agents commonly used in PAI were inorganic agents and organic agents [68]. Inorganic agents such as gold nanoparticles, carbon nanotubes, and quantum dots possess unique optical and physical properties that make them promising candidates for PAI. On the other hand, organic agents including organic dyes, porphyrins, and polymers offer advantages such as biocompatibility and biodegradability [53].

6.2.1 Inorganic Contrast Agents

Inorganic nanomaterials including metallic and carbon-based nanomaterials have been widely developed as PA contrast agents for both therapeutics and diagnostics through molecular research. They can be easily tuned for optical and magnetic properties by tailoring their composition, structure, size, and shape and by conjugating with various functional groups within a rapid synthesis time.

6.2.1.1 Gold Nanomaterials

The remarkable light absorption characteristics and efficient photothermal conversion of metallic nanomaterials can be attributed to the phenomenon known as localized surface plasmon resonance (LSPR). LSPR refers to collective oscillation of free electrons on the surface of metallic nanomaterials when excited by an electromagnetic field at a resonant frequency, resulting in enhanced light extinction. Importantly, the resonant frequency of metallic nanomaterials can be controlled by adjusting factors such as size, composition, geometry, structure, and sharpness of nanoparticles. This unique property allows metallic nanomaterials to possess distinct resonance peaks, making them excellent candidates for PA contrast agents. For example, Lu has pointed out that the optical absorption peak of gold nanosphere is similar to that of hemoglobin, making it difficult to use as a PA agent [69]. Lu et al. overcame this problem by surface conjugating with thiolated polyethylene gylcol-coated (PEGylated) hollow gold nanosphere (PEG-HAuNS) [69]. They tuned the light absorption peak to 800 nm using PEG-HAuNS with an outer diameter of 45 nm and successfully obtained PA efficiencies much higher than that of blood (Fig. 6.4a). Another notable research study has focused on enhancing absorption in the NIR-II window by manipulating the size of gold nanorods, which typically exhibit absorption peaks in the NIR-I window. Chen et al. have conducted a study to synthesize miniaturized gold nanorods (8×49 nm, width $\times$ length) that are 5–11 times smaller than conventional nanorods while maintaining their aspect ratio [70]. These miniaturized gold nanorods demonstrated peak optical absorption at 1050 nm (Fig. 6.4b). In addition, gold nanocages have garnered attention due to

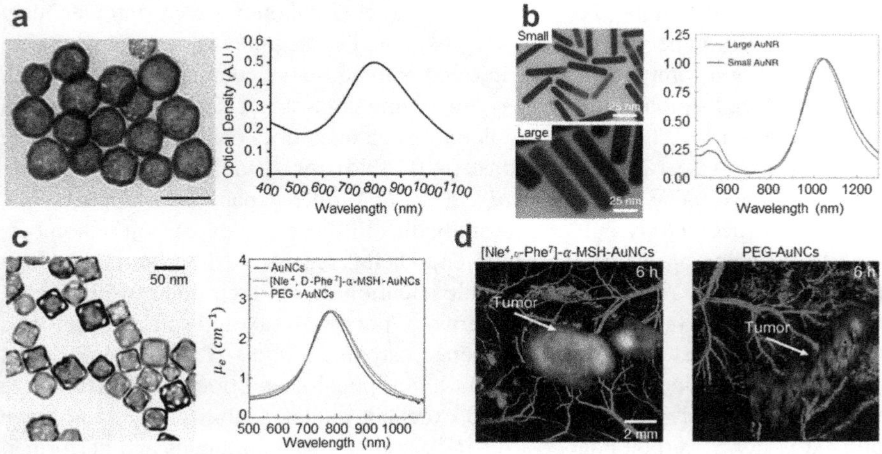

Fig. 6.4 Gold nanomaterial-based photoacoustic imaging. (**a**) TEM image of PEG-HAuNS and their optical absorbance spectrum in water which peaked at 800 nm. Reproduced with permission from [69]. (**b**) TEM images of as-synthesized small and large AuNRs and their normalized extinction spectra. Reproduced with permission from [70]. (**c**) TEM image of [Nle^4,D-Phe^7]-α-MSH-AuNCs and their extinction coefficient spectra. (**d**) In vivo blood vessels and tumor PA images with injection of [Nle^4,D-Phe^7]-α-MSH-AuNCs and PEG-AuNCs (as a control), respectively. (Reproduced with permission from [65]. TEM transmission electron microscopy; HAuNS hollow gold nanosphere; AuNR gold nanorod; PA photoacoustic)

their small size, relatively easy synthesis, high reproducibility, and excellent encapsulation abilities. Kim et al. have synthesized melanocyte-stimulating hormone gold nanocages ([Nle^4,D-Phe^7]-α-MSH-AuNCs) with a peak optical absorption at 790 nm specifically for enhancing contrast in PAI (Fig. 6.4c) [65]. When these nanocages were administered intravenously into mice with melanoma to facilitate active targeting, subsequent PA signals observed were amplified by 300% in comparison with signals produced by PEGylated gold nanocages (Fig. 6.4d).

6.2.1.2 Carbon-Based Nanomaterials

Carbon-based materials have captured the attention of numerous researchers due to their outstanding properties such as facile fabrication, long-term stability, distinctive absorption in the near-infrared (NIR) region, and high efficiency in converting light into heat through photothermal conversion. Despite their broad absorption spectra without distinct peaks, carbon-based materials can enhance the intensity of PA signals by incorporating plasmonic metal nanomaterials or small molecules within their carbon structure. This synergistic combination allows for improved PAI capabilities and opens up promising avenues for advanced biomedical applications. For example, Zerda et al. have demonstrated that targeted single-walled carbon nanotubes conjugated with cyclic Arg-Gly-Asp (RGD) peptides can serve as a contrast agent for PAI of tumors [71]. When these targeted nanotubes were intravenously administered to mice with tumors, they exhibited eight times greater PA signal in the tumor region compared to mice injected with non-targeted nanotubes. Kim et al. have introduced a pioneering strategy employing gold-plated carbon nanotubes as a contrast agent for PAI [72]. These golden carbon nanotubes exhibited a remarkable enhancement in near-infrared contrast (~102-fold) for visualization of lymphatic vessels in mice using just a few mJ cm^{-2}. To further expand their applications, researchers successfully conjugated antibodies to these golden carbon nanotubes, facilitating mapping of the lymphatic endothelial receptor. Although photoactive nanoparticles have received considerable attention due to their contrast-enhancing abilities, they still face challenges in terms of photostability and safety. To address these issues, Lee et al. have used N-doped carbon nanodots [73]. These N-doped carbon nanodots demonstrated not only high optical absorption in the NIR range but also remarkable photostability and excellent biodegradability. Lee et al. have utilized N-doped carbon nanodots for PA mapping of lymph nodes and performed cancer ablation through photothermal therapy.

6.2.2 Organic Contrast Agents

Metallic nanomaterials have achieved great success as PA contrast agents. However, their toxicity and low biodegradability are still issues. As alternatives, organic nanomaterials containing small molecules and semiconducting polymer particles are attracting attention as PA agents.

6.2.2.1 Small Organic Molecules

Small molecules are being increasingly used as contrast agents in PAI-based drug delivery systems because they are cost-effective and biocompatible with favorable pharmacokinetic properties. Additionally, small molecules can serve as strategic foundations for combination therapies in drug delivery systems because they can penetrate cell membranes and other physiological barriers to reach intracellular targets in different dosage forms. Representative small molecules such as PAI agents and porphysome nanovesicles are formed by self-assembled porphyrin bilayers. They exhibit large and tunable absorption coefficients, unique photothermal properties, and PA characteristics. Lovell et al. have developed porphysome nanovesicles with enzymatic biodegradability and potential to serve as PA agents with an intrinsic optical activity (Fig. 6.5a, b) [66]. Wang et al. have introduced a novel type of nanoparticles known as covalent organic framework NPs based on porphyrin molecules [74]. These NPs possess a large surface area and porous characteristics, enhancing their ability to harvest light and adsorb oxygen. However, one limitation was that the optical absorption peak of porphyrin-based materials was below 700 nm, leading to insufficient photostability during prolonged exposure to

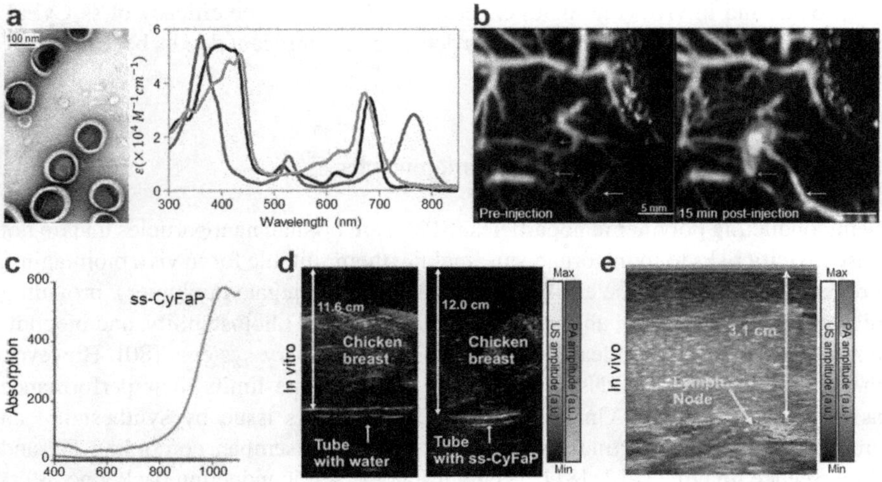

Fig. 6.5 Small organic molecule-based photoacoustic imaging. (**a**) Electron microscope image of negatively stained porphysomes and their absorbance. (**b**) Photoacoustic imaging of a rat lymphatic system before and after injection of porphysomes. Secondary lymph vessels are highlighted in cyan, the lymph node in red, and the inflowing lymph vessel in yellow. Reproduced with permission from [66]. (**c**) Absorbance of ss-CyFaP. (**d**) In vitro PA image (color scale) with overlaid ultrasound (US) image (grayscale) of a tube containing water (left) or ss-CyFaP micelles (right). PA signals are successfully detected at depths of up to 12 cm. (**e**) In vivo PA image of a lymph node, overlaid with a US image (grayscale), and captured at depths of up to 3.1 cm. Reproduced with permission from [79]

light. To overcome this challenge, Wu et al. have successfully modified the optical absorption peak by red-shifting it to 807 nm using porphyrin-diketopyrrolopyrrole NPs [75]. Cyanine dye, another prominent small molecule agent, has gained attention due to its strong absorption coefficient, easy optical tunability, and excellent biodegradability and biocompatibility [76]. Cyanine dyes have limited use because they degrade easily when exposed to light. To overcome this, researchers have tried encapsulation methods. One study successfully combined iridium complexes with a cyanine dye and then encapsulated them in hydrophobic PPO blocks of F127 to create iridium(III)-cyanine complex nanoparticles [77]. These nanoparticles have high water solubility. They remain stable when irradiated with 808 nm light. Other cyanine-based agents such as phthalocyanine (Pc) and naphthalocyanine (Nc) nanoagents are being explored as PA agents due to their strong light absorption in the near-infrared (NIR) region [78]. A strategy has been established for superior deep tissue imaging using NIR-II window lights that can highly penetrate into tissue. Chitgupi et al. have proposed the use of surfactant-stripped micelles, ss-CyFaP, made from cyanine fluoroalkylphosphate salt dye, which show high absorption at NIR-II window [79]. These micelles loaded with a hydrophobic dye using a temperature-sensitive Pluronic F127 surfactant demonstrated improved yield and storage stability, generating a peak absorption at 1064 nm (Fig. 6.5c). In vitro

(Fig. 6.5d) and in vivo (Fig. 6.5e) experiments confirmed the efficacy of ss-CyFaP micelles for PAI, with strong PA signals detected at significant depths beneath tissue layers up to 12 cm.

6.2.2.2 Semiconducting Polymer Nanoparticles

Semiconducting polymer nanoparticles (SPNs) are organic nanoparticles that do not pose toxicity risks to living organisms, making them suitable for in vivo bioimaging. Properties of SPNs can be easily adjusted using π-conjugated polymers, providing high structural flexibility, absorption in the NIR region, photostability, and biocompatibility, making them ideal PA agents for drug delivery systems [80]. However, the poor solubility of SPNs in biological environments limits their performance as PA contrast agents. Yin et al. have addressed this issue by synthesizing an amphiphilic semiconducting oligomer that can self-assemble, combining PA and fluorescence imaging (FLI) [81]. The hydrophobic semiconducting backbone offers a signal source with high photothermal efficiency, while PEG chains can improve water solubility, thus enhancing imaging capabilities of SPNs. Ding et al. have combined the surfactant-stripped micelles approach with semiconducting polymers (SPs) [58] for deep tissue imaging using NIR-II window. They developed semiconducting polymeric nanoparticles (NPs) by encapsulating a specially designed thiadiazoloquinoxaline-based polymer with F127 as a carrier, resulting in improved water solubility and biocompatibility. These newly designed NPs encapsulated in the surfactant-stripped micelles demonstrated a broad absorption range of 1000–1400 nm (Fig. 6.6a) and showed tumor PA signals peaking at 24 hours after injection in vivo. Their study confirmed deep tissue PAI capabilities of these surfactant-stripped semiconducting micelles by detecting PA signals at depths up to 5 cm beneath chicken tissue in experiments involving a rat intestine in vivo (Fig. 6.6b). Furthermore, rat intestines were observed at high resolution for monitoring the digestive process of these agents within the intestine over time (Fig. 6.6c and Table 6.1).

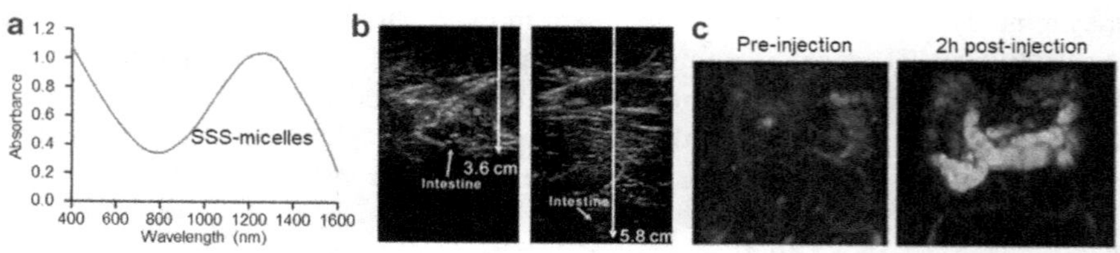

Fig. 6.6 Semiconducting polymer nanoparticle-based photoacoustic imaging. (**a**) Absorption spectrum of semiconducting polymer in SSS-micelles aqueous solution. (**b**) PA imaging of intestine in rats with chicken breast tissues stacked on rats after injecting SSS-micelles at different depths. (**c**) Depth-encoded PA images of rat intestine after gavage of concentrated SSS-micelles (56 mg mL − 1, 20 μL) at different time points. Reproduced with permission from [58]

Table 6.1 Summary of representative studies on inorganic and organic PA contrast agents

Ref	Type		PA agent	Peak absorption	Application
Lu [69]	Inorganic contrast agent	Gold nanomaterials	PEG-HAuNS	800 nm	Brain vasculature imaging
Chen [70]			Miniaturized AuNR	1050 nm	Tumor imaging
Kim [65]			PEGylated AuNC	790 nm	Tumor imaging
Zerda [71]		Carbon nanomaterials	Carbon nanotubes	690 nm	Tumor imaging
Kim [72]			Carbon nanotubes	525 nm	Lymphatic system imaging
Lee [73]			Carbon nanodots	–	Lymphatic system imaging
Lovell [66]	Organic contrast agent	Small organic molecule	Porphysome nanovesicles	680 nm	Lymphatic system imaging
Wang [74]			Porphyrin-based nanoparticles	425 nm	Tumor imaging
Wu [75]			Porphyrin-based nanoparticles	745 nm	Tumor imaging
Yang [77]			Iridium(III)–cyanine complex nanoparticles	820 nm	Tumor imaging
Chitgupi [79]			Surfactant-stripped micelles	1064 nm	Deep tissue imaging
Yin [81]		Semiconducting olymer nanoparticle	Amphiphilic semiconducting oligomer	585 nm	Tumor imaging
Ding [58]			Surfactant-stripped semiconductor polymeric micelles	1255 nm	Deep tissue imaging

HAuNS hollow gold nanospheres; AuNR gold nanorod; AuNC gold nanocage; PA photoacoustic

6.3 Reversibly Switchable PA Contrast Agent

In this section, we feature recently developed reversibly switchable PA contrast agents categorized by their key constituents as (1) bacteriophytochromes (BPh), (2) reversibly photoswitchable fluorescent proteins (RSFP), and (3) photochromic/thermochromic molecules.

6.3.1 Bacteriophytochrome

BPhs are light-sensitive proteins having bacterial or botanical origination. Under a red- or NIR- wavelength radiation, a methine bridge between tetrapyrrole C and D rings reversibly switches back and forth in cis/trans isomeric state, which is commonly noted as on state (Pfr) and off state (Pr) [63]. BPh is the most widely exploited category of switchable PA contrast agents due to its numerous favorable advantages. The absorption peak and switching wavelength of BPh fall in a red-/NIR range, which are favored spectra for deep tissue PAI. BPh is also a non-fluorescent chromophore, resulting in utmost PA yield due to a complementary relationship in energy transfer between PA and fluorescence. Furthermore, BPhs are passive chromophores that gain PA contrast only when binding to biliverdin, an endogenous chromophore abundantly found in mammalian cells and organs, and do not require an exogenous administration of chromophores. Instead, it is possible to functionalize live cells to genetically express BPh through transgene engineering, opening a new capability to unravel complex biological processes such as cell growth, carcinogenesis, and protein-protein interactions (PPI).

The concept of exploiting BPh as a reversibly switchable PA contrast agent was originally devised by Yao et al. [82]. BphP1 possesses two distinct optical absorption spectra, turning on with 630-nm illumination and turning off with 780-nm illumination (Fig. 6.7a). Under a relay of wavelength switching, the on-state extinction coefficient at 780 nm was ~ fourfold higher than that of off-state (Fig. 6.7b). Considering the photoswitching cycle of BphP1, a PAI time sequence was designed: 3.2 s of on-state acquisition under both 630 and 780 nm illumination, followed by 3.2 s of off-state acquisition under 780 nm only (Fig. 6.7c). Acquired PA images were subtracted to exclusively remain the differential contrast from BPh. To use as a live contrast agent for an in vivo experiment, U87 cells were genetically modified to express BphP1 and cultured. These cells were inoculated onto a rat's left kidney. After 1 week, a clear margin of a clustered tumor was observed from the differential PACT image (Fig. 6.7d). From this image, signals from blood-enriched internal organs were muted and spared only the BphP1 contrast, resulting in approximately 50-fold contrast-to-noise (CNR) improvement. Furthermore, tumor metastasis in a mouse liver was longitudinally monitored for 30 days after inoculation (Fig. 6.7e). After proliferating at the primary site on the right lobe of the liver, subsequent time points revealed spread of the tumor to other

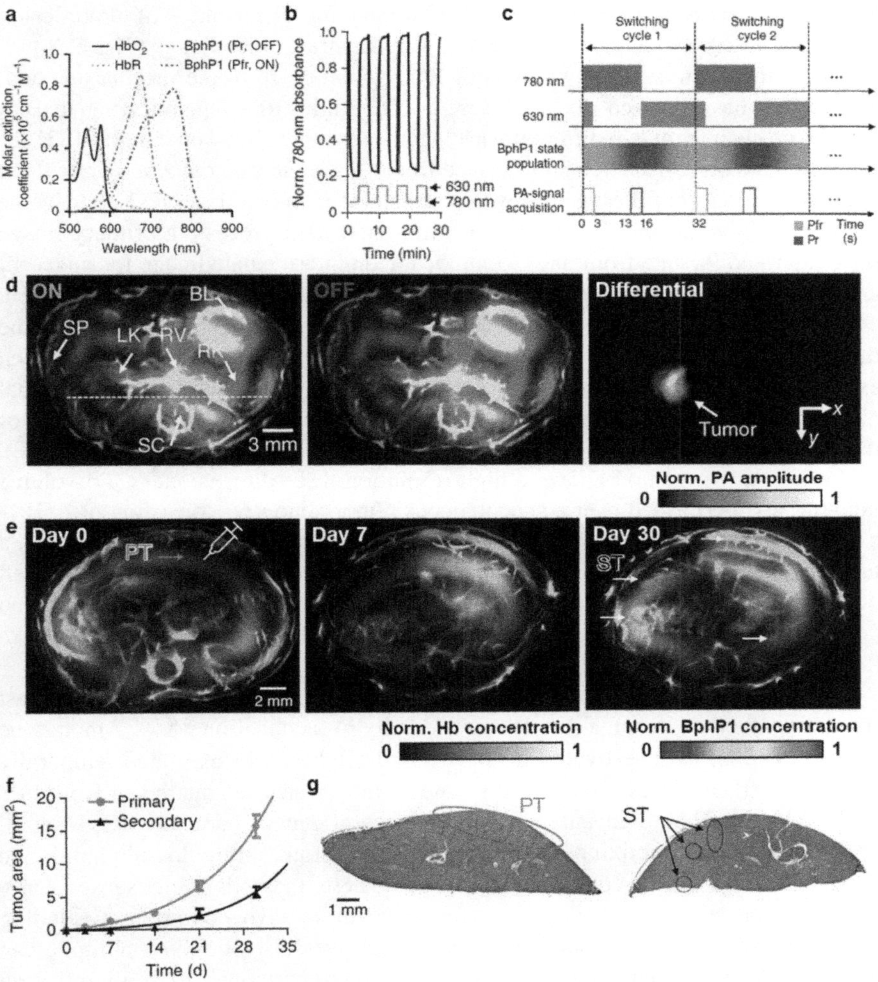

Fig. 6.7 Reversibly switchable PAI using bacteriophytochrome-expressing U87 cancer cells. (**a**) Optical spectrum of BphP1exhibiting photoconversion between red (Pr, OFF) and far-red (Pfr, ON). (**b**) Photoswitching stability between interleaved wavelengths. (**c**) Timing sequences of dual-wavelength PAI using BPhP1. (**d**) Background-suppressed differential imaging procedure from in vivo whole-body PACT images of a mouse injected with BphP1-expressing U87 cells into the left kidney. (**e**) Longitudinal PACT monitoring of BPhP1-expressing U87 cancer metastasis in a mouse liver. (**f**) Metastatic cancer proliferation identified from exponential increase in primary and secondary tumor area by time. (**g**) H&E-stained histological sections of mouse liver lobes with a primary tumor (PT) and secondary tumors (ST). (These images are reproduced with permission from [82])

liver lobes. The tumor area represented exponential growth trends of both primary and secondary tumors, with the latter exhibiting a delayed expansion (Fig. 6.7f).

Metastasis was histologically confirmed after imaging. The location of tumor lesion well agreed with the relative location of BphP1 contrast in PA image (Fig. 6.7h).

Extending from a single-wavelength PA acquisition from the original method, Märk et al. have devised dual-wavelength differential imaging, referring to BPh absorption changes at two different wavelengths for more abundant contrast [83]. To switch on and off an AGP1, a BPh derived from *Agrobacterium tumefaciens*, 670 nm (signal) and 755 nm (idler) beams were emitted from an identical OPO laser. Three PA images were acquired in a row for differential PAI process: a primary image for a saturated Pr state from the 755 nm beam, an intermediate image for mixed Pr and Pfr population from both 670 nm and 755 nm beams, and the last image for a saturated Pfr state from the 670 nm beam (Fig. 6.8a). In the intermediate state, the fluence ratio of the signal beam versus idler beam was set at 1:2, which has been experimentally confirmed to be an optimal ratio that yields the largest PA signal difference (Fig. 6.8b). Subtraction of intermediate image from the summation of the other two PA images yields a dual-wavelength differential image (Fig. 6.8c). Following to the method above, a high-resolution three-dimensional (3D) volume imaging was performed over a subcutaneous tumor xenograft consisting of AGP1-expressing HT29 cells. As a result, AGP1 contrast clearly margins the boundary of a tumor mass, while intertwining angiogenetic microvessels around the tumor are well described from background signal (Fig. 6.8d). Throughout 28-day post-inoculation period, different images clearly exhibited dynamic expansion of the tumor in both lateral and depth directions (Fig. 6.8e).

Instead of using multiple wavelengths, Li et al. have contrived a multi-contrast separating method from a single wavelength PA acquisition [84]. Among the two types of BPhs investigated in the study, DrBphP-PCM exhibited temporally slower PA signal decay from the Pfr state to the Pr state compared to RpBphP1 (Figs. 6.9a, b). Taking advantage of such temporal decay characteristics, Li et al. designed temporal unmixing parameters as demonstrated in the lock-in amplitude (Fig. 6.9c) and decay constant (Fig. 6.9d). These methods were subsequently validated for their differentiating performance in in vivo conditions, including an incubated tumor surrounded by brain cortical vessels. Lock-in amplitude can measure PA signal intensity temporally modulated to harmonics of an illumination frequency. As a result, it exhibited approximately 2.4 times higher CNR than conventional differential imaging, demonstrating a superior background suppression effect (Fig. 6.9e). The two BPhs were also distinguishable based on decay constant, with RpBphP1 exhibiting ~threefold higher value than DrBphP-PCM (Fig. 6.9f). Employing aforementioned contrast parameters, they also pioneered the first observation of protein-protein interactions using a split phytochrome complex. In more detail, they genetically separated DrBphP-PCM into DrPAS and PrGAF-PHY segments (referred to as 'DrSplit') and bound each segment with FRB protein and FKBP protein, respectively (Fig. 6.9g). In the presence of rapamycin, these two segments were reconstituted back into functional DrBphP-PCM, regaining its photoswitchable PA contrast. A step further from validation using tumor cells, they successfully identified the hydrodynamic transfection of DrSplit-encoded plasmid

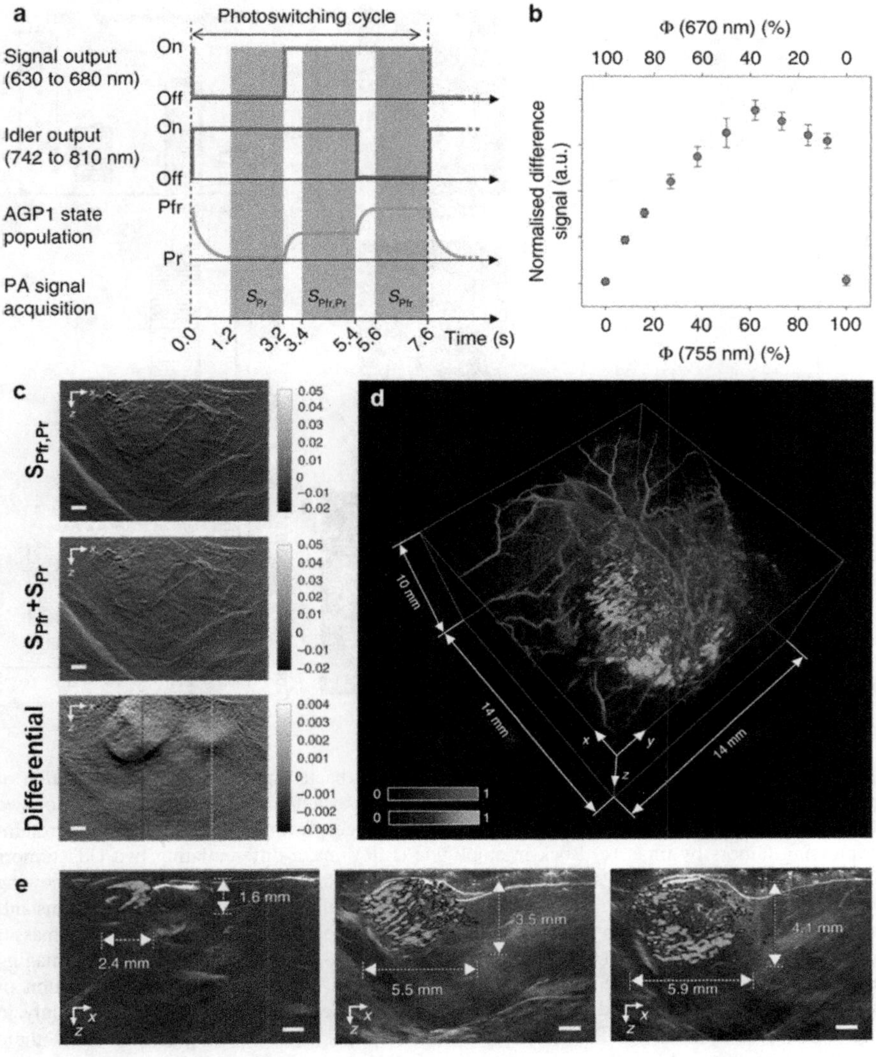

Fig. 6.8 Dual-wavelength 3D photoacoustic imaging of a photoswitchable phytochrome reporter protein using a Fabry-Perot scanner. (**a**) Timing sequences of dual-wavelength excitation and subsequent PA signal acquisition. (**b**) Normalized PA difference signal amplitudes to the following 650 nm/755 nm laser relative fluences. (**c**) Cross-sectional PA images of simultaneous irradiation ($S_{Pfr,Pr}$), the sum of individual irradiation ($S_{Pfr} + S_{Pr}$), and the subsequent background-suppressed differential image. (**d**) 3D volume rendering of differential AGP1-expressing HT29 cells (green) and non-switching hemoglobin map describing neighboring microvascular bed (red). (**e**) 28-day longitudinal X-Z MIPs of differential images attributing to AGP1-expressing tumor overlaid to anatomical images. (These images are reproduced with permission from [83])

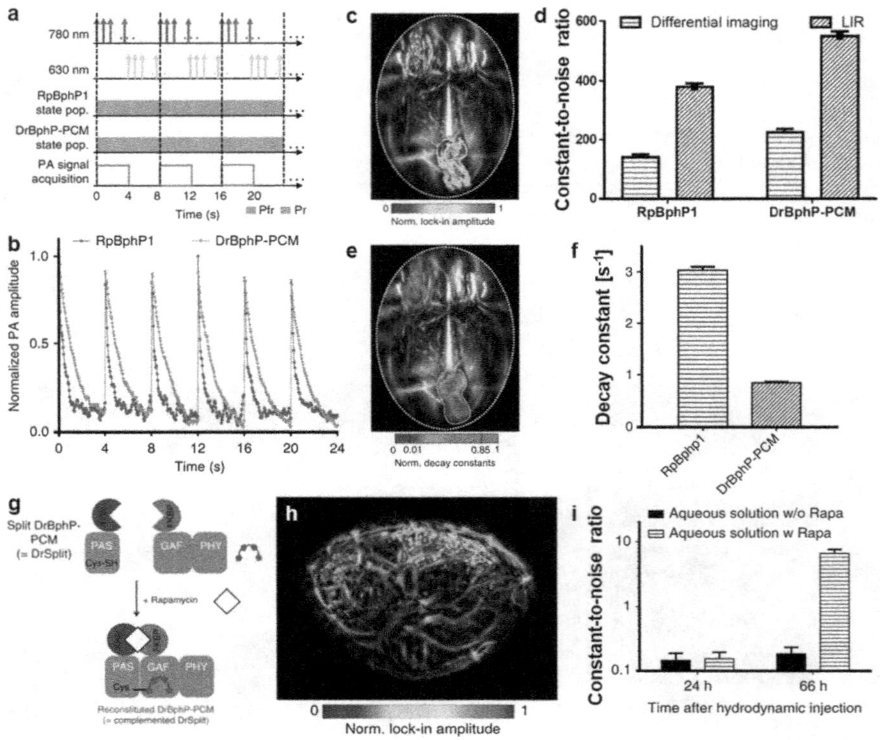

Fig. 6.9 Multi-contrast in vivo imaging of near-infrared bacteriophytochrome and detection of protein-protein interactions. (**a**) Timing sequences of photoswitching and PA acquisition of two BphPs (RpBphP1, DrBphP-PCM). (**b**) PA signal changes from two distinct chromoprotein-expressing tumors by time. (**c**) Lock-in amplitude (LIR) image differentiating two U87 tumors expressing two distinct BphPs. (**d**) Comparative constant-to-noise ratio enhancement showing from LIR images compared to the conventional differential imaging method. (**e**) Decay constant-encoded image separating two tumor masses. (**f**) Quantitative differentiation of two tumor masses from (**e**) by decay constant. (**g**) Concept illustration of protein-protein interaction (PPI) PA imaging using rapamycin-induced DrSplit recombining mechanism. (**h**) Hydrodynamic transfection of DrSplit to the mouse liver validated from LIR image of the mouse liver at 42 hours after rapamycin injection. (**i**) Comparative CNR quantification of LIR images in the liver region with and without rapamycin injection. (These images are reproduced with permission from [84])

into mouse liver cells (Fig. 6.9h). Quantitatively, the plasmid-injected mouse showed ~43 times CNR enhancement after rapamycin injection (Fig. 6.9i).

To differentiate three types of reversibly switchable BPh reporters (rsOAP) present simultaneously, Mishra et al. have employed a machine learning algorithm [85]. Each of ReBphP-PCM, RpBphP1-PCM, and DrBphP-PCM exhibited distinctive PA spectra (Fig. 6.10a) and PA temporal decay responses (Fig. 6.10b). From acquired PA images, multiple parameters were used for training the algorithm, including exponential fit (exponential coefficient, coefficient of determination),

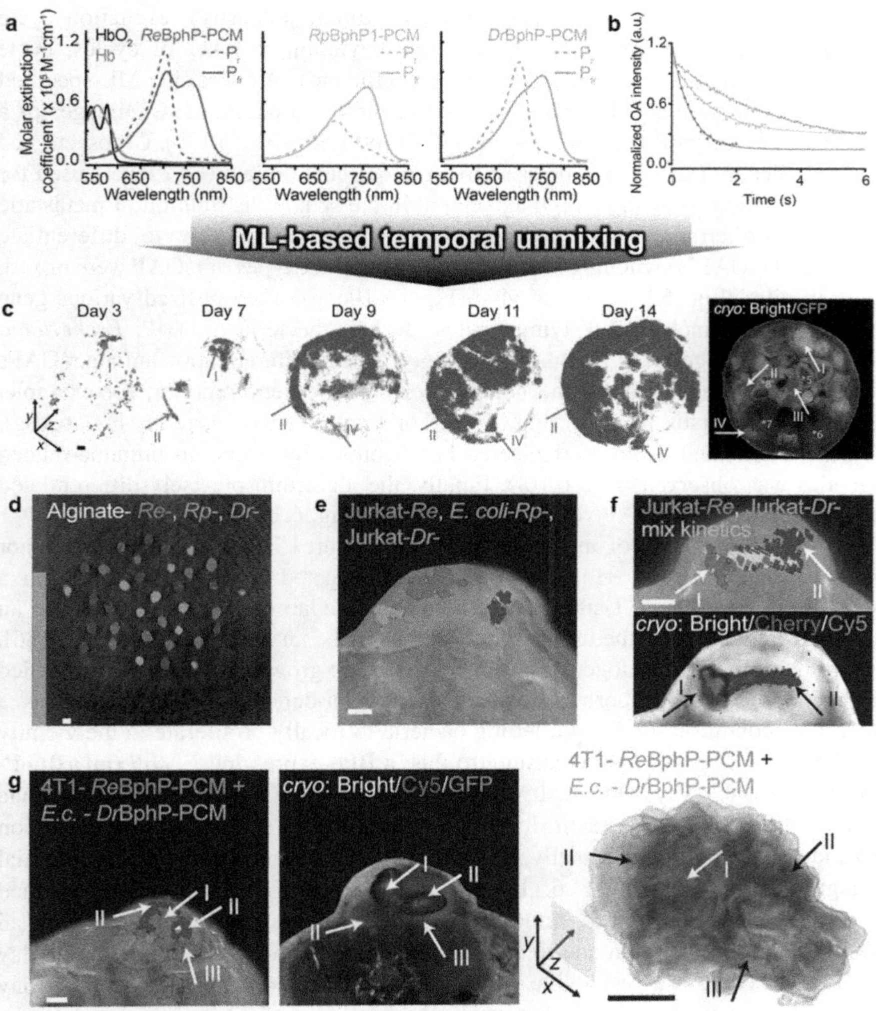

Fig. 6.10 Machine learning (ML)-based multiplexed differentiation of reversibly switchable chromoproteins. (**a**) Optical absorption spectra of three distinct reversibly switchable photoacoustic proteins (rsOAPs; ReBphP-PCM, RpBphP1-PCM, DrBphP-PCM). (**b**) PA decay curves of the three bacteriophytochromes. (**c**) Metastatic spreading of ReBphP-PCM-expressing HCT116 tumor cells in mouse liver labeled from ML-based analysis. High correlation between the rsOAP label and a coregistered fluorescence label from the cyroimage section was confirmed. Parallel visualization of (**d**) alginate bead phantom containing rsOAP-expressing *E. coli*, and (**e**) rsOAP-expressing Jurkat T cells and rsOAP-expressing *E. coli* injected into a 4T1 tumor xenograft. (**f**) Two engineered Jurkat T cells in a 4T1 tumor show mixing kinetics. (**g**) Spatial separation between rsOAP-expressing 4T1 tumor cells and *E. coli*. (These images are reproduced with permission from [85])

measured PA intensity (minimum, mean, maximum intensity), excitation cycle (minimum and maximum median, standard deviation, number of cycles, cycle length), and frequency response (Fourier coefficient). As a result, ML-specified labels effectively marked metastases observed in longitudinal MSOT images of a transplanted ReBphP-PCM-expressing HCT116 tumor (Fig. 6.10c). Cryosectioned fluorescence images of co-expressed green fluorescent protein (GFP) were used for cross-validation, showing a high agreement in the spatial distribution of metastatic lesions. Furthermore, they investigated the model's capability to differentiate individual rsOAP's switching kinetics from where three types of rsOAP were mixed, both in vitro (Fig. 6.10d) and in vivo (Figs. 6.10e–g). They utilized various gene reporter hosts, including T lymphocytes (Jurkat), bacteria (*E. coli*; *Escherichia coli*), and tumor cells (4T1). Correspondingly, a clear differentiation among rsOAPs was observed regardless of the heterogeneity of the gene reporter, for example, lymphocytes versus bacteria (Fig. 6.10e) or tumor versus bacteria (Fig. 6.10g). On the other hand, from co-registered homogenous reporters, an immune-related kinetics was observed (Fig. 6.10f). Finally, the algorithm precisely differentiated two heterogeneous rsOAP volumes in proximity (Fig. 6.10g).

While the majority of introduced research features BPh expression via tumor inoculation or passive delivery so far, tumor-targeted delivery has not been a widely excavated area. Using *E. coli* as a carrier, Gao et al. have figured out an active BPh delivery to the tumor, exploiting its niche anaerobic characteristic [86]. Along with rapid carcinogenesis, a hypermetabolic growth of tumor accompanied ischemia in the neighboring tissue. This oxygen-deficient environment was a favorable condition for *E. coli*, letting bacteria to focally proliferate in the vicinity of the tumor (Fig. 6.11a). According to this, a BPh-expressing *E. coli* (mDrBphP-PCMm/F469W) was cultured. It was then infused into a tumor-bearing mouse via tail vein injection. As a result, local colonization of *E. Coli* around tumor region was identified and longitudinally monitored with 7-day longitudinal PA differential imaging of the mouse (Fig. 6.11b). An increment in *E. coli* PA signals implied a local colonization around tumor. This time, the mouse was followed with an identical *E. coli* injection and treated with antibiotic shots of streptomycin on day 4 (Fig. 6.11c). As a result, the *E. coli* PA signal decreased by time from day 6, implying a post-antibiotic bacterial inhibition (Fig. 6.11d). Compared to the nontreated group, treated group exhibited a ~6.7-fold PA signal decrease from day 7 (Fig. 6.11e).

Thanks to great evolution in transgene technology, it is possible to genetically grant a mammalian animal to express photochromic proteins. Kasatkina et al. have bred an NIR transgenic mouse model (loxP-BphP1) that is genetically edited to express both NIR optogenetic manipulation and reversibly switchable PAI properties [87]. Specifically, BphP1-mCherry-TetR gene chain was inserted onto the loxP site of loxP-BphP1 knock-in mouse gene using Cre-recombinase. The crossing between a male loxP-BphP1 mouse with a female Vasa-Cre mouse led to a common expression of the BphP1-mCherry-TetR in their offspring (Fig. 6.12a). The lower abdomen of a pregnant Vasa-Cre mouse was noninvasively imaged with hybrid PA and US imaging. As a result, all seven embryos (BphP1-Cre$_{vasa}$) inside

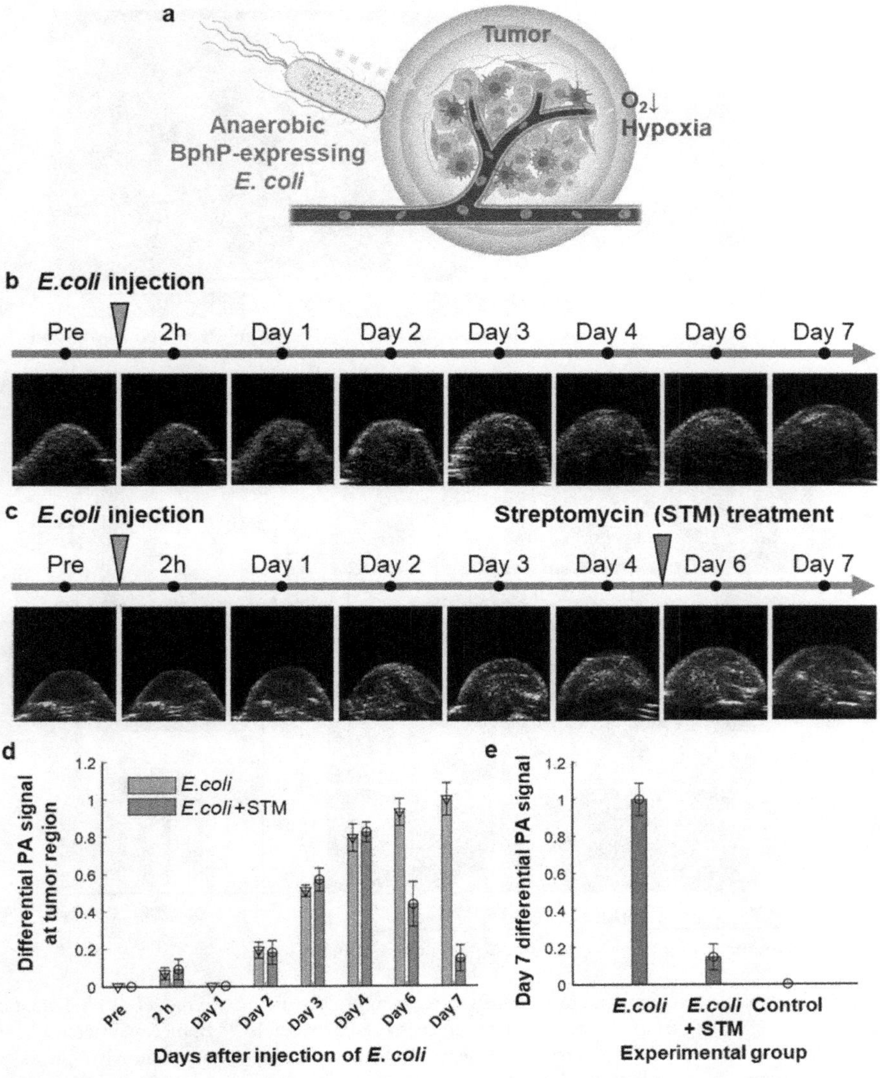

Fig. 6.11 Tumor-targeted switchable photoacoustic imaging using anaerobic bacterial carriers. (**a**) Schematic illustration of chromoprotein-expressing *E. coli* (mDrBphP-PCMm/F469W) actively delivered to a hypoxic tumor environment. (**b**) Longitudinal differential images of implanted tumor after intravenous injection of bacterial carrier. (**c**) Longitudinal differential images of implanted tumor repeated as (**b**) included with antibiotic treatment injection after day 4. (**d**) Relative differential PA signal change describing local colonization of *E. coli* in tumor regions and comparative *E. coli* apoptosis following antibiotic injection. (**e**) Comparison of differential PA signal at day 7 between study groups and a control group. (These images are reproduced with permission from [86])

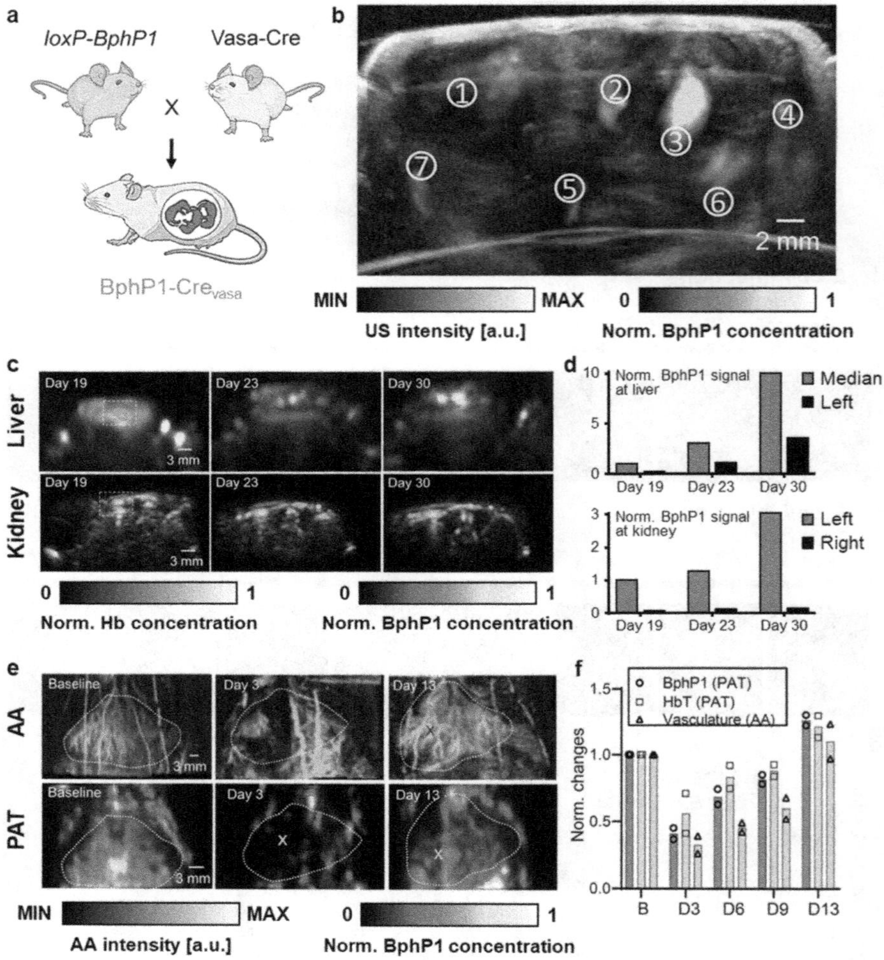

Fig. 6.12 Photoacoustic imaging of a BphP1-expressing transgenic mouse model. (**a**) Schematic illustration describing BphP1-Cre$_{vasa}$ generation crossbred between loxP-BphP1 male mouse and Vasa-Cre female mouse. (**b**) Overlaid image of BphP1-expressing embryos in vivo inside a maternal uterus over ultrasound (US) B-mode images. (**c**) Longitudinal BphP1 signal images of a loxP-BphP1 mouse after topical Cre-expressing viral infection in the liver and kidney. (**d**) Comparative quantification of mean BphP1 signals at injected region and control region in liver and kidneys. (**e**) Parallel acoustic angiography (AA) and PAT images of liver regeneration of BphP1-Cre$_{vasa}$ mouse after partial hepatectomy. (**f**) Multiparametric analysis of BphP1 signal, total hemoglobin (HbT) from PAT, and vascular volume from AA by time after hepatectomy. (These images are reproduced with permission from [87])

a maternal uterus expressed strong BphP1 signals (Fig. 6.12b). Next, a loxP-BphP1 mouse was examined to determine whether Cre-dependent BphP1 contrast was expressed by imaging after topical injection of Cre-expressing adeno-associated virus (AAV) into its liver median lobe and left kidney. According to viral infection over time, BphP1 contrast became broadened from its original injection site (Fig. 6.12c). At 30 days after injection, BphP1 signals in the liver and kidney were increased tenfold and threefold, respectively (Fig. 6.12d). Furthermore, there was a clear difference in BphP1 expression between injected sites (liver median lobe, left kidney) and control sites (liver left lobe, right kidney). Finally, to validate whether BphP1 expression was genetically inherited by a descendant BphP1-Cre_{vasa} mouse, a liver cell regeneration of the mouse was monitored after partial hepatectomy (Fig. 6.12e). Quantification of BphP1, total hemoglobin (HbT), and vascular volume validated a consistent liver regeneration by time (Fig. 6.12f). Although major losses of vascularity and BphP1 signal were observed after resection, restoration of both at day 13 demonstrated that BphP1 contrast was maintained.

6.3.2 Reversibly Switchable Fluorescent Proteins

Reversibly switchable fluorescent proteins (RSFP) are actively exploited following the advent of super-resolution fluorescence microscopy techniques such as stimulated emission depletion (STED) and reversible saturable/switchable optical linear fluorescence transitions (RESOLFT). Photoswitchability of RSFPs is achieved through cis/trans isomerization of the methine bridge linking hydroxybenzyl and imidazole rings anchored to the central β-barrel of the protein. Although fluorophores might reduce PA contrast compared to nonfluorescent chromophores, RSFPs have a huge influence in the field of confocal microscopy. They have become a compelling reason for implementation in PA imaging.

The first implementation of RSFPs into the PAI field was done by Stiel et al. through temporal unmixing of Dronpa and Dronpa-M159T contrast from background absorbers [88]. Their green fluorescence is switched off at 488 nm and switched back on at 420 nm excitation. Dronpa and Dronpa-M159T demonstrated different temporal PA responses, with the latter having faster signal decay and approximately fourfold higher change in PA amplitude upon switching than the former. Focusing on this difference, two proteins could be temporally differentiated. However, PA spectra of RSFPs did not match well with their optical absorption spectra. Technically, PAI is different from FLI as it involves pulsed illumination and relies on nonradiative energy transfer for signal generation. As a consequence, RSFPs require another optimized PA characterization other than FLI [89]. To meet such demand, Vetschcera et al. have followed up on ten commercially available RSFPs of their PA characteristics, employing an identical temporal unmixing imaging sequence [90].

Despite that, adopting RSFPs is still advantageous because they are genetically encodable. The well-established probing mechanism in fluorescent imaging can also be adopted in PAI. Specifically, it allows for probing dynamics of target chemical compositions or cellular microenvironments such as metabolite concentration, voltage, and pH. For example, Ca^{2+} sensing with a genetically encoded calcium indicator (GCaMP) is the nominal technique widely applied in neuronal activity or signaling pathway activation. To overcome high background absorption from adjacent tissues, Mishra et al. have used reversibly switchable GCaMP (rsGCaMP) triggered by 405/488 nm light irradiation [91]. In the presence of Ca^{2+}, the distal end of M13 and calmodulin can bind together and close the GCaMP chain loop, activating photoswitchability (Fig. 6.13a). The PA decay time was also relevant to Ca^{2+} concentration, in addition to illuminated laser intensity and wavelength (Fig. 6.13b). Such finding reinvents temporal unmixing of rsGCaMP as a molecular probe (Fig. 6.13c). To prove the applicability of rsGCaMP in vivo, PACT imaging was performed over the back of a mouse implanted with rsGCaMP1.1-expressing HeLa tumor cells (Fig. 6.13d). Despite less tissue penetration of blue light, photoswitching PA contrast specified cells. Furthermore, differences in photoswitching kinetics were identified for Ca^{2+}-varied implants treated with and without ionomycin (Fig. 6.13e).

6.3.3 Photo- and Thermo-Chromic Molecules

In this section, we will introduce chromic molecules behind chromoproteins deployed in PA imaging. Diarylethenes are commonly explored photochromic molecules discolored in ultraviolet (UV) irradiation. While UV light cannot penetrate deep tissue, Liu et al. have used upconverting nanoparticles (NaYF4:Yb/Tm@NaYF4) capable of converting irradiated NIR light into UV light [92]. This enabled triggering of photochromic molecule discoloration with NIR illumination. Consequently, the nanocomplex is reversibly switched on and off under NIR illumination of 980 nm and 640 nm, respectively. CNR of a differential PA image varied by HeLa cell population density, resulting in 20.9 for 10^6 cells to 1.5 for 2×10^4 cells. To overcome the shortcoming of the contrast from fewer cell numbers, they averaged sequential differential images from several cycles, attaining a fivefold CNR improvement from 2×10^4 cell population with five cycles of summation (Fig. 6.14d). After applying HeLa cell xenograft onto mouse skin, the tumor was clearly highlighted from averaged differential image with a CNR of 7.5.

A reversibly switching chromophore can undergo discoloration in response to various physical or chemical stimuli, not limited to light. For instance, thermochromic materials exhibit temperature-dependent color change due to alterations in crystal structure or stereoisomeric morphological conversion. Ma et al. have repurposed two types of thermochrome microparticles, named red ReST and black ReST referring to their original color, as reversibly switchable PA contrast agents (Fig. 6.14a) [93]. Water solubility and biocompatibility were improved through microencapsulation with hydrophilic polysaccharide alginate. ReST became colorless when heated above 37 °C and regained its original red color when cooled down below 25 °C (Fig. 6.14b). In variance of temperature, absorption coefficient was reduced by 57% for red ReST and 53% for black ReST at its absorption peak of 525 nm and 460/590 nm, respectively. Next, a 'D'-shaped red ReST phantom

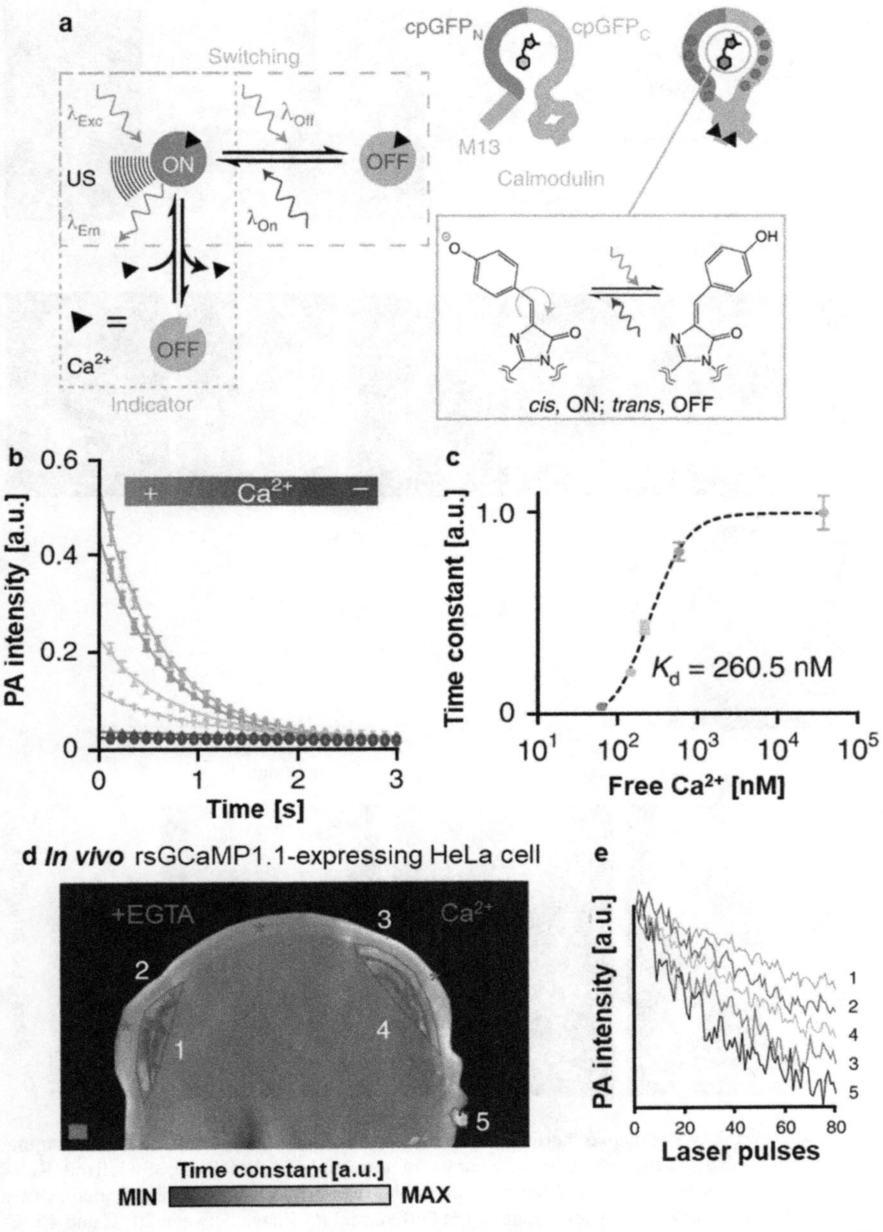

Fig. 6.13 Characterizing reversibly switchable genetically encoded calcium indicator (rsGCaMP) into Ca^{2+}-responsive PA probe. (**a**) Concept schematic of rsGCaMP selectively generating PA signal in the presence of Ca^{2+} ion. (**b**) Changes in temporal PA intensity decay from on state to off state depending on Ca^{2+} concentration. (**c**) Quantified mean time constant of exponential fits was obtained from (**b**). (**d**) Cross-sectional PACT images of three subcutaneous rsGCaMP1.1-expressing HeLa tumors in vivo untreated (left), 5 mM EGTA-treated (middle), and 5 μM Ionomcin and 10 mM Ca^{2+}-treated. (**e**) Photoswitching kinetics of numbered regions from (**d**). (These images are reproduced with permission from [91])

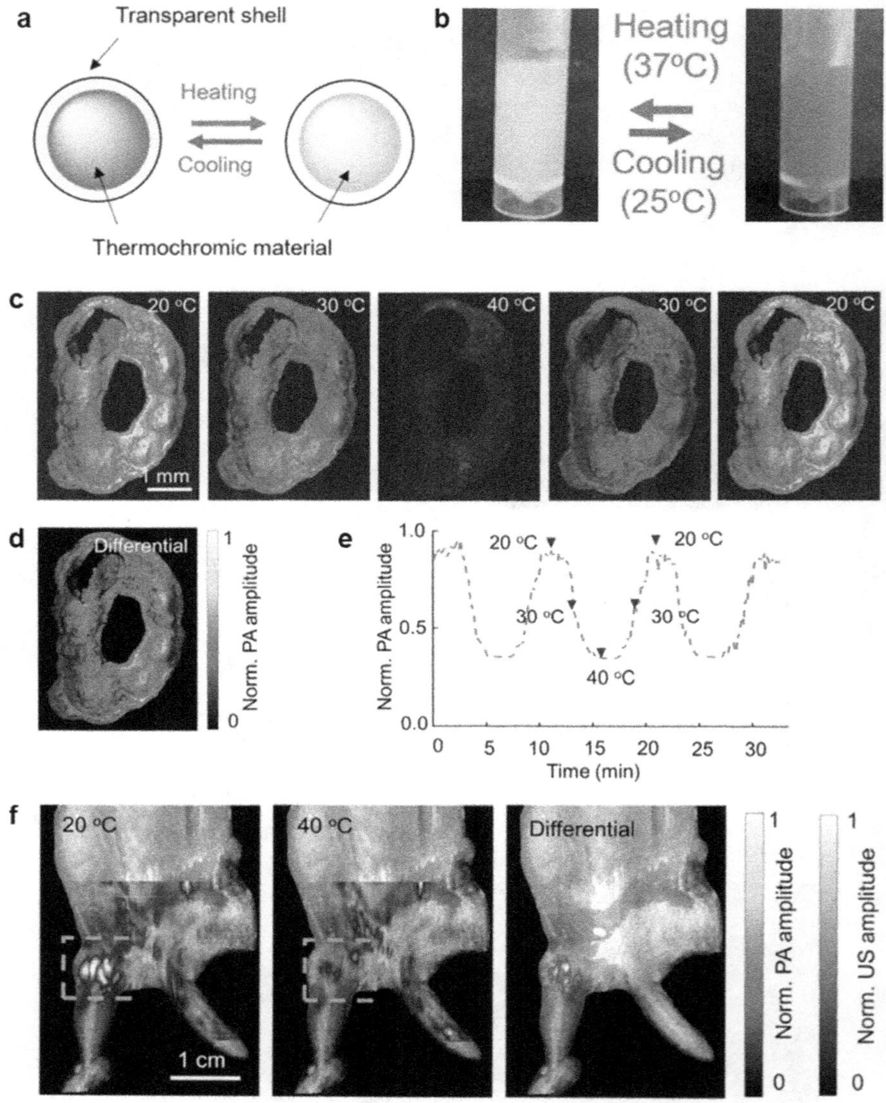

Fig. 6.14 Reversibly switchable thermochromics (ReST) for multiscale differential PA imaging. (**a**) Reversible thermal-responsive color change of encapsulated ReST. (**b**) Stabilized ReST aqueous suspension under repeated temperature cycle. (**c**) Serial changes in PAM images of a D-shaped red ReST phantom by temperature. (**d**) Differential PA image between 20 °C and 40 °C. (**e**) Temperature-dependent PA kinetics during temperature changes of red ReST. (**f**) In vivo PACT images of red ReST-mixed Matrigel implant in a mouse leg. (These images are reproduced with permission from [93])

was continuously imaged with high-speed PAM under three thermoswitching cycles between 20 °C and 40 °C (Fig. 6.14c). Similar to differential images from chromoproteins, the differential image between 20 °C and 40 °C showed decent background suppression (Fig. 6.14d). The dynamic change in PA amplitude to a PA switching ratio between two temperatures was ~3.3 (Fig. 6.14e). The ReST reversibility was also validated under in vivo conditions by imaging a subcutaneously implanted tumor-mimicking ReST Matrigel on a mouse (Fig. 6.14f). Temperature change was induced by controlling water temperature. Relevant PA amplitude changes were monitored from PACT images while anatomical structures were acquired from simultaneous US images. As a result, PA signal amplitude from a ReST implant decreased by ~25% between two temperatures, exclusively remaining differential contrast over the implant region (Table 6.2).

6.4 Discussion and Outlook

As a hybrid optical and acoustical imaging modality, PAI can image photochemical contrast from both endogenous and exogenous chromophores over depth limitation of conventional optical imaging techniques. Various organic and inorganic nanoparticles with superior photothermal properties have been extensively exploited as exogenous PA contrast agents. To overcome the imaging depth constraint due to background tissue signals, novel approaches utilizing reversibly switchable PA contrast agents allow exclusive control of the exogenous contrast by turning the optical contrast on and off. We further subcategorized studies under our scope based on their driving principle of contrast switchability.

We highlighted several key features observed in recent reversibly switchable contrast agents. Firstly, most studies were focused on genetic expression of switchable PA properties using biologically derived chromoproteins such as BPhs and RSFPs. This means that exogenous contrast can be introduced into an organism without needing direct injection, which is a significant departure from conventional contrast agents. Based on our review, we found that transduction of genetic expression for PA contrast covered a wide range, from cellular level including bacteria and mammalian tumor cells to complete organisms such as rodents.

Second, we observed successful biomedical applications of industry-level photochromic and thermochromic compounds, which have been reformulated to possess favorable properties for in vivo PAI. An example of this is the combination of UV-triggered photochrome with a UV-emitting upconversion nanoparticle, alternating to a switching via NIR wavelength suitable for deep tissue imaging. In addition, isolating compounds with biocompatible amphiphilic layers significantly enhanced their water solubility, a common challenge faced by NIR organic dyes.

In conclusion, we discovered a number of studies addressing reversibly switchable PA contrast agents, enhancing exogenous contrast to an undiscovered level by suppressing PA signals from endogenous chromophores enriched in biological tissues. We anticipate that these PAI approaches can facilitate new biomedical

Table 6.2 Summary of representative studies for reversibly switchable PA contrast agents

Ref	Switching scheme	Chromophore	Switching		PA Excitation	Optical Extinction ratio	PA Switching ratio	Differentiating method	Contrast-to-noise ratio
			On to off	Off to on					
Yao [82]	Bacterio-phytochrome	RpBphP1	780 nm	630 nm	780 nm	4.1	4.2	Differential	20
Märk [83]		AGP1	755 nm	670 nm	670/755 nm	2.25	3.1	Dual-wavelength differential	75
Li [84]		DrBphP-PCM, RpBphP1	780 nm	630 nm	780 um	9.9	8.6	LIR/decay constant	80
Mishra [85]		ReBphP-PCM	780 nm	630 nm	680 /770 nm	47	4.7	DL-based temporal unmixing	N/A
Gao [86]		mDrBphP-PCM	780 nm	630 nm	780 nm	16	17.2	Differential	12.7
Kasatkina [87]		RpBphP1-QPAS1 OT	790 nm	635 nm	750 nm	8	5	Differential	10 (in vitro)
Stiel [88]	RSFP	Dronpa, Dronpa-M159T	488 nm	420 nm	488 nm	N/A	10	Temporal unmixing	30 (in vitro)

Vetschera [90]		9 RSFPs (Dronpa, rsFastLime, Dronpa-M159T /Dronpa-2, Padron, rsEGFP2, rsCherry, rsCherryRev, rsCherryRev1.4, rsTagRFP)	420–570 nm	420–550 nm	N/A	N/A	0.28–0.91 (switching dynamic range)	Decay constant	N/A
Mishra [91]		rsGCaMP1.1, rsGCaMP1.4-ER	488 nm	405 nm	420/488 nm	9.5	40	LIR/decay constant	N/A
Liu [92]	Photochrome & upconversion	3ThacacH	680 nm	980 nm	680 nm	N/A	10	Differential	7.5
Ma [93]	Thermochrome	ReST	40 °C	20 °C	532 nm	5	4	Differential	40 (in vitro)

discoveries by probing the microenvironment in vivo and unraveling mechanisms of interactions at the cellular level.

Acknowledgments This work was supported by grants (2021M3C1C3097624, 2020R1A6A1A03 047902, 2023R1A2C3004880) of the National Research Foundation (NRF), grants (1711195277, RS-2020-KD000008) of Korea Medical Device Development Fund, a grant (2019-0-01906, Artificial Intelligence Graduate School Program (POSTECH)) of the Institute of Information & communications Technology Planning & Evaluation (IITP), and BK21 FOUR projects (POSTECH) funded by the Korean government (the Ministry of Science and ICT, the Ministry of Education, the Ministry of Trade, Industry and Energy, the Ministry of Health and Welfare, and the Ministry of Food and Drug Safety).

References

1. L. V. Wang and S. Hu, "Photoacoustic tomography: in vivo imaging from organelles to organs," *Science,* vol. 335, no. 6075, pp. 1458–1462, 2012.
2. S. Zackrisson, S. Van De Ven, and S. Gambhir, "Light in and sound out: emerging translational strategies for photoacoustic imaging," *Cancer research,* vol. 74, no. 4, pp. 979–1004, 2014.
3. L. V. Wang and H.-I. Wu, *Biomedical optics: principles and imaging*. John Wiley & Sons, 2012.
4. S. Manohar and D. Razansky, "Photoacoustics: a historical review," *Advances in optics and photonics,* vol. 8, no. 4, pp. 586–617, 2016.
5. L. V. Wang and J. Yao, "A practical guide to photoacoustic tomography in the life sciences," *Nature methods,* vol. 13, no. 8, pp. 627–638, 2016.
6. J. Kim, J. Y. Kim, S. Jeon, J. W. Baik, S. H. Cho, and C. Kim, "Super-resolution localization photoacoustic microscopy using intrinsic red blood cells as contrast absorbers," *Light: Science & Applications,* vol. 8, no. 1, p. 103, 2019.
7. J. Park *et al.*, "Quadruple ultrasound, photoacoustic, optical coherence, and fluorescence fusion imaging with a transparent ultrasound transducer," *Proceedings of the National Academy of Sciences,* vol. 118, no. 11, p. e1920879118, 2021.
8. W. Choi, E.-Y. Park, S. Jeon, and C. Kim, "Clinical photoacoustic imaging platforms," *Biomedical engineering letters,* vol. 8, pp. 139–155, 2018.
9. L. V. Wang, "Prospects of photoacoustic tomography," *Medical physics,* vol. 35, no. 12, pp. 5758–5767, 2008.
10. B. Park, S. Park, J. Kim, and C. Kim, "Listening to drug delivery and responses via photoacoustic imaging," *Advanced Drug Delivery Reviews,* vol. 184, p. 114235, 2022.
11. B. Park, D. Oh, J. Kim, and C. Kim, "Functional photoacoustic imaging: from nano-and micro-to macro-scale," *Nano Convergence,* vol. 10, no. 1, p. 29, 2023.
12. S. Jeon, J. Kim, D. Lee, J. W. Baik, and C. Kim, "Review on practical photoacoustic microscopy," *Photoacoustics,* vol. 15, p. 100141, 2019.
13. S. Hu and L. V. Wang, "Optical-resolution photoacoustic microscopy: auscultation of biological systems at the cellular level," *Biophysical journal,* vol. 105, no. 4, pp. 841–847, 2013.
14. J. Ahn, J. Y. Kim, W. Choi, and C. Kim, "High-resolution functional photoacoustic monitoring of vascular dynamics in human fingers," *Photoacoustics,* vol. 23, p. 100282, 2021/09/01/ 2021, doi: https://doi.org/10.1016/j.pacs.2021.100282.
15. J. Ahn *et al.*, "Fully integrated photoacoustic microscopy and photoplethysmography of human in vivo," *Photoacoustics,* vol. 27, p. 100374, 2022/09/01/ 2022, doi: https://doi.org/10.1016/j.pacs.2022.100374.
16. D. Kim, J. Ahn, E. Park, J. Y. Kim, and C. Kim, "In vivo quantitative photoacoustic monitoring of corticosteroid-induced vasoconstriction," *Journal of Biomedical Optics,* vol. 28, no. 8, pp. 082805–082805, 2023.

17. J. Ahn *et al.*, "In vivo photoacoustic monitoring of vasoconstriction induced by acute hyperglycemia," *Photoacoustics,* vol. 30, p. 100485, 2023/04/01/ 2023, doi: https://doi.org/10.1016/j.pacs.2023.100485.
18. W. Zhang *et al.*, "High-resolution, in vivo multimodal photoacoustic microscopy, optical coherence tomography, and fluorescence microscopy imaging of rabbit retinal neovascularization," *Light: Science & Applications,* vol. 7, no. 1, p. 103, 2018.
19. W. Liu and H. F. Zhang, "Photoacoustic imaging of the eye: a mini review," *Photoacoustics,* vol. 4, no. 3, pp. 112–123, 2016.
20. E. Park, Y.-J. Lee, C. Kim, and T. J. Eom, "Azimuth mapping of fibrous tissue in linear dichroism-sensitive photoacoustic microscopy," *Photoacoustics,* vol. 31, p. 100510, 2023/06/01/ 2023, doi: https://doi.org/10.1016/j.pacs.2023.100510.
21. J. W. Baik *et al.*, "Intraoperative Label-Free Photoacoustic Histopathology of Clinical Specimens," *Laser & Photonics Reviews,* vol. 15, no. 10, p. 2100124, 2021, doi: https://doi.org/10.1002/lpor.202100124.
22. J. Kim, J. Ahn, G. Kang, J. H. Hwang, and C. Kim, "High-resolution photoacoustic/ultrasound imaging of the porcine stomach wall: an ex vivo feasibility study," *Biomed. Opt. Express,* vol. 12, no. 11, pp. 6717–6729, 2021/11/01 2021, doi: https://doi.org/10.1364/BOE.441241.
23. S. Park, C. Lee, J. Kim, and C. Kim, "Acoustic resolution photoacoustic microscopy," *Biomedical Engineering Letters,* vol. 4, pp. 213–222, 2014.
24. J. Yao and L. V. Wang, "Photoacoustic microscopy," *Laser & photonics reviews,* vol. 7, no. 5, pp. 758–778, 2013.
25. J. W. Baik, J. Y. Kim, S. Cho, S. Choi, J. Kim, and C. Kim, "Super Wide-Field Photoacoustic Microscopy of Animals and Humans In Vivo," *IEEE Transactions on Medical Imaging,* vol. 39, no. 4, pp. 975–984, 2020, doi: https://doi.org/10.1109/TMI.2019.2938518.
26. W. Choi, D. Oh, and C. Kim, "Practical photoacoustic tomography: realistic limitations and technical solutions," *Journal of Applied Physics,* vol. 127, no. 23, 2020.
27. J. Yang, S. Choi, and C. Kim, "Practical review on photoacoustic computed tomography using curved ultrasound array transducer," *Biomedical Engineering Letters,* vol. 12, no. 1, pp. 19–35, 2022/02/01 2022, doi: https://doi.org/10.1007/s13534-021-00214-8.
28. S. Choi *et al.*, "Deep Learning Enhances Multiparametric Dynamic Volumetric Photoacoustic Computed Tomography In Vivo (DL-PACT)," *Advanced Science,* vol. 10, no. 1, p. 2202089, 2023, doi: https://doi.org/10.1002/advs.202202089.
29. W. Choi *et al.*, "Three-dimensional Multistructural Quantitative Photoacoustic and US Imaging of Human Feet in Vivo," *Radiology,* vol. 303, no. 2, pp. 467–473, 2022, doi: https://doi.org/10.1148/radiol.211029.
30. C. Lee, W. Choi, J. Kim, and C. Kim, "Three-dimensional clinical handheld photoacoustic/ultrasound scanner," *Photoacoustics,* vol. 18, p. 100173, 2020/06/01/ 2020, doi: https://doi.org/10.1016/j.pacs.2020.100173.
31. J. Kim *et al.*, "Multiparametric Photoacoustic Analysis of Human Thyroid Cancers In Vivo," *Cancer Research,* vol. 81, no. 18, pp. 4849–4860, 2021, doi: https://doi.org/10.1158/0008-5472.Can-20-3334.
32. Z. Chen *et al.*, "Multimodal Noninvasive Functional Neurophotonic Imaging of Murine Brain-Wide Sensory Responses," *Advanced Science,* vol. 9, no. 24, p. 2105588, 2022.
33. S. Na, Y. Zhang, and L. V. Wang, "Cross-Ray Ultrasound Tomography and Photoacoustic Tomography of Cerebral Hemodynamics in Rodents," *Advanced Science,* vol. 9, no. 25, p. 2201104, 2022.
34. H. He *et al.*, "Fast raster-scan optoacoustic mesoscopy enables assessment of human melanoma microvasculature in vivo," *Nature Communications,* vol. 13, no. 1, p. 2803, 2022.
35. D. J. Lawrence, M. E. Escott, L. Myers, S. Intapad, S. H. Lindsey, and C. L. Bayer, "Spectral photoacoustic imaging to estimate in vivo placental oxygenation during preeclampsia," *Scientific reports,* vol. 9, no. 1, p. 558, 2019.
36. I. Steinberg, D. M. Huland, O. Vermesh, H. E. Frostig, W. S. Tummers, and S. S. Gambhir, "Photoacoustic clinical imaging," *Photoacoustics,* vol. 14, pp. 77–98, 2019.

37. X. Tong *et al.*, "Non-Invasive 3D Photoacoustic Tomography of Angiographic Anatomy and Hemodynamics of Fatty Livers in Rats," *Advanced Science,* vol. 10, no. 2, p. 2205759, 2023.
38. X. Zhu *et al.*, "Real-time whole-brain imaging of hemodynamics and oxygenation at micro-vessel resolution with ultrafast wide-field photoacoustic microscopy," *Light: Science & Applications,* vol. 11, no. 1, p. 138, 2022.
39. C. Lee *et al.*, "Panoramic volumetric clinical handheld photoacoustic and ultrasound imaging," *Photoacoustics,* vol. 31, p. 100512, 2023/06/01/ 2023, doi: https://doi.org/10.1016/j.pacs.2023.100512.
40. B. Park *et al.*, "3D wide-field multispectral photoacoustic imaging of human melanomas in vivo: a pilot study," *Journal of the European Academy of Dermatology and Venereology,* vol. 35, no. 3, pp. 669–676, 2021, doi: https://doi.org/10.1111/jdv.16985.
41. N.-A. Fasoula *et al.*, "Non-invasive multispectral optoacoustic tomography resolves intrahepatic lipids in patients with hepatic steatosis," *Photoacoustics,* vol. 29, p. 100454, 2023.
42. K. Jansen, A. F. W. van der Steen, H. M. M. van Beusekom, J. W. Oosterhuis, and G. van Soest, "Intravascular photoacoustic imaging of human coronary atherosclerosis," *Optics Letters,* vol. 36, no. 5, pp. 597–599, 2011/03/01 2011, doi: https://doi.org/10.1364/OL.36.000597.
43. J. Yao *et al.*, "High-speed label-free functional photoacoustic microscopy of mouse brain in action," *Nature methods,* vol. 12, no. 5, pp. 407–410, 2015.
44. L. Lin and L. V. Wang, "The emerging role of photoacoustic imaging in clinical oncology," *Nature Reviews Clinical Oncology,* vol. 19, no. 6, pp. 365–384, 2022.
45. M. Yang *et al.*, "Synovial oxygenation at photoacoustic imaging to assess rheumatoid arthritis disease activity," *Radiology,* vol. 306, no. 1, pp. 220–228, 2023.
46. Y. Wang, Y. Zhan, L. M. Harris, S. Khan, and J. Xia, "A portable three-dimensional photoacoustic tomography system for imaging of chronic foot ulcers," *Quantitative imaging in medicine and surgery,* vol. 9, no. 5, p. 799, 2019.
47. J. Lv *et al.*, "In vivo photoacoustic imaging dynamically monitors the structural and functional changes of ischemic stroke at a very early stage," *Theranostics,* vol. 10, no. 2, p. 816, 2020.
48. G. Godefroy, B. Arnal, and E. Bossy, "Full-visibility 3D imaging of oxygenation and blood flow by simultaneous multispectral photoacoustic fluctuation imaging (MS-PAFI) and ultrasound Doppler," *Scientific Reports,* vol. 13, no. 1, p. 2961, 2023.
49. P. Hai *et al.*, "Label-free high-throughput photoacoustic tomography of suspected circulating melanoma tumor cells in patients in vivo," *Journal of biomedical optics,* vol. 25, no. 3, pp. 036002–036002, 2020.
50. L. Kang, X. Li, Y. Zhang, and T. T. Wong, "Deep learning enables ultraviolet photoacoustic microscopy based histological imaging with near real-time virtual staining," *Photoacoustics,* vol. 25, p. 100308, 2022.
51. R. Cao *et al.*, "Label-free intraoperative histology of bone tissue via deep-learning-assisted ultraviolet photoacoustic microscopy," *Nature Biomedical Engineering,* vol. 7, no. 2, pp. 124–134, 2023.
52. J. Shi *et al.*, "High-resolution, high-contrast mid-infrared imaging of fresh biological samples with ultraviolet-localized photoacoustic microscopy," *Nature photonics,* vol. 13, no. 9, pp. 609–615, 2019.
53. W. Choi, B. Park, S. Choi, D. Oh, J. Kim, and C. Kim, "Recent advances in contrast-enhanced photoacoustic imaging: overcoming the physical and practical challenges," *Chemical Reviews,* 2023.
54. M. Han, W. Choi, J. Ahn, H. Ryu, Y. Seo, and C. Kim, "In Vivo Dual-Modal Photoacoustic and Ultrasound Imaging of Sentinel Lymph Nodes Using a Solid-State Dye Laser System," *Sensors,* vol. 20, no. 13, p. 3714, 2020. [Online]. Available: https://www.mdpi.com/1424-8220/20/13/3714.
55. M. G. Cha *et al.*, "A dual modal silver bumpy nanoprobe for photoacoustic imaging and SERS multiplexed identification of in vivo lymph nodes," *Nanoscale,* 10.1039/C7NR03742B vol. 9, no. 34, pp. 12556–12564, 2017, doi: https://doi.org/10.1039/C7NR03742B.
56. J. Park *et al.*, "Bi-modal near-infrared fluorescence and ultrasound imaging via a transparent ultrasound transducer for sentinel lymph node localization," *Optics Letters,* vol. 47, no. 2, pp. 393–396, 2022/01/15 2022, doi: https://doi.org/10.1364/OL.446041.

57. W. Huang *et al.*, "In vivo quantitative photoacoustic diagnosis of gastric and intestinal dysfunctions with a broad pH-responsive sensor," *ACS nano,* vol. 13, no. 8, pp. 9561–9570, 2019.
58. Y. Ding *et al.*, "Surfactant-stripped semiconducting polymer micelles for tumor theranostics and deep tissue imaging in the NIR-II window," *Small,* vol. 18, no. 6, p. 2104132, 2022.
59. H. Xu, P. She, Z. Zhao, B. Ma, G. Li, and Y. Wang, "Duplex Responsive Nanoplatform with Cascade Targeting for Atherosclerosis Photoacoustic Diagnosis and Multichannel Combination Therapy," *Advanced Materials,* p. 2300439, 2023.
60. M. Gifani *et al.*, "Ultraselective carbon nanotubes for photoacoustic imaging of inflamed atherosclerotic plaques," *Advanced Functional Materials,* vol. 31, no. 37, p. 2101005, 2021.
61. Q. Shen *et al.*, "Stimuli-Responsive Organic Near-Infrared Photoacoustic Probes," *Advanced Functional Materials,* vol. 33, no. 21, p. 2300023, 2023, doi: https://doi.org/10.1002/adfm.202300023.
62. C. Kim, C. Favazza, and L. V. Wang, "In vivo photoacoustic tomography of chemicals: high-resolution functional and molecular optical imaging at new depths," *Chemical reviews,* vol. 110, no. 5, pp. 2756–2782, 2010.
63. D. M. Shcherbakova, M. Baloban, and V. V. Verkhusha, "Near-infrared fluorescent proteins engineered from bacterial phytochromes," *Current opinion in chemical biology,* vol. 27, pp. 52–63, 2015.
64. K. Mishra, J. P. Fuenzalida-Werner, V. Ntziachristos, and A. C. Stiel, "Photocontrollable proteins for optoacoustic imaging," *Analytical chemistry,* vol. 91, no. 9, pp. 5470–5477, 2019.
65. C. Kim *et al.*, "In vivo molecular photoacoustic tomography of melanomas targeted by bioconjugated gold nanocages," *ACS nano,* vol. 4, no. 8, pp. 4559–4564, 2010.
66. J. F. Lovell *et al.*, "Porphysome nanovesicles generated by porphyrin bilayers for use as multimodal biophotonic contrast agents," *Nature materials,* vol. 10, no. 4, pp. 324–332, 2011.
67. H. Takala *et al.*, "Signal amplification and transduction in phytochrome photosensors," *Nature,* vol. 509, no. 7499, pp. 245–248, 2014.
68. J. Weber, P. C. Beard, and S. E. Bohndiek, "Contrast agents for molecular photoacoustic imaging," *Nature methods,* vol. 13, no. 8, pp. 639–650, 2016.
69. W. Lu *et al.*, "Photoacoustic imaging of living mouse brain vasculature using hollow gold nanospheres," *Biomaterials,* vol. 31, no. 9, pp. 2617–2626, 2010.
70. Y.-S. Chen, Y. Zhao, S. J. Yoon, S. S. Gambhir, and S. Emelianov, "Miniature gold nanorods for photoacoustic molecular imaging in the second near-infrared optical window," *Nature nanotechnology,* vol. 14, no. 5, pp. 465–472, 2019.
71. A. De La Zerda *et al.*, "Carbon nanotubes as photoacoustic molecular imaging agents in living mice," *Nature nanotechnology,* vol. 3, no. 9, pp. 557–562, 2008.
72. J.-W. Kim, E. I. Galanzha, E. V. Shashkov, H.-M. Moon, and V. P. Zharov, "Golden carbon nanotubes as multimodal photoacoustic and photothermal high-contrast molecular agents," *Nature nanotechnology,* vol. 4, no. 10, pp. 688–694, 2009.
73. C. Lee *et al.*, "Biodegradable nitrogen-doped carbon nanodots for non-invasive photoacoustic imaging and photothermal therapy," *Theranostics,* vol. 6, no. 12, p. 2196, 2016.
74. D. Wang *et al.*, "Porphyrin-based covalent organic framework nanoparticles for photoacoustic imaging-guided photodynamic and photothermal combination cancer therapy," *Biomaterials,* vol. 223, p. 119459, 2019.
75. F. Wu *et al.*, "Small-molecule porphyrin-based organic nanoparticles with remarkable photothermal conversion efficiency for in vivo photoacoustic imaging and photothermal therapy," *ACS applied materials & interfaces,* vol. 11, no. 24, pp. 21408–21416, 2019.
76. D. Maji *et al.*, "Copper-catalyzed covalent dimerization of near-infrared fluorescent cyanine dyes: synergistic enhancement of photoacoustic signals for molecular imaging of tumors," *Analysis & Sensing,* vol. 2, no. 1, p. e202100045, 2022.
77. Q. Yang, H. Jin, Y. Gao, J. Lin, H. Yang, and S. Yang, "Photostable iridium (III)–cyanine complex nanoparticles for photoacoustic imaging guided near-infrared photodynamic therapy in vivo," *ACS applied materials & interfaces,* vol. 11, no. 17, pp. 15417–15425, 2019.

78. E.-Y. Park, D. Oh, S. Park, W. Kim, and C. Kim, "New contrast agents for photoacoustic imaging and theranostics: Recent 5-year overview on phthalocyanine/naphthalocyanine-based nanoparticles," *APL bioengineering,* vol. 5, no. 3, 2021.
79. U. Chitgupi *et al.*, "Surfactant-stripped micelles for NIR-II photoacoustic imaging through 12 cm of breast tissue and whole human breasts," *Advanced Materials,* vol. 31, no. 40, p. 1902279, 2019.
80. K. Pu *et al.*, "Semiconducting polymer nanoparticles as photoacoustic molecular imaging probes in living mice," *Nature nanotechnology,* vol. 9, no. 3, pp. 233–239, 2014.
81. C. Yin *et al.*, "Amphiphilic semiconducting oligomer for near-infrared photoacoustic and fluorescence imaging," *Acs Applied Materials & Interfaces,* vol. 9, no. 14, pp. 12332–12339, 2017.
82. J. Yao *et al.*, "Multiscale photoacoustic tomography using reversibly switchable bacterial phytochrome as a near-infrared photochromic probe," *Nature methods,* vol. 13, no. 1, pp. 67–73, 2016.
83. J. Märk *et al.*, "Dual-wavelength 3D photoacoustic imaging of mammalian cells using a photoswitchable phytochrome reporter protein," *Communications Physics,* vol. 1, no. 1, pp. 1–10, 2018.
84. L. Li *et al.*, "Small near-infrared photochromic protein for photoacoustic multi-contrast imaging and detection of protein interactions in vivo," *Nature communications,* vol. 9, no. 1, p. 2734, 2018.
85. K. Mishra *et al.*, "Multiplexed whole-animal imaging with reversibly switchable optoacoustic proteins," *Science Advances,* vol. 6, no. 24, p. eaaz6293, 2020.
86. R. Gao *et al.*, "Background-suppressed tumor-targeted photoacoustic imaging using bacterial carriers," *Proceedings of the National Academy of Sciences,* vol. 119, no. 8, p. e2121982119, 2022.
87. L. A. Kasatkina *et al.*, "Optogenetic manipulation and photoacoustic imaging using a near-infrared transgenic mouse model," *Nature communications,* vol. 13, no. 1, p. 2813, 2022.
88. A. C. Stiel, X. L. Deán-Ben, Y. Jiang, V. Ntziachristos, D. Razansky, and G. G. Westmeyer, "High-contrast imaging of reversibly switchable fluorescent proteins via temporally unmixed multispectral optoacoustic tomography," *Optics letters,* vol. 40, no. 3, pp. 367–370, 2015.
89. J. P. Fuenzalida Werner *et al.*, "Challenging a preconception: optoacoustic spectrum differs from the optical absorption spectrum of proteins and dyes for molecular imaging," *Analytical Chemistry,* vol. 92, no. 15, pp. 10717–10724, 2020.
90. P. Vetschera, K. Mishra, J. P. Fuenzalida-Werner, A. Chmyrov, V. Ntziachristos, and A. C. Stiel, "Characterization of reversibly switchable fluorescent proteins in optoacoustic imaging," *Analytical chemistry,* vol. 90, no. 17, pp. 10527–10535, 2018.
91. K. Mishra *et al.*, "Genetically encoded photo-switchable molecular sensors for optoacoustic and super-resolution imaging," *Nature Biotechnology,* vol. 40, no. 4, pp. 598–605, 2022.
92. C. Liu *et al.*, "Reversibly Photoswitching Upconversion Nanoparticles for Super-Sensitive Photoacoustic Molecular Imaging," *Angewandte Chemie,* vol. 134, no. 19, p. e202116802, 2022.
93. C. Ma *et al.*, "Multiscale photoacoustic tomography using reversibly switchable thermochromics," *Journal of Biomedical Optics,* vol. 28, no. 8, pp. 082804–082804, 2023.

Chapter 7
Technical Validation of Photoacoustic Imaging Systems Using Phantoms

Lina Hacker and James Joseph

Abstract Technical validation studies are paramount for objective performance evaluation and calibration of novel imaging devices, aiding their development and effective translation into clinical practice. Such studies usually rely on the use of tissue-mimicking test objects termed “phantoms.” This chapter outlines recommendations and best practices for technical validation studies in photoacoustic imaging (PAI) and provides an overview on current standardization efforts in the field. Specifically, it expands on the role of phantoms in the technical validation process, highlights key recommendations on properties and characterization methods for PAI phantoms, and provides a framework for technical validation studies of PAI systems.

Keywords Phantom · Standardization · Photoacoustic imaging · Technical validation

7.1 The Role of Phantoms in the Technical Validation of Medical Imaging Systems

As novel imaging technologies advance along the translational pipeline, thorough characterization, calibration, and technical device validation are required [1, 2] to establish accuracy, repeatability, reproducibility, and lack of bias of a biomarker measurement [3]. Advances in photoacoustic imaging (PAI) technology have resulted in the evolution of a myriad of imaging approaches, with differences in system configurations (ranging from tomographic, mesoscopic, and microscopic applications), image reconstruction and spectral unmixing techniques, and data quantifi-

L. Hacker
Department of Oncology, University of Oxford, Oxford, UK
e-mail: Lina.hacker@oncology.ox.ac.uk

J. Joseph (✉)
School of Science and Engineering, University of Dundee, Dundee, UK
e-mail: JJoseph001@dundee.ac.uk

W. Xia (ed.), *Biomedical Photoacoustics*,
https://doi.org/10.1007/978-3-031-61411-8_7

cation methodologies [4]. However, a standardized and robust technical framework for the validation and performance evaluation of PAI systems is still lacking. Usually, such frameworks rely on the use of standardized test objects that can provide repeatable and reproducible signal readouts. Whilst a variety of imaging test objects exists, tissue-mimicking phantoms (TMP, from ancient Greek φάντασμα (phántasma) = apparition, ghost or mirage) are considered to be the most relevant class of test objects for the clinical translation of PAI systems. TMPs are heavily deployed for conducting standardized tests in more matured imaging modalities such as magnetic resonance imaging (MRI), computed tomography (CT), or ultrasound [5]. TMPs enable the derivation of specific quantitative image quality metrics allowing objective device performance evaluation and inter-comparisons [6].

PAI offers unique scalability in both spatial resolution and penetration depth, providing not only structural but also functional information. The multi-functional capabilities of PAI impede the design of an all-encompassing phantom that fulfils the needs of every optical subspeciality. Therefore, a wide array of TMPs have been developed which differ widely in shape and complexity [7]. Broadly, these phantoms can be classified into two categories: (1) physical phantoms and (2) numerical (computational) phantoms. Physical phantoms are material objects that can be either of inorganic or organic origin [8]. Inorganic or engineered phantoms can be simple, homogenously composed or more complex, heterogeneously composed. Organic or biological phantoms are created of living matter and cover engineered and ex vivo tissues. Numerical phantoms cover in silico frameworks that are used for simulating complex systems or objects (digital twins) and conducting virtual experiments. Both physical and numerical types of phantoms play crucial roles in PAI research and technology development.

7.2 Tasks of Phantoms Along the Translational Pipeline

Phantoms fulfil various tasks along the full translational pipeline of an imaging modality [1, 9] (Fig. 7.1). These tasks include:

- Quantitatively and objectively assessing the performance of technologies across multiple stages of their development.
- Assessing drift in accuracy and precision of the imaging device over time ("routine quality performance").
- Facilitating system design optimization (hardware or software); quantifying/assessing the impact of system upgrades.
- Comparison between different devices and/or diagnostic modalities.
- Performing standardized biomarker readouts from multiple and distinct instruments.
- User training.
- Marketing.
- Technical demonstrations.
- Testing to support regulatory evaluation; Benchmarking.

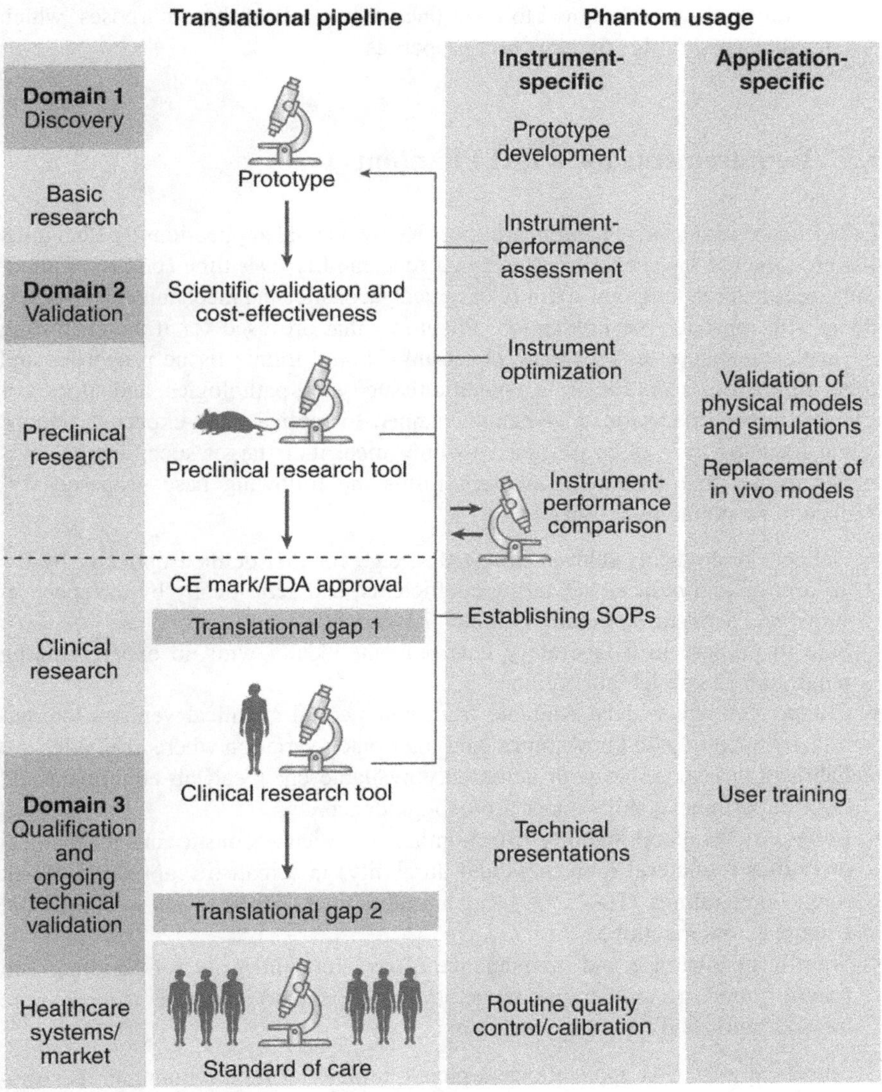

Fig. 7.1 Tasks of phantoms along the translational pipeline. Instrument- and application-specific tasks of phantoms are presented along the translational pipeline for optical imaging devices. Translational gap 1 refers to the step of transferring a research tool from a preclinical into a clinical research environment. Translational gap 2 refers to the step of integrating a tool from clinical research into clinical standard-of-care and routine patient use (SOPs = standard operating procedures). (Taken from [9])

Phantoms are often designed to meet one (or several) of these purposes, which dictates their final design and intrinsic properties.

7.3 Requirements for a PAI Phantom

The ideal properties of a phantom are specified by its final application [9]. Phantoms that are targeted toward testing the signal repeatability over time (precision phantoms) require high temporal stability of optical, acoustic, and mechanical properties, along with reproducible fabrication. Phantoms that are used for training, testing or verification purposes (accuracy phantoms) should mimic tissue properties and target geometries accurately (e.g., specific tissue types, pathologies, and species of interest) either in a static or dynamic manner, to replicate the expected signals. While phantoms are often designed to only meet their target specification, it is recommended that their bulk medium fulfils the following base properties [9] (ranked in no preferential order):

- Defined, biologically relevant properties, e.g., for PAI defined optically by the absorption and reduced scattering coefficients; and acoustically by the speed of sound and acoustic attenuation (Tables 7.1 and 7.2).
- Safe to prepare in a laboratory environment; ideally with no expert training required for material fabrication.
- Components are widely available from commercial chemical vendors internationally, ideally with known intra- and inter-batch variation where available.
- Fabrication is possible with commonly available chemical lab equipment and basic experimental skills under protocol guidance.
- Long-term temporal stability (>6 months) of optical, acoustic and mechanical properties (structural robustness and durability) in a realistic range of ambient room temperatures (18–25 °C) and humidities (30–80%) to allow convenient storage across the globe.
- Short-term tolerance and maintenance of structural integrity for handling and transportation in a temperature range between 4° and 40 °C based on appropriate transportation and handling procedures.

Specifically for PAI applications, a phantom material should maintain its structural and material integrity when in contact with an aqueous medium as water-based solutions are often used as acoustic coupling agents during signal acquisition. Additionally, the material should be photostable at the visible and near-infrared (NIR) wavelength range (532–1064 nm) or any other wavelengths of choice under safe exposure limits as encountered during imaging, handling and storage. A material should further allow the embedding of target inclusions without degradation to enable quantitative assessment of image quality metrics for specific applications. Ideally, the material should include targets made out of the same material type and targets made out of different material types (e.g., microspheres, wires, etc.).

Table 7.1 Definition of relevant material properties parameters in optical and acoustic imaging applications

	Parameter	Symbol	Unit	Definition	Ref
Optical	Optical absorption coefficient	μ_a	cm^{-1}	Probability of photon absorption per unit length travelled by the photon	[10, 11]
	Refractive index	n	–	Ratio of the velocity of light in a vacuum to the velocity of light in a medium	[10, 11]
	Anisotropy factor	g	–	Average of the cosine of the scattering angle which represents the effects of directionally dependent scattering	[10, 11]
	Reduced scattering coefficient	μ_s‘	cm^{-1}	Probability of photon scattering per unit length travelled by the photon	[10, 11]
Acoustic	Acoustic attenuation coefficient	α	$dB{\cdot}cm^{-1}{\cdot}MHz^{-y}$	Extent of reduction of an acoustic wave when propagating within a medium	[12]
	Speed of sound	c	$m{\cdot}s^{-1}$	Velocity of acoustic wave propagation within a medium	[12]
	Backscattering coefficient	μ_{bs}	$m^{-1}\ sr^{-1}$	Differential scattering cross section per unit volume at a scattering angle of 180°	[13]
Mechanical	Density	ρ	$kg{\cdot}m^{-3}$	Measure of mass per unit volume	[12]
	Young’s Modulus	E	$N{\cdot}m^{-2}$	Ratio of stress applied to a material to the strain that results from the stress	[12]
Thermoelastic	Grüneisen Parameter	Γ	–	Measure of thermoelastic efficiency	[14, 15]

The soft tissue values are just representative values. The properties may vary with the experimental conditions (e.g., temperature). See literature for values that pertain to specific samples and controls. Modified from [9]

Table 7.2 Overview of representative acoustic and optical properties found in soft tissues

	Acoustic properties		Optical properties		
Tissue type	c ($m \cdot s^{-1}$)	α ($dB \cdot cm^{-1}$) at f(MHz)	μ_a (cm^{-1})	μ_s' (cm^{-1})	Ref
Soft tissue	1450–1575	0.5–30 at 1–10 MHz	0.1–0.5	10–20	[16, 17]
Breast fat	1430–1480	1–18 at 1–10 MHz	0.05–0.4	3–8	[16, 17]
Breast parenchyma	1460–1520	2–25 at 1–10 MHz	0.1–0.3	5–15	[16]
Blood	1560–1570	0.1–0.2 $dB \cdot cm^{-1} \cdot MHz^{-1}$	2.0–10.0	10–15	[11, 18, 19]
Brain	1550	0.6 at 1 MHz	0.2–9	8–90	[19–21]
Liver	1510–1590	0.5–0.9 $dB \cdot cm^{-1} \cdot MHz^{-1}$	1.15–1.56	22–30	[11, 18, 22]
Prostate	1614	1.86 $dB \cdot cm^{-1} \cdot MHz^{-1}$	0.05–0.72	1–40	[11, 19]
Skin	~1600	2–4 $dB \cdot cm^{-1} \cdot MHz^{-1}$	0.05–1.11	2–21	[11, 19, 22]
Muscle	1540–1580	1.3–3.3 at 1 MHz	0.05–0.17	6–10	[11, 18]
Tendon	1670	4.7 $dB \cdot cm^{-1} \cdot MHz^{-1}$	a		[19]
Water	1480	0.0022 at 1 MHz	0.006–0.07	0.003	[19, 23]

Optical properties cover a spectrum from 600 to 900 nm. For values outside this wavelength range, please refer to the literature [11]
[a]No specific reference found

Importantly, an ideal material should be accessible to everyone in the scientific community. If it cannot be procured in a "ready to manufacture" state at more than one standard scientific material supplier, its ingredients (including chemical abstract service (CAS) numbers) and detailed manufacturing process should be openly published. Reproducible fabrication should be evidenced by a multicenter study to ensure that the material has broad accessibility, achieving desired properties within acceptable uncertainty limitations.

7.4 Optical and Acoustic Properties of a PAI Phantom

PAI phantoms require both biologically relevant optical and acoustic properties. In terms of optical properties, the background material should be characterized by low extinction values to allow tunability for a variety of tissue types. For specific applications, these values should be increased to match the average optical values of the biological tissue/condition of interest (Table 7.2). Target inclusions should exhibit a photoacoustic response at a wavelength relevant to the target system/application. Phantoms should ideally not exhibit any fluorescence or other non-linear optical behaviors that reduce photoacoustic conversion efficiency. Other biologically relevant optical properties include the anisotropy factor g, which is accounted for by the reduced scattering coefficient ($\mu_s' = \mu_s(1 - g)$) and the refractive index, n. Ideally, a PAI phantom material should exhibit forward scattering comparable to the tissue of interest, resulting in values of $g > 0$. The refractive index n should also mimic soft tissues if possible (Table 7.2).

An ideal material should also mimic the acoustic properties of tissues, particularly the tissue-specific speed of sound and acoustic attenuation values (Table 7.2). If quantification of other acoustic or mechanical properties can be performed, such as the ultrasound non-linearity parameter (B/A), acoustic back-scattering coefficient, echo reduction, density or Grüneisen parameter, it is recommended that the values approximate the values of the target tissue of interest as well. All chosen values should be repeatable with high precision.

7.5 PAI Phantom Materials

Many phantom materials have been tested for PAI, including hydrogels [24–26], PVA [27, 28], silicone [29, 30], PVCP [16, 31–35] and copolymer-in-oil materials [36–38] (Table 7.3). Inclusions have been made by embedding fluid-filled channels [16, 39] or solid absorbers such as wires or filaments [40], or by creating simple [31, 35, 41] or more complex vessel-mimicking [34, 42] inserts. PVCP [43] and copolymer-in-oil materials [38] are promising options for long-term stability phantoms due to their long shelf life and excellent matching of both optical and acoustic properties in the tissue-mimicking range. Hydrogel phantoms (e.g., agar-based) only offer limited stability, but benefit from an easy and cost-effective preparation procedure [44]. A brief overview of the reproducible fabrication of a copolymer-and-oil phantom [34, 36] (Fig. 7.2a) and an agar phantom [44] (Fig. 7.2b) for PAI applications can be found in Fig. 7.2, with detailed protocols published elsewhere [38, 44, 45].

A standardized phantom for PAI does not yet exist. The first efforts in this direction have been driven by the community-led consensus-based effort IPASC: the International Photoacoustic Standardisation Consortium [46]. Founded in 2018, IPASC combines stakeholders from academia, industry and government laboratories aiming to reach consensus on standardized PAI performance test methods, test objects and data management for improving the quality of preclinical studies and reinforcing efforts in clinical translation [4, 46]. The Test Objects and Methods theme is currently evaluating the use of a copolymer-mineral oil phantom for multicenter studies, where all constituent materials have defined chemical abstract service (CAS) numbers and are available from commercial chemistry suppliers, enabling local fabrication and testing.

7.6 Acoustic and Optical Characterization of PAI Phantoms

Phantoms require detailed characterization of their intrinsic properties. This requires specialist equipment that is regularly calibrated. Equipment hosted in research and industrial laboratories should ideally be cross-referenced to reference (e.g., metrology) institutes to determine the accuracy with which local characterization

Table 7.3 Overview of materials used in PAI phantoms

Material	TMM Properties			Handling			Fabrication				References
	Optical	Acoustic	Tuna-bility	Temporal Stability	Mecha-nical Stability	Storage/ Trans-port	Architec-tural Flexibility	Com-plexity	Safety	Biocompatibility	
Aqueous suspension	++	-	+	- -	- -	- -	-	++	++	Yes	[19]
Agar/Gelatin	++	++	++	-	-	-	++	++	++	Yes	[24], [25], [64]
Polyacrylamide	++	++	++	-	+	-	++	+	-	Yes	[26]
PVA	++	++	++	+	+	-	++	-	++	Yes	[27], [28], [65]
Co-polymer in oil	++	+	+	++	++	++	++	+	++	No	[38], [41], [42], [45], [66]–[68]
PVCP	++	+	+	++	++	++	+	+	+	No	[16], [31], [32], [69]
Silicone	++	-	+	++	++	++	++	+	++	No	[29], [70]
Polyurethane	++	+	+	++	++	++	++	+	+	No	[71]
Ex vivo tissues	++	++	- -	-	++	-	- -	++	++	Yes	[72]

Legend: green (++) = excellent performance; light green (+) = above-average performance; light red (–) = below-average performance; red (—) = poor performance in respective category.) Modified from [9]

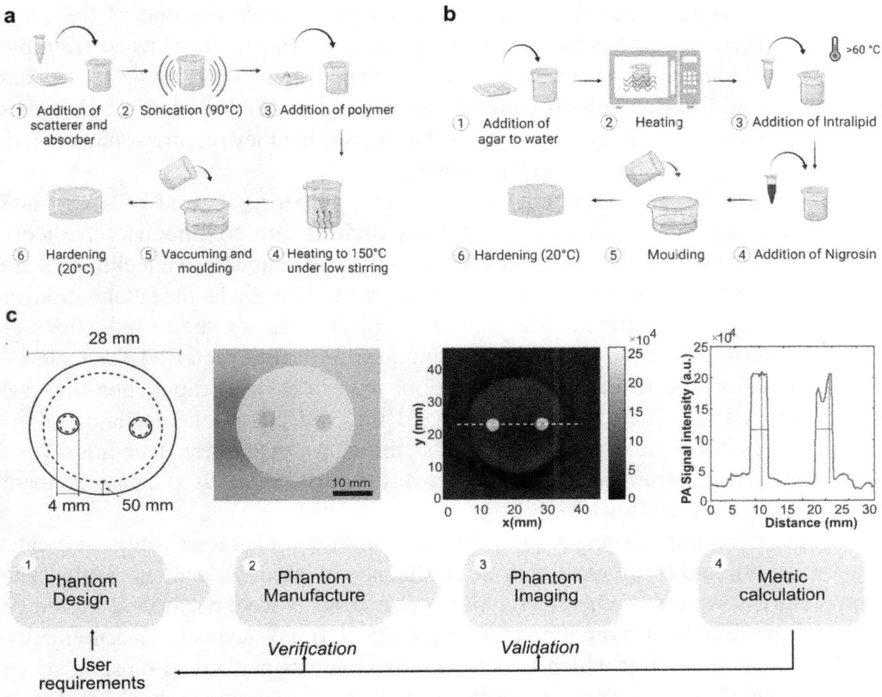

Fig. 7.2 Fabrication of PAI TMPs. (**a**) Fabrication of a copolymer-in-oil TMP material. (1) Materials for optical scattering and absorption are added to mineral oil and (2) sonicated at 90 °C until dissolved. (3) Polymer(s) and stabilizer are added and (4) the mixture is heated up in an oil bath to 150 °C under low stirring. (5) Upon dissolution of all components, the sample is poured in a suitable phantom mould and (6) left to harden at room temperature. (**b**) Fabrication of an agar TMP material. (1) Agar is added to deionized water and (2) heated until it fully dissolves. After cooling down to 60 °C, (3) pre-warmed intralipid and (4) an absorbing dye (e.g., Nigrosin) are added to the solution. (5) The sample is poured in a suitable phantom mould and (6) left to harden at room temperature. (**c**) Technical validation pipeline: (1) The user defines a phantom design according to their requirements (pictured schematic drawing of a tomographic phantom). (2) The phantom is manufactured (pictured a photograph of a cylindrical co-polymer in oil phantom). (3) The phantom is imaged using PAI equipment (pictured PAI cross-sectional image at 800 nm). (4) The metric of interest is calculated (pictured the intensity line profile of the white dotted line is shown (signal = blue, prominence = orange, width of signal peak at half maximum = grey).) (Created with BioRender. Figure 7.2a, c (modified) is reproduced from [38] under CC-BY 4.0 licence)

can be undertaken for a given material. Ideally, traceability to gold-standard metrology, such as those supported by the National Institute of Standards and Technology (NIST), should be achieved [9].

Techniques to characterize optical properties include steady-state domain systems [47, 48], time domain systems [49], time-frequency domain systems [50], spatial domain systems [51], and spatial frequency domain systems [52]. Steady-state domain systems based on integrating sphere systems are one of the most commonly used approaches for optical characterization due to their low cost, simple setup, short acquisition times, ability to characterize a broad range of optical properties, and detailed how-to-guides being readily available [47, 53, 54]. Time-domain optical approaches offer the highest accuracy, but they require sophisticated, costly equipment and long acquisition times.

Approaches for the measurement of acoustic properties (speed of sound and acoustic attenuation) [55–58] can be broadly divided into continuous wave techniques [59] and broadband pulse techniques [60]. Continuous wave techniques are highly accurate and beneficial for detecting small changes in the attenuation or sound velocity but are time-consuming and subject to errors due to reflections or other interfering signals [61]. Pulsed techniques are often preferred for material characterization due to their easy operation, lower cost and rapid, non-invasive measurement [62]. They can be categorized into the pulse-echo technique (one transducer as transmitter-receiver) and the through-transmission technique (two transducers as receiver/transducer respectively). Further details on measurement procedures are given elsewhere [63].

For the measurement technique of choice, sufficient characterization and calibration data should be given to provide confidence in the measurement results. The temperature at which the characterization measurements were performed should be reported (preferably at room temperature [18–25 °C]). For acoustic measurements, the frequency range over which the characterization was performed must be stated for reporting of measurements. Measurements from all techniques should include a robust uncertainty analysis, ideally with type A and type B uncertainty estimations.

7.7 Technical Validation Framework for PAI Systems

Technical validation studies are crucial to evaluate and compare the performance of novel medical imaging devices. Execution of a technical validation study demands careful evaluation of the device type(s) to be tested, testing method to be applied, the performance metrics to be evaluated, and minimum acceptance criteria, if applicable [6]. Specifically for phantom-based test methods, careful planning is required with respect to (1) phantom design, (2) phantom manufacture and validation, (3) data acquisition, and (4) data analysis (Fig. 7.2c).

Phantom geometry, including size and pattern of embedded targets, and material properties should be tailored towards the application and system of interest. Prior to their usage, phantoms should be rigorously characterized to ensure they meet desired specifications. The test methodology should be "pre-specified" to permit deviating from a specified protocol to produce more favorable outcomes. For data acquisition, the test object should be placed in a repeatable and stable position, e.g., by using a rigid holder, to prevent measurement drifts due to positioning inaccuracies. As temperature can impact photoacoustic signals, the test object should be temperature-stabilized under the experimental conditions of interest prior to data acquisition. During data acquisition, it is recommended to take multiple ($n \geq 5$) measurements over time and space of the target of interest. These measurements may extend over multiple wavelengths depending on the system type and application of interest. During the usage of a stable phantom material is highly recommended to avoid drifts in longitudinal measurements and ensure reproducibility over imaging sessions, different operators and/or instruments. Temporal variation can be assessed by the repeated measurements over minutes, hours, days, and months, with or without replacement of the test object.

For data quantification, raw radio frequency (RF) data may be employed, but generally, the usage of pre-defined image-based metrics is recommended. Such metrics may include (in-plane or out-of-plane) spatial resolution, geometric accuracy, image uniformity, depth of visualization, sensitivity, low-contrast detectability, and artifacts [6]. Test protocols should specify dimensions and locations of regions of interest (ROIs) and/or intensity line profiles and should include explicit formulas for computing image quality metrics. Adequate care should be taken for image acquisition and processing as any variations can greatly impact the measurement accuracy.

7.8 Conclusion

Thorough technical validation of novel imaging technologies such as PAI is paramount to support their development, facilitate their regulatory approval, and achieve clinical translation [9]. Technical standards for PAI do not yet exist but are pushed forward by the community [4, 6]. Such standards should cover (1) definitions of basic terminology; (2) performance metrics and test parameters; (3) phantom material and device-appropriate geometric designs; (4) standard operating procedures (SOPs) for data acquisition and analysis procedures [9]. Production of phantoms should ideally be scaled for manufacturing by a commercial vendor so that identical test objects with well-controlled properties and geometries complying with pre-specified recommendations can be produced. Whilst usage of standard basic performance assessment phantoms is important, focus should also be laid on the development of biomimetic, anthropomorphic phantoms [34] to provide more clinically relevant image quality assessment approaches.

In more mature imaging technologies, such as CT, X-ray mammography, ultrasound or MRI, standards have been established through standards organizations such as the International Electrotechnical Commission (IEC), International Organization for Standardization (ISO), and National Electrical Manufacturers Association (NEMA). Additionally, commercially available phantoms exist that are rigorously characterized by the manufacturer to ensure conformity to standards during acceptance testing, quality control, and maintenance/repairs. Similar developments should be pushed forward in PAI, requiring clear communication, commitment, and collaboration from all relevant stakeholders [9]. Only by developing and aligning robust testing frameworks, the translation of new technologies, such as PAI, can be accelerated and their full potential be exploited.

References

1. J. Hwang, J. C. Ramella-Roman, and R. Nordstrom, "Introduction: Feature Issue on Phantoms for the Performance Evaluation and Validation of Optical Medical Imaging Devices," *Biomed. Opt. Express*, vol. 3, no. 6, p. 1399, Jun. 2012, doi: https://doi.org/10.1364/boe.3.001399.
2. M. B. Aldrich *et al.*, "Seeing it through: translational validation of new medical imaging modalities," *Biomed. Opt. Express*, vol. 3, no. 4, p. 764, Apr. 2012, doi: https://doi.org/10.1364/BOE.3.000764.
3. J. P. B. O'Connor *et al.*, "Imaging biomarker roadmap for cancer studies," *Nat. Rev. Clin. Oncol.*, vol. 14, no. 3, pp. 169–186, Mar. 2017, doi: https://doi.org/10.1038/nrclinonc.2016.162.
4. H. Assi *et al.*, "A review of a strategic roadmapping exercise to advance clinical translation of photoacoustic imaging: From current barriers to future adoption," *Photoacoustics*, vol. 32, p. 100539, Aug. 2023, doi: https://doi.org/10.1016/J.PACS.2023.100539.
5. J. Palma-Chavez, K. A. Wear, Y. Mantri, J. V. Jokerst, and W. C. Vogt, "Photoacoustic imaging phantoms for assessment of object detectability and boundary buildup artifacts," *Photoacoustics*, vol. 26, p. 100348, Jun. 2022, doi: https://doi.org/10.1016/j.pacs.2022.100348.
6. J. Palma-Chavez, T. J. Pfefer, A. Agrawal, J. V. Jokerst, and W. C. Vogt, "Review of consensus test methods in medical imaging and current practices in photoacoustic image quality assessment," *J. Biomed. Opt.*, vol. 26, no. 09, Sep. 2021, doi: https://doi.org/10.1117/1.JBO.26.9.090901.
7. L. B. Christie *et al.*, "Review of imaging test phantoms," *J. Biomed. Opt.*, vol. 28, no. 08, Aug. 2023, doi: https://doi.org/10.1117/1.JBO.28.8.080903.
8. E. Fieremans and H. H. Lee, "Physical and numerical phantoms for the validation of brain microstructural MRI: A cookbook," *NeuroImage*, vol. 182. Academic Press Inc., pp. 39–61, Nov. 15, 2018, doi: https://doi.org/10.1016/j.neuroimage.2018.06.046.
9. L. Hacker, H. Wabnitz, A. Pifferi, T. J. Pfefer, B. W. Pogue, and S. E. Bohndiek, "Criteria for the design of tissue-mimicking phantoms for the standardization of biophotonic instrumentation," *Nat. Biomed. Eng.*, vol. 6, no. 5, pp. 541–558, May 2022, doi: https://doi.org/10.1038/s41551-022-00890-6.
10. I. J. Bigio and S. Fantini, *Quantitative Biomedical Optics*. Cambridge University Press, 2016.
11. S. L. Jacques, "Optical properties of biological tissues: a review," *Phys. Med. Biol.*, vol. 58, no. 11, pp. R37–R61, Jun. 2013, doi: https://doi.org/10.1088/0031-9155/58/11/R37.
12. F. A. Duck and Institute of Physics and Engineering in Medicine (Great Britain), *Physical properties of tissue : a comprehensive reference book.* .
13. D. Nicholas, "Evaluation of backscattering coefficients for excised human tissues: results, interpretation and associated measurements," *Ultrasound Med. Biol.*, vol. 8, no. 1, pp. 17–28, Jan. 1982, doi: https://doi.org/10.1016/0301-5629(82)90065-5.
14. E. Grüneisen, "Theorie des festen Zustandes einatomiger Elemente," *Ann. Phys.*, vol. 344, no. 12, pp. 257–306, 1912, doi: https://doi.org/10.1002/andp.19123441202.
15. Y. Villanueva, E. Hondebrink, W. Petersen, and W. Steenbergen, "Photoacoustic measurement of the Grüneisen parameter using an integrating sphere," *Rev. Sci. Instrum.*, vol. 85, no. 7, p. 074904, Jul. 2014, doi: https://doi.org/10.1063/1.4890666.
16. W. C. Vogt, C. Jia, K. A. Wear, B. S. Garra, and T. Joshua Pfefer, "Biologically relevant photoacoustic imaging phantoms with tunable optical and acoustic properties," *J. Biomed. Opt.*, vol. 21, no. 10, p. 101405, Oct. 2016, doi: https://doi.org/10.1117/1.JBO.21.10.101405.
17. IEC, "IEC 60601-2-37:2007 Particular requirements for the basic safety and essential performance of ultrasonic medical diagnostic and monitoring equipment," 2007. [Online]. Available: https://webstore.iec.ch/publication/2652.
18. T. D. Mast, "Empirical relationships between acoustic parameters in human soft tissues," *Acoust. Res. Lett. Online*, vol. 1, no. 2, pp. 37–42, Nov. 2000, doi: https://doi.org/10.1121/1.1336896.

19. M. O. Culjat, D. Goldenberg, P. Tewari, and R. S. Singh, "A Review of Tissue Substitutes for Ultrasound Imaging," *Ultrasound Med. Biol.*, vol. 36, no. 6, pp. 861–873, Jun. 2010, doi: https://doi.org/10.1016/j.ultrasmedbio.2010.02.012.
20. A. N. Yaroslavsky, P. C. Schulze, I. V Yaroslavsky, R. Schober, F. Ulrich, and H.-J. Schwarzmaier, "Optical properties of selected native and coagulated human brain tissues in vitro in the visible and near infrared spectral range," *Phys. Med. Biol.*, vol. 47, no. 12, p. 305, Jun. 2002, doi: https://doi.org/10.1088/0031-9155/47/12/305.
21. H. Azhari, "Appendix A: Typical Acoustic Properties of Tissues," in *Basics of Biomedical Ultrasound for Engineers*, Hoboken, NJ, USA: John Wiley & Sons, Inc., 2010, pp. 313–314.
22. T. L. Szabo, "Diagnostic Ultrasound Imaging - Inside out," in *Diagnostic Ultrasound Imaging*, Elsevier, 2004, pp. 243–295.
23. G. M. Hale and M. R. Querry, "Optical Constants of Water in the 200-nm to 200-μm Wavelength Region," *Appl. Opt.*, vol. 12, no. 3, p. 555, Mar. 1973, doi: https://doi.org/10.1364/ao.12.000555.
24. J. Laufer, E. Zhang, and P. Beard, "Evaluation of Absorbing Chromophores Used in Tissue Phantoms for Quantitative Photoacoustic Spectroscopy and Imaging," *IEEE J. Sel. Top. Quantum Electron.*, vol. 16, no. 3, pp. 600–607, 2010, doi: https://doi.org/10.1109/JSTQE.2009.2032513.
25. J. R. Cook, R. R. Bouchard, and S. Y. Emelianov, "Tissue-mimicking phantoms for photoacoustic and ultrasonic imaging," *Biomed. Opt. Express*, vol. 2, no. 11, p. 3193, Nov. 2011, doi: https://doi.org/10.1364/BOE.2.003193.
26. A. Hariri, J. Palma-Chavez, K. A. Wear, T. J. Pfefer, J. V. Jokerst, and W. C. Vogt, "Polyacrylamide hydrogel phantoms for performance evaluation of multispectral photoacoustic imaging systems," *Photoacoustics*, vol. 22, p. 100245, Jun. 2021, doi: https://doi.org/10.1016/j.pacs.2021.100245.
27. W. Xia, D. Piras, M. Heijblom, W. Steenbergen, T. G. Van Leeuwen, and S. Manohar, "Poly(vinyl alcohol) gels as photoacoustic breast phantoms revisited," *J. Biomed. Opt.*, vol. 16, no. 7, p. 075002, 2011, doi: https://doi.org/10.1117/1.3597616.
28. A. Kharine *et al.*, "Poly(vinyl alcohol) gels for use as tissue phantoms in photoacoustic mammography," vol. 48, no. 3, pp. 357–370, Feb. 2003, doi: https://doi.org/10.1088/0031-9155/48/3/306.
29. C. Avigo *et al.*, "Organosilicon phantom for photoacoustic imaging," *J. Biomed. Opt.*, vol. 20, no. 4, p. 046008, Apr. 2015, doi: https://doi.org/10.1117/1.JBO.20.4.046008.
30. F. Ratto *et al.*, "Hybrid organosilicon/polyol phantom for photoacoustic imaging.," *Biomed. Opt. Express*, vol. 10, no. 8, pp. 3719–3730, Aug. 2019, doi: https://doi.org/10.1364/BOE.10.003719.
31. S. E. Bohndiek, S. Bodapati, D. Van De Sompel, S.-R. R. Kothapalli, and S. S. Gambhir, "Development and Application of Stable Phantoms for the Evaluation of Photoacoustic Imaging Instruments," *PLoS One*, vol. 8, no. 9, pp. 1–14, Sep. 2013, doi: https://doi.org/10.1371/journal.pone.0075533.
32. M. Fonseca, B. Zeqiri, P. Beard, and B. Cox, "Characterisation of a PVCP-based tissue-mimicking phantom for quantitative photoacoustic imaging," vol. 9539, no. 0, p. 953911, 2015, doi: https://doi.org/10.1117/12.2183684.
33. E. Jeong *et al.*, "Fabrication and characterization of PVCP human breast tissue-mimicking phantom for photoacoustic imaging," *BioChip J.*, vol. 11, no. 1, pp. 67–75, Mar. 2017, doi: https://doi.org/10.1007/s13206-016-1109-4.
34. M. Dantuma, R. van Dommelen, and S. Manohar, "Semi-anthropomorphic photoacoustic breast phantom.," *Biomed. Opt. Express*, vol. 10, no. 11, pp. 5921–5939, Nov. 2019, doi: https:/ /doi.org/10.1364/BOE.10.005921.
35. M. Fonseca, B. Zeqiri, P. C. Beard, and B. T. Cox, "Characterisation of a phantom for multiwavelength quantitative photoacoustic imaging," *Phys. Med. Biol.*, vol. 61, no. 13, pp. 4950–4973, 2016, doi: https://doi.org/10.1088/0031-9155/61/13/4950.
36. L. C. Cabrelli, F. W. Grillo, D. R. T. Sampaio, A. A. O. Carneiro, and T. Z. Pavan, "Acoustic and Elastic Properties of Glycerol in Oil-Based Gel Phantoms," *Ultrasound Med. Biol.*, vol. 43, no. 9, pp. 2086–2094, Sep. 2017, doi: https://doi.org/10.1016/j.ultrasmedbio.2017.05.010.

37. F. W. Grillo, L. C. Cabrelli, D. R. T. Sampaio, A. A. O. Carneiro, and T. Z. Pavan, "Glycerol in oil-based phantom with improved performance for photoacoustic imaging," *IEEE Int. Ultrason. Symp. IUS*, pp. 1–4, Sep. 2017, doi: https://doi.org/10.1109/ULTSYM.2017.8091705.
38. L. Hacker *et al.*, "A copolymer-in-oil tissue-mimicking material with tuneable acoustic and optical characteristics for photoacoustic imaging phantoms," *IEEE Trans. Med. Imaging*, pp. 1–1, Jun. 2021, doi: https://doi.org/10.1109/TMI.2021.3090857.
39. M. Gehrung, S. E. Bohndiek, and J. Brunker, "Development of a blood oxygenation phantom for photoacoustic tomography combined with online pO2 detection and flow spectrometry," *J. Biomed. Opt.*, vol. 24, no. 12, p. 1, Oct. 2019, doi: https://doi.org/10.1117/1.JBO.24.12.121908.
40. W. C. Vogt *et al.*, "Phantom-based image quality test methods for photoacoustic imaging systems," *J. Biomed. Opt.*, vol. 22, no. 09, p. 1, 2017, doi: https://doi.org/10.1117/1.JBO.22.9.095002.
41. E. Maneas *et al.*, "Gel wax-based tissue-mimicking phantoms for multispectral photoacoustic imaging," *Biomed. Opt. Express*, vol. 9, no. 3, p. 1151, Mar. 2018, doi: https://doi.org/10.1364/BOE.9.001151.
42. E. Maneas *et al.*, "Anatomically realistic ultrasound phantoms using gel wax with 3D printed moulds," *Phys. Med. Biol.*, vol. 63, no. 1, p. 015033, Jan. 2018, doi: https://doi.org/10.1088/1361-6560/aa9e2c.
43. FDA, "Tissue-mimicking polymer test phantoms for optical-acoustic medical imaging | FDA." https://www.fda.gov/science-research/licensing-and-collaboration-opportunities/tissue-mimicking-polymer-test-phantoms-optical-acoustic-medical-imaging (accessed Sep. 29, 2020).
44. J. Joseph, M. R. Tomaszewski, I. Quiros-Gonzalez, J. Weber, J. Brunker, and S. E. Bohndiek, "Evaluation of Precision in Optoacoustic Tomography for Preclinical Imaging in Living Subjects.," *J. Nucl. Med.*, vol. 58, no. 5, pp. 807–814, May 2017, doi: https://doi.org/10.2967/jnumed.116.182311.
45. L. Hacker *et al.*, "A Stable Phantom Material for Optical and Acoustic Imaging," *J. Vis. Exp.*, no. 196, Jun. 2023, doi: https://doi.org/10.3791/65475.
46. S. Bohndiek, "Addressing photoacoustics standards," *Nature Photonics*, vol. 13, no. 5. Nature Publishing Group, p. 298, May 01, 2019, doi: 10.1038/s41566-019-0417-3.
47. P. Lemaillet, J.-P. Bouchard, J. Hwang, and D. W. Allen, "Double-integrating-sphere system at the National Institute of Standards and Technology in support of measurement standards for the determination of optical properties of tissue-mimicking phantoms," *J. Biomed. Opt.*, vol. 20, no. 12, p. 121310, Oct. 2015, doi: https://doi.org/10.1117/1.jbo.20.12.121310.
48. J. W. Pickering, S. A. Prahl, N. van Wieringen, J. F. Beek, H. J. C. M. Sterenborg, and M. J. C. van Gemert, "Double-integrating-sphere system for measuring the optical properties of tissue," *Appl. Opt.*, vol. 32, no. 4, p. 399, Feb. 1993, doi: https://doi.org/10.1364/ao.32.000399.
49. L. Spinelli *et al.*, "Calibration of scattering and absorption properties of a liquid diffusive medium at NIR wavelengths. Time-resolved method," *Opt. Express*, vol. 15, no. 11, p. 6589, 2007, doi: https://doi.org/10.1364/OE.15.006589.
50. B. Cletus, R. Künnemeyer, P. Martinsen, A. McGlone, and R. Jordan, "Characterizing liquid turbid media by frequency-domain photon-migration spectroscopy," *J. Biomed. Opt.*, vol. 14, no. 2, p. 024041, 2009, doi: https://doi.org/10.1117/1.3119282.
51. R. M. P. Doornbos, R. Lang, M. C. Aalders, F. W. Cross, and H. J. C. M. Sterenborg, "The determination of in vivo human tissue optical properties and absolute chromophore concentrations using spatially resolved steady-state diffuse reflectance spectroscopy," *Phys. Med. Biol.*, vol. 44, no. 4, pp. 967–981, Apr. 1999, doi: https://doi.org/10.1088/0031-9155/44/4/012.
52. G. Zaccanti, S. Del Bianco, and F. Martelli, "Measurements of optical properties of high-density media," *Appl. Opt.*, vol. 42, no. 19, p. 4023, Jul. 2003, doi: https://doi.org/10.1364/AO.42.004023.
53. D. Hu, R. Lu, Y. Huang, Y. Ying, and X. Fu, "Effects of optical variables in a single integrating sphere system on estimation of scattering properties of turbid media," *Biosyst. Eng.*, vol. 194, pp. 82–98, Jun. 2020, doi: https://doi.org/10.1016/j.biosystemseng.2020.03.012.

54. E. * Oregon Tech and A. Sw Parkway, "Everything I think you should know about Inverse Adding-Doubling," 2011. Accessed: Apr. 28, 2019. [Online]. Available: http://omlc.org.
55. B. Zeqiri, "An intercomparison of discrete-frequency and broadband techniques for the determination of ultrasonic attenuation," *Phys. Med. Ultrasound*, pp. 27–35, 1988.
56. J. C. Bamber, "Attenuation and absorption.," *Phys. Princ. Med. Ultrason.*, vol. 2, pp. 93–166, 1986.
57. W. Sachse and Y. H. Pao, "On the determination of phase and group velocities of dispersive waves in solids," *J. Appl. Phys.*, vol. 49, no. 8, p. 4320, Aug. 2008, doi: https://doi.org/10.1063/1.325484.
58. R. A. Kline, "Measurement of attenuation and dispersion using an ultrasonic spectroscopy technique," *J. Acoust. Soc. Am.*, vol. 76, no. 2, p. 498, Jun. 1998, doi: https://doi.org/10.1121/1.391591.
59. D. I. Bolef and J. De Klerk, "Some Continuous-Wave Techniques for the Measurement of Velocity and Attenuation of Ultrasonic Waves Between 1 and 1000 Mc," *IEEE Trans. Ultrason. Eng.*, vol. 10, no. 1, pp. 19–26, Jul. 1963, doi: https://doi.org/10.1109/T-UE.1963.29296.
60. H. J. McSkimin, "Ultrasonic Methods for Measuring the Mechanical Properties of Liquids and Solids," *Phys. Acoust.*, pp. 271–334, Jan. 1964, doi: https://doi.org/10.1016/B978-1-4832-2857-0.50010-1.
61. R. W. B. Stephens, "Ultrasonic spectral analysis for non-destructive evaluation," *Ultrasonics*, vol. 21, no. 6, p. 284, Nov. 1983, doi: https://doi.org/10.1016/0041-624X(83)90063-X.
62. M. J. W. (Malcolm J. W. Povey, *Ultrasonic techniques for fluids characterization*. Academic Press, 1997.
63. "IEC TS 63081:2019 | IEC Webstore." https://webstore.iec.ch/publication/32282 (accessed Jan. 21, 2021).
64. E. L. Madsen, M. A. Hobson, H. Shi, T. Varghese, and G. R. Frank, "Stability of heterogeneous elastography phantoms made from oil dispersions in aqueous gels," *Ultrasound Med. Biol.*, vol. 32, no. 2, pp. 261–270, Feb. 2006, doi: https://doi.org/10.1016/j.ultrasmedbio.2005.10.009.
65. E. Blumenröther, O. Melchert, M. Wollweber, and B. Roth, "Detection, numerical simulation and approximate inversion of optoacoustic signals generated in multi-layered PVA hydrogel based tissue phantoms," *Photoacoustics*, vol. 4, no. 4, pp. 125–132, Dec. 2016, doi: https://doi.org/10.1016/j.pacs.2016.10.002.
66. L. C. Cabrelli, F. W. Grillo, A. A. O. Carneiro, and T. Z. Pavan, "Copolymer-in-oil tissue-mimicking material with tunable acoustic properties," in *2016 IEEE International Ultrasonics Symposium (IUS)*, Sep. 2016, pp. 1–4, doi: 10.1109/ULTSYM.2016.7728859.
67. L. C. Cabrelli *et al.*, "Stable phantom materials for ultrasound and optical imaging," *Phys. Med. Biol.*, vol. 62, no. 2, pp. 432–447, Jan. 2017, doi: https://doi.org/10.1088/1361-6560/62/2/432.
68. L. C. Cabrelli, P. I. B. G. B. Pelissari, L. P. Aggarwal, A. M. Deana, A. A. O. Carneiro, and T. Z. Pavan, "Oil-based gel phantom for ultrasound and optical imaging," Jun. 2015, p. 95310L, doi: https://doi.org/10.1117/12.2180642.
69. G. Spirou, A. Oraevsky, I. Vitkin, and W. Whelan, "Optical and acoustic properties at 1064 nm of polyvinyl chloride-plastisol for use as a tissue phantom in biomedical optoacoustics," *Phys Med Biol*, vol. 50, 2005.
70. I. Feder, M. Wróbel, H. Duadi, M. Jędrzejewska-Szczerska, and D. Fixler, "Experimental results of full scattering profile from finger tissue-like phantom," *Biomed. Opt. Express*, vol. 7, no. 11, p. 4695, Nov. 2016, doi: https://doi.org/10.1364/BOE.7.004695.
71. C. F. Morsink *et al.*, "Design and characterization of color printed polyurethane films as biomedical phantom layers," *Biomed. Opt. Express*, vol. 14, no. 9, p. 4485, Sep. 2023, doi: https://doi.org/10.1364/BOE.491695.
72. B. W. Pogue and M. S. Patterson, "Review of tissue simulating phantoms for optical spectroscopy, imaging and dosimetry," *J. Biomed. Opt.*, vol. 11, no. 4, p. 041102, 2006, doi: https://doi.org/10.1117/1.2335429.

Part IV
Emerging Techniques in Photoacoustics

Chapter 8
Optical Wavefront Shaping in Biomedical Photoacoustics

Tianrui Zhao and Wenfeng Xia

Abstract Optical wavefront shaping is an emerging technique that controls light transportation in disordered media. It has attracted significant interest in the photoacoustics community in recent years. On one hand, photoacoustic signals can provide feedback on local light intensity to guide optical wavefront shaping non-invasively for focusing light through optically diffusive media. On the other hand, the wavefront shaping technique can be used to focus light through multimode fibres to achieve ultrathin photoacoustic endomicroscopy probes. This chapter reviews the latest research that lies at the intersection of wavefront shaping and photoacoustics, encompassing photoacoustic-guided wavefront shaping and multimode fibre-based photoacoustic endomicroscopy techniques.

Keywords Photoacoustic imaging · Wavefront shaping · Photoacoustic guidestar · Photoacoustic endomicroscopy · Ultrathin endoscope

8.1 Introduction

Optical wavefront shaping is an emerging technique that has shown potential to overcome optical scattering in biological tissue [1–4]. By modulating the incident wavefront to compensate for scattering, it is possible to shape the transmitted light into desired configurations within or through disordered media, enabling focusing of scattered light at target positions and providing great potential for deep-tissue microscopy imaging. In the past decade, advances in photoacoustics have contributed significantly to the development of optical wavefront shaping, and in return, recent advances in optical wavefront shaping have broadened the landscape of photoacoustic imaging (PAI) [5–8]. PAI is a hybrid modality that has been widely studied in biomedical sciences. By exciting ultrasound from

T. Zhao · W. Xia (✉)
School of Biomedical Engineering & Imaging Sciences, King's College London, London, UK
e-mail: tianrui.zhao@kcl.ac.uk; wenfeng.xia@kcl.ac.uk

W. Xia (ed.), *Biomedical Photoacoustics*,
https://doi.org/10.1007/978-3-031-61411-8_8

endogenous tissue chromophores or exogenous contrast agents with pulsed or intensity-modulated light, PAI encodes rich optical contrast into ultrasound signals to provide depth-resolved information of biological tissues with acoustically defined spatial resolution. Photoacoustic amplitude is proportional to the local light fluence at a specific spatial location, providing valuable feedback to guide wavefront shaping in turbid tissue non-invasively. The enhanced light intensity improves the signal-to-noise ratio of optically excited ultrasound waves. In optical imaging that relies on raster-scanning a laser beam for acquiring images in a pixel-by-pixel manner, wavefront shaping could also enhance the spatial resolution by focusing the laser beam into a tightly focused spot (e.g. from acoustically defined spatial resolution to optically defined spatial resolution). Furthermore, wavefront shaping has been explored for developing minimally invasive PAI tools that enable in-situ examinations of internal tissue from inside the body [9–11]. This technique allows the use of multimode fibres (MMFs) for forward-viewing ultrathin photoacoustic endoscopes, which offer substantially smaller probe sizes, significantly higher spatial resolutions, and much lower costs than previous devices that used coherent fibre bundles for excitation laser delivery.

This chapter aims to provide a comprehensive review of the recent developments in combining photoacoustics and optical wavefront shaping techniques for biomedical research. The chapter is organised as follows: Section 8.2 introduces the principles of the optical wavefront technique and discusses conventional methods for focusing laser light through disordered media. Section 8.3 presents the use of photoacoustic guidestars in wavefront shaping for effectively focusing laser light through tissue. Section 8.4 reviews the advancements in photoacoustic endomicroscopy based on MMFs.

8.2 Principle of Optical Wavefront Shaping

Photons propagating through biological tissues follow different trajectories (channels) with different optical path lengths due to optical scattering, which is caused by the heterogeneous spatial distribution of the refractive index. As a result, the incident wavefront is scrambled into a seemingly random output speckle pattern, as illustrated by Fig. 8.1a. The transmitted light (output) behind the media is the interferential result of photons from all the channels. The light field at any particular output mode, which is usually defined by a spatial position, can be computationally modelled as:

$$E_{\text{out}}^{m} = \sum_{n=1}^{N} t_{mn} E_{\text{in}}^{n} \tag{8.1}$$

where E_{out}^{m} is the output light field at the mth output mode E_{in}^{n} is input light field at the nth input mode, N is the total number of input modes, $t_{mn} = A_{mn}e^{i\theta_{mn}}$ is a

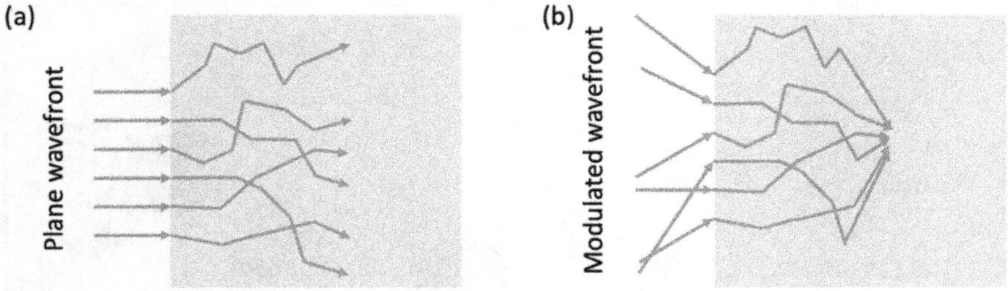

Fig. 8.1 Wavefront shaping for light focusing into an optical diffuser. (**a**) Natural light diffusion in a diffusive medium. (**b**) Light is focused into a spot in the diffusive medium with a modulated incident wavefront

transmission constant representing both the amplitude (A) and phase (θ) changes from the nth input mode to the mth output mode. As such, modulating the light phases at all input modes can compensate for the phase distortions and allow for constructive interference of different channels to achieve light focusing. To modulate the incident wavefronts, various spatial light modulators (SLM) have been explored including liquid-crystal-based SLM (LC-SLM), digital micromirror device (DMD), micro-electromechanical-systems-based SLM (MEMS-SLM), and deformable mirrors (DM). Illustrative diagrams showing light focusing inside an optically diffusive medium with wavefront shaping is presented in Fig. 8.1b. The enhancement of light intensity at the focused spot is often characterised by the peak-to-background ratio (PBR), which is the ratio of the light intensity at the focal position over the average background intensity.

Various methods have been developed for focusing light through disordered media. Early studies relied on iterative algorithms [12, 13], which used the light intensity at the target position as feedback to optimise the incident wavefront. As shown in Fig. 8.2a, one implementation of this approach is a continuous sequential algorithm that fixes an unchanged input mode for a static output light field and sequentially modulates other input modes to enhance light intensity at the output mode. This process can be iterated multiple times to maximise the PBR of the focus. Another commonly used iterative algorithm for wavefront shaping is genetic algorithms (GA) [14]. In GA, several random input wavefronts are used as the original generation population, each pattern is considered as the chromosome of an individual, and the phases and amplitudes coded in the input wavefront are considered chromosome codes. Output speckles are recorded as the input wavefronts are projected on the disordered media, and ranked according to the pre-defined feedback, usually the PBR values. Then individuals with larger PBRs were assigned larger probabilities to be selected as parents to produce the next generation population by crossing the parent chromosomes with a constant cross rate. Mutation can be introduced by randomly switching a small number of chromosome codes with a mutate rate to avoid locally optimal solutions. The producing and ranking phases are repeated many times to gradually enhance the feedback signal.

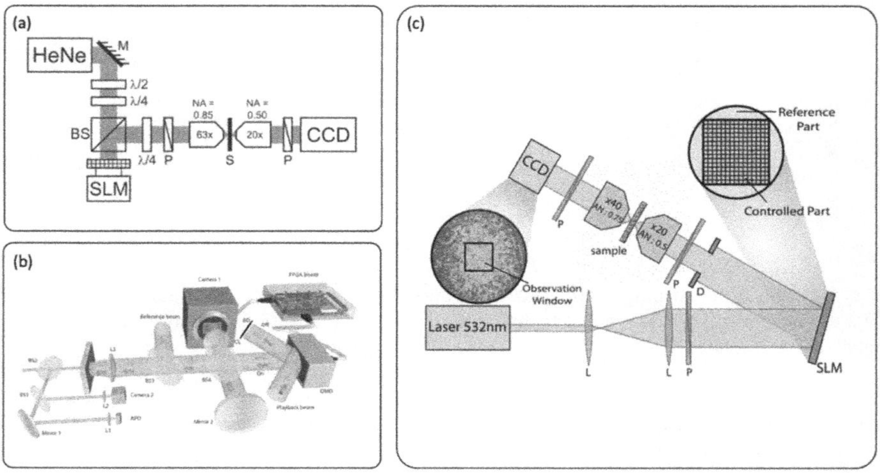

Fig. 8.2 Typical experimental configurations of wavefront shaping with (**a**) iterative algorithm (image adapted from [12] with permission), (**b**) digital optical phase conjugation (image adapted from [15] with permission), and (**c**) transmission matrix methods (image adapted from [19] with permission)

Digital optical phase conjugation (DOPC) is another approach for light focusing via wavefront shaping (Fig. 8.2b) [15–18]. It allows single-shot measurement of transmission constants (t_{mn}) linking all input modes to one output mode. In this approach, light is firstly focused at a specific output position and the transmitted light wavefront through the media is holographically recorded by a camera. Since light transport through the same media is reversible, light focusing could be recovered by projecting the phase-conjugated wavefront back with a spatial light modulator that is placed at the same optical plane as the camera. Compared to other methods, the single-shot nature enables ultrafast implementations of light focusing with the DOPC method.

Different from iterative algorithms and the DOPC approach, another commonly used approach characterises the disordered light transport between all input and output modes with a transmission matrix (TM) as:

$$E_{\text{out}} = \text{TM} \cdot E_{\text{in}} \tag{8.2}$$

where E_{out} is the output light field and E_{in} is the input light field that covers all input modes. The complex-valued TM [19–23] and its variants, such as binary transmission matrix [24–26] and real-valued intensity transmission matrix [27, 28], can be empirically measured from a series of input and output pairs. The light field measurement usually includes complex information of both phase and amplitude, which enables direct calculation of the TM value via simple matrix manipulation. An external reference arm that interferes with the output light field at the camera

Table 8.1 Theoretical peak-to-background ratio with various wavefront modulation strategies

Wavefront modulation methods	Phase and amplitude	Phase-only	Binary phase	Binary amplitude
Peak-to-background ratio	N	$\pi N/4$	N/π	$N/2\pi$

N is the number of independently controlled input modes

can be used to holographically capture the phase and amplitude information. The internal reference is another choice for TM characterisation, in which a fraction of the spatial light modulator was kept unchanged to project a static wavefront onto diffusers. A typical configuration of TM characterisation of a diffuser with internal reference is shown in Fig. 8.2c. While the central region of the modulator was modulated to change the light wavefront, the other region of the modulator was kept static, which provided a constant component in the light field as the internal reference. Computationally modelling the complex-valued TM with intensity-only light measurements has also been demonstrated via algorithms such as phase-shift algorithms [19, 29–32] and phase retrieval algorithms [26, 33–35]. With TM characterisation, the output light field could be tailored into a desired pattern by shaping the optimal input wavefront that is calculated via the following equation:

$$E_{\text{in}} = \text{TM}^{-1} E_{\text{out}}. \tag{8.3}$$

The output field is usually shaped into a tightly focused light spot that is scanned for microscopy imaging. While complex TM characterisation allows for amplitude-phase modulation for achieving nearly perfect light focusing [36, 37], the phase-only modulation, which offers high-quality light focusing with simpler incident light modulation methods, is commonly used in wavefront shaping. In the phase-only implementation, the spatial light modulator compensates for the phase latency indicated by the TM at each input mode, leading to significant constructive interference among light fields with the same phase at the target output mode. In addition, binary amplitude [15, 26, 28] and binary phase [38] modulation methods are also alternatives for light focusing but with degraded focusing performance. The theoretical PBRs with different modulation strategies are shown in Table 8.1.

8.3 Photoacoustics-Guided Wavefront Shaping

8.3.1 Guidestar-Assisted Light Focusing

Most biological tissues are turbid due to spatially varying refractive index that scatters light randomly. This limits the working range of imaging modalities that rely on optical focusing, as photons can only travel about 1 mm in biological tissue before significantly changing their directions. Wavefront shaping allows

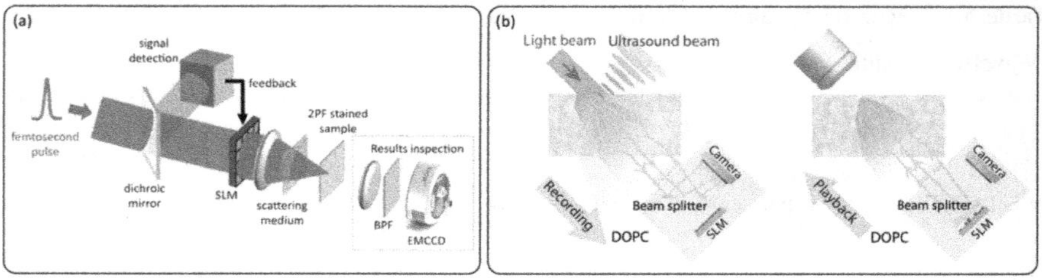

Fig. 8.3 Typical configurations of guidestar-assisted wavefront shaping. (**a**) Two-photons feedback (image adapted from [42] with permission). (**b**) Acoustic-optic feedback (image adapted from [45] with permission)

the correction of scrambled light transport through disordered media as presented in the previous section. However, these methods typically require a camera to capture the output light field information, which limits the usability of wavefront shaping for in vivo applications due to the invasive nature of these approaches. To address this, various research groups have been seeking alternative signals for use as guidestars that are closely associated with the local light intensity and can be measured non-invasively for directing light focusing inside biological tissues, including fluorescence [39–41], two-photons [42, 43], acoustic-optics [44–50], photoacoustics [5–7] and moving particles [51–53]. Typical configurations of fluorescence and acoustic-optic feedbacks are shown in Fig. 8.3. In general, fluorescence and two-photon guidestars require specific dyes (fluorophores) for generating light signals at a different wavelength from the excitation laser, while acoustic-optic guidestar is more flexible as the location and size of the optical focus are defined by the spatial location and size of the acoustic focus. Compared with feedback that relies on optical signals, photoacoustic guidestar is based on strongly absorptive tissue as it converts the excitation light into acoustic signals that can be measured externally. It is also well suited for improving the tissue penetration depth of photoacoustic imaging by enhancing the signal-to-noise ratios of the photoacoustic signals. Thus, it has attracted extensive research interest. For those who are interested in this topic, virtual guidestar-based wavefront shaping methods have been reviewed by Horstmeyer et al. in detail [54]. The next section (Sect. 8.3.2) will focus on photoacoustic guidestars that have shown promising results for improving photoacoustic imaging.

8.3.2 Photoacoustic Guidestars for Light Focusing

The photoacoustic signal amplitude that is optically excited from an absorber is proportional to the local optical fluence as:

$$P = aS\Gamma\mu_a F \tag{8.4}$$

where a is a constant characterising the ultrasound signal loss during propagation through the medium, *S* is the sensitivity of the ultrasound sensor, *Γ* is the Grüneisen coefficient, μ_a is the optical absorption coefficient of the absorber, and *F* is the local optical fluence. Since the excitation laser has a constant illumination duration *T*, the local optical fluence can be expressed as the integration of the light intensity during the light pulse over the illumination area $F = (I \cdot T)/A$, where I is the average light intensity over *T*. So, the photoacoustic amplitude can be further expressed as:

$$P = \frac{aS\Gamma\mu_a I}{AT}. \tag{8.5}$$

Assuming that the acoustic properties of the medium, the optical and thermoelastic properties of the absorber, and the acoustic sensitivity of the ultrasound detector remain unchanged in a common configuration, the photoacoustic signal amplitude is linearly proportional to the local light intensity. So, by measuring the photoacoustic signal amplitude non-invasively with external ultrasound sensors, non-invasive monitoring of the internal light intensity within a tissue can be realised. Compared to other optical guidestars, such as acousto-optic [45] and two-photon [42] based methods, wavefront shaping with a photoacoustic guidestar benefits from a relatively simple setup. Further, there are abundant endogenous absorbing chromophores available in biological tissue including DNA/RNA, water, lipid, and haemoglobin, etc., which ease the need for an exogenous contrast agent that is required in fluorescence-based methods. As a result of enhanced local light fluence, the enhanced photoacoustic amplitude could directly improve signal-to-noise ratios of the resulting photoacoustic images. The photoacoustic amplitude enhancement is closely associated with the light intensity enhancement achieved with wavefront shaping, which is determined by the wavefront modulation strategy and the number of controllable input modes (Table 8.1). The area for light focusing is usually larger than the sizes achievable with the optical-diffraction-limit, which allows it to cover multiple individual output modes inside the focal area. The resulting photoacoustic amplitude enhancement is therefore also influenced by the number of individual output modes covered in the focusing area. For instance, the resulting photoacoustic amplitude enhancement achieved by wavefront shaping using a DMD with binary amplitude modulation can be expressed as:

$$\eta_{PA} = \frac{N}{2\pi M} \tag{8.6}$$

where *M* is the number of output modes within the focal area.

Typically, a focused ultrasound transducer is used to guide the light convergence into the acoustic focus using iterative optimisation algorithms, and the wavefront-shaping-based light focusing can be directed to different locations by moving the transducer. A linear array ultrasound probe has also been used for guiding light focusing at a specific spatial location in photoacoustic images [55]. Various modulators, including deformable mirrors, LC-SLMs, and DMDs, have been used to shape the incident wavefront. In 2010, Kong et al. demonstrated photoacoustic-guided convergence of light through an optical diffuser using iterative wavefront modulations [56]. They used a deformable mirror and a continuous sequential algorithm to guide the laser beam behind the diffuser into the focus of the ultrasound transducer or onto small absorbers like graphite particles (Fig. 8.4a). It took 20 min to optimise the incident wavefront in 3 iterations, with each pixel of the deformable mirror completing 40 steps of phase increment from 0 to 2π.

Light focusing within the acoustically defined local area has been achieved using LC-SLM with a genetic algorithm in a series of studies. The diameter of the light focus, which is considered the point spread function in microscopy imaging, is defined by the size of the acoustic focus that is dependent on the frequency of the ultrasound transducer. Several methods were proposed to further reduce the size of the optical focal spot. In 2015, Conkey et al. [7] focused light into a ~ 13 μm-in-diameter spot with 24 times enhancement in light intensity, by using the Gaussian spatial sensitivity in the ultrasound focal area (~38 μm-in-diameter). The Gaussian spatial sensitivity gave higher weighting to ultrasound generated closer to the centre and thus prioritised focusing light at the centre location (Fig. 8.4b). In the same year, Lai et al. [6] reported an optical diffraction-limited focusing with a nonlinear, dual-pulse excitation photoacoustic feedback based on the Grüneisen relaxation effect. Different from previous studies that relied on linear photoacoustic feedback, they fired two sequential pulses to generate two photoacoustic waves with one incident wavefront. Whilst the temperature rise caused by the first pulse led to a lingering change in the Grüneisen parameter of the absorber, the second pulse firing within the thermal confinement time excited a stronger ultrasound wave. The difference between the two ultrasound amplitudes approximately increases in inverse proportion to the number of bright speckle grains covered by the acoustic focus. The method comprised two stages; first, a linear stage was based on a genetic algorithm with 800 iterations to focus light into the acoustic focus. Second, a non-linear stage further focused the laser beam into one grain (5–7 μm-in-diameter) with a light intensity enhancement of ~6000 times after 1600 iterations (Fig. 8.4c). In 2016, Tzang et al. reported nonlinear photoacoustic feedback based on lock-in detection of photoacoustic signals to focus light through scattering media [57]. Since nonlinear feedback can generate a single diffraction-limited focus even if the speckle field is non-localised, this feedback improved the light enhancement from 9 (with linear photoacoustic feedback) to 16 and decreased the number of modes in the focal region by ~1.4 times. In 2020, Inzunza-Ibarra et al. reported a sub-acoustic resolution optical focusing using flowing absorbers [58], which led to photoacoustic response variations due to different particle distributions. This variation was characterised by taking the standard deviation of the photoacoustic signal amplitude at each point in time over multiple measurements, and the decline in illumination volume increases the standard deviation of photoacoustic signals nonlinearly. As such, maximising this standard deviation with a genetic algorithm resulted in light focusing into a 12.7 μm-in-diameter spot (more than 16 times smaller than the size of the acoustic spot) with a 115 times optical intensity enhancement (Fig. 8.4d).

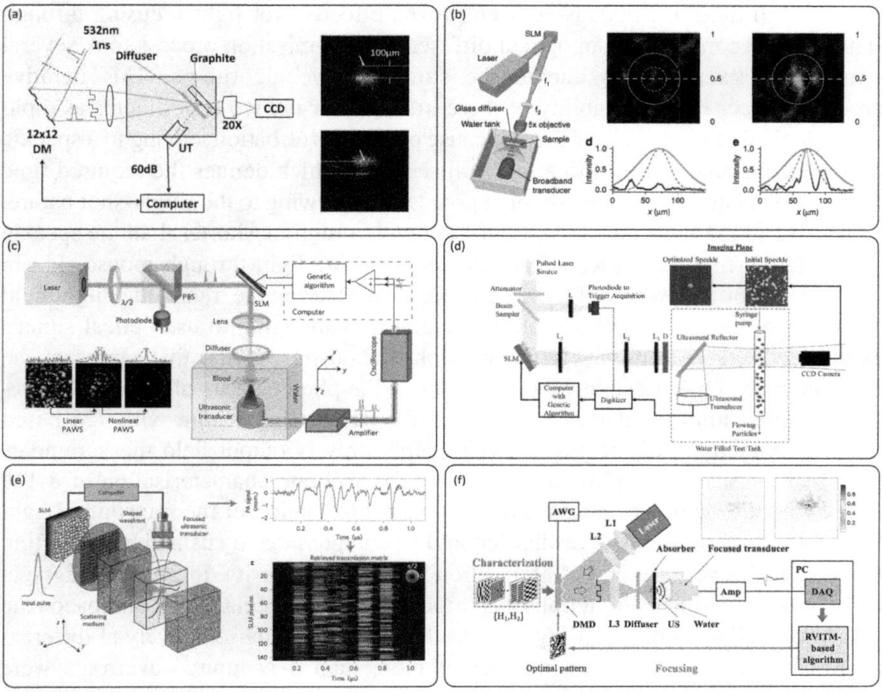

Fig. 8.4 Wavefront shaping with photoacoustic guidestars. (**a**) Laser light convergence through a carbon tape by using a deformable mirror modulating the incident wavefront and a focused ultrasound transducer for defining focusing positions. The inset images show that the laser beam was guided to the acoustic focus (marked by an arrow) with a diameter of 100 μm. Adapted from [56] with permission. UT, ultrasound transducer; DM, deformable mirror; CCD, camera. (**b**) Sub-acoustic light focusing. Based on the Gaussian spatial sensitivity in the ultrasound focal area, continuous optimisation of the incident wavefront focused light into the red dash circle which was smaller than the transducer focal region (red solid circle) [7]. (**c**) Dual-stage implementation of wavefront shaping with linear and non-linear photoacoustic signals as the feedback. In stage 1, the linear implementation focused randomly scattered light (left inset) into the acoustic focal region (middle inset). In stage 2, the use of nonlinear feedback further focused light onto a single-speckle grain (right inset). The acoustic focal area was marked by blue dashed circles. The profiles of light beams are illustrated by green lines. PBS, polarised beam splitter; SLM, spatial light modulator; λ/2, half-wave plate. Adapted from [6] with permission. (**d**) Sub-acoustic optical focusing through scattering using photoacoustic fluctuation feedback. The laser light was illuminated onto flowing particles. The standard deviation of the photoacoustic amplitude at each point in time over multiple measurements was used as the feedback. Increasing the standard deviation by using an SLM with a genetic algorithm led to laser light focusing inside a sub-acoustic resolution area. Adapted from [58] with permission. (**e**) Photoacoustic transmission matrix. A set of input (SLM pattern)-output (ultrasound signal) pairs were measured to calculate the transmission matrix with a phase-shift algorithm. Then, optimal input wavefronts were achieved for maximising the ultrasound amplitude at a targeted position [5]. (**f**) High-speed photoacoustic-guided wavefront shaping. A set of binary patterns were used to characterise the light transport through a diffuser with a real-valued intensity transmission matrix. The light speckles at the acoustic focal plane before and after wavefront shaping are shown in the insets. Amp, amplifier; L1–L3, convex lenses; DMD, digital micromirror device; AWG, arbitrary waveform generator. Adapted from [61] with permission

Although these methods have been proven effective for light focusing through static diffusive media like an optical diffuser, the optimisation process took several minutes to several hours due to the slow iterative algorithms. This iterative process reduces the applicability of these methods for in vivo experiments as rapid physiological motions in living tissue cause photon perturbation leading to a speckle decorrelation time of down to a few milliseconds, which defines the required time frame for the entire wavefront shaping process [59]. Owing to the single-shot nature, DOPC-based methods have shown promise with a runtime shorter than the speckle decorrelation time in tissue, which enabled focusing light through mouse skin in vivo [15, 50, 60]. However, the DOPC relied on measuring the transmitted light field at the target location using a camera, making it challenging to use optical signals as the feedback non-invasively (such as photoacoustic signals) that represent the light intensity. One possible way to accelerate the photoacoustically guided wavefront shaping implementation is replacing the iterative algorithm with TM-based methods, which usually require approximately only N output field measurements with both phase and amplitude information for accurate characterisation of a TM with N input modes. The absence of the phase information of the transmitted light field with the photoacoustic feedback could be compensated by using phase-shifting algorithms that retrieve phase information from light intensity. In 2013, Chaigne et al. [5] characterised the light transport through diffusive media with a photoacoustic TM, which allowed the focusing of light through the media selectively at different positions (Fig. 8.4e). A total number of more than 3 N input wavefronts were used for the TM characterisation. In this study, a total acquisition time of ~40 min was taken to achieve light focusing, which was mainly limited by the 10 Hz laser repetition rate. The slow spatial light modulators such as LC-SLMs that run at ~100 Hz also limited the speed of wavefront shaping. In 2021, we reported a high-speed photoacoustic-guided wavefront shaping method and demonstrated light focusing through an optical diffuser by using a DMD modulating the incident wavefront [61], as shown in Fig. 8.4f. The use of DMD significantly accelerated the characterisation implementation as it ran at 47 kHz. Furthermore, a real-valued intensity transmission matrix (RVITM), which linked the incident light intensity with the measured photoacoustic signal amplitude at the target location, was used to characterise the light transport. This algorithm reduced the required number of input-output pairs to 2 N for guiding binary amplitude modulations with N input modes. For experimental validation, a set of binary patterns was displayed on a DMD whilst the photoacoustic signals were detected by a focused ultrasound transducer. The computation for the RVITM involved only simple matrix manipulation. As a result, the total runtime, including the characterisation process and displaying the optimal DMD pattern for laser focusing, was ~300 ms, which is a significant step forward towards in vivo applications. Performance comparisons of the various photoacoustic-guided wavefront shaping methods are presented in Table 8.2.

Table 8.2 Wavefront shaping through optical diffusers with photoacoustic guidance

Reference	Modulator	Characterisation	Feedback	Input modes	Enhancement	Focus size (μm)	Runtime	RPM[a]
Kong et al. [56]	DM	Continuous sequential algorithm	Photoacoustic amplitude	140	–	–	20 min	8.6 s
Lai et al. [6]	LC-SLM	Genetic algorithm	Photoacoustic amplitude/dual-pulse	20,736	6000	5–7	12 h	0.14 s
Inzunza-Ibarra et al. [58]	LC-SLM	Genetic algorithm	Photoacoustic fluctuation	1024	115	12.7	–	–
Caravaca-Aguirre et al. [63]	LC-SLM	Genetic algorithm	Photoacoustic amplitude	804	10	–	15 min	1.12 s
Sun et al. [64]	DMD	Genetic algorithm	Photoacoustic amplitude	–	7.83	150 × μm	30 mins	–
Chaigne et al. [5]	LC-SLM	TM	Photoacoustic amplitude	160	–	–	40 min	66.67 μs
Tzang et al. [62]	LC-SLM	TM	Photoacoustic amplitude	256	16	300 μm	~ 1 h	14 s
Zhao et al. [61]	DMD	RVITM	Photoacoustic amplitude	4096	6.89	56 μm	306 ms	75 μs

[a]*RPM* Runtime per (input) mode

8.3.3 *Future Perspectives*

Since its first demonstration by Vellekoop et al. in 2007 [12], wavefront shaping for focusing light through scattering media has attracted immense interest. Despite advances in improving focusing performance, reducing invasiveness, increasing speed, and maintaining performance for deep tissue in vivo remains highly challenging due to short speckle decorrelation times in living tissues—a key technological bottleneck for in vivo applications. A recent study revealed that the tissue dynamic-induced light distortion can be decomposed into a stable component that remains constant between two breaths and a fast decorrelation component caused by rapid tissue motions such as blood flow; whilst the latter could be suppressed by averaging sequential light field measurements [60]. As such, a digital phase conjugation (DPC) wavefront shaping approach was iteratively implemented for high-performance light focusing through a mouse ear in vivo within a breath interval. While the DPC approach is not suitable for in vivo applications due to its invasiveness, the two-component concept could be applicable in photoacoustic-guided wavefront shaping for non-invasive light focusing. Another promising way is to employ ultrafast light modulators and modulation algorithms to overcome the speckle decorrelation due to significant tissue motions. In 2019, Tzang et al. reported the use of a grating light valve with thousands of free-standing silicon nitride ribbons on a silicon chip, to focus light through dynamic media such as diffusive phantoms and multimode fibres (MMFs) [62]. This modulator allowed 350 kHz continuous phase modulation and allowed a light-focusing implementation with a three-phase TM characterisation algorithm within several ms, which could be combined with a non-invasive photoacoustic guidestar for focusing light through living tissues.

8.4 Wavefront Shaping-Assisted Photoacoustic Endomicroscopy

In the past decade, various side-viewing photoacoustic probes have been developed to visualise tissue within hollow structures by rotational scanning of the probe [65], which could be particularly useful for applications in intravascular imaging [66, 67] and gastrointestinal tract imaging [68–70]. Recently, forward-viewing photoacoustic endoscopy probes have attracted significant interest, as they could be useful for guiding minimally invasive procedures by visualising critical tissue structures in front of the probe, such as needle biopsy, fetal surgery, and neurosurgery.

Early implementations relied on coherent fibre bundles based on the principles of either optical-resolution photoacoustic microscopy (OR-PAM) or photoacoustic computed tomography. With the microscopy approach, a focused laser beam could be scanned by a MEMS or a galvo mirror across the proximal facet of a fibre bundle to deliver light through the individual cores of the bundle to illuminate tissue at the distal facet, whilst the optically generated ultrasound could be received by an extracorporeal ultrasound transducer [71, 72]. The spatial resolution is defined by the size of the inter-core distance of the bundle. The tomographic approach used a bichromatic Fabry-Perot planar sensor integrated at the tip of a rigid fibre bundle, which was interrogated by raster-scanning a laser beam through the bundle. In this case, each core of the bundle represented a miniature and broadband ultrasound sensing element, leading to high-quality images of animals and human vasculatures [73] with a much larger penetration depth than that of OR-PAM. However, although the bundle-based microscopy enabled a ~ 7 μm spatial resolution, it is limited by gaps between individual cores within the fibre bundle.

8.4.1 Multimode Fibre-Based Photoacoustic Endomicroscopy

With recent advances in wavefront shaping, MMFs have shown the potential to revolutionise medical endoscopy for visualising hard-to-reach areas in the body owing to a large number of transverse optical modes [18, 22, 29, 74–76]. Compared to fibre bundles, MMFs are extremely cost-effective (2–3 orders of magnitude lower costs), and can achieve 1–2 orders of magnitude higher spatial resolution. Further, MMF-based lensless endoscopes provide greater flexibility as the focal spot diameter, shape, and focal plane can be adjusted in contrast to CFB-based implementations. Coherent light travels in thousands of modes in an MMF and the mode dispersion scrambles the incident wavefronts into seemingly random speckle patterns. Similar to light transportation within scattering media, the output of an MMF could be shaped into a tightly focused spot by modulating the incident wavefront. In 2013, Papadopoulos et al. reported the concept of OR-PAM imaging through an MMF with a black nylon wire [77]. Optical-diffraction-limited focusing was achieved with the DOPC method and the optically excited ultrasound was detected by an external piezoelectric transducer (Fig. 8.5a). In 2015, Stasio et al. developed a capillary waveguide that had a hollow core within a 330 μm diameter silica ring for dual-modal fluorescence and photoacoustic imaging [78]. After the waveguide characterisation with the DOPC method, the excited fluorescence light from the sample was collected through the silica ring and detected by a camera at the proximal probe tip, whilst the hollow core, filled with water, served as an ultrasound waveguide to deliver the excited ultrasound signal to the proximal probe tip and detected by a piezoelectric transducer. A black nylon wire placed at the distal probe tip was visualised by photoacoustic imaging. A photoacoustic microscopy image usually comprises several to tens of thousands of pixels which require the same number of scanning positions, which can lead to a long image acquisition time. Although the DOPC method and iterative algorithms allowed light focusing through MMFs, they involved either a time-consuming optimisation process or a complex optical system requiring the accurate placement of a camera and an SLM at the same optical plane.

Recent advances in fibre-optic ultrasound sensors have been key to the miniaturisation of MMF-based OR-PAM towards ultrathin endomicroscopy. In 2020, Mezil

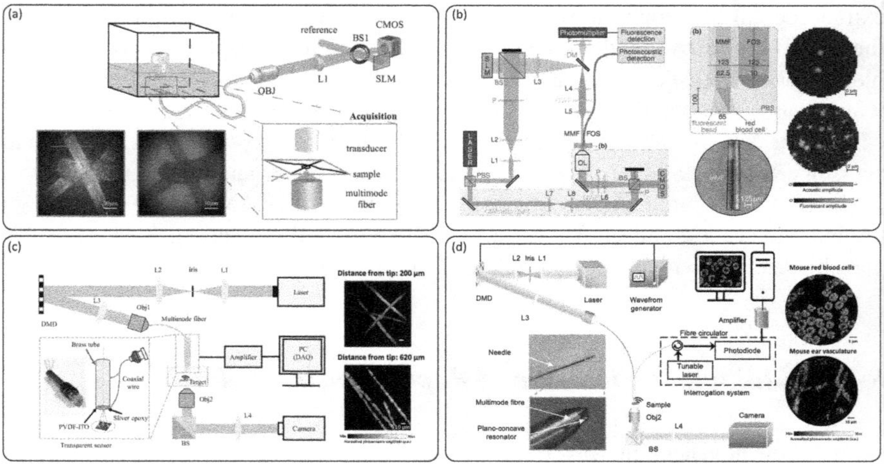

Fig. 8.5 Wavefront shaping-assisted photoacoustic endomicroscopy imaging systems. (**a**) Multimode fibre-based imaging probe with an external piezoelectric ultrasound transducer. Optical-resolution photoacoustic microscopy images of Nylon wires are shown in insets. The fibre was characterised by the digital optical phase conjugate method. Adapted from [77] with permission. (**b**) Multimode fibre-based imaging probe with a fibre-optic ultrasound sensor. Dual-modal photoacoustic and fluorescence imaging on red blood cells and fluorescent beads was achieved through the same probe. Adapted from [10] with permission. (**c**) Multimode fibre-based imaging probe with a miniature transparent piezoelectric transducer. Carbon fibres at various distances from the probe tip were achieved as shown in the insets. Adapted from [85] with permission. (**d**) High-speed photoacoustic endomicroscopy imaging system with a DMD and a fibre-optic ultrasound sensor. The multimode fibre and the sensor fibre were placed inside a needle as shown in the insets. High-speed photoacoustic microscopy imaging at ~3 fps was achieved by using a DMD modulating the incident wavefront. Imaging results of mouse red blood smear and ear micro-vasculature were shown in insets. Adapted from [9] with permission

et al. reported a dual-modal photoacoustic/fluorescence probe based on an MMF [10]. Excited fluorescence light was collected by the MMF whilst the photoacoustic signals were received by a fibre-optic ultrasound sensor (Fig. 8.5b). The probe had a 250 μm by 125 μm footprint and provided a field-of-view covering a circular area with a diameter of 50 μm. Validation was performed by imaging red blood cells. However, it took 30 s for single-frame acquisition due to the use of a slow LC-SLM. In 2021, we reported a dual-modal photoacoustic and fluorescence imaging probe through a multimode fibre [79]. A digital micromirror device (DMD) was employed to modulate the incident optical wavefront, and a real-valued intensity transmission matrix (RVITM) algorithm developed in our group [28] was used to characterise the light transport through the multimode fibre. The RVITM was then subsequently used to generate optimal DMD patterns for raster-scanning of a focused laser beam for imaging. We demonstrated that the spatiotemoral resolution of the imaging system was scalable with the laser scanning parameters. With a high-speed DMD running at 47 kHz, video-rate imaging was achieved and high-fidelity images were obtained from mouse red blood cells. In 2022, we miniaturised this probe by replacing the piezoelectric transducer with a fibre-optic ultrasound sensor that was based on a plano-concave micro-resonator at the tip of a single-mode fibre [80] (Fig. 8.5d). The MMF and the fibre-optic sensor were parallelly placed within the cannula of a 20-gauge needle. High-fidelity photoacoustic images of mouse red blood cells and ear vasculature ex vivo were achieved at a spatial resolution of 1.2 μm. Acquisition of images covering an area of 100 μm in diameter was achieved with the needle probe at a rate of ~3 frames per second, allowing the use of a video mosaicking program to automatically stitch sequential frames during the movement of the needle probe to enlarge the field of view. In addition to using MMFs for the delivery of the excitation laser, wavefront shaping could also be used to develop a photoacoustic computed tomography probe based on an MMF, by scanning a focused laser beam to interrogate a planar Fabry-Perot resonator at the MMF tip [9].

Transparent piezoelectric transducers [81–84] have been used in forward-viewing photoacoustic endoscopy probes. In contrast to a fibre-optic sensor that was placed adjacent to an MMF, a miniature transparent piezoelectric transducer can be placed directly at the fibre tip. The concept of using a miniature transparent piezoelectric transducer in photoacoustic endomicroscopy imaging was demonstrated in our recent work [85]. An ITO-coated PVDF thin film was cropped into 2 mm in diameter and coated onto the tip of a brass tube. An MMF was placed inside the tube with the piezoelectric thin film attached to the fibre tip. Photoacoustic imaging of carbon fibre samples at various distances from the probe tip was achieved as shown in Fig. 8.5c. The performance of different imaging probes is compared in Table 8.3. Although transparent ultrasound sensors benefit from a simple manufacturing process, high stability, and a low cost compared to optical ultrasound sensors, the sensitivity is compromised due to the small footprint.

8.4.2 Future Perspectives

The use of MMFs involves the challenges of fibre bending-induced system instability as changes in the fibre geometry can lead to substantial changes in the light transmission characteristics, leading to degraded focusing performance. As demonstrated in a prior study [86], gradient-index MMFs showed high robustness to fibre bending, which allowed raster-scan-based imaging even when the fibre was bent with large curvatures. For example, with mosaicking imaging in Ref. [9], although the needle probe was mechanically translated, the imaging performance was not substantially degraded, suggesting that this graded-index fibre was resistant to modest fibre bending. However, as a step towards clinical translation, future studies are required to investigate the impact of the MMF with complex shape changes and semi-rigid configurations (e.g. probe integrated within a rigid needle) on the imaging performance. Several approaches have been reported to address the bending issue, such as TM reconstruction based on the known fibre geometry [29] and GPU-accelerated real-time characterisation [87]. However, achieving a fully

Table 8.3 Multimode fibre-based photoacoustic endoscopy systems

Reference	Imaging modality	Resolution	Pixel acquisition rate	Ultrasound detection	Fibre calibration	Modulator	Fibre	Imaging targets
Papadopoulos et al. [77]	Photoacoustic	1.5 μm	60 Hz	External, piezoelectric	DOPC	LC-SLM	ø220 μm, 0.53 NA	Nylon wires
Stasio et al. [78]	Photoacoustic + fluorescence	1.5 μm	60 Hz	Proximal, piezoelectric	DOPC	LC-SLM	Ø330 μm capillary	Nylon wires, fluorescent beads
Mezil et al. [10]	Photoacoustic + fluorescence	1.5 μm	60 Hz	Fibre-optic sensor	TM	LC-SLM	Ø125 μm, 0.275 NA	Red blood cells + fluorescent beads
Zhao et al. [79]	Photoacoustic + fluorescence	1.7–3 μm	47 kHz	External, piezoelectric	RVITM	DMD	Ø140 μm, 0.29 NA	Mouse blood cells + fluorescent beads
Zhao et al. [9]	Photoacoustic	1.2–1.4 μm	22.7 kHz	Fibre-optic	RVITM	DMD	Ø140 μm, 0.29 NA	Mouse blood cells + vasculature
Zhao et al. [85]	Photoacoustic	2–3.7 μm	22.7 kHz	Transparent piezoelectric	RVITM	DMD	Ø140 μm, 0.29 NA	Carbon fibres

flexible MMF-based imaging probe with wavefront shaping remains challenging. Apart from MMFs, several novel optical fibres have been investigated for higher resistance to geometry changes of the fibre. A conformationally invariant multi-core fibre with twisted cores was reported, which permitted light focusing during fibre bending for confocal two-photon imaging [88]. A disordered glass-air Anderson localisation optical fibre was also demonstrated to be resistant to fibre bending for image transmission applications using deep learning [89].

Another area of recent developments has been the improvement of the imaging speed towards real-time imaging, crucial for use as intraoperative guidance. As the current DMD has a frame rate of up to 47 kHz, the highest acquisition rate for 100×100 pixels images is only 4.7 frames per second. Advanced image processing algorithms could be used to reduce the number of scanning points without sacrificing their spatial resolution. In 2021, we developed a deep learning-based approach to improve the imaging speed of the photoacoustic endomicroscopy system, by sub-sampling the physical scanning points and compensating its spatial resolution with a deep convolutional neural network (DCNN) [90, 91]. Demonstrated by imaging a carbon fibre phantom, this method increased the image acquisition rate by 16 times without significantly affecting the image quality. Another way is to employ faster hardware in the imaging system. For example, in 2018, Feldkhun et al. utilised a programmable acoustic optic deflector (AOD) as an alternative to traditional spatial light modulators [92]. It was based on a phase control technique using programmable AODs that are orders of magnitude faster than existing wavefront shaping methods. As such an array of RF-encoded beams were sent through the medium and measuring the phases of all scattered beams simultaneously with a fast single-pixel detector was performed in just 10 μs. In 2019, Tzang et al. reported light focusing via wavefront shaping using a grating light valve at 350 kHz [62]. A grating light valve is a device consisting of numerous silicon nitride ribbons suspended on a silicon chip. By electronically manipulating the deflection of these ribbons, the grating light valve serves as a programmable 1D phase modulator.

To exploit the potential of photoacoustic microscopy via an MMF, one of the central challenges is the lack of appropriate excitation light sources for multispectral imaging to obtain spatial distributions of the abundant absorbing chromophores including DNA/RNA, haemoglobin, lipids, water, and collagen, etc. Q-switched diode-pumped lasers and Ti: sapphire lasers are widely used for PAM which restrict the available wavelengths to only a few choices. Optical parametric oscillators can provide wavelength turnability over a wide range, but the low tuning speeds and pulse repetition rates prohibit them from real-time applications. Moreover, these light sources are usually bulky and expensive. As potential alternatives, multiple diode lasers with different wavelengths [93, 94] and Q-Switch lasers based on simulated Raman scattering [95] have shown potential for optical-resolution photoacoustic microscopy of blood and lymphatic vessels. Furthermore, since the light at various wavelengths suffers from different modal dispersions, multispectral MMF characterisation is highly desired for the development of multispectral PAI endomicroscopy probes through MMFs.

Acknowledgements This work was supported by the Academy of Medical Sciences/the Wellcome Trust/the Government Department of Business, Energy and Industrial Strategy/the British Heart Foundation/Diabetes UK Springboard Award [SBF006/1136]; Wellcome Trust, United Kingdom (203148/Z/16/Z); the Engineering and Physical Sciences Research Council, United Kingdom (NS/A000049/1); and the Royal Academy of Engineering under the Research Fellowship Scheme (RF2223/22/170). For the purpose of open access, the authors have applied a CC BY public copyright license to any author-accepted manuscript version arising from this submission.

References

1. Mosk, Allard P., et al. "Controlling waves in space and time for imaging and focusing in complex media." Nature photonics 6.5 (2012): 283–292.
2. Yoon, Seokchan, et al. "Deep optical imaging within complex scattering media." Nature Reviews Physics 2.3 (2020): 141–158.
3. Gigan, Sylvain, et al. "Roadmap on wavefront shaping and deep imaging in complex media." Journal of Physics: Photonics 4.4 (2022): 042501.
4. Yu, Zhipeng, et al. "Wavefront shaping: a versatile tool to conquer multiple scattering in multidisciplinary fields." The Innovation 3.5 (2022).
5. Chaigne, Thomas, et al. "Controlling light in scattering media non-invasively using the photoacoustic transmission matrix." Nature Photonics 8.1 (2014): 58–64.
6. Lai, Puxiang, et al. "Photoacoustically guided wavefront shaping for enhanced optical focusing in scattering media." Nature photonics 9.2 (2015): 126–132.
7. Conkey, Donald B., et al. "Super-resolution photoacoustic imaging through a scattering wall." Nature communications 6.1 (2015): 7902.
8. Bossy, Emmanuel, and Sylvain Gigan. "Photoacoustics with coherent light." Photoacoustics 4.1 (2016): 22–35.
9. Zhao, Tianrui, et al. "Ultrathin, high-speed, all-optical photoacoustic endomicroscopy probe for guiding minimally invasive surgery." Biomedical optics express 13.8 (2022): 4414–4428.
10. Mezil, Sylvain, et al. "Single-shot hybrid photoacoustic-fluorescent microendoscopy through a multimode fiber with wavefront shaping." Biomedical Optics Express 11.10 (2020): 5717–5727.
11. Cao, Hui, et al. "Controlling light propagation in multimode fibers for imaging, spectroscopy, and beyond." Advances in Optics and Photonics 15.2 (2023): 524–612.
12. Vellekoop, Ivo M., and A. P. Mosk. "Focusing coherent light through opaque strongly scattering media." Optics letters 32.16 (2007): 2309–2311.
13. Vellekoop, Ivo M. "Feedback-based wavefront shaping." Optics express 23.9 (2015): 12189–12206.
14. Conkey, Donald B., et al. "Genetic algorithm optimization for focusing through turbid media in noisy environments." Optics express 20.5 (2012): 4840–4849.
15. Wang, Daifa, et al. "Focusing through dynamic tissue with millisecond digital optical phase conjugation." Optica 2.8 (2015): 728–735.
16. Yaqoob, Zahid, et al. "Optical phase conjugation for turbidity suppression in biological samples." Nature photonics 2.2 (2008): 110–115.
17. Ayoub, Ahmed B., and Demetri Psaltis. "High speed, complex wavefront shaping using the digital micro-mirror device." Scientific Reports 11.1 (2021): 18837.
18. Papadopoulos, Ioannis N., et al. "High-resolution, lensless endoscope based on digital scanning through a multimode optical fiber." Biomedical optics express 4.2 (2013): 260–270.
19. Popoff, Sébastien M., et al. "Measuring the transmission matrix in optics: an approach to the study and control of light propagation in disordered media." Physical review letters 104.10 (2010): 100601.

20. Popoff, Sébastien, et al. "Image transmission through an opaque material." Nature communications 1.1 (2010): 81.
21. Kim, Moonseok, et al. "Transmission matrix of a scattering medium and its applications in biophotonics." Optics express 23.10 (2015): 12648–12668.
22. Choi, Youngwoon, et al. "Scanner-free and wide-field endoscopic imaging by using a single multimode optical fiber." Physical review letters 109.20 (2012): 203901.
23. Loterie, Damien, et al. "Digital confocal microscopy through a multimode fiber." Optics express 23.18 (2015): 23845–23858.
24. Tao, Xiaodong, et al. "High-speed scanning interferometric focusing by fast measurement of binary transmission matrix for channel demixing." Optics express 23.11 (2015): 14168–14187.
25. Zhao, Tianrui, et al. "Bayes' theorem-based binary algorithm for fast reference-less calibration of a multimode fiber." Optics express 26.16 (2018): 20368–20378.
26. Drémeau, Angélique, et al. "Reference-less measurement of the transmission matrix of a highly scattering material using a DMD and phase retrieval techniques." Optics express 23.9 (2015): 11898–11911.
27. Zhao, Tianrui, et al. "Focusing light through multimode fibres using a digital micromirror device: a comparison study of non-holographic approaches." Optics express 29.10 (2021): 14269–14281.
28. Zhao, Tianrui, et al. "Seeing through multimode fibers with real-valued intensity transmission matrices." Optics express 28.14 (2020): 20978–20991.
29. Plöschner, Martin, Tomáš Tyc, and Tomáš Čižmár. "Seeing through chaos in multimode fibres." Nature Photonics 9.8 (2015): 529–535.
30. Čižmár, Tomáš, and Kishan Dholakia. "Exploiting multimode waveguides for pure fibre-based imaging." Nature communications 3.1 (2012): 1027.
31. Čižmár, Tomáš, Michael Mazilu, and Kishan Dholakia. "In situ wavefront correction and its application to micromanipulation." Nature Photonics 4.6 (2010): 388–394.
32. Conkey, Donald B., Antonio M. Caravaca-Aguirre, and Rafael Piestun. "High-speed scattering medium characterization with application to focusing light through turbid media." Optics express 20.2 (2012): 1733–1740.
33. N'Gom, Moussa, et al. "Controlling light transmission through highly scattering media using semi-definite programming as a phase retrieval computation method." Scientific reports 7.1 (2017): 2518.
34. Huang, Guoqiang, et al. "Retrieving the optical transmission matrix of a multimode fiber using the extended Kalman filter." Optics express 28.7 (2020): 9487–9500.
35. Huang, Guoqiang, et al. "Generalizing the Gerchberg–Saxton algorithm for retrieving complex optical transmission matrices." Photonics Research 9.1 (2021): 34–42.
36. Gomes, André D., et al. "Near perfect focusing through multimode fibres." Optics Express 30.7 (2022): 10645–10663.
37. Xu, Jian, et al. "Focusing light through scattering media by transmission matrix inversion." Optics express 25.22 (2017): 27234–27246.
38. Liu, Yan, et al. "Focusing light inside dynamic scattering media with millisecond digital optical phase conjugation." Optica 4.2 (2017): 280–288.
39. Boniface, Antoine, Jonathan Dong, and Sylvain Gigan. "Non-invasive focusing and imaging in scattering media with a fluorescence-based transmission matrix." Nature communications 11.1 (2020): 6154.
40. Vellekoop, Ivo M., Meng Cui, and Changhuei Yang. "Digital optical phase conjugation of fluorescence in turbid tissue." Applied physics letters 101.8 (2012).
41. Van Putten, Elbert G., A. Lagendijk, and A. P. Mosk. "Optimal concentration of light in turbid materials." JOSA B 28.5 (2011): 1200–1203.
42. Katz, Ori, et al. "Noninvasive nonlinear focusing and imaging through strongly scattering turbid layers." Optica 1.3 (2014): 170–174.
43. Tang, Jianyong, Ronald N. Germain, and Meng Cui. "Superpenetration optical microscopy by iterative multiphoton adaptive compensation technique." Proceedings of the National Academy of Sciences 109.22 (2012): 8434–8439.

44. Xu, Xiao, Honglin Liu, and Lihong V. Wang. "Time-reversed ultrasonically encoded optical focusing into scattering media." Nature photonics 5.3 (2011): 154–157.
45. Ruan, Haowen, et al. "Deep tissue optical focusing and optogenetic modulation with time-reversed ultrasonically encoded light." Science advances 3.12 (2017): eaao5520.
46. Si, Ke, Reto Fiolka, and Meng Cui. "Fluorescence imaging beyond the ballistic regime by ultrasound-pulse-guided digital phase conjugation." Nature photonics 6.10 (2012): 657–661.
47. Wang, Ying Min, et al. "Deep-tissue focal fluorescence imaging with digitally time-reversed ultrasound-encoded light." Nature communications 3.1 (2012): 928.
48. Suzuki, Yuta, et al. "Continuous scanning of a time-reversed ultrasonically encoded optical focus by reflection-mode digital phase conjugation." Optics letters 39.12 (2014): 3441–3444.
49. Judkewitz, Benjamin, et al. "Speckle-scale focusing in the diffusive regime with time reversal of variance-encoded light (TROVE)." Nature photonics 7.4 (2013): 300–305.
50. Liu, Yan, et al. "Optical focusing deep inside dynamic scattering media with near-infrared time-reversed ultrasonically encoded (TRUE) light." Nature communications 6.1 (2015): 5904.
51. Ma, Cheng, et al. "Time-reversed adapted-perturbation (TRAP) optical focusing onto dynamic objects inside scattering media." Nature photonics 8.12 (2014): 931–936.
52. Zhou, Edward Haojiang, et al. "Focusing on moving targets through scattering samples." Optica 1.4 (2014): 227–232.
53. Ruan, Haowen, et al. "Focusing light inside scattering media with magnetic-particle-guided wavefront shaping." Optica 4.11 (2017): 1337–1343.
54. Horstmeyer, Roarke, Haowen Ruan, and Changhuei Yang. "Guidestar-assisted wavefront-shaping methods for focusing light into biological tissue." Nature photonics 9.9 (2015): 563–571.
55. Chaigne, Thomas, et al. "Light focusing and two-dimensional imaging through scattering media using the photoacoustic transmission matrix with an ultrasound array." Optics letters 39.9 (2014): 2664–2667.
56. Kong, Fanting, et al. "Photoacoustic-guided convergence of light through optically diffusive media." Optics letters 36.11 (2011): 2053–2055.
57. Tzang, Omer, and Rafael Piestun. "Lock-in detection of photoacoustic feedback signal for focusing through scattering media using wave-front shaping." Optics Express 24.24 (2016): 28122–28130.
58. Inzunza-Ibarra, Marco A., et al. "Sub-acoustic resolution optical focusing through scattering using photoacoustic fluctuation guided wavefront shaping." Optics Express 28.7 (2020): 9823–9832.
59. Brake, Joshua, Mooseok Jang, and Changhuei Yang. "Analyzing the relationship between decorrelation time and tissue thickness in acute rat brain slices using multispeckle diffusing wave spectroscopy." JOSA A 33.2 (2016): 270–275.
60. Yang, Jiamiao, et al. "Fighting against fast speckle decorrelation for light focusing inside live tissue by photon frequency shifting." ACS photonics 7.3 (2020): 837–844.
61. Zhao, Tianrui, et al. "High-speed photoacoustic-guided wavefront shaping for focusing light in scattering media." Optics Letters 46.5 (2021): 1165–1168.
62. Tzang, Omer, et al. "Wavefront shaping in complex media with a 350 kHz modulator via a 1D-to-2D transform." Nature Photonics 13.11 (2019): 788–793.
63. Caravaca-Aguirre, Antonio M., et al. "High contrast three-dimensional photoacoustic imaging through scattering media by localized optical fluence enhancement." Optics express 21.22 (2013): 26671–26676.
64. Sun, Jialin, et al. "Photoacoustic wavefront shaping with high signal to noise ratio for light focusing through scattering media." Scientific reports 9.1 (2019): 4328.
65. Zhao, Tianrui, et al. "Minimally invasive photoacoustic imaging: Current status and future perspectives." Photoacoustics 16 (2019): 100146.
66. Jansen, Krista, et al. "Intravascular photoacoustic imaging of human coronary atherosclerosis." Optics letters 36.5 (2011): 597–599.
67. Abran, Maxime, et al. "Development of a photoacoustic, ultrasound and fluorescence imaging catheter for the study of atherosclerotic plaque." IEEE transactions on biomedical circuits and systems 8.5 (2014): 696–703.

68. Yang, Joon-Mo, et al. "Simultaneous functional photoacoustic and ultrasonic endoscopy of internal organs in vivo." Nature medicine 18.8 (2012): 1297–1302.
69. Liu, Ning, Sihua Yang, and Da Xing. "Photoacoustic and hyperspectral dual-modality endoscope." Optics letters 43.1 (2018): 138–141.
70. Li, Yan, et al. "High-speed integrated endoscopic photoacoustic and ultrasound imaging system." IEEE Journal of selected topics in quantum electronics 25.1 (2018): 1–5.
71. Shao, Peng, et al. "Integrated micro-endoscopy system for simultaneous fluorescence and optical-resolution photoacoustic imaging." Journal of Biomedical Optics 17.7 (2012): 076024–076024.
72. Hajireza, Parsin, Wei Shi, and Roger Zemp. "Label-free in vivo GRIN-lens optical resolution photoacoustic micro-endoscopy." Laser Physics Letters 10.5 (2013): 055603.
73. Ansari, Rehman, et al. "All-optical forward-viewing photoacoustic probe for high-resolution 3D endoscopy." Light: Science & Applications 7.1 (2018): 75.
74. Turtaev, Sergey, et al. "High-fidelity multimode fibre-based endoscopy for deep brain in vivo imaging." Light: Science & Applications 7.1 (2018): 92.
75. Leite, Ivo T., et al. "Three-dimensional holographic optical manipulation through a high-numerical-aperture soft-glass multimode fibre." Nature Photonics 12.1 (2018): 33–39.
76. Ohayon, Shay, et al. "Minimally invasive multimode optical fiber microendoscope for deep brain fluorescence imaging." Biomedical optics express 9.4 (2018): 1492–1509.
77. Papadopoulos, Ioannis N., et al. "Optical-resolution photoacoustic microscopy by use of a multimode fiber." Applied Physics Letters 102.21 (2013).
78. Stasio, Nicolino, et al. "Towards new applications using capillary waveguides." Biomedical optics express 6.12 (2015): 4619–4631.
79. Zhao, Tianrui, et al. "Video-rate dual-modal photoacoustic and fluorescence imaging through a multimode fibre towards forward-viewing endomicroscopy." Photoacoustics 25 (2022): 100323.
80. Guggenheim, James A., et al. "Ultrasensitive plano-concave optical microresonators for ultrasound sensing." Nature Photonics 11.11 (2017): 714–719.
81. Dangi, Ajay, Sumit Agrawal, and Sri-Rajasekhar Kothapalli. "Lithium niobate-based transparent ultrasound transducers for photoacoustic imaging." Optics letters 44.21 (2019): 5326–5329.
82. Fang, Cheng, and Jun Zou. "Acoustic-resolution photoacoustic microscopy based on an optically transparent focused transducer with a high numerical aperture." Optics letters46.13 (2021): 3280–3283.
83. Park, Jeongwoo, et al. "Quadruple ultrasound, photoacoustic, optical coherence, and fluorescence fusion imaging with a transparent ultrasound transducer." Proceedings of the National Academy of Sciences 118.11 (2021): e1920879118.
84. Liu, Yu-Hang, et al. "Sensitive ultrawideband transparent PVDF-ITO ultrasound detector for optoacoustic microscopy." Optics Letters 47.16 (2022): 4163–4166.
85. Zhao, Tianrui, et al. "Wavefront Shaping-Assisted Forward-Viewing Photoacoustic Endomicroscopy Based on a Transparent Ultrasound Sensor." Applied Sciences 12.24 (2022): 12619.
86. Flaes, Dirk E. Boonzajer, et al. "Robustness of light-transport processes to bending deformations in graded-index multimode waveguides." Physical review letters 120.23 (2018): 233901.
87. Plöschner, Martin, and Tomáš Čižmár. "Compact multimode fiber beam-shaping system based on GPU accelerated digital holography." Optics letters 40.2 (2015): 197–200.
88. Tsvirkun, Victor, et al. "Flexible lensless endoscope with a conformationally invariant multicore fiber." Optica 6.9 (2019): 1185–1189.
89. Hu, Xiaowen, et al. "Unsupervised full-color cellular image reconstruction through disordered optical fiber." Light: Science & Applications 12.1 (2023): 125.
90. Zhao, Tianrui, et al. "Ai-enabled high-speed photoacoustic endomicroscopy through a multimode fibre." Photons Plus Ultrasound: Imaging and Sensing 2022. Vol. 11960. SPIE, 2022.
91. Zhao, Tianrui, et al. "Deep learning boosts the imaging speed of photoacoustic endomicroscopy." Photons Plus Ultrasound: Imaging and Sensing 2023. Vol. 12379. SPIE, 2023.

92. Feldkhun, Daniel, et al. "Focusing and scanning through scattering media in microseconds." Optica 6.1 (2019): 72–75.
93. Stylogiannis, Antonios, et al. "Frequency wavelength multiplexed optoacoustic tomography." Nature communications 13.1 (2022): 4448.
94. Jeng, Geng-Shi, et al. "Real-time interleaved spectroscopic photoacoustic and ultrasound (PAUS) scanning with simultaneous fluence compensation and motion correction." Nature communications 12.1 (2021): 716.
95. Liu, Chao, et al. "Five-wavelength optical-resolution photoacoustic microscopy of blood and lymphatic vessels." Advanced Photonics 3.1 (2021): 016002–016002.

Chapter 9
Optical Ultrasound Imaging Device Development and Characterisation: A Literature Review

Semyon Bodian, Sacha Noimark, Adrien E. Desjardins, and Ivan P. Parkin

Abstract The development of minimally invasive surgical interventions has drastically reduced patients' recovery periods and superficial scarring, lowering the risk of post-operative infections. However, using narrower incisions into the body has reduced the clinician's view and dexterity of their instruments, necessitating improved image guidance technology. Optical ultrasound (OpUS) generation is emerging as a promising route to generate ultrasound with broadband frequencies at high amplitudes minimally invasively. Here, through the use of optically absorbing films or coatings and pulsed laser excitation, high-resolution, real-time ultrasound imaging at clinically significant tissue depths can be achieved using a miniaturised device. Many carbonaceous and nanoparticulate-based thin films or composites have been examined for optimal OpUS generation. Further, complex patient-specific phantoms formed from tissue-mimicking materials have been developed

S. Bodian (✉)
Department of Chemistry, University College London, London, UK

Department of Medical Physics and Biomedical Engineering, University College London, London, UK

S. Noimark
Department of Chemistry, University College London, London, UK

Department of Medical Physics and Biomedical Engineering, University College London, London, UK

Wellcome EPSRC Institute of Surgical Science, University College London, London, UK
e-mail: s.noimark@ucl.ac.uk

A. E. Desjardins
Department of Medical Physics and Biomedical Engineering, University College London, London, UK

Wellcome EPSRC Institute of Surgical Science, University College London, London, UK
e-mail: a.desjardins@ucl.ac.uk

I. P. Parkin
Department of Chemistry, University College London, London, UK
e-mail: i.p.parkin@ucl.ac.uk

W. Xia (ed.), *Biomedical Photoacoustics*,
https://doi.org/10.1007/978-3-031-61411-8_9

to characterise OpUS imaging devices, mimicking clinical scenarios and providing training models for clinicians.

Keywords Optical Ultrasound Generation · Ultrasound Imaging · Minimally-Invasive Surgery · Multimodality Imaging · Nanocomposites · Fibre-Optic

9.1 Introduction

9.1.1 Minimally Invasive Surgery

Minimally invasive surgery (MIS) has revolutionised the field of medicine, utilising smaller incisions than conventional open surgery, leading to many benefits to both the patient and clinicians [1]. MIS interventions are commonplace across various medical disciplines, such as refractive surgery in ophthalmology [2, 3], stent placement in cardiology [4–6] and tumour biopsy for gastroenterology [7, 8]. The benefits of MIS include lower rates of post-operative infection, less scarring and shorter convalescence periods for the patient [9–13]. Overall, these outcomes result in lower costs for hospital admissions [9]. However, significant complexities are associated with MIS techniques [11, 12]. Unlike conventional open surgery in which the clinician can handle the patient's tissue through gloved hands, feeling its shape, texture, and temperature, in MIS, the clinician has no direct view, and interaction with patient tissue is through narrow incisions using external imaging and specialised instrumentation [14, 15]. As a result, the clinician's field of view, manual dexterity and range of motion are all limited, highlighting the need for improved imaging modalities [11]. Many clinical tools have since been developed to guide the clinician during MIS interventions, and one of the most critical is improved image guidance [16–20].

9.1.2 Medical Imaging for Minimally Invasive Surgery

To accurately diagnose and treat patients' diseases during surgery, clinicians require imaging modalities that provide high-resolution spatial information, utilising readily accessible and affordable equipment. X-ray imaging is widely used in clinical settings [21, 22] and provides critical pre-operative and intra-operative imaging, providing tissue-based contrast, dependent on the tissue's density and elemental composition. In X-ray imaging, high-energy photons are generated by a cathode, and these are directed at high velocity towards a metal plate, resulting in X-ray generation upon collision. When X-rays enter a patient's body, their energy is absorbed or scattered according to a tissue's density [23]. More dense body material, such as bone, absorbs more X-ray photons than softer tissues like fat, muscle and ligaments since the former comprises elements with higher atomic masses than the latter. One limitation of X-ray imaging is video-rate imaging cannot be achieved. However, when combined with a fluorescent contrast agent, video-rate imaging is possible in the form of fluoroscopy. Here, the contrast agent absorbs large portions of the incident X-ray photons and generates fluorescence, which is detected by a screen. For example, contrast dyes are used in coronary angiography, where the dye's flow is monitored through the body to evaluate the heart's function or locate blocked blood vessels [24]. A key drawback with X-ray fluoroscopy is that both the patient and the clinician are exposed to multiple high doses of X-ray radiation, posing concerns over radiation poisoning, particularly for the clinician. Another modality that utilises X-ray radiation is computed tomography (CT) imaging, where the X-ray source and detector are placed in a gantry and moved in a three-dimensional arc around a prone patient [25–28]. Like fluoroscopy, a contrast agent such as a dye is injected into or consumed by the patient, highlighting specific vasculature as part of a three-dimensional scan of the patient's body [25, 29, 30]. Like fluoroscopy, CT X-rays can damage tissues' DNA; thus, patients' and clinicians' exposure to such techniques must be limited [28, 29, 31], leading to intermittent use. It is advantageous if medical imaging modalities that are employed on a regular basis can be achieved without the use of ionising radiation.

One commonly used modality, magnetic resonance imaging (MRI), does not use ionising radiation, instead uses powerful magnets and electromagnetic radio pulses [32–34]. The corresponding MRI image contrast relies upon the different chemical environments of the protons in each tissue, derived from hydrogen atoms [35–37] within water molecules. MRI images can highly resolve soft tissues such as fat [35, 36, 38, 39] and muscle [35, 38, 39] due to their high water content in contrast to more dense tissues like bone [40]. However, MRI scans display low temporal resolutions [41] that are incapable of providing high spatial resolution in real time [42]. MRI scanners can also be costly [43–45] both to procure and to maintain, require the patient to remain stationary for long periods under claustrophobic conditions and may be unsuitable for patients with some implanted devices.

9.1.3 Medical Ultrasound Imaging

Ultrasound imaging is a non-invasive imaging modality that describes the transmission of high-frequency sound waves (above 20 kHz) towards an object [46, 47]. The ultrasound waves undergo reflection at the interfaces of media with differing acoustic impedances (the resistance a sound wave experiences passing through a material) [48]. An ultrasound sensor then receives the reflected ultrasound waves. This principle is described as the pulse-echo regime. The more significant the difference in acoustic impedance at an interface, the larger the proportion of incident ultrasound waves that undergo reflection at the interface [47–49]. The tissue area imaged can be reconstructed by recording the peak-to-peak amplitude (pressure) and time-of-flight of each ultrasound signal's arrival along with the speed of sound of the medium through which the ultrasound waves passed through [50–52]. Although widely used for medical imaging, the applications of ultrasound imaging range from non-destructive material characterisation [53–55] to motion sensors and range finding (Sonar) [56, 57].

The use of ultrasound imaging is widespread within health-care settings due to its highly affordable and portable apparatus. Medical ultrasound imaging provides real-time high-resolution imaging of bodily tissue, using differences in tissues' acoustic impedances to form image contrast. Such a modality provides structural information about tissue, such as its size, shape and density [58–60]. Inhomogeneous soft tissues comprising varying densities are well suited for ultrasound imaging. Thus, it is often used to image foetuses to monitor their development in the womb, measuring the growth and shape of their limbs [61]. In addition, ultrasound imaging is commonly used in oncological procedures, including the diagnosis of breast cancer, as well as to guide tissue biopsies. Ultrasound is widely used within clinical settings for medical imaging as it is non-invasive and considered safe because it employs non-ionising radiation. At ultrasound pressures used for clinical imaging, neither its transmission through tissue nor its reflection cause any change to the tissue [62]. However, ultrasound can be used for therapeutic applications using high-pressure ultrasound waves or focused ultrasound as a therapy. Unlike diagnostic ultrasound, focused ultrasound therapy generates larger ultrasound pressures, including the positive and negative phases of the ultrasound wave, creating a low enough density in the wave media during the negative phase that leads to cavitation forming a bubble. Upon the bubble's implosion, damage can be caused to surrounding tissue.

In ultrasound imaging, for each dimension represented within an ultrasound image, there is a corresponding resolution. Specifically, each resolution states the minimum distance that two objects within that dimension can be separated and still resolved as separate objects. For example, in two-dimensional ultrasound imaging, both axial and lateral resolutions are considered (Fig. 9.1). The axial resolution of an ultrasound image is mainly dependent on the frequency of the generated ultrasound waves, with higher frequencies able to resolve objects of lower spatial separation within the axial imaging plane [63]. However, higher ultrasound frequencies undergo greater acoustic attenuation when passing through a medium [64]. As a result, the higher the frequency of an ultrasound wave, the shorter the distance that it can penetrate a medium (before being fully absorbed). On the other hand, the lateral resolution of an ultrasound image is dependent on several factors. It predominantly relies upon the lateral dimensions of the ultrasound imaging probe as well as the error in its position when moved along a specific direction as part of an imaging scan [65]. The narrower the width of the imaging probe and the smaller the error in its position, the higher the resulting lateral resolution of the recorded image. One metric that can be used to classify an ultrasound image's quality is the signal-to-noise ratio (SNR), calculated by dividing the amplitude of the ultrasound signal reflected off tissue by the signal caused by background noise [66]. SNR is measured in decibels (dB), where an image with a higher SNR will feature less noise.

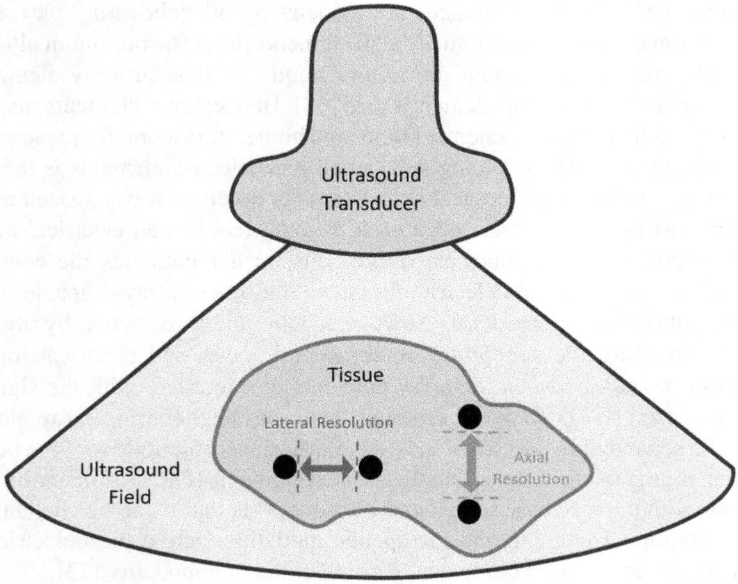

Fig. 9.1 Schematic illustration of lateral and axial resolutions of ultrasound image

9.1.4 Piezoelectric Ultrasound Transducers

Ultrasound imaging transducers often consist of piezoelectric crystals, which possess a lattice structure capable of deformation when an electrical voltage is applied across its structure [67]. The electronic charge within the crystal structure instigates the accumulation of mechanical stress, leading to the deformation of the crystal structure. The reciprocal linear relationship between the crystal's deformation and the applied voltage allows the same piezoelectric crystal to act as both an ultrasound wave generator and a sensor. Thus, applying a set voltage produces ultrasound waves of a particular pressure and frequency. In turn, the distortion of a piezoelectric crystal's lattice by a passing ultrasound wave is proportional to both its pressure and frequency and reflected in the corresponding yielded voltage. Piezoelectric ultrasound transducers are frequently arranged into two-dimensional arrays where each array element is connected to an electrical circuit. Each element produces an ultrasound signal that interferes with signals generated by surrounding elements, forming an ultrasound field. The greater the number of piezoelectric crystal elements within an array, the higher the lateral resolution of the transducer. Further, larger arrays can generate a greater extent of ultrasound pressures and bandwidths than smaller arrays, along with a more extensive range of directions that the ultrasound can be directed towards.

To use piezoelectric ultrasound transducers for in vivo ultrasound imaging in MIS interventions, these crystal arrays must be miniaturised to sub-millimetre-sized dimensions [68]. Various challenges are associated with fabricating piezoelectric ultrasound transducers featuring such small elements [69]. The minimum ultrasound wavelength, and thus maximum ultrasound frequency that an array element can produce, is equal to twice the element's size [63]. Hence, array elements must be of micron-scale dimensions to generate higher and higher ultrasound frequencies of the order of megahertz (MHz). Dicing such small piezoelectric elements is incredibly challenging, as is forming electrical connections between narrowly spaced adjacent array elements [70]. Moreover, since each element requires an electrical impulse, creating electrodes of micron-scale dimensions further increases the complexity of producing miniature piezoelectric ultrasonic transducer arrays capable of high-frequency ultrasound generation. Stray magnetic fields detected by the array elements can cause the generation of ultrasound waves and electrical impulses, leading to crosstalk between the array elements that interfere with the signal and image processing [71]. To avoid crosstalk between neighbouring array elements, they can be separated by electrical insulative materials [72]; however, this becomes more challenging to fabricate for smaller-scaled elements [68]. One major limitation associated with piezoelectric ultrasound transducers is that it can be challenging to integrate additional imaging and therapeutic modalities into a piezoelectric-based ultrasound device, limiting clinicians to a single imaging modality [73].

9.1.5 Optical Ultrasound Imaging

Optical ultrasound (OpUS) generation has provided a promising alternative to piezoelectric ultrasound imaging. Here, pulsed laser irradiation is absorbed by a medium, which converts the acquired optical energy into heat through the Photoacoustic (PA) effect, generating ultrasound [74–77]. The medium (OpUS layer) is responsible for the optical absorption of the laser pulses and its subsequent cyclical expansion and contraction, leading to pressure changes in the external environment resulting in emitted ultrasound waves. There are several advantages associated with OpUS generation compared to piezoelectric ultrasound generation. OpUS generation enables the miniaturisation of ultrasound transmitters using facile fabrication techniques without compromising on the resultant generated ultrasound pressures and bandwidths [74]. In addition, OpUS transmitters and sensors are immune to electromagnetic interference and are compatible with MRI [74, 78]. Furthermore, OpUS transmitters can accommodate additional modalities enabling multimodal imaging or concurrent therapies integrated using a single imaging device.

A key advantage of OpUS generation is the relationship between the laser pulse temporal width and the temporal width of the ultrasound pulse, and hence its frequency [79]. This occurs because the shape of the generated ultrasound pulse mirrors that of the laser pulse profile [80]. Consequently, ultrasound bandwidths have grown with the shortening of laser pulses culminating in the development of nanosecond laser pulses [79]. This relationship relies upon the assumption that the irradiating laser pulse length is less than the thermal conduction and stress propagation timeframes of the optically absorbing region in the irradiated medium [75, 81]. Depending on which is shortest, this region is referred to as either the optical penetration depth or the thickness of the optically absorbing medium.

The thermal confinement condition is represented as follows:

$$t_{th} = \frac{d_c^2}{4D_T} \tag{9.1}$$

Where t_{th}, d_c and D_T are the thermal conduction timeframe, the optically absorbing region's thickness and the thermal diffusivity rate, respectively. The thermal confinement assumption describes that the medium cannot dissipate the acquired thermal energy over the time length that the laser pulse is absorbed [82–85]. This maximises the temperature that can be reached in a medium leading to larger generated ultrasound pressures.

The stress confinement assumption assumes that the medium does not expand and maintains the same volume during the accruement of thermal energy from the absorbed laser pulse [82–85]. It is described as follows:

$$t_{st} = \frac{d_c}{C_s} \tag{9.2}$$

Where t_{st} is the stress propagation timeframe and C_s is the speed of sound in a medium.

When creating a medium for OpUS generation, it is crucial to consider a balance between generating high ultrasound pressures for imaging at deeper tissue depths and wide ultrasound bandwidths for high-resolution imaging. From the standpoint of ultrasound generation, it is preferable that a medium has a thickness similar to the optical penetration depth of the input laser wavelength, thereby minimising attenuation of the generated ultrasound signal, particularly at higher ultrasound frequencies. Thus, a maximal ultrasound bandwidth is generated, especially in high-frequency ultrasound. Two possible outcomes may be encountered if the medium thickness is decreased further. Firstly, if the thickness is decreased below the optical penetration depth without increasing its optical absorption coefficient, the medium will exhibit decreased optical absorbance and transmit some input laser light. This may lead to a lower accrual of optical energy and, thus, heat, generating decreased ultrasound pressures [75]. Secondly, if the optical absorption of the medium is maintained, however, the optical energy is accrued within a narrower space, leading to higher thermal densities, this can exceed the thermal damage threshold, causing the medium to disintegrate and impact on ultrasound generation [75]. Additionally, if the medium thickness becomes too thin, the stress and thermal timeframes can become shorter than the laser pulse length invalidating the stress and thermal confinement conditions [75].

The following relationship can be used to describe the pressure of the generated ultrasound wave [86]:

$$P\,\alpha\,\left(\mu_a * \Gamma\right) \tag{9.3}$$

P, μ_a and Γ are the generated ultrasound pressure, the optical absorption coefficient and the Grüneisen parameter, respectively. The Grüneisen parameter can be represented as follows [70]:

$$\Gamma = \frac{\beta * C_s^2}{C_p} \tag{9.4}$$

Where β, C_s and C_p are the volumetric thermal expansion coefficient, the speed of sound and the heat capacity at constant pressure, respectively. Thus, the equation describing the generated ultrasound pressure can be simplified to:

$$P \, \alpha \, \frac{\mu_\alpha * \beta * C_s^2}{C_p} \tag{9.5}$$

To maximise the generated ultrasound pressures, media are selected for their high μ_a coefficient at the laser wavelength employed, as well as their high β coefficients. A material's μ_a coefficient is the reciprocal of the distance a given electromagnetic wave can travel through a medium before being fully absorbed. Whilst the β coefficient measures how much a material's volume expands at a given temperature [70, 75]. To produce ultrasound pressures through the OpUS generation mechanism, the generating medium must undergo significant expansion and contraction to create sizeable disparities in the surrounding air pressure, forming ultrasound waves. As a result, higher ultrasound pressures are generated using materials with larger β coefficients. This quantity characterises the size of a material's volumetric expansion when it absorbs an input of thermal energy. Moreover, a high μ_a coefficient is a desirable property for OpUS generation as it enables more optical energy to be absorbed from the input laser pulse. This then leads to a more significant accumulation of heat within the material, bringing about a more sizeable volumetric expansion and higher yielded ultrasound pressures. The greater the ultrasound pressures that an imaging probe can produce, the deeper the tissue that the device can resolve. This is important as many clinical MIS interventions involve internal surgical procedures or removing tissue (for example, for biopsies) from within the patient. Thus, the high ultrasound frequencies required to provide high-resolution imaging are increasingly attenuated with the distance they travel through a tissue meaning those with higher pressures will reach the ultrasound detector. OpUS probes can be inserted within surgical instruments, minimising the distance between the ultrasound source and the imaged tissue.

9.2 Macroscale Composites

9.2.1 Metal Films

Various materials have been investigated for OpUS generation, referred to in earlier studies as thermoelastic ultrasound generation. Here, metal thin films were amongst the most widely employed as ultrasound-generating media. Such metals included chromium (Cr), molybdenum (Mo) and aluminium (Al) [87, 88]. Due to their simple structures, thin metal films were easy to fabricate, manipulate and model mathematically. These properties enabled many researchers to gain a deeper understanding of the mechanism of thermoelastic ultrasound generation, upon which more sophisticated composite films (composites) were designed [89–93]. The first models to describe the generation of ultrasound waves from pulsed heating of a surface by irradiation with electron/laser pulses were proposed by White et al., who also considered whether the irradiated surface was evenly heated or under different forms of strain, such as constraint [87]. The pulse repetition rate of the laser pulses, and hence the heat flux applied to the metal films, was found to be directly proportional to the emitted ultrasound pressures. Wavelengths within the visible spectrum were most strongly absorbed by the metal films forming the largest ultrasound pressures. In addition to greater ultrasound pressures, smaller radiation pressures were also formed at the film's surface. This trend proved that thermoelastic ultrasound generation resulted solely from the film's absorption of the pulse's energy rather than the radiation pressure exerted on its surface.

Several other regimes of ultrasound generation that use higher irradiating power densities were also explored, for example, the ablation (plasma) mechanism. The plasma regime describes the evaporation of a material's surface under the irradiation of a pulsed laser beam forming plasma, which expands, exerting radiation pressure on the film's surface, prompting the film to generate ultrasound [79]. The plasma regime is often unfavourable for many practical applications. Ultrasound waves manifest as three acoustic modes within a solid body: longitudinal, transverse and Rayleigh waves [92, 94, 95]. In longitudinal waves, the wave propagation and the particle displacement are parallel to one another. In contrast, the particle displacement in a transverse wave is perpendicular to its direction of propagation (Fig. 9.2). Both acoustic modes occur within the bulk of a body, whereas Rayleigh waves exist only on a body's surface, manifesting as longitudinal or transverse waves. Dewhurst et al. further detailed the workings of the thermoelastic and plasma regimes. The study recorded the ultrasound waveforms and pressures at the surface of aluminium and steel cylindrical discs using a Q-switched Nd: YAG laser [96]. The laser power density was gradually raised, moving from the thermoelastic into the plasma regime. The predominant acoustic modes measured at lower laser power densities (thermoelastic regime) were transverse waves with the lateral forces at the metal surface dominant [96]. In contrast, those recorded at higher laser power densities (plasma regime) were found to be longitudinal waves since forces perpendicular to the metal surface dominated [96]. These findings consolidated those of White et al., confirming that at lower laser power densities, the transient heating of the film's surface yielded ultrasound in the thermoelastic regime [87]. At higher laser power densities, the recoiling effect of the plasma generation on a film's surface in the plasma regime was responsible for the ultrasound generation, as characterised by their respective dominant acoustic modes. At constant laser pulse energy, an increase in the laser power density within the thermoelastic regime caused no change in the longitudinal and transverse waves' amplitudes. However, Dewhurst et al. (1982) demonstrated that once the laser power density rose to such a level as

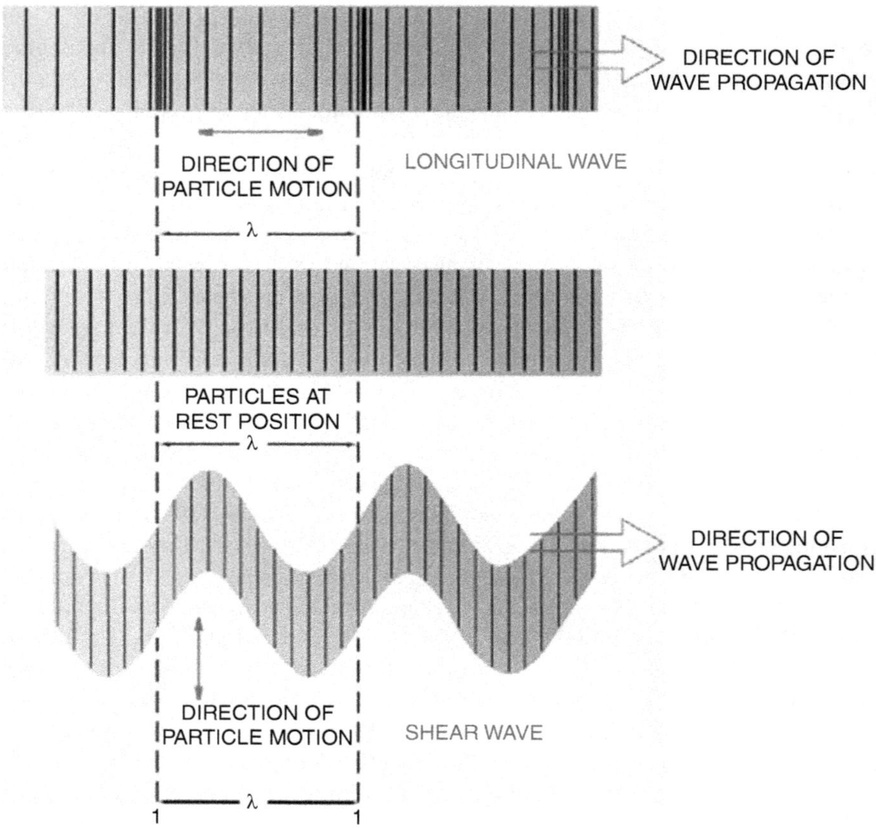

Fig. 9.2 Illustration of particle displacement in longitudinal and shear (compression) waves travelling through a medium. (Figure adapted from Messer et al. [165])

to cause ablation to the metal surface, the amplitudes of the longitudinal waves increased, whilst the amplitudes of transverse waves decreased.

The properties of the media surrounding the metal film and their effects on the yielded ultrasound signal have also been investigated, such as the thermoelastic ultrasound generation at solid/liquid interfaces [88]. Since most ultrasound experiments occur within water, studying ultrasound generation at solid-liquid interfaces is of particular interest [97]. Von Gutfeld et al. determined that the ultrasound pressures produced at a solid/liquid interface were directly proportional to a liquid's β coefficient whilst inversely proportional to the film thickness [88]. A molybdenum metal film sandwiched between a glass substrate and a liquid acted as the ultrasound-generating medium, which was irradiated with a pulsed dye laser. The three media were contained in a mylar polymer cell, separating them from the water bath hosting the panametric ultrasound transducer. Four liquids of differing β values were chosen as the liquid media of the cell (in order of decreasing β):

acetone, methanol, ethanol and water. The trend in ultrasound pressures mirrored that of the β coefficients, with the Mo/acetone interface generating the highest ultrasound pressures, whilst the Mo/water interface formed the lowest ultrasound pressures [88]. The relationship between the emitted ultrasound signal and the irradiating laser pulse length makes OpUS generation such a versatile technique. Acoustic harmonic frequencies with sub-nanosecond rise times were produced by Brienza et al., who irradiated a $LiNbO_3$ piezoelectric crystal with picosecond laser pulses, yielded by a Q-switched Nd: glass laser [93]. The harmonic ultrasound frequencies were created from a modulated laser's fundamental frequency that was modified by changing the length of the laser cavity. Frequencies as high as 2 GHz (the tenth harmonic, meaning ten times the fundamental frequency, 200 MHz) displayed SNRs exceeding 40 dB.

Through a thorough investigation of OpUS generation, it emerged that metal films were limited in their generated ultrasound pressures and bandwidths. This can be partly attributed to the metal films' high optical reflectivity and hence their low μ_a coefficients. Precisely, their sub-optimal β and μ_a coefficients restricted additional advances in generating higher ultrasound pressures and wider bandwidths [98]. To overcome the low β and μ_a coefficients of thin metal films, a move away from single-component films led to the development of multi-component composites. However, it should be highlighted that metal films' ultrasound pressures and efficiencies are still used as standards for subsequent media used for OpUS generation. Early metal films deposited for OpUS generation comprised a single component responsible for both absorbing the optical radiation and the film's subsequent volumetric expansion. However, it soon became clear that it might be advantageous to create composite materials for OpUS generation. The role of the first component (optical absorber) of the composite is to absorb the input laser radiation. Examples of optical absorbers include various nanoparticles [99], dye molecules [75] and even metal [100] films. The second component, often an elastic polymer (elastomer), undergoes pulsed expansion induced by the absorption of the heat emanated by the optical absorbers. This second component usually has a high β coefficient exhibited by its large expansion induced by a given heat input. Polydimethylsiloxane (PDMS) has been highlighted in the literature as an efficacious optical absorber host material for OpUS generation, facilitating the optical generation of high-frequency ultrasound [75]. Further, the more uniform the distribution of optical absorbers within the PDMS matrix, the shorter the heat diffusion timeframe, where the thermal energy accrued within the absorbers is transferred to the PDMS. The optical absorbers should therefore be surrounded by as many neighbouring PDMS molecules as possible. Note that aggregates of optical absorbers lead to heat sinks within the medium which can lead to thermal damage to the medium.

Composite materials have been created with monolayer or bilayer structures. With the latter, often the aim is for one layer to infiltrate into the other creating a composite region, rather than having a distinct boundary between the layers. Such integration often relies upon the density of the optical absorber region being low enough to enable the larger-sized PDMS polymer chains of fixed density

(0.965 g/cm^3) to pass between the optical absorber molecules [101]. A more integrated composite will likely be more mechanically and thermally robust. One such novel composite design was proposed by Lee et al., who fabricated a metal-polymer sandwich composite [100]. Based on a glass substrate, the composite consisted of a layered PDMS/Cr/PDMS/Al structure where an optical resonance cavity formed between the Cr and Al metal layers. These metal layers were deployed onto their surfaces through sputtering whilst the PDMS layers were spin-coated. The 10 nm thick Cr film acted as the ultrasound-generating medium, efficiently transferring the generated heat to the surrounding PDMS layers. However, the Al layer displayed a small μ_a coefficient resulting from its low thickness. Hence, the function of the 50 nm Cr film was to reflect any unabsorbed optical radiation back towards the Al film, increasing the composite's overall optical absorption. The optical resonance cavity raised the optical absorbance of the Cr layer to 90%. Lee et al. (2017) focussed on maximising both the composite's heat transfer rate and the optical absorption to improve the ultrasound performance rather than solely optimising the latter, the predominant aim of many past studies [102, 103]. Ultrasound pressures of over 1.82 MPa were measured from the PDMS/Cr/PDMS/Al film under a laser fluence of 2.35 mJ cm^{-2}. The sandwich composite design prevented the loss of heat generated in the metal layer to the substrate below, as seen in bilayer composites. Instead, Lee et al. (2017) calculated that over 70% of the input optical energy was transferred from the Al film to the adjacent PDMS layers. Equivalent composites of the same design using Parylene layers rather than PDMS were fabricated and produced lower generated ultrasound pressures and narrower bandwidths. This was explained by the higher PDMS β value compared to Parylene. In addition, PDMS has other highly desirable properties such as its optical transparency, simplicity to process and biological inertness [104]. The last property is especially advantageous for potential clinical applications such as minimally invasive OpUS image-guidance [105].

Despite the sandwich design of the metal-PDMS composite created by Lee et al. (2017), composites are often fabricated according to three designs: (i) bottom-up, (ii) top-down and (iii) all-in-one [75] (Fig. 9.3). In the bottom-up method, the nanoparticle layer is first deployed onto the substrate before the elastomer is used to create an overcoat. The opposite sequence occurs in the top-down method, where the elastomer is first deposited onto the substrate and cured. The nanoparticles are then incorporated within the elastomer using swell-encapsulation methods. Both the bottom-up and top-down methods can be used to create bilayer composites. In the all-in-one method, the elastomer and the nanoparticles are directly mixed to create a single dip-coating formulation prior to their deposition onto the substrate, creating an integrated composite. Using these methods, new materials for OpUS were created, replacing traditional thin metal films, with composites comprising metallic nanoparticles or carbonaceous absorbers. The thermal and optical properties of these composites can be tailored and optimised to a greater degree than the metal films, for example, by tuning nanoparticle size or shape [76].

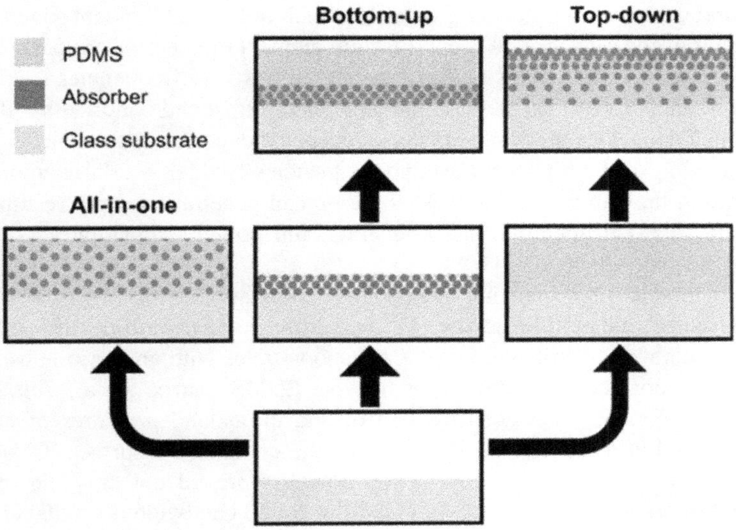

Fig. 9.3 Scheme illustrating the different designs of the 'all-in-one', 'bottom-up' and 'top-down' fabrication routes. (Reprinted from Noimark et al. [75])

9.2.2 Carbon Black

Early non-metallic composites fabricated for OpUS generation used carbonaceous optical absorbers. One example, developed by Buma et al., comprised a carbon black (CB) and PDMS composite [106]. Here, a formulation comprising CB nanoparticles, uncured PDMS and toluene was developed and deposited onto a microscope slide using spin-coating methods (all-in-one synthesis). The resultant films had a greater thickness than some metal-based generators which may contribute to the greater SNR observed in the ultrasound measurements. For example, the developed CB-PDMS coating was 25 μm thick (SNR: 38 dB), whereas a deposited Cr film had a thickness of 0.15 μm (SNR: 14 dB) [107]. The highly optically absorbing CB nanoparticles with their nano-sized dimensions facilitated a greater degree of heat transfer compared to thin-metal-PDMS composites. However, the ultrasound bandwidth of the 25 μm thick CB-PDMS composite was limited by the composite's optical penetration depth, estimated at 1 μm. Here, 24 μm of the CB-PDMS composite attenuated the generated ultrasound, in particular, the higher frequencies (>50 MHz) [107]. Buma et al. eliminated the Leaky Rayleigh Waves [95] (LRWs) responsible for broadening the ultrasound source element measured in their original CB-PDMS composite [106]. Previously, Buma et al. had suggested that these LRWs broadened the acoustic source leading to a corruption of the generated ultrasound field [106]. Further research involved using a PDMS substrate rather than glass to better-match the acoustic impedance of the substrate and composite coating. The PDMS base acted as a pressure release baffle, absorbing any backward projected ultrasound waves, rather than a rigid baffle, which reflects all incident sound waves. As a result of the reduction in the ultrasound element diameter, from 35 to 20 μm, wider ultrasound bandwidths and higher peak ultrasound frequencies (>75 MHz) were produced. To further increase the generated ultrasound bandwidths, Buma et al. concluded that the composite thickness must be reduced more to minimise acoustic attenuation of the higher ultrasound frequencies [95]. Hou et al. demonstrated that limiting the composite thickness to its optical penetration depth (estimated to be around 1 μm) resulted in broader generated ultrasound bandwidths as a result of reduced acoustic attenuation [108].

Using identical spin-coating methods for film application [106], CB-PDMS coatings with smaller thicknesses (11 μm) were applied onto a glass substrate. This was achieved by utilising larger CB nanoparticles with greater oil absorption, which were more evenly dispersed within the PDMS matrix. As a result, the PA conversion efficiency increased by 10 dB, and ultrasound pressures of 800 kPa were recorded at a distance of 10 mm from the composite's surface. In addition, the use of shorter laser pulses (5 ns vs 10 ns) improved not only the recorded ultrasound pressures but also the resultant ultrasound bandwidth (−6 dB) and peak frequency from 25.2 MHz and 45 MHz to 48 MHz and 60 MHz, respectively. These results highlight the relationship between the yielded ultrasound and the laser temporal profile. In addition, the results suggest that the use of shorter laser pulses (i.e. picosecond laser pulses) may further improve generated ultrasound bandwidths and ultrasound frequencies (>50 MHz), leading to higher ultrasound image resolutions. Nevertheless, this conclusion assumes that the laser pulse width, and not the composite thickness, was the limiting factor for the generation of higher ultrasound frequencies.

To compare the performance of macroscale OpUS transmitters with the state-of-the-art piezoelectric ultrasound transducers, Buma et al. compared the generated ultrasound signal produced by a CB-PDMS composite against a standard piezoelectric transducer. Here they demonstrated that CB-PDMS composites had superior ultrasound-generating efficiencies at higher ultrasound frequencies (>20 MHz) than those of standard piezoelectric ultrasound transducers used for interventional imaging [109]. At lower ultrasound frequencies (20 MHz), the CB-PDMS composite fabricated by Buma et al. displayed comparable efficiencies to a standard piezoelectric transducer [106]. Buma et al. theorised that the photoacoustic conversion efficiency of a thermoelastic ultrasound source was 'directly proportional to the laser power' and 'inversely proportional to the square of the laser pulse duration' [109]. Thus, it was deduced that the photoacoustic conversion efficiency of the CB-PDMS composite would increase at higher ultrasound frequencies. A 2D high-frequency OpUS array formed from a laser beam scanning a thermoelastic composite is a far simpler design than the piezoelectric equivalent, which comprises diced piezoelectric crystals. Therefore, thermoelastic all-optical ultrasound transducers were deemed as more suitable when devising high-frequency ultrasound 2D arrays.

For guidance of MIS interventions, the development of all-optical technology to achieve miniature transmit-receive OpUS devices is highly desirable to further facilitate translation of this technology from benchtop to clinic [79, 110]. Many popular OpUS detection techniques rely on interferometry [71, 111], where a resonant laser wavelength is reflected between two mirrors. The passage of an ultrasound wave through this structure then alters the separation of these two mirrors and changes the resonance wavelength, lowering its intensity [112]. The decrease in the intensity corresponds to the pressure of the passing ultrasound wave. This technique has been incorporated into various devices, both macroscale and microscale. For example, macroscale interferometric OpUS detectors comprise Fabry-Pérot (FP) etalons [113–115], whereas microscale interferometric OpUS detectors include plano-concave microresonators [116]. These hydrophones contain two polished mirrors deposited at the distal end of an optical fibre and overcoated with Parylene, waterproofing and protecting the underlying mirrors from mechanical abrasion [117]. Such interferometric OpUS detectors can be integrated into the same devices as OpUS transmitters [118]. An example is the all-optical ultrasound transducer devised by Hou et al. (2008), who integrated a 5.9-μm-thick etalon between the 11-μm-thick CB-PDMS composite produced by Hou et al. (2007) and the underlying glass substrate [115]. The transducer carried out 3D pulse-echo ultrasound imaging of metal wires, generating ultrasound bandwidths above 40 MHz. Since the CB-PDMS composite was deposited on top of the etalon, a hole (2 mm in diameter) was left in the middle of the etalon to allow the laser pulses to irradiate the composite to generate ultrasound. To increase the reflectivity within the etalon, both its inner surfaces were coated with a 30 nm gold layer. The resonance wavelength (1511.5 nm) reflected within the etalon was supplied by a continuous wave laser irradiating a 20 μm spot on the etalon.

Despite the development of the generated ultrasound signal and imaging performance of CB-PDMS composites, there are some drawbacks and challenges associated with using these composites. The extent to which solvents such as toluene can thin the CB-PDMS composite is limited and, therefore, using the fabrication methods explored, the 11-μm-thick CB-PDMS composite produced by Hou et al. was considered to be the thinnest CB-PDMS composite achievable [108]. To minimise film thickness and maintain the optical absorption coefficient and hence generated ultrasound pressures of the CB-PDMS composites, the concentration of CB nanoparticles integrated into PDMS must be increased. However, finding a solvent that is miscible with both the CB nanoparticles and PDMS is challenging since toluene is a non-polar solvent, whereas the CB nanoparticles are highly polarised [119]. Thus, increased use of a non-polar solvent can cause aggregation between the CB nanoparticles in the CB-PDMS formulations at high concentrations [120], preventing CB dispersion within the PDMS. As a result, large amounts of acoustic attenuation occur within these composites minimising both the ultrasound pressures (maximum 1.6 MPa) and bandwidths (12 MHz) they produce [75]. This impacts on high-frequency ultrasound generation (>50 MHz), restricting the achievable imaging resolution and the possibility of 3D imaging [79, 121].

9.2.3 Gold Nanoparticles

Gold nanoparticles exhibit a narrow absorption profile due to their surface plasmons, absorbing strongly at a particular resonance wavelength, whilst transmitting longer wavelengths that can be used to interrogate OpUS detectors such as FP etalons [75]. The size and shape of the gold nanoparticles determine the position of the resonance wavelength, and thus the absorption profiles are adjustable by fine-tuning their syntheses [71, 102, 103, 122–124].

To increase high-frequency ultrasound generation, Hou et al. fabricated a gold-PDMS (Au-PDMS) composite by nanoimprinting a two-dimensional gold array and overcoating it with a 4.5-μm-thick spin-coated PDMS layer [102]. The resultant Au-PDMS composite demonstrated a narrower optical absorption profile than broadband optically absorbing CB-PDMS composites, with wavelength-selective optical absorption exhibited. Further, these Au-PDMS composites achieved a 5 dB increase in ultrasound pressures at higher frequencies (>50 MHz) compared to a CB-PDMS composite fabricated by O'Donnell et al. (2008). Further, due to the smaller composite thickness, the Au-PDMS composite achieved wider ultrasound bandwidths resulting in a lower degree of acoustic attenuation compared to the CB-PDMS-based ultrasound transmitters.

To create an integrated all-optical transmit-receive OpUS device, a bilayer Au-PDMS composite-based ultrasound transmitter [102] was combined with an etalon developed by O'Donnell et al. [107]. This was achieved by fabricating a thin polymer etalon on top of the Au-PDMS composite, such that ultrasound detection could be achieved using an etalon interrogation wavelength of 1550 nm. The ultrasound performance of this device was compared to that of an 11-μm-thick CB-PDMS composite fabricated using the method reported by Hou et al. [108]. The produced ultrasound pressures were measured as a function of the laser wavelength in the 700–1000 nm range. The greatest ultrasound pressures were generated at 700 nm and corresponded with the optical absorbance maximum, with decreasing optical absorbance corresponding to reduced ultrasound pressures, with the lowest recorded pressures generated at the absorption minimum (1000 nm). The ratio between the maximum and minimum ultrasound pressures was 7.5:1. The disparity in the yielded ultrasound signals at different wavelengths highlights the potential of Au-PDMS composites for multimodality imaging or sensing. The generated ultrasound bandwidth (−6 dB) was measured at 16.5 MHz with a central frequency of 15 MHz. These values are smaller than those measured by Hou et al. [108]. This may be due to the increased all-optical ultrasound transducer source diameter of 400 μm compared to the 25 μm source used by Hou et al. (2007). This larger transducer source diameter corresponds to a resulting lower laser fluence and reduces the generated ultrasound signal.

A similar multi-layered all-optical ultrasound transducer, which generated high ultrasound pressures with wide frequency bandwidths, was fabricated by Hou et al. [111]. The device comprised a three-layered structure. A two-dimensional gold nanoparticle nanostructure array was deposited onto a glass substrate and over-

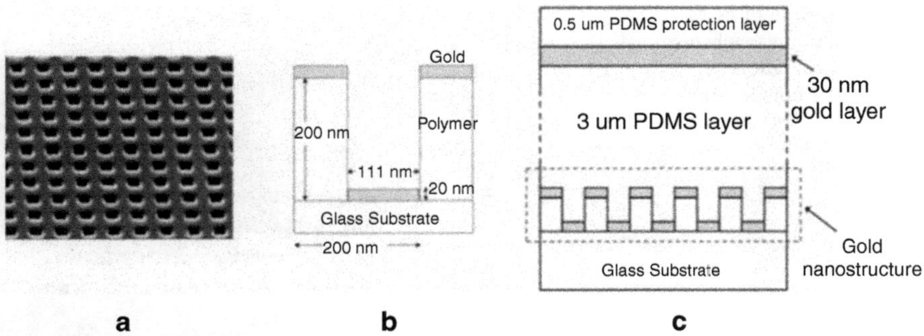

Fig. 9.4 Illustrations of Au-nanostructure. (**a**) SEM of Au nanostructure, (**b**) 2D Au nanostructure and (**c**) Au-PDMS film structure. (Reprinted from Hou et al. [71])

coated with a 3-μm-thick PDMS layer, whilst a 30-nm gold coating was deposited onto the PDMS layer (Fig. 9.4). Nanoimprint lithography and laser interference lithography were employed to fabricate the 2D-gold array using a SiO_2 mould. The two PDMS layers were spin-coated to sandwich the 30-nm gold layer deposited through 'gold evaporation'. The gold nanostructure partially absorbed the modulated laser input at frequencies in resonance with the gold nanostructure surface plasmons (700–800 nm), generating ultrasound. The transmitted wavelengths were reflected by the thicker gold layer and recirculated within the optical cavity formed between the gold nanostructure and the gold layer. Both exhibited high optical reflectivity coefficients, creating an etalon that enabled OpUS detection. The Au-PDMS all-optical ultrasound transducer recorded wide bandwidths of 57 MHz (−6 dB) with a central frequency of 40 MHz. Ultrasound pressures of 67 kPa were recorded 10 mm from the composite's surface using a 5 ns laser pulse at a laser fluence of 26 mJ cm^{-2}. The laser fluence was then raised to 5.1 J cm^{-2}, equalling the composite's damage threshold, generating ultrasound pressures of 1 MPa measured 3 mm in the far field. This was reportedly the first all-optical ultrasound transducer to utilise the wavelength-selective properties of gold nanoparticles, enabling high-frequency ultrasound imaging within one integrated device. This transducer performed ultrasound imaging, a first for such a device from an Au-PDMS multi-layered composite, imaging a wire and a human hair. It should be noted that the main difference between this transducer and that fabricated by O'Donnell et al. (2008) was that the former demonstrated higher ultrasound frequency generation and a larger theoretical SNR (>40 dB).

To simplify the fabrication of the Au-PDMS composites, Wu et al. explored the use of an 'in situ' synthesis, reducing the bulk gold reagent to nanoparticles within the PDMS matrix [103]. This 'one-pot' method ensured an even distribution of gold nanoparticles within the PDMS, with efficient heat transfer between the two components and provided greater control over the composite's optical absorption and its thickness. Here, gold (III) chloride trihydrate ($HAuCl_4 \cdot 3H_2O$) salt was mixed with a PDMS polymer solution, reducing the gold reagent and resulting in

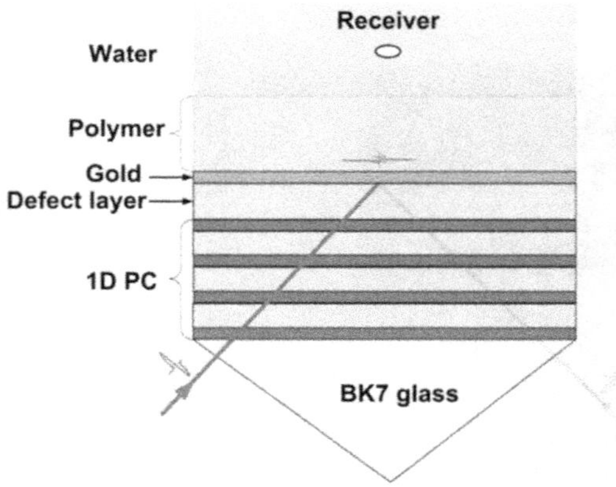

Fig. 9.5 Layout of PCM structure and the mechanism of OpUS generation. (Reprinted from Guo et al. [98])

the in situ formation of gold nanoparticles. The resultant polymer solution showed a characteristic ruby red colouration. The concentration of gold salt in the PDMS formulation did not affect the subsequent nanoparticle size, with formulations of increasing salt concentration achieving gold nanoparticles measuring 20 nm in diameter, corresponding to an optical absorption peak at 520 nm. Note that composites fabricated with a 3.75 wt. % demonstrated poor polymer curing; however, it was suggested that there is scope for fabrication optimisation. For a 450-μm-thick Au-PDMS composite, with a gold concentration of 1.79 wt%, the photoacoustic conversion efficiency was measured at 2.2×10^{-5}, which is three orders of magnitude greater than that measured for a standard Al film. The same composite generated ultrasound pressures of 189.49 kPa, measured 1.8 mm from the composite's surface, under a laser energy density of 13 mJ cm^{-2} and a pulse duration of 150 ns. Wu et al. (2012) reasoned that employing a shorter laser pulse could contribute towards widening the ultrasound bandwidths achieved from 3.1 MHz, since this value is far below those recorded by Hou et al. [102] and [111]. The examined Au-PDMS composite had a thickness of 450 μm and a gold nanoparticle concentration of 1.79 wt% [103]. Both the composite's gold concentration and thickness were varied to examine their effects on the resultant ultrasound performance. Increasing the gold salt concentration within the composite solution was found to result in Au-PDMS composites with higher optical absorption leading to more considerable ultrasound pressures. In addition, the generated acoustic pressures increased with composite thickness until an optimal thickness was achieved, after which point, increased composite thickness led to acoustic attenuation and decreased generated ultrasound pressures.

Guo et al. proposed a novel photonic crystal-metallic (PCM) structure design to overcome challenges with optical absorption and composite thickness. They fabricated an all-optical ultrasound transducer comprising an Au-PDMS composite deposited on top of a PCM structure deployed onto a prism substrate [98]. The PCM was formed from a series of dielectric layers (TiO_2/SiO_2) capped by a defect layer (SiO_2). The structure was designed such that 100% of specific wavelengths of the optical energy input was absorbed. Input light reflected off the Au-PDMS composite was circulated within the PCM structure through total internal reflection (TIR), acting as an FP cavity and enabling the 15-nm gold layer to absorb over 90% of the laser pulse (Fig. 9.5). This enabled greater absorption of the input laser light compared to other Au-PDMS transducer structures proposed, some of which absorbed up to 30% [111]. The transmitted light acted as an interrogation beam for the FP cavity. Using a 10-ns laser pulse ($\lambda = 532$ nm), ultrasound pressures of 28 kPa (10 mm from the composite surface) were generated with corresponding −6 dB bandwidths of 78 MHz. These pressures generated by the PCM Au-PDMS composite were over 10 times greater than those measured by the 2D Au-PDMS composite developed by Hou et al. [102]. The resulting photoacoustic conversion efficiency was determined to be 3×10^{-6}.

There are various advantages associated with creating all-optical ultrasound transducers fabricated with a PCM structure. Firstly, since the laser pulse is directed at an angle and not perpendicular to the Au-PDMS surface, higher laser pulse fluences can be adopted since any light not absorbed by the surface is internally reflected within the PCM structure. Thus, the imaged object would not be irradiated with high-energy laser light and thermally damaged. Utilising larger laser pulse fluences would theoretically enable higher ultrasound pressures to be generated. Additionally, the composite can be fabricated using simple deposition methods such as spin-coating to produce the PDMS coatings [98].

9.2.4 Carbonaceous Nanoparticles

Carbonaceous particles have been widely investigated as optical absorbers to create composites for OpUS generation. In particular, carbon nanotubes (CNTs) are well suited for the development of composites for OpUS due to their nanoscale dimensions, large aspect ratios, high thermal conductivities and large heat transfer rates. These properties facilitate the generation of higher thermal gradients, such that the resultant composites produced large ultrasound pressures compared to composites including Au-PDMS. In addition, CNT-containing composites often exhibit significant thermal damage thresholds owing to the ability of the elastomer to infiltrate between the CNT strands and adhere firmly to the underlying substrate. CNT-composites fabricated for applications in OpUS typically comprise multi-wall CNTs (MWCNTs). Here, the CNTs are concentrically stacked within one another. MWCNTs can be grown directly onto surfaces through use of a catalyst. Baac et al. designed a bilayer MWCNT-PDMS planar composite, growing the MWCNTs on a fused-silica substrate coated with an iron catalyst [120] (Fig. 9.6a). A PDMS layer was then spin-coated on top of the homogenously dense MWCNT underlayer. Composites with denser MWCNT layers were also fabricated, which generated higher ultrasound pressures; however, their increased composite thicknesses led to narrower ultrasound bandwidths. The resulting MWCNT-PDMS composite displayed an extinction coefficient of 80%. The nanoscale diameters and high aspect ratios of the MWCNTs provided efficient heat transfer from the MWCNTs to the surrounding PDMS. Seeded with a 6-ns 532-nm laser pulse,

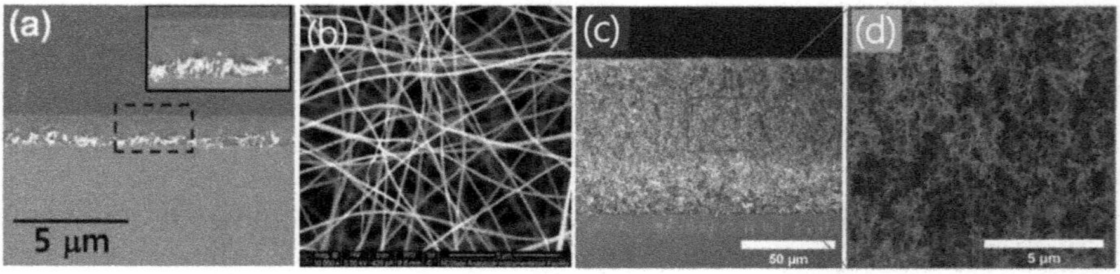

Fig. 9.6 SEM images of MWCNTs, CNFs and CSNPs. (**a**) MWCNTs grown on Fe-seeded glass by Baac et al. [120], (**b**) CNFs fabricated by electrospinning by Hsieh et al. [126] and (**c**) structure of CSNPs along with a magnified (**d**) view of the CSNP nanostructure by Bodian et al. [138]

the composite yielded ultrasound pressures 14 dB greater than those generated by the Au-PDMS composite and 25 dB greater than a standard 100 nm Cr film [120]. Observing the power spectra, the increase in ultrasound pressures of the MWCNT-PDMS composite against an Au-PDMS composite, formed from a gold-array overcoated with PDMS and a Cr film, respectively, can be seen at frequencies as high as 120 MHz. Moreover, when the disparity in the optical absorbances of Au-PDMS and MWCNT-PDMS composites was accounted for, the latter generated ultrasound pressures more than twice as high as the former. In particular, the MWCNT-PDMS power spectrum closely resembled that of the input laser pulse, demonstrating the dependency of the acoustic frequency generation mechanism on the input laser pulse.

Carbon nanofibres (CNFs), an allotrope to MWCNTs, have also been explored for applications in OpUS generation. Rather than forming multiple cylindrical graphene sheets nested within one another, here, the carbon comprises stacks of cones of graphene sheets. Motivation for using CNFs includes their lower cost and greater ease of handling than MWCNTs. Electrospinning is the most common method of deploying CNFs onto surfaces, ranging from larger planar surfaces to smaller optical fibre distal ends [125, 126]. This deposition technique can be used to effectively control the thickness of the CNF layer, minimising the degree of acoustic attenuation experienced by the generated ultrasound [86].

Electrospinning can be quite challenging as a deposition technique due to its complicated operating mechanism; however, it is well suited for applications in synthesising high aspect-ratio nanoparticles such as CNFs [127]. A planar electrospun CNF-PDMS composite was fabricated by Hsieh et al. (2015) using the following method. A solution of polyacrylonitrile (PAN) and N, N-dimethylformamide was electrospun through a needle syringe to form the 24-μm-thick CNF layer carbonised using nitrogen gas followed by a spin-coated PDMS overcoat (Fig. 9.6c) [127]. The composite was irradiated using a 532-nm 4-ns laser pulses with a laser fluence of 3.7 mJ cm^{-2}. The planar electrospun CNF-PDMS composite fabricated produced ultrasound pressures of 12.15 MPa, at a distance of 3.65 mm from the composite's surface, with corresponding frequency bandwidths (−6 dB) of 7.63 MHz [126]. These generated ultrasound pressures were amongst the highest reported to be produced by a planar, non-focussed OpUS transmitter. It should be noted that the ultrasound pressures decreased as one progressed outward from the composite's surface due to the attenuation of the higher ultrasound frequencies located at the front of the ultrasound wave pulse [126]. Despite this decay and fluctuations in the laser power, the 58-μm-thick CNF-PDMS composite yielded comparatively large ultrasound pressures at all measured axial distances than a standard CB-PDMS composite [126].

Candle-soot nanoparticles (CSNPs) have recently gained interest as new optical absorbers for composites for OpUS generation due to their unique hierarchical structures and low-cost simple synthesis [127]. Using CSNPs, high optical absorbances can be achieved within small thicknesses due to the 3D structure of the CSNPs. These CSNP coatings can be fabricated by positioning a desired substrate at a specified distance of a few millimetres above the wick of a burning candle (Fig. 9.6b). This rapid synthetic route facilitates the creation of CSNP coatings with thicknesses that can be modified by controlling the substrate-flame exposure period. Using a direct transfer method, Chang et al. (2015) developed a bilayer CSNP-PDMS composite ($d = 18.44$ μm). The direct transfer method used involved holding one slide above a candle flame, creating a CSNP coating, whilst a second slide was coated with an anti-adhesive and spin-coated with PDMS. The two slides were subsequently pressed together and then withdrawn, leaving an 18.44-μm-thick CSNP-PDMS composite that demonstrated an optical absorbance of 96% at 532 nm. The two layers were highly integrated, leading to a robust composite. The synthesised CSNPs possessed an intricate hierarchical structure with a high surface-area-to-volume ratio [70]. This led to a very low interfacial thermal resistance between the CSNPs and the surrounding PDMS, raising the heat transfer rate responsible for the generated ultrasound bandwidths. Upon excitation using a laser fluence of 3.57 mJ cm^{-2} (6-ns laser pulses), ultrasound pressures as high as 4.8 MPa were measured at a distance of 4.2 mm into the far field. The ultrasound pressures measured were twice as high as those generated by CNF-PDMS composites using the same excitation source. Furthermore, 21 MHz and 10 MHz were recorded as the ultrasound bandwidth (−6 dB) and the centre frequency, respectively. These generated ultrasound pressures were twice as high as those achieved by CNF-PDMS composites. The CSNP-PDMS composite fabricated exhibited a particularly promising photoacoustic conversion efficiency (4.41×10^{-3}). This was nearly 3 times greater than efficiencies achieved by composites comprising CNT-PDMS and nearly 24 times greater than those measured using Au-PDMS composites [127].

Although fewer studies have investigated the use of graphite as an optical absorber for macroscale ultrasound generating composites, these have shown that they demonstrate wide-band optical absorption and high resilience to extreme temperatures and pressures. Menichelli et al. generated ultrasound from a pure graphite film and a graphite-epoxy (Gr-Ep) composite [105]. An epoxy resin was employed as the elastomer matrix instead of the conventional PDMS. To simplify the experimental setup, the composites were applied onto a 1-mm-thick glass substrate to mimic fibre-optic OpUS probes, which were the target substrates of this study. The composite thickness and graphite concentration were varied and optimised through a trial-and-error method by creating a variety of different composites. The Gr-Ep composites displayed composite thicknesses varying from

10 to 500 μm, whilst the graphite concentration ranged between a fractional weight (FW) % of 0.4–11%. The photoacoustic conversion efficiencies rose with increasing graphite concentration in the composites as well as decreasing composite thicknesses down to the optical penetration depth. The Gr-Ep composite achieved a photoacoustic conversion efficiency two orders of magnitude greater than that of a 100-nm thick Al film. Illuminated by a Q-switched Nd: YAG laser ($\lambda = 1064$ nm), 20 kPa acoustic signals with a 30 MHz bandwidth (-3 dB) were generated by a 10-nm-thick Gr-Ep with a graphite concentration of 11% (FW). In addition, the ultrasound produced by both the pure graphite film and the Gr-Ep composite was similar despite their different acoustic properties.

With graphite displaying limited thermal conductivity, reduced graphene oxide (rGO) was investigated due to its higher thermal conductivity, larger surface area and enhanced mechanical robustness [128]. Hwan Lee et al. created an OpUS transmitter by combining rGO with a thin layer of Al demonstrating the suitability of rGO as an optical absorber in composites [129]. The rGO layer was synthesised using graphite flakes in the Hummers method to engineer graphene oxide that was then annealed at high temperatures and spin-coated onto Pyrex glass. The 100-nm-thick Al overcoat was applied using an 'E-gun evaporator'. The use of rGO in composites was reasoned attractive since it is a cheap form of graphene whose unique structural properties (high thermal conductance and structural strength) are desirable for OpUS generation [70]. The high optical absorbance of the rGO layer prevented any optical reflectance off the Al thermal expansion layer. Peak-to-peak ultrasound pressures of 7.5 MPa were recorded under a 532-nm 5-ns laser pulse with a laser pulse density of 43.28 mJ cm^{-2}. Since the coefficient of thermal expansion (CTE) of rGO is negative, the combination of rGO and Al created a bending motion during the generation of ultrasound, which was suggested by Hwan Lee et al. to enhance the emitted ultrasound pressures [129].

Lee et al. further explored rGO as an optical absorber for OpUS by creating composites with PDMS. The rGO layer was formed from spin-coating GO onto a Pyrex glass substrate, followed by annealing to form the 100-nm-thick rGO layer. A 5.5:1 PDMS: CB mixture was spin-coated over the bottom rGO layer. A 5-ns 532-nm Nd: YAG laser was used for the OpUS generation, producing a pulse energy of 25 mJ and a laser fluence of 35.66 mJ cm^{-2}. The uneven and lumpy rGO surface generated a series of interfering spherical ultrasound waves differing from the regular planar ultrasound waves emitted by many other OpUS transmitters. As a result, a non-uniform ultrasound field could be produced, leading to destructive interference, resulting in a narrow bandwidth. These rGO-PDMS composites generated ultrasound pressures of over 4.5 MPa, 2.2 times greater than that of the rGO-Al composite [130]. Furthermore, a 10 dB increase in ultrasound signal was observed across all ultrasound frequencies from the rGO-PDMS composite compared to the rGO-Al composite. This is in part due to the greater CTE of PDMS compared to Al; however, the high thermal conductivity of rGO significantly decreased the response time of the rGO-PDMS OpUS transmitter. Further, the ultrasound frequencies produced (>50 MHz) were so high that real-time ultrasound imaging was deemed feasible with the rGO-PDMS composite.

9.3 Fibre-Optic Composites

Macroscopic planar composites have limited applications in in vivo medical imaging due to their large dimensions. Innovations in materials technology were translated to the development of fibre-optic OpUS probes. With metre-scale longitudinal dimensions and minute lateral dimensions, these devices are well suited to applications in minimally invasive imaging, for example, intravascular ultrasound. Moreover, by changing the absorber component, the optical absorption of the coating can be tuned to enable multimodal imaging and therapy, for example, enabling combined ultrasound and photoacoustic imaging.

The application of micron-thickness composite coatings onto submillimetre-sized substrates for OpUS can be particularly challenging. Deposition techniques well suited for planar composite fabrication, such as spin-coating and chemical vapour deposition, are unsuitable for the application of coatings onto the distal ends of optical fibres [74, 75]. Instead, deposition techniques such as dip-coating and electrospinning [74, 104] have been explored [74, 104]. Many of these application methods have enabled researchers to integrate both OpUS detectors and transmitters into one fibre-optic device [122].

9.3.1 Gold Nanoparticles

Gold nanoparticles are well suited to the development of coatings for OpUS transmitters due to their wavelength-selective properties and high optical absorption coefficients. Examples include multi-layered Au-PDMS composites that comprise two layers of gold nanoparticles forming an FP cavity [71, 80, 131]. Wu et al. proposed an OpUS transducer design comprising repeating layers of a gold-polyethyleneimine (Au-PEI) composite applied onto the distal end of an 8.2 μm diameter single-mode optical fibre [99]. The multi-layered composite design was an innovative approach, compared to previous designs reported in the literature. Each composite layer comprised a PEI coating (positively charged) overcoated with gold nanoparticles (negatively charged). The gold nanoparticles were synthesised by reacting gold (III) chloride trihydrate ($HAuCl_4 \cdot 3H_2O$) with trisodium citrate ($Na_3C_6H_5O_7$), producing red gold nanoparticles ($\text{Ø} = 20$ nm). In parallel to the experimental studies, a simulation was run using identical conditions (optical irradiance laser power and composite thickness). The ten-layered fibre-optic composite coating achieved ultrasound pressures of up to 2.44 kPa, measured at a distance of 1.16 mm from the composite surface, using a pulse energy of 1.2 μJ. The ultrasound pressures were calculated to be twice the amplitude of those generated using a Gr-PEI composite. This was one of the first miniaturised OpUS fibre-optic transmitters designed for applications in in vivo medical imaging.

Exploiting the different peak absorption wavelengths of gold nanoparticles of varying sizes, Wu et al. coated a single-mode optical fibre with a multi-layered gold nanostructure, fabricated using a layer-by-layer technique to form a fibre-optic all-optical ultrasound transducer [132]. In total, five layers of gold nanoparticles were stacked on top of one another, forming an array. An FP cavity and a diaphragm were integrated into the fibre optic device such as OpUS detectors. Gold nanoparticles of different diameters were then separated within the fibre end-face. Those with a larger diameter exhibited a lower-energy maximum absorption peak corresponding to a smaller plasmonic resonance frequency. Ultrasound pressures of up to 1.56 kPa (measured 2.3 mm from the composite's surface) were generated by the transducer following the absorption of a 150-ns 527-nm laser pulse with a laser fluence of 1.77 J cm^{-2}. Compared to other Au-PDMS composites, these ultrasound pressures were limited; however, the ultrasound beam could be directed via selective excitation of the composite using a range of laser wavelengths [75, 99, 133, 134]. This feature removed the need for a mechanical scanning of the all-optical ultrasound transducer, often employed by other 2D OpUS arrays [106, 109]. Each laser wavelength excited gold nanoparticles with a specific diameter and hence generated ultrasound at that area of the composite. Specific ultrasound interference patterns were created by modifying the phase delay between each laser wavelength; hence, the ultrasound signal could be directed in a particular direction.

The one-pot protocol used for fabricating macroscale Au-PDMS composites [103] was applied to fabricating fibre-optic counterparts to form an all-optical ultrasound probe by Wu et al. [133]. The Au-PDMS composite was synthesised by mixing gold (III) chloride trihydrate ($HAuCl_4 \cdot 3H_2O$) into uncured PDMS. The PDMS reduced the gold salt into the nanoparticles. The synthesised gold nanoparticles had an average diameter of 20 nm corresponding to a narrow absorption profile centred at 520 nm. A 400 μm core multimode fibre was dip-coated with a 7.58 wt% Au-PDMS solution. This formed a 150-μm-thick domed composite, which was dried overnight. The measured ultrasound pressures and −6 dB bandwidth were 36.96 kPa (measured 1.8 mm from the composite's surface) and 2.1 MHz, respectively. This ultrasound performance demonstrated that OpUS generation from coated optical fibres was feasible; however, the ultrasound pressures and bandwidth were still inadequate for ultrasound imaging. The large degree of acoustic attenuation resulting from the thick composite minimised the ultrasound pressures and bandwidth. In addition, the wide laser pulse used ($\Delta t = 150$ ns) contributed to the small ultrasound bandwidth. Nevertheless, the photoacoustic conversion efficiency was calculated at 10^{-7}, an order of magnitude greater than that reported for standard Al films [133].

Zou et al. fabricated an Au-PDMS fibre-optic OpUS probe that generated sufficiently high ultrasound pressures and bandwidths to yield C-mode ultrasound images of swine tissue [134]. The gold nanoparticles were reduced within the PDMS by mixing a gold (III) chloride trihydrate ($HAuCl_4 \cdot 3H_2O$) salt in PDMS, forming a red colour in the same manner as by Wu et al. [103]. The solution underwent ultrasonication to prevent agglomeration. The tip of the 400-μm-diameter core multimode fibre was then dipped into the Au-PDMS solution forming a dip-coated Au-PDMS fibre-optic composite. Upon excitation with 532 nm light (6 ns pulse length), the Au-PDMS composite generated ultrasound pressures of 0.64 MPa

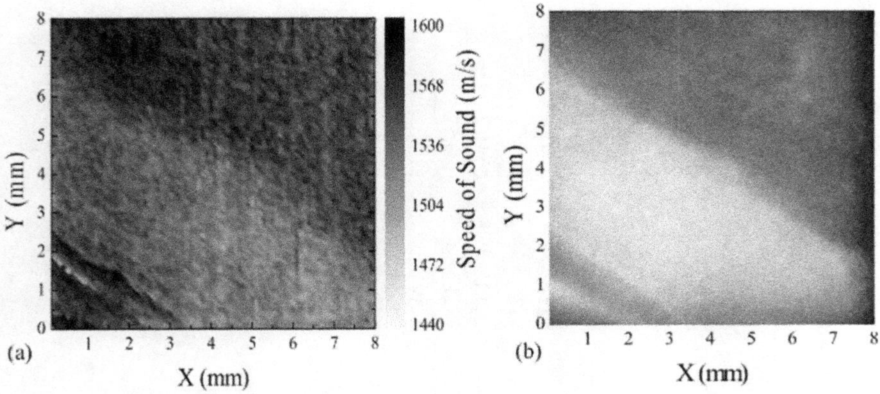

Fig. 9.7 Comparisons of ultrasound image and photograph of swine tissue. (**a**) Ultrasound image of a slice of swine tissue recorded by US transducer made up of Au-PDMS fibre-optic transmitter and a hydrophone. (**b**) Photograph of the slice of swine tissue. (Reprinted with permission from Zou et al. (2014) © The Optical Society [134])

measured at a distance of 1 mm from the surface. In addition, bandwidths of over 20 MHz were recorded, a significant improvement in the ultrasound performance demonstrated by Wu et al. [133]. The photoacoustic conversion efficiency was over five orders of magnitude greater than the standard Al metal film and three orders of magnitude greater than the Gr-Ep composite fabricated by Menichelli et al. [105]. However, the efficiency was an order of magnitude lower than that of a standard MWCNT-PDMS composite. These results were one of the first demonstrations that ultrasound imaging in MIS was feasible using fibre-optic OpUS transmitters resulting from the high-resolution ultrasound images of ex vivo animal tissue. The Au-PDMS-coated OpUS probe, coupled with a hydrophone, achieved resolutions of 200 μm. There was good registration between the ultrasound images and the photographs of the tissue specimen (Fig. 9.7, Table 9.1).

Using a slightly different synthetic route, a 'one-pot protocol', a 105-μm-thick Au-PDMS composite ([Au] = 7.58 wt%) was created on the distal end of a 400-μm multimode fibre. This composite had a smaller thickness (150 μm) than the Au-PDMS composite fabricated by Wu et al., despite similar syntheses and coating methods [133]. Zou et al. reported that the dip-coating method provided little control over the composite's thickness creating acoustic attenuation of the higher ultrasound frequencies due to the excess composite [134]. Focussed ion beam (FIB) milling was suggested as an alternative strategy to fabricate composites with smaller thicknesses [134]. Tian et al. used FIB to develop ultrasound probes with transmitter materials comprising micro-scale gold nanostructures and demonstrated that this method can be used to reduce the overall composite thickness and thus maximise the yielded ultrasound bandwidths [122] (Fig. 9.8). The gold nanopores were engraved into the gold nanostructure measuring 60 × 80 × 200 nm (height, diameter, spacing) by selecting a particular FIB current. Tian et al. discovered that regular gold

Table 9.1 Tabulated values of each material used for OpUS generation with the respective pressures and bandwidths (−6 dB). The distance between the material and the ultrasound receiver is stated along with the laser fluence used during the ultrasound generation. (Reprinted with permission from Noimark et al. [75]).

Author	Optically absorbing material	Coated substrate	Measured distance [mm]	Measured pressure [MPa]	–6 dB bandwidth [MHz]	Laser fluence [mJ cm^{-2}]
Baac et al. [166]	MWCNT	Concave lens	5.5	32	~20	42.4
Baac et al. [167]	AuNP	Glass slide	–	–	~80	3 mW cm^{-2}*
	MWCNT				~80	
Lee et al. [168]	MWCNT	Concave lens	9.2	70	25	9.6
Colchester et al. [135]	MWCNT	105 μm core optical fiber	2	0.45	12	41.6
		200 μm core optical fiber		0.9	15	36.3
Noimark et al. [74]	MWCNT	200 μm core optical fiber	3	1.34–4.5	23.15–39.8	16.2–87.9
Chang et al. [127]	CSNP	Glass slide	4.2	4.8	~21	3.57
	CNF			2.3	~10	
	CB			0.8	~12	
	Cr			0.4	~13	
Wu et al. [80]	AuNP	Glass slide	1.8	0.01–0.189	~4	3.67–13
Zou et al. [134]	AuNP	400 μm core optical fiber	1	0.64	~8	8.75
Hou et al. [102]	AuNP	Glass slide	10	0.002	~65	20.4
Buma et al. [109]	CB	Glass slide	10	0.150	–	10.4
Poduval et al. [125]	MWCNT	200 μm core optical fiber	1.5	0.2–1.59	29	35
Hsieh et al. [126]	CNF	Glass slide	3.65	12.15	7.63	3.71
	CB			1.60	7.84	
Lee and Cuo [100]	Cr	Planar glass	–	~0.6–1.82	–	2.35
	Ti			~0.5–1.0		

Fig. 9.8 SEM images of FIB-milled Au nanostructure on (**a**) fibre core, (**b**) magnified image of FIB-milled area, (**c**) Au nanopores (Ø = 80 nm) [122]. (Figure reprinted from Tian et al. [122])

nanostructures exhibit higher α values than irregular gold nanostructures [122]. The produced ultrasound signal pressure and bandwidth (−3 dB) were recorded at 2.73 kPa and 7 MHz, respectively. This ultrasound performance was a slight improvement on that generated by Wu et al. (2012), perhaps due to the FIB-milled gold nanostructure, whose smaller thickness led to an enhanced ultrasound signal.

9.3.2 Carbonaceous Nanoparticles

Electrospinning has been reported (Hsieh et al. [126]) as a niche method to fabricate CNFs for OpUS generation [126]. More recently, Poduval et al. successfully used this synthetic route to create CNF-PDMS bilayer composites on the distal end of a 200-μm optical fibres [125]. Due to the mesh-like structure of the electrospun CNF coating, the over-coated PDMS was able to infiltrate creating a well-integrated composite [125]. The CNF layer was created by electrospinning a PVA/MWCNT solution onto a fibre end-face. The PDMS overcoat was then applied using dip-coating methods. The CNFs were aligned along the fibre axis and Poduval et al. (2017) postulated that the alignment of the CNFs might influence the optical absorbance. Fibres aligned parallel to the polarisation of the laser pulse provided higher optical absorbances than those unaligned. Here, the CNF-PDMS composites exhibited high optical absorbances exceeding 97%. Several composites of differing CNF thicknesses were synthesised by varying the electrospinning time, a feature that created composites with customised CNF layer thicknesses and thus controllable overall composite thicknesses. Compared to a dip-coated CNF-PDMS standard film ($d = 60$ μm), a 13.7-μm-thick electrospun composite (without a PDMS overcoat) generated an ultrasound bandwidth of 29 MHz, far wider than the 8 MHz recorded by study's dip-coated equivalent. The broader bandwidth demonstrated the effect of reduced acoustic attenuation resulting from a thinner composite. The good integration between the CNF and PDMS resulted in good thermal conduction between the two constituents, translating to peak-to-peak ultrasound pressures of 1.59 MPa, over five times greater than the dip-coated standard CNF-PDMS fibre-optic composite (0.31 MPa) under a laser pulse energy of 11 μJ.

One particularly promising optical absorber for OpUS generation is MWCNT and the literature reports that macroscale planar MWCNT-PDMS composites have generated high ultrasound pressures (see Sect. 9.2.4). MWCNT-PDMS composites were created on the distal ends of optical fibres with a maximum dome thickness of less than 20 μm (Colchester et al. [135]). One key challenge of creating these composites was synthesising the MWCNT solvent dispersion that was compatible with PDMS. Here, stable MWCNT solutions were created by functionalising the MWCNTs using oleylamine-functionalised pyrene ligands that aided their dissolution in the xylene solvent. The MWCNT-PDMS formulation was subsequently dip-coated onto both 105-μm and 200-μm core multimode fibres. Xylene ensured the even distribution of the MWCNTs within the all-in-one MWCNT-PDMS solution producing homogenous ultrasound fields. A pulsed Nd: YAG laser ($\lambda = 1064$ nm, $\Delta t = 2$ ns) with pulse energies of 3.6 and 11.4 μJ (105 / 200 μm) generated laser fluences of 41.6 and 36.3 mJ cm^{-2} (105 / 200 μm), respectively, which irradiated the MWCNT-PDMS composites.

A hydrophone recorded ultrasound pressures of 0.44 and 0.89 MPa, respectively, generated at a distance of 2 mm from the composite surface. Additionally, −6 dB ultrasound bandwidths of 12 and 15 MHz (105/200 μm) were produced, respectively. These ultrasound pressures are higher than those generated by other fibre-optic composites mentioned in Sect. 9.3 [134]. Colchester et al. outlined that to achieve greater ultrasound bandwidths for higher ultrasound image resolution, a narrower laser pulse duration and a thinner composite were needed. Note that the latter option was cautiously suggested since a thinner composite could lead to lower ultrasound pressures [135].

In a follow-up study, Colchester et al. (2015) fabricated an all-optical ultrasound transducer comprising the MWCNT-PDMS-coated 200 μm-core multimode fibre, acting as an ultrasound source, whilst a FP cavity deposited on a single-mode fibre (fabricated by Beard et al. [136]) functioned as the ultrasound detector [77]. Upon excitation by a Q-switched Nd: YAG laser ($\lambda = 1064$ nm), the MWCNT-PDMS coated fibre generated ultrasound pressures in excess of 4 MPa at the fibre end-face with an ultrasound bandwidth of 15 MHz and a central frequency of 18 MHz.

Both optical fibres were housed within a metal tube and secured with heat shrink, with a total device diameter of less than 0.84 mm. Using this transmit-receive all-optical ultrasound transducer probe and raster scanning it across the tissue, high-resolution 2D B-mode ultrasound images were achieved with maximum axial and lateral resolutions of 64 μm and 88 μm, respectively, of a swine aorta and carotid artery. However, the narrow diameter of the MWCNT-PDMS-coated multimode fibre limited the generated ultrasound pressures. The ultrasound images exhibited well-resolved distinctions between different soft tissue types, including sub-structures such as lymph nodes and side branches visible in the swine aorta specimen. This study was one of the first to demonstrate that a fibre-optic all-optical ultrasound transducer small enough for endoscopic imaging could produce the required imaging resolutions to image vascular tissue accurately [135].

Noimark et al. [74] optimised the ultrasound generation performance of these MWCNT-PDMS-coated fibres by optimising the MWCNT-based coating formulations [74]. The MWCNTs were functionalised using an oleylamine-functionalised pyrene ligand as seen in [77, 135] and the nanotube loading was optimised. Further, using the functionalised MWCNTs, an organogel coating formulation was synthesised. They developed three different strategies for fabricating fibre-optic MWCNT-PDMS composites and undertook a consistency study that examined the fibre surface morphologies, optical absorption and generated ultrasound signals. One key challenge encountered was the formation of coffee-ring effects on the MWCNT-coated fibre tip. These effects are well documented in fields such as inkjet printing, when volatile solvents are used to create coatings. In the study by Noimark et al., fibre-optic OpUS transmitters (fibre diameter: 200 μm) with three types of MWCNT-PDMS composite coatings were created, with maximum dome thicknesses of less than 20 μm. The first composite (MWCNT-xylene composite) was created by dip-coating a fibre with a MWCNT-xylene solution forming a MWCNT coating with distinct coffee-ring drying effects. This coated fibre was subsequently overcoated with PDMS to form a bilayer bottom-up composite. Note that here, the highly optically absorbing MWCNT coating absorbed >97% of incident light within a coating with a thickness of less than 1 micron. The second composite (organogel composite) also comprised a bilayer structure, but here, an MWCNT organogel formulation was used to create the MWCNT coating, resulting in uniform coverage across the fibre end face. Subsequently the MWCNT-coated fibre was dip-coated with PDMS to form a composite. Lastly, an all-in-one MWCNT-PDMS formulation was created and dip-coated onto a fibre to form a third, 'integrated' composite coating. The two bottom-up coating methods minimised wastage of expensive MWCNTs compared to the integrated composite, whilst facilitating good integration between the MWCNTs and PDMS. The latter ensured a high rate of heat diffusion corresponding to a high-frequency ultrasound generation, as well as an enhanced level of structural protection for the MWCNTs being fully enveloped by the PDMS, as confirmed by SEM imaging. All three composites were highly optically absorbing with optical absorptions of greater than 94%. Moreover, the authors demonstrated that the fabrication technology developed here was translatable to create macroscale MWCT-PDMS composites on planar substrates.

The three types of MWCNT-PDMS composites on optical fibres achieved high ultrasound pressures and wide frequency bandwidths. Peak-to-peak ultrasound pressures reaching 1.36 MPa at a distance of 3 mm from the composite's surface

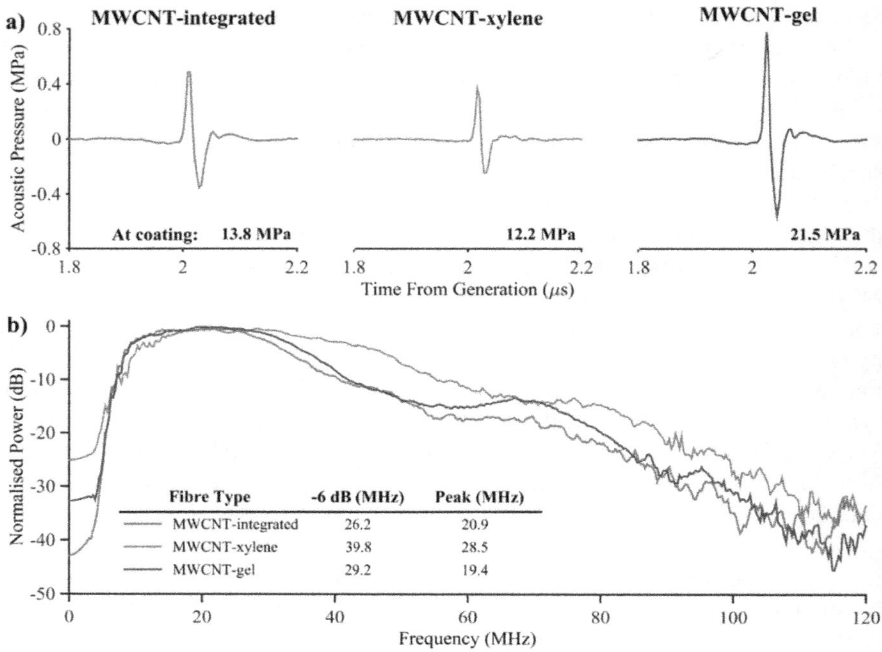

Fig. 9.9 Ultrasound measurements of MWCNT-PDMS fibre-optic composites. (**a**) Ultrasound pulse traces of MWCNT-integrated, MWCNT-xylene and MWCNT-gel composites along with their maximum ultrasound pressures measured at the composite surface. (**b**) The ultrasound bandwidths of each of the respective composites in a power spectrum MWCNT-integrated, MWCNT-xylene and MWCNT-gel are represented by the red, green, and blue lines with the ultrasound bandwidth (−6 dB) and peak ultrasound frequencies listed for each composite. (Figure taken from Noimark et al. [74])

were recorded by the organogel composite, followed by 0.91 MPa and 0.68 MPa recorded from the MWCNT-integrated and the MWCNT-xylene composites under the same experimental conditions, respectively [74]. The corresponding PA conversion efficiencies indicated that the bilayer composite demonstrated superior performance with calculated efficiencies of 0.65 MPa mJ^{-1} cm^{-2} (organogel) and 0.42 MPa mJ^{-1} cm^{-2} (MWCNT-xylene) compared to 0.37 MPa mJ^{-1} cm^{-2} achieved by the integrated composite. The trend in the ultrasound bandwidths was further evidence of the greater ultrasound performance of the bilayer composites in this study, with recorded −6 dB bandwidths of 39.8, 29.2 and 26.2 MHz for the MWCNT-xylene, organogel and integrated composites, respectively (Fig. 9.9). To test the coating stability, the ultrasound performance of all three composites was measured over a time period of 4 hours of continuous exposure to laser fluence and was found to remain unchanged. Moreover, the OpUS probes were stable over a period of 6 months with frequent use.

Using an organogel-coated OpUS transmitter and a fibre-based FP ultrasound transmitter, an all-optical transmit-receive device was fabricated. The device raster scanned across two pieces of swine aorta tissue, with each scan taking place on one piece of tissue at a time. High-resolution ultrasound images were achieved showing a well-delineated side-branch and depth variations in the tissue morphology, for example showing clear distinctions between the tunica and the adventitia. These fibre-optic MWCNT-PDMS nanocomposites were also used in the first in vivo imaging study using a fibre-optic all-optical ultrasound transducer, performed by Finlay et al. [118]. A highly miniature fibre-optic ultrasound imaging probe was assembled, comprising a fibre-optic MWCNT-PDMS composite produced by Noimark et al. [74], paired with a fibre-optic hydrophone and inserted within a clinical needle. The probe facilitated in vivo real-time imaging of a transeptal puncture within a swine heart, examining the atrial chambers and the septum separating them. These findings demonstrated the suitability of fibre-optic OpUS transmitters for potential clinical applications such as image-guided minimally invasive interventions.

CSNPs are a promising material for creating OpUS transmitters due to their unique, porous structure that allows for fast heat diffusion, aiding the high-frequency ultrasound generation necessary for high-resolution clinical imaging. Studies have shown that CSNP composites produced as free-standing membranes or applied onto planar and concave macroscale substrates exhibit high ultrasound generation performance and clinically relevant detail. However, less research has been devoted to applying CSNP-PDMS composites onto miniaturised substrates such as optical fibre end-faces.

Jiang et al. developed a method of coating a 600 μm-diameter optical fibre by holding it 2 cm above a candle flame to deposit a CSNP coating on its distal end [137]. The fibre was then overcoated with PDMS through dip-coating, forming a bilayer film [137]. The bilayer composite recorded ultrasound pressures of 750 kPa 7.5 mm from the fibre's surface under irradiation from a 532-nm 6-ns laser pulse with a laser pulse density of 1 mJ/pulse [137].

Despite this initial research, a need remained for more development and investigation into the performance of fibre-optic OpUS transmitters featuring CSNP-PDMS composites to achieve large tissue penetration depths and high spatial resolution ultrasound imaging. To address this need, Bodian et al. compared two facile and reproducible fabrication methods, a novel one-step fabrication method, 'all-in-one' (AiO), in addition to the fabrication strategy used by Jiang et al., termed the 'Direct Deposition' (DD) method [138] (Fig. 9.10). The AiO method involved mixing CSNPs with PDMS and dip-coating a 200-μm-diameter fibre to form an integrated CSNP-PDMS coating with an average thickness of 54.7 ± 1.71 μm. The DD method comprised holding the distal end of the fibre 2.5 mm from the candle flame's wick for 4 seconds before dip-coating it with PDMS to form a bilayer composite with an average thickness of 51.8 ± 0.95 μm. Both types of nanocomposite exhibited a smooth, black-domed structure and displayed broadband optical absorption (>98% between 500 and 1400 nm). They also generated high peak-to-peak ultrasound pressures (>3 MPa) and wide bandwidths (>29 MHz). The

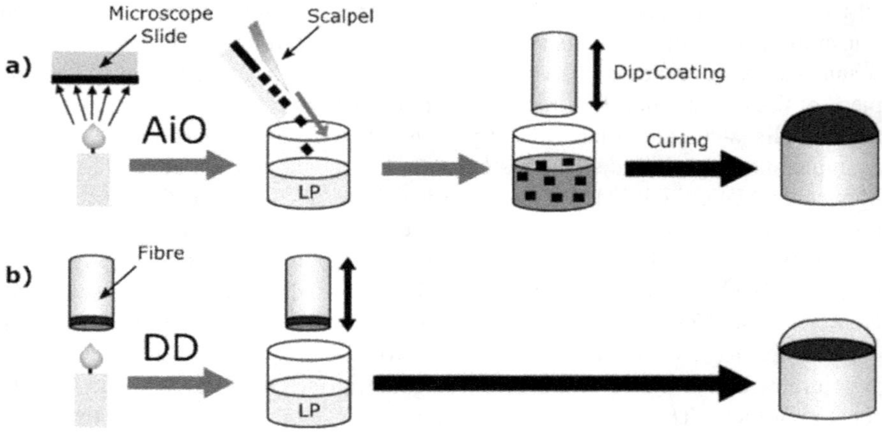

Fig. 9.10 Schematic of the two CSNP-PDMS nanocomposite fabrication techniques compared by Bodian et al. [138]. (**a**) The all-in-one (AiO) scheme comprises of CSNPs collected on a glass slide positioned above a candle flame, which are deposited with a set amount of uncured PDMS and mixed before an optical fibre is dip-coated with the mixture to form an integrated composite. (**b**) The direct deposition (DD) method derived from Jiang et al. [137], where a fibre's distal end is held above a candle flame, leaving a CSNP coating before being dip-coated with an overcoat of PDMS, forming a bilayer composite

ultrasound signals produced by both nanocomposites enabled B-mode ultrasound imaging of a wire phantom and ex vivo lamb brain cerebellum and cerebrum tissue, with imaging depths of over 8 mm. Moreover, high-resolution (<40-μm axial resolution) B-mode ultrasound imaging of ex vivo lamb brain tissue demonstrated how CSNP-PDMS OpUS transmitters can allow for high-fidelity, minimally invasive imaging of biological tissues. This study was the first to demonstrate ex vivo imaging of brain tissue using all-optical ultrasound transducers, demonstrating their potential for integration with devices used in neurological interventions to guide surgeons and provide them with structural information.

9.3.3 Quantum Dots

Recent innovations in the development of coatings for OpUS had led to the use of quantum dots (QDs) for OpUS generation [75, 76]. QDs are nanoscale structures with unique electronic properties, making them valuable in various scientific and technological applications [139–142]. Through modifying synthetic protocol, it is possible to tune their bandgaps, allowing customisation of their light absorption and emission profiles [143–145]. QDs are used in LEDs and liquid-crystal displays, including flat-screen TVs [146]. Some QDs containing heavy metals can accumulate in biological tissues, posing toxicity concerns and leading to bans in household electronics [147, 148]. Biomedical applications prefer heavy-metal-free QDs like $CuInS_2$ (CIS) QDs [149–153], which are used in biological imaging for cancer diagnoses [154] and therapy [155]. The surfaces of QDs can be functionalised for diverse applications including gene delivery [156] and multimodality imaging [76, 142, 155, 157, 158]. QD performance depends on core size, composition, and synthesis method, with the heat-up method being suitable for large-scale production [159–163]. QDs with low photoluminescence quantum yields are preferred for efficient OpUS generation, as they convert absorbed optical energy into heat for ultrasound rather than emitting light [164].

Bodian et al. fabricated the first QD-PDMS composite-coated optical fibre for ultrasound imaging. The QD coating comprised a bilayer structure featuring CIS QDs overcoated with PDMS, using a two-layer dip-coating strategy [76]. A cleaved and cleaned 200-μm-core multimodal fibre was dipped into a concentrated CIS QD-xylene paste to form a CIS QD coating, followed by a PDMS (Med-1000) overcoat.

These composites showed great promise for multimodality imaging. They demonstrated high optical absorption (>90%) at the Nd:YAG laser wavelength of 532 nm employed for OpUS generation and low optical absorption (<5%) at near-infrared wavelengths beyond 700 nm, enabling concurrent photoacoustic imaging. Ultrasound pressures of up to 3.7 MPa upon pulsed excitation at 532 nm and bandwidths (−6 dB) of 18 MHz were recorded by the fibre-optic CIS-PDMS composite at a distance of 1.5 mm from the composite's surface.

The fibre-optic composite was paired with a fibre-optic FP hydrophone (fabricated by Beard et al. in the same design as Beard et al. [116]) to form an all-optical ultrasound transducer, performing co-registered ultrasound and photoacoustic imaging of a gel-wax ink-filled phantom. Here, a 1064 nm pulsed excitation source was used for photoacoustic imaging. The phantom comprised polymethylpentene tubes filled with three different concentrations of CB in water positioned within a plastic dish filled with gel-wax. The CB/water mixtures acted as contrast for the photoacoustic imaging, whilst the plastic tubing and dish acted as ultrasound contrast. When both sets of images (ultrasound and photoacoustic) were overlaid on top of one another, the outline of each of the three tubes was resolvable with bright orange colouration observable within each tube indicating the CB/water ink solutions (Fig. 9.11). The composite was able to perform co-registered laser-generated ultrasound and photoacoustic imaging of a gel-wax ink-filled phantom.

In summary, QDs hold great potential as optical absorbers to produce wavelength-selective elastomeric composites, which can aid in dual-mode OpUS imaging. Their tailored optical absorption profiles make them a suitable choice for meeting specific requirements for a given modality or optical set-up. However, it is important to note that research is being conducted to identify QD species that are free of heavy metals due to concerns about their toxicities. As a result, it is expected that more QD species will be employed in composites for OpUS imaging, facilitating different combinations of imaging and therapeutic modalities.

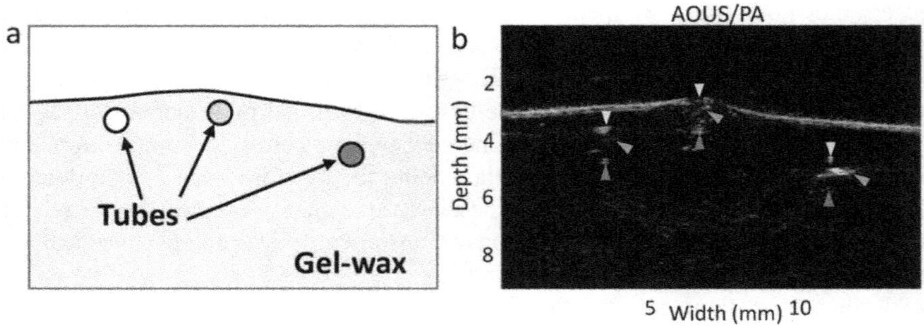

Fig. 9.11 Combined ultrasound and photoacoustic imaging system and results. (**a**) A cross-sectional view of a gel-wax ink-filled phantom with three tubes containing carbon black ink at varying concentrations. (**b**) An overlaid image of ultrasound and photoacoustic images of the phantom cross-section from panel (a), with the ultrasound signal shown in white and denoted by yellow and blue arrows indicating the top and bottom of the phantom's plastic tubes, while the photoacoustic signal is represented in red/yellow and highlighted by a green arrow. (Reprinted from Bodian et al. [76])

9.4 Conclusion

This chapter discusses how MIS techniques have transformed medicine, resulting in reduced infection rates and shorter recovery periods. However, limited visibility and interaction with patient tissue present challenges, which necessitate improved imaging modalities. Medical imaging plays a crucial role, with X-ray and CT providing valuable pre-operative and intra-operative insights, albeit with radiation concerns. While MRI is radiation-free, it has limitations in real-time imaging. Ultrasound emerges as a non-invasive, cost-effective and safe modality, particularly suitable for real-time imaging. Piezoelectric ultrasound transducers face challenges in miniaturisation, leading to the exploration of OpUS imaging. OpUS exhibits promise for advancing high-resolution ultrasound imaging in medical applications along with other advantages, including miniaturisation advantages, multimodal capabilities and compatibility with MRI.

Various thin metal films, including chromium (Cr), molybdenum (Mo) and aluminium (Al), were widely employed for OpUS generation due to their simple structures and ease of fabrication. Thermoelastic ultrasound generation mechanisms were explored, with studies into how pulse repetition rate and laser wavelengths affect the generated ultrasound signal. Metal films' limitations in pressure and bandwidth led to the development of sophisticated composite films, such as PDMS/Cr/PDMS/Al, enhancing heat transfer and absorption. Studies on solid/liquid interfaces provided insights, showing ultrasound pressures are proportional to liquid's β and inversely proportional to film thickness. Different composite fabrication methods, including bottom-up, top-down, and all-in-one, offer opportunities for tailoring OpUS materials.

This chapter also explores OpUS transducers, focusing on carbonaceous and gold nanoparticles. CB-PDMS composites, with varying thicknesses, demonstrate improved ultrasound efficiency over metallic counterparts. With a narrower absorption profile, gold-PDMS composites outperform CB-PDMS at higher frequencies. CNTs, CNFs and CSNPs offer promising OpUS performance with unique advantages. Using rGO in composites enhances ultrasound pressures, making it suitable for real-time imaging.

Innovative materials and technology have propelled the development of fibre-optic OpUS probes, revolutionising minimally invasive medical imaging. Overcoming challenges faced by macroscopic composites, fibre-optic probes, incorporating gold nanoparticles, carbonaceous nanoparticles (CNFs, MWCNTs and CSNPs) and QDs demonstrate versatility. Gold nanoparticle-coated probes enable multimodal imaging, while carbonaceous nanoparticles, especially MWCNTs, provide high ultrasound pressures and wide ultrasound bandwidths. CNFs, integrated into PDMS, exhibit superior ultrasound bandwidths. CSNPs with a unique porous structure facilitate high-frequency ultrasound generation. QDs offer tuneable bandgaps, and CIS QDs present a biocompatible option. The first QD-PDMS composite-coated optical fibres show high optical absorption, enabling co-registered ultrasound and photoacoustic imaging. Ongoing research explores diverse materials and fabrication methods, enhancing the performance and versatility of fibre-optic OpUS probes in medical imaging, promising breakthroughs in multimodal applications.

Acknowledgements This work was supported by the Engineering and Physical Sciences Research Council (EPSRC) (Grant Nos. EP/S513726/1 and NS/A000050/1), EPSRC UK Research and Innovation (UKRI) Innovation Fellowship (EP/S001506/1), Johnson Matthey PLC (Grant No. JM11372) and the Wellcome Trust (Grant Nos. WT101957 and 203145Z/16/Z). AED owns shares in Echopoint Medical Ltd. For the purpose of open access, the authors have applied a CC BY public copyright license to any author-accepted manuscript version arising from this submission.

Bibliography

1. van der Putten, E. P. W.; Goossens, R. H. M.; Jakimowicz, J. J.; Dankelman, J. Haptics in Minimally Invasive Surgery - A Review. *Minimally Invasive Therapy and Allied Technologies*. 2008, pp 3–16. https://doi.org/10.1080/13645700701820242.
2. Kim, T.-I.; Alió Del Barrio, J. L.; Wilkins, M.; Cochener, B.; Ang, M. *Refractive Surgery*; 2019; Vol. 393. www.thelancet.com.
3. Fine, I. H.; Mojon, D. S. *Minimally Invasive Ophthalmic Surgery*; Springer Berlin Heidelberg, 2010. https://doi.org/10.1007/978-3-642-02602-7.
4. Liekweg, W. G.; Misra, R. Minimally Invasive Direct Coronary Artery Bypass, Percutaneous Transluminal Coronary Angioplasty, and Stent Placement for Left Main Stenosis. *J Thorac Cardiovasc Surg* **1997**, *113* (2), 411–412.
5. Bacha, E. A.; Hijazi, Z. M.; Cao, Q. L.; Abdulla, R.; Starr, J. P.; Quinones, J.; Koenig, P.; Agarwala, B. Hybrid Pediatric Cardiac Surgery. *Pediatr Cardiol* **2005**, *26* (4), 315–322. https://doi.org/10.1007/s00246-005-8648-0.
6. Mack, M. J. Minimally Invasive Cardiac Surgery. *Surgical Endoscopy and Other Interventional Techniques*. April 2006. https://doi.org/10.1007/s00464-006-0110-8.

7. Kitagawa, Y.; Fujii, H.; Mukai, M.; Kubota, T.; Ando, N.; Watanabe, M.; Ohgami, M.; Otani, Y.; Ozawa, S.; Hasegawa, H.; Furukawa, T.; Kumai, K.; Ikeda, T.; Nakahara, T.; Kubo, A.; Kitajima, M. The Role of the Sentinel Lymph Node in Gastrointestinal Cancer. *Surgical Clinics of North America* **2000**, *80* (6), 1799–1809.
8. Malkan, A. D.; Loh, A. H. P.; Sandoval, J. A. Minimally Invasive Surgery in the Management of Abdominal Tumors in Children. *Journal of Pediatric Surgery*. W.B. Saunders 2014, pp 1171–1176. https://doi.org/10.1016/j.jpedsurg.2014.04.010.
9. Jaffray, B. Minimally Invasive Surgery. *Arch Dis Child* **2005**, *90* (5), 537–542. https://doi.org/10.1136/adc.2004.062760.
10. Matern, U.; Waller, P. New Technology Instruments for Minimally Invasive Surgery Principles of Ergonomic Handles. *Surg Endosc* **1999**, *13*, 174–182.
11. van der Putten; E. P.Westebring; Goossens, R. H. M.; Jakimowicz, J. J.; Dankelman, J. Haptics in Minimally Invasive Surgery - A Review. *Minimally Invasive Therapy and Allied Technologies*. 2008, pp 3–16. https://doi.org/10.1080/13645700701820242.
12. Smith, Z. A.; Fessler, R. G. Paradigm Changes in Spine Surgery-Evolution of Minimally Invasive Techniques. *Nature Reviews Neurology*. August 2012, pp 443–450. https://doi.org/10.1038/nrneurol.2012.110.
13. Dogangil, G.; Davies, B. L.; Rodriguez Y Baena, F. A Review of Medical Robotics for Minimally Invasive Soft Tissue Surgery. *Proceedings of the Institution of Mechanical Engineers, Part H: Journal of Engineering in Medicine*. May 1, 2010, pp 653–679. https://doi.org/10.1243/09544119JEIM591.
14. Våpenstad, C.; Hofstad, E. F.; Langø, T.; Mårvik, R.; Chmarra, M. K. Perceiving Haptic Feedback in Virtual Reality Simulators. *Surg Endosc* **2013**, *27* (7), 2391–2397. https://doi.org/10.1007/s00464-012-2745-y.
15. el Rassi, I.; el Rassi, J. M. A Review of Haptic Feedback in Tele-Operated Robotic Surgery. *Journal of Medical Engineering and Technology*. Taylor and Francis Ltd July 3, 2020, pp 247–254. https://doi.org/10.1080/03091902.2020.1772391.
16. Bauzano, E.; Muñoz, V. F.; Garcia-Morales, I. Auto-Guided Movements on Minimally Invasive Surgery for Surgeon Assistance. In *IEEE/RSJ 2010 International Conference on Intelligent Robots and Systems, IROS 2010 - Conference Proceedings*; 2010; pp 1843–1848. https://doi.org/10.1109/IROS.2010.5650875.
17. Burschka, D.; Corso, J. J.; Dewan, M.; Lau, W.; Li, M.; Lin, H.; Marayong, P.; Ramey, N.; Hager, G. D.; Hoffman, B.; Larkin, D.; Hasser, C. Navigating Inner Space: 3-D Assistance for Minimally Invasive Surgery. In *Robotics and Autonomous Systems*; 2005; Vol. 52, pp 5–26. https://doi.org/10.1016/j.robot.2005.03.013.
18. Boppart, S. A.; Deutsch, T. F.; Rattner, D. W. Optical Imaging Technology in Minimally Invasive Surgery: Current Status and Future Directions. *Surg Endosc* **1999**, *13*, 718–722.
19. Schols, R. M.; Connell, N. J.; Stassen, L. P. S. Near-Infrared Fluorescence Imaging for Real-Time Intraoperative Anatomical Guidance in Minimally Invasive Surgery: A Systematic Review of the Literature. *World Journal of Surgery*. Springer New York LLC May 1, 2015, pp 1069–1079. https://doi.org/10.1007/s00268-014-2911-6.
20. Vajsbaher, T.; Ziemer, T.; Schultheis, H. A Multi-Modal Approach to Cognitive Training and Assistance in Minimally Invasive Surgery. *Cogn Syst Res* **2020**, *64*, 57–72. https://doi.org/10.1016/j.cogsys.2020.07.005.
21. Park, H.; Han, K. N.; Choi, B. H.; Yoon, H.; An, H. J.; Lee, J. S.; Kim, H. K. Ultra-Low-Dose Intraoperative X-Ray Imager for Minimally Invasive Surgery: A Pilot Imaging Study. *Transl Lung Cancer Res* **2022**, *11* (4), 588–599. https://doi.org/10.21037/tlcr-21-909.
22. Cysewska-Sobusiak, A. R.; Sowier, A.; Skrzywanek, P. Application of Combined Methods of Imaging in Minimally Invasive Surgery. In *Annual International Conference of the IEEE Engineering in Medicine and Biology - Proceedings*; 2003; Vol. 2, pp 1043–1046. https://doi.org/10.1109/iembs.2003.1279423.
23. Seibert, J. A. X-Ray Imaging Physics for Nuclear Medicine Technologists. Part 1: Basic Principles of X-Ray Production*. *J Nucl Med Technol* **2004**, *32* (3), 139–147.

24. Briguori, C.; Manganelli, F.; Scarpato, P.; Elia, P. P.; Golia, B.; Riviezzo, G.; Lepore, S.; Librera, M.; Villari, B.; Colombo, A.; Ricciardelli, B. Acetylcysteine and Contrast Agent-Associated Nephrotoxicity. *J Am Coll Cardiol* **2002**, *40* (2), 298–303. https://doi.org/10.1016/S0735-1097(02)01958-7.
25. Bibb, R.; Winder, J. A Review of the Issues Surrounding Three-Dimensional Computed Tomography for Medical Modelling Using Rapid Prototyping Techniques. *Radiography* **2010**, *16* (1), 78–83. https://doi.org/10.1016/j.radi.2009.10.005.
26. Zanzonico, P.; Dauer, L.; St Germain, J. Operational Radiation Safety for PET-CT, SPECT-CT, and Cyclotron Facilities. *Health Phys* **2008**, *95* (5), 554–570.
27. Koshevaya, E.; Nazarovskaia, D.; Simakov, M.; Belousov, A.; Morozov, V.; Gandalipov, E.; Krivoshapkina, E.; Krivoshapkin, P. Surfactant-Free Tantalum Oxide Nanoparticles: Synthesis, Colloidal Properties, and Application as a Contrast Agent for Computed Tomography. *J Mater Chem B* **2020**, *8* (36), 8337–8345. https://doi.org/10.1039/d0tb01204a.
28. Diederich, S.; Lenzen, H. Radiation Exposure Associated with Imaging of the Chest: Comparison of Different Radiographic and Computed Tomography Techniques. In *Cancer*; John Wiley and Sons Inc., 2000; Vol. 89, pp 2457–2460. https://doi.org/10.1002/1097-0142(20001201)89:11+<2457::aid-cncr22>3.0.co;2-7.
29. Brenner, D. J.; Hall, E. J.; Phil, D. *Computed Tomography-An Increasing Source of Radiation Exposure*; 2007; Vol. 357. www.nejm.org.
30. Kawamata, A.; Ariji, Y.; Langlais, R. P. Three-Dimensional Computed Tomography Imaging in Dentistry. *Applications of Digital Imaging Modalities for Dentistry* **2000**, *44* (2), 395–410.
31. Pathe, C.; Eble, K.; Schmitz-Beuting, D.; Keil, B.; Kaestner, B.; Voelker, M.; Kleb, B.; Klose, K. J.; Heverhagen, J. T. The Presence of Iodinated Contrast Agents Amplifies DNA Radiation Damage in Computed Tomography. *Contrast Media Mol Imaging* **2011**, *6* (6), 507–513. https://doi.org/10.1002/cmmi.453.
32. Lopez, H. S.; Liu, F.; Weber, E.; Crazier, S. Passive Shim Design and a Shimming Approach for Biplanar Permanent Magnetic Resonance Imaging Magnets. *IEEE Trans Magn* **2008**, *44* (3), 394–402. https://doi.org/10.1109/TMAG.2007.914770.
33. Vogel, M. W.; Guridi, R. P.; Su, J.; Vegh, V.; Reutens, D. C. 3D-Spatial Encoding with Permanent Magnets for Ultra-Low Field Magnetic Resonance Imaging. *Sci Rep* **2019**, *9* (1). https://doi.org/10.1038/s41598-018-37953-1.
34. Zhao, H.; Crozier, S.; Doddrell, D. M. A Hybrid, Inverse Approach to the Design of Magnetic Resonance Imaging Magnets. *Med Phys* **2000**, *27* (3), 599–607. https://doi.org/10.1118/1.598899.
35. Arshiya Ara, S.; Katti, G.; Shireen, A. Magnetic Resonance Imaging (MRI)-A Review. *International Journal of Dental Clinics* **2011**, *3* (1), 65–70.
36. Lee, N.; Hyeon, T. Designed Synthesis of Uniformly Sized Iron Oxide Nanoparticles for Efficient Magnetic Resonance Imaging Contrast Agents. *Chem Soc Rev* **2012**, *41* (7), 2575–2589. https://doi.org/10.1039/c1cs15248c.
37. Yan, G. P.; Robinson, L.; Hogg, P. Magnetic Resonance Imaging Contrast Agents: Overview and Perspectives. *Radiography*. December 2007. https://doi.org/10.1016/j.radi.2006.07.005.
38. Höger, S. A.; Gast, L. v.; Marty, B.; Hotfiel, T.; Bickelhaupt, S.; Uder, M.; Heiss, R.; Nagel, A. M. Sodium and Quantitative Hydrogen Parameter Changes in Muscle Tissue after Eccentric Exercise and in Delayed-onset Muscle Soreness Assessed with Magnetic Resonance Imaging. *NMR Biomed* **2022**. https://doi.org/10.1002/nbm.4840.
39. Carlier, P. G.; Marty, B.; Scheidegger, O.; Loureiro De Sousa, P.; Baudin, P. Y.; Snezhko, E.; Vlodavets, D. Skeletal Muscle Quantitative Nuclear Magnetic Resonance Imaging and Spectroscopy as an Outcome Measure for Clinical Trials. *Journal of Neuromuscular Diseases*. IOS Press 2016, pp 1–28. https://doi.org/https://doi.org/10.3233/JND-160145.
40. Hindley, J.; Regan, L.; Stewart, E.; Tempany, C.; Hynnen, K.; Inbar, Y.; Kim, K.; Geschwind, J.; Hesley, G.; Gostout, B.; Sklair-Levy, M. MRI Guidance of Focused Ultrasound Therapy of Uterine Fibroids: Early Results. *American Journal of Roentgenology* **2004**, *183* (December), 1713–1719.

41. Santini, F.; Pansini, M.; Hrabak-Paar, M.; Yates, D.; Langenickel, T. H.; Bremerich, J.; Bieri, O.; Schubert, T. On the Optimal Temporal Resolution for Phase Contrast Cardiovascular Magnetic Resonance Imaging: Establishment of Baseline Values. *Journal of Cardiovascular Magnetic Resonance* **2020**, *22* (1). https://doi.org/10.1186/s12968-020-00669-1.
42. Nayak, K. S.; Lim, Y.; Campbell-Washburn, A. E.; Steeden, J. Real-Time Magnetic Resonance Imaging. *Journal of Magnetic Resonance Imaging*. John Wiley and Sons Inc January 1, 2022, pp 81–99. https://doi.org/10.1002/jmri.27411.
43. Ben-Galim, P.; Steinberg, E. L.; Amir, H.; Ash, N.; Dekel, S.; Arbel, R. Accuracy of Magnetic Resonance Imaging of the Knee and Unjustified Surgery. *Clin Orthop Relat Res* **2006**, *447*, 100–104. https://doi.org/10.1097/01.blo.0000203471.50808.b7.
44. Bisset, R. A. L.; Thomas, N. B.; Turnbull, I. W.; Lee, S. Postmortem Examinations Using Magnetic Resonance Imaging: Four Year Review of a Working Service. *Br Med J* **2002**, *324* (7351), 1423–1424. https://doi.org/10.1136/bmj.324.7351.1423.
45. Noone, T. C.; Semelka, R. C.; Chaney, D. M.; Reinhold, C. Abdominal Imaging Studies: Comparison of Diagnostic Accuracies Resulting from Ultrasound, Computed Tomography, and Magnetic Resonance Imaging in the Same Individual. *Magn Reson Imaging* **2004**, *22* (1), 19–24. https://doi.org/10.1016/j.mri.2003.01.001.
46. Jeong, J. J.; Choi, H. An Impedance Measurement System for Piezoelectric Array Element Transducers. *Measurement (Lond)* **2017**, *97*, 138–144. https://doi.org/10.1016/j.measurement.2016.10.053.
47. Suresh, K. S.; Greeshma, S. L. Study on Ultrasound Detection of Arteries. *IJISET-International Journal of Innovative Science, Engineering & Technology* **2014**, *1*.
48. Aldrich, J. E. Basic Physics of Ultrasound Imaging. *Critical Care Medicine*. May 2007. https://doi.org/10.1097/01.CCM.0000260624.99430.22.
49. Reda, R.; Zanza, A.; Cicconetti, A.; Bhandi, S.; Miccoli, G.; Gambarini, G.; di Nardo, D. Ultrasound Imaging in Dentistry: A Literature Overview. *Journal of Imaging*. MDPI November 1, 2021. https://doi.org/10.3390/jimaging7110238.
50. Bernhardt, M.; Vishnevskiy, V.; Rau, R.; Goksel, O. Training Variational Networks with Multidomain Simulations: Speed-of-Sound Image Reconstruction. *IEEE Trans Ultrason Ferroelectr Freq Control* **2020**, *67* (12), 2584–2594. https://doi.org/10.1109/TUFFC.2020.3010186.
51. Zhang, J.; Wang, K.; Yang, Y.; Anastasio, M. A. Simultaneous Reconstruction of Speed-of-Sound and Optical Absorption Properties in Photoacoustic Tomography via a Time-Domain Iterative Algorithm. In *Photons Plus Ultrasound: Imaging and Sensing 2008: The Ninth Conference on Biomedical Thermoacoustics, Optoacoustics, and Acousto-optics*; SPIE, 2008; Vol. 6856, p 68561F. https://doi.org/10.1117/12.764171.
52. Anis, F.; Su, R.; Nadvoretsky, V. v.; Conjusteau, A.; Ermilov, S. A.; Oraevsky, A. A.; Anastasio, M. A. Image Reconstruction and System Optimization for Three-Dimensional Speed of Sound Tomography Using Laser-Induced Ultrasound. In *Photons Plus Ultrasound: Imaging and Sensing 2013*; SPIE, 2013; Vol. 8581, p 85814T. https://doi.org/10.1117/12.2008813.
53. le Jeune, L.; Robert, S.; Lopez Villaverde, E.; Prada, C. Plane Wave Imaging for Ultrasonic Non-Destructive Testing: Generalization to Multimodal Imaging. *Ultrasonics* **2016**, *64*, 128–138. https://doi.org/10.1016/j.ultras.2015.08.008.
54. Marmonier, M.; Robert, S.; Laurent, J.; Prada, C. Real-Time 3D Imaging with Fourier-Domain Algorithms and Matrix Arrays Applied to Non-Destructive Testing. *Ultrasonics* **2022**, *124*. https://doi.org/10.1016/j.ultras.2022.106708.
55. Pérez, E.; Kirchhof, J.; Krieg, F.; Römer, F. Subsampling Approaches for Compressed Sensing with Ultrasound Arrays in Non-Destructive Testing. *Sensors (Switzerland)* **2020**, *20* (23), 1–26. https://doi.org/10.3390/s20236734.
56. Lanza, G. M. Ultrasound Imaging: Something Old or Something New? *Investigative Radiology*. Lippincott Williams and Wilkins September 1, 2020, pp 573–577. https://doi.org/10.1097/RLI.0000000000000679.

57. Leighton, T. G. What Is Ultrasound? *Progress in Biophysics and Molecular Biology*. January 2007, pp 3–83. https://doi.org/10.1016/j.pbiomolbio.2006.07.026.
58. Johnstone, A. J.; Beggs, I. Ultrasound Imaging of Soft-Tissue Masses in the Extremities. *J Bone Joint Surg* **1994**, *76* (5), 688–689.
59. Li, Y.; Lu, G.; Zhou, Q.; Chen, Z. Advances in Endoscopic Photoacoustic Imaging. *Photonics* **2021**, *8* (7). https://doi.org/10.3390/photonics8070281.
60. Jacobson, J. A.; Middleton, W. D.; Allison, S. J.; Dahiya, N.; Lee, K. S.; Levine, B. D.; Lucas, D. R.; Murphey, M. D.; Nazarian, L. N.; Siegel, G. W.; Wagner, J. M. Ultrasonography of Superficial Soft-Tissue Masses: Society of Radiologists in Ultrasound Consensus Conference Statement. *Radiology* **2022**, *304* (1), 18–30. https://doi.org/10.1148/radiol.211101.
61. Seo, J.; Kim, Y. sun. Ultrasound Imaging and beyond: Recent Advances in Medical Ultrasound. *Biomedical Engineering Letters*. Springer Verlag May 1, 2017, pp 57–58. https://doi.org/10.1007/s13534-017-0030-7.
62. Bradley, M.; Morgan, D.; Pentlow, B.; Roe, A. The Groin Hernia – an Ultrasound Diagnosis? *Ann R Coll Surg Engl* **2003**, *85*, 178–180.
63. Ng, A.; Swanevelder, J. Resolution in Ultrasound Imaging. *Continuing Education in Anaesthesia, Critical Care and Pain* **2011**, *11* (5), 186–192. https://doi.org/10.1093/bjaceaccp/mkr030.
64. Shung, K. K. High Frequency Ultrasonic Imaging. *Journal of Medical Ultrasound*. Elsevier (Singapore) Pte Ltd 2009, pp 25–30. https://doi.org/10.1016/S0929-6441(09)60012-6.
65. Zhao, T.; Desjardins, A. E.; Ourselin, S.; Vercauteren, T.; Xia, W. Minimally Invasive Photoacoustic Imaging: Current Status and Future Perspectives. *Photoacoustics* **2019**, *16* (October), 100146. https://doi.org/10.1016/j.pacs.2019.100146.
66. Xu, M.; Yang, X.; Ding, M.; Yuchi, M. Spatio-Temporally Smoothed Coherence Factor for Ultrasound Imaging. *IEEE Trans Ultrason Ferroelectr Freq Control* **2014**, *61* (1), 182–190. https://doi.org/10.1109/TUFFc.2014.2889.
67. La, T. G.; Le, L. H. Flexible and Wearable Ultrasound Device for Medical Applications: A Review on Materials, Structural Designs, and Current Challenges. *Advanced Materials Technologies*. John Wiley and Sons Inc March 1, 2022. https://doi.org/10.1002/admt.202100798.
68. Wissmeyer, G.; Pleitez, M. A.; Rosenthal, A.; Ntziachristos, V. Looking at Sound: Optoacoustics with All-Optical Ultrasound Detection. *Light: Science and Applications*. Nature Publishing Group December 1, 2018. https://doi.org/10.1038/s41377-018-0036-7.
69. Jung, J.; Lee, W.; Kang, W.; Shin, E.; Ryu, J.; Choi, H. Review of Piezoelectric Micromachined Ultrasonic Transducers and Their Applications. *Journal of Micromechanics and Microengineering*. Institute of Physics Publishing September 22, 2017. https://doi.org/10.1088/1361-6439/aa851b.
70. Lee, T.; Baac, H. W.; Ok, J. G.; Jay Guo, L. Polymer-Nanomaterial Composites for Optoacoustic Conversion. *Functional Organic and Hybrid Nanostructured Materials* **2018**, *14*, 519–546. https://doi.org/10.1002/9783527807369.ch14.
71. Hou, Y.; Kim, J. S.; Ashkenazi, S.; Huang, S. W.; Guo, L. J.; O'Donnell, M. Broadband All-Optical Ultrasound Transducers. *Appl Phys Lett* **2007**, *91* (7). https://doi.org/10.1063/1.2771058.
72. Ritter, T. A.; Shrout, T. R.; Tutwiler, R.; Shung, K. K. A 30-MHz Piezo-Composite Ultrasound Array for Medical Imaging Applications. *IEEE Trans Ultrason Ferroelectr Freq Control* **2002**, *49* (2), 217–230. https://doi.org/10.1109/58.985706.
73. Zhao, T.; Su, L.; Xia, W. Optical Ultrasound Generation and Detection for Intravascular Imaging: A Review. *J Healthc Eng* **2018**, *2018*, 1–14. https://doi.org/10.1155/2018/3182483.
74. Noimark, S.; Colchester, R. J.; Blackburn, B. J.; Zhang, E. Z.; Alles, E. J.; Ourselin, S.; Beard, P. C.; Papakonstantinou, I.; Parkin, I. P.; Desjardins, A. E. Carbon-Nanotube–PDMS Composite Coatings on Optical Fibers for All-Optical Ultrasound Imaging. *Adv Funct Mater* **2016**, *26* (46), 8390–8396. https://doi.org/10.1002/adfm.201601337.
75. Noimark, S.; Colchester, R. J.; Poduval, R. K.; Maneas, E.; Alles, E. J.; Zhao, T.; Zhang, E. Z.; Ashworth, M.; Tsolaki, E.; Chester, A. H.; Latif, N.; Bertazzo, S.; David, A. L.; Ourselin, S.; Beard, P. C.; Parkin, I. P.; Papakonstantinou, I.; Desjardins, A. E. Polydimethylsiloxane

Composites for Optical Ultrasound Generation and Multimodality Imaging. *Adv Funct Mater* **2018**, *28* (9), 1–16. https://doi.org/10.1002/adfm.201704919.
76. Bodian, S.; Colchester, R. J.; Macdonald, T. J.; Ambroz, F.; Briceno de Gutierrez, M.; Mathews, S. J.; Fong, Y. M. M.; Maneas, E.; Welsby, K. A.; Gordon, R. J.; Collier, P.; Zhang, E. Z.; Beard, P. C.; Parkin, I. P.; Desjardins, A. E.; Noimark, S. CuInS2 Quantum Dot and Polydimethylsiloxane Nanocomposites for All-Optical Ultrasound and Photoacoustic Imaging. *Adv Mater Interfaces* **2021**, *8* (20). https://doi.org/10.1002/admi.202100518.
77. Colchester, R. J.; Zhang, E. Z.; Mosse, C. A.; Beard, P. C.; Papakonstantinou, I.; Desjardins, A. E. Broadband Miniature Optical Ultrasound Probe for High Resolution Vascular Tissue Imaging. *Biomed Opt Express* **2015**, *6* (4), 1502. https://doi.org/10.1364/BOE.6.001502.
78. Alles, E. J.; Noimark, S.; Maneas, E.; Zhang, E. Z.; Beard, P. C.; Desjardins, A. E.; Parkin, I. P. Video-Rate All-Optical Ultrasound Imaging. *Biomed Opt Express* **2018**, *9* (8), 3481–3494. https://doi.org/10.1364/BOE.9.003481.
79. Chen, S.-L. Review of Laser-Generated Ultrasound Transmitters and Their Applications to All-Optical Ultrasound Transducers and Imaging. *Applied Sciences* **2016**, *7* (1), 25. https://doi.org/10.3390/app7010025.
80. Wu, N.; Zou, X.; Zhou, J.; Wang, X. Fiber Optic Ultrasound Transmitters and Their Applications. *Measurement (Lond)* **2016**, *79*, 164–171. https://doi.org/10.1016/j.measurement.2015.10.002.
81. Rajagopal, S.; Cox, B. T. Modelling Laser Ultrasound Waveforms: The Effect of Varying Pulse Duration and Material Properties. *J Acoust Soc Am* **2021**, *149* (3), 2040–2054. https://doi.org/10.1121/10.0003558.
82. Ruiz-Veloz, M.; Martínez-Ponce, G.; Fernández-Ayala, R. I.; Castro-Beltrán, R.; Polo-Parada, L.; Reyes-Ramírez, B.; Gutiérrez-Juárez, G. Thermally Corrected Solutions of the One-Dimensional Wave Equation for the Laser-Induced Ultrasound. *J Appl Phys* **2021**, *130* (2). https://doi.org/10.1063/5.0050895.
83. Nikitichev, D. I.; Xia, W.; Hill, E.; Mosse, C. A.; Perkins, T.; Konyn, K.; Ourselin, S.; Desjardins, A. E.; Vercauteren, T. Music-of-Light Stethoscope: A Demonstration of the Photoacoustic Effect. *Phys Educ* **2016**, *51*, 1–9.
84. Silva, A. D.; Henriques, C. A.; Malva, D. v.; Calvete, M. J. F.; Pereira, M. M.; Serpa, C.; Arnaut, L. G. Photoacoustic Generation of Intense and Broadband Ultrasound Pulses with Functionalized Carbon Nanotubes. *Nanoscale* **2020**, *12* (40), 20831–20839. https://doi.org/10.1039/d0nr04986g.
85. Oraevsky, A. A.; Jacques, S. L.; Tittel, F. K. Measurement of Tissue Optical Properties by Time-Resolved Detection of Laser-Induced Transient Stress. *Appl Opt* **1997**, *36* (1), 402–415.
86. Lee, T.; Baac, H. W.; Li, Q.; Guo, L. J. Efficient Photoacoustic Conversion in Optical Nanomaterials and Composites. *Adv Opt Mater* **2018**, *6* (24), 1–30. https://doi.org/10.1002/adom.201800491.
87. White, R. M. Generation of Elastic Waves by Transient Surface Heating. *J Appl Phys* **1963**, *34* (12), 3559–3567. https://doi.org/10.1063/1.1729258.
88. Von Gutfeld, R. J.; Budd, H. F. Laser-Generated MHz Elastic Waves from Metallic-Liquid Interfaces. *Appl Phys Lett* **1979**, *34* (10), 617–619. https://doi.org/10.1063/1.90637.
89. Lee, R. E.; White, R. M. Excitation of Surface Elastic Waves by Transient Surface Heating. *Appl Phys Lett* **1968**, *12* (1), 12–14. https://doi.org/10.1063/1.1651832.
90. Carome, E. F.; Clark, N. A.; Moeller, C. E. Generation of Acoustic Signals in Liquids by Ruby Laser-Induced Thermal Stress Transients. *Appl Phys Lett* **1964**, *4* (6), 95–97. https://doi.org/10.1063/1.1753985.
91. Carome, E. F.; Clark, N. A.; Moeller, C. E. Generation of Acoustic Signals in Liquids by Ruby Laser-Induced Thermal Stress Transients. *Appl Phys Lett* **1964**, *4* (6), 95–97. https://doi.org/10.1063/1.1753985.
92. Arzt, R. M.; Dransfeld, K. Excitation of Rayleigh Waves at High Frequencies and at Low Temperatures. *Appl Phys Lett* **1965**, *7* (6), 156–158. https://doi.org/10.1063/1.1754354.

93. Brienza, M. J.; Demaria, A. J. Laser-Induced Microwave Sound by Surface Heating. *Appl Phys Lett* **1967**, *11* (2), 44–46. https://doi.org/10.1063/1.1755021.
94. Davies, S. J.; Edwards, C.; Taylor, G. S.; Palmer, S. B. Laser-Generated Ultrasound: Its Properties, Mechanisms and Multifarious Applications. *J Phys D Appl Phys* **1993**, *26* (3), 329–348. https://doi.org/10.1088/0022-3727/26/3/001.
95. Buma, T.; Spisar, M.; O'Donnell, M. A High-Frequency, 2-D Array Element Using Thermoelastic Expansion in PDMS. *IEEE Trans Ultrason Ferroelectr Freq Control* **2003**, *50* (9), 1161–1176. https://doi.org/10.1109/TUFFC.2003.1235327.
96. Dewhurst, R. J.; Hutchins, D. A.; Palmer, S. B.; Scruby, C. B. Quantitative Measurements of Laser-Generated Acoustic Waveforms. *J Appl Phys* **1982**, *53* (6), 4064–4071. https://doi.org/10.1063/1.331270.
97. Yeh, H.-C.; Wolf, B. S. A Simple Portable Water Bath for Superficial Ultrasonography. *Am J Roentgen Ray Society* **1978**, *130*, 275–278.
98. Guo, Y.; Baac, H. W.; Chen, S.-L.; Norris, T. B.; Guo, L. J. Broad-Band High-Efficiency Optoacoustic Generation Using a Novel Photonic Crystal-Metallic Structure. *Photons Plus Ultrasound: Imaging and Sensing 2011* **2011**, *7899* (February 2011), 78992C. https://doi.org/10.1117/12.876012.
99. Wu, N.; Sun, K.; Wang, X. Fiber Optics Photoacoustic Generation Using Gold Nanoparticles as Target. *Sensors and Smart Structures Technologies for Civil, Mechanical, and Aerospace Systems 2011* **2011**, *7981* (April 2011), 798118. https://doi.org/10.1117/12.880446.
100. Lee, T.; Guo, L. J. Highly Efficient Photoacoustic Conversion by Facilitated Heat Transfer in Ultrathin Metal Film Sandwiched by Polymer Layers. *Adv Opt Mater* **2017**, *5* (2), 1–9. https://doi.org/10.1002/adom.201600421.
101. Cai, L. H.; Kodger, T. E.; Guerra, R. E.; Pegoraro, A. F.; Rubinstein, M.; Weitz, D. A. Soft Poly(Dimethylsiloxane) Elastomers from Architecture-Driven Entanglement Free Design. *Advanced Materials* **2015**, *27* (35), 5132–5140. https://doi.org/10.1002/adma.201502771.
102. Hou, Y.; Kim, J.-S.; Ashkenazi, S.; O'Donnell, M.; Guo, L. J. Optical Generation of High Frequency Ultrasound Using Two-Dimensional Gold Nanostructure. *Appl Phys Lett* **2006**, *89* (9), 093901. https://doi.org/10.1063/1.2344929.
103. Wu, N.; Tian, Y.; Zou, X.; Silva, V.; Chery, A.; Wang, X. High-Efficiency Optical Ultrasound Generation Using One-Pot Synthesized Polydimethylsiloxane-Gold Nanoparticle Nanocomposite. *Journal of the Optical Society of America B* **2012**, *29* (8), 2016. https://doi.org/10.1364/JOSAB.29.002016.
104. Zhao, T.; Su, L.; Xia, W. Optical Ultrasound Generation and Detection for Intravascular Imaging: A Review. *J Healthc Eng* **2018**, *2018*, 1–14. https://doi.org/10.1155/2018/3182483.
105. Biagi, E.; Margheri, F.; Menichelli, D. Efficient Laser-Ultrasound Generation by Using Heavily Absorbing Films as Targets. *IEEE Trans Ultrason Ferroelectr Freq Control* **2001**, *48* (6), 1669–1680. https://doi.org/10.1109/58.971720.
106. Buma, T.; Spisar, M.; O'Donnell, M. High-Frequency Ultrasound Array Element Using Thermoelastic Expansion in an Elastomeric Film. *Appl Phys Lett* **2001**, *79* (4), 548–550. https://doi.org/10.1063/1.1388027.
107. O'Donnell, M.; Hou, Y.; Kim, J. S.; Ashkenazi, S.; Huang, S. W.; Guo, L. J. Optoacoustic Generation of High Frequency Sound for 3-D Ultrasonic Imaging in Medicine. *European Physical Journal: Special Topics* **2008**, *153* (1), 53–58. https://doi.org/10.1140/epjst/e2008-00392-9.
108. Hou, Y.; Ashkenazi, S.; Huang, S. W.; O'Donnell, M. Improvements in Optical Generation of High-Frequency Ultrasound. *IEEE Trans Ultrason Ferroelectr Freq Control* **2007**, *54* (3), 682–686. https://doi.org/10.1109/TUFFC.2007.292.
109. Buma, T.; Spisar, M.; O'Donnell, M. Thermoelastic Expansion vs. Piezoelectricity for High-Frequency, 2-D Arrays. *IEEE Trans Ultrason Ferroelectr Freq Control* **2003**, *50* (8), 1065–1068. https://doi.org/10.1109/TUFFC.2003.1226551.
110. Hsieh, B. Y.; Chen, S. L.; Ling, T.; Guo, L. J.; Li, P. C. All-Optical Transducer for Ultrasound and Photoacoustic Imaging by Dichroic Filtering. *IEEE International Ultrasonics Symposium, IUS* **2012**, *20* (2), 1410–1413. https://doi.org/10.1109/ULTSYM.2012.0352.

111. Hou, Y.; Kim, J. S.; Huang, S. W.; Ashkenazi, S.; Guo, L. J.; O'Donnell, M. Characterization of a Broadband All-Optical Ultrasound Transducer - From Optical and Acoustical Properties to Imaging. *IEEE Trans Ultrason Ferroelectr Freq Control* **2008**, *55* (8), 1867–1877. https://doi.org/10.1109/TUFFC.2008.870.
112. Alessandra Cutolo, M.; Breglio, G. Interferometric Fabry-Perot Sensors for Ultrasound Detection on the Tip of an Optical Fiber. *Results in Optics* **2022**, *6*. https://doi.org/10.1016/j.rio.2021.100209.
113. Hamilton, J. D. High Frequency Optoacoustic Arrays Using Etalon Detection. *IEEE Trans Ultrason Ferroelectr Freq Control* **2000**, *47* (1), 160–169. https://doi.org/10.1109/58.818758.
114. Hou, Y.; Ashkenazi, S.; Huang, S. W.; O'Donnell, M. An Integrated Optoacoustic Transducer Combining Etalon and Black PDMS Structures. *IEEE Trans Ultrason Ferroelectr Freq Control* **2008**, *55* (12), 2719–2725. https://doi.org/10.1109/TUFFC.2008.988.
115. Huang, S. W.; Hou, Y.; Ashkenazi, S.; O'Donnell, M. High-Resolution Ultrasonic Imaging Using an Etalon Detector Array. *Appl Phys Lett* **2008**, *93* (11), 1–4. https://doi.org/10.1063/1.2982584.
116. Zhang, E. Z.; Beard, P. C. A Miniature All-Optical Photoacoustic Imaging Probe. In *Photons Plus Ultrasound: Imaging and Sensing 2011*; SPIE, 2011; Vol. 7899, p 78991F. https://doi.org/10.1117/12.874883.
117. Zhang, E.; Beard, P. Broadband Ultrasound Field Mapping System Using a Wavelength Tuned, Optically Scanned Focused Laser Beam to Address a Fabry Perot Polymer Film Sensor. *IEEE Trans Ultrason Ferroelectr Freq Control* **2006**, *53* (7), 1330–1338. https://doi.org/10.1109/TUFFC.2006.1665081.
118. Finlay, M. C.; Mosse, C. A.; Colchester, R. J.; Noimark, S.; Zhang, E. Z.; Ourselin, S.; Beard, P. C.; Schillin, R. J.; Parkin, I. P.; Papakonstantinou, I.; Desjardins, A. E. Through-Needle All-Optical Ultrasound Imaging in Vivo: A Preclinical Swine Study. *Light Sci Appl* **2017**, *6* (12), e17103-7. https://doi.org/10.1038/lsa.2017.103.
119. Zhang, S.; Zhong, R.; Hong, R.; Hui, D. On Factors Affecting Surface Free Energy of Carbon Black for Reinforcing Rubber. *Nanotechnol Rev* **2020**, *9* (1), 170–181. https://doi.org/10.1515/ntrev-2020-0015.
120. Won Baac, H.; Ok, J. G.; Park, H. J.; Ling, T.; Chen, S. L.; Hart, A. J.; Guo, L. J. Carbon Nanotube Composite Optoacoustic Transmitters for Strong and High Frequency Ultrasound Generation. *Appl Phys Lett* **2010**, *97* (23), 1–4. https://doi.org/10.1063/1.3522833.
121. Bost, W.; Stracke, F.; Weib, E. C.; Narasimhan, S.; Kolios, M. C.; Lemor, R. High Frequency Optoacoustic Microscopy. In *Proceedings of the 31st Annual International Conference of the IEEE Engineering in Medicine and Biology Society: Engineering the Future of Biomedicine, EMBC 2009*; IEEE Computer Society, 2009; pp. 5883–5886. https://doi.org/10.1109/IEMBS.2009.5334452.
122. Tian, Y.; Wu, N.; Zou, X.; Felemban, H.; Cao, C.; Wang, X. Fiber-Optic Ultrasound Generator Using Periodic Gold Nanopores Fabricated by a Focused Ion Beam. *Optical Engineering* **2013**, *52* (6), 065005. https://doi.org/10.1117/1.oe.52.6.065005.
123. Ryu, D.; Loh, K. J.; Ireland, R.; Karimzada, M.; Yaghmaie, F.; Gusman, A. M. In Situ Reduction of Gold Nanoparticles in PDMS Matrices and Applications for Large Strain Sensing. *Smart Struct Syst* **2011**, *8* (5), 471–486. https://doi.org/10.12989/sss.2011.8.5.471.
124. Sharifi, M.; Attar, F.; Saboury, A. A.; Akhtari, K.; Hooshmand, N.; Hasan, A.; El-Sayed, M. A.; Falahati, M. Plasmonic Gold Nanoparticles: Optical Manipulation, Imaging, Drug Delivery and Therapy. *Journal of Controlled Release*. Elsevier B.V. October 1, 2019, pp 170–189. https://doi.org/10.1016/j.jconrel.2019.08.032.
125. Poduval, R. K.; Noimark, S.; Colchester, R. J.; Macdonald, T. J.; Parkin, I. P.; Desjardins, A. E.; Papakonstantinou, I. Optical Fiber Ultrasound Transmitter with Electrospun Carbon Nanotube-Polymer Composite. *Appl Phys Lett* **2017**, *110* (22), 223700. https://doi.org/10.1063/1.4984838.
126. Bao-Yu Hsieh, Jinwook Kim, Jiadeng Zhu, Sibo Li, Xiangwu Zhang, and X. J. A Laser Ultrasound Transducer Using Carbon Nanofibers—Polydimethylsiloxane Composite Thin Film. *Appl Phys Lett* **2015**, *021902* (December 2014). https://doi.org/10.1063/1.4905659.

127. Wei-Yi Chang, Wenbin Huang, Jinwook Kim, S. L. and X. J. Candle Soot Nanoparticles-Polydimethylsiloxane Composites for Laser Ultrasound Transducers. *Appl Phys Lett* **2015**, *107* (July), 161903–161907. https://doi.org/10.1063/1.4934587.
128. Hidayah, N. M. S.; Liu, W. W.; Lai, C. W.; Noriman, N. Z.; Khe, C. S.; Hashim, U.; Lee, H. C. Comparison on Graphite, Graphene Oxide and Reduced Graphene Oxide: Synthesis and Characterization. In *AIP Conference Proceedings*; American Institute of Physics Inc., 2017; Vol. 1892. https://doi.org/10.1063/1.5005764.
129. Hwan Lee, S.; Park, M.; Yoh, J. J.; Song, H.; Yun Jang, E.; Hyup Kim, Y.; Kang, S.; Seop Yoon, Y. Reduced Graphene Oxide Coated Thin Aluminum Film as an Optoacoustic Transmitter for High Pressure and High Frequency Ultrasound Generation Reduced Graphene Oxide Coated Thin Aluminum Film as an Optoacoustic Transmitter for High Pressure and High Frequency. *Appl Phys Lett* **2012**, *101* (24), 241909. https://doi.org/10.1063/1.4772498.
130. Lee, S. H.; Lee, Y.; Yoh, J. J. Reduced Graphene Oxide Coated Polydimethylsiloxane Film as an Optoacoustic Transmitter for High Pressure and High Frequency Ultrasound Generation. *Appl Phys Lett* **2015**, *106* (8), 1–5. https://doi.org/10.1063/1.4913970.
131. Morris, P.; Hurrell, A.; Shaw, A.; Zhang, E.; Beard, P. A Fabry–Pérot Fiber-Optic Ultrasonic Hydrophone for the Simultaneous Measurement of Temperature and Acoustic Pressure. *J Acoust Soc Am* **2009**, *125* (6), 3611–3622. https://doi.org/10.1121/1.3117437.
132. Wu, N.; Tian, Y.; Zou, X.; Wang, X. Study of the Compact Fiber Optic Photoacoustic Ultrasonic Transducer. *Sensors and Smart Structures Technologies for Civil, Mechanical, and Aerospace Systems 2012* **2012**, *8345* (April 2012), 83453Z. https://doi.org/10.1117/12.915243.
133. Wu, N.; Tian, Y.; Zou, X.; Wang, X. Fiber Optic Photoacoustic Ultrasound Generator Based on Gold Nanocomposite. **2013**, No. April 2013, 86940Q. https://doi.org/10.1117/12.2012241.
134. Zou, X.; Wu, N.; Tian, Y.; Wang, X. Broadband Miniature Fiber Optic Ultrasound Generator. *Opt Express* **2014**, *22* (15), 18119. https://doi.org/10.1364/OE.22.018119.
135. Colchester, R. J.; Mosse, C. A.; Bhachu, D. S.; Bear, J. C.; Carmalt, C. J.; Parkin, I. P.; Treeby, B. E.; Papakonstantinou, I.; Desjardins, A. E.; Colchester, R. J.; Mosse, C. A.; Bhachu, D. S.; Bear, J. C.; Carmalt, C. J.; Parkin, I. P.; Treeby, B. E.; Papakonstantinou, I.; Desjardins, A. E. Laser-Generated Ultrasound with Optical Fibres Using Functionalised Carbon Nanotube Composite Coatings. *Appl Phys Lett* **2014**, *104* (17), 173502–173505. https://doi.org/10.1063/1.4873678.
136. Zhang, E. Z.; Beard, P. C. Characteristics of Optimized Fibre-Optic Ultrasound Receivers for Minimally Invasive Photoacoustic Detection. *Photons Plus Ultrasound: Imaging and Sensing 2015* **2015**, *9323*, 932311. https://doi.org/10.1117/12.2081904.
137. Chang, W. Y.; Jiang, X. A Fiber Optic Laser Ultrasound Transducer Using Candle Soot Nanoparticles/PDMS Composites. *Proceedings of the IEEE Conference on Nanotechnology* **2018**, *2018*, 1–2. https://doi.org/10.1109/NANO.2018.8626346.
138. Bodian, S.; Aytac-Kipergil, E.; Zhang, S.; Lewis-Thompson, I.; Sathasivam, S.; Mathews, S. J.; Alles, E. J.; Zhang, E. Z.; Beard, P. C.; Gordon, R. J.; Collier, P.; Parkin, I. P.; Desjardins, A. E.; Colchester, R. J.; Noimark, S. Comparison of Fabrication Methods for Fiber-Optic Ultrasound Transmitters Using Candle-Soot Nanoparticles. *Adv Mater Interfaces* **2023**, *10* (9). https://doi.org/10.1002/admi.202201792.
139. Moradlou, O.; Rabiei, Z.; Banazadeh, A.; Warzywoda, J.; Zirak, M. Carbon Quantum Dots as Nano-Scaffolds for A-Fe2O3growth: Preparation of Ti/CQD@A-Fe2O3photoanode for Water Splitting under Visible Light Irradiation. *Appl Catal B* **2018**, *227* (January), 178–189. https://doi.org/10.1016/j.apcatb.2018.01.016.
140. Kuang, H.; Zhao, Y.; Ma, W.; Xu, L.; Wang, L.; Xu, C. Recent Developments in Analytical Applications of Quantum Dots. *TrAC - Trends in Analytical Chemistry* **2011**, *30* (10), 1620–1636. https://doi.org/10.1016/j.trac.2011.04.022.
141. Raphael, E.; Jara, D. H.; Schiavon, M. A. Optimizing Photovoltaic Performance in CuInS2and CdS Quantum Dot-Sensitized Solar Cells by Using an Agar-Based Gel Polymer Electrolyte. *RSC Adv* **2017**, *7* (11), 6492–6500. https://doi.org/10.1039/c6ra27635k.

142. Zhou, J.; Yang, Y.; Zhang, C. Y. Toward Biocompatible Semiconductor Quantum Dots: From Biosynthesis and Bioconjugation to Biomedical Application. *Chem Rev* **2015**, *115* (21), 11669–11717. https://doi.org/10.1021/acs.chemrev.5b00049.
143. Drbohlavova, J.; Adam, V.; Kizek, R.; Hubalek, J. Quantum Dots - Characterization, Preparation and Usage in Biological Systems. *Int J Mol Sci* **2009**, *10* (2), 656–673. https://doi.org/10.3390/ijms10020656.
144. Vasudevan, D.; Gaddam, R. R.; Trinchi, A.; Cole, I. Core-Shell Quantum Dots: Properties and Applications. *J Alloys Compd* **2015**, *636*, 395–404. https://doi.org/10.1016/j.jallcom.2015.02.102.
145. Karakoti, A. S.; Shukla, R.; Shanker, R.; Singh, S. Surface Functionalization of Quantum Dots for Biological Applications. *Adv Colloid Interface Sci* **2015**, *215*, 28–45. https://doi.org/10.1016/j.cis.2014.11.004.
146. Shirasaki, Y.; Supran, G. J.; Bawendi, M. G.; Bulović, V. Emergence of Colloidal Quantum-Dot Light-Emitting Technologies. *Nat Photonics* **2013**, *7* (1), 13–23. https://doi.org/10.1038/nphoton.2012.328.
147. Hauck, T. S.; Anderson, R. E.; Fischer, H. C.; Newbigging, S.; Chan, W. C. W. In Vivo Quantum-Dot Toxicity Assessment. *Small* **2010**, *6* (1), 138–144. https://doi.org/10.1002/smll.200900626.
148. Nakamura, H.; Kato, W.; Uehara, M.; Nose, K.; Omata, T.; Otsuka-Yao-Matsuo, S.; Miyazaki, M.; Maeda, H. Tunable Photoluminescence Wavelength of Chalcopyrite CuInS 2-Based Semiconductor Nanocrystals Synthesized in a Colloidal System. *Chemistry of Materials* **2006**, *18* (14), 3330–3335. https://doi.org/10.1021/cm0518022.
149. Leach, A. D. P.; Macdonald, J. E. Optoelectronic Properties of CuInS2Nanocrystals and Their Origin. *Journal of Physical Chemistry Letters* **2016**, *7* (3), 572–583. https://doi.org/10.1021/acs.jpclett.5b02211.
150. Berends, A. C.; Rabouw, F. T.; Spoor, F. C. M.; Bladt, E.; Grozema, F. C.; Houtepen, A. J.; Siebbeles, L. D. A.; De Mello Donegá, C. Radiative and Nonradiative Recombination in CuInS2Nanocrystals and CuInS2-Based Core/Shell Nanocrystals. *Journal of Physical Chemistry Letters* **2016**, *7* (17), 3503–3509. https://doi.org/10.1021/acs.jpclett.6b01668.
151. Shei, S. C.; Chiang, W. J.; Chang, S. J. Synthesis of CuInS2 Quantum Dots Using Polyetheramine as Solvent. *Nanoscale Res Lett* **2015**, *10* (1). https://doi.org/10.1186/s11671-015-0789-3.
152. Castro, S. L.; Bailey, S. G.; Raffaelle, R. P.; Banger, K. K.; Hepp, A. F. Synthesis and Characterization of Colloidal CuInS 2 Nanoparticles from a Molecular Single-Source Precursor. *Journal of Physical Chemistry B* **2004**, *108* (33), 12429–12435. https://doi.org/10.1021/jp049107p.
153. Kolny-Olesiak, J.; Weller, H. Synthesis and Application of Colloidal CuInS2 Semiconductor Nanocrystals. *ACS Appl Mater Interfaces* **2013**, *5* (23), 12221–12237. https://doi.org/10.1021/am404084d.
154. Xiong, W. W.; Yang, G. H.; Wu, X. C.; Zhu, J. J. Aqueous Synthesis of Color-Tunable CuInS2/ZnS Nanocrystals for the Detection of Human Interleukin 6. *ACS Appl Mater Interfaces* **2013**, *5* (16), 8210–8216. https://doi.org/10.1021/am402328t.
155. Fang, M.; Peng, C. W.; Pang, D. W.; Li, Y. Quantum Dots for Cancer Research: Current Status, Remaining Issues, and Future Perspectives. *Cancer Biol Med* **2012**, *9* (3), 151–163. https://doi.org/10.7497/j.issn.2095-3941.2012.03.001.
156. Ghafary, S. M.; Nikkhah, M.; Hatamie, S.; Hosseinkhani, S. Simultaneous Gene Delivery and Tracking through Preparation of Photo-Luminescent Nanoparticles Based on Graphene Quantum Dots and Chimeric Peptides. *Sci Rep* **2017**, *7* (1), 1–14. https://doi.org/10.1038/s41598-017-09890-y.
157. Zhu, Y.; Hong, H.; Xu, Z.; Li, Z.; Cai, W. Quantum Dot-Based Nanoprobes for In Vivo Targeted Imaging. *Curr Mol Med* **2013**, *13* (10), 1549–1567. https://doi.org/10.2174/1566524013666131111121733.

158. Mahajan, K. D.; Fan, Q.; Dorcéna, J.; Ruan, G.; Winter, J. O. Magnetic Quantum Dots in Biotechnology - Synthesis and Applications. *Biotechnol J* **2013**, *8* (12), 1424–1434. https://doi.org/10.1002/biot.201300038.
159. Schulze, A. S.; Tavernaro, I.; Machka, F.; Dakischew, O.; Lips, K. S.; Wickleder, M. S. Tuning Optical Properties of Water-Soluble CdTe Quantum Dots for Biological Applications. *Journal of Nanoparticle Research* **2017**, *19* (2). https://doi.org/10.1007/s11051-017-3757-2.
160. Zhong, H.; Lo, S. S.; Mirkovic, T.; Li, Y.; Ding, Y.; Li, Y.; Scholes, G. D. Noninjection Gram-Scale Synthesis of Monodisperse Pyramidal CuInS2 Nanocrystals and Their Size-Dependent Properties. *ACS Nano* **2010**, *4* (9), 5253–5262.
161. Zhong, H.; Zhou, Y.; Ye, M.; He, Y.; Ye, J.; He, C.; Yang, C.; Li, Y. Controlled Synthesis and Optical Properties of Colloidal Ternary Chalcogenide CuInS2 Nanocrystals. *Chemistry of Materials* **2008**, *20* (20), 6434–6443.
162. Petryayeva, E.; Algar, W. R. Single-Step Bioassays in Serum and Whole Blood with a Smartphone, Quantum Dots and Paper-in-PDMS Chips. **2015**, *140* (12). https://doi.org/10.1039/c5an00475f.
163. Bera, D.; Qian, L.; Tseng, T. K.; Holloway, P. H. Quantum Dots and Their Multimodal Applications: A Review. *Materials* **2010**, *3* (4), 2260–2345. https://doi.org/10.3390/ma3042260.
164. Yao, J.; Wang, L. V. Perspective on Fast-Evolving Photoacoustic Tomography. *J Biomed Opt* **2021**, *26* (06), 1–19. https://doi.org/10.1117/1.jbo.26.6.060602.
165. Messer, M. Pulsed Ultrasonic Doppler Velocimetry for Measurement of Velocity Profiles in Small Channels and Cappilaries, 2005. https://www.researchgate.net/publication/27524353.
166. Baac, H. W.; Ok, J. G.; Maxwell, A.; Lee, K.-T.; Chen, Y.-C.; Hart, A. J.; Xu, Z.; Yoon, E.; Guo, L. *J Sci Rep* **2012**, *2*, 989
167. Baac, H. W.; Ok, J. G.; Park, H. J.; Ling, T.; Chen, S.-L.; Hart, A. J.; Guo, L. *J Appl Phys Lett* **2010**, *97*, 234104
168. Lee, T.; Ok, J. G.; Guo, L. J.; Baac, H. W. *Appl Phys Lett* **2016**, *108*, 104102.

Chapter 10
Optical Ultrasound Sensors for Biomedical Photoacoustics

Yuecheng Shen and Jun Ma

Abstract Recent developments in optical ultrasound sensors bring new opportunities for photoacoustic imaging. Compared to conventional piezoelectric transducers, these optical sensors generally hold the advantages of high sensitivity, broad bandwidth, wide acceptance, and miniaturized footprint. In this chapter, we review the historical developments of optical ultrasound sensors and briefly introduce three important types of miniaturized sensors and the ways of parallel interrogation. Using these state-of-the-art sensors, representative imaging applications on a variety of live animals are also described here. We envision that the development of optical ultrasound sensors is promising in advancing deeper and higher-resolution photoacoustic imaging.

Keywords Photoacoustic imaging · Optical ultrasound sensor · On-chip device · Fabry-Pérot interferometers · Fiber Bragg grating · Micro-ring resonator

10.1 Ultrasound Sensors for Photoacoustic Imaging

With the increasing need for deep and high-resolution photoacoustic imaging (PAI), there is a growing demand for developing ultrasound sensors with high sensitivity, broad bandwidth, and wide acceptance angles. For both photoacoustic microscopy (PAM) and photoacoustic computed tomography (PACT), commercialized piezoelectric transducers are the most widely used ultrasound sensors. Historically, piezoelectric transducers were developed for ultrasonography, in which sending and receiving ultrasonic signals are required for pulse-echo ultrasound imaging.

Y. Shen (✉)
State Key Laboratory of Precision Spectroscopy, East China Normal University, Shanghai, China
e-mail: ycshen@lps.ecnu.edu.cn

J. Ma
Guangdong Provincial Key Laboratory of Optical Fiber Sensing and Communications, Institute of Photonics Technology, Jinan University, Guangzhou, China
e-mail: jun.ma@jnu.edu.cn

W. Xia (ed.), *Biomedical Photoacoustics*,
https://doi.org/10.1007/978-3-031-61411-8_10

Since the imaging scenario of PAI is different from that of medical ultrasound imaging, piezoelectric transducers may not be the perfect choice for PAI where the ultrasound waves are commonly generated by optical or microwave pulses [1]. The opacity of the piezoelectric transducer can block the optical path in PAI, which increases the complexity of optical alignment and induces acoustic loss if an opto-acoustic combiner is used. Moreover, the miniaturization of ultrasound sensors is particularly valuable for endoscopy and certain tomographic applications. Unfortunately, directly reducing the size of piezoelectric transducers causes many difficulties in practice, as the sensitivity dramatically degrades proportionally to the square root of the sensing area [2]. To mitigate these difficulties, transparent materials have been considered [3] and miniaturized capacitive micromachined ultrasonic transducers have been developed [4, 5]. However, the demonstrated performance of these devices so far, in terms of sensitivity and bandwidth, cannot fully meet the demanding performance in PAI yet.

Optical detection is long considered to have the advantage of high precision and high sensitivity. Thus, for decades, optical ultrasound sensors have been actively researched as alternatives for piezoelectrical transducers in PAI. Compared to commercialized piezoelectric transducers, optical ultrasound sensors have been demonstrated with broad bandwidth and a nearly size-independent high sensitivity. Due to these advantages, optical ultrasound sensors present great promise in advancing PAI. Up to now, there exists a variety of ways for optical sensing on ultrasonic waves, and some good review articles on optical ultrasound sensors can be found in Refs. [1, 6]. In this chapter, we restricted ourselves to describing interferometric optical ultrasound sensors using miniaturized optical resonators. This type of optical ultrasound sensor has been demonstrated with high sensitivity while maintaining small form factors, showing great potential as a new generation of sensitive and broadband ultrasound sensors. Then we describe the state-of-the-art means of implementing parallel interrogation to optical ultrasound sensor arrays. Finally, we summarize recent progress in the applications of PAI with these optical ultrasound sensors.

10.2 Optical Ultrasound Sensors with Optical Microresonators

A variety of optical ultrasound sensors based on different sensing principles have been actively developed in recent years. In principle, any change in the fundamental parameters of the optical waves due to the presence of the ultrasound wave can be exploited as the basis for optical ultrasound sensors. One example is to record the intensity fluctuations reflected from the interface of two different media caused by the ultrasound wave that passes through the interface [7]. To increase the sensitivity, metal-dielectric coating at the interface can be introduced to excite the surface plasmon resonance for enhanced light-ultrasound interaction [8, 9]. This type of optical ultrasound sensor can cover a large bandwidth of up to 180 MPa [10, 11], which benefits the axial resolution in PAI. Without relying on a reflective interface, the deflection of the probing beam at the existence of the acoustic wave can be exploited as well [12, 13]. This deflection, which is commonly seen with an acousto-optic deflector, can be quantified by a position-sensitive detector. Alternatively, except for detecting ultrasound-induced position change, ultrasound-induced phase change or even polarization change [14] can be detected as well, which serves as an effective tool to characterize the acoustic wave [15, 16].

To improve detection sensitivity, optical interferometric systems are adopted to interrogate ultrasound-induced phase change in optical paths. In early works, Michelson interferometers and Mach-Zehnder interferometers were demonstrated with the capability to detect ultrasound waves [17, 18]. Later on, optical resonators, together with fiber-coupling systems, are emerging as promising candidates for optical ultrasound sensors [19–21]. Optical resonators can confine light to a small volume to enhance light-sound interaction, thereby increasing the detecting sensitivity. Moreover, recent developments in nanofabrication techniques make micro-scale or even nano-scale optical resonators achievable. Therefore, this type of optical ultrasound sensor relying on micro-resonators and interferometric design generally holds the advantage of high sensitivity while maintaining a miniaturized footprint. In the following, we will focus on describing three frequently used optical ultrasound sensors: Fabry-Pérot interferometers [22–27], π-phase-shifted Bragg gratings [28–34] (π-BGs), and micro-ring resonators [35–37] (MRRs).

10.2.1 Fabry-Pérot Interferometers

Early designs of Fabry-Pérot interferometers trap the probing beam between two flat mirrors. In this way, the stress induced by the incident ultrasound wave modulates the optical path length of the cavity, thus changing its resonant frequency. By recording the transient intensity variation of the transmitted or reflected probing beam, this type of optical interferometer ultrasound sensor achieves a sensitivity down to 50 Pa and a bandwidth of up to 40 MHz [24]. Although such a design can bring convenience in fabricating dense sensor arrays [23, 38], it suffers from the problem that the probing beam may walk off laterally upon each round trip. This problem causes difficulties in achieving high optical confinement in interferometers, degrading the optical finesse of the cavity and thus the sensitivity [27].

A recent design with a plano-concave optical microcavity proves to considerably boost the detecting sensitivity by using a wavefront-matched concave mirror to strengthen the optical confinement of the light beam bouncing forth and back in the cavity [27]. A schematic illustration of the optical ultrasound sensor is depicted in Fig. 10.1a, which comprises a solid plano-concave polymer microcavity formed between two highly reflective mirrors. To create an acoustically homogeneous planar structure, the cavity is encapsulated within a layer of the same polymer. To maximize the sensitivity of the sensor, the divergence of the interrogation beam

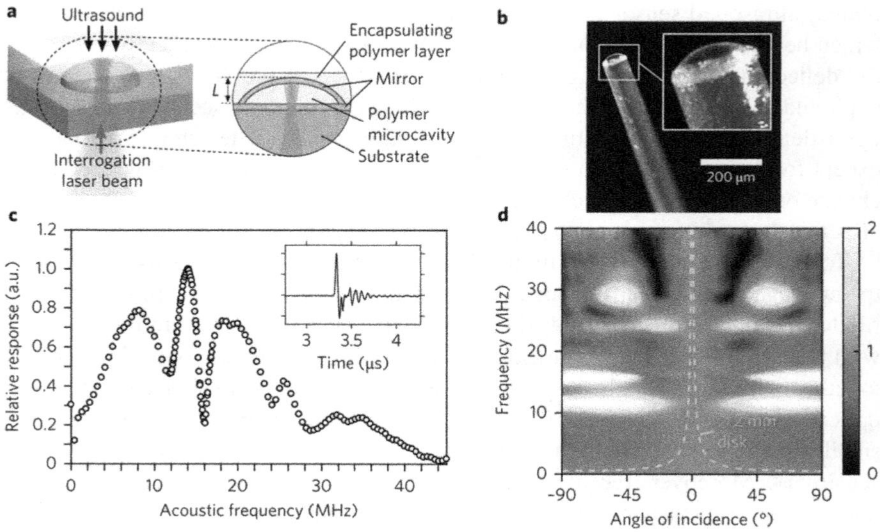

Fig. 10.1 Plano-concave optical microresonator optical ultrasound sensor. (**a**) A schematic illustration of the sensor (L = cavity thickness). The sensor comprises a plano-concave polymer microcavity encapsulated in a planar polymer layer. (**b**) Photograph of the microcavity after being attached to the tip of a single-mode optical fiber, enabling a highly miniaturized flexible probe-type ultrasound sensor. (**c**) Frequency response of a typical example of the sensor with a 16-μm thickness. (**d**) Angular response of a typical example of the sensor with a 16-μm thickness (normalized to $\theta = 0°$). As a comparison, the contour line shows the 50% cut-off for the modeled response of a disk-shaped purely spatially averaging piezoelectric sensor of a diameter of 2 mm. (Reprinted by permission from Springer Nature: Nature Photonics, Ultrasensitive plano-concave optical microresonators for ultrasound sensing, Guggenheim, J. A. et al., © 2017)

and the curvature of the top mirror should be perfectly matched. The frequency of the interrogation laser beam is tuned to have a wavelength close to the resonant frequency of the cavity. To implement the sensor for imaging applications, the microcavity is attached to the tip of a single-mode optical fiber, as shown in the photograph displayed in Fig. 10.1b. A typical example of the frequency response for a cavity with a thickness of 16 μm is illustrated in Fig. 10.1c, in which the noise-equivalent pressure (NEP) at 3.5 MHz is 9.3 Pa over a 20 MHz measurement bandwidth (2.1 mPaHz$^{-1/2}$). Moreover, the measured angular response is shown in Fig. 10.1d, indicating that the sensor is effectively omnidirectional with a high sensitivity of up to ±90° for most frequencies up to 40 MHz. By choosing different cavity thicknesses, this optical ultrasound sensor encounters a counterbalance between the sensitivity and detecting bandwidth. Generally, it is found that the microcavity with a smaller thickness supports a wider bandwidth but has a lower sensitivity.

10.2.2 π-BGs

Optical ultrasound sensors based on π-BGs are essentially one-dimensional cavities fabricated on optical fibers or waveguides. In these microcavities, light is confined within a dimension that is even smaller than the physical size of the Bragg gratings. Early designs of this type of sensor support sensitivities down to 100 Pa and bandwidths up to 77 MHz [30, 31]. However, these sensors are line detectors, which require less straightforward imaging reconstruction than point detectors.

Recently, nearly ideal point-like sensor geometry was reported with miniaturized π-BGs in optical waveguides based on silicon-on-insulator technology [33]. The sensor, which was named silicon waveguide-etalon detector (SWED), has a sensing area of only 220 nm × 500 nm, providing per-area sensitivity that is 100,000,000 times better than that of piezoelectric detectors. A schematic of the detector, as well as its operating principle, is described in Fig. 10.2a.

The SWED contains a single continuous silicon waveguide with four sections: a silver layer, a spacer, a cavity, and a Bragg grating. A continuous-wave laser pump light is confined within the cavity, with a wavelength tuned off-resonance to the point of the steepest slope. When the incident ultrasound wave induces resonance change of the cavity, the variations in the reflected optical intensity, which is subsequently measured by a photodiode, can be recorded to represent time-varying ultrasound signals. Four SWEDs were manufactured and reported in this work, with

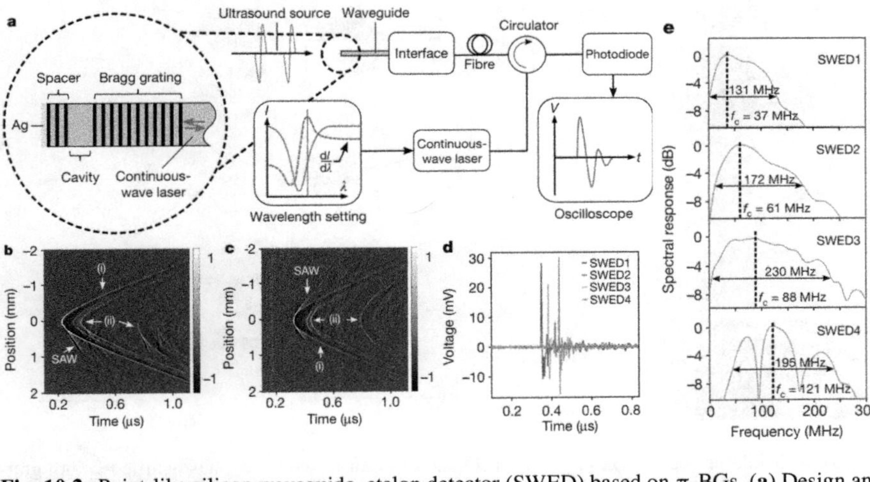

Fig. 10.2 Point-like silicon waveguide–etalon detector (SWED) based on π-BGs. (**a**) Design and operating principles of the SWED. (**b**, **c**) Spatial responses of SWED3 acquired by scanning the SWED linearly over a broadband ultrasound point source. The scanning path was either parallel to the short dimension of the chip facet (**b**) or perpendicular to the long dimension of the chip facet (**c**). SAW: surface acoustic waves. (**d**) Temporal responses of all four SWEDs. (**e**) Frequency responses of all four SWEDs. (Reprinted by permission from Springer Nature: Nature, A submicrometre silicon-on-insulator resonator for ultrasound detection, Shnaiderman, R. et al., © 2020)

different spacer lengths of 26 μm, 14 μm, 9 μm, and 3.5 μm. As a typical example, Fig. 10.2b and c show the spatial response of SWED3 characterized by scanning linearly over a broadband ultrasound point source along different directions, resulting in characteristic curved profiles. To further characterize the detection bandwidth, the temporal responses of all four SWEDs are measured and shown in Fig. 10.2d. By taking Fourier transformations to the temporal responses, respective frequency responses are shown in Fig. 10.2e. Notably, the central frequency increases with the decreased spacer length. Moreover, an ultrawide detection bandwidth of up to 230 MHz was reported for SWED3. The authors further investigated the sensitivity of the SWED using a calibrated needle hydrophone and found that the NEP for SWED1 was 45 Pa (9 mPaHz$^{-1/2}$) over a 25-MHz bandwidth around the central frequency at 37 MHz [33].

Using the same silicon-photonics platform, a miniaturized π-BGs-based optical ultrasound sensor with a wide bandwidth of over 200 MHz was developed as illustrated in Fig. 10.3a-c [34]. The ultrasound impinging onto the silicon π-BG induced modulation on the phase difference between the optical beams in the resonator and the reference beam of a Mach-Zehnder interferometer (MZI) as shown in Fig. 10.3d-g. By reading out the differential output voltage signal from the

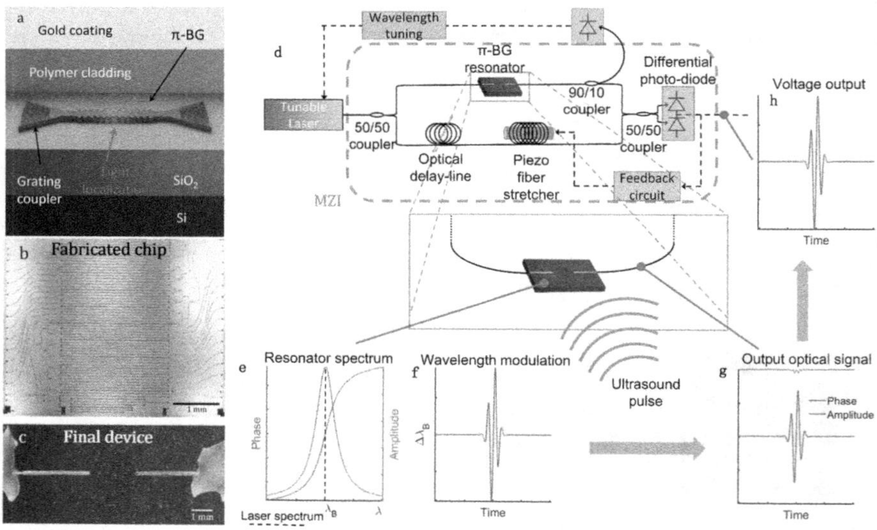

Fig. 10.3 Miniaturized silicon-photonics acoustic detector based on π-phase-shifted Bragg grating (π-BG) with grating couplers. (**a**) Schematic of the silicon-photonics detector. Photograph of (**b**) the fabricated silicon-photonics chip and (**c**) the final assembled devices. (**d**) Schematic of the optical demodulation system. (**e**) The spectrum of the resonator. (**f**) Wavelength modulation induced by ultrasound. (**g**) Corresponding optical output from the resonator (**h**) Differential voltage signal. (Reprinted by permission from Springer Nature: Nature, Silicon-photonics acoustic detector for optoacoustic micro-tomography, Hazan, Y. et al., © 2022)

stabilized MZI (Fig. 10.3h), the NEP of the sensor was reduced to 2.2 mPaHz$^{-1/2}$ due to the cancellation of phase noise from the laser.

10.2.3 Micro-Ring Resonator

Micro-ring resonators have the advantages of high-quality factor, miniaturized form factor, and optical transparency [35, 36, 39–41]. To date, optical ultrasound sensors that employ either polymer-based [36, 39], silicon-based [42], or even chalcogenide-based [43] micro-rings have been reported to maintain a high-quality factor of up to 10^5 with a small footprint [44, 45], an ultrabroad frequency bandwidth of up to 350 MHz [46], high sensitivity with a noise-equivalent pressure (NEP) down to 1.3 mPaHz$^{-1/2}$ [42], and a wide acceptance angle over ±30° [35]. All these advantages make micro-ring resonators attractive for PAI.

A representative optical ultrasound sensor based on the micro-ring resonator reported recently with superior performances was manufactured in silicon photonics [42]. This sensor is about 20 μm in size, leading to a NEP below 1.3 mPaHz$^{-1/2}$ within a frequency range of 3–30 MHz. A principle diagram of the micro-ring resonator-based optical ultrasound sensor is shown in Fig. 10.4a. The incident ultrasound wave causes a vibration to an acoustically resonant membrane. A split-rib waveguide has one part on the movable membrane and one part on a fixed substrate, which is separated by a tiny 15 nm gap, as shown in Fig. 10.4b. This waveguide was fabricated with a ring shape, which effectively forms a micro-ring resonator and is interrogated by a laser. Due to the strong optical field concentration in the gap, a small ultrasound-induced change in the height of the gap leads to a big change in the effective refractive index of the waveguide, thus modifying the resonant frequency of the micro-ring resonator. The use of the membrane here is to enhance the sensitivity of the sensor, as silicon is relatively rigid. A typical example of the sensor with a membrane diameter of 20 μm was characterized by exciting a broadband ultrasound pulse and the measured temporal response is shown in Fig. 10.4c. After a series of calculations, the sensitivity of the sensor is computed and shown as the left axis of Fig. 10.4d, after comparing it to a calibrated hydrophone. The noise amplitude spectral density of the sensor was also calculated from a recorded time trace without applying ultrasound, which is shown as the right axis of Fig. 10.4d. By dividing these two curves, the NEP spectral density can be calculated, which is below 1.3 mPaHz$^{-1/2}$ throughout the frequency range from 3 MHz to 30 MHz. Moreover, the authors demonstrate an array that contains ten sensors with a fine pitch of only 30 μm. They argued that by tuning these micro-ring sensors to operate at slightly different resonant frequencies, parallel interrogation could be realized by using only two optical fibers using on-chip photonic multiplexing.

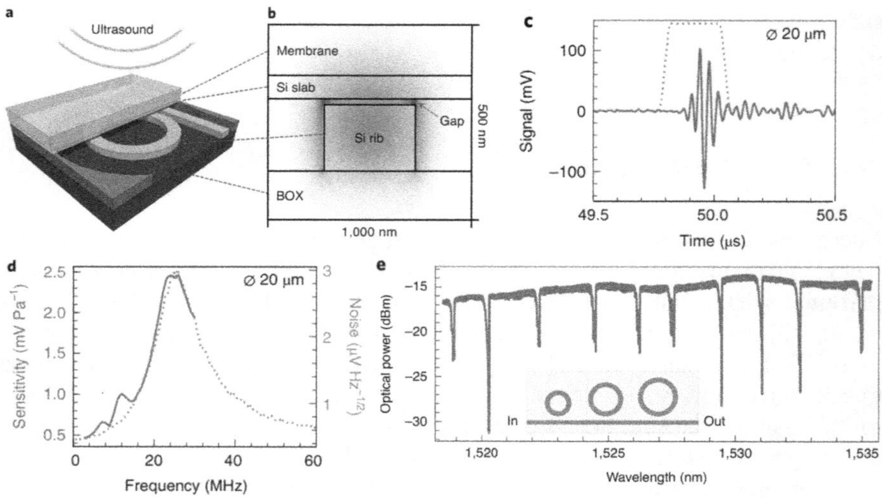

Fig. 10.4 Optical ultrasound sensor based on the micro-ring resonator. (**a**, **b**) Schematic of the micro-ring sensor (**a**) and of the split-rib waveguide with a tiny gap (**b**) showing the silicon dioxide lower cladding (BOX) and upper cladding (membrane). Overlaid is a plot of the simulated electric field intensity of the waveguide mode, showing the high intensity in the gap. (**c**) Temporal response of the sensor. A time window (red dotted line) selects the first arrival of the pressure pulse and excludes unwanted reflections from the back surface of the substrate. (**d**) Sensor sensitivity (left axis) and noise amplitude spectral density (right axis) of the sensor. (**e**) Optical transmission spectrum of the sensor array that contains ten micro-ring resonators. These micro-ring resonators have slightly different radii, enabling roughly equally spaced resonant wavelengths within the free spectral range. The inset shows an example of three micro-ring resonators. (Reprinted by permission from Springer Nature: Nature Photonics, Sensitive, small, broadband and scalable optomechanical ultrasound sensor in silicon photonics, Westerveld, W. J. et al., © 2021)

10.2.4 Parallel Interrogation for the Sensor Array

Most optical ultrasound sensors based on micro-resonators reported were demonstrated with a single element. In practice, a sensor array with many elements can tremendously facilitate imaging speed and largely benefit imaging applications. The Fabry-Pérot interferometer is naturally a two-dimensional sensor array [23, 38]. However, parallel interrogation of this type of sensor array generally involves photodetector arrays, which increase electrical complexity and suffer from considerable crosstalk. Similarly, fiber-optic detector arrays also require a rather complicated imaging system [47–50]. The integrated on-chip sensor array was first reported in 2008, which contains four micro-rings [36]. However, the quality factors of the micro-resonators fabricated at that time are relatively low, exhibiting considerable spectrum overlapping between adjacent resonant frequencies. Therefore, it remains a challenge to enable all four elements to function simultaneously. In 2021, micro-rings with high-quality factors were reported using silicon photonics. In that work, a one-dimensional sensor array containing ten micro-ring sensors was shown to have a well-separated spectrum for adjacent resonant frequencies [42]. However, this system had only one laser and one photoreceiver, permitting only one micro-ring sensor to work at a time. The authors envision that by utilizing wavelength division multiplexing, parallel interrogation could be realized in the future [46]. Such an interrogation strategy is rather cumbersome, as it either requires a sufficient number of source-detector pairs [42] or a frequency-swept source [48]. Alternatively, the employment of a phase-modulated pulse allows simultaneous multi-channel detection for ultrasound waves by using up to four elements [50]. However, one photoreceiver per detecting channel is still required, which inevitably increases system complexity.

Recent progress on the parallel interrogation to the optical ultrasound sensor array was made through the employment of a digital optical frequency comb (DOFC) [43]. The DOFC exhibits a fine comb tooth in the frequency domain, serving as an efficient tool for high-precision measurement, which has been demonstrated previously for ultrafine spectral measurement (0.01 pm resolution) [51] and low-frequency ultrasound detection (165 kHz) [52]. With this tool, the transmission spectrum of the sensor array can be probed with the advantages of stability, flexibility, and tunability. For a fiber-based resonator, similar comb structures were created optically with pulsed light and two fiber Bragg gratings for ultrasonic detection [30]. In this work, the authors fabricated an optical ultrasound array that contains 15 micro-ring resonators and coupled this sensor array to a single bus waveguide using chalcogenide materials. These micro-ring sensors were characterized with a frequency bandwidth up to 175 MHz (−6 dB), a NEP down to $2.2\,\text{mPaHz}^{-1/2}$ or 7.1 Pa within a 20-MHz range, an acceptance angle of ±30° (−3 dB at 25 MHz), and a small footprint of $0.85 \times 40 \times 40\ \mu\text{m}^3$ [43]. The experimental setup to realize parallel interrogation for the sensor array is schematically shown in Fig. 10.5a. A continuous-wave narrow-linewidth tunable laser was chosen as the optical source. The generation process of the DOFC started with a digital signal with a multi-carrier bandwidth of 40 GHz and a sequence length of 1536, which was synthesized through the orthogonal frequency division multiplexing method. To demodulate the DOFC, the measured signal was Fourier transformed into the frequency domain, leading to the determination of the transmission spectrum of the sensor array. Figure 10.5b plots the measured transmission spectrum as a function of time for all 15 micro-rings. When no ultrasound is present, the resonant frequency of each micro-ring remains as a flat line, indicating a stable detecting environment. When the laser-induced ultra-sound signal is present to modulate the optical properties of these micro-rings, Fig. 10.5c plots the measured transmission spectrum as a function of time. By

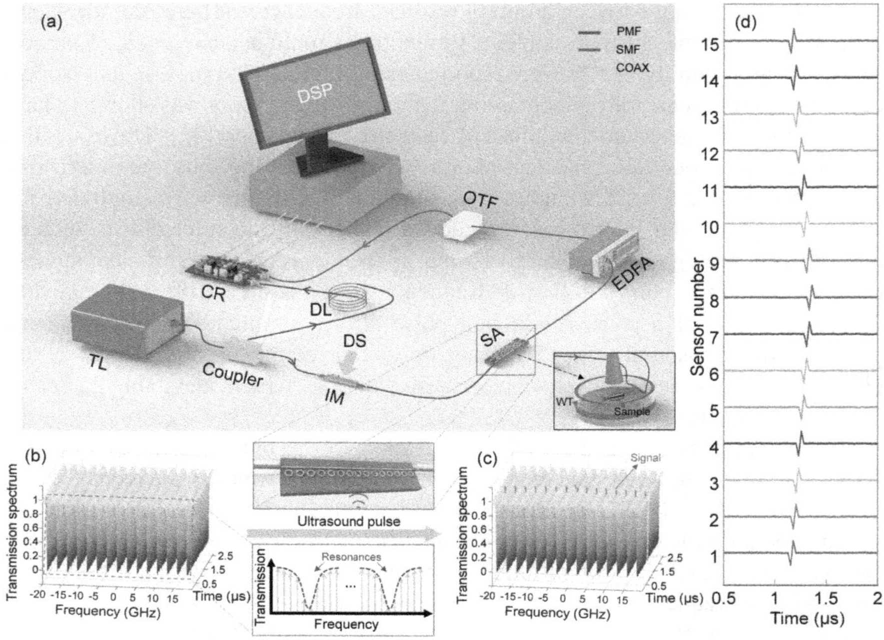

Fig. 10.5 Principle diagram of parallel interrogation to the sensor array. (**a**) Experimental setup of the imaging system. The two insets show the generation and receiving processes of the PA signal. TL tunable laser, IM intensity modulator, DS digital signal, SA sensor array, WT water tank, EDFA erbium-doped fiber amplifier, OTF optical tunable filter, DL optical delay line, CR coherent receiver, DSP digital signal processing, PMF polarization maintaining fiber, SMF single mode fiber, COAX coaxial cable. (**b**) The measured transmission spectrum of the sensor array as a function of time in the null case. Inset: a schematic diagram that illustrates how the DOFC samples the transmission spectrum of the sensor array. DOFC digital optical frequency comb. (**c**) The measured transmission spectrum of the sensor array when ultrasound is present to modulate the sensor array. (**d**) The reconstructed ultrasound signal is a function of time for each micro-ring sensor. (Reprinted by permission from Springer Nature: Nature Communications, Parallel interrogation of the chalcogenide-based micro-ring sensor array for photoacoustic tomography, Pan, J. et al., © 2021)

scrutinizing the temporal change in the transmission spectrum, time-dependent acoustic signals arrived at each micro-ring can be determined, which is illustrated in Fig. 10.5d.

10.3 Imaging Applications by Using Optical Ultrasound Sensors

Due to the superior performance, various applications involving PAI have been demonstrated through optical ultrasound sensors, including laser-scanning endoscopy [53, 54] and microscopy on single cells [55], mouse ears [45], and mouse brains [56]. Recently, three-dimensional PAT for a variety of biological tissue with an imaging depth of up to 8 mm has also been reported [57]. Due to the limited space, in this section, we describe two representative imaging applications that are particularly suitable for optical ultrasound sensors with miniaturized size.

10.3.1 Miniature Forward-Viewing Endoscopic Probe

To date, there is a substantial effort in clinal PAI in advancing the development of non-invasive imaging instruments [58, 59]. Among them, there is increasing interest in developing miniature endoscopic photoacoustic devices for minimally invasive applications such as the assessment of prostate and colorectal cancers and coronary artery disease and guiding interventional procedures such as laparoscopic surgery [54]. To satisfy this need, the authors describe a miniature forward-viewing three-dimensional photoacoustic probe, which comprises a coherent fiber bundle with a Fabry-Pérot ultrasound sensor [24, 60, 61]. In particular, this probe consists of a rigid 76 mm long, 3.2 mm diameter fiber-optic bundle with 50,000 cores, which can be interrogated through point-by-point scanning.

Using this endoscopic probe, an ex vivo duck embryo was imaged using an excitation wavelength at 590 nm, which is schematically shown in Fig. 10.6a. The maximum intensity projections of the three-dimensional image data set are shown in Fig. 10.6b-g. Both the fine capillary network of the chorioallantoic membrane and vessels with diameters of approximately 50 μm can be visualized and identified.

The endoscopic probe was further employed to image the microvasculature in a 5-week-old BALB/C mouse skin on the abdominal side. The maximum intensity projections of the three-dimensional image data set along the x–y and x–z at two distinct areas are shown in Fig. 10.7. The blood vessels from the surrounding tissue to a depth of approximately 2 mm can be identified. Specifically, these images show the blood vessels, which range in diameter from 53 μm to 180 μm, in the dermis and hypodermis layers of the abdominal skin.

Compared to the fiber bundle with a diameter of several millimeters, multi-mode fibres (MMFs) with a smaller diameter have attracted increasing interest in developing miniature endoscopy probes. By using a 140 μm-diameter MMF to deliver the excitation light and a highly sensitive fibre-optic microresonator ultrasound sensor to detect the PA signals, an ultrathin size and low-cost MMF-based PA endomicroscopy probe was integrated within a 20 gauge medical needle [62]. Through wavefront shaping with a digital micromirror device, rapid raster-scanning of a focused light spot at the distal end of the MMF was used to acquire PA image of mouse red blood cells over a region of 100 μm in diameter with a fast imaging speed of ~3 frames per second and an optical diffraction-limited resolution of ~1.2 μm. The probe can potentially provide sub-cellular spatial resolution images

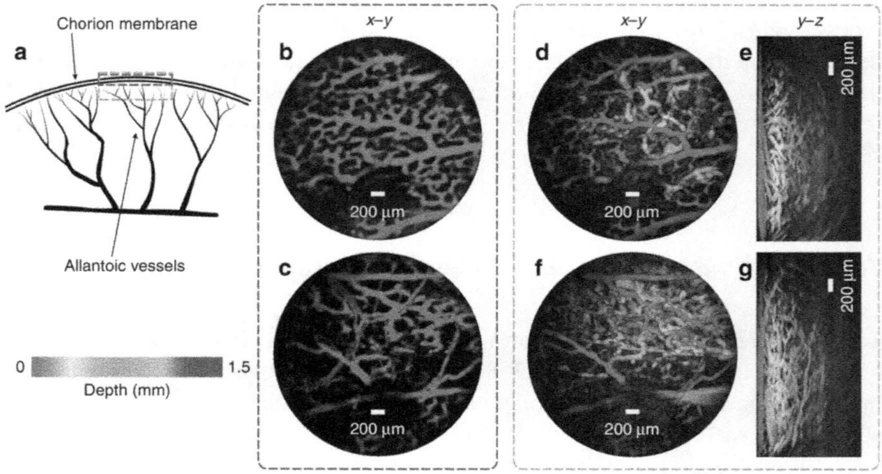

Fig. 10.6 Imaging results of an ex vivo duck embryo. (**a**) A schematic diagram of an avian embryonic vasculature. (**b**, **c**) x–y projections for depth range z = 0–200 µm of two regions of the same embryo, showing the microvascular anatomy of the chorioallantoic membrane. (**d**, **f**) x–y projections for the depth range z = 0–1.5 mm for the same two regions as in (b) and (c), respectively. (**e**, **g**) y–z projections. The colormap that corresponds to the imaging depth is provided at the left bottom corner. (Reprinted by permission from Springer Nature: Light: Science & Applications, All-optical forward-viewing photoacoustic probe for high-resolution 3D endoscopy, Ansari, R. et al., © 2018)

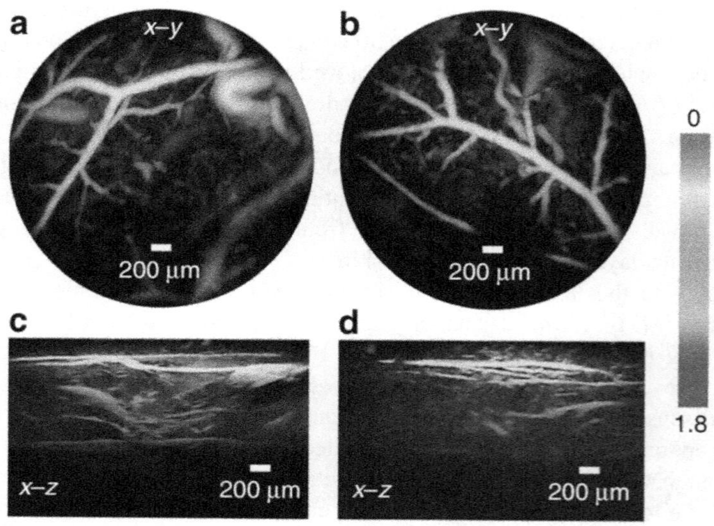

Fig. 10.7 Imaging results of an ex vivo mouse abdominal skin microvasculature. (**a**, **b**) x–y projections for depth range z = 0–2 mm of two regions. (**c**, **d**) x–z projections for the same region. The colormap that corresponds to the imaging depth is provided on the right. (Reprinted by permission from Springer Nature: Light: Science & Applications, All-optical forward-viewing photoacoustic probe for high-resolution 3D endoscopy, Ansari, R. et al., © 2018)

with functional, molecular, and microstructural information of tissue for minimally invasive surgery guiding in real-time.

10.3.2 *In Vivo Cortical Imaging with Implanted Optical Ultrasound Sensors*

The miniaturized optical ultrasound sensor brings advantages when being mounted or even implanted in the brains of live animals. A recent work developed a low-cost soft nanoimprint lithography method to fabricate disposable optical ultrasound sensors based on micro-ring resonators. By integrating the sensor to the inner surface of the chronic cranial window (CCW), ultrasound sensing CCW (usCCW) was demonstrated and PAM for cortical vascular structure in a live mouse was monitored for over 28 days [56].

As illustrated in Fig. 10.8a, the authors performed a craniotomy and surgically implanted the usCCW onto the skull. The miniaturized sensor fitted well inside the narrow space between the dura and glass window, as shown in Fig. 10.8b. During the imaging process, the excitation light illuminates the imaging area through the usCCW, as shown in Fig. 10.8c. A bright-field optical microscope image was provided in Fig. 10.8d to show the image area, varying the depth-coded PAM image shown in Fig. 10.8e. Notably, the PAM image shows much finer details of small capillaries. A three-dimensional visualization of the area is provided in Fig. 10.8f. In addition to imaging blood vessels, temporary hemorrhage was also visualized and highlighted in Figs. 10.8g-i. Such an accomplishment is attributed to the capability of segmentation of PAM, which was further strengthened due to the high-axial resolution offered by the optical ultrasound sensor.

Long-term intravital imaging using usCCW was also reported in Fig. 10.9. The usCCW in open skull configuration was implanted on the mouse's forehead following craniotomy. Since only two optical fibers were attached, the existence of the sensor imposes a minimal constraint on the motion of the mouse, as shown in Fig. 10.9a. The mouse was only fixed to the animal holder during the imaging experiment (Fig. 10.9b and c) and was realized in the cage for the rest of the time. In this study, the authors performed PAM imaging over 28 days and marginal reductions in the quality factors of the micro-ring sensors were observed (Fig. 10.9d), confirming the stability of usCCW for in vivo applications. The maximum intensity projections of PAM images over 28 days are shown in Fig. 10.9e, manifesting the evolution of the cortical vasculature in the same area for selected days. Medium postsurgical bleeding occurs on Day 3 and the hemorrhage is gradually cleared on Day 4. Aggressive neovascularization is also clearly observed in the following days.

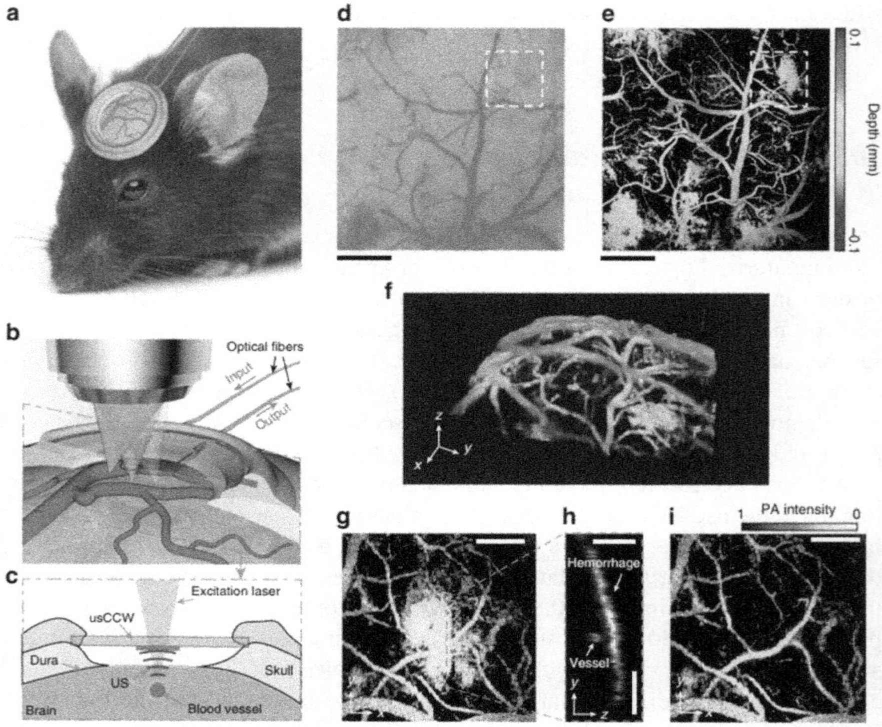

Fig. 10.8 In vivo PAM using a usCCW. (**a**) Schematic illustration of implanting the usCCW on the mouse skull. (**b**) Illustration of optical scanning through the usCCW. (**c**) Optical excitation and ultrasonic detection geometry. (**d**) Brightfield image of the cortical region captured using an optical microscope. (**e**) Depth-coded maximum intensity projection image of the same region. (**f**) Three-dimensional visualization of the imaging area. (**g**) PAM image of the highlighted region using the white dashed box in (e). (**h**) B-scan image of the highlighted green dashed line in (g). (**i**) The hidden vessels beneath the hemorrhage area. Scale bars, (a)-(b) 0.5 mm; (g)-(i) 200 μm. (Reprinted by permission from Springer Nature: Nature Communications, Disposable ultrasound-sensing chronic cranial window by soft nanoimprinting lithography, Li, H. et al., © 2019)

10.4 Conclusions

PAI presents challenges for ultrasound detection, which brings great interest in optical ultrasound detection schemes. Compared to piezoelectric transducers, optical ultrasound sensors based on micro-resonators can maintain miniaturized size without compromising sensitivity, thus presenting great promise in advancing PAT to deeper and higher-resolution imaging.

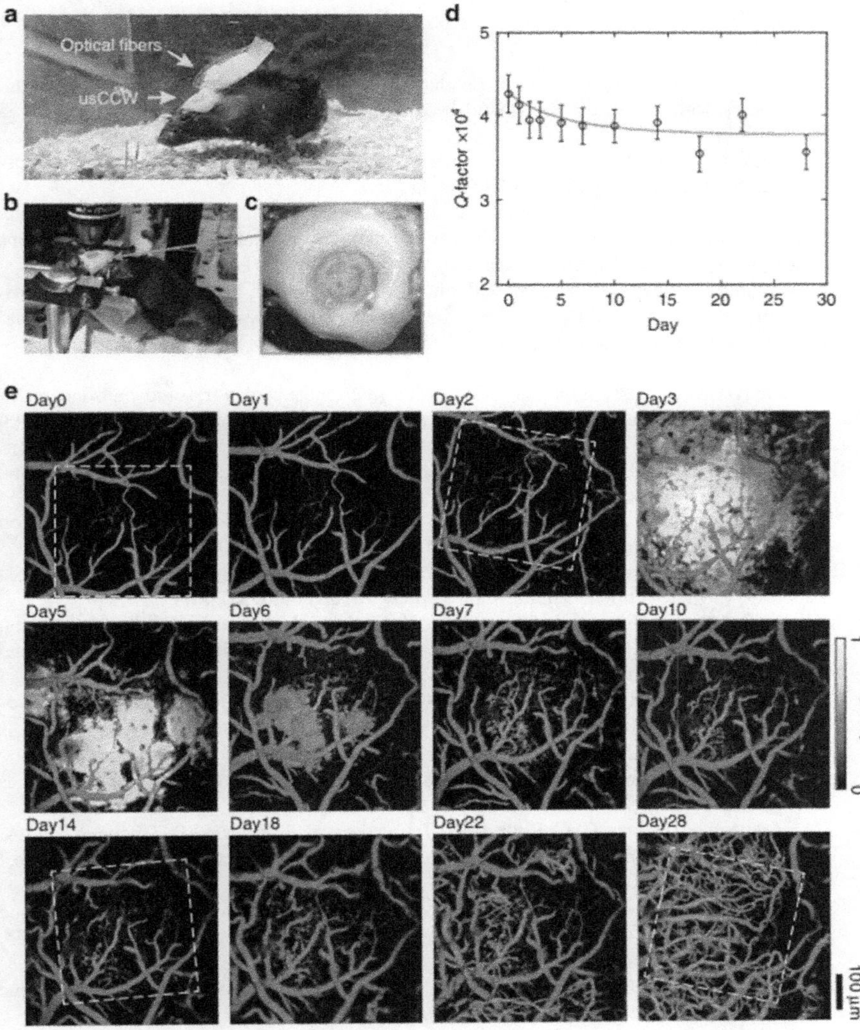

Fig. 10.9 Long-term intravital PAM imaging of live mouse brains. (**a**) Photograph of a free-moving mouse in the breeding cage after the implantation. (**b**) Photograph of the mouse mounted under the microscope during the imaging. (**c**) A magnified optical image shows the implanted transparent usCCW and the clear optical view of the cerebral cortex. (**d**) Quality factors measured over the 28 days, showing a marginal reduction. Error bars show the standard deviation of measured results. (**e**) Maximum amplitude projections of PAM images show cortical vasculature in the same area over 28 days. (Reprinted by permission from Springer Nature: Nature Communications, Disposable ultrasound-sensing chronic cranial window by soft nanoimprinting lithography, Li, H. et al., © 2019)

References

1. G. Wissmeyer, M. A. Pleitez, A. Rosenthal, and V. Ntziachristos, "Looking at sound: optoacoustics with all-optical ultrasound detection," Light: Science & Applications **7**, 53 (2018).
2. M. W. Amy, I. M. Konstantin, and V. W. Lihong, "Noise-equivalent sensitivity of photoacoustics," Journal of Biomedical Optics **18**, 097003 (2013).
3. G. W. J. Brodie, Y. Qiu, S. Cochran, G. C. Spalding, and M. P. Macdonald, "Optically transparent piezoelectric transducer for ultrasonic particle manipulation," IEEE Transactions on Ultrasonics, Ferroelectrics, and Frequency Control **61**, 389–391 (2014).
4. I. O. Wygant, X. Zhuang, D. T. Yeh, O. Oralkan, A. S. Ergun, M. Karaman, and B. T. Khuri-yakub, "Integration of 2D CMUT arrays with front-end electronics for volumetric ultrasound imaging," IEEE Transactions on Ultrasonics, Ferroelectrics, and Frequency Control **55**, 327–342 (2008).
5. B. T. Khuri-Yakub and Ö. Oralkan, "Capacitive micromachined ultrasonic transducers for medical imaging and therapy," Journal of Micromechanics and Microengineering **21**, 054004 (2011).
6. B. Dong, C. Sun, and H. F. Zhang, "Optical Detection of Ultrasound in Photoacoustic Imaging," IEEE Transactions on Biomedical Engineering **64**, 4–15 (2017).
7. X. Zhu, Z. Huang, G. Wang, W. Li, D. Zou, and C. Li, "Ultrasonic detection based on polarization-dependent optical reflection," Opt. Lett. **42**, 439–441 (2017).
8. R. Nuster, G. Paltauf, and P. Burgholzer, "Comparison of surface plasmon resonance devices for acoustic wave detection in liquid," Opt. Express **15**, 6087–6095 (2007).
9. T. Wang, R. Cao, B. Ning, A. J. Dixon, J. A. Hossack, A. L. Klibanov, Q. Zhou, A. Wang, and S. Hu, "All-optical photoacoustic microscopy based on plasmonic detection of broadband ultrasound," Applied Physics Letters **107**, 153702 (2015).
10. G. Paltauf, H. Schmidt-Kloiber, K. P. Köstli, and M. Frenz, "Optical method for two-dimensional ultrasonic detection," Applied Physics Letters **75**, 1048–1050 (1999).
11. J. E. Parsons, C. A. Cain, and J. B. Fowlkes, "Cost-effective assembly of a basic fiber-optic hydrophone for measurement of high-amplitude therapeutic ultrasound fields," The Journal of the Acoustical Society of America **119**, 1432–1440 (2006).
12. S. M. Maswadi, B. L. Ibey, C. C. Roth, D. A. Tsyboulski, H. T. Beier, R. D. Glickman, and A. A. Oraevsky, "All-optical optoacoustic microscopy based on probe beam deflection technique," Photoacoustics **4**, 91–101 (2016).
13. R. A. Barnes, S. Maswadi, R. Glickman, and M. Shadaram, "Probe beam deflection technique as acoustic emission directionality sensor with photoacoustic emission source," Appl. Opt. **53**, 511–519 (2014).
14. P. Zhang, Y. Miao, Y. Ma, P. Niu, L. Zhang, L. Zhang, and F. Gao, "All-optical ultrasonic detector based on differential interference," Opt. Lett. **47**, 4790–4793 (2022).
15. C. I. Zanelli and S. M. Howard, "Schlieren metrology for high frequency medical ultrasound," Ultrasonics **44**, e105–e107 (2006).
16. R. Nuster, P. Slezak, and G. Paltauf, "High resolution three-dimensional photoacoustic tomography with CCD-camera based ultrasound detection," Biomed. Opt. Express **5**, 2635–2647 (2014).
17. H. A. Deferrari, R. A. Darby, F. A. Andrews, "Vibrational displacement and mode-shape measurement by a laser interferometer," J. Acoust. Soc. Am. **42**, 982–990 (1967).
18. J. Bauer-Marschallinger, A. Höllinger, B. Jakoby, P. Burgholzer, and T. Berer, "Fiber-optic annular detector array for large depth of field photoacoustic macroscopy," Photoacoustics, **5**, 1–9 (2017).
19. J. Ma, J. Zhao, H. Chen, L. P. Sun, J. Li, and B. O. Guan, "Transparent microfiber Fabry-Perot ultrasound sensor with needle-shaped focus for multiscale photoacoustic imaging," Photoacoustics, **30**, 100482 (2023).

20. X. Bai, J. Ma, X. Li, L. Jin, and B.-O. Guan, "Focus-tunable fiber-laser ultrasound sensor for high-resolution linear-scanning photoacoustic computed tomography," Appl. Phys. Lett. **116**, 153701 (2020).
21. J. Ma, Y. He, X. Bai, L.-P. Sun, K. Chen, K. Oh, and B.-O. Guan, "Flexible microbubble-based Fabry–Pérot cavity for sensitive ultrasound detection and wide-view photoacoustic imaging," Photon. Res. **8**, 1558–1565 (2020).
22. S. Ashkenazi, Y. Hou, T. Buma, and M. O'Donnell, "Optoacoustic imaging using thin polymer étalon," Applied Physics Letters **86**, 134102 (2005).
23. S.-W. Huang, Y. Hou, S. Ashkenazi, and M. O'Donnell, "High-resolution ultrasonic imaging using an etalon detector array," Applied Physics Letters **93**, 113501 (2008).
24. E. Zhang, J. Laufer, and P. Beard, "Backward-mode multiwavelength photoacoustic scanner using a planar Fabry-Perot polymer film ultrasound sensor for high-resolution three-dimensional imaging of biological tissues," Appl. Opt. **47**, 561–577 (2008).
25. P. Hajireza, K. Krause, M. Brett, and R. Zemp, "Glancing angle deposited nanostructured film Fabry-Perot etalons for optical detection of ultrasound," Opt. Express **21**, 6391–6400 (2013).
26. S. Preisser, W. Rohringer, M. Liu, C. Kollmann, S. Zotter, B. Fischer, and W. Drexler, "All-optical highly sensitive akinetic sensor for ultrasound detection and photoacoustic imaging," Biomed. Opt. Express **7**, 4171–4186 (2016).
27. J. A. Guggenheim, J. Li, T. J. Allen, R. J. Colchester, S. Noimark, O. Ogunlade, I. P. Parkin, I. Papakonstantinou, A. E. Desjardins, E. Z. Zhang, and P. C. Beard, "Ultrasensitive plano-concave optical microresonators for ultrasound sensing," Nature Photonics **11**, 714–719 (2017).
28. A. Rosenthal, D. Razansky, and V. Ntziachristos, "High-sensitivity compact ultrasonic detector based on a pi-phase-shifted fiber Bragg grating," Opt. Lett. **36**, 1833–1835 (2011).
29. Q. Wu and Y. Okabe, "High-sensitivity ultrasonic phase-shifted fiber Bragg grating balanced sensing system," Opt. Express **20**, 28353–28362 (2012).
30. A. Rosenthal, S. Kellnberger, D. Bozhko, A. Chekkoury, M. Omar, D. Razansky, and V. Ntziachristos, "Sensitive interferometric detection of ultrasound for minimally invasive clinical imaging applications," Laser & Photonics Reviews **8**, 450–457 (2014).
31. G. Wissmeyer, D. Soliman, R. Shnaiderman, A. Rosenthal, and V. Ntziachristos, "All-optical optoacoustic microscope based on wideband pulse interferometry," Opt. Lett. **41**, 1953–1956 (2016).
32. R. Shnaiderman, G. Wissmeyer, M. Seeger, D. Soliman, H. Estrada, D. Razansky, A. Rosenthal, and V. Ntziachristos, "Fiber interferometer for hybrid optical and optoacoustic intravital microscopy," Optica **4**, 1180–1187 (2017).
33. R. Shnaiderman, G. Wissmeyer, O. Ülgen, Q. Mustafa, A. Chmyrov, and V. Ntziachristos, "A submicrometre silicon-on-insulator resonator for ultrasound detection," Nature **585**, 372–378 (2020).
34. Y. Hazan, A. Levi, M. Nagli, and A. Rosenthal, "Silicon-photonics acoustic detector for optoacoustic micro-tomography," Nature Communications **13**, 1488 (2022).
35. H. Li, B. Dong, Z. Zhang, H. F. Zhang, and C. Sun, "A transparent broadband ultrasonic detector based on an optical micro-ring resonator for photoacoustic microscopy," Scientific Reports **4**, 4496 (2014).
36. A. Maxwell, S. W. Huang, T. Ling, J. S. Kim, S. Ashkenazi, and L. J. Guo, "Polymer Microring Resonators for High-Frequency Ultrasound Detection and Imaging," IEEE Journal of Selected Topics in Quantum Electronics **14**, 191–197 (2008).
37. C. y. Chao, S. Ashkenazi, S. w. Huang, M. O. Donnell, and L. J. Guo, "High-frequency ultrasound sensors using polymer microring resonators," IEEE Transactions on Ultrasonics, Ferroelectrics, and Frequency Control **54**, 957–965 (2007).
38. J. D. Hamilton, T. Buma, M. Spisar, and M. O. Donnell, "High frequency optoacoustic arrays using etalon detection," IEEE Transactions on Ultrasonics, Ferroelectrics, and Frequency Control **47**, 160–169 (2000).
39. S.-W. Huang, S.-L. Chen, T. Ling, A. Maxwell, M. O'Donnell, L. J. Guo, and S. Ashkenazi, "Low-noise wideband ultrasound detection using polymer microring resonators," Applied Physics Letters **92**, 193509 (2008).

40. T. Ling, S.-L. Chen, and L. J. Guo, “High-sensitivity and wide-directivity ultrasound detection using high Q polymer microring resonators,” Applied Physics Letters **98**, 204103 (2011).
41. C. Zhang, S. L. Chen, T. Ling, and L. J. Guo, “Review of Imprinted Polymer Microrings as Ultrasound Detectors: Design, Fabrication, and Characterization,” IEEE Sensors Journal **15**, 3241–3248 (2015).
42. W. J. Westerveld, M. Mahmud-Ul-Hasan, R. Shnaiderman, V. Ntziachristos, X. Rottenberg, S. Severi, and V. Rochus, “Sensitive, small, broadband and scalable optomechanical ultrasound sensor in silicon photonics,” Nature Photonics **15**, 341–345 (2021).
43. J. Pan, Q. Li, Y. Feng, R. Zhong, Z. Fu, S. Yang, W. Sun, B. Zhang, Q. Sui, J. Chen, Y. Shen, and Z. Li, “Parallel interrogation of the chalcogenide-based micro-ring sensor array for photoacoustic tomography,” Nat. Commun. **14**, 3250 (2023).
44. T. Ling, S.-L. Chen, and L. J. Guo, “Fabrication and characterization of High Q polymer micro-ring resonator and its application as a sensitive ultrasonic detector,” Opt. Express **19**, 861–869 (2011).
45. Z. Xie, S.-L. Chen, T. Ling, L. J. Guo, P. L. Carson, and X. Wang, “Pure optical photoacoustic microscopy,” Opt. Express **19**, 9027–9034 (2011).
46. C. Zhang, T. Ling, S.-L. Chen, and L. J. Guo, “Ultrabroad bandwidth and highly sensitive optical ultrasonic detector for photoacoustic imaging,” Acs Photonics **1**, 1093–1098 (2014).
47. G. A. Cranch, P. J. Nash, and C. K. Kirkendall, “Large-scale remotely interrogated arrays of fiber-optic interferometric sensors for underwater acoustic applications,” IEEE Sensors Journal **3**, 19–30 (2003).
48. H. Gabai, I. Steinberg, and A. Eyal, “Multiplexing of fiber-optic ultrasound sensors via swept frequency interferometry,” Opt. Express **23**, 18915–18924 (2015).
49. J. Bauer-Marschallinger, K. Felbermayer, and T. Berer, “All-optical photoacoustic projection imaging,” Biomed. Opt. Express **8**, 3938–3951 (2017).
50. Y. Hazan and A. Rosenthal, “Simultaneous multi-channel ultrasound detection via phase modulated pulse interferometry,” Opt. Express **27**, 28844–28854 (2019).
51. Y. Bao, X. Yi, Z. Li, Q. Chen, J. Li, X. Fan, and X. Zhang, “A digitally generated ultrafine optical frequency comb for spectral measurements with 0.01-pm resolution and 0.7-μs response time,” Light: Science & Applications **4**, e300–e300 (2015).
52. J. Pan, B. Zhang, Z. Liu, J. Zhao, Y. Feng, L. Wan, and Z. Li, “Microbubble resonators combined with a digital optical frequency comb for high-precision air-coupled ultrasound detectors,” Photon. Res. **8**, 303–310 (2020).
53. B. Dong, S. Chen, Z. Zhang, C. Sun, and H. F. Zhang, “Photoacoustic probe using a microring resonator ultrasonic sensor for endoscopic applications,” Opt. Lett. **39**, 4372–4375 (2014).
54. R. Ansari, E. Z. Zhang, A. E. Desjardins, and P. C. Beard, “All-optical forward-viewing photoacoustic probe for high-resolution 3D endoscopy,” Light: Science & Applications **7**, 75 (2018).
55. B. Dong, H. Li, Z. Zhang, K. Zhang, S. Chen, C. Sun, and H. F. Zhang, “Isometric multimodal photoacoustic microscopy based on optically transparent micro-ring ultrasonic detection,” Optica **2**, 169–176 (2015).
56. H. Li, B. Dong, X. Zhang, X. Shu, X. Chen, R. Hai, D. A. Czaplewski, H. F. Zhang, and C. Sun, “Disposable ultrasound-sensing chronic cranial window by soft nanoimprinting lithography,” Nature Communications **10**, 4277 (2019).
57. Q. Rong, Y. Lee, Y. Tang, T. Vu, C. Taboada, W. Zheng, J. Xia, D. A. Czaplewski, H. F. Zhang, C. Sun, and J. Yao, “High-Frequency 3D Photoacoustic Computed Tomography Using an Optical Microring Resonator,” BME Frontiers **2022**(2022).
58. K. S. Valluru, K. E. Wilson, and J. K. Willmann, “Photoacoustic imaging in oncology: translational preclinical and early clinical experience,” Radiology **280**, 332–349 (2016).
59. S. Zackrisson, S. Van De Ven, and S. Gambhir, “Light in and Sound Out: Emerging Translational Strategies for Photoacoustic Imagingtranslational Strategies for Photoacoustic Imaging,” Cancer research **74**, 979–1004 (2014).

60. B. Zabihian, J. Weingast, M. Liu, E. Zhang, P. Beard, H. Pehamberger, W. Drexler, and B. Hermann, "In vivo dual-modality photoacoustic and optical coherence tomography imaging of human dermatological pathologies," Biomed. Opt. Express **6**, 3163–3178 (2015).
61. A. P. Jathoul, J. Laufer, O. Ogunlade, B. Treeby, B. Cox, E. Zhang, P. Johnson, A. R. Pizzey, B. Philip, T. Marafioti, M. F. Lythgoe, R. B. Pedley, M. A. Pule, and P. Beard, "Deep in vivo photoacoustic imaging of mammalian tissues using a tyrosinase-based genetic reporter," Nat. Photon. **9**, 239–246 (2015).
62. T. Zhao, T. T. Pham, C. Baker, M. T. Ma, S. Ourselin, T. Vercauteren, E. Zhang, P. C. Beard, and W. Xia, "Ultrathin, high-speed, all-optical photoacoustic endomicroscopy probe for guiding minimally invasive surgery," Biomed. Opt. Express, **13**, 4414–4428 (2022).

Chapter 11
Protoacoustics and Its Applications in Radiation Oncology

Kaitlyn Kim and Shawn (Liangzhong) Xiang

Abstract Protoacoustics, also known as ionoacoustics, is a rapidly evolving area in medical imaging technology, combining thermoacoustic principles with proton therapy. This technology has attracted significant interest in the scientific and clinical communities. Protoacoustics relies on the interaction between protons and tissues during radiation therapy, generating acoustic waves that can be captured and transformed into images. These images provide valuable data on the proton beam–tissue interaction, enabling real-time range verification during proton therapy. This chapter discusses the theoretical foundations, instrumentation development, software advancements, and potential applications of protoacoustics, such as Bragg peak localization and in vivo dosimetry. Despite some technical and regulatory challenges, protoacoustics holds great promise for revolutionizing cancer imaging and therapy with improved precision and effectiveness.

Keywords Protoacoustic imaging · Radiation oncology · Bragg peak localization · In Vivo dosimetry · FLASH therapy

11.1 Introduction

Protoacoustics, also known as ionoacoustics, is a novel and rapidly evolving area in medical imaging technology. Recognized for its unique capabilities, protoacoustics unites the principles of thermoacoustic imaging with cutting-edge proton therapy.

K. Kim
Department of Biomedical Engineering, University of California, Irvine, CA, USA

S. (Liangzhong) Xiang (✉)
Department of Biomedical Engineering, University of California, Irvine, CA, USA

Beckman Laser Institute & Medical Clinic, University of California, Irvine, CA, USA

Department of Radiological Sciences, University of California, Irvine, CA, USA
e-mail: liangzhx@hs.uci.edu

W. Xia (ed.), *Biomedical Photoacoustics*,
https://doi.org/10.1007/978-3-031-61411-8_11

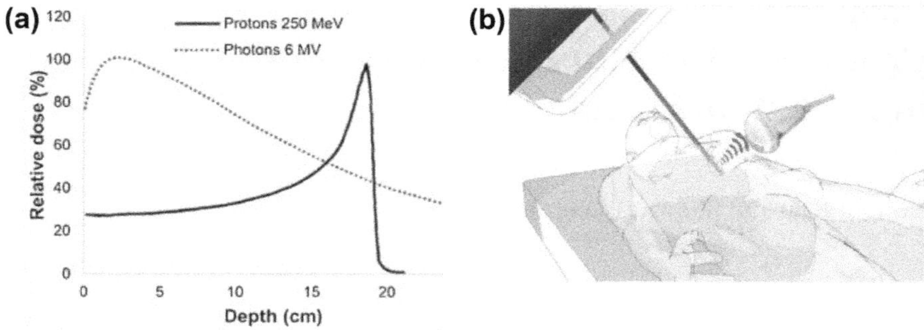

Fig. 11.1 Schematic figures of proton therapy and how protoacoustic imaging can be implemented in clinic settings. (**a**) Proton dose deposition curve compared to a photon beam. Compared to photons, proton beams have a distinct peak known as the Bragg peak, where a majority of the radiation dose stops just after the peak. (**b**) General schematic of how protoacoustics can be applied in a clinical setting. The proton beam irradiating the patient emits acoustic waves, which are detected by the ultrasound transducer. (Figure courtesy of Caron et al. [4])

The potential for this technology is attracting a burgeoning interest within the scientific community and clinical practice.

As proton beams travel and reach the end of their range, they produce a defined peak, better known as the Bragg peak (Fig. 11.1a). If the peak is aligned to the target tumor location, the maximum amount of dose can reach the tumor without severely affecting surrounding organs. After reaching this Bragg peak, the proton dose drastically decreases instantly, also known as dose fall-off, and this allows for potential reduction in toxicity to surrounding organs [10]. Although the Bragg peak allows for higher dose delivery directly to the tumor, this elusive trait is barricaded by insufficient knowledge of its location [11]. Thus, there is a need for modalities that are able to locate the Bragg peak and the amount of dose deposited into the location to ensure treatment accuracy.

The operational principle of protoacoustics hinges on the interaction between protons and human tissues during radiation therapy. When protons are propelled into the body, they decelerate and deposit their energy, prompting the tissues to heat and subsequently expand [15, 29]. The resultant expansion triggers the emission of acoustic waves that can be captured by ultrasound sensors and transformed into images (Fig. 11.1b). Such images deliver invaluable data on the location and extent of the interaction between the proton beam and the tissues, effectively serving as real-time range verification during proton therapy. This imaging approach holds immense significance in cancer therapies to guarantee accurate targeting and to minimize damage to healthy tissues.

Historically, efforts to harness protoacoustics for clinical application have been ongoing since 1979 [21]. However, after some initial advancements in the 1980s, the field observed a period of stagnation until a renewed wave of interest emerged around 2015 [15, 22]. This resurgence was stimulated by both the technological advancements in ultrasound transducer technology and the growing sophistication of pulsed proton sources. As a result, protoacoustics is now poised on the brink of the clinical realm, promising a new era in cancer imaging and therapy.

11.1.1 Theory Behind Protoacoustics

The fundamentals of protoacoustics can be described by the thermoacoustic effect, a principle that states that when an energy source irradiates a target, pressure waves are emitted. In the case of proton beam irradiation, this generated pulse is more significant in the Bragg peak location. The propagation of the protoacoustic waves can be described by the thermoacoustic wave equation [28]:

$$\nabla^2 p(\vec{r},t) - \frac{1}{{v_s}^2}\frac{\partial^2}{\partial t^2}p(\vec{r},t) = -\frac{\beta}{C_p}\frac{\partial}{\partial t}H(\vec{r},t) \tag{11.1}$$

where *H* ($r^{\rightarrow}$,*t*) is the heat deposition at point $\vec{r}$ and time t, v_s is the speed of sound in the medium, β is the thermal expansion coefficient, and C_p is the specific heat capacity. This pressure wave equation can be further simplified by assuming that each individual proton pulse will deposit energy instantaneously. The solution to Eq. (11.1) due to delta pulse excitation is given as:

$$P_G(\vec{r},t) = \frac{\beta}{4\pi C_p}\frac{\partial}{\partial t}\left(\frac{1}{ct}\int_{S(\vec{r},t)} H\left(\vec{r}'\right)dS'(t)\right); \left|\vec{r}-\vec{r}'\right| = vt, (\vec{r},t) \tag{11.2}$$

where $S'(t)$ represents a time-dependent spherical surface at the center of the detector location $(\vec{r})$ such that $\left|\vec{r}-\vec{r}'\right| = ct$. The time-of-flight information with proton-induced acoustic pulses can be used to localize the Bragg Peak, which can improve the precision of proton therapy for cancer patients.

For each individual proton pulse, the dose can be related to the acoustic pressure through the following equation:

$$P_0 = \Gamma \rho D \tag{11.3}$$

where Γ represents the Gruneisen coefficient, a dimensionless parameter, and ρ is the density of water. This relationship is a key characteristic of measured protoacoustic amplitude *P* and the radiation dose deposition, *D,* allowing for dose measurements in vivo on patients [1, 5].

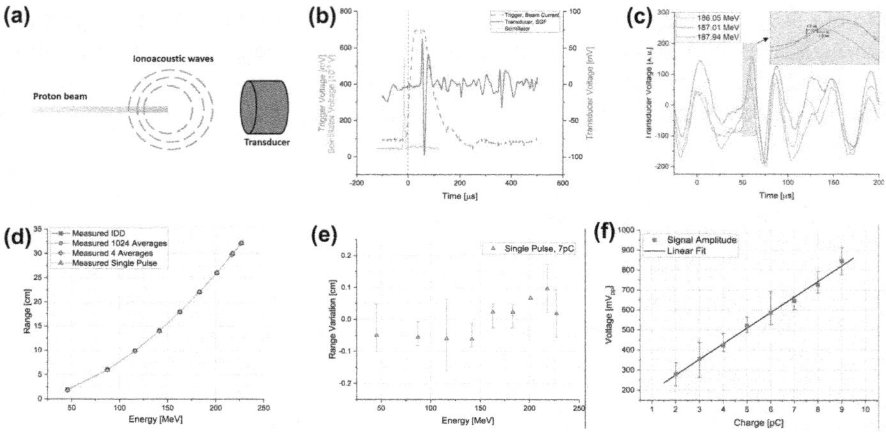

Fig. 11.2 A-line protoacoustic signal measurement can be used for localizing the Bragg peak. (**a**) General setup for protoacoustic measurements using a single-point transducer. (**b**) Typical waveform seen from ionoacoustic measurements with a single-point transducer alongside the trigger signal and the beam current trigger from the scintillator. (**c**) Resolution of the single point measurements with varying beam energies. (**d**) Comparison of various measurement averages against the measured ionization depth dose (IDD) when observing the relationship between energy and range. (**e**) Deviation in the single pulse measurements across various energies. (**f**) Display of the linear relationship between the single voltage charge depositions. (Figure courtesy of Caron et al. [4])

11.2 Instrumentation Development

The initial research into protoacoustic measurements was conducted using single ultrasound detectors (Fig. 11.2a), primarily due to their affordability and ease of use [22]. Mechanical scanning of the transducer was necessary to construct the required images (Fig. 11.4a, b). As the field advanced, ultrasound arrays were introduced into the domain of protoacoustic imaging, allowing for a more comprehensive capture of acoustic data [12, 16, 30]. Several systems employed a linear array configuration, a method frequently adopted in clinical settings due to its simplicity and efficiency. More advanced systems, however, utilized a matrix array to enable three-dimensional imaging. The use of matrix arrays marks a significant progression, offering a more detailed and multidimensional perspective of the area being studied.

11.2.1 Single Transducer-Based Detection System

Typically, a singular acoustic transducer is employed to detect A-line protoacoustic signals (Fig. 11.2b). One of the earliest experiments to verify the detectability of acoustic signals from an ion beam source was conducted by Sulak et al. in 1979 using a 150 MeV proton beam [21]. Acoustic signals were detected within a water medium by deploying two hydrophones. One hydrophone captured signals ranging from 1 kHz to 150 kHz, while the other extended its detection up to 1 MHz. This groundbreaking study confirmed the feasibility of acoustic signal detection from a proton source, laying the foundation for future research. Again, in 1991, the successful detection of acoustic signals from a proton source was demonstrated using a 300 kHz hydrophone [22]. In these cases, hydrophones served as the primary tools for acoustic signal measurements, becoming a kind of benchmark in the field [13]. However, these tools require precise alignment for high-signal capture and often suffer from a low signal-to-noise ratio (SNR). Sueysa et al. [20] further refined this technology by developing an optical hydrophone that leverages laser interferometry to detect low-pressure signals from the proton beam [20]. This innovative approach addressed the issue of signal reduction due to misalignment and improved SNR when compared to the conventional hydrophone. Conventional ultrasound transducers, usually composed of either lead zirconate titanate (PZT) or polyvinylidene fluoride (PVDF), are also being used for protoacoustic signal detection. Protoacoustic signals can be utilized to localize the Bragg peak at different proton energies (Fig. 11.2c). In the study by Assmann et al. [2], PZT-based transducers were used to measure single proton pulses at 3.5 MHz and 10 MHz frequencies [2]. However, a low SNR remains a challenge with this technique. Meanwhile, Lascaud et al. utilized PVDF transducers to detect pulsed proton signals as low as 100 kHz, with tungsten serving as the backing of the transducer to enhance sensitivity [9]. Comparatively, PZT transducers exhibit a higher detection sensitivity than PVDF-based ones, though bandwidth becomes a limiting factor for the former. In contrast, Lascaud et al.'s [8] research showcases that capacitive micromachined ultrasonic transducers (CMUTs) are capable of a wide bandwidth without degrading or attenuating the proton signal [8]. Their 12 MHz CMUT prototype could detect signals within a range of several hundred kHz to a few MHz. It is clear that continued research is necessary to identify an acoustic transducer type that amalgamates both a broad bandwidth and high detection sensitivity, making significant strides toward optimal protoacoustic detection.

11.2.2 Array Transducers

The detection methodologies discussed so far have primarily relied on single-point measurements. While these detectors carry certain advantages, their use is somewhat constrained without a translational stage for mechanical movement. This limitation hinders the creation of images, confining them to single A-line measurements and impeding real-time 3D in vivo dosimetry with a single transducer. Recognizing this shortcoming, there has been a growing momentum toward employing transducer arrays for detecting protoacoustic waves (Fig. 11.4d). By leveraging an array, real-time in vivo proton dosimetry becomes feasible, and a 3D image of the beam's dose deposition can be reconstructed (Fig. 11.4e–g). Ionizing Radiation

Acoustic Imaging (iRAI), an approach that mirrors the principles of protoacoustics by detecting thermoacoustic waves generated from an electron or x-ray source, has experimentally succeeded in using a 2D matrix array [30]. In the realm of protoacoustics, experimental studies utilizing array transducers have not yet been published. However, numerous simulation studies have hinted at encouraging prospects. One such study showcased a 3D protoacoustic imaging workflow using a matrix array [17]. In this case, dose reconstruction was achieved using the array, and frequency characteristics from the protoacoustic signals were identified. With the advancements in transducer array technology, protoacoustic imaging is poised to take significant leaps in its journey toward 3D imaging and real-time in vivo dosimetry, which could prove revolutionary in cancer treatment.

11.3 Software Development

Given the low signal-to-noise ratio (SNR) inherent in protoacoustic measurements, signal processing plays a crucial role in noise reduction. Various techniques have been investigated to enhance the SNR, including the use of low-pass filters and wavelet denoising filters, among others. In addition, the development of image reconstruction algorithms has been pivotal for not only localizing the Bragg peak but also quantifying the radiation dose. These advancements in signal processing and image reconstruction are instrumental in harnessing the full potential of protoacoustic imaging, thus improving the precision and effectiveness of radiation therapy.

11.3.1 Signal Processing

The processing of protoacoustic signals necessitates specific procedures like averaging and filtering, which are essential to enhancing the signal-to-noise ratio (SNR). Commonly, digital filters such as band-pass filters are employed for this purpose [26]. Wavelet filtering has also been utilized as an effective tool for boosting the SNR (Fig. 11.3a) [19, 24]. Some studies have highlighted its usefulness in the context of protoacoustic signal processing, showcasing its potential in the field. Moreover, some research has incorporated the use of Savitzky-Golay filters [4]. The application of these filters has proven to significantly elevate the SNR, particularly for single pulse protoacoustic measurements. The advantages brought about by this technique underline the need for further exploration of diverse filtering methods to maximize signal clarity and accuracy in protoacoustic imaging.

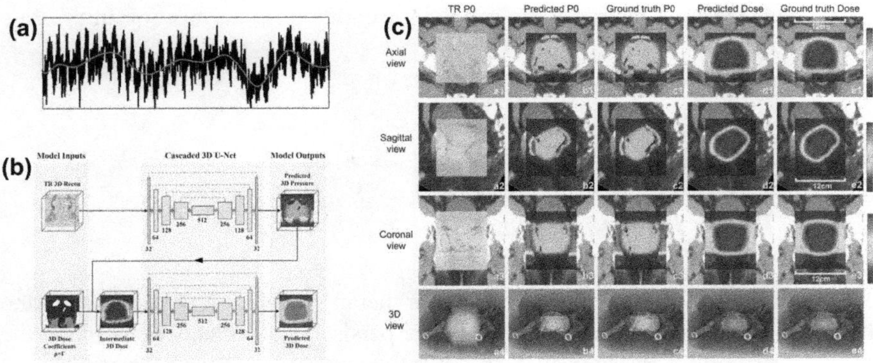

Fig. 11.3 Signal processing and image reconstruction are vital components of protoacoustic imaging. (**a**) Wavelet filtering has been used to improve the SNR of the protoacoustic signals. Figure courtesy of Bergen et al. [3] (**b**) Workflow for the proposed deep cascaded convolutional neural network (CNN) to reconstruct pressure and dose maps using a matrix ultrasound array. Numbers represent the data channels. Figure courtesy of Jiang et al. (**c**) Various slices of the pressure and dose maps from different views. Figure courtesy of Jiang et al. [6]

11.3.2 Reconstruction Algorithms

In protoacoustics, the visualization of the Bragg peak location and associated dose information is reliant on the vital process of reconstruction. Various reconstruction algorithms have been developed, drawing inspiration from photoacoustic reconstruction [27]. Kellenberger et al. [7] employed a modified backprojection algorithm and model-based reconstruction to obtain protoacoustic images, effectively showcasing the maximum intensity of the proton beam [7]. Furthermore, advancements in 3D dose distribution reconstruction algorithms tailored for specific types of machines, such as isochronous cyclotrons and synchrocyclotrons, have been made [25]. An innovative approach involves numerical time-reversal, which uses transducer arrays to approximate thermoacoustic waves and reconstruct the acoustic source amplitude [12]. However, most of the reconstruction algorithms for protoacoustic imaging are limited by the ultrasound transducer's restricted viewing angle. To address this issue, a deep learning-based method utilizing a deep cascaded convolutional neural network (DC-CNN) has been developed (Fig. 11.3b) [6]. Validation of the DC-CNN method involved 81 prostate cancer patients' treatment plans, where dose calculations and PA simulations were performed using commercial software and the open-source k-wave package, respectively. The proposed method was trained and tested on patient samples, with the reconstructed pressures and dose maps compared to ground truth data using various metrics (Fig. 11.3c). Results demonstrated significant improvements in limited-view PA image quality, with clear and accurate pressure structures and high agreement between derived doses and ground truth data. This approach aims to reconstruct high-quality radiation-induced pressures and derive precise 3D dosimetry from PA signals detected by a matrix array, enabling accurate dose verification in prostate proton therapy.

11.4 Application

11.4.1 Bragg Peak Localization

One of the main advantages of proton beam therapy lies in its unique feature, the Bragg peak. However, a significant challenge persists in accurately pinpointing the exact location of this Bragg peak. Consequently, numerous groups have endeavored to localize the Bragg peak, aiming to confirm its position within a sub-millimeter range. Assmann and colleagues, for instance, successfully gauged the Bragg peak position of a 20 MeV proton beam to an accuracy of less than 100 micrometers [2]. This achievement was made possible by chopping the proton beam into pulse widths of 8 ns–4.3 μs and measuring the results with a 3.5 MHz PZT transducer. Several other studies have reported the submillimeter range verification of higher energy proton beams reaching up to 220 MeV (Fig. 11.2d, e) [18]. This task typically presents difficulties due to the low SNR associated with higher energy proton beams. One solution to this challenge involves filtering the signals through simulated templates. There is also an ongoing exploration into the application of Spherical Ionoacoustic waves with Resonant frequency (SPIREs) [23]. Experiments have shown that spherical gold markers in tumors can emit these SPIRE waves. The amplitude of these waves can be linearly correlated with the residual beam range of the marker position. This remarkable characteristic presents the opportunity to diminish range uncertainty, thereby offering a more accurate determination of the Bragg peak's location.

11.4.2 In Vivo Dosimetry

In radiation therapy, in vivo dosimetry is a crucial aspect that clinicians often prioritize [14]. A significant finding in this regard was the measurement of the peak-to-peak voltage of various signals, followed by calculating the mean and standard deviation for each charge deposition set (Fig. 11.2f) [4]. The data were then fitted with a linear curve, resulting in an R-squared value of 0.994, indicating a strong fit. This dosimetric data suggests a linear relationship between the delivered charge and the signal's amplitude. This implies that it is possible to assess the deposited dose in the medium by relating it to the deposited charge. This characteristic offers an advantage over other dosimetry techniques as it allows for monitoring the actual dose delivered to the patient. Consequently, treatment plans can be adjusted based on these results, leading to a more personalized and efficient approach to treatment. Protoacoustics, which utilizes thermoacoustic waves emitted from the proton beam during therapy, proves to be effective in facilitating in vivo dosimetry. The implementation of this advanced technique has the potential to usher in a new era in radiation therapy, enhancing treatment precision and effectiveness.

11.5 Outlook

As a promising imaging modality, protoacoustics holds significant potential to transform the landscape of biomedicine, especially in the domain of cancer therapeutics. One exciting prospect is the possibility of realizing three-dimensional (3D) in vivo dosimetry using a two-dimensional (2D) matrix array. This could significantly improve the precision of proton therapy by providing real-time monitoring and verification of the radiation dose being delivered to the patient's tumor.

Furthermore, the development of a dual-modal imaging system would enable protoacoustic imaging to be integrated with other imaging modalities, such as planning CT scans. The incorporation of planning CT scans would provide a crucial means to correct the sound speed in heterogeneous tissues, thus refining the accuracy of protoacoustic imaging. With the integration of pulse-echo ultrasound (Fig. 11.4c), real-time tumor motion management can be achieved, akin to how it is done with MRI, thereby enhancing the adaptability and efficacy of proton therapy.

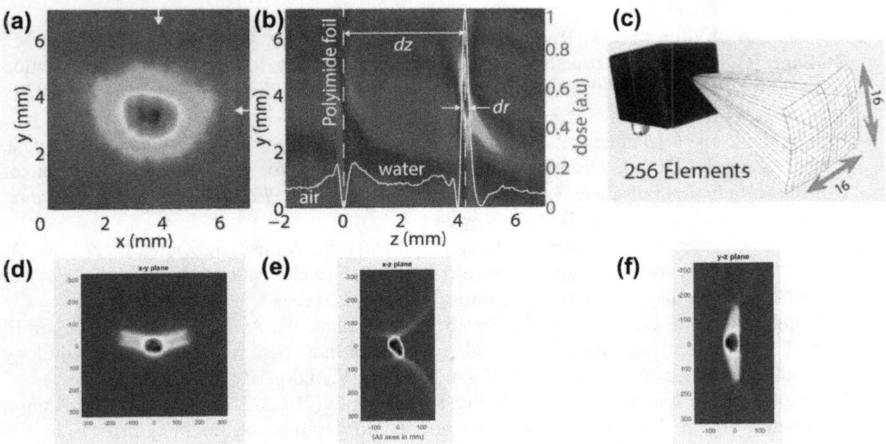

Fig. 11.4 Protoacoustic imaging can be used for in vivo range verification during proton therapy. (**a**) 3D reconstruction of a proton beam, displayed at the maximum intensity projection (MIP). (Figure courtesy of Kellenberger et al. [8]). (**b**) YZ-slice of the reconstruction with the MIP shown. (**c**) Setup of a 16 × 16 transducer array that is capable of 3D protoacoustic imaging. (Figure courtesy of Samant et al). (**d**) Reconstruction of a simulated proton source using the 16 × 16 planar array from Fig. 4d along the x–y plane, (**e**) x–z plane, and (**f**) y–z plane. (Figure courtesy of Samant et al. [18])

However, the integration of protoacoustic imaging into clinical practice would require navigating through several technical and regulatory hurdles. From a technical standpoint, the hardware used in protoacoustic imaging systems would need to be optimized to provide high-quality, real-time images that can guide clinical decision-making. Similarly, software algorithms used for image reconstruction and signal processing would need to be fine-tuned to handle the complexities of real-world clinical data.

From a regulatory perspective, protoacoustic imaging systems would need to undergo rigorous testing to demonstrate their safety, efficacy, and reliability in a variety of clinical scenarios. Moreover, the costs associated with developing, implementing, and maintaining these systems would need to be justified by demonstrating significant improvements in patient outcomes.

Nevertheless, with ongoing advancements in technology and increasing clinical recognition of the potential benefits of proton therapy, the prospects for the integration of protoacoustic imaging into clinical practice remain bright. Future research will undoubtedly focus on addressing these challenges and exploring the full range of possibilities that protoacoustics can bring to the field of biomedicine.

References

1. Ahmad, M., Xiang, L., Yousefi, S., & Xing, L. (2015). Theoretical detection threshold of the proton-acoustic range verification technique. *Medical Physics*, *42*(10), 5735–5744. https://doi.org/10.1118/1.4929939
2. Assmann, W., Kellnberger, S., Reinhardt, S., Lehrack, S., Edlich, A., Thirolf, P. G., Moser, M., Dollinger, G., Omar, M., Ntziachristos, V., & Parodi, K. (2015). Ionoacoustic characterization of the proton Bragg peak with submillimeter accuracy. *Medical Physics*, *42*(2), 567–574. https://doi.org/10.1118/1.4905047
3. Bergen, R., Sun, L., Pandey, P., Wang, S., Bjegovic, K., Gonzalez, G., Chen, Y., Lopata, R., & Xiang, L. (2023). Discrete Wavelet Transformation for the Sensitive Detection of Ultrashort Radiation Pulse with Radiation-induced Acoustics. *IEEE Transactions on Radiation and Plasma Medical Sciences* [Underr Review]
4. Caron, J., Gonzalez, G., Pandey, P. K., Wang, S., Prather, K., Ahmad, S., Xiang, L., & Chen, Y. (2023). Single pulse protoacoustic range verification using a clinical synchrocyclotron. *Physics in Medicine & Biology*, *68*(4), 045011. https://doi.org/10.1088/1361-6560/ACB2AE
5. Hickling, S., Xiang, L., Jones, K. C., Parodi, K., Assmann, W., Avery, S., Hobson, M., & El Naqa, I. (2018). Ionizing radiation-induced acoustics for radiotherapy and diagnostic radiology applications. *Medical Physics*, *45*(7), e707–e721. https://doi.org/10.1002/MP.12929
6. Jiang, Z., Sun, L., Yao, W., Wu, Q. J., Xiang, L., & Ren, L. (2022). 3D in vivo dose verification in prostate proton therapy with deep learning-based proton-acoustic imaging. *Physics in Medicine & Biology*, *67*(21), 215012. https://doi.org/10.1088/1361-6560/AC9881
7. Kellnberger, S., Assmann, W., Lehrack, S., Reinhardt, S., Thirolf, P., Queirós, D., Sergiadis, G., Dollinger, G., Parodi, K., & Ntziachristos, V. (2016). Ionoacoustic tomography of the proton Bragg peak in combination with ultrasound and optoacoustic imaging. *Scientific Reports 2016 6:1*, *6*(1), 1–7. https://doi.org/10.1038/srep29305
8. Lascaud, J., Kalunga, R., Lehrack, S., Wieser, H. P., Englbrecht, F. S., Wurl, M., Assmann, W., Savoia, A. S., & Parodi, K. (2019). Applicability of Capacitive Micromachined Ultrasonic Transducers for the detection of proton-induced thermoacoustic waves. *IEEE*

International Ultrasonics Symposium, IUS, 2019-October, 143–146. https://doi.org/10.1109/ULTSYM.2019.8926023

9. Lascaud, J., Kowalewski, R., Wollant, B., Carmigniani, H., Schnurle, K., Dash, P., Wieser, H. P., Bortfeldt, J., Kalunga, R., Rouffaud, R., Gerard, A., Vidal, M., Herault, J., Certon, D., & Parodi, K. (2021). Optimization of the backing material of a low frequency PVDF detector for ion beam monitoring during small animal proton irradiation. *IEEE International Ultrasonics Symposium, IUS*. https://doi.org/10.1109/IUS52206.2021.9593703
10. Liu, H., & Chang, J. Y. (2011). Proton therapy in clinical practice. *Chinese Journal of Cancer*, *30*(5), 315. https://doi.org/10.5732/CJC.010.10529
11. Lomax, A. J. (2020). Proton therapy special feature: Review article myths and realities of range uncertainty. *British Journal of Radiology*, *93*(1107). https://doi.org/10.1259/BJR.20190582/ASSET/IMAGES/LARGE/BJR.20190582.G008.JPEG
12. Mast, T. D., Johnstone, D. A., Dumoulin, C. L., Lamba, M. A., & Patch, S. K. (2023). Reconstruction of thermoacoustic emission sources induced by proton irradiation using numerical time reversal. *Physics in Medicine & Biology*, *68*(2), 025003. https://doi.org/10.1088/1361-6560/ACABFC
13. Nie, W., Jones, K. C., Petro, S., Kassaee, A., Sehgal, C. M., & Avery, S. (2018). Proton range verification in homogeneous materials through acoustic measurements. *Physics in Medicine & Biology*, *63*(2), 025036. https://doi.org/10.1088/1361-6560/AA9C1F
14. Olaciregui-Ruiz, I., Beddar, S., Greer, P., Jornet, N., McCurdy, B., Paiva-Fonseca, G., Mijnheer, B., & Verhaegen, F. (2020). In vivo dosimetry in external beam photon radiotherapy: Requirements and future directions for research, development, and clinical practice. *Physics and Imaging in Radiation Oncology*, *15*, 108. https://doi.org/10.1016/J.PHRO.2020.08.003
15. Parodi, K., & Assmann, W. (2015). Ionoacoustics: A new direct method for range verification. *30*(17), 1540025. https://doi.org/10.1142/S0217732315400258
16. Patch, S. K., Nguyen, C., Dominguez-Ramirez, D., Labarbe, R., Janssens, G., Cammarano, D., Lister, J., Finch, C., Lambert, J., Pandey, J., Bennett, C., Porteous, E., Chirvase, C., Cohilis, M., Souris, K., Ono, S., & Lynch, T. (2021). Thermoacoustic range verification during pencil beam delivery of a clinical plan to an abdominal imaging phantom. *Radiotherapy and Oncology*, *159*, 224–230. https://doi.org/10.1016/J.RADONC.2021.03.027
17. Samant, P., Trevisi, L. M., Chen, Y., Zwart, T., & Xiang, L. (2022). 3-D Protoacoustic Imaging Through a Planar Ultrasound Array: A Simulation Workflow. *IEEE Transactions on Radiation and Plasma Medical Sciences*, *7*(1), 83–95. https://doi.org/10.1109/TRPMS.2022.3177236
18. Schauer, J., Wieser, H. P., Huang, Y., Ruser, H., Lascaud, J., Würl, M., Chmyrov, A., Vidal, M., Herault, J., Ntziachristos, V., Assmann, W., Parodi, K., & Dollinger, G. (2022). Proton beam range verification by means of ionoacoustic measurements at clinically relevant doses using a correlation-based evaluation. *Frontiers in Oncology*, *12*, 925542. https://doi.org/10.3389/FONC.2022.925542/BIBTEX
19. Sohn, J., Nie, W., Avery, S., He, X., Zhou, J., James Sohn, al, Avery, S. M., Yang, X., Liu, T., & Sohn, J. J. (2020). Wavelet-based protoacoustic signal denoising for proton range verification. *11319*(16), 14–19. https://doi.org/10.1117/12.2550768
20. Sueyasu, S., Takayanagi, T., Miyazaki, K., Kuriyama, Y., Ishi, Y., Uesugi, T., Unlu, M. B., Kudo, N., Chen, Y., Kasamatsu, K., Fujii, M., Kobayashi, M., Rohringer, W., & Matsuura, T. (2023). Ionoacoustic application of an optical hydrophone to detect proton beam range in water. *Medical Physics*, *50*(4), 2438–2449. https://doi.org/10.1002/MP.16189
21. Sulak, L., Armstrong, T., Baranger, H., Bregman, M., Levi, M., Mael, D., Strait, J., Bowen, T., Pifer, A. E., Polakos, P. A., Bradner, H., Parvulescu, A., Jones, W. V., & Learned, J. (1979). Experimental studies of the acoustic signature of proton beams traversing fluid media. *Nuclear Instruments and Methods*, *161*(2), 203–217. https://doi.org/10.1016/0029-554X(79)90386-0
22. Tada, J., Hayakawa, Y., Hosono, K., & Inada, T. (1991). Time resolved properties of acoustic pulses generated in water and in soft tissue by pulsed proton beam irradiation—A possibility of doses distribution monitoring in proton radiation therapy. *Medical Physics*, *18*(6), 1100–1104. https://doi.org/10.1118/1.596618

23. Takayanagi, T., Uesaka, T., Nakamura, Y., Unlu, M. B., Kuriyama, Y., Uesugi, T., Ishi, Y., Kudo, N., Kobayashi, M., Umegaki, K., Tomioka, S., & Matsuura, T. (2020). On-line range verification for proton beam therapy using spherical ionoacoustic waves with resonant frequency. *Scientific Reports 2020 10:1*, *10*(1), 1–10. https://doi.org/10.1038/s41598-020-77422-2
24. Vallicelli, E. A., Corona, M., Dell'acqua, M., Baschirotto, A., & Matteis, M. De. (2021). Denoising for Enhancing Signal-to-Noise Ratio in Proton Sound Detectors. *BioCAS 2021—IEEE Biomedical Circuits and Systems Conference, Proceedings*. https://doi.org/10.1109/BIOCAS49922.2021.9644987
25. Van Dongen, K. W. A., De Blécourt, A. J., Lens, E., Schaart, D. R., & Vos, F. M. (2019). Reconstructing 3D proton dose distribution using ionoacoustics. *Physics in Medicine & Biology*, *64*(22), 225005. https://doi.org/10.1088/1361-6560/AB4CD5
26. Wang, J., Sohn, J. J., Lei, Y., Nie, W., Zhou, J., Avery, S., Liu, T., & Yang, X. (2023). Deep learning-based protoacoustic signal denoising for proton range verification. *Biomedical Physics & Engineering Express*, *9*(4), 045006. https://doi.org/10.1088/2057-1976/ACD257
27. Xu, M., & Wang, L. V. (2005). Universal back-projection algorithm for photoacoustic computed tomography. *Physical Review E—Statistical, Nonlinear, and Soft Matter Physics*, *71*(1), 016706. https://doi.org/10.1103/PHYSREVE.71.016706/FIGURES/5/MEDIUM
28. Xu, M., & Wang, L. V. (2006). Photoacoustic imaging in biomedicine. *Review of Scientific Instruments*, *77*(4), 41101. https://doi.org/10.1063/1.2195024/913562
29. Yu, Y., Li, Z., Zhang, D., Xing, L., & Peng, H. (2019). Simulation studies of time reversal-based protoacoustic reconstruction for range and dose verification in proton therapy. *Medical Physics*, *46*(8), 3649–3662. https://doi.org/10.1002/MP.13661
30. Zhang, W., Oraiqat, I., Litzenberg, D., Chang, K. W., Hadley, S., Sunbul, N. B., Matuszak, M. M., Tichacek, C. J., Moros, E. G., Carson, P. L., Cuneo, K. C., Wang, X., & El Naqa, I. (2023). Real-time, volumetric imaging of radiation dose delivery deep into the liver during cancer treatment. *Nature Biotechnology 2023*, 1–8. https://doi.org/10.1038/s41587-022-01593-8

Part V
Photoacoustic Imaging Combined with Other Modalities

Chapter 12
Hybrid Photoacoustic and Laser-Induced Ultrasound Computed Tomography

David Thompson, Michael Jaeger, Damien Gasteau, and Srirang Manohar

Abstract We present a lab-built prototype for a 3D hybrid photoacoustic/laser-induced ultrasound computed tomography device for breast phantom imaging. The device, inspired by a hemispherical geometry, allows some flexibility of combining laser-induced ultrasound transmitters, light sources for photoacoustic imaging and sensitive, purpose-built ultrasound detectors in standardised mounting points. Our initial results employ tailored 1 MHz centre frequency laser-induced ultrasound transmitters to achieve sound speed mapping in a breast-sized phantom. The phantom, a two-layer breast analogue constructed using everyday lab supplies, consists of polyvinyl chloride plastisol and agarose-based tissue mimicking materials. Using transmission-mode and reflection-mode acquisition protocols, respectively, 2D maps of the sound speed and ultrasound reflectivity are reconstructed. Both sets of data compare well with the ground truth of the phantom. Simultaneously, conventional photoacoustic imaging of the phantom is also performed. A comparison of image quality in photoacoustic images using three different sound speed models, a single-speed, two-speed and full sound speed map model, shows the advantages of measuring the sound speed map, especially when imaging more complex objects such as the human breast. This proof-of-concept work shows promise for implementation with a higher density of ultrasound detection elements, where the benefits of laser-induced ultrasound in otherwise cross-talk-prone situations may become clear.

Keywords Photoacoustic tomography · Laser-induced ultrasound CT · Sound speed map · Phantom and aberration correction

D. Thompson (✉) · D. Gasteau · S. Manohar
Multi-Modality Medical Imaging group (M3I), Technical Medical Centre, University of Twente, Enschede, The Netherlands
e-mail: d.thompson@utwente.nl

M. Jaeger
Optoacoustic Imaging Group, University of Bern, Bern, Switzerland

W. Xia (ed.), *Biomedical Photoacoustics*,
https://doi.org/10.1007/978-3-031-61411-8_12

12.1 Introduction

Photoacoustic tomography (PAT) is an area of increasing interest in the biomedical imaging field in recent decades [1–5]. An important area of focus for PAT [6–11], and ultrasound computed tomography (USCT), is imaging of the human breast for the screening and diagnosis of breast cancer. The absence of bone and air, combined with the possibility of 360-degree access, makes the breast a prime target for these techniques [12, 13].

Imaging using PAT is achieved by sending short (ns) light pulses into the tissue, where they are absorbed predominantly by blood. Upon absorption of the light, the tissue heats up. Due to the thermal expansion, an ultrasonic pulse is generated, which is then detected. The promise of using photoacoustic imaging is that it can use the advantages of light such as relatively low cost and the spectroscopic potential to perform molecular imaging. The image resolution is better than in purely optical methods as the acoustic signals undergo significantly less scattering than light in biological tissues. The combination of high resolution at depth and the chemical sensitivity of the optical excitation is a powerful one, however PAT still carries some challenges. To reconstruct a photoacoustic image, the arrival times of the detected signals are assigned to a spatial location via the sound speed, which in most cases is assumed depending on the organ of interest. This assumed sound speed does not take into account the fact that different tissue types have different sound speeds, nor is the physical structure of the target organ considered. Combining with USCT can make a crucial contribution to a PAT system by enabling through-transmission sound speed mapping as well as reflection mode structural imaging [14–17]. Specifically for breasts, which can exhibit considerable variations in tissue types and structure, an accurate sound speed map may be of added value. Sound speed maps can be used in image reconstruction to calculate more accurate back projection paths, improving the PAT image reconstruction and signal-to-noise ratio (SNR).

Various breast PAT and USCT imaging prototypes have been developed and are still under development [6–13, 18] using conventional piezoelectric transducers for photoacoustic detection and ultrasound transmission and detection. An alternative way to generate ultrasound is by what is called laser-induced ultrasound (LIUS). The method uses pulsed laser light and absorption by a film or layer [19] to produce thermo-elastic expansion, in other words, the photoacoustic effect, to generate ultrasound. This ultrasound is generated to insonify the subject under investigation for ultrasound imaging. The advantages of using this approach are the ability to change the spatial characteristics of the ultrasound emission simply by changing the shape of the illumination spot, as well as to change the frequency characteristics of the ultrasound transmission conveniently by changing the optical properties of the absorbing layer [20]. The film or layer is usually made of inexpensive and abundant materials, thus giving a degree of flexibility which is more difficult to achieve using other means.

A few systems have used LIUS for ultrasound imaging in the past [21–26]. However, these have been developed for imaging relatively small structures for the application of imaging studies on mouse models of disease. In this work, we have designed and built a 3D lab prototype PAT/LIUS-CT system for large breast-sized objects. We present initial, 2D, phantom imaging results from a single imaging plane. The phantom was a 12 cm diameter, 3.2 cm height cylinder, containing 2 different materials with differing acoustic and optical properties. We imaged the phantom using PAT and LIUS-CT, then used the sound speed maps to reconstruct the PAT images. We compare the quality of PAT images using a single-speed, 2-speed, and fully mapped reconstruction, as well as looking at the sound speed images for their merit. B-mode reflection LIUS-CT images are also reconstructed to inspect the structural properties of the phantom.

12.2 Materials and Methods

12.2.1 Concept

The imaging system, depicted in Fig. 12.1, is based upon a design for a hemispherical geometry clinical breast imaging system [18]. As such, all scanning protocols described below are derived from the source and detector positioning in the clinical system. It consists of a pair of curved wedges held on two arms which are mounted on a two-axis rotation stage. The two curved wedges (30 cm radius of curvature) can be rotated independently of each other to synthesise a hemisphere, reminiscent of a hemispherical imaging bowl as used in PAT [9, 18] except inverted (Fig. 12.1a). The wedges have holes to accommodate LIUS transmitters, ultrasound detectors or optical fibre bundles for photoacoustic excitation as the case may be. The scanning arms can rotate over the full 360° range, as well as relative to one another over a 329° range limited only by the need to not collide with one another. This allows the emulation of a full circular or hemispherical array, depending on the placement of transmitters and receivers.

In this demonstration study, we used only a 2D geometry where one arm holds a single detector and the other a single LIUS transmitter. A close-up of the inner surface of one of the arms containing a single detector at an arbitrary position is shown in Fig. 12.1b, with an arrow indicating the eventual location of the transmitter or receiver in this work. Optical fibre bundles can be mounted separately to allow the addition of photoacoustic imaging. In 2D PAT configuration, only a single arm with a single detector is mounted as shown in Fig. 12.1c.

12.2.2 Scanning Protocols

Three scanning protocols were followed, one each for transmission and reflection LIUS-CT and one for PAT. Fig. 12.2a shows an overview of the transmission

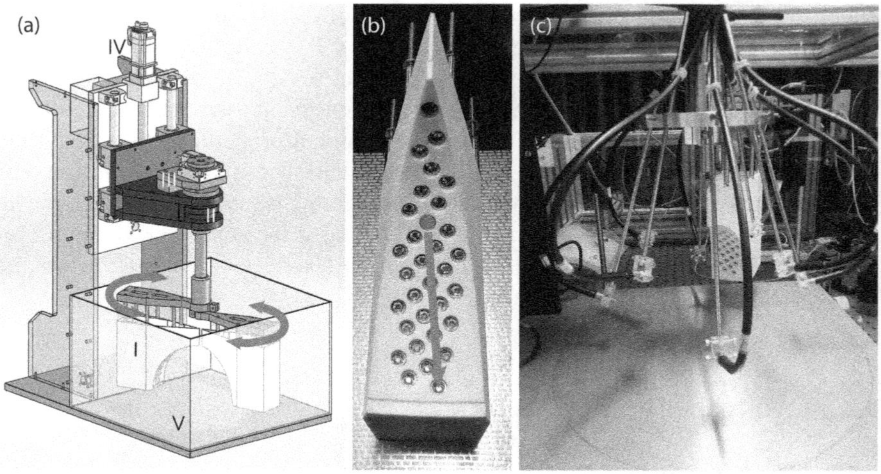

Fig. 12.1 (**a**) SolidWorks® drawing of the set-up, showing the two scanning arms (I) mounted on a two-axis rotation stage within a water tank (V). The rotation stages (II and III) are mounted on a translation stage (IV) to allow vertical motion as well. (**b**) Close-up of the inner face of one of the scanning arms, showing dummies mounted in all the holes except one, which contains a detector. The arrow indicates the mounting location eventually used for both LIUS transmitter and detector in this work. (**c**) Photograph of the set-up in the PAT mode, without the water tank for clarity. The fibre bundles are evenly spaced around the centre of the scanning volume to have illumination as uniform as possible with this bundle

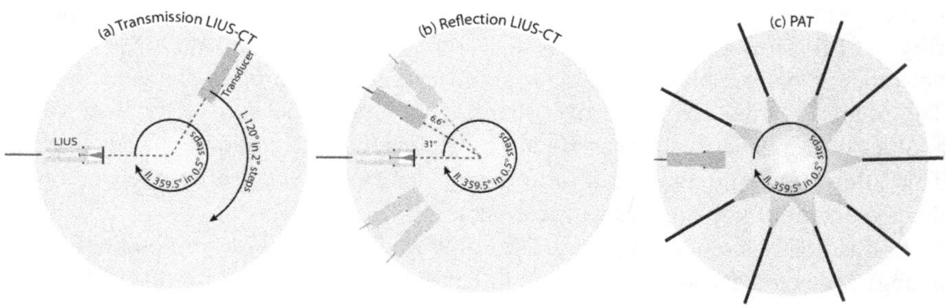

Fig. 12.2 (**a**) Transmission LIUS-CT protocol, showing the 120° scanning arc of the detector and the 359.5° scanning range of the transmitter arm. (**b**) Reflection mode LIUS-CT protocol, showing the four positions the detector takes up for a single projection and the angles between them and again the full scanning range of 359.5° is also indicated. (**c**) PAT scanning protocol, showing the illumination geometry and the scanning range

LIUS-CT acquisition protocol. For a single projection, the LIUS transmitter is kept stationary and the detector is scanned over a 120° arc on the opposite side of the phantom in 2° steps. A total of 719 such projections with a 0.5° step size are taken to cover the full 360° range. Fig. 12.2b shows the reflection mode LIUS-CT scanning protocol, where each projection sees the detector in four positions, two on either side of the LIUS transmitter. Due to the size of the arms, the minimum angular distance the transmitter can have from the detector is 31° on either side. The distance between adjacent detector positions is chosen to be 6.6° for a detector spacing of 15 mm, similar to what it would be in a full imaging bowl. Again, the angular step between projections is 0.5°, giving a total of 719 projections. The LIUS transmitters have a wide, cylindrically symmetric, emission profile which is roughly matched to the sensitivity profile of the detectors. In a 2D reflection mode configuration as used here, this leads to out-of-plane clutter due to backscattered signals originating above or below the imaging plane. To alleviate this, reflection mode scans were taken at six additional elevational positions, two below and four above the intended imaging plane, at 1.5 mm intervals covering most of the phantom. The choice to perform two scans above and four below the imaging plane was out of necessity, as the plane containing the features of interest in the phantom was located nearer to the top of the phantom than to the bottom. Taking three slices above and three below would mean the topmost slice would be mostly in water, adding little information for the reconstruction of the image. Fig. 12.2c depicts the PAT imaging protocol, where the entire assembly of the detector and nine fibre bundles are rotated over 359.5° in 0.5° steps. Out-of-plane signals were less of a problem in the PAT imaging case, as all of the strong absorbers placed in the phantom as photoacoustic sources are placed in a single plane.

12.2.3 Hardware

The main components of the set-up, in both LIUS-CT and PAT modes, are shown in the triggering schematics in Fig. 12.3. In either mode, a PC with LabVIEW (National Instruments, USA) first positions the stage or stages prior to sending the command to start an acquisition. The acquisition command is sent to a DAC system specialised for photoacoustic mammoscopy [8] (PA Imaging R&D B.V., the Netherlands). This DAC allows the triggering of a maximum of 20 acquisitions per second, where each acquisition is performed with a sampling rate of 25 MHz and a maximum duration of 163.84 μs. In the LIUS-CT mode, a 20 Hz, 5 V TTL signal is supplied to the DAC by a function generator (AFG 3102, Tektronix, USA), synchronised to the 100 Hz trigger signal required for optimal stability by the 1064 nm, 4-ns pulsed diode-pumped LIUS laser (M-Nano PR147, Montfort Laser, Austria). The 20 Hz detection trigger is delayed by 220 μs relative to the 100 Hz illumination trigger to account for the laser's trigger and Q-switch delays. The laser is coupled with the fibre of the lab-built LIUS transmitter [20], which emits an acoustic pulse with an opening angle of 45.8°. The pulse traverses the sample and is picked up by the ultrasound detector (Imasonic SAS, France). One of two detectors is used in each acquisition, one with a flat surface or one with a diverging acoustic lens built in. The flat detector with a 3 mm active element diameter has a minimum detectable pressure (MDP) of 0.23 Pa, a 1 MHz centre frequency with a 100% bandwidth, and a $30° e^{-1}$ opening angle. The lens-equipped detector has a larger

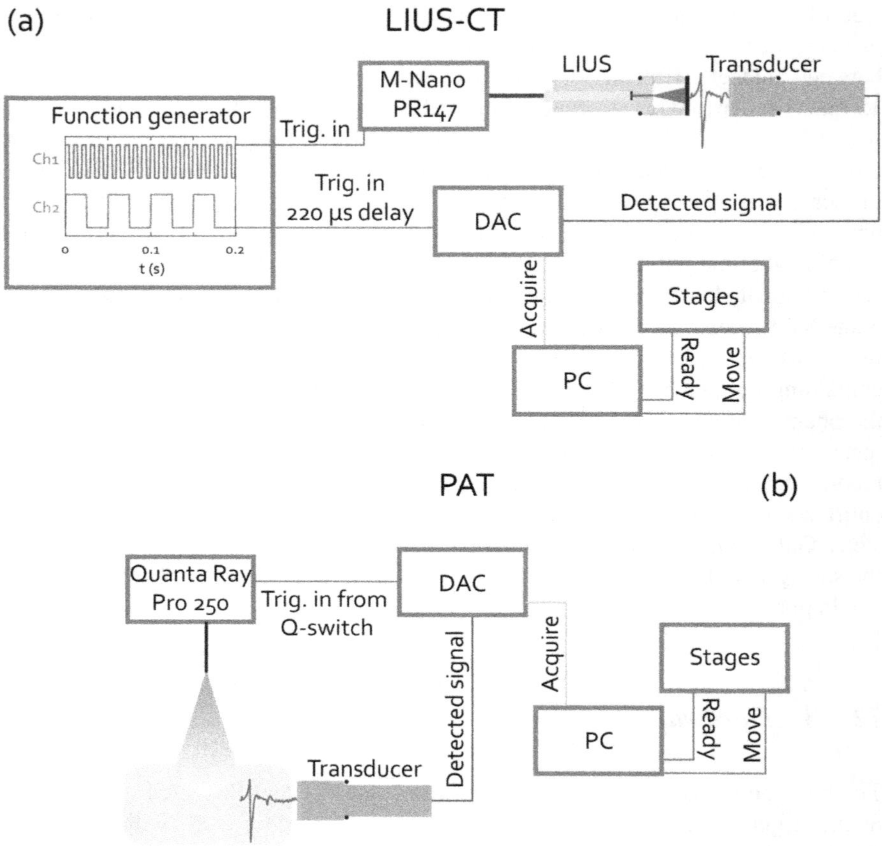

Fig. 12.3 (**a**) Triggering schematic for the LIUS-CT measurements. A function generator continuously supplies a TTL signal at 100 Hz to the LIUS laser, while a 20 Hz trigger, delayed by 220 μs to account for trigger and Q-switch delays, is supplied to the DAC. (**b**) Triggering schematic for the PAT measurements. The PAT laser, firing at a 10 Hz repetition rate, supplies a trigger pulse to the DAC each time the Q-switch activates. (**a**) and (**b**) The PC moves the stages and waits for them to finish, then sends the acquisition command to the DAC which will take the desired number of averages on arrival of the next trigger pulse

opening angle of 58°, though a higher MDP of 0.5 Pa due to insertion losses in the lens. The frequency response of the lens-equipped detector is centred at 0.78 MHz with a 134% bandwidth. The signal from the detector is stored in local memory on the DAC system and transferred to the PC for display in LabVIEW, following which it is saved in a .txt file for later processing.

In the PAT mode, the role of the PC, rotation stage and DAC remain unchanged. The detection is now triggered by the 10 Hz Q-switch-synchronised trigger pulse from the 532 nm, 10 ns pulse PAT laser (Quanta Ray pro 250, Spectra-Physics, USA). The light from the laser is coupled into a fibre bundle with nine separate outputs which are mounted evenly spread around the phantom. The excited PA signals are detected and again transferred to the PC and saved for later processing.

12.2.4 LIUS Transmitters

The laser-induced ultrasound transmitters are based on the absorption of ns-pulsed laser light in polydimethylsiloxane (PDMS) (Sylgard 184, Dow Corning, USA) doped with optically absorbing carbon black (Printex 60, Palmer Holland, USA), which causes the emission of a photoacoustically generated ultrasound pulse. The transmitters are described in more detail in a previous work [20], but will also be briefly discussed here. Figure 12.4 shows a diagram and a photograph of the LIUS transmitter, as well as plots showing the temporal and spectral properties of the LIUS signals. The optically absorbing part of the LIUS transmitter consists of a 10 mm diameter, 1 mm thick PDMS disc containing 0.075% by weight carbon black. The active element size is determined by the illumination spot diameter, which is approximately 3.6 mm in this work. A 100 μm thick, 5 mm diameter

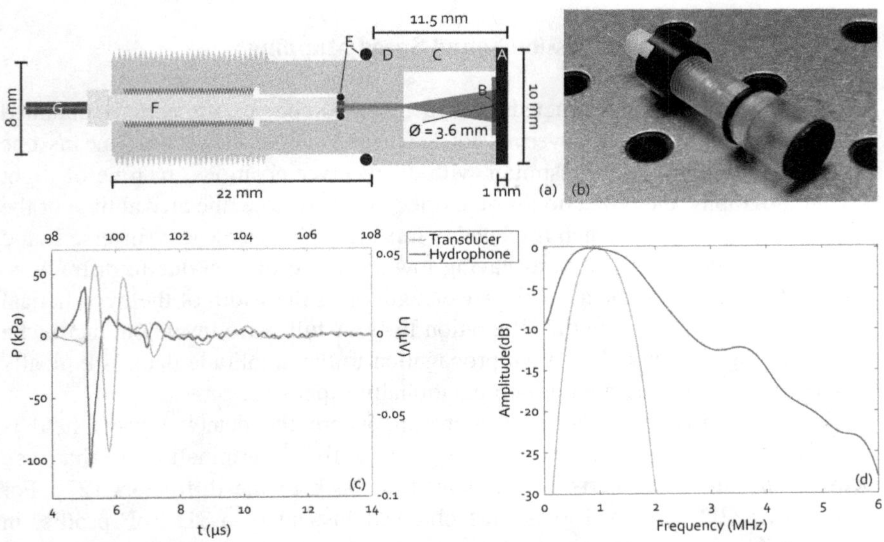

Fig. 12.4 (**a**) Schematic and photograph of the LIUS transmitter (top), indicating the various components: (A) optically absorbing active element, (B) high-impedance backing (glass coverslip), (C) tube segment enclosing air cavity, (D) threaded mounting piece with fibre passage, (E) outer and inner O-rings for water-tightness and fibre clamping (inner only), (F) fibre clamping chuck, (G) optical fibre for light delivery. (**b**) Photograph of LIUS transmitter. (**c**) Waveform from the LIUS transmitter as measured by a hydrophone and the detector in the imaging apparatus. (**d**) Frequency spectra of both signals show the narrower bandwidth of the detector, with a good overlap of the maxima. (Adapted from Ref. [20])

borosilicate glass microscopy cover slip (Agar Scientific, UK) is bonded to the back of the PDMS with cyanoacrylate adhesive (401, Loctite, Germany) to increase the positive pressure content in the signal. The glass-backed PDMS disc is bonded in the same way to a PMMA casing containing an air pocket to prevent reflections from within the casing to mix with the LIUS pulse. The casing contains a multimode optical fibre (FT600UMT, Thorlabs, USA) to supply the light. The casing can be mounted in the scanning arms by inserting it from the front into one of the mounting holes and fastening it with a specially made nut. The plots in the bottom half of Fig. 12.4 show the temporal and frequency response as measured with a fibre-optic hydrophone with a 0–30 MHz −6dB bandwidth (Precision Acoustics, UK), and as measured with the custom detector which is also used in this work. The transmitter has a 0.94 MHz centre frequency with a 200% −6 dB bandwidth, showing a large overlap with the more narrowband detector. Pressures of up to 200 kPa can be generated when illuminating with 1 mJ laser pulse energy, without adverse effects on the transmitter. The LIUS transmitter has a conical emission pattern with a far-field opening angle of 45.8°.

12.2.5 LIUS-CT Image Reconstructions

12.2.5.1 Through-Transmission Sound Speed Mapping

The sound speed is reconstructed from a large number of projections obtained with different view angles covering 360°, where a single view angle means one rotational position of the transmitter with all receiver positions. In time-of-flight (ToF) tomography, the projection is quantified as the ToF, i.e. the arrival time of the US pulse at the receiver, which is related to line integrals of slowness (inverse sound speed). ToF is commonly seen as having low spatial resolution due to diffraction, that is the finite size of the acoustic wavelength blurs the width of the hypothetical acoustic rays that constitute the integration lines. A full-wave inversion, that is the iterative fitting of a model of wave propagation to the amplitude data, is typically used to avoid this problem but is computationally expensive.

We have chosen an alternative approach, where the detected wave field is virtually back-migrated into the sample prior to ToF determination, in that way focusing the diffracted parts of the wavefront back to the diffractors [27]. For each view, a 2D map of ToF is thus obtained instead of a 1D ToF profile, in which the ToF increment imprinted by a diffractor onto the projection is resolved with (ideally) diffraction-limited resolution at the location of the diffractor. In the present study, a back-migrated view is calculated for each rotational position of the transmitter, using the paraxial approximation/angular spectrum technique. Given the quite large 2° angle step between receiver positions (corresponding to 4.5 mm), the back-migrated wave field contained a high level of grating artefacts. To mitigate these artefacts, back-migrated views were coherently compounded over a range of rotational positions (5° around each view angle) prior to ToF determination, thereby virtually increasing the spatial sampling density of the receivers per view angle. The view angles were chosen identical to the transmitter rotation angles (i.e. 0° to 359.5° in 0.5° steps). As in ref. [27], a modified filtered back projection is then performed to reconstruct the 2D distribution of slowness: the 2D ToF maps of all views are averaged after scan conversion to lab coordinates, and the average is ramp-filtered in the frequency domain.

12.2.5.2 Reflection B-Mode Imaging

For reflection B-mode imaging, a complex radio-frequency (CRF)-mode image is reconstructed for each rotational position of the transmitter from the Hilbert-transformed signals, using a delay-and-sum algorithm. Again, coherent compounding of the CRF-mode images over a 30° range of rotational positions is used, to virtually increase the spatial sampling density of the transmitter–receiver pairs and thus reduce grating artefacts. Given the isotropic angular sensitivity of the transmitter and receiver, such CRF-mode images have a strong contribution from out-of-plane echoes. In order to focus on elevation (direction z perpendicular to the imaging plane), CRF-mode images from the same view angle but from multiple vertical positions were coherently averaged. Such images were generated for view angles covering the 360° in 15° steps. All images were then incoherently averaged (i.e. after taking the square modulus) to yield the final B-mode image.

12.2.6 Photoacoustic Image Reconstruction

The photoacoustic images were reconstructed using the k-Wave software package in Matlab® [28], using time reversal. In k-Wave, time reversal is implemented by propagating the detected signals back into the reconstruction region, taking the user-defined acoustic properties of the medium into account. The reconstruction was performed on a 2D grid of 1024 × 1024 pixels, and the sound speed can be defined separately for each pixel, if desired. Three reconstructions were performed, using three different sound speed maps. The first reconstruction assumes a homogeneous sound speed throughout the image, which is optimised by the eye to achieve a sharp image. The second reconstruction uses a 2-speed model, with 1485 m/s outside the phantom and 1441 m/s inside the phantom, the average speed of the phantom materials. The final reconstruction incorporates the measured sound speed map from the LIUS-CT data to also differentiate between the sound speed in different regions within the phantom.

12.2.7 Phantom

A cylindrical 2-layer phantom was developed to provide a regular outer surface of the phantom, but an irregular interface between two layers, in a simple analogy to the fat and glandular tissues of the human breast rendered down to a 2D geometry. The tissue-mimicking materials (TMMs) were prepared according to Dantuma et al. [29]. The outer layer of the phantom was made of polyvinyl chloride plastisol (PVCP, Lure Flex Firm, Lure Factors, UK) with 25 mg/mL glass beads (G4649, Sigma Aldrich, USA) with diameters $\leq 106\,\mu m$ added to provide acoustic scattering, as well as 6.4 mg/mL TiO_2 powder (Sigma Aldrich, USA) to provide optical scattering. The inner layer is made from a gel of 1% agarose (Sigma Life Sciences, USA) in water, with 105.5 mg/mL glass beads and 3.5 mg/mL TiO_2

The outer layer of the phantom was created by pouring the PVCP-based TMM mixture into a 12 cm diameter springform pan to a depth of 3 cm in two stages. After pouring the first layer, two ink-coated stainless steel microspheres (100 μm diameter) were placed at two locations 2–3 cm from the edge of the pan, to serve as photoacoustic point sources. The second layer was poured on top and once it had set and cooled, an irregularly shaped opening was cut into the centre of the PVCP cylinder, using a duck-shaped cookie cutter. Three black surgical sutures of diameter 100 μm (NYL1CN, VetSuture SAS, France) were threaded through the PVCP and inner cavity as shown in Fig. 12.5a to also provide photoacoustic sources which extend through both TMMs. After placement of the sutures, the outer wall of the spring form pan is replaced and the agar-based TMM is poured into the inner cavity and allowed to cool and set. Figure 12.5b, c show the finished phantom placed inside the imaging apparatus in LIUS-CT and PAT modes, respectively.

By using something a bit more complex than a simple geometry of nested cylinders with different sound speeds, or singular point or wire absorbers placed in water, we are hoping to bridge a gap between simple test objects and more complex tissue-mimicking phantoms [29, 30] to perform a preliminary characterisation of most performance metrics of an imaging system such as the one presented here. The presence of a simple cylindrical outer layer, a complex inner interface in the shape of a duck, a pair of point absorbers and sub-resolution wires in a non-straight alignment, is meant to give a variety of targets in a single phantom. Having a phantom with such a variety of characteristics can also help to make up for the currently time-intensive scanning procedure.

For characterisation of the material sound speeds and the acoustic attenuation, a block measuring $60 \times 40 \times 30\ mm^3$ was cast from both TMMs. These two blocks were then characterised using the modified insertion method described by Dantuma et al. [29], using the differences in arrival time and amplitude of signals transmitted through the 60 mm side and the 40 mm side and a water sound speed determined by the method of Marczak [31] from the measured water temperature. The water temperature is measured using a thermocouple (USB-TC01, National Instruments, USA). The signals are produced from a single-element PZT transducer (Olympus NDT, USA) and detected using a 1 mm needle hydrophone (Precision Acoustics,

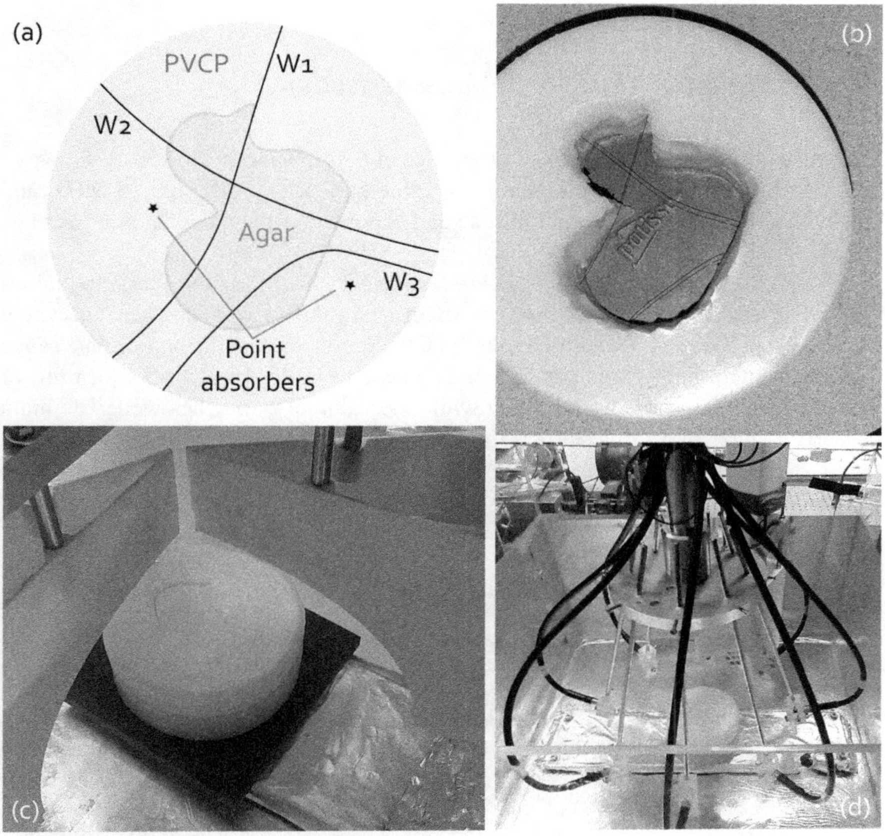

Fig. 12.5 (**a**) Schematic representation of the phantom in a top view, indicating the 2 TMM layers, point absorbers as they were approximately placed and the sutures labelled W1 through W3. (**b**) Photograph of the PCVP-based TMM outer layer of the phantom placed on the base of the spring form pan, with three surgical sutures threaded through. (**c**) Phantom with a cavity filled in with agarose gel-based TMM, placed in the imaging device in the USCT mode. (**d**) Full phantom in the set-up in the PAT mode, with nine fibre bundles evenly spaced around

UK). Transducers with centre frequencies of 1, 2.25, 5, 7.5 and 15 MHz were used. A power law describing the frequency-dependent acoustic attenuation is fitted to the data:

$$\alpha(f) = \alpha_0 f^b \qquad (12.1)$$

Here, α_0 is in dB/cm/MHz^{-b}, f is in MHz and b is a dimensionless exponent governing the frequency dependence.

12.3 Results and Discussion

12.3.1 Phantom Materials' Characterisation

The values of the relevant acoustic properties of both phantom TMMs are shown in Table 12.1. The sound speed is the mean value between 1 MHz and 15 MHz, and the acoustic attenuation is the amplitude and exponent of the attenuation power law (Eq. 12.1).

Although the sound speed values are somewhat lower than the literature values for breast tissue, the contrast between them is similar to what is found in actual breasts. The acoustic attenuation in the PVCP outer layer is in the appropriate range for human mammary glandular tissue. The agar has a low α_0 of about a third of the literature value for adipose tissue while the value of b matches well. The main property of interest for this study is the sound speed contrast, which is in the correct range to simulate breast tissue. With the acoustic attenuation properties of the PVCP layer matching well with literature values for glandular tissue, the acoustic properties of the phantom were deemed sufficient to proceed with the imaging study.

12.3.2 SoS Map and B-Mode US Images

Figure 12.6a shows the reconstructed sound speed map, clearly showing the three regions, water, the outer phantom layer and the inner duck-shaped layer. The sound speed has mean values of 1484.6 m/s in water, based on the mean measured water temperature of 20.6 °C, 1408.5 m/s in PVCP and 1465.1 m/s in agar. The mean sound speed value in the agar layer is lower than the result in Table 12.1. The sound speed distribution is also highly non-uniform, with higher sound speed values near the boundary of the duck and varying between 1470 and 1500 m/s, while the central region contains values between 1440 and 1460 m/s. The non-uniformity is an artefact, probably caused by view angle-dependent acoustic refraction at the irregular interface between the agar and PVCP compartments. Based on the single sound speed measurement for agar in Table 12.1, actual sound speed fluctuations within the material itself also cannot be entirely excluded.

Table 12.1 Measured values of the acoustic properties for the phantom TMMs, and literature values for the same properties for adipose and glandular breast tissues

	Acoustic properties		
Property	v_s (m/s)	α_0 (dB/cm/MHz^{-6})	b
PVCP	1410	0.8112	1.404
Agar	1476	0.0539	1.77
Breast (glandular) [32, 33]	1515	0.87	1.5
Breast (adipose) [33, 34]	1436–1481	0.16	1.7

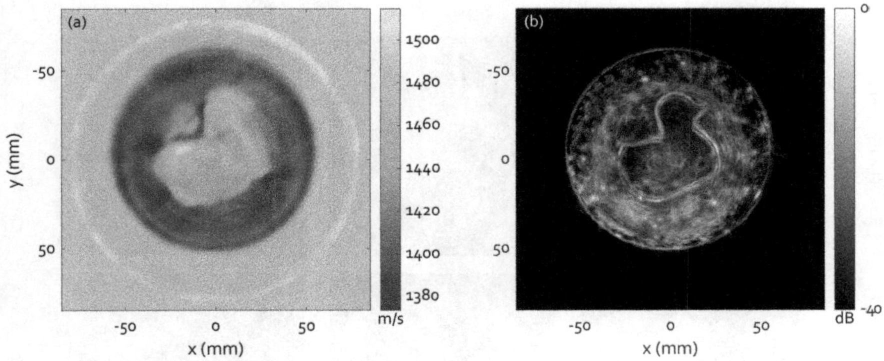

Fig. 12.6 (**a**) Sound speed map of the phantom. (**b**) Vertically compounded B-mode image, reconstructed without the use of the sound speed map. The sound speed used to reconstruct was 1486 m/s based on the average water temperature during the measurement (20.3 °C)

The resolution of the material boundaries, determined by taking perpendicular cross sections across each boundary in 10 locations, is consistently 6 mm for the water–PVCP boundary, while it fluctuates between 6 and 10 mm for the PVCP–agar boundary. As with the non-uniformity in the sound speed, the variations in the PVCP–agar boundary resolution may also be influenced by view-angle-dependent acoustic refraction at the irregular shape, in contrast to the regularly cylindrical outer boundary where refraction did not depend on view angle.

The B-mode image in Fig. 12.6b clearly shows the duck-shaped outline of the agar layer of the phantom, including a difference in the spatial distribution of echoes. The contrast between the inner and outer phantom layers is −4.3 dB, as determined by the difference between the mean normalised dB values of the PVCP and agar layers in the image. The clear specular reflection at the boundary assists in separating the two regions. The boundary appears as a double line because of the sound speed mismatch between the PVCP and the agar, as the sound speed map was not used in this image reconstruction. Looking at cross sections of the brightest point scatterers and the material interfaces in the image, FHWM values between 2 and 3 mm are observed.

12.3.3 PAT Reconstructions

Figure 12.7 shows a comparison of six PAT reconstructions, one for each combination of the three sound speed maps and two detectors. In the single-speed reconstructions, an optimal sound speed was chosen empirically to visually maximise the sharpness of the wires in the inner layer. The optimum sound speed was 1438 m/s for the images taken with the flat detector and 1416 m/s for those taken with the lens-equipped detector. Depending on the choice of a single sound speed,

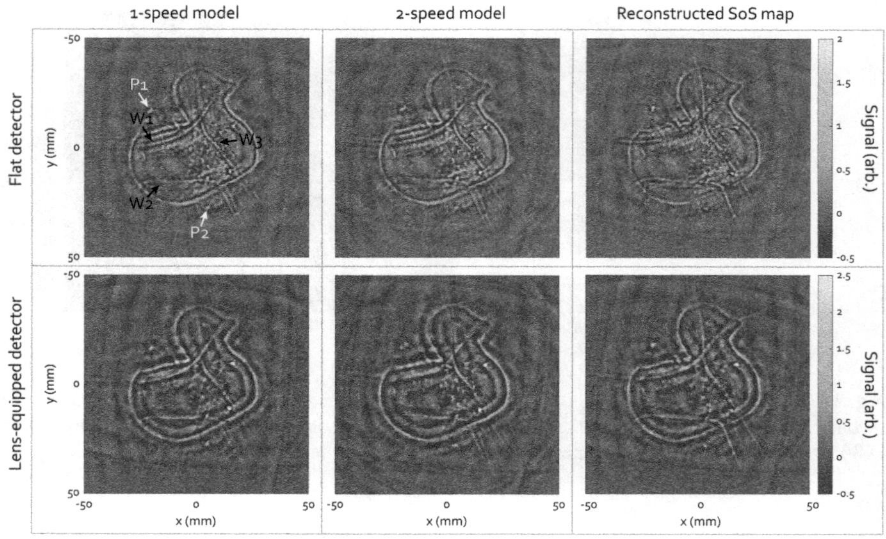

Fig. 12.7 Comparison of PAT images taken with both detector types, reconstructed with either a single-speed model (1438 m/s with a flat detector, 1416 with lens-equipped), 2-speed model (1485.5 m/s outside, 1441 m/s inside) or the measured SoS map. The 2-speed model offers a substantial improvement over the single-speed model, though structures further from the centre still appear double. The full sound speed map gives the best reconstruction, with a consistent PSF throughout the imaging region and higher SNR (see also Table 12.2)

the quality of the reconstruction is optimal for a different part of the image. With the sound speed optimised for the inner layer, the wires appear to split outside of the boundary of the duck, whereas if one were to optimise the sound speed to maximise the sharpness of the outer point sources, some of the inner wires would be too blurred to be clearly seen, while others would appear split.

The 2-speed model employs the water sound speed of 1484.6 m/s based on the measured water temperature outside of the phantom region. The inside of the phantom, the location of which can be identified in the B-mode image, is given a sound speed of 1443 m/s, the mean of the phantom materials (see also Table 12.1). The 2-speed reconstructions look similar to the single-speed images, with the wires appearing well focused in the central region, less so towards the outside of the phantom. This can be seen in the wires, when following them from the inside outwards they appear to gradually split and fade in contrast. The wires' splitting can be understood by considering the difference in arrival time in signals measured from opposite sides of the phantom. On the near side, a signal from the edge propagates mostly through water, whereas a signal from the same location must traverse the entire phantom to reach a detector on the other side. The same holds true for the point absorbers, marked P1 and P2 in Fig. 12.7. Signals of the same object detected on opposite sides of the phantom will thus experience different sound speeds on the

Table 12.2 Size (FWHM, mm) and SNR values (unitless) for the three wires (W1–W3 in top left Fig. 12.7) and the known point absorbers (P1 and P2)

Model	1-speed				2-speed				Reconstructed SoS Map			
Detector	Flat		Lens		Flat		Lens		Flat		Lens	
	FWHM	SNR	FWHM	SNR	FWHM	SNR	FWHM	SNR	FWHM	SNR	FWHM	SNR
W1	0.76	0.65	0.82	1.37	0.75	0.12	1.15	0.31	0.72	2.41	0.89	3.63
W2	0.72	0.64	0.76	0.49	0.72	0.78	0.89	0.42	0.67	2.75	0.88	2.40
W3	0.73	1.59	0.71	1.66	0.87	1.72	0.80	1.61	0.69	2.58	1.04	2.91
P1	1.56	5.69	1.38	6.29	2.03	4.53	2.38[a]	4.32	0.98	7.82	1.15	6.29
P2	1.24	3.35	1.38	3.31	1.30	3.27	1.47	2.97	0.79	3.75	1.07	2.80

[a]Split into two distinct parts, the value denotes the distance between peaks

way to the detector, leading to them finally being reconstructed in slightly different locations.

In the reconstructions employing the reconstructed sound speed map, the splitting of the wires is no longer visible. The size of the point absorbers is more consistent, and they appear more point-like. To quantify the resolution of the images, the mean full width at half maximum (FWHM) in mm of the 2-point sources associated with those placed by us as indicated in Fig. 12.5a and the wires (indicated as P1, P2 and W1-W3 in Fig. 12.7, respectively) is listed in Table 12.2 for each sub-image of Fig. 12.7. For the point sources the mean FWHM is the mean value of the maximum and minimum FWHM, for the wires it is the mean of the minimum FWHM, only evaluated where the wire is clearly a single, non-split structure. Other point sources than those actually placed in the phantom can be seen on the inside of the duck region, these are most likely motes of dust or other contaminations that made their way into the phantom material.

Table 12.2 also lists the signal-to-noise ratios (SNRs) of the same three wires and two point absorbers. The SNR, taken in hand-defined regions of interest containing a feature and a surrounding margin of background, is defined to be

$$\mathrm{SNR} = \frac{\mu_{ft}}{\sigma_{bg}}, \tag{12.2}$$

where μ_{ft} is the mean value of the amplitude of a feature and σ_{bg} is the standard deviation of the background amplitude directly surrounding the feature. For the wires, the ROI to be evaluated is based on the wires as seen in the reconstructed images using the sound speed map, making the effect of the splitting visible.

The FWHM values of the wires do not change more than 50 μm between the different sound speed maps, largely because only the well-reconstructed part of the wire was considered each time. As can be seen easily by the eye in Fig. 12.7, the wires are well reconstructed over their entire visible length when using the reconstructed sound speed map, and only partially in the other cases, mostly inside the duck-shaped inclusion. The splitting of the wires away from the image centre appears somewhat worse in the 2-speed reconstruction compared to the single-speed images. The point sources are more similar in size to one another in the images employing the reconstructed sound speed map compared to the other two methods.

For the flat detector, the SNRs of W2 and W3 increase a little going from the 1-speed model to the 2-speed model, while that of W1 decreases. For the lens-equipped detector, the SNR barely changes for W2 and W3 between the 1-speed and 2-speed models, while W1 again decreases by a factor of 4.4. For both detector types, however, the SNR of all wires in the sound speed-mapped images is superior to the other models.

While similar image qualities can be achieved locally for each reconstruction approach, the map-based reconstruction shows a more consistent quality throughout the image. Comparing the results from the flat detector and the lens-equipped detector, the FWHM of all features is smaller in the images taken with the former, in keeping with the lower centre frequency and lower bandwidth of the latter, as discussed in Sect. 12.2.3.

12.3.4 General Discussion

In this work, we developed instrumentation for the first time to image large-volume (breast-sized) samples using hybrid photoacoustic and USCT imaging. The ultrasound for USCT is generated using specially made LIUS sources. The LIUS sources have similar frequency content as the expected PA signals, to better match a single, optimised detector. The work demonstrates the use of LIUS transmitters specially fabricated for the purpose and is the culmination of steps taken where we first showed the potential for improved contrast in linear array detection geometries using plane wave [35] and synthetic aperture [19] approaches using LIUS; and where we provided a detailed description [20] of the means to fabricate the LIUS transmitters used in the current work.

Since the imaging device in its current form is a technology demonstrator and not optimised for application, it takes considerable time to collect enough data for a high-quality image, as much as 20 h for either type of LIUS-CT image. The PAT images can be obtained within 30 min, and are thus less problematic. This limitation means that for the present work, the phantom was required to be stable in both position and composition for the data acquisition time. While this was not a problem for the PVCP phantom, we cannot discount that the properties of the agar changed slightly over time, as the sample made (1% w/v agarose in water) constitutes a weak gel and can absorb water. Further, agarose gel is also vulnerable to microbial infection which may change the optical properties slightly.

The addition of more detectors in a curvilinear array for a 2D configuration will drastically decrease the imaging time as fewer positions for the detection arm per projection will be needed. Using a laser with a higher repetition rate and a DAC system with a correspondingly higher acquisition rate will enable faster PAT imaging as well, possibly even allowing for continuous rotation of the scanner. For the purpose of phantom imaging, however, the current acquisition speed is adequate, if somewhat cumbersome.

In future work, it may be of interest to perform 3D imaging with both types of detector used for the PAT images. For an optimal resolution, it would be best to use the flat detector, though weaker features may then be obscured where the larger acceptance angle of the lens-equipped detector could render them visible. Meanwhile, to further optimise 2D LIUS-CT imaging it might be interesting to confine the emitted LIUS field to the imaging plane of interest, rather than having an axisymmetric source. This may be achieved by changing the shape of the illumination spot on the LIUS transmitter to a more linear or elliptical shape [19], or by changing the shape of the absorber itself. This will reduce out-of-plane reflections and could remove or reduce the need for axial scanning for the reflection mode acquisition.

12.4 Conclusion

We have presented a lab-based reconfigurable acoustic tomography device for transmission and reflection LIUS-CT and PAT, and demonstrated its use on a multimodal 2D phantom. The transmission sound speed images with a 120° scanning range per projection and 719 projections show sound speed values which correspond well to the independently measured values, enough to improve the PA images. The reflection mode LIUS-CT image shows a −4.3 dB contrast between the different materials in the phantom, with a resolution of a few millimetres. The effect of the choice of sound speed model is evident in the quality of the PAT reconstructions, with the superior result coming from the use of the measured sound speed map. The 2D configuration of this system, combined with a simple phantom, shows potential for continued development into a fully 3D system for assessment of different measurement configurations for PAT and LIUS-CT of more complex breast phantoms.

Acknowledgements The authors would like to express their thanks to Gerard Oude Meijers for his indispensable role in the refurbishment and integration of the translation and rotation stages. Thanks are also due to Johan van Hespen for his part in the design of the scanning arms, as well as his assistance in the programming of the scanning and acquisition software. We would also like to thank Maura Dantuma for her assistance in the fabrication of the tissue-mimicking materials used in the phantom. This work was funded as a part of the European Horizon 2020 PAMMOTH project under Grant Agreement No. 732411, an initiative of the Photonics Public-Private Partnership.

Disclosures The authors declare no conflicts of interest.

References

1. A. Özbek, X. L. Deán-Ben, and D. Razansky, "Optoacoustic imaging at kilohertz volumetric frame rates," *Optica***5**, 857–863 (Optical Society of America, 2018).

2. S. A. Ermilov, T. Khamapirad, A. Conjusteau, M. H. Leonard, R. Lacewell, K. Mehta, T. Miller, and A. A. Oraevsky, "Laser optoacoustic imaging system for detection of breast cancer," *Journal of biomedical optics* **14**, 024007 (International Society for Optics; Photonics, 2009).
3. G. L. Menezes, R. M. Pijnappel, C. Meeuwis, R. Bisschops, J. Veltman, P. T. Lavin, M. J. Van De Vijver, and R. M. Mann, "Downgrading of breast masses suspicious for cancer by using optoacoustic breast imaging," *Radiology* **288**, 355–365 (Radiological Society of North America, 2018).
4. J.-W. Muller, R. van Hees, M. van Sambeek, P. Boutouyrie, M. Rutten, P. Brands, M. Wu, and R. Lopata, "Towards in vivo photoacoustic imaging of vulnerable plaques in the carotid artery," *Biomedical Optics Express* **12**, 4207–4218 (Optical Society of America, 2021).
5. J. J. Niederhauser, M. Jaeger, R. Lemor, P. Weber, and M. Frenz, "Combined ultrasound and optoacoustic system for real-time high-contrast vascular imaging in vivo," *IEEE transactions on medical imaging* **24**, 436–440 (IEEE, 2005).
6. S. Manohar and M. Dantuma, "Current and future trends in photoacoustic breast imaging," *Photoacoustics* **16**, 100134 (Elsevier, 2019).
7. S. Manohar, A. Kharine, J. C. van Hespen, W. Steenbergen, and T. G. van Leeuwen, "The twente photoacoustic mammoscope: System overview and performance," *Physics in Medicine & Biology* **50**, 2543 (IOP Publishing, 2005).
8. S. M. Schoustra, D. Piras, R. Huijink, T. J. Op't Root, L. Alink, W. M. F. Kobold, W. Steenbergen, and S. Manohar, "Twente photoacoustic mammoscope 2: System overview and three-dimensional vascular network images in healthy breasts," *Journal of biomedical optics* **24**, 121909 (International Society for Optics; Photonics, 2019).
9. R. A. Kruger, C. M. Kuzmiak, R. B. Lam, D. R. Reinecke, S. P. Del Rio, and D. Steed, "Dedicated 3D photoacoustic breast imaging," *Medical physics* **40**, 113301 (Wiley Online Library, 2013).
10. L. Lin, P. Hu, J. Shi, C. M. Appleton, K. Maslov, L. Li, R. Zhang, and L. V. Wang, "Single-breath-hold photoacoustic computed tomography of the breast," *Nature communications* **9**, 1–9 (Nature Publishing Group, 2018).
11. M. Toi, Y. Asao, Y. Matsumoto, H. Sekiguchi, A. Yoshikawa, M. Takada, M. Kataoka, T. Endo, N. Kawaguchi-Sakita, et al., "Visualization of tumor-related blood vessels in human breast by photoacoustic imaging system with a hemispherical detector array," *Scientific reports* **7**, 1–11 (Nature Publishing Group, 2017).
12. H. Gemmeke, L. Berger, T. Hopp, M. Zapf, W. Tan, R. Blanco, R. Leys, I. Peric, and N. V. Ruiter, *The new generation of the KIT 3D USCT* (KIT Scientific Publishing, 2018).
13. B. Schäfgen, M. Juskic, M. Radicke, M. Hertel, R. Barr, A. Pfob, R. Togawa, J. Nees, A. von Au, et al., "Evaluation of the FUSION-x-US-II prototype to combine automated breast ultrasound and tomosynthesis," *European Radiology* **31**, 3712–3720 (Springer, 2021).
14. J. Jose, R. G. Willemink, W. Steenbergen, C. H. Slump, T. G. van Leeuwen, and S. Manohar, "Speed-of-sound compensated photoacoustic tomography for accurate imaging," *Medical physics* **39**, 7262–7271 (Wiley Online Library, 2012).
15. S. Ermilov, R. Su, A. Conjusteau, F. Anis, V. Nadvoretskiy, M. Anastasio, and A. Oraevsky, "Three-dimensional optoacoustic and laser-induced ultrasound tomography system for preclinical research in mice: Design and phantom validation," *Ultrasonic imaging* **38**, 77–95 (SAGE Publications Sage CA: Los Angeles, CA, 2016).
16. M. Jaeger, G. Held, S. Peeters, S. Preisser, M. Grünig, and M. Frenz, "Computed ultrasound tomography in echo mode for imaging speed of sound using pulse-echo sonography: Proof of principle," *Ultrasound in medicine & biology* **41**, 235–250 (Elsevier, 2015).
17. M. Pérez-Liva, J. Herraiz, J. Udias, E. Miller, B. Cox, and B. Treeby, "Time domain reconstruction of sound speed and attenuation in ultrasound computed tomography using full wave inversion," *The Journal of the Acoustical Society of America* **141**, 1595–1604 (Acoustical Society of America, 2017).
18. M. Dantuma, F. Lucka, S. Kruitwagen, A. Javaherian, L. Alink, R. van Meerdervoort, M. Nanninga, T. Root, B. De Santi, et al., "Fully three-dimensional sound speed-corrected multi-wavelength photoacoustic breast tomography," *arXiv preprint arXiv:2308.06754* (2023).

19. D. Thompson, H. Kruit, D. Gasteau, and S. Manohar, “Laser-induced synthetic aperture ultrasound imaging,” *Journal of Applied Physics* **128**, 163105 (AIP Publishing LLC, 2020).
20. D. Thompson, J. Nagel, D. Gasteau, and S. Manohar, “Laser-induced ultrasound transmitters for large-volume ultrasound tomography,” *Photoacoustics*, 100312 (Elsevier, 2021).
21. S. Manohar, R. G. Willemink, F. van der Heijden, C. H. Slump, and T. G. van Leeuwen, “Concomitant speed-of-sound tomography in photoacoustic imaging,” *Applied physics letters* **91** (AIP Publishing, 2007).
22. S. Resink, J. Jose, R. G. Willemink, C. H. Slump, W. Steenbergen, T. G. van Leeuwen, and S. Manohar, “Multiple passive element enriched photoacoustic computed tomography,” *Optics letters* **36**, 2809–2811 (Optica Publishing Group, 2011).
23. J. Jose, R. G. Willemink, S. Resink, D. Piras, J. G. van Hespen, C. H. Slump, W. Steenbergen, T. G. van Leeuwen, and S. Manohar, “Passive element enriched photoacoustic computed tomography (PER PACT) for simultaneous imaging of acoustic propagation properties and light absorption,” *Optics express* **19**, 2093–2104 (Optica Publishing Group, 2011).
24. J. Xia, C. Huang, K. Maslov, M. A. Anastasio, and L. V. Wang, “Enhancement of photoacoustic tomography by ultrasonic computed tomography based on optical excitation of elements of a full-ring transducer array,” *Optics letters* **38**, 3140–3143 (Optica Publishing Group, 2013).
25. A. Bychkov, V. Zarubin, A. Karabutov, V. Simonova, and E. Cherepetskaya, “On the use of an optoacoustic and laser ultrasonic imaging system for assessing peripheral intravenous access,” *Photoacoustics* **5**, 10–16 (Elsevier, 2017).
26. A. Bychkov, V. Simonova, V. Zarubin, E. Cherepetskaya, and A. Karabutov, “The progress in photoacoustic and laser ultrasonic tomographic imaging for biomedicine and industry: A review,” *Applied Sciences* **8**, 1931 (MDPI, 2018).
27. M. Jaeger, T. Schweizer, C. Etter, P. Stähli, and M. Frenz, “Diffraction-limited spatial resolution using synthetic focus time-of-flight ultrasound tomography,” in *2019 IEEE international ultrasonics symposium (IUS)*, (IEEE, 2019).
28. B. E. Treeby and B. T. Cox, “K-wave: MATLAB toolbox for the simulation and reconstruction of photoacoustic wave fields,” *Journal of biomedical optics* **15**, 021314 (International Society for Optics; Photonics, 2010).
29. M. Dantuma, R. van Dommelen, and S. Manohar, “Semi-anthropomorphic photoacoustic breast phantom,” *Biomedical optics express* **10**, 5921 (Optical Society of America, 2019).
30. M. Dantuma, S. C. Kruitwagen, M. J. Weggemans, T. J. Op't Root, and S. Manohar, “Suite of 3D test objects for performance assessment of hybrid photoacoustic-ultrasound breast imaging systems,” *Journal of biomedical optics* **27**, 074709–074709 (Society of Photo-Optical Instrumentation Engineers, 2022).
31. W. Marczak, “Water as a standard in the measurements of speed of sound in liquids,” *the Journal of the Acoustical Society of America* **102**, 2776–2779 (Acoustical Society of America, 1997).
32. C. Li, N. Duric, P. Littrup, and L. Huang, “In vivo breast sound-speed imaging with ultrasound tomography,” *Ultrasound in medicine & biology* **35**, 1615–1628 (Elsevier, 2009).
33. F. d'Astous and F. Foster, “Frequency dependence of ultrasound attenuation and backscatter in breast tissue,” *Ultrasound in medicine & biology* **12**, 795–808 (Elsevier, 1986).
34. S. Yongchen, D. Yanwu, T. Jie, and T. Zhensheng, “Ultrasonic propagation parameters in human tissues,” in *IEEE 1986 ultrasonics symposium*, (IEEE, 1986).
35. D. Thompson, D. Gasteau, and S. Manohar, “Spatially compounded plane wave imaging using a laser-induced ultrasound source,” *Photoacoustics* **18**, 100154 (Elsevier, 2020).

Chapter 13
Multimodal Optoacoustic Imaging

Zhenyue Chen, Irmak Gezginer, Quanyu Zhou, and Daniel Razansky

Abstract Optoacoustic (OA) or photoacoustic imaging is an emerging technique synergistically combining optical excitation with ultrasonic detection to acquire diverse functional, anatomical, molecular, and metabolic information from living tissues across various scales, from organelles and cells to whole organs and organisms. Despite its powerful capabilities, the technique is hindered with limited specificity, penetration depth, and primarily blood-related contrast in mammalian tissues. Multi-modal approaches combining OA imaging with other imaging techniques have been proposed to overcome these limitations, thus fully exploiting the unique capabilities and complementarity of OA imaging. This chapter summarizes the current state-of-the-art in multimodal optoacoustic imaging systems and their applications, including combinations with ultrasound, fluorescence, and magnetic resonance imaging as well as ionizing methods, such as X-ray computed tomography, single-photon-emission computed tomography, and positron emission tomography.

Keywords Optoacoustic imaging · Photoacoustic imaging · Multimodal imaging · Hybrid system · Dual-modal · Contrast agent · Ultrasound (US) · Fluorescence · Magnetic resonance imaging (MRI) · X-ray computed tomography

Z. Chen · I. Gezginer · Q. Zhou
Institute for Biomedical Engineering and Institute of Pharmacology and Toxicology, Faculty of Medicine, University of Zurich, Zurich, Switzerland

Institute for Biomedical Engineering, Department of Information Technology and Electrical Engineering, ETH Zurich, Zurich, Switzerland

D. Razansky (✉)
Institute for Biomedical Engineering and Institute of Pharmacology and Toxicology, Faculty of Medicine, University of Zurich, Zurich, Switzerland

Institute for Biomedical Engineering, Department of Information Technology and Electrical Engineering, ETH Zurich, Zurich, Switzerland

Zurich Neuroscience Center (ZNZ), Zurich, Switzerland
e-mail: daniel.razansky@uzh.ch

W. Xia (ed.), *Biomedical Photoacoustics*,
https://doi.org/10.1007/978-3-031-61411-8_13

(CT) · Single-photon-emission computed tomography (SPECT) · Positron emission tomography (PET)

13.1 Introduction

Optoacoustic (OA) imaging, also known as photoacoustics, is an emerging technique combining optical excitation with ultrasonic detection that can render rich optical contrast from living tissues with high spatial and temporal resolution. OA has gained increasing attention of the biomedical research community due to its unique hybrid properties allowing to visualize, with high sensitivity, the biodistribution of endogenous chromophores as well as exogenous optical absorbers, such as nanoparticles, organic dyes, or fluorescent proteins [1–3]. In comparison to optical microscopy techniques, OA can achieve greater (centimeter-scale) penetration into living tissues by exploiting the low scattering of ultrasound waves, thus breaking through the barriers imposed by optical diffusion [4]. Different embodiments of OA have attained imaging at multiple scales ranging from subcellular organelles to organs with the same optical contrast [4, 5].

On the microscopic scale, optical resolution optoacoustic microscopy (OAM) approaches are suitable for high-resolution imaging at depths within the mean free path of photon transport, i.e., ~1 mm in biological tissues [6]. On the other hand, OA tomography can resolve the optical absorption distribution with ultrasonic resolution in the 10–100 s micrometers range at centimeter-scale depths, where the excitation light beam is fully diffuse [7]. State-of-the-art tomography systems based on multi-element and matrix arrays capitalize on simultaneous excitation of the entire imaged volume, rapid wavelength-tuning lasers, parallel data acquisition, and ultrafast computations on graphics processing units (GPUs) to accelerate the acquisition and visualization of volumetric spectrally resolved data in real time [8, 9].

The versatility of OA systems has been demonstrated in label-free anatomical [10], functional [11–14], and molecular imaging with exogenous contrast agents [15–17]. A detailed description of OA methods and applications can be found in recently published reviews [7, 18–20]. Despite its powerful technical capabilities, OA suffers from several limitations, including penetration depth, inevitable trade-offs between penetration and spatial resolution, as well as limited sensitivity and specificity. More importantly, OA contrast primarily stems from optical absorption, making it insensitive to other contrast mechanisms related to scattering, fluorescence, acoustic, and magnetic properties of tissue. To overcome such limitations, it is common practice to combine multiple modalities to seek complementary information for a more comprehensive understanding of biological processes.

Due to the inherent hybrid optical and ultrasound nature of OA, multi-modality efforts have been mainly focused on hybridization with ultrasound (US) [21–23], fluorescence (FL) [14, 24–26], and other optical imaging modalities [27–29]. Recent progress on multimodal OA has also seen valuable combinations with magnetic resonance imaging (MRI) [30–34], computed tomography (CT), and positron emission tomography (PET) to exploit entirely different contrast to enhance anatomical, functional, and molecular information. This chapter thus focuses on hybridization efforts of OA with other imaging modalities, as well as their applications in preclinical research and clinical diagnostics and treatment monitoring.

13.2 Hybridization Between OA and US

OA and US imaging share similar signal acquisition and processing pipelines and share other common advantages, such as the use of non-ionizing radiation, real-time imaging capability, high spatial resolution, and portability [35]. Combining the two modalities may thus result in a great complementary value as US enables easy anatomical navigation and localization of structures, while OA can provide additional functional information, thus enhancing clinical applicability [22]. Moreover, transmission-mode US imaging can render additional acoustic tissues properties, such as the speed of sound (SoS) and acoustic attenuation (AA), which can be used for improving accuracy of tomographic OA reconstructions [36–38]. For hybridization with reflection-mode US, linear array [39–44] and concave arrays [45–48] are commonly employed in conjunction with broad-beam illumination through a fiber bundle or mirrors [40] fixed on the side of the transducer array (Fig. 13.1a).

It should be noted that data acquisition and image formation strategies of B-mode US fundamentally differ from the tomographic OA imaging approach commonly employed to render optimal image quality. For B-mode US imaging, array transducers with relatively small inter-element pitch below acoustic wavelength are typically used to avoid image quality deterioration due to the side lobes and resolution loss. In contrast, large element size and angular tomographic coverage are preferred in OA imaging to attain higher sensitivity in detecting weak signals and mitigate limited-view image artifacts in tomographic reconstructions. To this end, multi-segment array was used for optimal OA-US hybridization consisting of a linear segment located between two surrounding concave segments [22]. The linear segment utilized a small inter-element pitch (1.25λ) for better performance in the US mode, while the concave segments with 3λ pitch were optimized for OA signal detection (Fig. 13.1a). OA and US images can simultaneously be obtained and further enhanced by image superimposition (Fig. 13.1b). The two modalities are highly synergistic since the OA contrast mainly emphasizes vascular structures, while US images provide complementary anatomical information on other soft tissues. In addition, US has been exploited for motion correction and light fluence correction in hybrid OA-US imaging [39].

Compared to reflection-mode US imaging, transmission-mode US can additionally provide the maps of SoS and AA distribution in the imaged tissue. A typical transmission reflection optoacoustic ultrasound (TROPUS) system is based on a full-ring array geometry and employed for hybrid OA, reflection-, and transmission-mode US imaging [37] (Fig. 13.1c). A dedicated trigger paradigm is employed to synchronize the excitation and detection of signals in a time-interleaved way. For OA signal excitation, the illumination light is coupled through fiber bundle

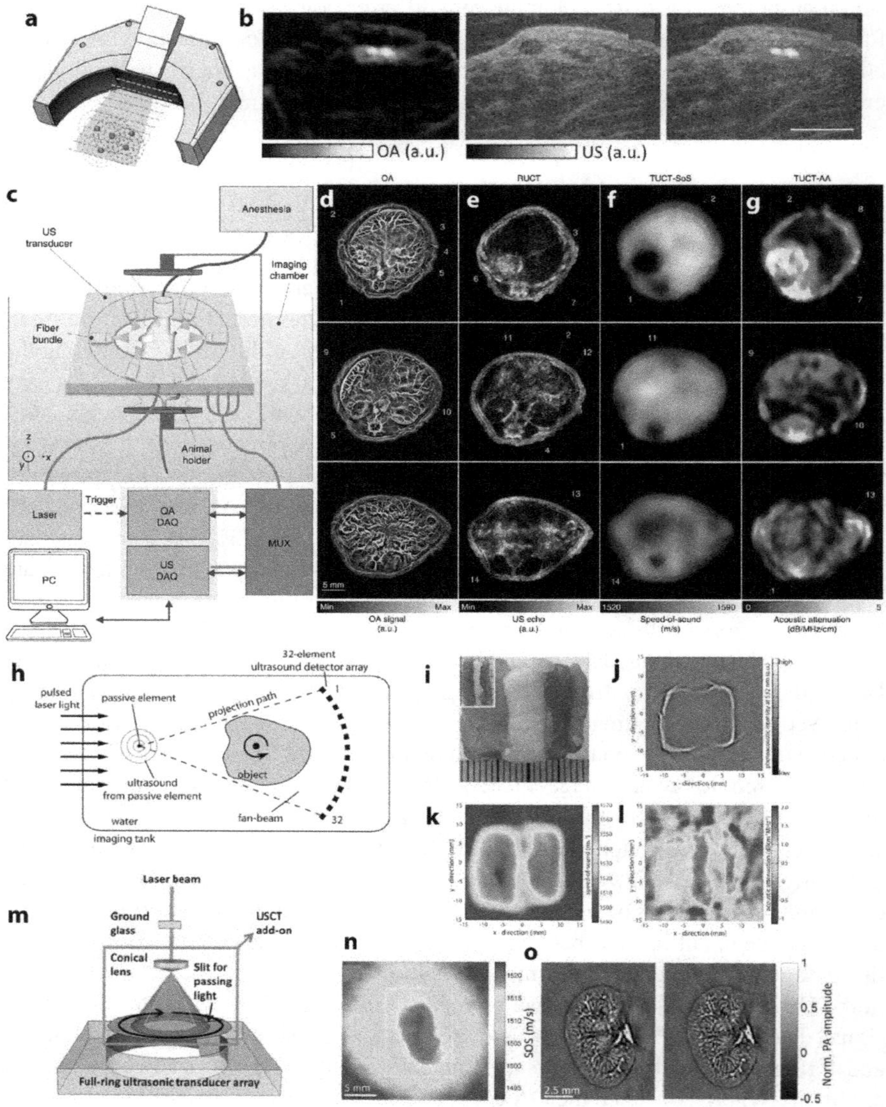

Fig. 13.1 Hybrid OA and reflection-mode US imaging. (**a**) A multi-segment array probe for optimal imaging performance in both modes. (**b**) OA and US images of the wrist of a healthy volunteer. From left to right: OA image at 800 nm; the corresponding US image; superimposed hybrid OA-US image. Scale bar: 10 mm. Reprinted with permission from Ref. [22]. (**c**) Transmission–reflection optoacoustic ultrasound (TROPUS) computed tomography system. (**d**) Representative cross-sections acquired in the OA mode. (**e**) The corresponding reflection-mode US images. (**f, g**) The corresponding transmission-mode US images showing the distribution of the SoS and AA, respectively. Annotations: 1: spinal cord; 2: liver; 3: vena porta; 4: vena cava; 5: aorta; 6: stomach; 7: ribs; 8: skin/fat layer; 9: spleen; 10: right kidney; 11: cecum; 12: pancreas; 13: intestines; 14: muscle. Reprinted with permission from [37]. (**h**) Schematics of laser-induced US based

and fixed at both sides of the ring-array transducer. In the US mode, synthetic transmit aperture technique is used, whereas neighboring channels were grouped together as active sub-apertures. The in vivo applicability of the technique has been demonstrated via noninvasive whole-body imaging of mice to render fully coregistered OA (Fig. 13.1d), reflection-mode US (Fig. 13.1e), SoS (Fig. 13.1f), and acoustic attenuation (Fig. 13.1g) maps. The system was subsequently used for noninvasive multi-parametric characterization of mammalian tumors [49] and fatty liver disease [50].

Simultaneous OA and US imaging is commonly hindered by conflicting requirements for signal amplification and synchronization of active transmissions. To avoid the use of additional transmit-receive systems, active US can instead be generated by illuminating an optical absorber placed next to the imaging subject [38, 51, 52] or direct excitation of the transducer surface with the pulsed light [53]. In the former case, a single or multiple strong light absorbers (e.g., horsetail hair), also called "passive" US element, were positioned outside the imaging object with a 32-element curvilinear array used as the signal receiver (Fig. 13.1h–l). The transducer array was used to collect signals from both modalities temporally separated by different time-of-flight (TOF). Alternatively, ring-shaped light and a rotating slit were employed to selectively activate different transducer elements in sequence (Fig. 13.1m–o). Laser ultrasound (LUS) imaging was further suggested where OA sources were generated on the tissue surface and combined with laser interferometric detection to image targets up to ~5 cm depth in a fully noncontact fashion, making it a promising approach for clinical applications such as nonintrusive patient monitoring and intraoperative imaging [54].

Besides preclinical research [37, 55], hybrid OA-US has also successfully been employed in clinical studies. Compared to standalone US, the OA mode in dual-modal systems provides high vascular contrast, which can improve the diagnostic specificity in diagnostics of melanoma [56], thyroid [42], or breast cancers [44, 48, 57]. Additionally, multispectral optoacoustic tomography (MSOT) measurements employing five distinct wavelengths have enabled retrieving the maps of oxygen saturation (sO_2) to differentiate the papillary thyroid cancer (PTC) nodules from benign cases, with PTC nodules showing much lower sO_2 values and skew angle of the sO_2 distribution [41]. In breast cancer patients, the dual-modal system detects sentinel lymph nodes located near the primary tumor. While in the US mode lymph nodes were mainly recognized based on their distinct structural features,

Fig. 13.1 (continued) on a single optical absorber (passive US element). (**i**) Photograph of the top view of the biological specimen. Inset shows the side view. Three layers are identified: from left to right—muscle, fat, muscle. A fourth layer composed of muscle mixed with fat is barely distinguishable. (**j**) OA image shows only surface of the muscle tissue. (**k, l**) SoS and acoustic attenuation images of the specimen. Reprinted with permission from Ref. [51]. (**m**) Schematic of laser-induced US system based on a ring array in conjunction with ring-shaped laser beam and rotating slit. (**n**) The SoS map of the mouse kidney. (**o**) Comparison of OA image rendered by assuming a constant SoS versus and an improved image achieved by considering a non-constant SoS map in (**n**). (Reprinted with permission from Ref. [53])

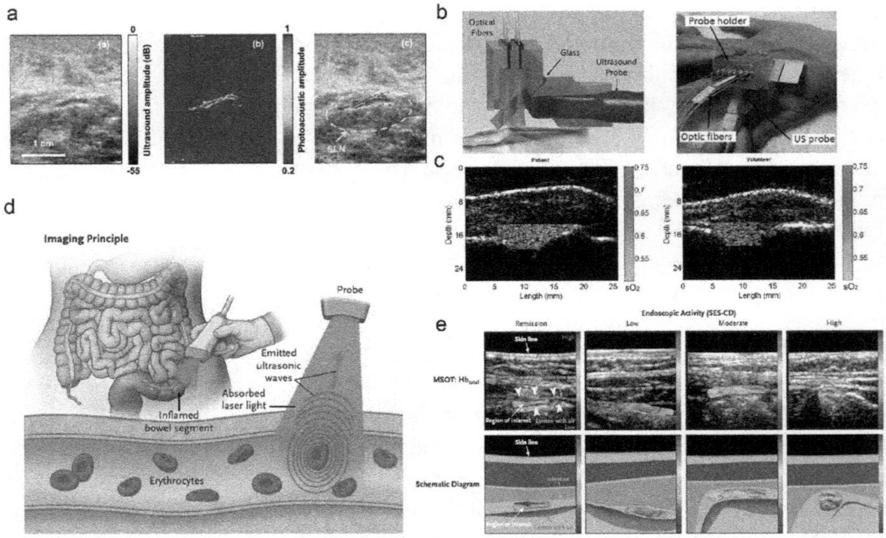

Fig. 13.2 Clinical applications of hybrid OA and US imaging. (**a**) US, OA, and overlaid OA-US images of a sentinel lymph node post methylene blue injection. Reprinted with permission from Ref. [43]. (**b**) Hybrid OA-US system for arthritis imaging in human fingers. (**c**) Color-encoded sO_2 map superimposed on gray-scale US image of inflammatory metacarpophalangeal (MCP) finger joint (left) and healthy MCP finger joint (right). Reprinted with permission from Ref. [58]. (**d**) Hybrid OA-US system for transabdominal imaging [60]. (**e**) Color-encoded total hemoglobin map overlaid on the gray-scale US image of intestinal wall in patients with different degrees of endoscopic inflammation [60]

OA imaging was used to identify subcutaneously injected methylene blue dye (Fig. 13.2a), thus enabling the lymph node staging and guiding cancer prognosis and treatment evaluation [43]. In addition to cancer imaging, the dual-modal system (Fig. 13.2b) was also applied in accessing hypoxia in the inflammatory regions of finger joints (Fig. 13.2c) in patients with inflammatory arthritis [58] or synovitis [59]. Furthermore, it was exploited to distinguish active disease from remission cases in Crohn's disease patients by assessing total hemoglobin total hemoglobin (HbT) levels in the intestinal wall overlaid on the structural US-based information (Fig. 13.2d, e) [60]. A comprehensive review on clinical applications of hybrid OA and US systems can be found elsewhere [61].

Dual-modal OA-US imaging can be also implemented in a scanning microscopy fashion employing motorized translational stages [62–65] or slider-crank scanners [66] where the data acquisition is sequentially switched between the OA and US modes. The anatomical structure provided by scanning acoustic microscopy was reported to extract the contour map of uneven surfaces, thus improving dynamic focusing [64]. The ultrasonic echo generated by acoustic microscopy can also be utilized to remove the multi-reflection artifacts in OA images by assuming an equivalent impulse function calculated from US and subsequently used in OA signal recovery [65]. Multimodal endoscopic [67, 68] and microscopic [69] imaging systems based on single-element transducers have also been reported and found wide applications in multi-parametric imaging of brain function [64], melanoma lesions [70], angiogenesis, and skull bone morphogenesis [62, 63] in mice.

13.3 Merging OA with Optical Imaging

FL imaging is a popular technique in biological discovery owing to its high sensitivity and specificity in detecting fluorochromes against the tissue background. The resolution of FL imaging greatly deteriorates with depth in scattering objects, which generally limits the effective penetration depth in fluorescence microscopy to ~1 mm in opaque living tissues. In contrast, OA imaging could preserve high resolution in deep tissues benefiting from the insignificant (typically reduced by several orders of magnitude) scattering and attenuation of ultrasound relative to light. However, OA suffers from relatively low tissue contrast as it is mainly sensitive to hemoglobin, which exhibits strong light absorption in the visible and near infrared wavebands. Thus, synergistic integration of OA and FL imaging is a popular solution to combine the high molecular sensitivity of FL imaging with the excellent hemodynamic contrast and scalable penetration of OA imaging.

Hybrid OA and FL microscopy systems commonly share the same excitation light path while differing in the detection path. To separate the collection of ultrasound and light signals, several types of system design have been proposed (Fig. 13.3a, d, g, j). A straightforward idea is to collect light and ultrasound signals from different directions (Fig. 13.3a), which can enable hybrid implementations integrating OAM with nonlinear optical microscopy such as two-photon microscopy (2 PM), second-harmonic generation (SHG), and third harmonic generation (THG) microscopy [71, 72]. Mechanical scanning is used to acquire two-dimensional (2D) OA image while optical scanning with galvo mirror is used in the nonlinear optical modalities. Another design has integrated OA with confocal fluorescence microscopy (CFM) and 2 PM based on a commercial FL microscope, facilitating sectioned retinal slice imaging in transgenic mice. The FL mode was used to detect yellow fluorescent protein (YFP)-labeled dipolar cells, while melanin in the retinal pigmented layer provided the absorption contrast in the OA images [73]. By combing OA mesoscopy, OAM and 2 PM working in label-free modes, the reported multi-photon and multi-scale OAM system (Fig. 13.3b) not only provides different contrast modes but also offers imaging at different geometrical scales, achieving uniquely broad resolution and a 1000-fold volume sampling increase compared to volumes scanned by optical microscopy and leading to new possibilities in biological visualization (Fig. 13.3c) [74].

Although collecting US and light signals from different directions represents a straightforward solution for dual-modality imaging, the application of this approach is restricted to low scattering and thin specimens, such as zebrafish and mouse ear. To address this limitation, a customized miniaturized ultrasound transducer with a

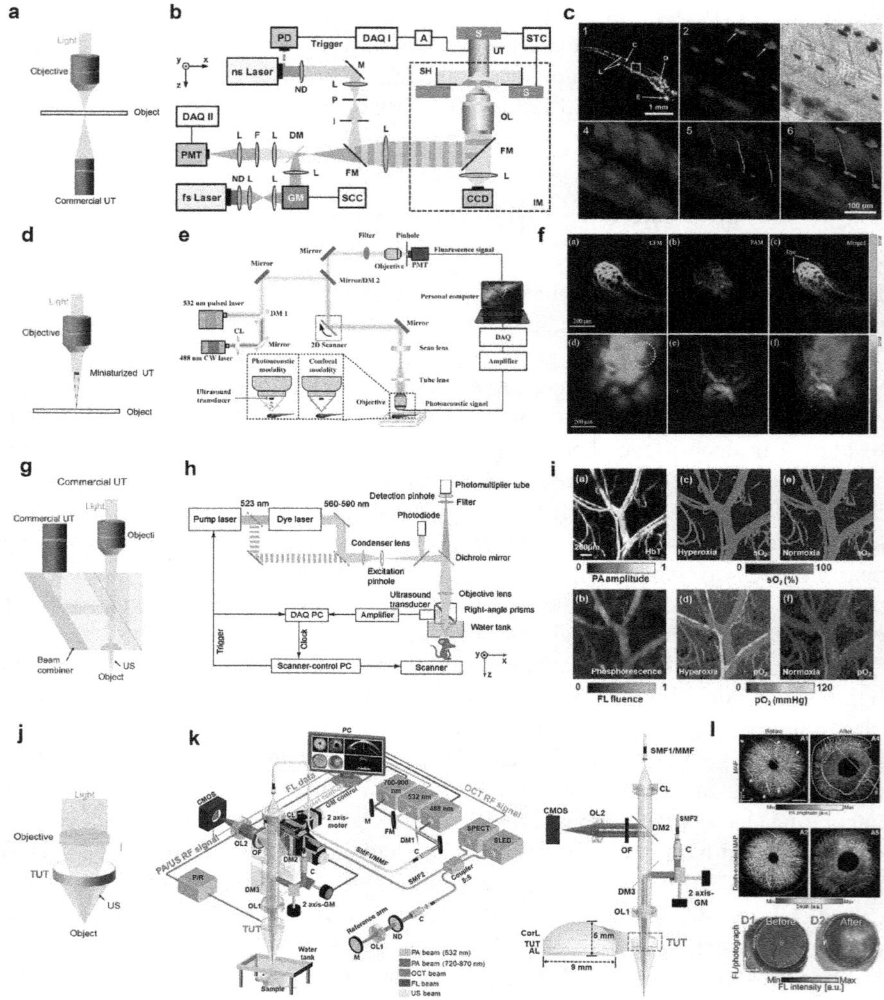

Fig. 13.3 Four common designs in dual-modality OA-FL microscopy and their corresponding experimental results. (**a**) Typical implementation involving light and ultrasound collection from opposite directions. (**b**) Example of an integrated multi-photon and multi-scale OAM system. (**c**) Hybrid label-free imaging of a wild-type zebrafish larva ex vivo. (1) Raster scanning OAM image of the zebrafish, visualizing lateral (L) and central (C) melanocyte stripes, inner organs (O), and the eyes (E). Microscopy images were obtained from the area indicated by the white box. Imaging was performed with the (2) OAM, (3) brightfield, (4) SHG, and (5) THG modalities. (6) Overlay of the OAM, SHG, and THG images. Reproduced with permission from Ref. [74]. (**d**) System design by positioning a miniaturized ultrasound transducer to collect light and ultrasound from the same direction. (**e**) Integrated OAM-CFM system with a miniature ultrasound transducer. (**f**) Whole-body zebrafish and xenograft tumor imaging in the mouse ear with the integrated OAM-CFM system. Green channel: GFP-labeled cells detected in FL mode.

size of 0.6 × 0.5 × 0.2 mm3 could be inserted between the objective and imaging objects to collect both light and US signals from the same side (Fig. 13.3d) [75]. The illumination for both modalities and the FL light collection could be performed through an external cone. The working distance for this miniature US transducer was ~1 mm. The compact design allowed the use of high numerical aperture (NA) objective, and the influence on FL excitation or collection was negligible due to the miniature size of transducer. The hybrid OAM-CFM system (Fig. 13.3e) has been successfully applied to imaging the zebrafish larvae and a xenograft 4T1 breast cancer model inoculated in the mouse ear, where fluorescent protein–labeled 4T1 breast cancer cells were detected in FL mode with surrounding vascular network depicted in the OA mode (Fig. 13.3f) [75].

Another type of design involves using an acoustic-optical beam splitter between the objective and imaging objects (Fig. 13.3g). The hybrid OAM-CFM system shares the same tunable nanosecond pulsed dye laser and objective [76]. An acoustic-optical beam splitter, consisting of two prisms and an oil–glass surface at the diagonal to reflect US signals, is used to reflect the generated OA waves while allowing the transmission of back-scattered FL [77]. Due to the requirement of a long working distance to position the acoustic-optical beam splitter, a low numerical aperture (NA, ~0.2) is used. 2D raster scanning is achieved with a mechanical stage, providing automatically coregistered OA and FL images. An example of a dual-modality system is shown in Fig. 13.3h, which was utilized for hemoglobin sO_2 and oxygen partial pressure (pO_2) mapping of mouse ear under normoxia and hyperoxia conditions [76, 78]. The sO_2 map was acquired with OA mode at 570/578 nm dual wavelengths through spectral unmixing (Fig. 13.3i, top). The pO_2 value was quantified by measuring the lifetime of oxygen-sensitive phosphorescent probe injected through the tail vein (Fig. 13.3i, bottom) using CFM. The variation of sO_2 and pO_2 correlated well under both conditions.

Utilization of a transparent ultrasound transducer (TUT) represents a promising avenue for multimodal imaging due to its on-axis excitation and detection capabilities (Fig. 13.3j), allowing for seamless integration of US imaging, OAM, OCT, and FL imaging [79]. A typical design was reported with lithium niobate (LNO)-based TUT incorporating spherical acoustic focusing lens. The customized TUT

Fig. 13.3 (continued) Red channel: vascular distribution in the OAM mode. Reprinted with permission from Ref. [75]. (**g**) System design by inserting an acoustic-optical beam splitter between the objective and imaged objects. (**h**) Representative integrated OAM-CFM system using an acoustic-optical beam splitter. (**i**) Multiparametric vascular imaging in the mouse ear under hyperoxia and normoxia using OAM-CFM system based on the acoustic-optical beam splitter. Top: OA signal intensity and spectrally unmixed sO_2 maps. Bottom: FL intensity and pO_2 maps. Reproduced with permission from Ref. [78]. (**j**) System design by incorporating a transparent ultrasound transducer. (**k**) Quadruple fusion system using lithium niobate (LNO) crystal-based transparent ultrasound transducer (TUT). (**l**) Vascular mapping and inflammation area quantification in rat's eye before and after alkali burns using quadruple fusion system with OA intensity map (top), color-encoded OA depth map (middle), and FL image (bottom). Reproduced with permission from Ref. [79]

was featured with an acoustic pressure field with 0.13 mm lateral full-width-at-half-maximum (FWHM), 0.89 mm axial FWHM, and ~70% optical transparency in the visible range. Multiple light sources, including two pulsed lasers for OAM, a continuous wave (CW) laser for FL imaging, and a superluminescent light-emitting diode for OCT, were coupled to the corresponding single- and multi-mode fibers to provide co-axial illumination focused by the objective through the TUT. Given the TUT's fixed position, mechanical scanning of the imaging head was required to cover the x-y plane for OA and US, while optical scanning with 2D galvo mirror was utilized for OCT (Fig. 13.3k). This quadruple fusion imaging system was exploited to provide multiparametric visualization of ophthalmologic injuries and oncologic diseases in rodents. Structural changes in the corneal neovascularization in the rat's eye were monitored with single-wavelength OAM while inflammation areas post ophthalmologic injuries were depicted in the FL mode (Fig. 13.3l). In addition to the LNO-based TUT, an optically transparent micro-ring resonator (MRR)-based ultrasound transducer has been developed [80]. This device consists of a bus waveguide and a ring-shaped waveguide with a low-dielectric gap, offering high ultrasound detection sensitivity, a wide angular detection angle, and submillimeter size. An additional narrow-band tunable laser was required to be coupled through the bus waveguide while the generated ultrasound could be detected by quoting the wavelength shift of the MRR resonance modes. Through fabrication of the MRR-based transducer directly on the microscope coverslip, a multimodal system incorporating OAM and CFM has been developed based on a commercial inverted microscope platform [81]. The system has been employed for cellular resolution imaging of ex vivo samples, such as mouse blood smear and retinal pigment epithelium samples.

Besides microscopic imaging, multi-modal OA and FL imaging can be implemented on a macroscopic scale, which however is not straightforward due to the limited physical space available for combining large contact US array transducers with light illumination and collection over a wide field-of-view (FOV). Simultaneous OA and epifluorescence imaging was proposed in transmission mode with a 60-element hemispherical transducer array and a CCD camera arranged on the opposite side of the sample [82]. This approach is limited to thin samples and the image quality is compromised due to the insufficient number of tomographic projections for OA. More recently, simultaneous OA and epifluorescence imaging in reflection mode was proposed using an electron multiplying charge-coupled device (EMCCD)-based fiberscope and a 512-element hemispherical transducer array (Fig. 13.4a) [83]. The customized fiberscope integrated a fiber optic image guide consisting of 100,000 fibers, co-axis illumination fiber bundle, an optical relay system, and an EMCCD for FL detection. The diameter of fiberscope's shaft matches the cylindrical opening of the ultrasound transducer to provide simultaneous OA and FL readings with a similar FOV while OA provided an additional depth dimension. Illumination of both readings was provided with a nanosecond pulsed laser that triggered both the EMCCD and ultrasound transducer simultaneously. The hybrid system has been employed to study indocyanine green (ICG) perfusion (Fig. 13.4b, c) [83], nanoparticles bio-distribution in solid tumors [26], and simultaneous neural

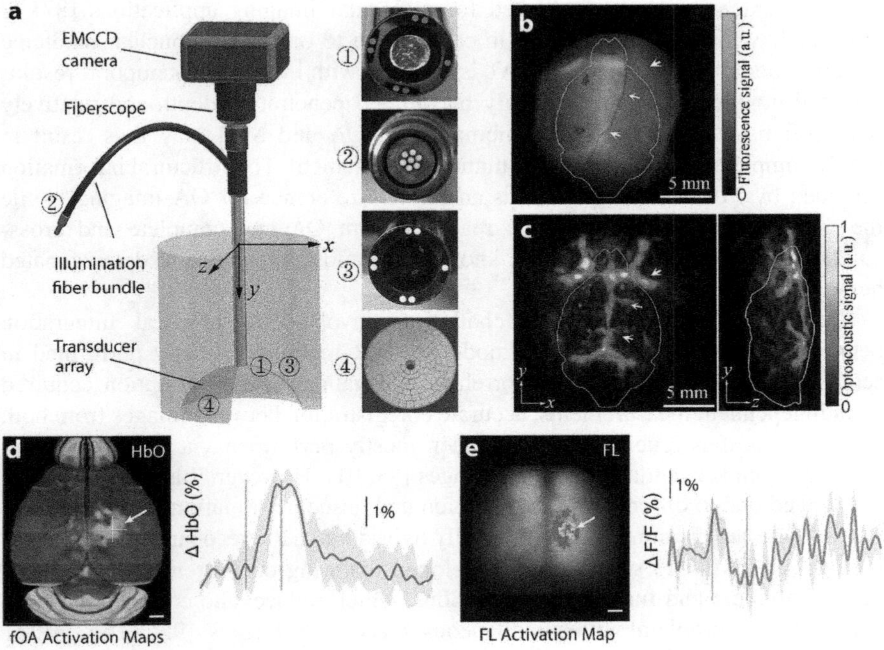

Fig. 13.4 Simultaneous OA tomography and epifluorescence imaging system and application for functional neuroimaging. (**a**) Schematic of the hybrid OA-FL system. (**b**, **c**) FL and volumetric OA images of the mouse brain after intravenous ICG injection. Reprinted with permission from Ref. [83]. (**d**, **e**) Activation map and fractional signal changes from oxygenated hemoglobin (HbO) and genetically encoded calcium indicator GCaMP6f post electrical hindpaw stimulation. Reprinted with permission from Ref. [84]

and hemodynamic activity recordings (Fig. 13.4d, e) [84] in the mouse brain in a noninvasive manner.

13.4 Hybrid OA and Magnetic Resonance Imaging

Magnetic resonance imaging (MRI) is featured with excellent soft tissue contrast and whole-body imaging capability, both in small animals and in humans. As a noninvasive, nonionizing modality, it has been used in various preclinical and clinical applications and currently is the imaging modality of choice for various indications [85]. For instance, functional MRI (fMRI) by measuring the blood oxygen level dependent (BOLD) signal has been the gold standard to study neuronal and brain dynamics derived via indirect hemodynamic activity measurements [86]. MRI is generally associated with very high equipment and maintenance costs and sub-optimal trade-offs between the spatial and temporal resolution. Even though

MRI has extensively been explored for molecular imaging applications [87], it also has low molecular sensitivity in comparison to optical and nuclear medicine imaging modalities. In contrast, OAT is featured with high spatio-temporal resolution and higher molecular sensitivity but limited penetration depth and relatively poor soft tissue contrast. The combination of OA and MRI may thus result in highly complementary features and mutual enhancement. The structural information provided by MRI can be utilized as anatomical reference for OA imaging, while the multiparametric hemodynamic readings from OA can complete and cross-validate the BOLD signals that are known to be mostly sensitive to deoxygenated hemoglobin (HbR).

Due to significant technical challenges involved in physical integration between the two modalities, dual-mode OA-MR imaging is vastly performed in sequence [88–90]. To combine, correlate, and validate the information acquired from independent measurements, accurate coregistration between images from both modalities is thus crucial which has been mostly performed via landmark-based methods from sequentially acquired images [88, 91]. However, this is significantly challenged due to differences in orientation and tissue deformations resulting from independent acquisitions. Recently, a fully hybrid magnetic resonance optoacoustic tomography (MROT) scanner was developed featuring concurrent dual-modal in vivo anatomical and functional acquisitions, which reduces the complexity of the coregistration problem with simultaneously acquired images [92]. The MROT scanner was composed of an OA module inserted into a 9.4 T preclinical MRI scanner. The module includes an MRI-compatible transducer array, fiber bundle, and radiofrequency (RF) coils (Fig. 13.5a). With the simultaneous 4D image data acquired with the hybrid scanner, corruption of OA signals due to RF pulses was observed and addressed with an algorithm automatically eliminating the corrupted frames of the data. Furthermore, field inhomogeneities in MRI images were observed when using water as the acoustic coupling medium which was eliminated by using deuterium oxide (heavy water) and adopting a reconstruction algorithm to appropriate the speed of sound difference between heavy water and biological tissue [92]. Hemodynamic responses induced by breathing oxygen challenge were successfully captured by both modalities in the entire mouse brain [92]. Additionally, dynamic brain responses under an electrical stimulation paradigm demonstrated synchronized changes in BOLD and OA hemodynamic components (oxygenated hemoglobin (HbO), HbR, HbT, and sO_2) (Fig. 13.5b) [93]. Stronger changes in the OA signal intensities were observed compared to BOLD, and the spatial analysis of the responses showed high correlation between BOLD and OA readings (Fig. 13.5c). The BOLD and HbR components were found less sensitive in detecting sensory responses in comparison to other hemodynamic components recorded by OA [93]. Owing to the excellent molecular sensitivity of OA, dual-mode OA-MR was employed in molecular imaging of glucose uptake in an orthotopic glioblastoma in mouse. OA signals from IRDye 800CW 2-DG were detected within the peripheries of the mouse brain within an hour post-administration, and MRI anatomy images can provide spatial information of the probe signals in relation to the tumor tissue in the cortex. Coregistered OA-MR images offered complementary

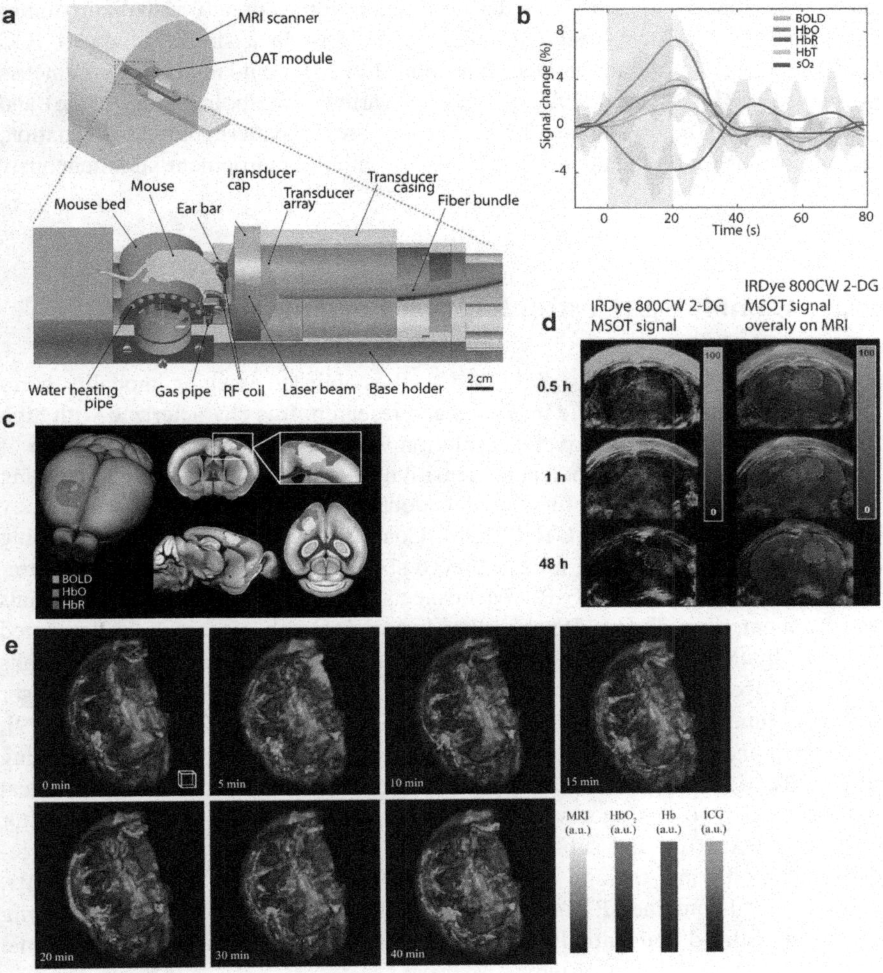

Fig. 13.5 OAT-MRI multimodal imaging. (**a**) The hybrid magnetic resonance and optoacoustic tomography (MROT) scanner layout enabling concurrent MRI and OAT readouts. Reproduced from Ref. [92]. (**b**) Averaged fractional signal intensity changes of BOLD and OAT hemodynamic components following the sensory stimulation. (**c**) Superposition of the BOLD, HbO, and HbR activation maps onto the mouse brain atlas. The white color indicates the overlapping region of the three components. Reproduced from Ref. [93]. (**d**). Left: IRDye 800CW 2-DG MSOT signals (green) in U87MG orthotopic glioma tumor model at various time points post-injection (20 nmol, 100 μL). Right: IRDye 800CW 2-DG optoacoustic signals were superimposed on MRI anatomical images. The red lines show the circumference of the tumor. Reprinted from Ref. [88]. (**e**) Temporal anatomical-molecular imaging of ICG perfusion inside 4T1 tumor. The HbR, HbO, ICG distributions on top of a co-registered MRI image of the 4T1 tumor microenvironment. OA images were acquired at a 5-min interval. Reprinted from Ref. [90]

anatomical, functional, and molecular information for accurately visualizing tumor hemodynamics and dye retention (Fig. 13.5d) [88]. In a similar study of ICG perfusion inside 4T1 tumor in mice, dual-mode OA-MR could render the volumetric images of MRI-T2, HbO, HbR, and ICG at various time points (Fig. 13.5e) and could reveal information on tumor size, border, vascular pattern, blood oxygenation, and molecular probe metabolism of the tumor micro-environment simultaneously [90].

13.5 Combining OA with Ionizing Imaging Modalities

X-ray computed tomography (CT) is a well-established imaging modality commonly employed in diagnostic medicine and research. It is characterized with high penetration depth and high spatial resolution with the drawbacks being the use of ionizing radiation and low molecular sensitivity [94]. OA-CT multimodal imaging can therefore incorporate the functional readouts from OAT with the 3D whole body structural information from CT [95, 96]. Contrast-enhanced in vivo CT imaging of 4T1 breast tumor–bearing mice achieved sixfold signal enhancement post intratumoral (i.t.) injection of WS_2-PEG nanosheets (Fig. 13.6a) and fivefold OA signals enhancement (Fig. 13.6b). On the other hand, single-photon-emission computed tomography (SPECT) allows for molecular imaging with high sensitivity by means of radioactive substances to the detriment of low spatial resolution and longer scanning time [97]. Nanoagents labeled with radionuclides, e.g., ^{125}I [98, 99], ^{131}I [100], and ^{99m}Tc [101], have been employed for OA-SPECT hybrid imaging (Fig. 13.6c, d). Multi-modal OAT-CT-SPECT probes have been used in preclinical settings as diagnostics agents for sentinel lymph node mapping [98] and tumor imaging [100, 102], and substantially as theranostics agents to perform image-guided photothermal therapy (PTT) [103–109], combined PTT/radiotherapy [95, 110–112], and combined PTT/chemotherapy [113], where the dual-modal imaging has been realized sequentially with preclinical OA scanners and small animal nuclear medicine imaging systems.

Another popular nuclear medicine imaging modality is positron emission tomography (PET), which is widely used in disease diagnosis, pharmacokinetic, and therapy response studies. While allowing for whole-body quantitative tracking of radionuclide-labeled contrast agents with high sensitivity, similarly to SPECT, PET suffers from inferior spatial resolution, slow imaging speed, absence of anatomical information, and the need for an onsite cyclotron [114]. In preclinical PET imaging, the low spatial resolution often leads to partial volume effects and thus inaccurate estimates of activity based on PET images [115]. The higher resolution tomographic data of OA may complement the excellent sensitivity and deep tissue quantitative imaging capacities of PET by providing additional functional readouts and aiding biodistribution studies with its superior temporal resolution. To this end, OA-PET dual-modal imaging has been applied for image-guided cancer therapy, gut imaging, Alzheimer's disease studies (Fig. 13.6e–g) [116–121].

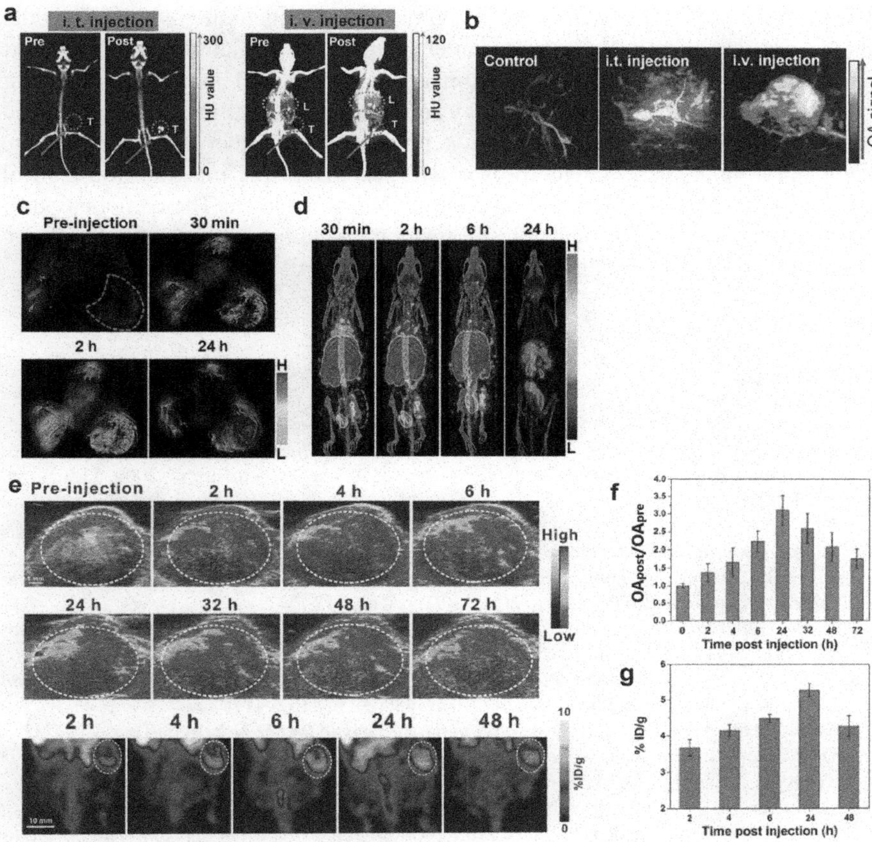

Fig. 13.6 Hybrid OA with ionizing imaging modalities. (**a**) CT images of mice pre- and post-i.t. and -i.v. injection of WS_2-PEG nanosheets. Red dashed circle indicates the tumor and green dashed circle indicates the mouse liver. (**b**) OA images of tumors on mice pre- and post- i.t. or -i.v. injection of WS_2-PEG. Reprinted from Ref. [96]. (**c**) In vivo dynamic OA imaging of osteosarcoma-bearing mice pre- and post-injection of Cy@Silk. Tumor region is indicated by the red dashed area. (**d**) In vivo dynamic SPECT imaging of osteosarcoma-bearing mice at 30 min and 2, 6, and 24 h post-injection of Cy@Silk-^{99m}Tc. Tumor region is indicated by the red dashed area. Reprinted from Ref. [122]. (**e**) OA and PET images of U87MG-tumor-bearing mice pre- and postinjection of PGM NPs. (**f**), (**g**) OA signal intensity changes and time-dependent tumor uptake of CPGM NPs. Reproduced from Ref. [117]

13.6 Conclusion and Outlook

In conclusion, OA imaging has successfully been combined with US, optical, MRI, and ionizing imaging modalities, thus delivering highly complementary anatomical, functional, and molecular readings and enabling cross-validation between modalities. OA has obvious benefits in the multi-modal setting owing to its portability and ability to render rich optical absorption contrast from living tissues with high spatial and temporal resolution. Some technical challenges and limitations of OA, such as limited penetration depth, lack of quantification due to wavelength-dependent light attenuation, image artifacts introduced by acoustic heterogeneities, can be alleviated by taking advantage of the information provided by another modality. Despite promising pilot studies showing prospective benefits of OA imaging in a clinical setting, so far it has mostly been used in preclinical research. The combination of OA with other well-established clinical imaging modalities can improve diagnostic capacities, thus greatly facilitating OA translation into the clinical space.

References

1. Cho, E.C., et al., *Measuring the Optical Absorption Cross Sections of Au-Ag Nanocages and Au Nanorods by Photoacoustic Imaging.* Journal of Physical Chemistry C, 2009. **113**(21): p. 9023–9028.
2. Kim, G., et al., *Indocyanine-green-embedded PEBBLEs as a contrast agent for photoacoustic imaging.* Journal of Biomedical Optics, 2007. **12**(4).
3. Razansky, D., et al., *Multispectral opto-acoustic tomography of deep-seated fluorescent proteins in vivo.* Nature Photonics, 2009. **3**(7): p. 412–417.
4. Wang, L.V., *Multiscale photoacoustic microscopy and computed tomography.* Nature Photonics, 2009. **3**(9): p. 503–509.
5. Razansky, D., J. Klohs, and R. Ni, *Multi-scale optoacoustic molecular imaging of brain diseases.* Eur J Nucl Med Mol Imaging, 2021. **48**(13): p. 4152–4170.
6. Shi, J., et al., *Advances in super-resolution photoacoustic imaging*, 2018. **8**(8): p. 724–732.
7. Deán-Ben, X.L., et al., *Advanced optoacoustic methods for multiscale imaging of in vivo dynamics.* Chemical Society Reviews, 2017. **46**(8): p. 2158–2198.
8. Deán-Ben, X.L., A. Ozbek, and D. Razansky, *Volumetric Real-Time Tracking of Peripheral Human Vasculature With GPU-Accelerated Three-Dimensional Optoacoustic Tomography.* IEEE Transactions on Medical Imaging, 2013. **32**(11): p. 2050–2055.
9. Özbek, A., X.L. Deán-Ben, and D. Razansky, *Optoacoustic imaging at kilohertz volumetric frame rates.* Optica, 2018. **5**(7): p. 857–863.
10. Nam, S.Y., et al., *Combined ultrasound and photoacoustic imaging to noninvasively assess burn injury and selectively monitor a regenerative tissue-engineered construct.* Tissue Eng Part C Methods, 2015. **21**(6): p. 557–66.
11. Wang, X., et al., *Noninvasive imaging of hemoglobin concentration and oxygenation in the rat brain using high-resolution photoacoustic tomography.* J Biomed Opt, 2006. **11**(2): p. 024015.
12. Wang, X.D., et al., *Noninvasive laser-induced photoacoustic tomography for structural and functional in vivo imaging of the brain.* Nature Biotechnology, 2003. **21**(7): p. 803–806.
13. Tomaszewski, M.R., et al., *Oxygen-Enhanced and Dynamic Contrast-Enhanced Optoacoustic Tomography Provide Surrogate Biomarkers of Tumor Vascular Function, Hypoxia, and Necrosis.* Cancer Research, 2018. **78**(20): p. 5980–5991.
14. Chen, Z., et al., *Multimodal Noninvasive Functional Neurophotonic Imaging of Murine Brain-Wide Sensory Responses.* Advanced Science. **n/a**(n/a): p. 2105588.
15. Weber, J., et al., *An activatable cancer-targeted hydrogen peroxide probe for photoacoustic and fluorescence imaging.* Cancer Res, 2019.
16. Sun, I.C., et al., *Photoacoustic imaging of cancer cells with glycol-chitosan-coated gold nanoparticles as contrast agents.* J Biomed Opt, 2019. **24**(12): p. 1–5.

17. Gargiulo, S., S. Albanese, and M. Mancini, *State-of-the-Art Preclinical Photoacoustic Imaging in Oncology: Recent Advances in Cancer Theranostics.* Contrast Media Mol Imaging, 2019. **2019**: p. 5080267.
18. Liu, W. and J. Yao, *Photoacoustic microscopy: principles and biomedical applications.* Biomedical Engineering Letters, 2018. **8**(2): p. 203–213.
19. Lin, L. and L.V.J.N.R.C.O. Wang, *The emerging role of photoacoustic imaging in clinical oncology.* 2022. **19**(6): p. 365–384.
20. Attia, A.B.E., et al., *A review of clinical photoacoustic imaging: Current and future trends.* 2019. **16**: p. 100144.
21. Rebling, J., et al., *Dual-wavelength hybrid optoacoustic-ultrasound biomicroscopy for functional imaging of large-scale cerebral vascular networks.* Journal of Biophotonics, 2018. **11**(9): p. e201800057.
22. Deán-Ben, X.L., E. Merčep, and D. Razansky, *Hybrid-array-based optoacoustic and ultrasound (OPUS) imaging of biological tissues.* Applied Physics Letters, 2017. **110**(20): p. 203703.
23. Lafci, B., et al., *Noninvasive multiparametric characterization of mammary tumors with transmission-reflection optoacoustic ultrasound.* Neoplasia, 2020. **22**(12): p. 770–777.
24. Estrada, H., et al., *High-resolution fluorescence-guided transcranial ultrasound mapping in the live mouse brain.* Science Advances, 2021. **7**(50): p. eabi5464.
25. Chen, Z., et al., *Hybrid system for in vivo epifluorescence and 4D optoacoustic imaging.* Optics Letters, 2017. **42**(22): p. 4577–4580.
26. Chen, Z., et al., *Concurrent fluorescence and volumetric optoacoustic tomography of nanoagent perfusion and bio-distribution in solid tumors.* Biomedical optics express, 2019. **10**(10): p. 5093–5102.
27. Berer, T., et al., *Multimodal noncontact photoacoustic and optical coherence tomography imaging using wavelength-division multiplexing.* J Biomed Opt, 2015. **20**(4): p. 46013.
28. Zhang, W., et al., *High-resolution, in vivo multimodal photoacoustic microscopy, optical coherence tomography, and fluorescence microscopy imaging of rabbit retinal neovascularization.* Light Sci Appl, 2018. **7**: p. 103.
29. Chen, Z., et al., *Non-invasive multimodal optical coherence and photoacoustic tomography for human skin imaging.* Sci Rep, 2017. **7**(1): p. 17975.
30. Zhang, S., et al., *MRI Information-Based Correction and Restoration of Photoacoustic Tomography.* IEEE Transactions on Medical Imaging, 2022. **41**(9): p. 2543–2555.
31. Qin, C., et al., *Tyrosinase as a multifunctional reporter gene for Photoacoustic/MRI/PET triple modality molecular imaging.* 2013. **3**(1): p. 1–8.
32. Zhang, S., et al., *MRI Information-Based Correction and Restoration of Photoacoustic Tomography.* 2022. **41**(9): p. 2543–2555.
33. Chen, Z., et al., *Hybrid magnetic resonance and optoacoustic tomography (MROT) for preclinical neuroimaging.* 2022. **11**(1): p. 332.
34. Chen, Z., et al., *Simultaneous Functional Magnetic Resonance and Optoacoustic Imaging of Brain-Wide Sensory Responses in Mice.* 2023. **10**(3): p. 2205191.
35. Deán-Ben, X.L. and D. Razansky, *Functional optoacoustic human angiography with handheld video rate three dimensional scanner.* Photoacoustics, 2013. **1**(3): p. 68–73.
36. Zhang, Y. and L. Wang, *Video-rate full-ring ultrasound and photoacoustic computed tomography with real-time sound speed optimization.* Biomedical Optics Express, 2022. **13**(8): p. 4398–4413.
37. Merčep, E., et al., *Transmission–reflection optoacoustic ultrasound (TROPUS) computed tomography of small animals.* Light: Science & Applications, 2019. **8**(1): p. 18.
38. Jose, J., et al., *Speed-of-sound compensated photoacoustic tomography for accurate imaging.* Medical physics, 2012. **39**(12): p. 7262–7271.
39. Jeng, G.-S., et al., *Real-time interleaved spectroscopic photoacoustic and ultrasound (PAUS) scanning with simultaneous fluence compensation and motion correction.* Nature communications, 2021. **12**(1): p. 716.

40. Niederhauser, J.J., et al., *Combined ultrasound and optoacoustic system for real-time high-contrast vascular imaging in vivo.* IEEE transactions on medical imaging, 2005. **24**(4): p. 436–440.
41. Kim, J., et al., *Multiparametric Photoacoustic Analysis of Human Thyroid Cancers in Vivophotoacoustic Analysis of Human Thyroid Cancers.* Cancer Research, 2021. **81**(18): p. 4849–4860.
42. Yang, M., et al., *Photoacoustic/ultrasound dual imaging of human thyroid cancers: an initial clinical study.* Biomedical optics express, 2017. **8**(7): p. 3449–3457.
43. Garcia-Uribe, A., et al., *Dual-modality photoacoustic and ultrasound imaging system for noninvasive sentinel lymph node detection in patients with breast cancer.* Scientific reports, 2015. **5**(1): p. 1–8.
44. Neuschler, E.I., et al., *A pivotal study of optoacoustic imaging to diagnose benign and malignant breast masses: a new evaluation tool for radiologists.* Radiology, 2018. **287**(2): p. 398–412.
45. Merčep, E., et al., *Hybrid optoacoustic tomography and pulse-echo ultrasonography using concave arrays.* IEEE transactions on ultrasonics, ferroelectrics, and frequency control, 2015. **62**(9): p. 1651–1661.
46. Merčep, E., et al., *Whole-body live mouse imaging by hybrid reflection-mode ultrasound and optoacoustic tomography.* Optics letters, 2015. **40**(20): p. 4643–4646.
47. Dima, A. and V. Ntziachristos, *In-vivo handheld optoacoustic tomography of the human thyroid.* Photoacoustics, 2016. **4**(2): p. 65–69.
48. Becker, A., et al., *Multispectral optoacoustic tomography of the human breast: characterisation of healthy tissue and malignant lesions using a hybrid ultrasound-optoacoustic approach.* European radiology, 2018. **28**: p. 602–609.
49. Lafci, B., et al., *Noninvasive multiparametric charac-terization of mammary tumors with transmission-reflection optoacoustic ultrasound.* Neoplasia, 2020. **22**(12): p. 770–777.
50. Lafci, B., et al., *Multimodal assessment of non-alcoholic fatty liver disease with transmission-reflection optoacoustic ultrasound.* Theranostics, 2023. **13**(12): p. 4217–4228.
51. Jose, J., et al., *Passive element enriched photoacoustic computed tomography (PER PACT) for simultaneous imaging of acoustic propagation properties and light absorption.* Optics express, 2011. **19**(3): p. 2093–2104.
52. Resink, S., et al., *Multiple passive element enriched photoacoustic computed tomography.* Optics letters, 2011. **36**(15): p. 2809–2811.
53. Xia, J., et al., *Enhancement of photoacoustic tomography by ultrasonic computed tomography based on optical excitation of elements of a full-ring transducer array.* Optics letters, 2013. **38**(16): p. 3140–3143.
54. Zhang, X., et al., *Full noncontact laser ultrasound: first human data.* Light: Science & Applications, 2019. **8**(1): p. 119.
55. Lafci, B., et al., *Multimodal Assessment of Non-Alcoholic Fatty Liver Disease with Transmission-Reflection Optoacoustic Ultrasound.* bioRxiv, 2022: p. 2022.08. 16.504139.
56. Park, B., et al., *3D wide-field multispectral photoacoustic imaging of human melanomas in vivo: a pilot study.* Journal of the European Academy of Dermatology and Venereology, 2021. **35**(3): p. 669–676.
57. Jose, J., et al., *Imaging of tumor vasculature using Twente photoacoustic systems.* Journal of biophotonics, 2009. **2**(12): p. 701–717.
58. Jo, J., et al., *A functional study of human inflammatory arthritis using photoacoustic imaging.* Scientific reports, 2017. **7**(1): p. 15026.
59. van den Berg, P.J., et al., *Feasibility of photoacoustic/ultrasound imaging of synovitis in finger joints using a point-of-care system.* Photoacoustics, 2017. **8**: p. 8–14.
60. Knieling, F., et al., *Multispectral optoacoustic tomography for assessment of Crohn's disease activity.* New England Journal of Medicine, 2017. **376**(13): p. 1292–1294.
61. Wen, Y., et al., *Clinical photoacoustic/ultrasound dual-modal imaging: Current status and future trends.* 2022. **13**.

62. Li, W., et al., *Tracking Strain-Specific Morphogenesis and Angiogenesis of Murine Calvaria with Large-Scale Optoacoustic and Ultrasound Microscopy.* Journal of Bone and Mineral Research, 2022. **37**(5): p. 1032–1043.
63. Estrada, H., et al., *Intravital optoacoustic and ultrasound bio-microscopy reveal radiation-inhibited skull angiogenesis.* Bone, 2020. **133**: p. 115251.
64. Ning, B., et al., *Ultrasound-aided multi-parametric photoacoustic microscopy of the mouse brain.* Scientific reports, 2015. **5**(1): p. 18775.
65. Chen, W., C. Tao, and X. Liu, *Artifact-free imaging through a bone-like layer by using an ultrasonic-guided photoacoustic microscopy.* Optics Letters, 2019. **44**(5): p. 1273–1276.
66. Ly, C.D., et al., *Development of fast photoacoustic and ultrasound imaging system based on slider-crank scanner for small animals and humans study.* Expert Systems with Applications, 2022. **206**: p. 117939.
67. Yang, J.-M., et al., *Simultaneous functional photoacoustic and ultrasonic endoscopy of internal organs in vivo.* Nature Medicine, 2012. **18**(8): p. 1297–1302.
68. Dai, X., et al., *Miniature Endoscope for Multimodal Imaging.* ACS Photonics, 2017. **4**(1): p. 174–180.
69. Estrada, H., et al., *Virtual craniotomy for high-resolution optoacoustic brain microscopy.* Scientific Reports, 2018. **8**(1): p. 1459.
70. Wang, Y., et al., *Toward in vivo biopsy of melanoma based on photoacoustic and ultrasound dual imaging with an integrated detector.* Biomedical optics express, 2016. **7**(2): p. 279–286.
71. Tserevelakis, G.J., et al., *Hybrid multiphoton and optoacoustic microscope.* Optics letters, 2014. **39**(7): p. 1819–1822.
72. Soliman, D., et al., *Combining microscopy with mesoscopy using optical and optoacoustic label-free modes.* Scientific reports, 2015. **5**(1): p. 1–9.
73. Rao, B., et al., *Integrated photoacoustic, confocal, and two-photon microscope.* Journal of biomedical optics, 2014. **19**(3): p. 036002–036002.
74. Soliman, D., et al., *Combining microscopy with mesoscopy using optical and optoacoustic label-free modes.* Sci Rep, 2015. **5**: p. 12902.
75. Liu, C., et al., *The integrated high-resolution reflection-mode photoacoustic and fluorescence confocal microscopy.* Photoacoustics, 2019. **14**: p. 12–18.
76. Wang, Y., et al., *Integrated photoacoustic and fluorescence confocal microscopy.* IEEE Trans Biomed Eng, 2010. **57**(10): p. 2576–8.
77. Hu, S., et al., *Intravital imaging of amyloid plaques in a transgenic mouse model using optical-resolution photoacoustic microscopy.* Optics letters, 2009. **34**(24): p. 3899–3901.
78. Wang, Y., et al., *In vivo integrated photoacoustic and confocal microscopy of hemoglobin oxygen saturation and oxygen partial pressure.* Optics letters, 2011. **36**(7): p. 1029–1031.
79. Park, J., et al., *Quadruple ultrasound, photoacoustic, optical coherence, and fluorescence fusion imaging with a transparent ultrasound transducer.* Proceedings of the National Academy of Sciences, 2021. **118**(11): p. e1920879118.
80. Li, H., et al., *A transparent broadband ultrasonic detector based on an optical micro-ring resonator for photoacoustic microscopy.* Scientific reports, 2014. **4**(1): p. 1–8.
81. Dong, B., et al., *Isometric multimodal photoacoustic microscopy based on optically transparent micro-ring ultrasonic detection.* Optica, 2015. **2**(2): p. 169–176.
82. Kosik, I. and J. Carson, *Combined 3D photoacoustic and 2D fluorescence imaging of indocyanine green contrast agent flow*. SPIE BiOS. Vol. 8581. 2013: SPIE.
83. Chen, Z., et al., *Hybrid system for in vivo epifluorescence and 4D optoacoustic imaging.* Opt Lett, 2017. **42**(22): p. 4577–4580.
84. Chen, Z., et al., *Multimodal Noninvasive Functional Neurophotonic Imaging of Murine Brain-Wide Sensory Responses.* Advanced Science, 2022. **9**(24): p. 2105588.
85. Glover, G.H., *Overview of functional magnetic resonance imaging.* Neurosurg Clin N Am, 2011. **22**(2): p. 133–9, vii.
86. Handwerker, D.A., et al., *The continuing challenge of understanding and modeling hemodynamic variation in fMRI.* Neuroimage, 2012. **62**(2): p. 1017–23.

87. Terreno, E., et al., *Challenges for molecular magnetic resonance imaging.* Chem Rev, 2010. **110**(5): p. 3019–42.
88. Attia, A.B.E., et al., *Multispectral optoacoustic and MRI coregistration for molecular imaging of orthotopic model of human glioblastoma.* Journal of Biophotonics, 2016. **9**(7): p. 701–708.
89. Park, S., et al., *Real-time Triple-modal Photoacoustic, Ultrasound, and Magnetic Resonance Fusion Imaging of Humans.* IEEE Trans Med Imaging, 2017. **36**(9): p. 1912–1921.
90. Zhang, S., et al., *In vivo co-registered hybrid-contrast imaging by successive photoacoustic tomography and magnetic resonance imaging.* Photoacoustics, 2023. **31**: p. 100506.
91. Gehrung, M., et al., *Co-registration of optoacoustic tomography and magnetic resonance imaging data from murine tumour models.* Photoacoustics, 2020. **18**: p. 100147.
92. Chen, Z., et al., *Hybrid magnetic resonance and optoacoustic tomography (MROT) for preclinical neuroimaging.* Light Sci Appl, 2022. **11**(1): p. 332.
93. Chen, Z., et al., *Simultaneous Functional Magnetic Resonance and Optoacoustic Imaging of Brain-Wide Sensory Responses in Mice.* Adv Sci (Weinh), 2023. **10**(3): p. e2205191.
94. Withers, P.J., et al., *X-ray computed tomography.* Nature Reviews Methods Primers, 2021. **1**(1): p. 18.
95. Tang, W., et al., *Multifunctional Two-Dimensional Core-Shell MXene@Gold Nanocomposites for Enhanced Photo-Radio Combined Therapy in the Second Biological Window.* ACS Nano, 2019. **13**(1): p. 284–294.
96. Cheng, L., et al., *PEGylated WS(2) nanosheets as a multifunctional theranostic agent for in vivo dual-modal CT/photoacoustic imaging guided photothermal therapy.* Adv Mater, 2014. **26**(12): p. 1886–93.
97. Khalil, M.M., et al., *Molecular SPECT Imaging: An Overview.* Int J Mol Imaging, 2011. **2011**: p. 796025.
98. Akers, W.J., et al., *Multimodal sentinel lymph node mapping with single-photon emission computed tomography (SPECT)/computed tomography (CT) and photoacoustic tomography.* Transl Res, 2012. **159**(3): p. 175–81.
99. Yao, M., et al., *Engineering of SPECT/Photoacoustic Imaging/Antioxidative Stress Triple-Function Nanoprobe for Advanced Mesenchymal Stem Cell Therapy of Cerebral Ischemia.* ACS Appl Mater Interfaces, 2020. **12**(34): p. 37885–37895.
100. Liu, Y., et al., *Aggregation-Induced Absorption Enhancement for Deep Near-Infrared II Photoacoustic Imaging of Brain Gliomas In Vivo.* Adv Sci (Weinh), 2019. **6**(8): p. 1801615.
101. Guo, Y., et al., *Magnetic-responsive and targeted cancer nanotheranostics by PA/MR bimodal imaging-guided photothermally triggered immunotherapy.* Biomaterials, 2019. **219**: p. 119370.
102. Shang, W., et al., *Core-Shell Gold Nanorod@Metal-Organic Framework Nanoprobes for Multimodality Diagnosis of Glioma.* Adv Mater, 2017. **29**(3).
103. Chen, M., et al., *Core-shell Pd@Au nanoplates as theranostic agents for in-vivo photoacoustic imaging, CT imaging, and photothermal therapy.* Adv Mater, 2014. **26**(48): p. 8210–6.
104. Jin, Y., et al., *Encapsulating tantalum oxide into polypyrrole nanoparticles for X-ray CT/photoacoustic bimodal imaging-guided photothermal ablation of cancer.* Biomaterials, 2014. **35**(22): p. 5795–804.
105. Jing, L., et al., *Prussian blue coated gold nanoparticles for simultaneous photoacoustic/CT bimodal imaging and photothermal ablation of cancer.* Biomaterials, 2014. **35**(22): p. 5814–21.
106. Li, Z., et al., *Multimodal Imaging-Guided Antitumor Photothermal Therapy and Drug Delivery Using Bismuth Selenide Spherical Sponge.* ACS Nano, 2016. **10**(10): p. 9646–9658.
107. Liang, X., et al., *Activatable near infrared dye conjugated hyaluronic acid based nanoparticles as a targeted theranostic agent for enhanced fluorescence/CT/photoacoustic imaging guided photothermal therapy.* Biomaterials, 2017. **132**: p. 72–84.
108. Wu, B., et al., *Gadolinium-chelate functionalized bismuth nanotheranostic agent for in vivo MRI/CT/PAI imaging-guided photothermal cancer therapy.* Biomaterials, 2018. **159**: p. 37–47.

109. Yang, S., et al., *Multifunctional Bi@PPy-PEG Core-Shell Nanohybrids for Dual-Modal Imaging and Photothermal Therapy.* ACS Appl Mater Interfaces, 2018. **10**(2): p. 1605–1615.
110. Yu, X., et al., *Ultrasmall Semimetal Nanoparticles of Bismuth for Dual-Modal Computed Tomography/Photoacoustic Imaging and Synergistic Thermoradiotherapy.* ACS Nano, 2017. **11**(4): p. 3990–4001.
111. Cheng, Y., et al., *Glutathione-Mediated Clearable Nanoparticles Based on Ultrasmall Gd(2)O(3) for MSOT/CT/MR Imaging Guided Photothermal/Radio Combination Cancer Therapy.* Mol Pharm, 2019. **16**(8): p. 3489–3501.
112. Wang, Y., et al., *Oxygen-supplementing mesoporous polydopamine nanosponges with WS(2) QDs-embedded for CT/MSOT/MR imaging and thermoradiotherapy of hypoxic cancer.* Biomaterials, 2019. **220**: p. 119405.
113. Men, X., et al., *Thermosensitive Polymer Dot Nanocomposites for Trimodal Computed Tomography/Photoacoustic/Fluorescence Imaging-Guided Synergistic Chemo-Photothermal Therapy.* ACS Appl Mater Interfaces, 2020. **12**(46): p. 51174–51184.
114. Shukla, A.K. and U. Kumar, *Positron emission tomography: An overview.* J Med Phys, 2006. **31**(1): p. 13–21.
115. Vaquero, J.J. and P. Kinahan, *Positron Emission Tomography: Current Challenges and Opportunities for Technological Advances in Clinical and Preclinical Imaging Systems.* Annu Rev Biomed Eng, 2015. **17**: p. 385–414.
116. Zhang, Y., et al., *Surfactant-Stripped Frozen Pheophytin Micelles for Multimodal Gut Imaging.* Adv Mater, 2016. **28**(38): p. 8524–8530.
117. Wang, S., et al., *Core-Satellite Polydopamine-Gadolinium-Metallofullerene Nanotheranostics for Multimodal Imaging Guided Combination Cancer Therapy.* Adv Mater, 2017. **29**(35).
118. Liu, Y., et al., *Highly specific noninvasive photoacoustic and positron emission tomography of brain plaque with functionalized croconium dye labeled by a radiotracer.* Chem Sci, 2017. **8**(4): p. 2710–2716.
119. Zhang, Y., et al., *Non-invasive multimodal functional imaging of the intestine with frozen micellar naphthalocyanines.* Nat Nanotechnol, 2014. **9**(8): p. 631–8.
120. Xu, C., et al., *Bacteria-like mesoporous silica-coated gold nanorods for positron emission tomography and photoacoustic imaging-guided chemo-photothermal combined therapy.* Biomaterials, 2018. **165**: p. 56–65.
121. Lin, J., et al., *Multimodal-Imaging-Guided Cancer Phototherapy by Versatile Biomimetic Theranostics with UV and γ-Irradiation Protection.* Adv Mater, 2016. **28**(17): p. 3273–9.
122. Wang, Y., et al., *In Vivo Photoacoustic/Single-Photon Emission Computed Tomography Imaging for Dynamic Monitoring of Aggregation-Enhanced Photothermal Nanoagents.* Analytical Chemistry, 2019. **91**(3): p. 2128–2134.

Part VI
Pre-clinical and Clinical Applications

Chapter 14
Multifaceted Potential of Photoacoustic Imaging for Preclinical Cancer Research

Deeksha Sankepalle, Allison Sweeney, and Srivalleesha Mallidi

Abstract Cancer, due to its intricate biology and heterogeneity, remains an insurmountable challenge at various molecular, cellular, and tissue levels. Substantial efforts are being dedicated to unraveling the complexities of the macro- and microenvironments of tumors that will eventually aid in developing innovative and precise treatment strategies for improving patient survival rates. In the quest to decipher tumors' biological, physical, and molecular attributes and their responses to therapy, animal cancer models have emerged as invaluable preclinical platforms. Noninvasive imaging of these preclinical cancer models enables researchers to visualize, quantify, and comprehend the disease processes, evaluate treatment effectiveness, and refine experimental designs. Photoacoustic (PA) imaging, a modality that stands at the intersection of optics and acoustics, has garnered significant attention in the realm of preclinical cancer research. Among the diverse array of medical imaging modalities available, PA imaging offers the unique feature of providing three-dimensional, high-resolution, functional, and molecular information with or without the use of exogenous contrast agents. In this chapter, we present the multifaceted potential of PA imaging for preclinical cancer research and delve into the advantages and limitations of this modality in providing insights into the tumor microenvironment and therapy response. By shedding light on the evolving landscape of PA imaging, we aim to provide preclinical researchers with a comprehensive summary of the role it can play in the relentless pursuit of more effective cancer diagnosis and treatment strategies.

Keywords Preclinical research · Murine models · Tumor vasculature · Blood oxygen saturation · Therapy monitoring

D. Sankepalle · A. Sweeney
Department of Biomedical Engineering, Tufts University, Medford, MA, USA

S. Mallidi (✉)
Department of Biomedical Engineering, Tufts University, Medford, MA, USA

Wellman Center for Photomedicine, Massachusetts General Hospital, Harvard Medical School, Boston, MA, USA
e-mail: Srivalleesha.Mallidi@tufts.edu

W. Xia (ed.), *Biomedical Photoacoustics*,
https://doi.org/10.1007/978-3-031-61411-8_14

Abbreviations

AR-PAM	Acoustic resolution photoacoustic microscopy
CNR	Contrast to noise ratio
DCIS	Ductal carcinoma in situ
GEMM	Genetically engineered mouse models
ICG	Indocyanine green
LED	Light-emitting diode
MRI	Magnetic resonance imaging
MSOT	Multispectral optoacoustic tomography
NIR	Near infrared
NSCLC	Nonsmall cell lung cancer
OE	Oxygen-enhanced
OR-PAM	Optical resolution photoacoustic microscopy
PA	Photoacoustic
PHANTOM	Photoacoustic Annotation Toolkit in MATLAB
PDT	Photodynamic therapy
PDX	Patient-derived xenograft
PET	Positron emission tomography
RSOM	Reflective-scanning optoacoustic microscopy
SNR	Signal-to-noise ratio
SPECT	Single photon emission computed tomography
US	Ultrasound
VEGF(R)	Vascular endothelial growth factor (receptor)
2DG	2-deoxyglucose

14.1 Introduction

Cancer is one of the leading causes of death worldwide. In fact, in the United States alone, the American Cancer Society projected around two million new cancer cases and about 600 thousand cancer-related deaths in 2023 [1]. Despite remarkable progress in various fields of medicine, from early diagnosis to targeted therapies, the battle against cancer is far from over. The primary reason for the ongoing challenge is due to the highly intricate and complex biological, physiological, and molecular heterogeneity expressed in tumors. This complexity underscores the critical need for preclinical research that serves as the essential bridge between laboratory discoveries and clinical trials involving human subjects. In the preclinical phase, researchers (from both the academic and industrial realms) can delve deep into the molecular and cellular intricacies of cancer in controlled environments, gaining invaluable insights into its biology, progression, and response to new anticancer interventions.

Preclinical cancer research is performed on species ranging from zebrafish [2] to large animals (canines, primates, etc.) [3], and these models remain indispensable for unraveling the complexities of cancer and accelerating the development of novel and more effective treatments [4]. Among the different animal models available, rodents are the most widely used in preclinical cancer research due to their cost-effectiveness while providing a dynamic tumor microenvironment [4]. The most cost-effective rodent model involves subcutaneously implanting human or mouse-derived cancer cell lines in mice. Orthotopic xenograft models go further by implanting cells directly into the organ of tumor origin. Patient-derived xenograft (PDX) models use freshly resected human tumor tissue implanted in immunosuppressed mice, and this is more complex and expensive methodology. Another widely used mouse model that revolutionized cancer research is the Genetically Engineered Mouse Model (GEMM) where specific human cancer aspects are mimicked by specifically altering mouse genetics [5]. In all of these preclinical models, high-resolution imaging is crucial to advance our understanding of the intricacies of cancer biology such as tumor growth, proliferation, senescence, molecular and genetic heterogeneity in the tumor microenvironment, metastasis, and overall response to various types of cancer treatments.

Various imaging techniques have emerged as powerful and precise tools for assessing preclinical tumor models and associated microenvironments [6]. Intravital microscopy, nuclear imaging techniques of Positron Emission Tomography (PET) and Single Photon Emission Computed Tomography (SPECT), nonionizing imaging techniques like magnetic resonance imaging (MRI), ultrasound (US), and optical imaging modalities such as fluorescence and bioluminescence imaging are currently used for obtaining information on tumor structure and function. In the last two decades, photoacoustic (PA) imaging has gained immense popularity and stands out amongst the imaging modalities specified above for its noninvasive, nonionizing, near real-time, deep-tissue, and 3D functional imaging capabilities of the tumor microenvironment, particularly tumor vasculature [7]. Photoacoustic imaging, also known as optoacoustic imaging, is an innovative biomedical imaging technique that harnesses the principles of light absorption and acoustic wave generation to produce high-resolution images of biological tissues. The fundamental principle of PA imaging begins with the delivery of short pulses of laser light into the tissue of interest. These laser pulses are typically in the near-infrared (NIR) range to ensure deeper tissue penetration [8]. Within the tissue, the laser light is absorbed by specific molecules known as chromophores. These chromophores can be naturally occurring, such as hemoglobin in blood vessels or they can be exogenous contrast agents designed to explicitly target cancer biomarkers [7, 9]. Upon absorption of the laser energy, the chromophores undergo rapid heating, causing localized and transient thermal expansion within the tissue. This sudden expansion generates ultrasonic or acoustic waves, a phenomenon termed the photoacoustic effect. These acoustic waves then propagate outward from the point of absorption in all directions [10]. Ultrasound detectors, usually placed on the tissue's surface, capture the resulting acoustic signals. By precisely measuring the time of arrival and intensity of these signals, a detailed 2D or 3D image of the tissue can be reconstructed. This reconstructed image represents the spatial distribution of the absorbed laser energy and, consequently, the distribution of the chromophores within the tissue [11].

Photoacoustic imaging offers a distinctive advantage in its deep tissue penetration capabilities, surpassing conventional optical imaging methods. Combining the strengths of both optical and US imaging enables the visualization of intricate structures within tumors and their microenvironments. PA imaging provides exceptional spatial resolution compared to MRI, whole-body fluorescence, or bioluminescence modalities used widely in preclinical cancer research, facilitating the meticulous examination of tissue structures at a microscopic level, essential for studying the characteristics of the tumor microenvironment [12]. The ability of PA imaging to be multiscale (the resolution can be varied by using various optical excitation and acoustic detection schemes), capturing images across organelles to entire organs, makes it a versatile modality for various preclinical research applications [13, 14]. Additionally, it enables temporal imaging, facilitating real-time monitoring of dynamic processes like changes in tumor blood flow, oxygenation, and responses to treatments [15]. Moreover, by relying on the tissue's optical absorption properties, PA imaging permits the simultaneous imaging of multiple chromophores or molecules (e.g., blood vessels, lipids, collagen, melanin) within the tumor using the same detection scheme but different illumination wavelengths [16]. As an alternative to endogenous chromophores, exogenous contrast agents such as dyes, metal/inorganic, and organic nanoparticles can also be used in PA imaging to garner information on drug accumulation and tumor biomarker expression [17, 18].

In this chapter, we delve into the multifaceted potential of photoacoustic imaging that enhances our understanding of tumor molecular and functional processes, making it a powerful tool in preclinical cancer research. Specifically, we discuss the features of currently available commercial preclinical photoacoustic imaging systems and explore advantages and limitations in providing insights into the tumor microenvironment using endogenous and exogenous contrast agents. Additionally, we examine how PA imaging, mainly using commercially available systems, can aid in monitoring the response of various traditional and emerging cancer therapeutics. Overall, by shedding light on the recent developments and the evolving landscape of this imaging modality, we aim to provide researchers with a comprehensive summary of the role PA imaging can play in preclinical cancer research, with the ultimate goal of developing more effective cancer diagnostic and therapeutic strategies.

14.2 Current Preclinical PA Imaging Systems

There are several types of PA imaging systems commercially available for preclinical research such as the Vevo LAZR-X (Visualsonics, FujiFilm, Canada), MSOT In-Vision (iTheraMedical, Germany), RSOM (Raster Scan Optoacoustic mesoscopy (iTheraMedical, Germany), LOIS 3D (Tomowave Systems, USA), Tritom (Photosound Technologies, USA), Photoacoustic microscope (Opticho Inc),

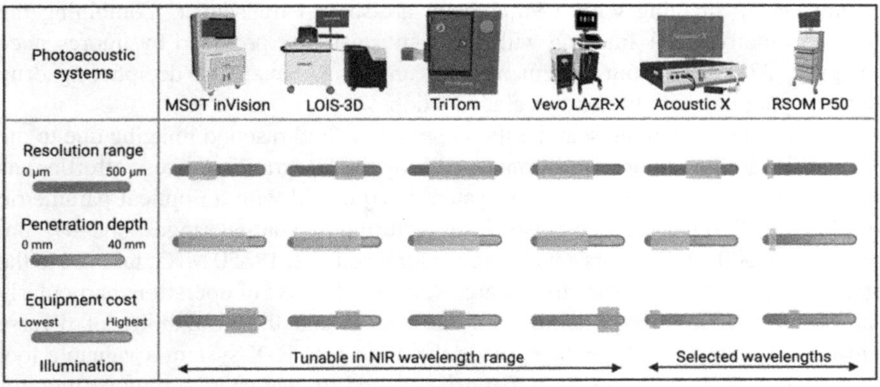

Fig. 14.1 A comparison of presently accessible photoacoustic systems for preclinical research. The resolution, penetration depth, and cost of each equipment are provided in a range, recognizing that each application may require distinct system configurations such as the center frequency of the transducer and the operating wavelength of the laser. (Schematic based on data from Ref. [28])

and Acoustic-X (Cyberdyne Inc) [19, 20]. The PA imaging systems are generally categorized into microscopy, computed tomography, endoscopy, and multimodality imaging systems. The MSOT, LOIS, and Tritom are tomography systems where cross-sections of the rodent body can be visualized. The Vevo LAZR-X and Acoustic-X systems are not tomography systems but function in the reflectance mode where a coaxial epi-illumination scheme with the light source and the linear array transducer on the same side of the tissue is employed. The RSOM and Opticho utilize high-frequency single-element transducers while other systems utilize either linear, curved, or spherical array transducers. Overall, the light and transducer configurations primarily determine the resolution, penetration depth, and, ultimately, the cost of these imaging systems, as schematically depicted in Fig. 14.1.

The tomography systems generally involve placing the rodent's body in a water bath where the transducer is rotated around the animal or the chamber holding the rodent is rotated while the transducer is fixed. For example, the LOIS-3D system is equipped with a 96-element broadband arc-shaped transducer and orthogonal optical fiber illumination with a 360-degree rotating stage for 3D photoacoustic tomography [21–23]. The MSOT inVision system features an angular coverage of 270 or 310° with a ring illumination and detector. The system also involves the translation of the stage on which the animal is placed and has been used for a wide range of preclinical cancer applications, such as detecting deeply located orthotopic pancreatic [24] and prostate tumors [25], monitoring infiltration of T cells in immunotherapy using multispectral PA imaging [26], and imaging molecularly targeted contrast agents. The inVision system is also equipped with a handheld linear array transducer to enable the imaging of large animal models. Tritom is a multimodality tomography system for live mice imaging where the system consists of an optical/photoacoustic orthogonal and epi-excitation for dual fluorescence and photoacoustic imaging with a 96-element arc-shaped transducer. Combining the high specificity of PA imaging with the high sensitivity provided by fluorescence imaging [27], the Tritom system offers a unique advantage of deciphering drug uptake dynamics with tumor vascular function.

Photoacoustic imaging seamlessly integrates with ultrasound imaging due to the similarities in the required electronics, making the integration process effortless as in the Vevo LAZR-X system. This system is equipped with an optical parametric oscillator (OPO) nanosecond pulsed laser featuring two output ranges: 680–970 nm and 1200–2000 nm. It offers a wide range of transducers, 18–50 MHz, tailored to the specific requirements of various research studies. The ease of operation, particularly the absence of a closed chamber for the rodent and the availability of diverse ultrasound imaging features have made the Vevo LAZR-X system a valuable tool in numerous studies. It has been extensively used to characterize tumors, monitor therapy progress, monitor and assess contrast agent uptake, and assess drug delivery efficacy [29–37].

An emerging alternative for nanosecond pulsed laser sources in PA imaging is light emitting diodes (LEDs) as used in the Acoustic-X [38]. Contrary to tunable lasers, this LED system has high-density high-power LED with fixed wavelengths of 690, 750, 850, and 940 nm that can operate at 1, 2, 3 or 4 kHz frequency with a selectable pulse width from 30 ns to 150 ns. This system comes equipped with a 7 MHz and a 10 MHz 128-element linear array transducer with a maximum imaging depth of 40 mm. Although the output energy by the LED arrays is low, studies have shown that decent vascular structure can be visualized with LED-based PA systems in humans and animals [39]. For example, Xavierselvan et al. attempted 3D in vivo imaging on subcutaneous xenograft mouse models with the Acoustic X system and were able to establish a good correlation with tumor vasculature immunofluorescence images [40]. To enhance the weak PA signals, Hariri et al. utilized a reactive oxygen and nitrogen species with a near-infrared absorbing small molecule called CyBA to produce a 3.2-fold increase in PA intensity [41]. More recently, several studies have utilized deep learning methodologies to improve the low signal-to-noise ratio (SNR) and contrast-to-noise ratio (CNR) in LED-based PA to overcome the requirement of high frame averaging, thereby facilitating real-time imaging [42, 43].

The raster-scan optoacoustic mesoscopy (RSOM) system has a nanosecond 532 nm pulsating laser that can generate a maximum energy of 1 mJ at a pulse repetition rate of 1 kHz. The photoacoustic signals are detected by a focused spherical single-element transducer with a 50 MHz center frequency. A combination of these configurations provides a 20 μm resolution at several mm depth [44]. Due to the high-resolution images provided by RSOM, beyond evaluating tumor vascular density, assessing vessel diameter, tortuosity, and branching is also possible, as recently demonstrated by Brown et al. in breast cancer PDX models [45]. More specifically employing various vessel segmentation algorithms on the tumor vascular network, the authors studied the correlation between immunohistochemistry markers of vasculature and a multitude of imaging-derived parameters, including classical descriptors such as vessel diameter and length to more complex descriptors of network topology such as number of edges, nodes, and loops within the vasculature. In another study by Haedicke et al., RSOM was used to quantify therapy-induced differences in tumor vascular density [46]. Studies such as these underscore the utility of PA imaging to noninvasively study vascular networks and their evolution during treatment without the use of exogenous contrast agents.

On the microscopic scale, the majority of the research groups build custom systems with variations in transducer type, transducer frequencies, and optical illumination schemes catering to their specific needs. In general photoacoustic microscopy (PAM) is categorized into optical resolution PAM (OR-PAM) or acoustic resolution PAM (AR-PAM) [47] where the main difference lies in the size of the focused laser spot on the Region of Interest (ROI). Since the lateral resolution is determined by the focused light spot size, AR-PAM employs tightly focused light and hence has superior resolution. OR-PAM has a collimated light beam and the resolution is determined by the numerical aperture of the focused ultrasonic transducer [48]. PAM has been employed from multiscale to functional imaging using endogenous biomolecules to molecular contrast agents. Baik et al. used an AR-PAM system that was modified by Opticho to achieve super-wide field PAM of animals in vivo with multilayered and multiplane whole-body imaging [49]. Imai et al. developed a PAM that provided real-time information on the development of hypoxic and low vascularization regions [50]. Recently, Kim et al. developed a deep learning algorithm for high-speed bidirectional raster scanning of mice ears using the OR-PAM developed by Baik et al. [51]. Lin et al. developed a custom OR-PAM equipped with a 75 MHz single-element transducer and 532 nm excitation laser to monitor the development of tumor angiogenesis of carcinoma [52]. Various advancements have been made to overcome the limitations of AR-PAM and OR-PAM [53–55], especially addressing the depth dependence of AR-PAM using the synthetic-aperture focusing technique by Gao et al. [56].

14.3 Selected Applications of Photoacoustic Imaging in Preclinical Cancer Research

The contrast observed in PA images reflects variations in optical absorptions within tissue, attributed to both endogenous chromophores and exogenous agents (such as dyes and nanoparticles). Endogenous chromophores, which are naturally occurring molecules within the tissue, primarily include hemoglobin, melanin, DNA, RNA, lipids, water, and collagen. Hemoglobin, in particular, serves as the primary light absorber in the visible and near-infrared (NIR) segments of the optical spectrum, enabling the acquisition of high-resolution images of tumor vasculature at greater depths through PA imaging. Notably, oxygenated and deoxygenated hemoglobin exhibit distinct spectral properties. These properties enable the calculation of total hemoglobin (HbT) content in the tumor and the blood oxygen saturation (StO_2) levels defined as the ratio of oxygenated hemoglobin to total hemoglobin. This is possible through the postprocessing of several PA images collected at multiple wavelengths in the NIR range to unmix the contribution of each chromophore to the PA signal. Furthermore, PA imaging has proven to be a valuable tool for monitoring alterations in tumor morphology and vascular heterogeneities at various stages of growth and during treatment (Sect. 14.3.1). The nanosecond pulsed lasers used in PA imaging are now equipped with fast tuning capabilities, enabling rapid switching between wavelengths without the need for additional modifications to the electronics. This advancement facilitates the acquisition of spectroscopic PA images. Particularly noteworthy is PA imaging's ability to simultaneously assess endogenous chromophores like lipids and collagen, which have distinct optical absorption properties compared to hemoglobin, within tumors alongside vascular oxygenation. This is accomplished through the utilization of multiwavelength spectroscopic PA imaging techniques (Sect. 14.3.2). Finally, we highlight the potential of PA imaging when combined with targeted contrast agents—a field propelled by advances in nanotechnology and bioconjugation (Sect. 14.3.3). This combination allows us to exemplify the multifaceted potential of PA imaging in obtaining multiparametric molecular and functional information on tumors (Table 14.1).

14.3.1 Photoacoustic Imaging of Tumor Vascular Structure and Function

14.3.1.1 Tumor Vascular Characterization

Assessing the tumor microenvironment goes beyond just visualizing its structural aspects; it necessitates the imaging of its physiological functions, including vascular density and functional properties like blood flow, perfusion, and oxygen levels. This holistic evaluation is vital for understanding tumor behavior. Photoacoustic imaging stands out as a powerful tool in this regard, as it enables the deep tissue visualization of hemoglobin, providing an effective means to monitor critical indicators of malignancy such as angiogenesis, increased vascular density, enhanced permeability, and tortuosity. These factors profoundly impact the supply of oxygen and nutrients to the neoplastic tissue, making PA imaging a valuable asset in comprehending the dynamic aspects of tumor biology (Fig. 14.2).

Excellent visual correlation between PA image of vasculature and immunofluore scence-derived vascular density in murine xenograft tumor models was observed in several studies [36, 40, 75]. For instance, when utilizing low-frequency transducers like the 7 MHz transducer, Xavierselvan et al. have illustrated a clear and robust qualitative and quantitative correlation between PA images obtained through LED-based PA excitation and immunofluorescence images obtained by staining for the type I integral membrane glycoprotein CD31, which is highly expressed on endothelial cells [40]. PA imaging also proves valuable in distinguishing inter- and intratumor variations in vascular characteristics. Quiroz-Gonzales et al. employed

Table 14.1 Specifications of various commercially available PA imaging systems. References for a few selected preclinical studies are also provided

System	Wavelengths (nm)	Transducers	Application	References
Vevo LAZR-X Ultrasound & Photoacoustic	680–970 nm and 1200–2000 nm	Linear array 15, 21, 30 and 40 MHz	HbT and StO_2	[29–31, 33–37, 57, 58]
			Lipids	[59]
			Ex Con	[41, 60]
Acoustic X Ultrasound & Photoacoustic	470, 520, 620, 660, 690, 750, 820, 850, 940, 970 nm and 750/850, 690/850, 820/940 nm LED arrays	Linear array 7 and 10 MHz	HbT and StO_2	[38, 40, 61]
			Ex Con	[38, 41, 42]
Tritom (Fluorescence + Photoacoustic)	660–1064 nm, extended 460–659 nm and 1065–1300 nm	Arch-shaped array 6 MHz	Ex Con	[27, 62]
MSOT inVision	710–950 nm	270° Cylindrical array	HbT and StO_2	[63–66]
			Ex Con	[24, 25, 67–69]
			Endogenous	[26, 70, 71]
LOIS 3D	750–850 nm	Arc-shaped array Ultrawide-band 50 kHz–8 MHz	HbT and StO_2 Endogenous	[22]
			Ex Con	[21, 23]
Optiocho microphotoa-coustics	559–532 nm	Single element 50 MHz	HbT and StO_2	[49–51]
RSOM	532 nm	Single element 50 and100 MHz	HbT and StO_2	[45, 46]
			Ex Con	[72–74]

Ex Con Exogenous Contrast agents

noninvasive PA assessment with a multispectral optoacoustic tomography (MSOT) imaging system to reveal that total hemoglobin, indicative of vascular density, is elevated in the tumor rim compared to the core in murine breast cancer tumor models, aligning with the prevailing understanding of tumors being hypoxic due to their extensive metabolic load, resulting in low vascular environments within the tumor core [65]. While tomography imaging systems may not offer intricate vascular details, the availability of high-resolution PA imaging systems like reflective-scanning optoacoustic microscopy (RSOM) in conjunction with advanced vessel segmentation algorithms facilitates the visualization and quantification of three-dimensional vessel network complexity. Recently, Brown et al. demonstrated the potential of statistical and topological analyses in characterizing tumor vascular networks using patient-derived xenograft (PDX) breast cancer models. They applied these analyses to provide comprehensive parameterization, ranging from general descriptors such as vessel diameter and length to more intricate descriptors involving

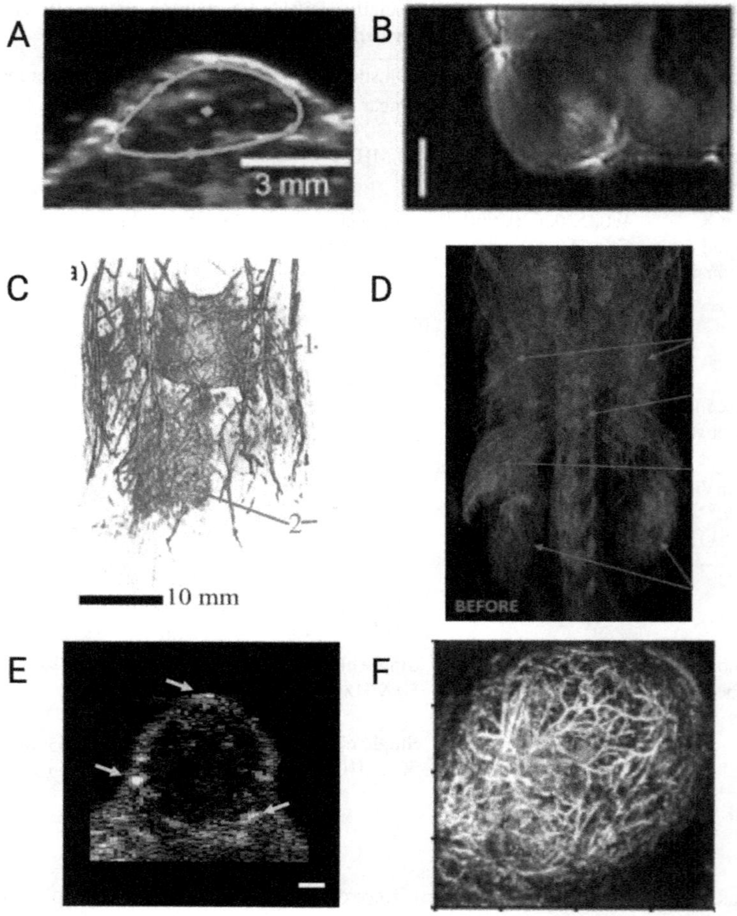

Fig. 14.2 Total hemoglobin (HbT) images acquired on photoacoustic systems mentioned in the above sections. (**a**) U87 tumor imaged at 750/850 nm and 21 MHz linear array transducer on Vevo LAZR [33]. (**b**) MDA-MB-468 tumor imaged at day 21 using a ring transducer MSOT at 800 nm [64]. (**c**) 3D rendering of PAT volumes using 532 nm illumination on Tritom [27]. (**d**) Maximum intensity projection of 800 nm PA image acquired on LOIS-3D [23]. (**e**) HbT map computed on Acoustic X using 800 nm LED laser source and 7 MHz transducer [40]. (**f**) RSOM MIP image of the vascular network of a PDX breast tumor acquired at 532 nm using a 50 MHz transducer [45]

vessel connectivity and loops, deciphered through various segmentation algorithms. These descriptors were subsequently correlated with immunohistochemical markers such as CD31 staining area (indicative of vessels), α-smooth muscle actin vessel coverage (a marker of pericyte/smooth muscle coverage and vessel maturity), and carbonic anhydrase-IX (a marker of hypoxia) to establish ex vivo biological validation of the in vivo descriptors [45]. In another study by Lin et al. [52], angiogenesis in a subcutaneous dorsal murine model of breast cancer was monitored using OR-PAM. The authors were able to quantify features of tumor angiogenesis, including the change of vascular tortuosity, the dilation of vessel diameters, and the increase of blood supply over several days, and observed the common feature of increased vascular density and tortuosity over time. All these studies point to the fact that PA imaging offers a noninvasive assessment of tumor vascular structure in preclinical cancer research without the use of exogenous contrast agents, a feature that was previously only available through terminal histology procedures. Another key advantage of utilizing PA imaging for assessing tumor vasculature is that it can be noninvasively assessed in three dimensions that cannot transparently be achieved through histological procedures.

14.3.1.2 Tumor Vascular Blood Oxygen Saturation (StO_2) and Perfusion

The studies mentioned above serve as clear illustrations of PA imaging's capabilities in comprehending the morphology of tumor vasculature. Beyond structural differences, understanding vascular function, particularly oxygen delivery to the tumor via the StO_2 parameter, a surrogate marker for tumor oxygenation status, is crucial. In line with the well-established fact that tumors tend to be hypoxic due to their high metabolic activity, PA imaging possesses the capability to identify lower StO_2 levels in tumors compared to healthy tissue. Several groups have extensively reviewed the role of PA imaging in probing the tumor microenvironment and tumor hypoxia for preclinical research [76–78]. The intricacies involved in instrumentation and mathematical models in deciphering the contribution of individual chromophores, i.e., the oxygenated and deoxygenated hemoglobin, have been provided by Li et al. [79].

One of the early milestones for PA imaging preclinical cancer research was demonstrated by Li et al. in a murine model of glioblastoma, where tumor vasculature exhibited lower StO_2 values compared to normal brain vasculature [80]. Importantly, StO_2 and total hemoglobin measurements in these models were performed without the need for craniotomy to remove the rodent skull. The study further revealed that hemoglobin content was higher in the tumor core than in adjacent healthy tissue, indicating elevated vascular development in tumors but with lower StO_2 values indicative of hypoxia formation. Over the past 15 years, numerous rodent tumor models have been assessed for variations in StO_2 compared to normal tissue using PA imaging, and significant technological advancements have been made to enhance its robustness and high-throughput capabilities. To exemplify these technological advancements, a recent study by Zhang et al. in glioblastoma serves as a noteworthy example in comparison to the study by Li et al. In the work by Zhang et al. dual-scale multiwavelength photoacoustic imaging approach was utilized to decipher the blood–brain-barrier (BBB) disruption during the tumor progression [81]. Their findings demonstrated spatial heterogeneity and distinct morphological features in the vascular distribution within the tumor region with tumor progression. Additionally, they employed the contrast agent Evans Blue to assess BBB perfusion capabilities, revealing higher accumulation in the tumor during advanced stages.

Along with fixed assessments of StO_2 and HbT obtainable with conventional PA imaging, oxygen-enchanced (OE) PA imaging allows dynamic assessment of StO_2 without the use of contrast agents. Parameters gathered from OE-PA serve as surrogate biomarkers for relative perfusion, rate of oxygen diffusion, and necrosis within a tumor [25, 82]. The information on perfusion gathered from OE-PA can also be utilized to assess if hypoxic regions are cyclic or acute or if a tumor is undergoing vascular normalization due to therapy [83, 84]. OE-PA is based on the clinically validated OE-MRI and evaluates the change in StO_2 after a shift in breathing from air (21% oxygen) to 100% oxygen, known as an oxygen challenge [85]. Compared to the static measurements of StO_2 and HbT, OE-PA markers have a good, repeatable correlation with histologic markers of vascular function [25]. Tomaszewski et al. [25] utilized PA imaging during an oxygen challenge and detected significantly different oxygenation dynamics between different murine models of prostate cancer that could not be differentiated with static PA imaging. Ron et al. [84] used a custom MSOT system to generate 3D oxygenation maps of mice with orthotopic breast cancer xenografts undergoing an oxygen challenge. Spatial StO_2 profiles and oxygen challenge responses revealed hemodynamic differences in the rim, hypoxic, and normoxic cores. Further evaluation of the hypoxic core showed cyclic hypoxia patterns had a 21% disparity between cycles measured via StO_2 and Hb variations. Ghosh et al. [83] used OE-MSOT to examine the treatment response of human nonsmall cell lung cancer (NSCLC) xenografts to heme-targeting agents. Static MSOT revealed that treated NSCLC tumors showed a reduction of HbT and an increase in StO_2, while OE-MSOT showed that treated tumors displayed a considerable and significant rise in StO_2 from the oxygen gas challenge compared to control tumors. These results indicate that treatment with heme-targeting agents greatly increased tumor perfusion and were also validated by immunohistochemical analysis.

14.3.1.3 PA Imaging as a Therapy Monitoring Tool

Cancer therapies can have significant effects on tumor vasculature and oxygenation. Radiation therapy can damage tumor blood vessels, leading to reduced blood flow and oxygen delivery, while some chemotherapy agents may target the abnormal vasculature within tumors. Additionally, therapies designed to normalize tumor blood vessels can improve oxygenation, potentially enhancing the overall effectiveness of treatment and reducing tumor hypoxia. The assessment of therapy-induced changes in tumor vasculature in preclinical research was primarily restricted to immunohistochemistry (a terminal procedure that cannot be used for longitudinal monitoring), intravital microscopy (which lacks deeper penetration depth), and MRI (which is expensive and has low resolution). The availability of PA imaging has enabled longitudinal monitoring of tumor vascular structure and function in 3D without the use of exogenous contrast agents. Below, we showcase a few examples where PA imaging was used as a therapy monitoring tool in various types of therapy.

Rich et al. demonstrated radiation therapy-induced alterations in tumor hemodynamics within patient-derived xenograft models of head and neck squamous cell carcinoma. Their findings revealed that PA imaging, conducted using the Vevo LAZR-X system, could detect changes in StO_2 levels at early time points, even before observable changes in tumor volume occurred, and these alterations exhibited a strong correlation with the radiation treatment dose [35]. Similarly, Costa et al. conducted a study using head and neck xenograft mouse models imaged with the MSOT system. They discovered that in the 10 Gy and 20 Gy groups, tumors with higher pretreatment blood StO_2 levels exhibited improved responses compared to tumors with lower baseline levels. Interestingly, the treatment response in the 30 Gy group appeared to be independent of blood StO_2 levels. Changes in StO_2, monitored at 24 h and 96 h following the 10 Gy and 20 Gy doses, revealed that resistant tumors had higher blood StO_2 levels at both time points compared to those that exhibited a decline in StO_2 following radiation therapy [86]. PA imaging can aid in monitoring vascular disruptions in radiotherapy-based combinatorial treatment, as demonstrated by Hysi et al., who employed acoustic resolution PA imaging to monitor vascular disruption in combined nanobubble and radiation therapy. Histopathological evaluation of tumor vascularity apoptosis showed good correlation with photoacoustic biomarkers of treatment response [36].

Photodynamic therapy (PDT) is a treatment is light-activated photo-cytotoxic treatment where reactive oxygen species generated by photosensitizers excited with light of a specific wavelength is used to destroy cancer cells. As with radiotherapy, the hypoxic nature of tumors severely hinders the effectiveness of PDT and intertumoral heterogeneity can lead to varying PDT outcomes [87]. Therefore, monitoring tumor oxygenation prior to, during, and after PDT is a valuable preliminary indication of the treatment's long-term efficacy [88]. The role of PA in guided PDT has been reviewed by Hester et al. [89]. Several studies utilized either commercially available PA imaging systems or custom-built devices to monitor tumor hemodynamics postPDT. For example, Neuschmelting et al. utilized handheld MSOT to monitor the effect of PDT with WST11 photosensitizer on a subcutaneous murine model of renal cell adenocarcinoma. MSOT images were gathered prior to photosensitizer or vehicle administration and collected for 1 h following light delivery and at 24 and 48 h posttreatment. Changes in StO_2 were most prominent during the first 20 min after PDT, and the treated tumors displayed a 60% StO_2 decrease from pretreatment values within the first hour, while the control tumors displayed no change. This reduction in StO_2 continued throughout the 24 and 48 h posttreatment time points in the PDT group [68]. In another study, Mai et al. demonstrate the feasibility of endogenous contrast PAM for monitoring the vasculature changes induced by PDT in melanoma lesions of the mouse ear. Particularly, a quantitative evaluation of the vasculature was conducted using Hessian filter segmentation. Before and after PDT treatment, 3D morphological, depth-resolved images were acquired and quantified using the vascular metrics of PA signal, vessel diameter, vessel density, perfused vessel density, and vessel complexity. These metrics were utilized to determine the effect of PDT. The quantitative and qualitative results showed PDT-induced changes in microvasculature structure, showing the potential of OR-PAM to evaluate PDT efficacy [90].

A groundbreaking and pioneering study by Mallidi et al. stands out as one of the first to illustrate that PA imaging is not only capable of monitoring PDT but can also predict the nonresponsive 3D regions within the tumor. In a murine model of glioblastoma, Mallidi et al. showcased that tumors responding to PDT exhibited an average StO_2 change of 85% within 24 h posttreatment, whereas the nonresponding group displayed no significant alteration in StO_2 levels. Through the analysis of 3D PA oxygen saturation maps of the tumor (Fig. 14.3a), they constructed treatment prediction maps that precisely correlated with the regions where tumor local recurrence had occurred. The prediction maps generated clearly showcased the entire tumor volume to be green (treated) or red (nontreated) for the responding and nonresponding tumors indicating a significant amount of vascular change or no change within the tumor volume, respectively (Fig. 14.3b). Furthermore for the tumors that regrew posttherapy, prediction maps were able to identify these precise regions of regrowth within just 24 h of treatment (bottom panel of Fig. 14.3b), which would not have been possible without the use of 3D PA imaging [33].

Extensive research has gone into the development of therapies to curb or normalize tumor vasculature [91, 92]. Agents like bevacizumab, DC101, and cabozantinib targeting vascular endothelial growth factor (VEGF) disrupt the formation of new blood vessels within tumors. Multikinase inhibitors such as cabozantinib also target several other tumor markers to reduce tumor blood supply, and hinder tumor growth and metastasis. The last few years saw a plethora of studies where various mouse models undergoing vascular-targeted therapies were longitudinally monitored using PA imaging. For example, on an orthotopic mouse model of ovarian cancer, Bohndiek et al. used photoacoustic tomography to assess vascular regression and normalization resulting from trebananib treatment [93]. In mice given the treatment, the photoacoustic signal of total hemoglobin decreased while the signal from oxyhemoglobin increased. Changes in serum biomarkers gathered before and after the cessation of treatment, along with ex vivo histology, confirmed normalization of remaining tumor vasculature and a 30% decrease in microvessel density. In another study, Liapis et al. monitored tumor hemodynamics in two types of breast cancer xenografts using the MSOT system and observed a significant decrease in tumor StO_2 and HbT concentrations immediately following treatment initiation [64]. This study emphasizes the importance of gathering data promptly following the initiation of therapy. In another study, Zhou et al. utilized OR-PAM to longitudinally monitor DC101, a VEGFR2 monoclonal antibody, effect on prostate cancer xenografts. Diminished vessel tortuosity, smaller vessel diameter, and homogenized intratumoral vessel distribution were observed between day 1 and 3 of therapy. The findings indicated vascular normalization, which was pathologically confirmed by increased fractional pericyte coverage, functional

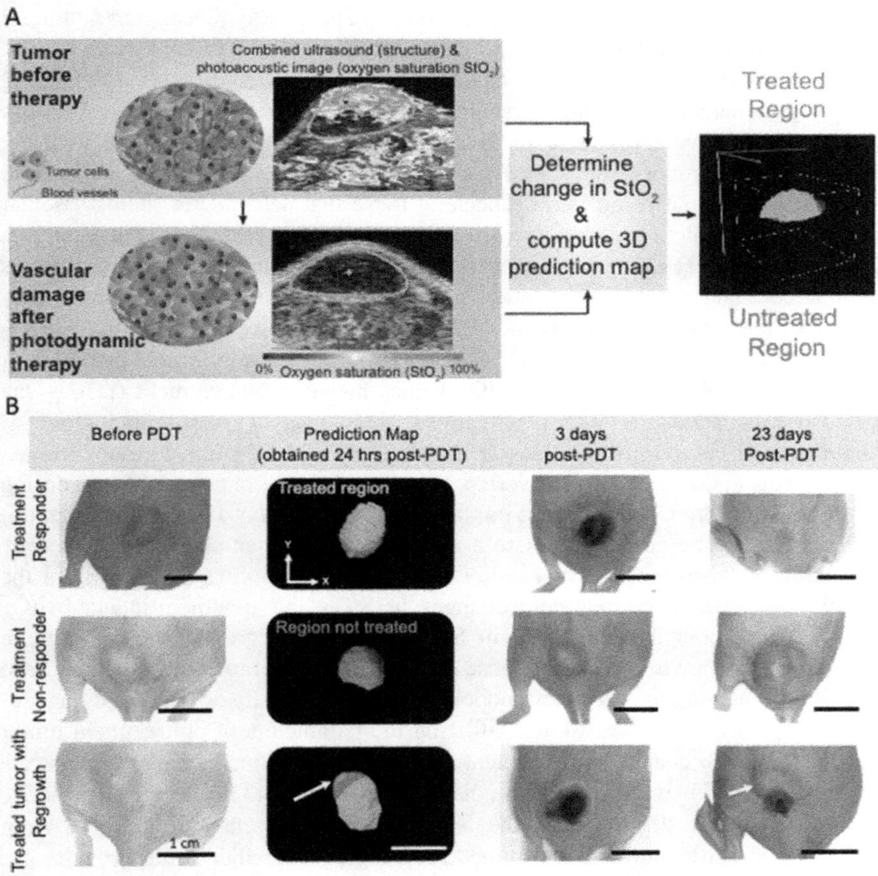

Fig. 14.3 (**a**) Schematic representation of calculating the prediction map, where 3D images before therapy are compared to images posttherapy to compute the treatment prediction map. Representative ultrasound images (grayscale) overlaid with photoacoustic oxygenation maps of a tumor before and after treatment are shown. Red represents highly oxygenated areas while blue represents hypoxic regions. (**b**) Treatment prediction map calculated 24 h postPDT from photoacoustic StO_2 maps in three different groups. Photographs of the same tumor are shown at different time points: before PDT, 3 days posttreatment, and 23 days posttreatment. The scale bars in the photographs represent 1 cm. The green region in the prediction map indicates treated regions and the red areas represent nontreated regions. The top panel represents the mouse treatment responder group that showed no visible tumor regrowth at 23 days postPDT. The center panel shows a mouse in the nonresponder group. This mouse did not show any necrosis at 3 days posttreatment and continued to grow. The prediction map clearly shows that this tumor was not treated. The bottom panel shows a representative mouse that had tumor regrowth. Three days posttherapy, the tumor appeared to have complete necrosis; however, the prediction map indicated a nontreated region in the top half of the tumor (indicated by a white arrow). Twenty three days post therapy we observe regrowth from the same region of the tumor as shown by the prediction map. (Images adapted from Ref. Mallidi et al. [33])

perfusion, and drug delivery. OR-PAM imaging subsequently revealed intratumoral vascular regression following four DC101 treatments [94].

Predicting treatment response is a complex task as it involves considering various parameters due to the heterogeneity within and between tumors. The task complexity further increases if the therapeutic dose is suboptimal. To effectively anticipate how a tumor will respond to a particular treatment regimen, it is crucial to incorporate multiple parameters, especially those that account for tumor vascular structural and functional changes pre- to posttreatment initiation. For example, Claus and Sweeney et al. gathered several PA imaging-based vascular parameters pretreatment and at several points posttreatment. Volumetric visualization of StO_2 from pre- and posttreatment revealed that chemotherapy-induced changes in StO_2 were visible 72 h posttreatment (D14—Day 14 postimplantation in Fig. 14.4a). A scatter bubble plot of changes in StO_2 values between pretreatment (D10—Day 10 postimplantation) and 72 h posttreatment (D14—Day 14 postimplantation) vs. tumor growth rate β (day^{-1}) shows that the treated and the control groups formed distinct clusters with minimal overlap (Fig. 14.4b). The change in StO_2 during tumorgenesis from D7 (day 7 post implantation) to D10 (day 10 post implantation) is indicated by the color bar where a decrease in oxygenation is represented by red, while an increase in oxygenation is represented by blue. This displays the correlation between pretreatment changes in StO_2 and treatment-induced StO_2 changes as tumors that increased in StO_2 during their growth showed a more significant decrease in StO_2 in response to cabozantinib therapy. The data were used to build a multivariate regression model to predict chemotherapy-induced changes in tumor growth rate (Fig. 14.4c). Utilizing the parameters of pretreatment tumor volume and StO_2 changes pretreatment and early posttreatment, tumor growth rate was predicted with 100% positive predictive power and 58.33% negative predictive power. The study demonstrates the feasibility of StO_2 changes gathered from endogenous multiparametric US-guided PA imaging to predict tumor growth rates in suboptimal therapies.

Photoacoustic imaging has also been used to monitor the therapeutic response of traditionally used chemotherapeutic agents such as doxorubicin and novel therapies such as T-cell therapy. For example, May et al. utilized US-guided PA imaging to examine alterations in tumors following treatment with heat-activated cytotoxic thermosensitive liposomes containing doxorubicin [37]. In another study on breast cancer models, Hysi et al. demonstrated a decrease in tumor StO_2 due to the administration of doxorubicin-encapsulated thermosensitive liposomes [95]. In a recent study, Kimm et al. retrieved multiscale PA data regarding T-cell therapy-induced therapeutic effects on EG7 tumors. Combined macro- (MSOT) and mesoscale (RSOM) characteristic changes across the tumor mass were monitored longitudinally to showcase vessel regression during cytotoxic T lymphocyte therapy [26].

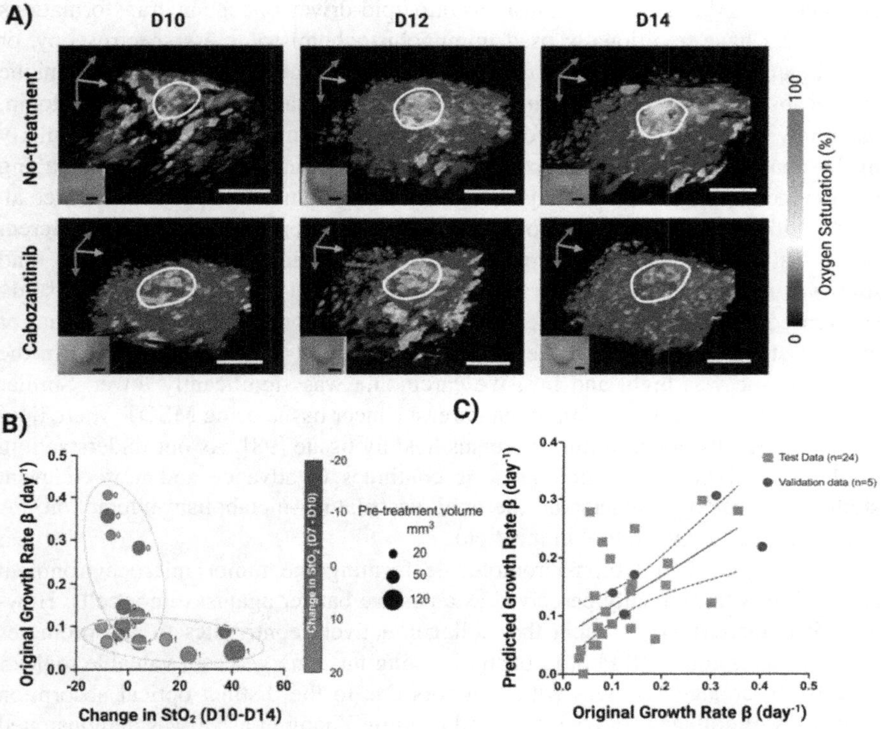

Fig. 14.4 (**a**) 3D rendered images of tumor regions from the day before the first administration of treatment (D10) immediately through the first 72 h after the start of treatment (D14). Insets shown in the lower left corner of 3D images are photographs of tumors taken immediately before the corresponding PA image acquisition. Both black and white scale bars = 5 mm. Green, blue, and red arrows indicate the x-, y-, and z-direction, respectively. (**b**) Scatter bubble plot of change in StO_2 values between Day 10 and Day 14 vs. tumor growth rate (β) values. The change in StO_2 from Day 7 to Day 10 is indicated by the color bar where a decrease in oxygenation is represented by red, while an increase in oxygenation is represented by blue. The size of the bubbles is representative of the pretreatment volume of the tumors on D10. The label 0 indicates tumors in the control group, while 1 corresponds to treated group. (**c**) Representative data from the k-fold cross-validation of the multivariate linear regression model developed from the pretreatment and change in oxygen saturation StO_2 values at various time points pre and post treatment. The training data set is represented by green squares ($n = 24$), the representative regression line in black, and the validation data set is represented by blue circles ($n = 5$)* Adapted from Ref. Claus and Sweeney et al. [34]

14.3.2 Photoacoustic Imaging of Tumors with Endogenous Contrast Agents Other Than Hemoglobin

There is increasing evidence that lipids, molecules that provide energy storage and are structural components of membranes, in excess can promote cancer progression and metastasis [96, 97]. To gain insights into lipid-driven oncogenic transformations researchers have traditionally used immunohistochemistry, mass spectroscopy, or PET imaging that are terminal or utilize exogenous contrast agents. The unique optical absorption properties of lipids, though weaker than melanin and hemoglobin, have now provided an avenue for PA imaging to examine their role, particularly in the context of vasculature, oxygen consumption, and metabolism. Leveraging the distinct optical absorption profiles of hemoglobin and lipids, Wilson et al. used multiwavelength spectroscopic imaging to differentiate among four different breast pathologies (normal, hyperplasia, ductal carcinoma in situ (DCIS), and invasive breast carcinoma) in a transgenic mouse model of breast cancer. The study showcased decreased lipid content in invasive carcinoma compared to normal or hyperplastic tissue while, on the other hand, the total hemoglobin content in the normal tissue was high, and invasive carcinoma was significantly lower. Similar studies were also performed in human breast cancer tissue using MSOT where lipid layers appear disrupted in tumors versus healthy tissue [98]. As our understanding of the role of lipids in carcinogenesis continues to advance and more relevant studies conducted to understand the lipid–vasculature–metabolism interaction, PA will emerge as a pivotal tool in this field.

Collagen, a pivotal biomacromolecule forming the tumor microenvironment scaffold, was traditionally perceived as a passive barrier against tumor cells. However, it is increasingly evident that collagen actively contributes to and promotes tumor progression [99]. Photoacoustic imaging has emerged as a valuable tool for visualizing collagen content within tumors due to the distinct optical absorption properties of collagen. In research conducted by Zhang et al., it was demonstrated that the lipid and collagen levels in tumors were significantly lower than those in normal tissue, and there was a statistically significant difference between squamous and basal cell carcinoma, aligning with histopathological findings [60]. In another study by Li et al., a clear differentiation in collagen content was observed among three types of breast cancer xenografts, with luminal-type breast cancer exhibiting higher collagen content compared to triple-negative tumors [75]. Although the imaging of collagen, particularly in the NIR-II regions where collagen exhibits greater absorption than hemoglobin, is a recent development, further research is essential to unlock the full potential of PA imaging in unraveling the role of collagen in tumor progression and therapy response.

Melanin, a highly absorbing endogenous chromophore, has been leveraged in PA imaging to assess the depth of melanomas in preclinical mouse models (reference). Its unique optical absorption properties also enable the detection of circulating metastatic tumor cells [19, 100]. Furthermore, Lauvadu et al. demonstrated the noninvasive capability of PA imaging in detecting melanoma brain metastases [29]. Specifically, they examined melanin, oxygenated, and deoxygenated hemoglobin content within metastatic tumors using an ultrasound-coupled PA imaging probe through the intact skull. This research revealed a distinct differentiation between control and metastatic brains, characterized by lower hemoglobin and melanoma cell signals in the control group. While melanin is not universally expressed in all cancer models, recent efforts have focused on developing melanin-like and melanin-based nanosystems to harness their strong absorption properties. PA imaging can capitalize on their high absorption coefficients for imaging their uptake within tumors [22, 101].

14.3.3 Contrast Agent-Enabled Molecular PA Imaging

Molecules or nano entities designed to have exceptional optical absorption properties and simultaneously convert the absorbed energy into acoustic signals are perfect contrast agents in PA imaging. These agents can be engineered to bind specifically to molecular targets of interest, making it possible to visualize disease-specific biomarkers or track the distribution of drugs within the tumor or the body. The ability to tailor contrast agents for specific applications has expanded the capabilities of photoacoustic imaging, enabling preclinical researchers and clinicians to gain molecular insights into complex biological processes at a level of detail and at greater penetration depth than was previously unattainable. In essence, the marriage of optical contrast agents with photoacoustic imaging has propelled this technology to the forefront of preclinical cancer research and is clearly evident with the exponentially growing research and review publications in the field [9, 102–105]. In the sections below, we briefly discuss the contrast agents and their utility in molecular PA imaging for preclinical cancer research.

14.3.3.1 Optical Dyes

Optical dyes such as Indocyanine Green (ICG) and IRDye800 have an absorption peak at around 780–800 nm that is leveraged by several researchers to image vascular perfusion and lymphatic structures and also map biomarkers in tumors by conjugating these dyes to various cancer-specific antibodies or molecules. For example, Taruttis et al. demonstrated the capability of MSOT to obtain pharmacokinetics and biodistribution of ICG (in its free form) in mouse models [106]. Woold et al. developed a tracer agent where ICG is liposomal encapsulated in their J-aggregate form that can specifically target and trace folate receptors in ovarian cancer xenograft models [107]. In another study, Wilson et al. conjugated ICG to an antibody targeted a novel B7-H3 biomarker overexpressed in breast cancer. Multiwavelength PA imaging showed a 3.01-fold increase in molecular B7-H3-ICG signal in tumors compared to control conditions. In another study by Attia et al., IRDye800 was conjugated to 2deoxyglucose (2DG). The probe was then used to image glucose uptake in an orthotopic glioblastoma mouse model using the MSOT system [108, 109] The study identified that the IRDye800-2DG molecule exhibited peak contrast at the tumor site 48 h postinjection concerning the surrounding normal tissue. Similarly in another study, Karlas et al. showcased the potential of MSOT to concurrently image tumor apoptosis and oxygenation, achieved through the utilization of both exogenous apoptosis-targeting dye, xPLORE-APOFL750©, and endogenous contrast from hemoglobin [69]. Mice with orthotopic 4T1 breast tumors were imaged both prior to and after the administration of doxorubicin treatment. At the 24-h mark following treatment, the PA signal of xPLORE-APOFL750© and deoxyhemoglobin exhibited an increase resulting from Doxorubicin treatment, and the apoptotic status of the tumor was validated using ex vivo fluorescence analysis. Extensive research is currently ongoing to develop contrast agents for photoacoustic imaging in the NIR-II (1000–1700 nm) to enable deep tissue PA imaging [110].

14.3.3.2 Organic and Inorganic Nanoparticles

The last two decades have seen a notable surge in the development of nanoparticle-based contrast agents for photoacoustic imaging. For instance, metal nanoparticles such as silver and gold nanoparticles are highly regarded for their biocompatibility and ease of functionalization with molecules that can specifically target cancer cells or transport therapeutic payloads and are excellent PA contrast agents. Beyond gold and silver, various other metal nanoparticles (such as copper and platinum nanoparticles), carbon nanotubes, liposomes, mesoporous silica-based nanoparticles, etc. have also garnered attention for their unique photoacoustic properties. These diverse nanoparticles can be employed individually or combined with therapeutic agents to fabricate multifunctional contrast agents that provide a comprehensive and detailed insight into the tumor microenvironment, while also inflicting damage on the tumors. By incorporating targeting ligands onto these nanoparticles, they can be directed to selectively bind to cancer cells or tumor-specific biomarkers, thus facilitating the precise localization and characterization of malignant tissue. Currently, several of these nanoparticles are in the development stage. Preclinical rodent cancer models and PA imaging aid researchers to better understand the pharmacokinetics and biodistribution of the nanoparticle-based contrast agents as they have distinct optical absorption properties than the endogenous chromophores. Numerous comprehensive reviews have thoroughly explored the intricacies of nanoparticle fabrication and their utility as photoacoustic contrast agents in deciphering the tumor microenvironment. Given the breadth and depth of these existing reviews, this chapter will not delve into those specific details. Instead, interested readers are encouraged to refer to the provided references for an in-depth exploration of this crucial aspect of nanoparticle-based preclinical cancer research [111].

14.4 Summary and Future Directions in PA Imaging

Over the past decade, PA imaging has garnered substantial attention in the field of preclinical cancer research due to its numerous advantages. This imaging modality is not only cost-effective but also capable of deeply penetrating tissues without the use of ionizing radiation. This makes it a noninvasive and safe option for obtaining high-resolution images of tumors located deep within the body. One of its primary strengths lies in its capacity to provide unparalleled three-dimensional insights into tumor vasculature and function. Additionally, it proves to be an excellent tool for monitoring the uptake of targeted contrast agents, further enhancing its utility in research. With the advent of genetic reporters [112], the potential applications of photoacoustic imaging in investigating various cellular communications at high resolution deep within tissue are becoming a reality. Furthermore, significant efforts are underway to develop endoscopic photoacoustic imaging systems that will enable the imaging of rodent cancer models with deeply seated tumors, such as orthotopic colon cancer models. For instance, Liang et al. and Kim et al. have developed endoscopic systems for imaging the rat rectum and colon, respectively [113, 114]. These developments will also contribute to a better understanding of tumor vasculature within its natural environment.

Despite these remarkable features, PA imaging does come with its share of limitations. First, it is a relatively low-throughput technique compared to preclinical fluorescence imaging systems since only one mouse can be imaged at a time on existing PA systems. Given the extensive use of multiple mice in preclinical cancer research experiments, there is a pressing need for high-throughput imaging systems. This will require extensive revamping of detection and light illumination schematics. With the development in transducer and laser technologies going at a rapid pace, it will not be a surprise if PA imaging systems that can monitor several mice become a reality.

A significant limitation of photoacoustic imaging is its operator dependence, akin to ultrasound imaging. This means that acquired images cannot be easily quantified, and operators must possess knowledge of photoacoustic system attributes, image optimization techniques, and common artifacts. Factors such as the presence of air bubbles in and around the imaging area can introduce artifacts that inexperienced users may mistakenly interpret as genuine signals. In an ideal scenario, a photoacoustic imaging system would possess an aperture of infinite size, an infinite bandwidth, a lossless medium, and a homogeneously known speed of sound. However, in practical scenarios, the limited bandwidth, finite size of the transducer, or the space/medium-dependent speed of sound can contribute to the loss of resolution. Additionally, quantifying photoacoustic signals requires a deep understanding of signal processing, noise reduction, and knowledge of laser–tissue interactions. For instance, background signals from other chromophores in the tissue can significantly affect the photoacoustic signal from the chromophore of interest [19]. Furthermore, signal attenuation due to tissue absorption becomes notably pronounced at greater depths, particularly in heterogeneous samples. To mitigate this signal loss, researchers have extensively explored optical fluence compensation methods. One such approach involves Monte Carlo simulations of light transport, considering the illumination geometry and acoustic detectors, to develop algorithms that enhance signals at deeper depths [115]. Additionally, user-friendly interfaces such as the Ultrasound-guided Photoacoustic Image Annotation Toolkit in MATLAB (PHANTOM) [116] have been developed to assist inexperienced users in performing fluence compensation on photoacoustic images—an approach previously limited to select photoacoustic imaging research groups. Finally, as the number of photoacoustic imaging users grows, standardization and open sharing of data will provide a strong foundation that will not only expand its user base but also facilitate integrated preclinical cancer research with clinical translation efforts [117]. Despite these minor drawbacks, photoacoustic imaging is making significant strides and is rapidly becoming an indispensable tool in preclinical cancer research. Its unique capabilities and potential to provide valuable insights into tumor biology make it an essential asset in the quest to advance our understanding of cancer and develop effective treatment strategies.

References

1. Siegel, R.L., et al., *Cancer statistics, 2023.* CA Cancer J Clin, 2023. **73**(1): p. 17–48.
2. Astell, K.R. and D. Sieger, *Zebrafish In Vivo Models of Cancer and Metastasis.* Cold Spring Harb Perspect Med, 2020. **10**(8).
3. Schachtschneider, K.M., et al., *Editorial: "Humanized" Large Animal Cancer Models: Accelerating Time and Effectiveness of Clinical Trials.* Front Oncol, 2019. **9**: p. 793.
4. Ireson, C.R., et al., *The role of mouse tumor models in the discovery and development of anticancer drugs.* Br J Cancer, 2019. **121**(2): p. 101–108.
5. Onaciu, A., et al., *Spontaneous and Induced Animal Models for Cancer Research.* Diagnostics (Basel), 2020. **10**(9).
6. Pulumati, A., et al., *Technological advancements in cancer diagnostics: Improvements and limitations.* Cancer Rep (Hoboken), 2023. **6**(2): p. e1764.
7. Mallidi, S., G.P. Luke, and S. Emelianov, *Photoacoustic imaging in cancer detection, diagnosis, and treatment guidance.* Trends in Biotechnology, 2011. **29**(5): p. 213–221.
8. Li, L. and L.V. Wang, *Recent Advances in Photoacoustic Tomography.* BME Frontiers, 2021. **2021**: p. 9823268.
9. Weber, J., P.C. Beard, and S.E. Bohndiek, *Contrast agents for molecular photoacoustic imaging.* Nature Methods, 2016. **13**(8): p. 639–650.
10. Choi, W., D. Oh, and C. Kim, *Practical photoacoustic tomography: Realistic limitations and technical solutions.* Journal of Applied Physics, 2020. **127**(23).
11. Tzoumas, S., et al., *Unmixing Molecular Agents From Absorbing Tissue in Multispectral Optoacoustic Tomography.* IEEE Transactions on Medical Imaging, 2014. **33**(1): p. 48–60.
12. Brown, E., J. Brunker, and S.E. Bohndiek, *Photoacoustic imaging as a tool to probe the tumour microenvironment.* Dis Model Mech, 2019. **12**(7).
13. Wang, L.V., *Multiscale photoacoustic microscopy and computed tomography.* Nat Photonics, 2009. **3**(9): p. 503–509.
14. Park, B., et al., *Functional photoacoustic imaging: from nano- and micro- to macro-scale.* Nano Converg, 2023. **10**(1): p. 29.
15. Taruttis, A. and V. Ntziachristos, *Advances in real-time multispectral optoacoustic imaging and its applications.* Nature Photonics, 2015. **9**(4): p. 219–227.
16. Yao, J., J. Xia, and L.V. Wang, *Multiscale Functional and Molecular Photoacoustic Tomography.* Ultrasonic Imaging, 2016. **38**(1): p. 44–62.
17. Choi, W., et al., *Recent Advances in Contrast-Enhanced Photoacoustic Imaging: Overcoming the Physical and Practical Challenges.* Chemical Reviews, 2023. **123**(11): p. 7379–7419.
18. Farooq, A., et al., *Exogenous Contrast Agents in Photoacoustic Imaging: An In Vivo Review for Tumor Imaging.* Nanomaterials (Basel), 2022. **12**(3).
19. Gargiulo, S., S. Albanese, and M. Mancini, *State-of-the-Art Preclinical Photoacoustic Imaging in Oncology: Recent Advances in Cancer Theranostics.* Contrast Media Mol Imaging, 2019. **2019**: p. 5080267.

20. Beard, P., *Biomedical photoacoustic imaging.* Interface Focus, 2011. **1**(4): p. 602–31.
21. Zhou, H.-C., et al., *Intravital NIR-II three-dimensional photoacoustic imaging of biomineralized copper sulfide nanoprobes.* Journal of Materials Chemistry B, 2021. **9**(13): p. 3005–3014.
22. Liopo, A., R. Su, and A.A. Oraevsky, *Melanin nanoparticles as a novel contrast agent for optoacoustic tomography.* Photoacoustics, 2015. **3**(1): p. 35–43.
23. Su, R., et al., *Small animal optoacoustic tomography system for molecular imaging of contrast agents.* SPIE BiOS. Vol. 9708. 2016: SPIE.
24. Napp, J., et al., *Fluorescence- and multispectral optoacoustic imaging for an optimized detection of deeply located tumors in an orthotopic mouse model of pancreatic carcinoma.* International Journal of Cancer, 2018. **142**(10): p. 2118–2129.
25. Tomaszewski, M.R., et al., *Oxygen Enhanced Optoacoustic Tomography (OE-OT) Reveals Vascular Dynamics in Murine Models of Prostate Cancer.* Theranostics, 2017. **7**(11): p. 2900–2913.
26. Kimm, M.A., et al., *Longitudinal imaging of T cell-based immunotherapy with multi-spectral, multi-scale optoacoustic tomography.* Scientific Reports, 2020. **10**(1): p. 4903.
27. Thompson, W.R., et al., *Characterizing a photoacoustic and fluorescence imaging platform for preclinical murine longitudinal studies.* Journal of Biomedical Optics, 2023. **28**(3): p. 036001.
28. Fatima, A., et al., *Review of cost reduction methods in photoacoustic computed tomography.* Photoacoustics, 2019. **15**: p. 100137.
29. Lavaud, J., et al., *Exploration of melanoma metastases in mice brains using endogenous contrast photoacoustic imaging.* International Journal of Pharmaceutics, 2017. **532**(2): p. 704–709.
30. Keša, P., et al., *Quantitative In Vivo Monitoring of Hypoxia and Vascularization of Patient-Derived Murine Xenografts of Mantle Cell Lymphoma Using Photoacoustic and Ultrasound Imaging.* Ultrasound Med Biol, 2021. **47**(4): p. 1099–1107.
31. Scheepbouwer, C., et al., *A Multimodal Imaging Approach for Longitudinal Evaluation of Bladder Tumor Development in an Orthotopic Murine Model.* PLOS ONE, 2016. **11**(8): p. e0161284.
32. Karmacharya, M.B., L.R. Sultan, and C.M. Sehgal, *Photoacoustic monitoring of oxygenation changes induced by therapeutic ultrasound in murine hepatocellular carcinoma.* Scientific Reports, 2021. **11**(1): p. 4100.
33. Mallidi, S., et al., *Prediction of Tumor Recurrence and Therapy Monitoring Using Ultrasound-Guided Photoacoustic Imaging.* Theranostics, 2015. **5**(3): p. 289–301.
34. Claus, A., et al., *3D Ultrasound-Guided Photoacoustic Imaging to Monitor the Effects of Suboptimal Tyrosine Kinase Inhibitor Therapy in Pancreatic Tumors.* Frontiers in Oncology, 2022. **12**.
35. Rich, L.J., et al., *Photoacoustic Imaging as an Early Biomarker of Radio Therapeutic Efficacy in Head and Neck Cancer.* Theranostics, 2018. **8**(8): p. 2064–2078.
36. Hysi, E., et al., *Photoacoustic imaging biomarkers for monitoring biophysical changes during nanobubble-mediated radiation treatment.* Photoacoustics, 2020. **20**: p. 100201.
37. May, J.P., et al., *Photoacoustic Imaging of Cancer Treatment Response: Early Detection of Therapeutic Effect from Thermosensitive Liposomes.* PLOS ONE, 2016. **11**(10): p. e0165345.
38. Joseph, J., et al., *Technical validation studies of a dual-wavelength LED-based photoacoustic and ultrasound imaging system.* Photoacoustics, 2021. **22**: p. 100267.
39. Agrawal, S., et al., *Photoacoustic Imaging of Human Vasculature Using LED versus Laser Illumination: A Comparison Study on Tissue Phantoms and In Vivo Humans.* Sensors, 2021. **21**(2): p. 424.
40. Xavierselvan, M., M.K.A. Singh, and S. Mallidi, *In Vivo Tumor Vascular Imaging with Light Emitting Diode-Based Photoacoustic Imaging System.* Sensors, 2020. **20**(16): p. 4503.
41. Hariri, A., et al., *Molecular imaging of oxidative stress using an LED-based photoacoustic imaging system.* Scientific Reports, 2019. **9**(1): p. 11378.

42. Hariri, A., et al., *Deep learning improves contrast in low-fluence photoacoustic imaging.* Biomedical Optics Express, 2020. **11**(6): p. 3360–3373.
43. Mallidi, S. and A. Paul, *U-Net enhanced real-time LED-based photoacoustic imaging*. 2023.
44. Omar, M., et al., *Pushing the optical imaging limits of cancer with multi-frequency-band raster-scan optoacoustic mesoscopy (RSOM).* Neoplasia, 2015. **17**(2): p. 208–14.
45. Brown, E.L., et al., *Quantification of vascular networks in photoacoustic mesoscopy.* Photoacoustics, 2022. **26**: p. 100357.
46. Haedicke, K., et al., *High-resolution optoacoustic imaging of tissue responses to vascular-targeted therapies.* Nature Biomedical Engineering, 2020. **4**(3): p. 286–297.
47. Yao, J. and L.V. Wang, *Photoacoustic Microscopy.* Laser Photon Rev, 2013. **7**(5).
48. Liu, W. and J. Yao, *Photoacoustic microscopy: principles and biomedical applications.* Biomed Eng Lett, 2018. **8**(2): p. 203–213.
49. Baik, J.W., et al., *Super Wide-Field Photoacoustic Microscopy of Animals and Humans In Vivo.* IEEE Transactions on Medical Imaging, 2020. **39**(4): p. 975–984.
50. Imai, T., et al., *Direct measurement of hypoxia in a xenograft multiple myeloma model by optical-resolution photoacoustic microscopy.* Cancer Biology & Therapy, 2017. **18**(2): p. 101–105.
51. Kim, J., et al., *Deep learning alignment of bidirectional raster scanning in high speed photoacoustic microscopy.* Sci Rep, 2022. **12**(1): p. 16238.
52. Lin, R., et al., *Longitudinal label-free optical-resolution photoacoustic microscopy of tumor angiogenesis in vivo.* Quant Imaging Med Surg, 2015. **5**(1): p. 23–9.
53. Huang, B. and T.T. Wong, *Review of low-cost light sources and miniaturized designs in photoacoustic microscopy.* Journal of Biomedical Optics, 2023. **29**(S1): p. S11503.
54. Yao, J., et al., *Editorial: Breaking the speed limits in photoacoustic microscopy.* Photoacoustics, 2023. **32**: p. 100541.
55. Park, B., et al., *Functional photoacoustic imaging: from nano- and micro- to macro-scale.* Nano Convergence, 2023. **10**(1): p. 29.
56. Gao, R., et al., *Achieving depth-independent lateral resolution in AR-PAM using the synthetic-aperture focusing technique.* Photoacoustics, 2022. **26**: p. 100328.
57. Dolet, A., et al., *In Vitro and In Vivo Multispectral Photoacoustic Imaging for the Evaluation of Chromophore Concentration.* Sensors (Basel), 2021. **21**(10).
58. Raes, F., et al., *High Resolution Ultrasound and Photoacoustic Imaging of Orthotopic Lung Cancer in Mice: New Perspectives for Onco-Pharmacology.* PLoS One, 2016. **11**(4): p. e0153532.
59. Wilson, K.E., et al., *Multiparametric spectroscopic photoacoustic imaging of breast cancer development in a transgenic mouse model.* Theranostics, 2014. **4**(11): p. 1062–71.
60. Zhang, M., et al., *Ultrasound-guided photoacoustic imaging for the selective detection of EGFR-expressing breast cancer and lymph node metastases.* Biomedical Optics Express, 2016. **7**(5): p. 1920–1931.
61. Bulsink, R., et al., *Oxygen Saturation Imaging Using LED-Based Photoacoustic System.* Sensors, 2021. **21**(1): p. 283.
62. Mokrousov, M.D., et al., *Indocyanine green dye based bimodal contrast agent tested by photoacoustic/fluorescence tomography setup.* Biomedical Optics Express, 2021. **12**(6): p. 3181–3195.
63. Shah, A., et al. *Combined dynamic contrast enhanced ultrasound and multispectral optoacoustic tomography for imaging tumour hypoxia.* in *2016 IEEE International Ultrasonics Symposium (IUS)*. 2016.
64. Liapis, E., et al., *Resolution of Spatial and Temporal Heterogeneity in Bevacizumab-Treated Breast Tumors by Eigenspectra Multispectral Optoacoustic Tomography.* Cancer Research, 2020. **80**(23): p. 5291–5304.
65. Quiros-Gonzalez, I., et al., *Optoacoustics delineates murine breast cancer models displaying angiogenesis and vascular mimicry.* Br J Cancer, 2018. **118**(8): p. 1098–1106.
66. He, H., et al., *Capsule optoacoustic endoscopy for esophageal imaging.* Journal of Biophotonics, 2019. **12**(10): p. e201800439.

67. Chen, S., et al., *Photoactivated Nanohybrid for Dual-Nuclei MR/US/PA Multimodal-Guided Photothermal Therapy.* Bioconjugate Chemistry, 2022. **33**(9): p. 1729–1740.
68. Neuschmelting, V., et al., *WST11 Vascular Targeted Photodynamic Therapy Effect Monitoring by Multispectral Optoacoustic Tomography (MSOT) in Mice.* Theranostics, 2018. **8**(3): p. 723–734.
69. Karlas, A., et al., *Multi-Aspect Optoacoustic Imaging of Breast Tumors under Chemotherapy with Exogenous and Endogenous Contrasts: Focus on Apoptosis and Hypoxia.* Biomedicines, 2021. **9**(11): p. 1696.
70. Lucero, M.Y. and J. Chan, *Photoacoustic imaging of elevated glutathione in models of lung cancer for companion diagnostic applications.* Nature Chemistry, 2021. **13**(12): p. 1248–1256.
71. Escudero Sanchez, L., et al., *Feasibility and sensitivity study of radiomic features in photoacoustic imaging of patient-derived xenografts.* Scientific Reports, 2022. **12**(1): p. 15142.
72. Haedicke, K., et al., *Sonophore labeled RGD: a targeted contrast agent for optoacoustic imaging.* Photoacoustics, 2017. **6**: p. 1–8.
73. Novoselova, M.V., et al., *Multifunctional nanostructured drug delivery carriers for cancer therapy: Multimodal imaging and ultrasound-induced drug release.* Colloids and Surfaces B: Biointerfaces, 2021. **200**: p. 111576.
74. Barmin, R.A., et al., *Optoacoustic/Fluorescent/Acoustic Imaging Probe Based on Air-Filled Bubbles Functionalized with Gold Nanorods and Fluorescein Isothiocyanate.* ACS Omega, 2021. **6**(5): p. 3809–3821.
75. Li, J., et al., *Spatial heterogeneity of oxygenation and haemodynamics in breast cancer resolved in vivo by conical multispectral optoacoustic mesoscopy.* Light Sci Appl, 2020. **9**: p. 57.
76. Nasri, D., et al., *Photoacoustic imaging for investigating tumor hypoxia: a strategic assessment.* Theranostics, 2023. **13**(10): p. 3346–3367.
77. Hu, S. and L.V. Wang, *Photoacoustic imaging and characterization of the microvasculature.* J Biomed Opt, 2010. **15**(1): p. 011101.
78. John, S., et al., *Niche preclinical and clinical applications of photoacoustic imaging with endogenous contrast.* Photoacoustics, 2023. **32**: p. 100533.
79. Li, M., Y. Tang, and J. Yao, *Photoacoustic tomography of blood oxygenation: A mini review.* Photoacoustics, 2018. **10**: p. 65–73.
80. Li, M.L., et al., *Simultaneous Molecular and Hypoxia Imaging of Brain Tumors In Vivo Using Spectroscopic Photoacoustic Tomography.* Proceedings of the IEEE, 2008. **96**(3): p. 481–489.
81. Zhang, J., et al., *In vivo characterization and analysis of glioblastoma at different stages using multiscale photoacoustic molecular imaging.* Photoacoustics, 2023. **30**: p. 100462.
82. Tomaszewski, M.R., et al., *Oxygen-Enhanced and Dynamic Contrast-Enhanced Optoacoustic Tomography Provide Surrogate Biomarkers of Tumor Vascular Function, Hypoxia, and Necrosis.* Cancer Research, 2018. **78**(20): p. 5980–5991.
83. Ghosh, P., et al., *Oxygen-Enhanced Optoacoustic Tomography Reveals the Effectiveness of Targeting Heme and Oxidative Phosphorylation at Normalizing Tumor Vascular Oxygenation.* Cancer Res, 2020. **80**(17): p. 3542–3555.
84. Ron, A., et al., *Volumetric Optoacoustic Imaging Unveils High-Resolution Patterns of Acute and Cyclic Hypoxia in a Murine Model of Breast Cancer.* Cancer Research, 2019. **79**(18): p. 4767–4775.
85. Mason, R.P., *Oxygen breathing challenge- the simplest theranostic.* Theranostics, 2017. **7**(16): p. 3873–3875.
86. Márcia Martinho, C., et al., *Photoacoustic imaging for the prediction and assessment of response to radiotherapy in vivo.* bioRxiv, 2018: p. 329516.
87. Gunaydin, G., M.E. Gedik, and S. Ayan, *Photodynamic Therapy for the Treatment and Diagnosis of Cancer-A Review of the Current Clinical Status.* Front Chem, 2021. **9**: p. 686303.

88. Hong, L., et al., *Recent Advances in Strategies for Addressing Hypoxia in Tumor Photodynamic Therapy.* Biomolecules, 2022. **12**(1).
89. Hester, S.C., et al., *Role of Ultrasound and Photoacoustic Imaging in Photodynamic Therapy for Cancer.* Photochem Photobiol, 2020. **96**(2): p. 260–279.
90. Mai, T.T., et al., *In Vivo Quantitative Vasculature Segmentation and Assessment for Photodynamic Therapy Process Monitoring Using Photoacoustic Microscopy.* Sensors (Basel), 2021. **21**(5).
91. Ellis, L.M. and D.J. Hicklin, *VEGF-targeted therapy: mechanisms of anti-tumour activity.* Nature Reviews Cancer, 2008. **8**(8): p. 579–591.
92. Jain, R.K., *Normalization of tumor vasculature: an emerging concept in antiangiogenic therapy.* Science, 2005. **307**(5706): p. 58–62.
93. Sarah, E.B., et al., *Photoacoustic Tomography Detects Early Vessel Regression and Normalization During Ovarian Tumor Response to the Antiangiogenic Therapy Trebananib.* Journal of Nuclear Medicine, 2015. **56**(12): p. 1942.
94. Zhou, H.C., et al., *Optical-resolution photoacoustic microscopy for monitoring vascular normalization during anti-angiogenic therapy.* Photoacoustics, 2019. **15**: p. 100143.
95. Hysi, E., et al., *Photoacoustic signal characterization of cancer treatment response: Correlation with changes in tumor oxygenation.* Photoacoustics, 2017. **5**: p. 25–35.
96. Martin-Perez, M., et al., *The role of lipids in cancer progression and metastasis.* Cell Metab, 2022. **34**(11): p. 1675–1699.
97. Butler, L.M., et al., *Lipids and cancer: Emerging roles in pathogenesis, diagnosis and therapeutic intervention.* Adv Drug Deliv Rev, 2020. **159**: p. 245–293.
98. Diot, G., et al., *Multispectral Optoacoustic Tomography (MSOT) of Human Breast Cancer.* Clin Cancer Res, 2017. **23**(22): p. 6912–6922.
99. Fang, M., et al., *Collagen as a double-edged sword in tumor progression.* Tumour Biol, 2014. **35**(4): p. 2871–82.
100. Deán-Ben, X.L., et al., *Deep tissue volumetric optoacoustic tracking of individual circulating tumor cells in an intracardially perfused mouse model.* Neoplasia, 2020. **22**(9): p. 441–446.
101. Fu, M., et al., *Biosynthesis of Melanin Nanoparticles for Photoacoustic Imaging Guided Photothermal Therapy.* Small, 2023. **19**(14): p. 2205343.
102. Lemaster, J.E. and J.V. Jokerst, *What is new in nanoparticle-based photoacoustic imaging?* Wiley Interdiscip Rev Nanomed Nanobiotechnol, 2017. **9**(1).
103. Chitgupi, U. and J.F. Lovell, *Naphthalocyanines as contrast agents for photoacoustic and multimodal imaging.* Biomed Eng Lett, 2018. **8**(2): p. 215–221.
104. Borg, R.E. and J. Rochford, *Molecular Photoacoustic Contrast Agents: Design Principles & Applications.* Photochem Photobiol, 2018. **94**(6): p. 1175–1209.
105. Li, M., et al., *Sound Out the Deep Colors: Photoacoustic Molecular Imaging at New Depths.* Mol Imaging, 2020. **19**: p. 1536012120981518.
106. Taruttis, A., et al., *Fast multispectral optoacoustic tomography (MSOT) for dynamic imaging of pharmacokinetics and biodistribution in multiple organs.* PLoS One, 2012. **7**(1): p. e30491.
107. Wood, C.A., et al., *Clinically translatable quantitative molecular photoacoustic imaging with liposome-encapsulated ICG J-aggregates.* Nature Communications, 2021. **12**(1): p. 5410.
108. Dogra, V., et al., *Photoacoustic imaging with an acoustic lens detects prostate cancer cells labeled with PSMA-targeting near-infrared dye-conjugates.* J Biomed Opt, 2016. **21**(6): p. 66019.
109. Attia, A.B.E., et al., *Multispectral optoacoustic and MRI coregistration for molecular imaging of orthotopic model of human glioblastoma.* Journal of Biophotonics, 2016. **9**(7): p. 701–708.
110. Jiang, S., J. Lin, and P. Huang, *Nanomaterials for NIR-II Photoacoustic Imaging.* Advanced Healthcare Materials, 2023. **12**(16): p. 2202208.
111. Upputuri, P.K. and M. Pramanik, *Recent advances in photoacoustic contrast agents for in vivo imaging.* WIREs Nanomedicine and Nanobiotechnology, 2020. **12**(4): p. e1618.
112. Brunker, J., et al., *Photoacoustic imaging using genetically encoded reporters: a review.* J Biomed Opt, 2017. **22**(7).

113. Kim, M., et al., *Intra-instrument channel workable, optical-resolution photoacoustic and ultrasonic mini-probe system for gastrointestinal endoscopy.* Photoacoustics, 2022. **26**: p. 100346.
114. Liang, Y., et al., *Optical-resolution functional gastrointestinal photoacoustic endoscopy based on optical heterodyne detection of ultrasound.* Nature Communications, 2022. **13**(1): p. 7604.
115. Kirillin, M., et al., *Fluence compensation in raster-scan optoacoustic angiography.* Photoacoustics, 2017. **8**: p. 59–67.
116. Sweeney, A., et al., *Ultrasound-guided Photoacoustic image Annotation Toolkit in MATLAB (PHANTOM) for preclinical applications.* bioRxiv, 2023: p. 2023.11.07.565885.
117. Gröhl, J., et al., *The IPASC data format: A consensus data format for photoacoustic imaging.* Photoacoustics, 2022. **26**: p. 100339.

Chapter 15
Ghost-Particle-Enhanced Photoacoustic Flow Cytometry for Identification of Bacterial Cells in Body Fluids

Robert H. Edgar and John A. Viator

Abstract Photoacoustic flow cytometry can detect small particles in flow by generating high-frequency ultrasonic waves in these particles given that there is optical contrast between the particles and the medium of flow. These ultrasonic waves indicate the presence of the particles and enable capture and further analysis. In the case of in vitro bacterial detection in body fluids for rapid identification of infection, we provide optical contrast using dyed bacteriophage that specifically targets bacterial cells. While these dyed and, hence, optically absorbing bacteriophages provide the optical contrast to bacteria for photoacoustic detection, the nature of bacteriophage results in lysis of the bacteria in an hour or so, thus precluding imaging or other analysis. In order to prevent lysis of the bacteria, but preserve the highly specific nature of bacteriophage to target bacterial species, we remove the capsids and the genetic material responsible for bacterial lysis and replace them with dyed latex microspheres. These engineered bacteriophage particles can therefore act as photoacoustic tags enabling detection and identification of bacterial species in body fluids while preserving the cells for imaging or molecular analysis.

Keywords Bacteria · Bacteriophage · Diagnostic · Optical · Optoacoustic

15.1 Introduction

Lasers provide many methods for improving human health in many areas of medicine, including dermatology, surgery, oncology, and other fields. We have used lasers for several applications, including laser ablation of blood clots, port wine

R. H. Edgar
Department of Bioengineering, University of Pittsburgh, Pittsburgh, PA, USA
e-mail: viatorj@duq.edu

J. A. Viator (✉)
Department of Bioengineering, University of Pittsburgh, Pittsburgh, PA, USA

Department of Engineering, Duquesne University, Pittsburgh, PA, USA

W. Xia (ed.), *Biomedical Photoacoustics*,
https://doi.org/10.1007/978-3-031-61411-8_15

stain measurements, and various methods of photoacoustic sensing [10–12, 16]. Photoacoustic generation occurs when light induces acoustic vibrations in matter. There are several mechanisms that can induce this effect, including thermoelastic expansion, where light absorption causes heating and subsequent expansion in the targeted medium [13]. This expansion results in a pressure wave that is typically in the megahertz regime in biological media. The laser generated acoustic wave is robust and does not suffer from optical scattering in complex media, as is the case in fluorescence and other purely optical methods.

Taking advantage of photoacoustic generation for biomedical purposes, we have developed photoacoustic flow cytometry to detect pathological cells under flow from body fluids. In photoacoustic flow cytometry, we irradiate a flowing sample, choosing the laser wavelength to ensure there is no optical absorption in the fluid or in healthy cells, though there should be high absorption in the pathological cells. For example, we exploited the strong optical absorber, melanin, in circulating melanoma cells to detect and enumerate them in human patient blood samples [6]. We were able to count the cancer cells and developed a simple predictive model for patients advancing to metastatic disease.

We have also used photoacoustic flow cytometry to rapidly identify bacterial species in human patient samples by pairing them with modified probes to ensure optical absorption. Our method uses a novel bacterial tag, optically modified bacteriophage, that binds irreversibly and specifically and does not require a bacterial culture step or DNA amplification like many clinical diagnostics. Bacteriophage can be modified by the addition of protein dyes to create optically absorbing bacterial tags. Bacteriophages present many advantages over antibodies or other types of tags. Bacteriophages have greater specificity than antibodies, are easier to produce, bind irreversibly, and are more stable [1, 3, 8, 15].

However, the life cycle of bacteriophage ultimately results in lysis of the bacterial cell [17]. Bacteriophages used in our technology follow the lytic life cycle with a half-life on the order of forty minutes [2, 7, 14]. For rapid identification of bacterial infection, it is clinically important to preserve the bacterial cell so that testing delays do not result in artificially depressing the detected cell number, since lysed bacterial cells will not be counted. Preventing natural lysis of the cell would obviate the need for restrictive timing in the clinical protocol and allow for imaging or downstream analysis that requires an intact or viable cell. Since cell lysis is due to the incorporation of the genetic material from bacteriophage into the bacterial cell, we create and use bacteriophage ghost particles in the photoacoustic process [4, 9]. These ghost particles are bacteriophages treated to remove the native DNA from the capsid, while keeping the bacteria-specific tail spikes intact. The procedure entails osmotic shock of bacteriophage suspensions in detergent solutions. This process causes the bacteriophage capsid structure to expand, making it more porous, and allowing the DNA molecules to diffuse into the surrounding medium. Thus, we can use bacteriophage to tag specifically tag bacteria for rapid identification without loss of the cells themselves, ultimately providing a method that can be used in research laboratories and in clinical settings.

15.2 Materials

We culture bacterial cells to pair with the bacteriophage ghost particles that will be used for photoacoustic sensing. After growing a bacterial strain, we dye the matching bacteriophage and then create the non-lysing ghost particles by osmotic shock.

15.2.1 Bacterial Culturing

1. Rich culture media (Luria Broth: Tryptone, Yeast extract, Sodium Chloride)
2. 37°C shaking water bath
3. Bacterial streak plate or frozen stock
4. LB agar plate
5. 37°C bacterial incubator
6. Culture sticks
7. Sterile loop
8. 16×125 mm glass culture tubes with aeration caps

15.2.2 Bacteriophage

1. Polyethylene glycol 8000 (PEG)
2. Direct Red 81 dye
3. Bacteriophage stock at $> 10^1 1$ plaque forming units (pfu) per milliliter to match bacterial species
4. 10 mM Tris, pH 7.5
5. 10 mM $MgCl_2$
6. 68 mM NaCl as bacteriophage buffer
7. Nanodrop 2300 spectrometer

15.2.3 Bacteriophage Tails and Microspheres

1. Cesium chloride
2. Bio-Rad HPLC
3. BioTek Synergy H1 with 96 well plate
4. Thermo Fisher/FEI T12 Spirit with Gatan US 1000 and Orius CCD camera
5. Streptavidin coated dyed polystyrene microspheres, 0.10 μm diameter
6. Spin-X concentrator
7. Slide-A-Lyzer, 2 kD dialysis cassettes

15.2.4 Photoacoustic Flow Cytometry

1. Photoacoustic flow cytometry system consisting of nanosecond laser, acoustic chamber, transducer, digitizer of 10 MHz or more, and data recording device
2. Mineral oil, light
3. 5 ml syringe
4. Sonotech LithoClear acoustic gel
5. Tegam 4040B amplifier
6. LabView software (National Instruments)
7. 1% Bovine serum albumin (BSA)

15.3 Methods

Photoacoustic flow cytometry is used to detect small particles under flow that have some optical absorption. Using cultured bacterial cells, we can induce optical absorption by tagging them with their corresponding bacteriophage. Once the bacterial cells are introduced to the photoacoustic system, they can be detected, enumerated, and captured for further study or analysis. Cultured cells can be added to blood samples or PBS for testing.

15.3.1 Bacterial Culturing

1. Using aseptic technique, remove freezer stock from −80°C freezer.
2. Using a sterile culture loop, remove a small portion of frozen stock and streak on LB agar plate using standard microbiology techniques.
3. Label LB agar plate and place in 37°C bacterial incubator for 16–24 hours.
4. Place 4 ml of LB into a 16×125 mm glass culture tube with aeration cap.
5. Using aseptic technique and a sterile inoculation loop, transfer one bacterial colony into the LB within the glass culture tube.
6. Place inoculated culture in 37°C shaking water bath for 16 hours.
7. Remove culture from shaking water bath and serially dilute to reach desired concentration.
8. Add diluted bacteria to processed blood and use for testing.

15.3.2 Bacteriophage

1. Obtain bacteriophage suspension with high titer concentration of at least 10^{11} pfu. Introduce to Direct Red 81 dye in order to bind dye molecules to bacteriophage, thus inducing optical absorption for 532 nm laser light.

2. Pellet and resuspend the bacteriophage in 10 mM Tris, pH 7.5, 10 mM $MgCl_2$, 68 mM NaCl, a bacteriophage buffer, to remove any unbound dye.
3. Using the Nanodrop 2300 spectrometer or other suitable devices, record absorbance spectrum to verify optical absorption at 532 nm.

15.3.3 Bacteriophage Tails and Microspheres

1. Purify bacteriophage with CsCl gradient purification by taking 1×10^{12} pfu/ ml of bacteriophage and add CsCl to increase density to 1.5 g/ ml.
2. Use osmotic shock to remove bacteriophage capsids, leaving only tails. Leave bacteriophage in CsCl overnight, and then introduce to phage buffer to cause rapid diffusion of CsCl, resulting in capsid separation.
3. Further purify bacteriophage tails using Bio-Rad HPLC and determine purity using spectrophotometer.
4. Determine protein purity and concentration using BioTek Synergy H1 by measuring optical absorbance at 260 and 280 nm.
5. Wash streptavidin coated microspheres in PBS four times in Spin-X concentrator, removing stabilizer and antimicrobial agents.
6. Prepare biotin in PBS and combine with purified bacteriophage tails and incubate on ice for two hours.
7. Remove excess biotin by dialysis using 2kD Slide-A-Lyzer cassettes.
8. Gently mix biotinylated tails with streptavidin for 30 minutes and wash 10 times to remove unbound tails.

15.3.4 Photoacoustic Flow Cytometry

1. Our photoacoustic flow cytometry system consists of an Nd:YAG laser coupled into a 1000 μm, 0.39 numerical aperture optical fiber. This is used to produce 532 nm laser light with 5 nanosecond pulses. Laser beam energy coupled through the optical fiber is maintained and measured from 1.9–2.1 mJ for tagged bacterial cell detection. Laser light is directed toward a quartz tube with 10 μm thick walls passing through a 3D-printed flow chamber, Fig. 15.1. The 10 μm thick walls allow for the propagation of ultrasonic waves, as well as provide an optically transparent pathway for the sample to flow through. The optical fiber is placed 5 mm from the quartz tube to create a detection volume of 0.04 μL. Laser beam shape is assumed Gaussian, and fluence is calculated to be 0.014 mJ/ cm^2. A 2.25 MHz transducer (immersion transducer, 0.50 in. element diameter F=0.80 in.) focused on the quartz sample tube is fitted to the base of the 3D-printed flow chamber. The transducer focal point is aligned to incorporate the quartz tube illuminated by the laser beam. The internal volume of the chamber is filled with Sonotech LithoClear acoustic gel to provide a

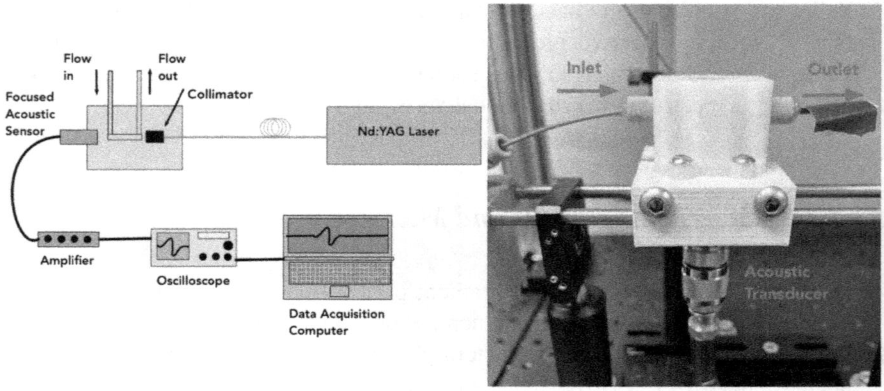

Fig. 15.1 (Left) Schematic showing photoacoustic setup used for testing S. aureus samples. (Right) The actual flow chamber is shown

medium for the propagation of acoustic waves. Syringe pumps create alternating flow of sample and mineral oil equal to a 60 μL/min flow rate. The introduction of sample and the immiscible mineral oil induces two-phase flow [12]. Two-phase flow is employed to allow for sample collection and further analysis while eliminating the possibility of samples becoming stuck or delayed inside the tubing. Signals are amplified with a gain of 50 using a Tegam 4040B amplifier and sent to a desktop computer running a customized LabView program.

2. Allow a 1% BSA solution to sit in the sample syringe, tubing, and the flow chamber for 1 hour prior to use to create a hydrophobic barrier inside the tubing/syringe.
3. Turn on the laser system and allow it to establish thermal equilibrium.
4. Layer 1 ml of mineral oil on top of sample in a 5 ml syringe.
5. Place sample syringe in first syringe pump and place second syringe with mineral oil in second syringe pump.
6. Turn on the syringe pumps and watch for air bubbles to pass through the Quartz tube in the flow chamber before beginning test.
7. Start the computer interface.
8. Begin the photoacoustic test once any air bubbles have cleared the flow chamber.
9. Set the digitizing data acquisition device to take a minimum of 4 averages to reduce random noise.
10. Set the amplifier on 32 dB amplification, with a 1 MHz high-pass filter.

15.4 Results

Figure 15.2 shows the results of phage tails attached to microspheres. Micrographs were taken on a Thermo Fisher/FEI T12 Spirit using a Gatan US 1000 and Orius

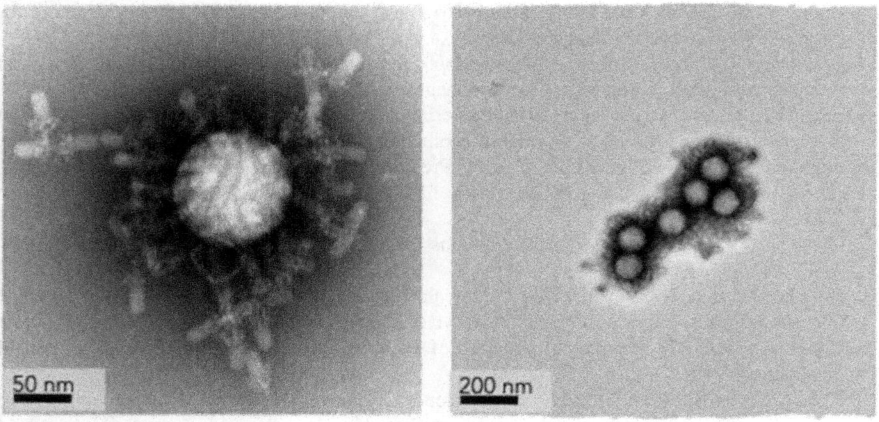

Fig. 15.2 (Left) A single microsphere with several phage tails attached. (Right) Several spheres at lower resolution showing numerous phage tails. These tails were attached in random orientation, much like the image at left

CCD camera (Hillsboro, Oregon). At left a single sphere is shown with several tails attached. The tails are randomly oriented. The image on the right shows six spheres with tails similarly oriented in their attachment.

15.5 Discussion

Successful depletion of genetic material from bacteriophage and attachment of a known optical absorber allow us to use the specific recognition proteins in natural bacteriophage for identification of bacterial species in liquid samples using photoacoustic flow cytometry. While we were able to rapidly identify bacteria before[5], using intact bacteriophage would cause cell lysis so that testing needed to be accomplished within an hour or so. Additionally, our use of dyed microspheres provides uniform optical absorption, allowing for more predictable photoacoustic analysis. These methods will enable optimal use of photoacoustic flow cytometry for detection and identification of bacterial infection in human patients.

References

1. Bartual SG, Otero JM, Garcia-Doval C, et al (2010) Structure of the bacteriophage t4 long tail fiber receptor-binding tip. Proceedings of the National Academy of Sciences 107(47):20287–20292
2. Bertani G (1953) Lysogenic versus lytic cycle of phage multiplication. In: Cold Spring Harbor symposia on quantitative biology, Cold Spring Harbor Laboratory Press, pp 65–70

3. Bonilla N, Rojas MI, Cruz GNF, et al (2016) Phage on tap–a quick and efficient protocol for the preparation of bacteriophage laboratory stocks. PeerJ 4:e2261
4. Boulanger P (2009) Purification of bacteriophages and sds-page analysis of phage structural proteins from ghost particles. In: Bacteriophages. Springer, p 227–238
5. Edgar RH, Cook J, Noel C, et al (2019) Bacteriophage-mediated identification of bacteria using photoacoustic flow cytometry. Journal of Biomedical Optics 24(11):115003–115003
6. Edgar RH, Tarhini A, Sander C, et al (2021) Predicting metastasis in melanoma by enumerating circulating tumor cells using photoacoustic flow cytometry. Lasers in surgery and medicine 53(4):578–586
7. Gandon S (2016) Why be temperate: lessons from bacteriophage λ. Trends in microbiology 24(5):356–365
8. Hodyra-Stefaniak K, Miernikiewicz P, Drapała J, et al (2015) Mammalian host-versus-phage immune response determines phage fate in vivo. Scientific reports 5:14802
9. Kaiser A (1966) On the internal structure of bacteriophage lambda. The Journal of general physiology 49(6):171–178
10. Li B, Majaron B, Viator JA, et al (2004) Accurate measurement of blood vessel depth in port wine stained human skin in vivo using pulsed photothermal radiometry. Journal of biomedical optics 9(5):961–966
11. O'Brien CM, Rood K, Sengupta S, et al (2011) Detection and isolation of circulating melanoma cells using photoacoustic flowmetry. JoVE (Journal of Visualized Experiments) (57):e3559
12. O'Brien CM, Rood KD, Bhattacharyya K, et al (2012) Capture of circulating tumor cells using photoacoustic flowmetry and two phase flow. Journal of biomedical optics 17(6):061221–061221
13. Sigrist MW (1986) Laser generation of acoustic waves in liquids and gases. J Appl Phys 60:R83–121
14. Storms ZJ, Brown T, Cooper DG, et al (2014) Impact of the cell life-cycle on bacteriophage t4 infection. FEMS microbiology letters 353(1):63–68
15. Thomas Jr GJ, Becka R, Sargent D, et al (1990) Conformational stability of p22 tailspike proteins carrying temperature-sensitive folding mutations. Biochemistry 29(17):4181–4187
16. Viator JA, Prahl SA (1999) Laser thrombolysis using long pulse frequency-doubled nd: Yag lasers. Lasers in Surgery and Medicine: The Official Journal of the American Society for Laser Medicine and Surgery 25(5):379–388
17. Young R (1992) Bacteriophage lysis: mechanism and regulation. Microbiological reviews 56(3):430–481

Chapter 16
Photoacoustic Spectrum Analysis for Soft Tissues

Yingna Chen, Shiying Wu, Mengjiao Zhang, Jiayan Li, Menglu Qian, and Qian Cheng

Abstract Biomedical photoacoustics has shown great potential for precise medical diagnosis because it can provide structural, physiological/pathological characteristics, and metabolic information of biological tissues noninvasively in vivo. Photoacoustic imaging has made great breakthroughs in many preclinical studies, including microvascular imaging, blood oxygen detection, and tumor detection. However, compared with photoacoustic imaging, the photoacoustic spectrum can provide more information, such as the rich molecular information in the optical spectrum and the rich microstructural information in the ultrasonic spectrum, which is closely related to the disease evolution process. Recently, photoacoustic spectrum analysis (PASA) has demonstrated the ability to quantitatively extract physicochemical information from biological tissues to distinguish between normal and diseased tissues, especially for classifying, grading, and staging cancer tissue, making it one of the most promising methods for noninvasive, accurate diagnosis of clinical diseases. In this chapter, we introduce the methods of PASA for the diagnosis of soft tissue diseases.

Keywords Photoacoustic spectrum analysis (PASA) · Tumor microenvironment · Tumor grading · Tumor identification · Molecular typing of tumors

---

Authors Yingna Chen, Shiying Wu, Mengjiao Zhang, and Jiayan Li have equally contributed to this chapter.

Y. Chen · S. Wu · M. Zhang · J. Li · M. Qian · Q. Cheng (✉)
Institute of Acoustics, School of Physics Science and Engineering, Tongji University, Shanghai, China
College of Science and Technology, Ningbo University, Zhejiang, China
e-mail: q.cheng@tongji.edu.cn

W. Xia (ed.), *Biomedical Photoacoustics*,
https://doi.org/10.1007/978-3-031-61411-8_16

16.1 Introduction

One of the problems worldwide is the cancer diagnosis and evaluation [1]. Therefore, the development of technologies to improve detection and diagnostic accuracy is necessary. In this order of ideas, it is also mandatory to develop technologies that evaluate the cancer grades or categories, guiding further treatment. However, human carcinogenesis is a complex evolutionary process that involves multiple levels of association and coupling from microgenes to macrotissues [2–4]. Genetic damage or mutation causes abnormal cell differentiation and affects the tissue microenvironment. Gene damage is difficult to predict and localize, even though it is not a requirement for tumor development [5], while abnormal cell differentiation often relies on invasive pathology testing, so heterogeneity of the tissue microenvironment becomes one of the main directions for early detection of cancer. Then, the development of technologies that support the evaluation of tissues becomes more relevant due to the higher requirement for detection. For example, positron emission tomography (PET), which has been rapidly developed in the last decade, takes advantage of the extremely high glucose consumption of cancer cells to achieve tumor detection [6]. However, the ionizing radiation of PET is still an unavoidable shortcoming.

Among the technologies for tissue analysis and determination of components, photoacoustic techniques stand out [7]. In particular, photoacoustic spectrum analysis (PASA) is considered by some researchers as a promising technology for disease diagnosis. PASA has some characteristics such as the fact that it uses nonionizing optical excitation, making it safe for detection; it has high intrinsic optical absorption contrast, making it free of additional injection; it has abundant physiochemical information with quantitative ability [8, 9]. Owing to all these advantages, the first two decades of the twenty-first century saw a quick development of the PASA technique, making it a convenient option to be incorporated in tissue detection and disease diagnosis. The PASA technique allows the obtaining of optical qualities of components in tissues, depending on their specific optical absorption, which is useful in disease diagnosis since the content of chemical components such as collagen, lipid, and hemoglobin is closely associated with the evolution of the diseases [10]. Besides, the unfolding of PASA on the ultrasound spectrum provides information on the microstructure of the tissue, which is also important for disease diagnosis [11, 12]. Therefore, this chapter is to perform our work on PASA in biological soft tissue research from methodology to application. In addition, considering the small size of clinical samples, we also present some machine-learning methods that can help extract features from the PASA data. In this way, it will be possible to know the proposed various embodiments of PASA, why they are done, and how they can be applied to real problems. The PASA experimental setup continues to be optimized and focused on the specific problems of real clinical applications where it can serve as a supportive technique to improve the probability and accuracy of disease diagnosis in its early stage and realize the evolution of the treatment guidance.

16.2 Various Embodiments of PASA

The occurrence and evolution of diseases are often accompanied by changes in the content of chemical components and microstructure characteristics, such as the reduction of collagen content and the enhancement of microstructure heterogeneity in prostate cancer (PCa) [13], which are the basis for the realization of disease diagnosis and quantitative evaluation by PASA. The photoacoustic spectrum contains both optical and acoustic dimensions. The optical dimension mainly utilizes the difference in light absorption of different chemical components to achieve specific detection, which makes photoacoustic more colorful than traditional ultrasound detection. Compared to optical absorption dimension, less attention has been paid to the ultrasound dimensions and does not take full advantage of ultrasound information on tissue heterogeneity. In this section, we focus on the multidimensional information that the ultrasound dimension of photoacoustics can bring us and elaborate on its effectiveness in disease detection.

16.2.1 Photoacoustic Power Spectrum

The frequency domain is the fundamental dimension of ultrasound signal unfolding. Compared to the time series, observing the frequency domain signal can provide us with more information about tissues from various angles, particularly structural information. It has been well studied that the ultrasonic spectrum is used to characterize structures, including target structure and size [14, 15], and it lays a foundation for the structural quantification study of photoacoustics based on the ultrasonic spectrum. One of the differences is that the ultrasound signals received by photoacoustics are excited and passively received by the laser, compared to the active emission and reception of conventional ultrasound detection. There is a difference in the noise and signal randomness, so the power spectrum analysis is required to analyze energy density at different frequencies.

The power spectral density in the actual application cannot be accurately calculated but estimated due to the limited points collected in a finite period. The classical periodogram method is adopted for power spectrum analysis, which means that N sampling points of random signals are treated as a signal with finite energy. The real power spectrum is estimated by dividing the square of amplitude by N after carrying out a Fourier transform. To reduce noise and improve the signal-to-noise ratio, the window function and sliding average method are combined on this basis.

The calculated photoacoustic power spectrum can indicate the internal size of tissue clusters. Take the most basic single spherical sound source as an example. The calculated sound pressure is related to spherical radius $\boldsymbol{r_0}$, and the smaller the r_0 is, the higher the frequency of the spectral peak (Fig. 16.1a, b).

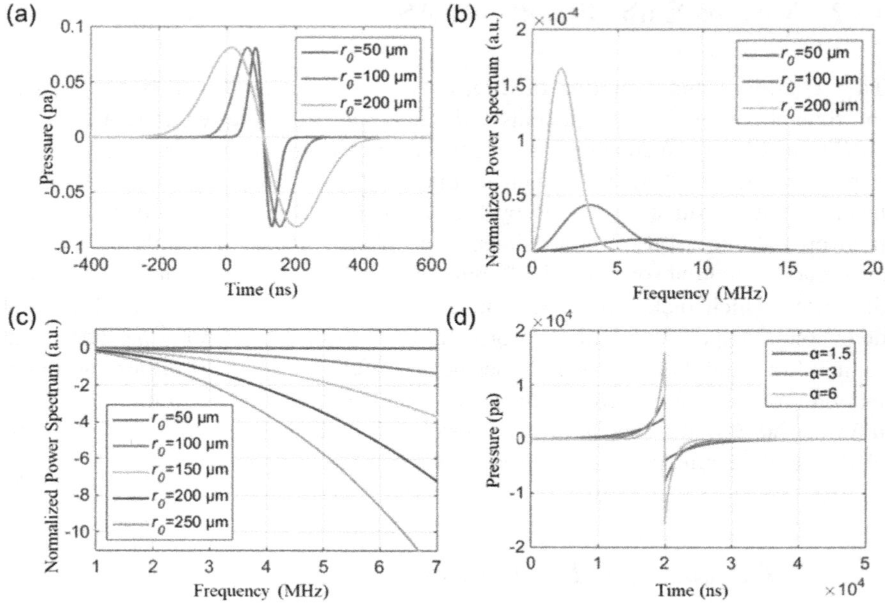

Fig. 16.1 Numerical study verifying the ability of power spectrum characterization of chemical component microstructure and content. (**a**) Photoacoustic time domain signals from a single spherical sound source. (**b**) Normalized power spectrum of a single spherical sound source (transducer receiving distance $r = 1.6 \times 10^{-2}$ cm). (**c**) Normalized power spectra of randomly distributed clusters of different sizes. (**d**) Photoacoustic signals in a semi-infinite ideal fluid with free boundary conditions (detection distance $r = 3$ cm)

$$p\,(r,t) = -\frac{\beta I_0 v_c^3}{2\pi^{3/2} C_p r_0^3}\frac{1}{r}\left(t - \frac{r}{v_c}\right)\exp\left\{-\left[\frac{v_c}{r_0}\left(t - \frac{r}{v_c}\right)\right]^2\right\} \tag{16.1}$$

Under the more complex stochastic distribution model of sound sources, the power spectrum is then rewritten based on the autocorrelation function of spatial distribution function $\boldsymbol{R_A}\left(\boldsymbol{\Delta\vec{r}}\,;\boldsymbol{d}\right)$ and transducer directivity function $\boldsymbol{R_D}\left(\boldsymbol{\Delta\vec{r}}\,,\boldsymbol{f}\right)$ [16]. To eliminate the system response, a differential power spectrum was proposed with a mathematical representation that is as follows:

$$S_d(f) = \frac{\int_{\Delta V} R_A\left(\Delta\vec{r}\,;d\right) R_D\left(\Delta\vec{r}\,,f\right)\,\mathrm{d}\Delta V}{\int_{\Delta V} R_A\left(\Delta\vec{r}\,;d_c\right) R_D\left(\Delta\vec{r}\,,f\right)\,\mathrm{d}\Delta V} \tag{16.2}$$

It is worth noting that the response frequency band of most commercial ultrasonic transducers is difficult to cover higher frequencies, such as over 20 MHz. However, it has been studied that there is still a difference in the lower bands of the photoacoustic power spectrum (Fig. 16.1c), which is useful in practical applications.

In addition to microstructure data, the photoacoustic power spectrum provides information about component content. According to Parseval's theorem, the power spectrum is directly related to the signal power in the time domain, which is influenced by the tissue's optical absorption (Figure 16.1d). Making the incident light energy is sufficient with free surface, the light absorption is proportional to the content of components in the tissue (Equation 16.3):

$$\lim_{\alpha z \gg 1} p(z,t) = \frac{v_c \beta I_0}{2C_p \tau} \begin{cases} \left(1 - e^{-\alpha v_c \tau}\right) e^{\alpha v_c t'}, t' < 0 \\ e^{-\alpha v_c t'} - e^{\alpha v_c (t' - \tau)}, 0 < t' < \tau \\ \left(1 - e^{\alpha v_c \tau}\right) e^{-\alpha v_c t'}, t' > \tau \end{cases} \tag{16.3}$$

where v_c is the sound velocity of the ideal fluid, β is the coefficient of linear thermal expansion, α is the light absorption coefficient, I_0 is the incident light intensity, C_p is the specific heat capacity at constant pressure, and τ is the laser pulse width.

16.2.2 Photoacoustic Power Azimuth Spectrum (PA-PAS)

The photoacoustic power spectrum has proven the ability to characterize the microstructure and component content. However, in complex biological environments, many factors affect the photoacoustic signal, including the mechanisms of diffraction, scattering, and absorption in tissues. Scattering, in particular, can cause the measured photoacoustic signal to have angular anisotropy. Furthermore, the limited perspective and directivity of a general hydrophone can result in a limited visual angle and missing information. This can lead to a significant deviation in the photoacoustic power spectrum of heterogeneous tissues. To address this issue, the photoacoustic power azimuth spectrum was developed. This supplements the photoacoustic signals under the full circumscribed angle by circumscribed acquisition. Additionally, it provides angular information that can further supplement the structural features of tissues.

This section introduces the PA-PAS method based on our research on the detection of vascular as an example [17]. Blood vessels, which often have a preferred growth direction, cause the signal to be anisotropic. Identifying such characteristic is critical for diagnosing and evaluating various diseases.

The blood vessel is simplified as a thin-walled round tube model, the sound propagation of the cylindrical sound source is directional, and the maximum signal is generated in the radial direction of the cylinder. Therefore, the photoacoustic field produced by blood vessels due to the light absorption of hemoglobin is also directivity, with a maximum signal in the radial direction of blood vessels or perpendicular to the direction of the blood vessel growth. For the actual vascular network, vessels with different growth directions may have different signal directivities. Therefore, through the analysis of acoustic signals in different directions, it is expected to obtain structural characteristics such as the growth direction or curvature of blood vessels.

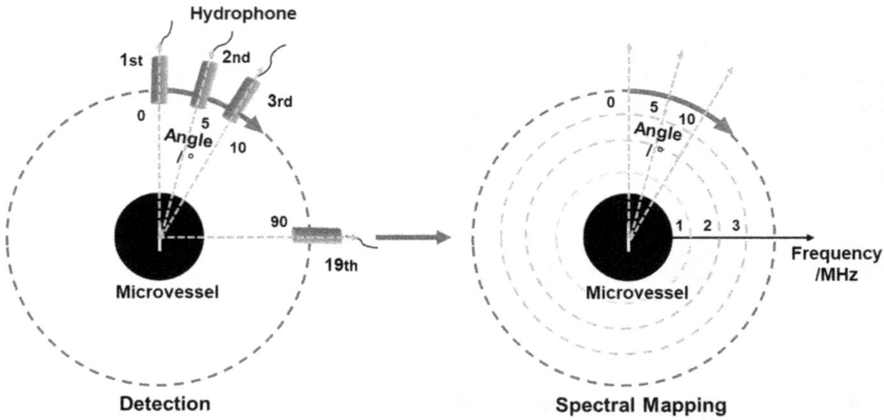

Fig. 16.2 Detection and spectral mapping processes used in the PA-PAS method. 1st, 2nd, 3rd, and 19th represent the numbers indicating the acoustic sensor locations; 0, 5, 10, and 90 represent the angles; 1, 2, and 3 represent the PA spectral frequencies [17]

We mapped the acoustic power spectrum of the PA signals along the azimuth direction. The power spectra of the PA signals were unfolded along the radius direction, and the spectra acquired at all the angles were combined along the angle axis to form a PA-PAS map, as shown in Fig. 16.2. The radius axis gives the ultrasonic frequency distribution and represents the structural size of the detected sample; the angle axis represents the detection angles, showing the structural growth direction of the detected sample. Owing to the differences in the structural growth direction and size of the detected sample, each detected sample should possess a unique PA-PAS map.

16.2.3 Photoacoustic Time-Frequency Spectrum (PA-TFS)

Although PA-PAS incorporates multiangle heterogeneity analysis, it is still rely on photoacoustic power spectrum analysis. When using the Fourier transform to calculate the power spectrum, we lose time information and are unable to obtain the frequency distribution for each moment. That is to say, what we get is an whole signal of the target tissue, but it does not show the variations in the physical and chemical properties of the tissue across time domain. A solution to the problem was proposed using photoacoustic time-frequency spectrum (PA-TFS), dual dimensional information in time and frequency domains.

The Fourier transform can only provide a general understanding of the significance of frequency components. However, time-frequency analysis techniques such as the short-time Fourier transform, continuous wavelet transform, Wigner–Ville distribution, and Hilbert Yellow transform can break down the entire signal into

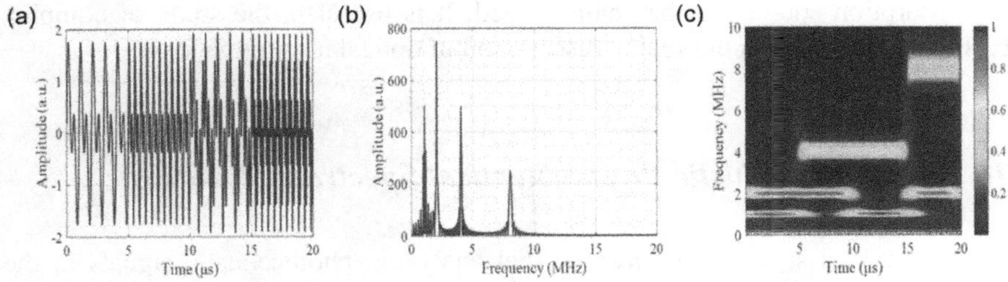

Fig. 16.3 Comparison between the power spectrum and time-frequency spectrum is conducted through simulation. (**a**) Nonstationary signal, (**b**) frequency spectrum of the nonstationary signal by Fourier transform, (**c**) time-frequency spectrum of the nonstationary signal by continuous wavelet transform

shorter time domain signals by applying windowing to the nonstationary signal. The windowed shorter time domain signals are analyzed to obtain the joint distribution information of both the time and frequency domains. The information on frequency change with time can be obtained effectively from nonstationary signals, which can be used in time-frequency filtering and time-varying signal research. The time-frequency analysis formula is as follows:

$$G(f, \tau) = \int_{-\infty}^{+\infty} f(t) w(t - \tau) h(t) \mathrm{d}t \qquad (16.4)$$

where $f(t)$ is the nonstationary time domain signal, $w(t)$ is the window function, and $h(t)$ is the kernel function of various transformations.

We customize a nonstationary signal composed of various time-varying frequencies (Fig. 16.3a). The frequency components of the signal in 0–5 s are 1 MHz and 2 MHz, the frequency components of the signal in 5–10 s are 2 MHz and 4 MHz, the frequency components of the signal in 10–15 s are 1 MHz and 4 MHz, and the frequency components of the signal in 15–20 s are 2 MHz and 8 MHz. Fourier transform is applied to the signal to obtain the frequency spectrum of the signal (Fig. 16.3b). It can be seen that the whole signal is mainly composed of the frequency components of 1 MHz, 2 MHz, 4 MHz, and 8 MHz, but the change information of the four main frequency components with time cannot be obtained. The time-frequency spectrum (Fig. 16.3c) is obtained by continuous wavelet transform of the nonstationary signal. The time spectrum provides the joint distribution information of the time domain and frequency domain of the nonstationary signal and clearly describes the frequency distribution at different times.

The time-frequency spectrum contains rich information in the time domain and frequency domain, which corresponds to the spatial distribution and microstructure information of biological tissues. Based on time-frequency spectral analysis of photoacoustic signals from the biological tissue, information on the frequency spectrum distribution at each moment in time can be obtained. Additionally, by extracting quantitative parameters from the PA-TFS, the space distribution of the absorption source can be characterized. It is useful in the study of complex biological tissues with the spatial distribution of optical absorption sources.

16.2.4 Photoacoustic Physiochemical Spectrum (PA-PCS)

In the previous sections, we learned that analyzing photoacoustic signals in the acoustic dimension can provide valuable insights into microstructures. To further quantify this information, we propose using photoacoustic power azimuth spectrum and time-frequency spectrum, taking into account the angular anisotropy and spatial distribution of the signals. By leveraging the sensitivity of photoacoustic detection to different chemical components in multiwavelength detection, we can create a map that synchronously combines optical and acoustic dimensions. This map contains a wealth of physical and chemical information, which we refer to as a photoacoustic physiochemical spectrum.

For our study, we used laser wavelengths ranging from 690 to 750 nm and 1200 to 1700 nm, covering the absorption bands of major biochemical components in soft tissues including hemoglobin, collagen, lipid, and water in the first and second infrared windows [18]. Figure 16.4 shows a photoacoustic physiochemical spectrum consisting of the optical and sound power spectrum. To create this spectrum, the signal under each wavelength is transformed into a optical spectrum, and the sound power spectrum are aligned as a matrix. This is then visualized as a pseudo-color map, with the colors representing the strength of the photoacoustic signals in different frequency ranges.

Compared to the photoacoustic power spectrum at a single wavelength that only detects one component in tissue, the physiochemical spectrum provides a more comprehensive analysis of various chemical components and their microstructure characteristics. This allows for the study of the interrelationships between these components, which is crucial in understanding the development of diseases. Addi-

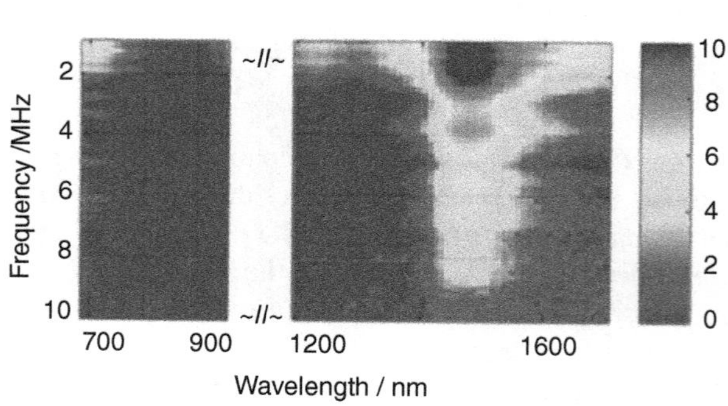

Fig. 16.4 General photoacoustic physiochemical spectrum. The x-coordinate is the wavelength, the y-coordinate is the power spectrum, and the pseudo color represents the signal intensity

tionally, the high-dimension physiochemical spectrum offers the potential for deep feature mining, which can extract disease diagnostic factors with greater robustness.

16.3 Applications of PASA

In the second section, various photoacoustic spectrum methods and their application backgrounds are introduced. These include the fundamental method of analyzing the photoacoustic power spectrum, the suggested method of PA-PAS based on tissue heterogeneity, and the PA-TFS method for examining tissue spatial distribution. In this section, we present the clinical feasibility results of photoacoustic spectrum analysis in soft tissues, along with the specific quantitative features used to characterize the disease. The application is presented in the same order as the photoacoustic spectroscopic methods presented.

16.3.1 Skin Cancer Diagnosis Using Photoacoustic Power Spectrum

The tumor microenvironment (TME) contains collagen and lipid, which play crucial roles in tumor development and invasion. PASA has been introduced, having the potential to be an effective tool in evaluating naturally occurring pigments in biological tissues for tumor diagnosis. Therefore, we propose the PASA for characterizing the tumor-related features of lipid and collagen contents in the TME to identify different types of tumors. Skin cancers are great research objects for photoacoustic detection as they are superficial, which contributes to a great signal-to-noise ratio and high safety. Meanwhile, clinical skin cancer detection continues to rely on invasive puncture, and new technological approaches are needed to enable cancer diagnosis and differentiation. Thus, experiments on skin cancer diagnosis through photoacoustics were conducted [19–21].

Ex vivo experiments were conducted using three types of human skin tissues: normal, squamous cell carcinoma (SCC), and basal cell carcinoma (BCC). The relative lipid and collagen contents in the TME were assessed based on the PASA parameters and compared with histology. Support vector machine (SVM), one of the simplest machine learning tools, was applied for automatic skin cancer type detection.

Based on the results of the PASA, it was found that the levels of lipid and collagen in tumors were significantly lower than those in normal tissues. There was also a statistical difference between SCC and BCC ($p < 0.05$), which was consistent with the histopathological results. Using SVM-based categorization, diagnostic accuracies of 91.7% (normal), 93.3% (SCC), and 91.7% (BCC) were achieved. The study verified that collagen and lipid in the TME can serve as biomarkers of tumor

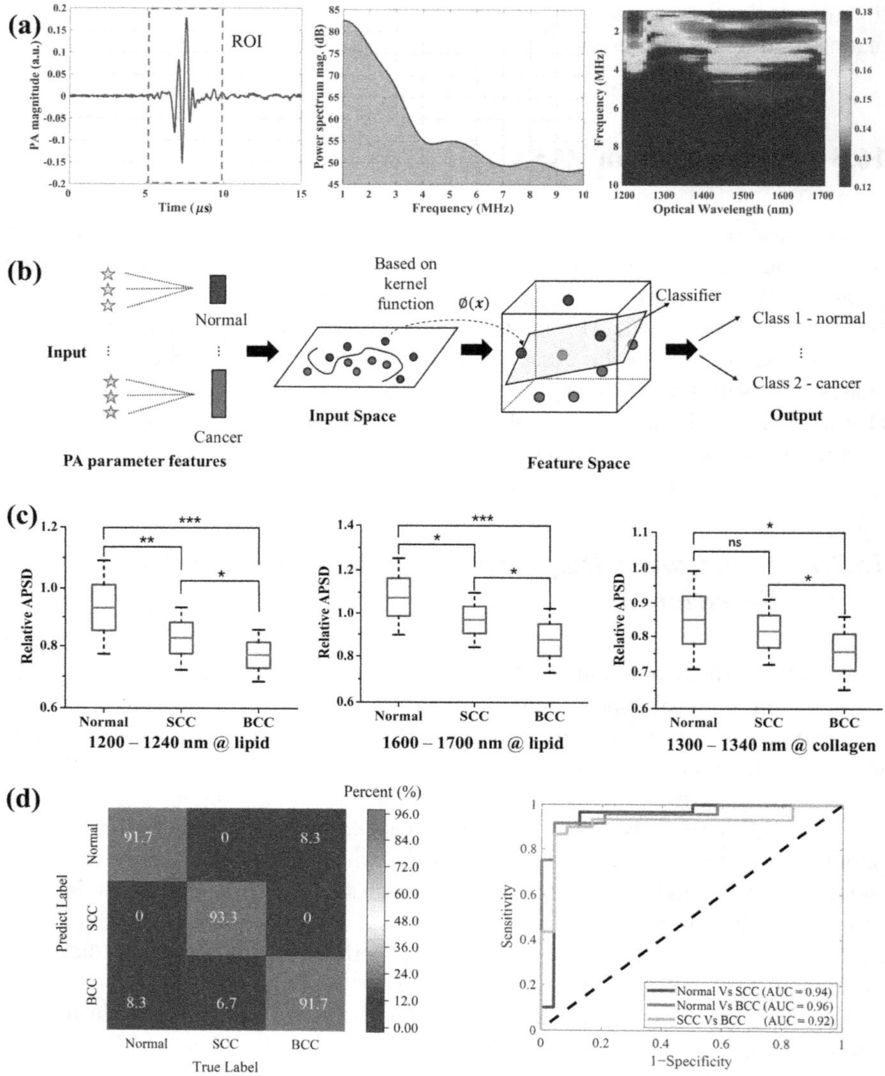

Fig. 16.5 (**a**) Photoacoustic power spectral analysis method. (**b**) Classification of normal and skin cancer tissues using an SVM classifier. (**c**) Statistical results of the PA-related APSD values for the lipid and collagen contents in different skin tissues. (**d**) SVM classification results [20]

diversity, and accurate tumor classification is possible based on the collagen and lipid content through PASA. This method presents a new way of diagnosing tumors. Since the PA technique is noninvasive and has greater detection depth than pure optical technology, this method has significant potential for noninvasive and more accurate diagnosis of tumors in vivo (Fig. 16.5).

16.3.2 Myocardial Infarction (MI) Boundary Demarcation Using Photoacoustic Power Spectrum

Myocardial infarction (MI) is a serious global health issue that causes illness and death. Recent therapeutic approaches such as stem cell, biomaterial, gene, and growth hormone therapies have proven effective in restoring and retaining the heart's contractile properties after MI by targeting the infarct border. However, conventional clinical imaging methods have their limitations. Ultrasound (US) has difficulty expressing soft tissue contrast, single-photon emission computed tomography (SPECT) and positron emission tomography (PET) have lower spatial resolution, and magnetic resonance imaging (MRI) is time-consuming. Finding a noninvasive, real-time imaging technique to accurately identify the MI border is crucial.

After a myocardial infarction (MI), activated fibroblasts expand into the damaged tissue, triggering a fibrotic response and collagen deposition. Molecular changes are known to occur after MI, making PA a promising tool for producing image contrast based on these differences. As a result, we are developing a new PA-based approach to detect the MI border by observing changes in the collagen content [22].

The purpose of this study was to test the feasibility of using dual-wavelength PA spectral analysis (DWPASA) to measure specific molecular characteristics associated with myocardial infarction (MI) and help identify the edges of the affected tissue. The study involved conducting ex vivo PA experiments on samples from a rat MI model. By analyzing broadband PA signals generated by exciting the samples with 1310 and 1450 nm wavelengths, we were able to define a new quantitative parameter called "the ratio of the areas of the power spectral densities (RAPSD)." This parameter helped to quantify specific molecular characteristics both before and after MI. To test the reliability of this method, we compared the results obtained with DWPASA to those obtained using standard Masson staining. Results showed that MI tissue contained more collagen than the normal tissue, resulting in higher RAPSD values ($p < 0.001$). Additionally, the cross-sectional length of the MI and the two-dimensional projection of the MI shape (area) determined using DWPASA were in substantial agreement with Masson staining ($\text{ICC} = 0.76$, $p < 0.001$, $\text{IoU} = 0.72$). The study shows that RAPSD is a more sensitive indicator of disease progression than the PA signal amplitude of individual molecules and thus can help distinguish between normal and pathological tissues (Fig. 16.6).

16.3.3 Vascular Morphological Estimation Using PA-PAS

The morphology and function of microvasculature are closely linked to the occurrence and development of numerous diseases, including changes in size and density. These quantitative changes can offer valuable information about the dynamics of tumor angiogenesis, which is crucial for clinicians in the diagnosis and treatment of cancer.

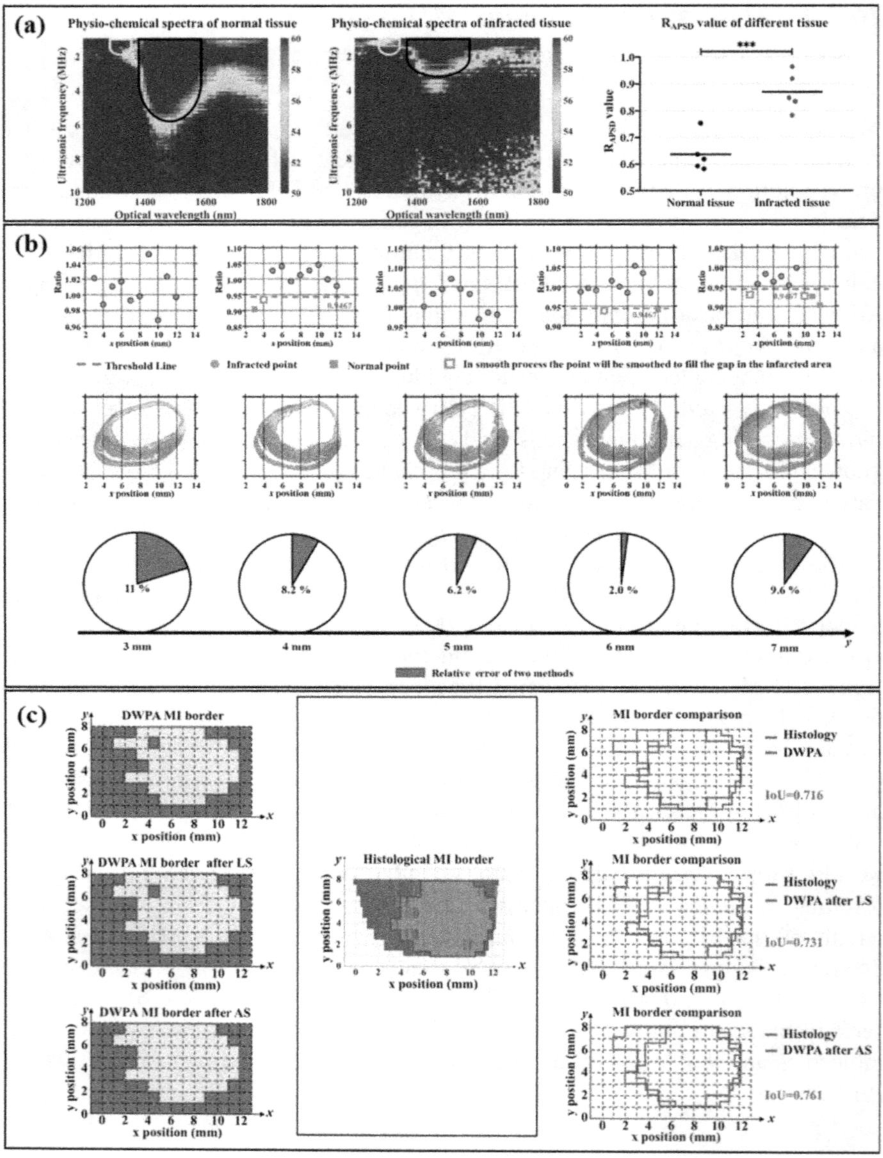

Fig. 16.6 (**a**) Physiochemical spectrograms of normal and infarcted tissues. The right figure is the RAPSD statistics, which can be used to differentiate between normal and infarcted tissues. (***statistically significant difference ($p < 0.001$)). (**b**) Typical cross-sectional myocardial infarction (MI) border determination results for each of the five slices of sample number 492. (**c**) Typical diagrams showing a two-dimensional MI area border (sample number 492) [22]

The use of PA technology has become widespread in the detection of blood vessels due to the significant contrast between hemoglobin in the blood and surrounding tissue. PASA has proven useful in detecting absorbers smaller than the system resolution, allowing for the analysis of microvascular structures. However, blood vessels have tubular structures and dendritic distributions, which cause the generated PA signals to be inhomogeneous and exhibit anisotropy. This results in significant differences at various receiving angles. Previous studies of the power spectra of blood vessels have ignored the differences in angles. To address this issue, we have developed a new method called PA-PAS [17], which combines power spectra with angles to more accurately evaluate microvessel structures. The PA-PAS method maps the acoustic power spectrum of the PA signals along the azimuthal direction, revealing microvessel characteristics such as the structural direction and size. Additionally, we calculate a PA spectral parameter known as power-weighted mean frequency (PWMF) to quantify microvessel sizes, which reflect the main frequency component of biological tissues.

According to the research findings, information about the PA signal of microvessels is mainly concentrated in the width direction. Ultrasonic frequencies that reflect the structural sizes of microvessels are more effective in this direction. The PA-PAS method is better than the PACT method in differentiating microvascular sizes, especially for smaller vessels. The PA-PAS maps show that the power spectra are more extended along the radius axis for smaller vessels. Furthermore, PWMF values along the structural direction can be used to measure microvascular structural sizes. The PWMF values decrease linearly with the size of the microvessel, which can accurately determine its size (Fig. 16.7).

16.3.4 Prostate Cancer Grading Using PA-TFS

Prostate cancer (PCa) has recently become the most frequently diagnosed cancer in men worldwide, making up 10% of all new male cancer cases. The "gold standard" for confirming a PCa diagnosis and determining tumor aggressiveness is currently magnetic resonance (MRI)-/ultrasound (US)-guided biopsy puncture pathology. During this procedure, 16- or 18-gauge biopsy needles are used to extract 15-mm biopsy needle strips from the prostate tissue. Each tissue sample is then graded and assigned a Gleason score (GS) based on the glandular cavity architecture. However, this grading system is subjective and can produce an approximate 80% accurate diagnosis. Additionally, the pathology process itself is time-consuming. While US and MRI methods have shown great potential in detecting PCa, they focus on overall tissue morphology and do not account for the distribution of glandular cavity structure information, which is crucial for accurately evaluating the Gleason system. As a result, there is an urgent need for a new method that can assess PCa aggressiveness more quickly, objectively and accurately.

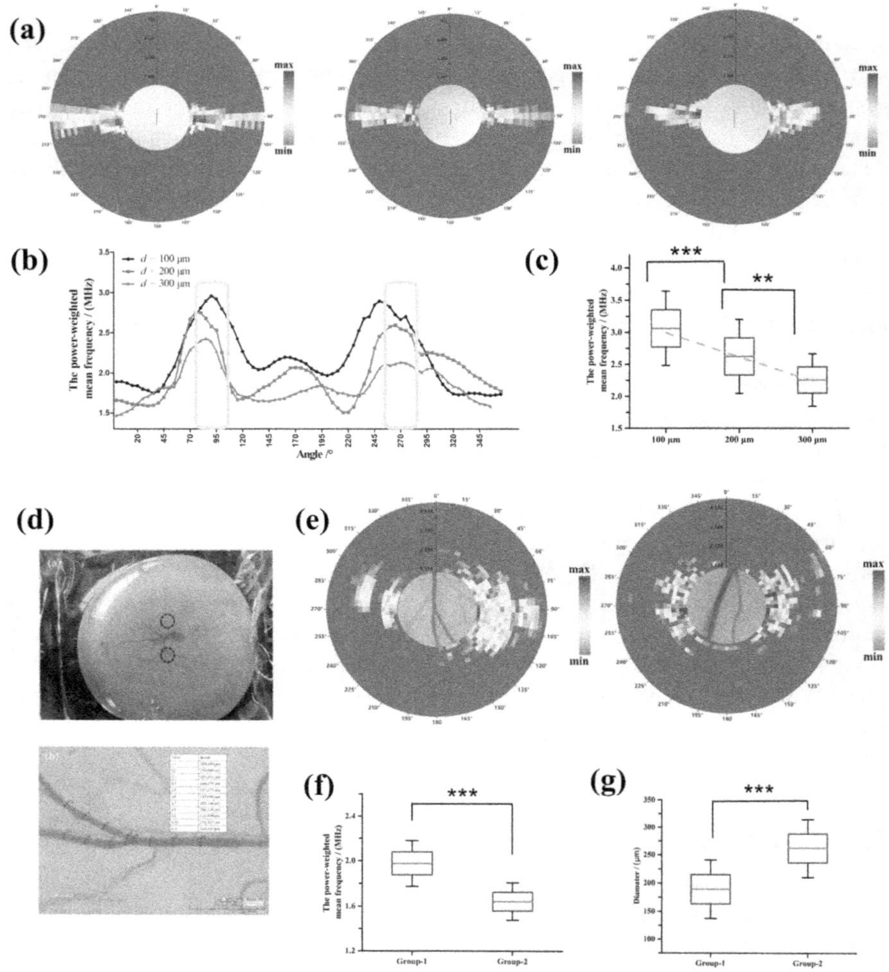

Fig. 16.7 (**a**) PA-PAS maps of different microvessels, whose diameters are 100, 200, and 300 μm. (**b**) PWMF-angle curves of all samples (the solid yellow boxes correspond to the width direction of the microvessels). (**c**) PWMF values obtained for different microvascular sizes along their width direction. The PWMF values decrease linearly with the increase in the vascular diameter, as indicated by the green dashed line. (**d**) Representative chick embryo model. (**e**) Representative PA-PAS maps of the detected microvessels in two groups of the chick embryo model. (**f**) Quantified PWMF values in the width direction of the microvessels in the two groups. (**g**) Sizes of the blood vessels in the two groups were measured microscopically [17]

In the field of PASA, the PA signal can be analyzed in the frequency domain to assess the size and microstructure of various chemical components in biological tissues. This method is consistent with the evaluation characteristics of the Gleason system. In a previous study, PASA was conducted on the entire effective signal, which is generated by whole biological tissue, and the PA spectral parameter slope reflects the proportion of high- and low-frequency components [23]. However, when analyzing prostate biopsy cores, which consist of a 15-mm biopsy needle strip, only 20–30% (approximately 4 mm) is the tumor region, while the rest is normal prostate tissue. In this case, the slope parameter can provide inaccurate results as it reflects the average structure size of the tissue, including the normal region. To obtain more accurate results, the tumor region must be selected first before PASA. The continuous wavelet transform (CWT) can be used to provide joint distribution information of the time and frequency domains of PA signals [24]. By using CWT, we can obtain the frequency distribution information along the 15-mm biopsy needle strip. As the difference between prostate cancer (PCa) and normal tissue lies in the structural size of the glandular cavity, the influence of normal tissues can be removed by utilizing the frequency difference between the structure of the normal tissue and PCa using CWT.

Convolutional neural networks (CNNs) are a type of deep feedforward neural network that uses convolution computation. They are one of the most popular algorithms in deep learning due to their ability to create rich representations, fast inference, and filter-sharing properties. CNNs have greatly improved the accuracy of image classification and super-resolution reconstruction. The deep residual network (ResNet) is a common type of CNN, which creates "shortcuts" (skip connections) between the front and back layers to better train a deeper network and achieve higher accuracy. ResNets (ResNet-18, ResNet-34, ResNet-50, ResNet-101, and ResNet-152) perform better than other CNN models for image classification in the ImageNet dataset because they can extract image features well. ResNet-18 performs similarly to other ResNets but can retain more low-scale features due to being shallow. We used ResNet-18 for the intelligent grading of prostate cancer.

Our study used PA-TFS to assess the aggressiveness of prostate cancer (PCa). We conducted an ex vivo experiment on human prostate biopsy needle strips with varying Gleason scores (GSs). Firstly, we performed CWT time-frequency analysis on each sample to identify the tumor region of the entire biopsy needle strip. Next, we conducted a linear fitting on the PA power spectrum curve of the tumor region to determine the quantified spectral parameter slope. The results showed that higher GSs led to smaller glandular cavity structures and higher heterogeneity, resulting in higher slopes at both 1210 nm and 1310 nm ($p < 0.01$). Furthermore, we used ResNet-18 on the PA-TFS images of the tumor region to directly evaluate the aggressiveness of PCa. The classification accuracy of the PA-TFS of the tumor region using ResNet-18 was 89% at 1210 nm and 92% at 1310 nm. The total testing time was less than 7 min, including 3 min for PA detection, 2 min for post-processing time, and 1 min for decision-making. Our objective was to grade different GSs by characterizing the microstructure of chemical content in the frequency spectrum. Our results showed that PA-TFS can rapidly and objectively identify PCa (Fig. 16.8).

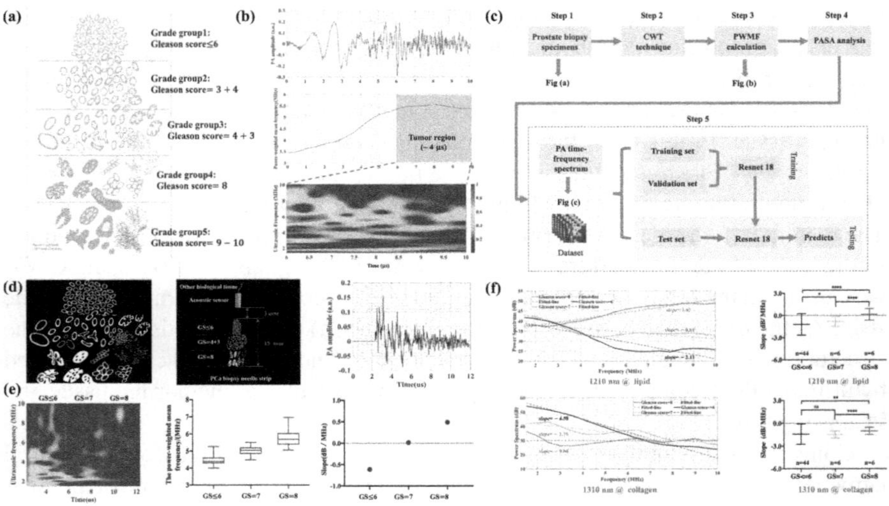

Fig. 16.8 (**a**) Newly modified Gleason grading diagram presented in the ISUP publication. (**b**) Typical PA time-domain signal, PWMF curve, and corresponding PA time-frequency spectrum (PA-TFS) of the human PCa biopsy needle strip. (**c**) Block diagram of the PA signal processing. (**d**) Simulation results of prostate cancer tissue. (**e**) Quantification results of numerical simulation. (**f**) Quantification results of ex vivo PASA of different GS tissues at two wavelengths of 1210 nm and 1310 nm [24]

16.3.5 Breast Cancer Evaluation Using PA-PCS (Multiwavelength Power Spectrum)

The extracellular matrix (ECM) is primarily made up of collagen, which goes through various changes during the onset and growth of cancer. Traditional methods of examining these changes, like histopathological examination, are invasive and limited to in-vitro tissues. However, utilizing PA-PCS can allow for in vivo, real-time detection of collagen, providing a new way to study the transformation of ECM during cancer progression.

PA-PCS was used to distinguish between different breast cancer molecular subtypes [25]. The study involved examining 50 breast cancer samples from nude mice, including three molecular subtypes known as HER2-enriched, luminal, and triple negative. We used a pulsed OPO laser with tunable wavelengths ranging from 1200 to 1700 nm and an energy density of 20 mJ/cm^2 that was lower than the ANSI laser safety standard. They extracted the spectrum of "area of power spectrum density (APSD)" from the PA power spectra to evaluate the content of biomacromolecules. We then used regression analysis to compare the APSD spectra pre-processed by the Savitzky–Golay 1st (S-G 1st) algorithm, against the Masson's trichrome staining results. They used the partial least-squares (PLS) algorithm with tenfold cross-validation and calculated the loadings and variable importance in projection (VIP) scores to evaluate the importance of each wavelength.

The results of the histological staining analysis revealed significant differences in collagen levels between the luminal subtype and the other two subtypes. The correlation coefficients between PASA and the staining results were 0.78 and 0.66 for the training and cross-validation sets, respectively. The wavelengths with the highest loadings and VIP scores were found in the absorption band of collagen, demonstrating the effectiveness of the analysis model. These findings suggest that PASA has potential as a noninvasive and semiquantitative method for detecting collagen in vivo. This could improve our understanding of ECM changes and potentially be used as a diagnostic tool for cancer subtype classification at a macromolecular level (Fig. 16.9).

16.4 Discussion and Conclusion

This chapter demonstrates how photoacoustic spectrum analysis can be used to quantify the features of soft tissue lesions to help characterize the diseases and meet clinical requirements. We introduce four different types of photoacoustic spectrum, including the original photoacoustic power spectrum, photoacoustic azimuth spectrum (PA-PAS), photoacoustic time-frequency spectrum (PA-TFS), and the high-dimensional photoacoustic physiochemical spectrum (PA-PCS). These spectrums are validated with experimental results. However, the robustness of quantitative features is a limitation of QPA in practical clinical requirements. To overcome this, we extracted specific quantitative parameters that can characterize the chemical component content or microstructure evolution of the tissue based on the evolution of different diseases. The features developed here have great potential in clinical applications. We have successfully demonstrated the applicability of our methods in skin cancer, blood vessels, myocardial infarction, prostate cancer, breast cancer, and osteoporosis, both in ex vivo samples and in vivo mice experiments. However, for the soft tissues with complex structures and biological compositions, ensuring the robustness of features extracted from complex PA signals is still a bottleneck, which will limit the further application of PASA methods in the clinical diagnosis of more complex soft tissues. Correct physical modeling of complex biological media is one of the keys to solving this problem [21]. In addition, although machine learning is introduced as a powerful method for feature extraction with good robustness, its internal black-box nature still prevents it from being truly applied in clinic. Therefore, we recommend the hybrid model of the physical model and data as a development direction to promote the PASA method.

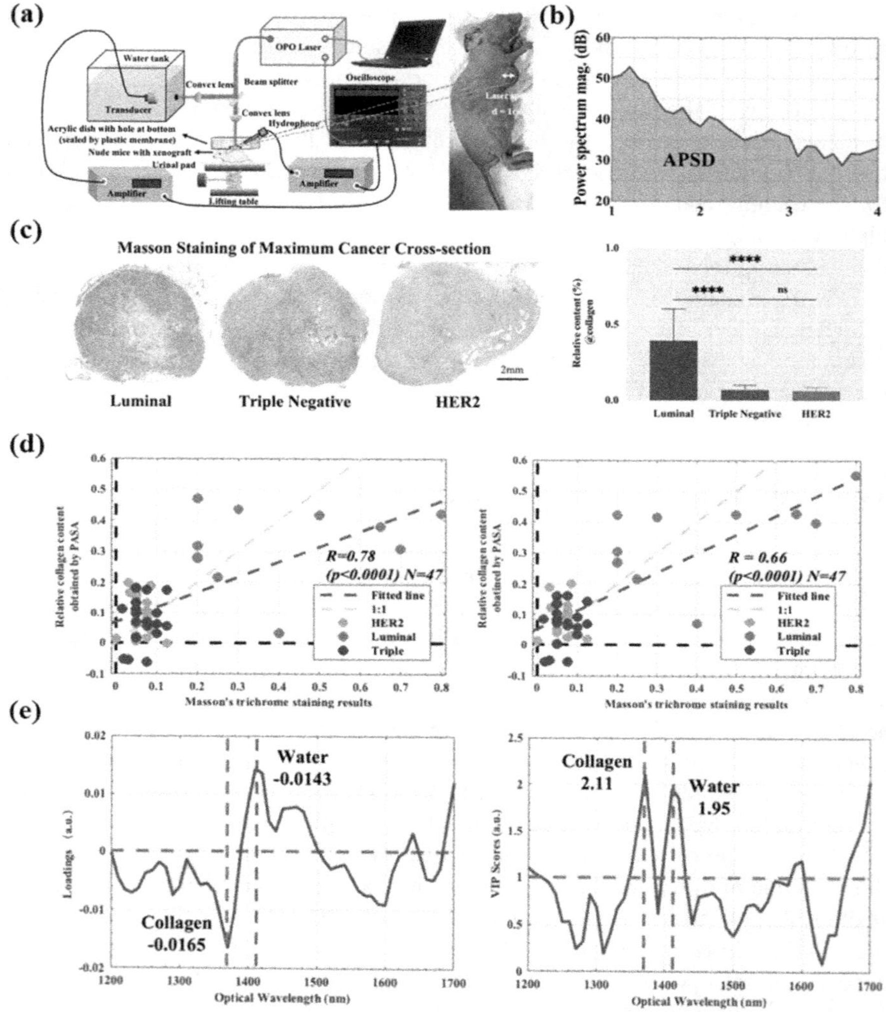

Fig. 16.9 (**a**) In vivo experimental setup for PA measurements. (**b**) Schematic diagram of photoacoustic power spectrum and parameter "area of power spectrum density (APSD)." (**c**) Histopathological photographs for the maximum cross section of the nude mice xenografts (left); statistics analysis and Tukey multiple comparison test results for the collagen contents of xenografts obtained from Masson's trichome staining (****$p < 0.0001$, ns: no statistical significance) (right). (**d**) Analysis of correlation between the relative collagen contents attained via PASA and histopathological method: training set (left) and cross-validation set (right). (**e**) Variable loadings (left) and VIP scores spectrum (right) for comparing the impact of each wavelength in the regression model. The absorption peaks for collagen and water were marked by the specific values in the figure [25]

References

1. Siegel RL, Miller KD, Wagle NS, Jemal A. Cancer statistics. CA: A Cancer J. Clin. 2023; 73: 17–48.
2. Lorusso G, Rüegg C. The tumor microenvironment and its contribution to tumor evolution toward metastasis. Histochem. Cell Biol. 2008; 130:1091–103.
3. Polyak K, Haviv I, Campbell IG. Co-evolution of tumor cells and their microenvironment. Trends Genet. 2009; 25:30–8.
4. Comings DE. A General Theory of Carcinogenesis. Proc. Natl. Acad. Sci. 1973; 70:3324–8.
5. Bodmer W, Bielas J, Beckman R. Genetic Instability Is Not a Requirement for Tumor Development. Cancer Res. 2008; 68:3558–60;
6. Bailey DL, Karp JS, Surti S. Physics and Instrumentation in PET. Positron Emission Tomography: Basic Sciences. London: Springer; 2005.
7. G BA. On the production and reproduction of sound by light. Proc. Am. Assoc. Adv. Sci. 1881; 29:115–36.
8. Xu G, Meng Z-X, Lin JD, Yuan J, Carson PL, Joshi B, et al. The Functional Pitch of an Organ: Quantification of Tissue Texture with Photoacoustic Spectrum Analysis. Radiology. 2014; 271:248–54.
9. Das D, Sharma A, Rajendran P, Pramanik M. Another decade of photoacoustic imaging. Phys. Med. Biol. 2021; 66:05TR01.
10. Rosencwaig A. Photoacoustic Spectroscopy. Annu. Rev. Biophys. Bioeng. 1980; 9:31–54.
11. Alizadeh AA, Aranda V, Bardelli A, Blanpain C, Bock C, Borowski C, et al. Toward understanding and exploiting tumor heterogeneity. Nat. Med. 2015; 21:846–53.
12. Marusyk A, Polyak K. Tumor heterogeneity: Causes and consequences. Biochim. Biophys. Acta (BBA) – Rev. Cancer. 2010; 1805:105–17.
13. Burns-cox N, Avery NC, Gingell JC, Bailey AJ. Changes in collagen metabolism in prostate cancer: a host response that may alter progression. J. Urol. 2001; 166:1698–701.
14. Povey MJW. Ultrasound particle sizing: A review. Particuology. 2013; 11:135–47.
15. Moore MP, Spencer T, Salter DM, Kearney PP, Shaw TRD, Starkey IR, et al. Characterisation of coronary atherosclerotic morphology by spectral analysis of radiofrequency signal: in vitro intravascular ultrasound study with histological and radiological validation. Heart. 1998; 79:459–67.
16. Wang S, Tao C, Wang X, Liu X. Quantitative detection of stochastic microstructure in turbid media by photoacoustic spectral matching. Appl. Phys. Lett. 2013; 102:114102.
17. Zhang M, Chen Y, Xie W, Wu S, Liao J, Cheng Q. Photoacoustic power azimuth spectrum for microvascular evaluation. Photoacoustics. 2021; 22:100260.
18. Wang H-W, Chai N, Wang P, Hu S, Dou W, Umulis D, et al. Label-Free Bond-Selective Imaging by Listening to Vibrationally Excited Molecules. Phys. Rev. Lett. 2011; 106:238106.
19. Zhang H, Zhang G, Zhang Y, Wen L, Zhang M, Pan J, et al. Quantitatively assessing port-wine stains using a photoacoustic imaging method: A pilot study. J. Am. Acad. Dermatol. 2021; 85:1613–6.
20. Zhang M, Wen L, Zhou C, Pan J, Wu S, Wang P, et al. Identification of different types of tumors based on photoacoustic spectral analysis: preclinical feasibility studies on skin tumors. J. Biomed. Opt. 2023; 28:065004.
21. Cheng Q, Qian M, Wang X, Zhang H, Wang P, Wen L, et al. Diagnosis and Treatment Monitoring of Port-Wine Stain Using LED-Based Photoacoustics: Theoretical Aspects and First In-Human Clinical Pilot Study. LED-Based Photoacoustic Imaging: From Bench to Bedside. Singapore: Springer; 2020; 351–77.
22. Ma K, Wu S, Huang S, Xie W, Zhang M, Chen Y, et al. Myocardial infarct border demarcation by dual-wavelength photoacoustic spectral analysis. Photoacoustics. 2022; 26:100344.
23. Chen Y, Xu C, Zhang Z, Zhu A, Xu X, Pan J, et al. Prostate cancer identification via photoacoustic spectroscopy and machine learning. Photoacoustics. 2021; 23:100280.

24. Wu S, Liu Y, Chen Y, et al. Quick identification of prostate cancer by wavelet transform-based photoacoustic power spectrum analysis. Photoacoustics, 2022, 25: 100327.
25. Li J, Chen Y, Ye W, Zhang M, Zhu J, Zhi W, et al. Molecular breast cancer subtype identification using photoacoustic spectral analysis and machine learning at the biomacromolecular level. Photoacoustics. 2023; 30:100483.

Chapter 17
Photoacoustic Techniques for Bone Characterization

Ting Feng, Weiya Xie, Wenyi Xu, Ya Gao, Teng Liu, Dean Ta, Menglu Qian, and Qian Cheng

Abstract Biomedical photoacoustics (PA) combines the advantages of selective optical absorption of biological tissues and high penetration depth of ultrasound, providing information on the macroscopic physiological structure and microscopic molecular level of biological tissues. Thus, it has great potential in the application of biomedical diagnosis. Among them, PA detection of bone is an important branch. In contrast to soft tissue, which can be regarded as a homogeneous medium, bone tissue is a two-phase complex medium (mineralized bone trabecular network in the solid phase and embedded bone marrow in the liquid phase), and this anisotropic and heterogeneous structure poses significant challenges to bone tissue characterization methods based on PA technology. This chapter introduces our work on PA detection for complex bone tissue, including the principle of PA wave generation and propagation, the methods of PA imaging and spectrum analysis, and their clinical applications. Finally, this chapter analyzes the current challenges in this field and anticipates future development trends based on the current research status of the PA technology.

Authors Ting Feng, Weiya Xie, Wenyi Xu, and Ya Gao have equally contributed to this chapter.

T. Feng
Institute of Acoustics, School of Physics Science and Engineering, Tongji University, Shanghai, China

Institute of Biomedical Engineering and Technology, Fudan University, Shanghai, China

W. Xie · W. Xu · Y. Gao · M. Qian · Q. Cheng (✉)
Institute of Acoustics, School of Physics Science and Engineering, Tongji University, Shanghai, China
e-mail: q.cheng@tongji.edu.cn

T. Liu
School of Electronic and Optical Engineering, Nanjing University of Science and Technology, Nanjing, China

D. Ta
Department of Electronic Engineering, Fudan University, Shanghai, China

W. Xia (ed.), *Biomedical Photoacoustics*,
https://doi.org/10.1007/978-3-031-61411-8_17

Keywords Photoacoustic spectrum analysis (PASA) · Photoacoustic bone imaging · Bone characterization · Bone microstructure · Bone metabolism

17.1 Introduction

The incidence rate of skeletal diseases ranks first among the five major diseases of the middle-aged and elderly. It is estimated that by the middle of the twenty-first century, the number of patients with skeletal diseases in China will reach 200 million, and the related medical prevention and treatment costs will reach 1800 billion yuan [1]. Early prevention and treatment of skeletal diseases can effectively circumvent the development of serious bone conditions, such as bone fractures and a series of complications. However, the current status of prevention and treatment of skeletal diseases is not optimistic. The main problem is the lack of a unified screening method with high accuracy and simplicity for high-risk populations. Therefore, in order to achieve early screening and diagnosis of skeletal diseases, it is urgent to develop bone assessment and disease diagnosis methods that can evaluate bone health and easy promotion.

The evaluation of bone tissue is mainly based on two types of factors: (1) physical properties, such as bone mineral density (BMD), bone microarchitecture (BMA), which are directly related to bone strength and (2) metabolic characteristics of molecules in bone tissue, such as the content of lipid and type I collagen, which can reflect bone turnover levels and predict fracture risk at an earlier stage compared to physical property testing of bone [2]. In recent years, most commonly used techniques for evaluating the physical properties of bone tissue in clinical practice have been based on X-ray or ultrasound (US), such as dual energy X-ray absorptiometry (DEXA), quantitative computed tomography (QCT), and quantitative ultrasound (QUS) techniques. DEXA can provide BMD information and is considered as the gold standard for the diagnosis of osteoporosis, but this type of method can only explain 60–80% of changes in bone strength [3]. QCT can provide BMA information of the spine and femur and achieve high-resolution three-dimensional imaging of tibia, radius, and other surrounding bones, as well as provide microstructure information such as cortical bone thickness, porosity, trabecular shape, size, number, and direction of cancellous bone. However, the radiation exposure generated by QCT is large and cannot provide organic chemical composition information related to bone blood flow and bone metabolism [4–6]. The radiation-free QUS technology provides a practical and economical alternative method for bone evaluation, which can provide physical information, such as BMD, BMA, and elasticity, but provides less information on chemical composition [7]. Magnetic resonance imaging (MRI) can be used to evaluate the chemical composition of bone (such as bone marrow adipose tissue), but it cannot evaluate the mineral and collagen content, which are highly related to bone metabolism [8]. Pure optical imaging methods can also be used to evaluate the chemical composition of bone tissue, but due to the strong optical attenuation caused by the strong scattering of light transmitting in bone tissue, the detection depth of traditional pure optical imaging methods is limited [9–11]. In summary, the bone assessment methods currently used in clinical practice cannot simultaneously perform noninvasive and radiation-free detection of bone microstructure and bone metabolism. Especially for the determination of biochemical indicators related to bone metabolism, invasive methods (bone biopsy or serum testing) are still mainly used in clinical practice. Therefore, developing a noninvasive and nonradiative in vivo detection method that can simultaneously evaluate microstructure and metabolic information using imaging, mass spectrometry, and other technologies is of great scientific significance and has high clinical value for the early diagnosis and treatment monitoring of bone tissue diseases.

Nowadays, a novel, noninvasive, and radiation-free medical imaging method called photoacoustic imaging (PAI), represented by L. V. Wang's research group, is thriving in the application field of medical disease diagnosis [12–14]. PAI is based on the principle of imaging the PA signals generated by biological tissues excited by pulsed laser. The amplitude and spectral characteristics of photoacoustic (PA) signals depend on the light source, optical, thermal, and mechanical properties of the irradiated biological tissue, as well as the acoustic properties of the biological tissue along the PA signal propagation path. Therefore, by reconstructing the PA images or analyzing PA signals, the physiological structure and functional information of biological tissues can be obtained [12]. First, PAI can provide physiological structural information of biological tissues. PA signals have been proven to carry macroscopic to microscopic structural features of biological tissues and have been used for imaging various disease models based on changes in the microstructure of biological tissues. Second, PAI can provide information on the chemical composition of biological tissues, as it reflects the light energy "absorbed" by biological tissues, and the absorption characteristics of different wavelengths of light by biological tissues are closely related to their internal chemical composition. In addition, PAI has smaller attenuation and higher resolution because the PA signal generated by biological tissue only transmits once in one direction in biological tissue, while the acoustic signal in ultrasonic imaging transmits twice in biological tissue. Therefore, compared with ultrasonic bone imaging, the PA signal in PAI of bone is less affected by the acoustic attenuation and scattering of bone tissue and has the potential to achieve a higher resolution imaging of bone microstructures [12]. At present, PAI has been used in biomedical imaging research of different soft tissues, such as brain function, melanoma, and breast tumor, and is considered to be a promising emerging clinical precision diagnosis technique [12].

PA detection of bone tissue diseases is an important branch of PAI, but compared with the rapid development of PA diagnosis for soft tissue diseases, the research progress of PA technology in the field of bone disease diagnosis is relatively slow. The main characteristics that distinguish bone tissue from soft tissue are the following: (1) The physical structure of bone tissue is complex, consisting of a nonuniform-layered curved structure composed of soft tissue, cortical bone, and cancellous bone. The cancellous bone is a solid–liquid two-phase porous network structure (the trabecular framework of the bone can be seen as a solid phase, and the bone marrow filled in the bone pores can be seen as a liquid phase). (2)

The chemical components of bone tissue are complex, including bone minerals, collagen, lipid, hemoglobin, and water. The physical structure and chemical composition characteristics of bone tissue pose key bottlenecks in the application of PA technology in bone tissue detection: both light and US generate strong scattering and attenuation during propagation, accompanied by factors such as spectral coloring, frequency dispersion, and waveform distortion. Specifically, it can be divided into the following two categories:

The Problem of PA Excitation and Propagation in Complex Media: PA sources of bone minerals are distributed network; the transmitted light pulse in bone tissue faces multiple boundary problems. The PA signal generation and propagation in bone tissue contains several propagation modes such as shear wave and longitudinal wave (which can also be divided into fast wave and slow wave), and there will be mode conversion between different propagation modes.

Nonlinear Problem: The scattering and attenuation of light and PA signal in bone tissue are severe, and there are various physical factors involved in the generation and propagation of PA signals. Among them, the most typical "spectral coloring" (also known as "spectral deformation") and "frequency dispersion" have a serious impact on the accuracy of bone parameter inversion based on PA spectroscopy.

The above impacts pose significant challenges in the transformation of relevant PA theories and methods toward clinical applications. This review focuses on the existing academic research progress and the problems in PA detection and imaging of complex bone tissues, including the theory of PA bone detection, PA bone analysis methods, PAI method of bone tissue, and their clinical applications, as shown in Fig. 17.1. Finally, prospects were made for future development trends of bone detection.

17.2 Theory of Photoacoustic Bone Detection

The theory of PA bone detection mainly includes four parts: light transmission, light absorption, PA excitation, and US propagation in complex bone tissues. The relevant physical processes are shown in Fig. 17.2.

17.2.1 Light Transmission Within Bone Tissue

Bone tissue is a complex biological medium, which can be regarded as composed of three basic biological media: soft tissue, cortical bone, and cancellous bone. It is usually assumed that each basic biological medium is isotropic. The transmission of light within various basic biological media and between different biological media is studied.

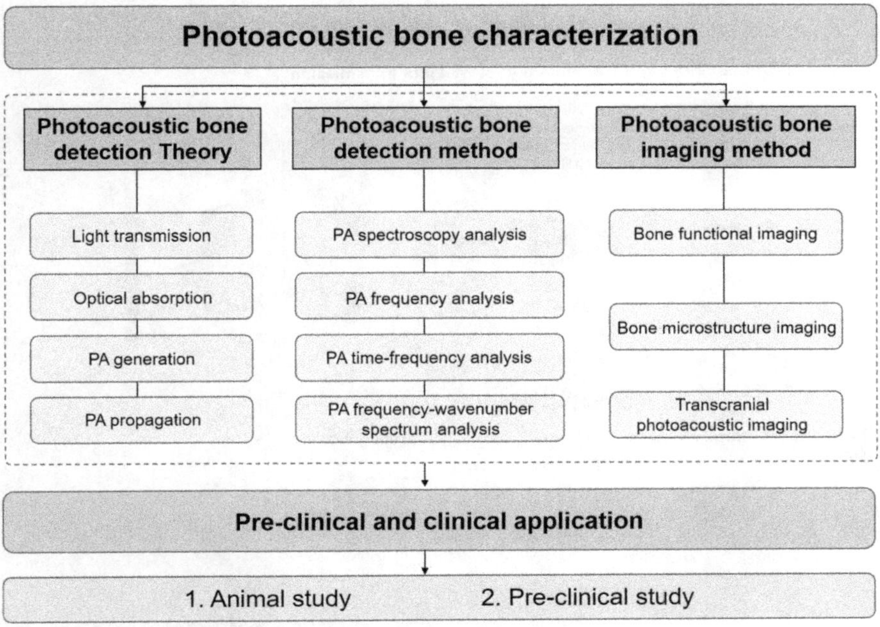

Fig. 17.1 Framework of photoacoustic bone detection technology

The physical process of light diffusion in biological tissues can be approximated by the following equation [15, 16]:

$$Q(\boldsymbol{r}) = \mu_a(\boldsymbol{r})\,\Phi(\boldsymbol{r}) - \nabla \cdot [D(\boldsymbol{r})\,\nabla(\boldsymbol{r})]. \tag{17.1}$$

In the equation, $Q(\boldsymbol{r})$ represents the light source, Φ represents the light distribution in biological tissues, μ_a represents the light absorption coefficient, and D represents the light diffusion coefficient, all of which are functions of spatial position $\boldsymbol{r}$.

When light is transmitted in a single biological medium, if the scale of optical heterogeneity is smaller than the mean free path of several photons, the optical flux is approximately smooth and can be characterized by a "macroscopic homogeneous" medium model. For macroscopic homogeneous medium, light absorption, light diffusion, and light scattering coefficient can be considered independent of the spatial position. When the light source is a point light source, the solution of luminous flux can be expressed as follows based on Green's function:

$$\Phi(\boldsymbol{r}) = \frac{1}{4\pi D\,|\boldsymbol{r} - \boldsymbol{r}'|} \exp\left(-\mu_{\text{eff}}\,|\boldsymbol{r} - \boldsymbol{r}'|\right). \tag{17.2}$$

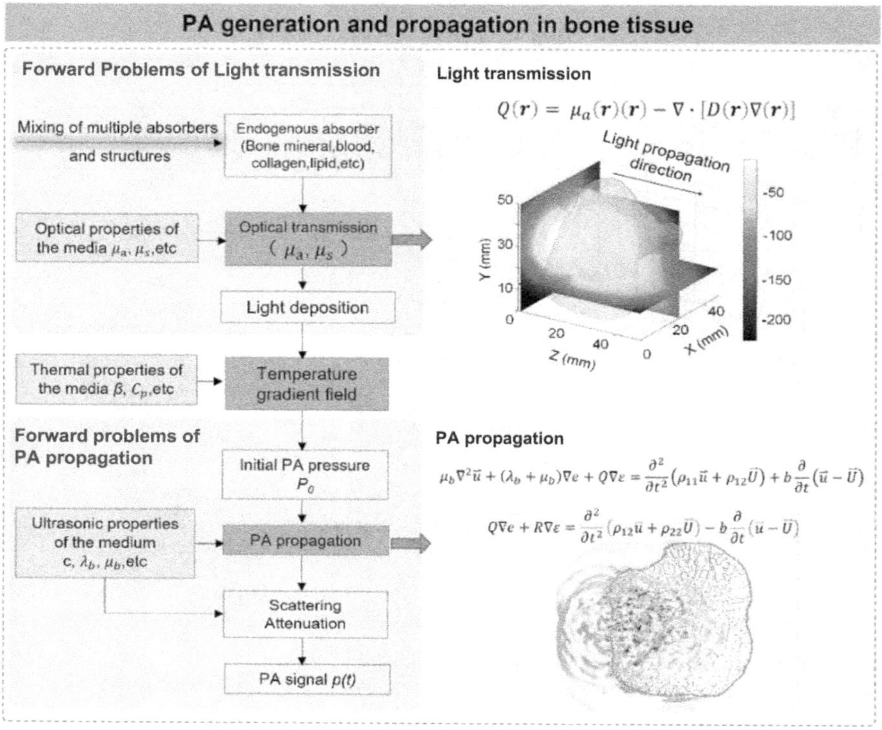

Fig. 17.2 Schematic diagram of the physical process of photoacoustic signal generation and propagation

In the formula, $\mu_{\text{eff}} = \sqrt{3\mu_a\left(\mu_a + \mu_s'\right)}$ is the effective attenuation coefficient, and $\boldsymbol{r}'$ is the spatial position of the point source. When the light reduced scattering coefficient μ_s' in biological media is much greater than the light absorption coefficient, the effective attenuation coefficient can be simplified as $\mu_{\text{eff}} \approx \sqrt{3\mu_a\mu_s'}$.

When light is transmitted in two different biological media, the boundary conditions of different basic biological media are different as the physical properties of different basic biological media are different. According to the refractive indices of two basic biological media, physical problems of light transmission can be divided into two categories: refractive index match and refractive index mismatch. When the refractive indices of adjacent two basic biological media match (such as skin and subcutaneous soft tissue can be approximately matched), the boundary condition is

$$\Phi(\boldsymbol{r}) - 2D\frac{\partial\Phi(\boldsymbol{r})}{\partial z} = 0. \tag{17.3}$$

When the refractive indices of two adjacent layers of basic biological media does not match (such as soft tissue and bone tissue does not match), the boundary conditions are

$$\begin{aligned} \Phi(\boldsymbol{r}) - 2D\tfrac{1+R_J}{1-R_\Phi}\tfrac{\partial\Phi(\boldsymbol{r})}{\partial z} &= 0 \\ R_\Phi = \textstyle\int_0^\pi 2\sin\theta\cos\theta R(\cos\theta)\,d\theta &\quad . \\ R_J = \textstyle\int_0^{\frac{\pi}{2}} 3\sin\theta(\cos\theta)^2 R(\cos\theta)\,d\theta & \end{aligned} \tag{17.4}$$

In summary, given the light source, the distribution light after transmission through the complex bone tissue composed of multiple basic biological media can be obtained. Based on the theory of light transmission in complex biological media, Feng et al. preliminarily explored the attenuation of light transmitting through the human calcaneus and the distribution of light energy deposition in the calcaneus based on the finite element analysis method, and proved that in the near-infrared range, light can penetrate the entire calcaneus (thickness $\approx$3 cm), with a penetration rate of about 1‰ [17]. However, the difference of BMD in different age groups has an impact on light energy transmission.

17.2.2 Absorption of Light in Bone Tissue

When bone tissue is radiated by light energy, chromophores (absorbers) in bone tissue absorb light energy, the light energy is converted into heat energy, and its temperature field meets the heat conduction equation:

$$\left(\nabla^2 - \frac{1}{\kappa}\frac{\partial}{\partial t}\right)\vartheta(\boldsymbol{r},t) = -\frac{\mu_a(\lambda)\,I_0(t)e^{\mu_a(\lambda)\boldsymbol{r}}}{K}, \ (-l < z < 0), \tag{17.5}$$

where κ is the thermal diffusivity, K is the thermal conductivity, $\vartheta(\boldsymbol{r},t)$ is the temperature distribution function, and I_0 is the light intensity. The light energy deposition distribution $\Phi(\boldsymbol{r})$ in the biological tissue can be solved by formulas (17.1)–(17.4); $\mu_a(\lambda)$ is the light absorption coefficient of biological medium, which is determined by various optical absorbers inside biological tissue and can be expressed as

$$\mu_a(\lambda,\boldsymbol{r}) = \sum_{l=1}^{L}\mu_{a_l}(\lambda,\boldsymbol{r})\,C_l, \tag{17.6}$$

where μ_a is the light absorption coefficient of the mixed biological medium, L is the type of absorber, μ_{a_l} and C_l are the light absorption coefficient and concentration coefficient of different substances in the biological medium, respectively. When the pulsed laser irradiates bone tissue, biological macromolecules within the bone tissue selectively absorb light energy of different wavelengths and generate PA signals.

The biochemical components that absorb light energy in bone tissues mainly include bone matrix (minerals and collagen fibers) and bone marrow (deoxyhemoglobin, oxygenated hemoglobin, water, lipid, and collagen). Within the window range of the near-infrared optical window regions I (690~950 nm) and II (1200~1800 nm), several major biomolecules in bone tissue exhibit significant characteristic optical absorption, including deoxyhemoglobin, oxygenated hemoglobin, lipid, collagen, and water. According to formula (17.6) and Lambert Beer's law, the light absorption coefficient of bone tissue at different wavelengths can be expressed as follows:

$$\begin{aligned}\mu_{atotal}(\lambda_1) =& \varepsilon_m(\lambda_1) C_m + \varepsilon_c(\lambda_1) C_c + \varepsilon_{Hb}(\lambda_1) C_{Hb} + \varepsilon_{HbO_2}(\lambda_1) C_{HbO_2} \\ &+ \varepsilon_l(\lambda_1) C_l + \varepsilon_w(\lambda_1) C_w \\ &\vdots \\ \mu_{atotal}(\lambda_6) =& \varepsilon_m(\lambda_6) c_m + \varepsilon_c(\lambda_6) c_c + \varepsilon_{Hb}(\lambda_6) c_{Hb} + \varepsilon_{HbO_2}(\lambda_6) C_{HbO_2} \\ &+ \varepsilon_l(\lambda_6) C_l + \varepsilon_w(\lambda_6) C_w\end{aligned} \quad (17.7)$$

Among them, $\lambda_1 \cdots \lambda_6$ are the characteristic wavelengths corresponding to minerals, collagen, deoxyhemoglobin, oxygenated hemoglobin, lipid, and water. ε is the optical attenuation coefficient (here, it is approximately the optical absorption coefficient $\alpha_l(\lambda)$ of each absorber), and C is the concentration coefficient of the six chemical components. Therefore, from the above formula, it can be seen that the biochemical composition information of bone tissue can be obtained by analyzing PA signals in the absorption spectrum dimension.

17.2.3 Excitation Theory of Optical Sound Source

When biological tissue is irradiated by a pulsed laser or an amplitude modulated laser, the chromophore in the biological tissue absorbs light energy and generates thermal gradient field, and the stress field is generated based on thermoelastic effect. According to the motion equation, the volume expansion equation and the heat conduction equation, the wave equation with a heat source in the ideal fluid can be derived as

$$\left(\nabla^2 - \frac{1}{c^2}\frac{\partial^2}{\partial t^2}\right) \boldsymbol{u}(\boldsymbol{r}, t) = -\beta \nabla \vartheta(\boldsymbol{r}, t), \quad (17.8)$$

or

$$\nabla^2 p - \frac{1}{c^2}\frac{\partial^2 p}{\partial t^2} = -\frac{B}{C_p}\frac{\partial H(\boldsymbol{r}, t)}{\partial t}, \quad (17.9)$$

where c is the propagation speed of acoustic wave, $\boldsymbol{u}(\boldsymbol{r}, t)$ is the displacement vector at time t and position $\boldsymbol{r}$, β is the coefficient of thermal expansion, p is the sound pressure, B is the adiabatic volume modulus, C_p is the specific heat capacity at constant pressure, and $H(\boldsymbol{r}, t)$ is the thermal power density function of the heat source.

When pulsed laser irradiates layered [17, 18], slender cylindrical and spherical biological tissues [19, 20], the temperature distribution function or thermal power density distribution function are, respectively, as follows:

$$\vartheta\left(z, t\right) = \frac{\mu_a\left(\lambda\right) I_0(t) \mathrm{e}^{\mu_a(\lambda) z}}{C_p \rho}, \left(-l < z < 0\right), \tag{17.10}$$

$$H\left(r, t\right) = \mu_a\left(\lambda\right) I_0(t) \mathrm{e}^{-\frac{r^2}{\tau^2}}, \left(0 < r < R_0\right), \tag{17.11}$$

$$\vartheta\left(r, t\right) = \frac{E_a}{\rho \sigma \left(4\pi \kappa t\right)^{\frac{3}{2}}} \mathrm{e}^{\frac{-r^2}{4\kappa t}}, \left(0 < r < r_0\right), \tag{17.12}$$

where ρ is the density of the medium, l is the thickness of the layered medium, R_0 is the radius of the fine cylinder, r_0 is the radius of the spherical point source, τ is the radius of the pulsed laser, E_a is the light energy absorbed by the optical absorption source, and σ is the specific heat of the fluid.

Given the active-source wave equation and temperature distribution function, the PA field generated by three basic PA sources can be calculated. The PA source of complex biological tissues can be regarded as the combination of these three basic PA sources, and then the PA field generated by complex biological tissues can be obtained.

Taking a spherical PA source as an example, based on the active-source wave equation, combined with the following initial conditions and boundary conditions:

$$\begin{array}{c} u\left(\infty, t\right) = 0 \\ u\left(r, 0\right) = 0 \\ \frac{\partial u(r,0)}{\partial t} = c\left(r, 0\right) = 0 \end{array} . \tag{17.13}$$

The frequency domain solution of the spherical PA source can be obtained:

$$P\left(r, \omega\right) = j \frac{\beta E_a c^2}{2\pi \sigma r_0^2} \frac{1}{r} \omega \mathrm{e}^{-\frac{\omega^2}{4}}, \tag{17.14}$$

where ω is the angular frequency.

It can be seen that the PA signal is directly related to the optical, thermal, mechanical properties, and size of the spherical PA source. The same is true for the frequency domain solutions of the PA fields produced by the other two basic PA sources. The frequency domain solution of complex medium is the combination of the PA field generated by the basic PA source, so the related physical characteristics of the PA source can be reflected from the frequency domain of the sound signal.

17.2.4 Theory of Sound Propagation in Bone

As mentioned in Sect. 17.2.1, bone tissue is a complex medium composed of soft tissue, cortical bone, and cancellous bone. Soft tissue and cortical bone can be regarded as relatively uniform media, and the acoustic propagation in continuous and uniform media can be described by acoustic transmission equation. Cancellous bone is a solid–liquid two-phase porous medium, and the liquid phase (bone marrow) is regarded as the load of the solid phase (bone trabecula), causing complicated acoustic propagation in it. Therefore, the difficulty of acoustic propagation in bone tissue is mainly focused on the acoustic propagation in cancellous bone.

17.2.4.1 Static Sound Field of Solid–Liquid Two-Phase Porous Cancellous Bone

Cancellous bone is a complex porous medium composed of interconnected trabecular bone forming a solid frame and liquid bone marrow filling the trabecular space. It can be regarded as a porous elastic medium composed of a solid frame and viscous fluid [21]. Biot and Willis established the isothermal quasi-static elastic theory of saturated liquid-filled pore elastic solids, described the propagation of elastic waves in saturated liquid-filled pore elastic media, simulated the elastic, inertial, and viscous coupling between liquids and solids, and has been widely used to describe the propagation of two P-waves and one S-wave in porous media [21–27]. Rice and Cleary further extended the Biot's theory, and the newly extended Biot's theory proved to be more convenient to solve relevant problems and explain the obtained solutions [28–31]. At present, the Biot's theory of elastic waves in porous media has been rapidly applied to nondestructive evaluation of bone [32–38].

Biot's theory is applicable to the case where the acoustic frequency is less than the critical frequency f_c of pore medium, f_c is defined as follows [39]:

$$f_c = \frac{\phi\eta}{2\pi\rho_f\mathcal{K}}, \tag{17.15}$$

where ϕ is the porosity, η is the fluid viscosity, ρ_f is the density of the liquid in the pore, and $\mathcal{K}$ is the permeability. For cancellous bone filled with bone marrow, the range of f_c is 1–10 kHz [40, 41]. When the frequency of acoustic wave propagating in the pore medium is $f < f_c$, the flow of liquid relative to solid through the pore can be considered as a Poseuille flow, that is, the inertia force of the fluid flowing in the capillary in the pore medium can be ignored compared with the viscous force it is subjected to.

17.2.4.1.1 Constitutive Equation of Saturated Liquid-Filled Porous Elastic Media

A saturated fluid-filled pore elastic medium is a solid–liquid two-phase medium with an elastic solid framework with porosity, and a liquid that can flow with respect to the solid framework, assuming that the fluid flow through the pore fulfills Poiseuille's law. The average porosity, the average size, and the number of pores are basically the same in one acoustic wave wavelength. The pore wall is impermeable and the thermoelastic effect is ignored.

A unit volume cube is considered to represent the solid–liquid phase cancellous bone. The constitutive equation of the solid skeleton of the cancellous bone is (the following are all using the Ines subscript summation convention)

$$\sigma_{ij} = 2\mu_b e_{ij} + \lambda_b e\delta_{ij} + Q\varepsilon\delta_{ij}. \tag{17.16}$$

And the constitutive equation of liquid filling in pores is

$$\tau = Qe + R\varepsilon, \tag{17.17}$$

where σ_{ij} is the stress component of a liquid-filled porous solid skeleton, μ_b and λ_b are the elastic constants of the frame, which depend not only on the material elasticity of the frame but also on its structural parameters such as porosity ϕ and tortuosity. R is another elastic constant, while Q is the Biot's elastic constant that describes the coupling of the liquid and solid frame. $e_{ij} = \frac{1}{2}\left(u_{i,j} + u_{j,i}\right)$ is the strain of the solid skeleton, $e = e_{11} + e_{22} + e_{33} = \nabla \cdot \boldsymbol{u}$ is the normal strain for solid, $\varepsilon_{ij} = \frac{1}{2}\left(U_{i,j} + U_{j,i}\right)\delta_{ij}$ is the strain of the liquid in the pore, and $\varepsilon = \varepsilon_{11} + \varepsilon_{22} + \eta_{33} = \nabla \cdot \boldsymbol{U}$ is the strain for fluid $\boldsymbol{u}$ is the solid skeleton displacement vector, $\boldsymbol{U}$ is the liquid skeleton displacement vector. There are four elastic constants μ_b, λ_b, R, and Q in the constitutive equation of saturated liquid-filled porous media. The stress component of the liquid–solid coupling containing Biot's elastic constant Q is added to the constitutive equations of solid skeleton and liquid, respectively.

17.2.4.1.2 Equation of Motion

The solid–liquid two-phase cube element is considered to be relatively small relative to the elastic wave wavelength and relatively large relative to the pore size. It is assumed that the liquid flows through the pores without viscosity. Since the liquid-filled pore medium is statistically isotropic, it is equivalent in x, y, and z directions, and there is no coupling in dynamics. Therefore, in the rectangular coordinate system, the kinetic energy T of the liquid-filled porous medium per unit volume can be expressed as

$$2T=\rho_{11}\left[\left(\frac{\partial u_x}{\partial t}\right)^2+\left(\frac{\partial u_y}{\partial t}\right)^2+\left(\frac{\partial u_z}{\partial t}\right)^2\right]+\rho_{22}\left[\left(\frac{\partial U_x}{\partial t}\right)^2+\left(\frac{\partial U_y}{\partial t}\right)^2+\left(\frac{\partial U_z}{\partial t}\right)^2\right]$$
$$+2\rho_{12}\left[\left(\frac{\partial u_x}{\partial t}\right)\left(\frac{\partial U_x}{\partial t}\right)+\left(\frac{\partial u_y}{\partial t}\right)\left(\frac{\partial U_y}{\partial t}\right)+\left(\frac{\partial u_z}{\partial t}\right)\left(\frac{\partial U_z}{\partial t}\right)\right]. \quad (17.18)$$

where ρ_{mn} is the mass coefficient related to the porosity ϕ and the mass densities ρ_s and ρ_f of the solid and fluid. ρ_{12} is the mass coupling parameter between the fluid phase and the solid phase.

Since porous media are dynamically equivalent in all directions, we first discuss the equation of motion of porous media along the x direction. Using Lagrange equation, the motion equation of skeleton and liquid particle in the x direction can be obtained:

$$f_x=\frac{\partial}{\partial t}\left(\frac{\partial T}{\partial \dot{u}_x}\right)=\frac{\partial^2}{\partial t^2}\left(\rho_{11}u_x+\rho_{12}U_x\right), \quad (17.19)$$

$$F_x=\frac{\partial}{\partial t}\left(\frac{\partial T}{\partial \dot{U}_x}\right)=\frac{\partial^2}{\partial t^2}\left(\rho_{12}u_x+\rho_{22}U_x\right). \quad (17.20)$$

where f_x and F_x is acting on the unit volume per unit volume in skeleton and pore fluid on the force along the x direction, $\dot{u}_x=\frac{\partial u_x}{\partial t}$ and $\dot{U}_x=\frac{\partial U_x}{\partial t}$ are skeleton and liquid particle vibration velocity in the x direction, respectively.

Since there is no external force acting, the internal force acting on the porous solid skeleton and liquid along the x direction is the gradient of stress:

$$f_x=\frac{\partial \sigma_{xx}}{\partial x}+\frac{\partial \sigma_{xy}}{\partial y}+\frac{\partial \sigma_{xz}}{\partial z}, \quad (17.21)$$

$$F_x=\frac{\partial \tau}{\partial x}. \quad (17.22)$$

By substituting constitutive Eqs. (17.16) and (17.17) of skeleton and liquid into Eqs. (17.21) and (17.22), kinetic equations of particles in solid skeleton and liquid in the x direction can be obtained:

$$\mu_b\nabla^2u_x+(\lambda_b+\mu_b)\frac{\partial e}{\partial x}+Q\nabla\varepsilon=\frac{\partial^2}{\partial t^2}\left(\rho_{11}u_x+\rho_{12}U_x\right), \quad (17.23)$$

$$Q\frac{\partial e}{\partial x}+R\frac{\partial \varepsilon}{\partial x}=\frac{\partial^2}{\partial t^2}\left(\rho_{12}u_x+\rho_{22}U_x\right). \quad (17.24)$$

As the porous cube on the dynamics are equivalent in all directions, the solid skeleton displacement vector $\boldsymbol{u}$ and liquid displacement vector $\boldsymbol{U}$ motion equation can be expressed as

$$\mu_b \nabla^2 \boldsymbol{u} + (\lambda_b + \mu_b)\,\nabla e + Q\nabla\varepsilon = \frac{\partial^2}{\partial t^2}\left(\rho_{11}\boldsymbol{u} + \rho_{12}\boldsymbol{U}\right), \tag{17.25}$$

$$Q\nabla e + R\nabla\varepsilon = \frac{\partial^2}{\partial t^2}\left(\rho_{12}\boldsymbol{u} + \rho_{22}\boldsymbol{U}\right). \tag{17.26}$$

The motion equation shows that the motion of the skeleton and the liquid is separated and coupled with each other. Equations (17.25) and (17.26) can be used to study the propagation characteristics of waves in solid–liquid two-phase porous media.

17.2.4.1.3 Propagation of Expansion and Rotation Waves

There are two types of P-wave and one type of S-wave in porous media. The first type of P-wave (fast wave) corresponds to the solid–liquid motion in the phase, and the second type of P-wave (slow wave) corresponds to the solid–liquid motion in the reverse phase. The propagation of these three waves will be analyzed below.

Because of the statistical isotropy of the material, the rotation wave and the expansion wave are not coupled and obey the independent propagation equation. To derive from the motion equation of displacement wave equation, the introduction of liquid in porous media of solid and liquid scalar and vector rotation makes $\nabla \cdot \boldsymbol{u} = e$, $\nabla \cdot \boldsymbol{U} = \varepsilon$, $\nabla \times \boldsymbol{u} = \boldsymbol{\omega}$, $\nabla \times \boldsymbol{U} = \boldsymbol{\Omega}$.

Applying the divergence to Eqs. (17.25) and (17.26) of motion, using $\nabla \cdot \nabla = \nabla^2$, the wave equation for the expansion of a solid skeleton and a liquid is obtained:

$$P\nabla^2 e + Q\nabla^2\varepsilon = \frac{\partial^2}{\partial t^2}\left(\rho_{11}e + \rho_{12}\varepsilon\right), \tag{17.27}$$

$$Q\nabla^2 e + R\nabla^2\varepsilon = \frac{\partial^2}{\partial t^2}\left(\rho_{12}e + \rho_{22}\varepsilon\right), \tag{17.28}$$

where Biot's elastic coefficient is $P = (\lambda_b + 2\mu_b)$.

Taking the curl of $(\nabla\times)$ on both sides of equations of motion (17.25) and (17.26), using $\nabla \times \nabla = 0$, the wave equation for the rotation vector of a solid skeleton and a liquid particle is obtained

$$\mu_b \nabla^2 \boldsymbol{\omega} = \frac{\partial^2}{\partial t^2}\left(\rho_{11}\boldsymbol{\omega} + \rho_{12}\boldsymbol{\Omega}\right), \tag{17.29}$$

$$\frac{\partial^2}{\partial t^2}\left(\rho_{12}\boldsymbol{\omega}+\rho_{22}\boldsymbol{\Omega}\right)=0. \tag{17.30}$$

Biot defines μ_b as the Biot's elastic coefficient $N = \mu_b$. Therefore, the four Biot elastic coefficients of saturated liquid-filled pore media are N, P, Q, and R.

By solving the above homogeneous wave equations of rotation and expansion, shear waves and compression waves propagating in saturated liquid porous media can be obtained.

1. *Propagation of Rotating Waves*

It can be obtained from Eq. (17.30) that $\boldsymbol{\Omega} = -\frac{\rho_{12}}{\rho_{22}}\boldsymbol{\omega}$, substituting Eq. (17.29) gives the wave equation of rotating wave $\boldsymbol{\omega}$

$$N\nabla^2\boldsymbol{\omega}=\left(\rho_{11}-\frac{\rho_{12}^2}{\rho_{22}}\right)\frac{\partial^2\boldsymbol{\omega}}{\partial t^2}. \tag{17.31}$$

The shear wave velocity of saturated liquid porous media is obtained by solving the wave equation

$$c_t^2=\frac{N}{\rho_c}=N/\left(\rho_{11}-\frac{\rho_{12}^2}{\rho_{22}}\right)=\frac{\rho_{22}}{\left(\rho_{11}\rho_{22}-\rho_{12}^2\right)}N, \tag{17.32}$$

where $\rho_c = \left(\rho_{11} - \frac{{\rho_{12}}^2}{\rho_{22}}\right)$, which corresponds to the quality of the porous medium involved in the propagation of acoustic waves.

The results show that rotation wave $\boldsymbol{\omega}$ in the liquid is generated by rotation wave $\boldsymbol{\Omega}$ in the solid through mass coupling, and both have the same shear velocity. Because $\rho_{12} < 0$, so the direction of $\boldsymbol{\Omega}$ and $\boldsymbol{\omega}$ is the same.

2. *Propagation of Expansion Waves*

First, considering the wave propagating along the x direction of the solid skeleton and the liquid in the pores in the saturated liquid porous medium, the wave equations of Eqs. (17.27) and (17.28) can be simplified as

$$P\frac{\partial^2 e}{\partial x^2}+Q\frac{\partial^2 \varepsilon}{\partial x^2}=\frac{\partial^2}{\partial t^2}\left(\rho_{11}e+\rho_{12}\varepsilon\right), \tag{17.33}$$

$$Q\frac{\partial^2 e}{\partial x^2}+R\frac{\partial^2 \varepsilon}{\partial x^2}=\frac{\partial^2}{\partial t^2}\left(\rho_{12}e+\rho_{22}\varepsilon\right). \tag{17.34}$$

Let the solution of the above wave equation be

$$e=A_s\mathrm{e}^{-\mathrm{i}(\varpi t-kx)}, \tag{17.35}$$

$$\varepsilon = A_f \mathrm{e}^{-\mathrm{i}(\varpi t - kx)}. \tag{17.36}$$

The solutions are substituted into wave Eqs. (17.33) and (17.34), respectively, and the homogeneous equations of coefficients A_s and A_f are obtained after collation:

$$\left(P - \rho_{11} c_L^2\right) A_s + \left(Q - \rho_{12} c_L^2\right) A_f = 0, \tag{17.37}$$

$$\left(Q - \rho_{12} c_L^2\right) A_s + \left(R - \rho_{22} c_L^2\right) A_f = 0. \tag{17.38}$$

In the above simultaneous equation, the sufficient and necessary condition for A_s and A_f to have nonzero solutions is that the determinant of the coefficients is equal to zero, namely,

$$\left(R - \rho_{22} c_L^2\right) \left(P - \rho_{11} c_L^2\right) - \left(Q - \rho_{12} c_L^2\right)^2 = 0. \tag{17.39}$$

By expanding the above equation, we can get the equation about the velocity of longitudinal waves:

$$\left(\rho_{11}\rho_{22} - \rho_{12}^2\right) \mathrm{c}_L^4 - (\rho_{11} R + \rho_{22} P - 2\rho_{12} Q)\, \mathrm{c}_\mathrm{L}^2 + \left(PR - Q^2\right) = 0. \tag{17.40}$$

To make the result representation more concise, let $(\rho_{11} R + \rho_{22} P - 2\rho_{12} Q) = M$, $(PR - Q^2) = L$, and $\rho_{11}\rho_{22} - {\rho_{12}}^2 = \rho_{22}\rho_c$.

Therefore, longitudinal wave velocity Eq. (17.40) is simplified to

$$\left(\rho_{11}\rho_{22} - \rho_{12}^2\right) c_L^4 - M c_L^2 + L = 0. \tag{17.41}$$

Equation (17.41) is a quadratic equation for ${c_L}^2$, and the solutions of its two P-wave velocities are

$$c_{Lj}^2 = \frac{M \pm \sqrt{M^2 - 4\left(\rho_{11}\rho_{22} - \rho_{12}^2\right) L}}{2\left(\rho_{11}\rho_{22} - \rho_{12}^2\right)}, (j = 1, 2)\,. \tag{17.42}$$

Because the Biot's elastic constants P, Q, R and the mass coefficient ρ_{11}, ρ_{22} are positive, while ρ_{12} is negative, $M^2 - 4\left(\rho_{11}\rho_{22} - \rho_{12}^2\right) L = (R\rho_{11} + P\rho_{22} - 2Q\rho_{12})^2 - 4\left(\rho_{11}\rho_{22} - \rho_{12}^2\right)\left(PR - Q^2\right) = (R\rho_{11} - P\rho_{22})^2 + 4\rho_{11}\rho_{22} Q^2 + 4PR\rho_{12}^2 - 4RQ\rho_{11}\rho_{22} - 4PQ\rho_{11}\rho_{22} > 0$. That is, there are two real P-wave velocities for expansion waves in saturated liquid-filled porous media:

$$c_{L1} = \left(\frac{M + \sqrt{M^2 - 4\left(\rho_{11}\rho_{22} - \rho_{12}^2\right)L}}{2\left(\rho_{11}\rho_{22} - \rho_{12}^2\right)} \right)^{1/2}, \tag{17.43}$$

$$c_{L2} = \left(\frac{M - \sqrt{M^2 - 4\left(\rho_{11}\rho_{22} - \rho_{12}^2\right)L}}{2\left(\rho_{11}\rho_{22} - \rho_{12}^2\right)} \right)^{1/2}. \tag{17.44}$$

Therefore, there are two elastic longitudinal waves in saturated liquid-filled porous media. The one with fast propagation speed is usually referred to as fast longitudinal wave and the one with slow propagation speed is referred to as slow longitudinal wave.

According to Eqs. (17.37) and (17.38), the amplitude ratio of fast P-wave ($j = 1$) and slow P-wave ($j = 2$) propagating in solid skeleton and liquid can be obtained, namely,

$$\frac{A_f^j}{A_s^j} = -\frac{\left(P - \rho_{11}c_{Lj}^2\right)}{\left(Q - \rho_{12}c_{Lj}^2\right)} = -\frac{\left(Pk_{Lj}^2 - \rho_{11}\varpi^2\right)}{\left(Qk_{Lj}^2 - \rho_{12}\varpi^2\right)} = \mathcal{L}_j, \; (j = 1, 2). \tag{17.45}$$

The p-wave in the solid skeleton and the liquid can be expressed as

$$e = A_s^{(1)} \mathrm{e}^{-\mathrm{i}(\varpi t - k_{L1}x)} + A_s^{(2)} \mathrm{e}^{-\mathrm{i}(\varpi t - k_{L2}x)}, \tag{17.46}$$

$$\begin{aligned} \varepsilon = &A_f^{(1)} \mathrm{e}^{-\mathrm{i}(\varpi t - k_{L1}x)} + A_f^{(2)} \mathrm{e}^{-\mathrm{i}(\varpi t - k_{L2}x)} = \mathcal{L}_1 A_s^{(1)} \mathrm{e}^{-\mathrm{i}(\varpi t - k_{L1}x)} \\ &+ \mathcal{L}_2 A_s^{(2)} \mathrm{e}^{-\mathrm{i}(\varpi t - k_{L2}x)}. \end{aligned} \tag{17.47}$$

Therefore, the fast and slow longitudinal waves in saturated liquid-filled porous media are, respectively,

$$e^{(1)} + \varepsilon^{(1)} = A_s^{(1)} \mathrm{e}^{-\mathrm{i}(\varpi t - k_{L1}x)} + A_f^{(1)} \mathrm{e}^{-\mathrm{i}(\varpi t - k_{L1}x)} = A_s^{(1)} \left(1 + \mathcal{L}_1\right) \mathrm{e}^{-\mathrm{i}(\varpi t - k_{L1}x)}, \tag{17.48}$$

$$e^{(2)} + \varepsilon^{(2)} = A_s^{(2)} \mathrm{e}^{-\mathrm{i}(\varpi t - k_{L2}x)} + A_f^{(2)} \mathrm{e}^{-\mathrm{i}(\varpi t - k_{L2}x)} = A_s^{(2)} \left(1 + \mathcal{L}_2\right) \mathrm{e}^{-\mathrm{i}(\varpi t - k_{L2}x)}. \tag{17.49}$$

17.2.4.2 Modified Biot's Theory

Sound velocity and attenuation are two main characteristics of elastic wave propagation in medium. When considering fluid viscosity, it is necessary to consider the energy dissipation caused by fluid-solid relative motion, especially when the frequency of sound wave is higher than critical frequency f_c of the porous medium, the liquid flow in the pores of the porous medium does not satisfy the laminar flow condition described by Poiseuille's law, and the motion equation needs to be corrected by the high frequency and viscosity.

The size of the trabecular bone of cancellous bone ranged from 50 to 200 μm, and the size of the trabecular space (bone marrow clusters) ranged from 0.2 to 3 mm [42]. The speed of sound of trabecular bone and bone marrow is 3200 m/s and 1435 m/s, respectively [43]. The frequencies of PA signals generated by both are much higher than f_c, which no longer satisfies the laminar flow condition described by Poiseuille's law. Therefore, it is necessary to modify the Biot's theory at high frequency. On the other hand, fluid bone marrow has viscosity, causing energy dissipation due to the relative motion of fluid and solid also, so the viscosity correction needs to be introduced.

When there is relative motion between liquid and solid particles in liquid-saturated porous media, energy loss D will occur due to viscosity in liquid:

$$D = \frac{b}{2}\left[\left(\dot{u}_x - \dot{U}_x\right)^2 + \left(\dot{u}_y - \dot{U}_y\right)^2 + \left(\dot{u}_z - \dot{U}_z\right)^2\right], \tag{17.50}$$

where $b = \frac{\eta\phi^2}{K}$ is the dissipation factor, which is a function of porosity ϕ.

The internal forces acting on the solid phase and on the liquid in the pores in the liquid-saturated porous medium along the x direction are determined by the Lagrange equation considering dissipation:

$$\frac{\partial}{\partial t}\left(\frac{\partial T}{\partial \dot{u}_x}\right) + \frac{\partial D}{\partial \dot{u}_x} = f_x, \tag{17.51}$$

$$\frac{\partial}{\partial t}\left(\frac{\partial T}{\partial \dot{U}_x}\right) + \frac{\partial D}{\partial \dot{U}_x} = F_x. \tag{17.52}$$

By substituting kinetic energy T and dissipated energy D into Eqs. (17.51) and (17.52), we can get

$$f_x = \frac{\partial}{\partial t}\left(\frac{\partial T}{\partial \dot{u}_x}\right) + \frac{\partial D}{\partial \dot{u}_x} = \frac{\partial^2}{\partial t^2}\left(\rho_{11}u_x + \rho_{12}U_x\right) + b\frac{\partial}{\partial t}\left(u_x - U_x\right), \tag{17.53}$$

$$F_x = \frac{\partial}{\partial t}\left(\frac{\partial T}{\partial \dot{U}_x}\right) + \frac{\partial D}{\partial \dot{U}_x} = \frac{\partial^2}{\partial t^2}\left(\rho_{12}u_x + \rho_{22}U_x\right) - b\frac{\partial}{\partial t}\left(u_x - U_x\right). \tag{17.54}$$

Substitute Eqs. (17.16), (17.17), (17.19), and (17.20) into Eqs. (17.53) and (17.54), then the equation of motion of solid skeleton displacement vector $\boldsymbol{u}$ and liquid displacement vector $\boldsymbol{U}$ can be represented as when considering the dissipation

$$\mu_b \nabla^2 \boldsymbol{u} + (\lambda_b + \mu_b) \nabla e + Q \nabla \varepsilon = \frac{\partial^2}{\partial t^2} (\rho_{11} \boldsymbol{u} + \rho_{12} \boldsymbol{U}) + b \frac{\partial}{\partial t} (\boldsymbol{u} - \boldsymbol{U}), \tag{17.55}$$

$$Q \nabla e + R \nabla \varepsilon = \frac{\partial^2}{\partial t^2} (\rho_{12} \boldsymbol{u} + \rho_{22} \boldsymbol{U}) - b \frac{\partial}{\partial t} (\boldsymbol{u} - \boldsymbol{U}). \tag{17.56}$$

The rotational wave and expansion wave with dissipation are solved according to Eqs. (17.55) and (17.56).

17.2.4.2.1 Propagation of Expansion and Rotation Waves

By performing curl operations on both sides of Eqs. (17.55) and (17.56), the motion equation of particle rotation in a viscous damping porous medium is obtained:

$$\mu_b \nabla^2 \boldsymbol{\omega} = \frac{\partial^2}{\partial t^2} (\rho_{11} \boldsymbol{\omega} + \rho_{12} \Omega) + b \frac{\partial}{\partial t} (\boldsymbol{\omega} - \Omega), \tag{17.57}$$

$$0 = \frac{\partial^2}{\partial t^2} (\rho_{12} \boldsymbol{\omega} + \rho_{22} \Omega) - b \frac{\partial}{\partial t} (\boldsymbol{\omega} - \Omega). \tag{17.58}$$

Suppose a rotating wave is an elastic wave traveling in the x direction. Apply the Fourier transform to the equation of motion $t \to \varpi$, $\omega \to \omega^*$, $\Omega \to \Omega^*$, we can obtain

$$\mu_b \frac{\partial^2 \omega^*}{\partial x^2} = -\varpi^2 \left(\rho_{11} \omega^* + \rho_{12} \Omega^*\right) + \mathrm{i} \varpi b \left(\omega^* - \Omega^*\right), \tag{17.59}$$

$$0 = -\varpi^2 \left(\rho_{12} \omega^* + \rho_{22} \Omega^*\right) - \mathrm{i} \varpi b \left(\omega^* - \Omega^*\right). \tag{17.60}$$

It can be obtained from Eq. (17.60) $\Omega^* = -\frac{(\varpi \rho_{12} + \mathrm{i} b)\omega^*}{(\varpi \rho_{22} - \mathrm{i} b)}$, by substituting in Eq. (17.59), the second order differential equation of ω^* is obtained

$$\mu_b \frac{\partial^2 \omega^*}{\partial x^2} = -\left[\frac{\left(\rho_{11} + \frac{b}{\mathrm{i}\varpi}\right)\left(\rho_{22} + \frac{b}{\mathrm{i}\varpi}\right) - \left(\rho_{12} - \frac{b}{\mathrm{i}\varpi}\right)^2}{\left(\rho_{22} + \frac{b}{\mathrm{i}\varpi}\right)} \right] \varpi^2 \omega^*. \tag{17.61}$$

Therefore, the shear wave velocity in porous media with viscous loss can be represented by elastic coefficient $N = \mu_b$:

$$\begin{aligned} c_{tD}^2 &= N \left(\rho_{22} + \tfrac{b}{\mathrm{i}\varpi}\right) / \left[\left(\rho_{11} + \tfrac{b}{\mathrm{i}\varpi}\right)\left(\rho_{22} + \tfrac{b}{\mathrm{i}\varpi}\right) - \left(\rho_{12} - \tfrac{b}{\mathrm{i}\varpi}\right)^2\right] \\ &= N \rho_{22}(\varpi) / \left[\rho_{11}(\varpi) \rho_{22}(\varpi) - \rho_{12}^2(\varpi)\right], \end{aligned} \tag{17.62}$$

where $\rho_{11}(\varpi) = \left(\rho_{11} + \frac{b}{\mathrm{i}\varpi}\right)$, $\rho_{12}(\varpi) = \left(\rho_{12} - \frac{b}{\mathrm{i}\varpi}\right)$, and $\rho_{22}(\varpi) = \left(\rho_{22} + \frac{b}{\mathrm{i}\varpi}\right)$.

Compare Eq. (17.62) and Eq. (17.32), it can be found that the viscous dissipation of pore medium makes quality coefficients ρ_{mn} becomes $\rho_{mn}(\varpi)$, namely the Biot's quality coefficient with viscosity loss of pore medium $\rho_{mn}(\varpi)$ can be represented as

$$\rho_{mn}(\varpi) = \rho_{mn} + (-1)^{m+n}\frac{b}{\mathrm{i}\varpi}, (m, n = 1, 2). \tag{17.63}$$

It makes the shear wave velocity complex, so when the shear wave propagates in the saturated liquid-filled porous medium with energy dissipation, the amplitude of the rotating wave gradually decreases with the increase of the propagation distance.

17.2.4.2.2 Propagation of Expansion and Rotation Waves

Similarly, the wave equation of expansion waves in a dissipated saturated liquid porous medium is as follows:

$$P\nabla^2 e + Q\nabla^2\varepsilon = \frac{\partial^2}{\partial t^2}(\rho_{11}e + \rho_{12}\varepsilon) + b\frac{\partial}{\partial t}(e - \varepsilon), \tag{17.64}$$

$$Q\nabla^2 e + R\nabla^2\varepsilon = \frac{\partial^2}{\partial t^2}(\rho_{12}e + \rho_{22}\varepsilon) - b\frac{\partial}{\partial t}(e - \varepsilon). \tag{17.65}$$

Assuming that expansion wave propagation along the x direction is plane wave with complex wave number k and circular frequency ϖ, its expression is as follows:

$$e = A_1\mathrm{e}^{\mathrm{i}(\varpi t - kx)}, \tag{17.66}$$

$$\varepsilon = A_2\mathrm{e}^{\mathrm{i}(\varpi t - kx)}. \tag{17.67}$$

Substitute the solution of the above expansion wave into one-dimensional wave Eqs. (17.68) and (17.69)

$$P\frac{\partial^2 e}{\partial x^2} + Q\frac{\partial^2\varepsilon}{\partial x^2} = \frac{\partial^2}{\partial t^2}(\rho_{11}e + \rho_{12}\varepsilon) + b\frac{\partial}{\partial t}(e - \varepsilon), \tag{17.68}$$

$$Q\frac{\partial^2 e}{\partial x^2} + R\frac{\partial^2\varepsilon}{\partial x^2} = \frac{\partial^2}{\partial t^2}(\rho_{12}e + \rho_{22}\varepsilon) - b\frac{\partial}{\partial t}(e - \varepsilon). \tag{17.69}$$

A system of linear equations for A_1 and A_2 can be obtained:

$$\left(\varpi^2\rho_{11} - k^2P - \mathrm{i}\varpi b\right)A_1 + \left(\varpi^2\rho_{12} - k^2Q + \mathrm{i}\varpi b\right)A_2 = 0, \tag{17.70}$$

$$\left(\varpi^2\rho_{12} - k^2Q + \mathrm{i}\varpi b\right)A_1 + \left(\varpi^2\rho_{22} - k^2R - \mathrm{i}\varpi b\right)A_2 = 0. \tag{17.71}$$

The sufficient and necessary condition for the existence of a solution where the amplitudes of the expansion waves A_1 and A_2 are nonzero is that the coefficient determinant of the above simultaneous equation is zero, namely

$$\left(\varpi^2\rho_{11} - k^2P - \mathrm{i}\varpi b\right)\left(\varpi^2\rho_{22} - k^2R - \mathrm{i}\varpi b\right) - \left(\varpi^2\rho_{12} - k^2Q + \mathrm{i}\varpi b\right)^2 = 0, \tag{17.72}$$

$$\begin{aligned}&\varpi^4\left(\rho_{11}\rho_{22} - \rho_{12}^2\right) - \varpi^2k^2\left(R\rho_{11} + P\rho_{22} - 2Q\rho_{12}\right) + k^4\left(PR - Q^2\right)\\ &- \mathrm{i}\varpi b\left[\varpi^2\left(\rho_{11} + \rho_{22} + 2\rho_{12}\right) - k^2\left(P + R + 2Q\right)\right] = 0.\end{aligned} \tag{17.73}$$

In order to simplify the expression, the quadratic equation of longitudinal wave velocity c_L^{*2} is obtained by using parameter $H = P + R + 2Q$ and relation $(PR - Q^2) = L$, $(\rho_{11}R + \rho_{22}P - 2\rho_{12}Q) = M$, $\rho_{11}\rho_{22} - {\rho_{12}}^2 = \rho_{22}\rho_c$, $\rho_{11} + \rho_{22} + 2\rho_{12} = \rho$, and $\varpi^2 = k^2{c_L}^2$:

$$\left(\rho_{11}\rho_{22} - \rho_{12}^2 + \frac{b}{\mathrm{i}\varpi}\rho\right)c_L^4 - \left(M + \frac{b}{\mathrm{i}\varpi}H\right)c_L^2 + L = 0. \tag{17.74}$$

The solution to the above equation of sound velocity in lossy saturated liquid-filled porous media is the complex longitudinal wave velocity:

$$c_{Lj}^{*2} = \frac{\left(M + \frac{b}{\mathrm{i}\varpi}H\right) \pm \sqrt{\left(M + \frac{b}{\mathrm{i}\varpi}H\right)^2 - 4L\left(\rho_{11}\rho_{22} - \rho_{12}^2 + \frac{b}{\mathrm{i}\varpi}\rho\right)}}{2\left(\rho_{11}\rho_{22} - \rho_{12}^2 + \frac{b}{\mathrm{i}\varpi}\rho\right)}, (j = 1, 2). \tag{17.75}$$

From $M(\varpi) = \left(M + \frac{b}{\mathrm{i}\varpi}H\right) = (R\rho_{11} + P\rho_{22} - 2Q\rho_{12}) + \frac{b}{\mathrm{i}\varpi}(P + R + 2Q) = R\left(\rho_{11} + \frac{b}{\mathrm{i}\varpi}\right) + P\left(\rho_{22} + \frac{b}{\mathrm{i}\varpi}\right) - 2Q\left(\rho_{12} - \frac{b}{\mathrm{i}\varpi}\right) = R\rho_{11}(\varpi) + P\rho_{22}(\varpi) - 2Q\rho_{12}(\varpi)$ and $\rho_{11}\rho_{22} - {\rho_{12}}^2 + \mathrm{i}\varpi b(\rho_{11} + \rho_{22} + 2\rho_{12}) = (\rho_{11} + \mathrm{i}\varpi b)(\rho_{22} + \mathrm{i}\varpi b) - (\rho_{12} - \mathrm{i}\varpi b)^2 = \rho_{11}(\varpi)\rho_{22}(\varpi) - {\rho_{12}}^2(\varpi)$, we can obtain

$$c_{Lj}^{*2} = \frac{M(\varpi) \pm \sqrt{M^2(\varpi) - 4L\left[\rho_{11}(\varpi)\rho_{22}(\varpi) - \rho_{12}^2(\varpi)\right]}}{2\left[\rho_{11}(\varpi)\rho_{22}(\varpi) - \rho_{12}^2(\varpi)\right]}, (j = 1, 2). \tag{17.76}$$

Comparing Eq. (17.76) and Eq. (17.42), the change in the speed of sound of P-wave is the same as that of S-wave. Fast and slow P-wave velocities of porous media also become complex wave velocity with the existence of viscous damping in saturated liquid-filled porous media, three corresponding Biot's quality coefficients ρ_{mn} change in complex number $\frac{b}{\mathrm{i}\varpi}$, which can be represented as

$$\rho_{mn}(\varpi) = \rho_{mn} + (-1)^{m+n}\frac{b}{\mathrm{i}\varpi}, (j = 1, 2). \tag{17.77}$$

Damping in liquid-filled porous media makes both fast and slow P-waves complex. The amplitudes of fast and slow P-waves decay with the increase in propagation distance. When acoustic waves propagating through saturated liquid-filled porous medium, damping loss causes the changes in quality coefficient ρ_{mn} in complex number $\frac{b}{\mathrm{i}\varpi}$. If the influence of high-frequency dissipation needs to be taken into account, Biot's mass coefficient ρ_{mn} need to be transformed as follows:

$$\rho_{mn}(\varpi) = \rho_{mn} + (-1)^{m+n}\frac{bF(v)}{\mathrm{i}\varpi}, (m, n = 1, 2), \tag{17.78}$$

where $F(v) = \frac{1}{3}\frac{\mathrm{i}^{\frac{1}{2}} v \tanh\left(\mathrm{i}^{\frac{1}{2}} v\right)}{1-\mathrm{i}^{\frac{1}{2}} v \tanh\left(\mathrm{i}^{\frac{1}{2}} v\right)}$ is the complex number, representing the deviation of friction with Poiseuille as frequency increases, reflecting the phase difference between velocity and friction. The frequency-dependent dissipation coefficient $v = a\left(\frac{2\pi f \rho_f}{\eta}\right)^{\frac{1}{2}}$, where a is the average pore size of the pore medium.

The attenuation coefficient is the imaginary part of the complex wave number

$$\alpha_{Lj}^* = \left|\mathrm{Im}\left(\frac{\varpi}{c_{Lj}^*}\right)\right|, (j = 1, 2). \tag{17.79}$$

The high-frequency attenuation coefficient is influenced by several factors, including porosity ϕ, average pore size a of the solid–liquid two-phase porous medium, frequency f of the acoustic wave, and fluid viscosity η. Therefore, porosity ϕ of cancellous bone can be inferred based on the frequency-dependent attenuation coefficient.

17.3 Photoacoustic Spectral Analysis Method

From the theoretical analysis in Sect. 17.2, it can be seen that the PA signal contains the chemical composition and physical structure information of the bone tissue. [44, 45]. The current PA spectral analysis methods used for bone quality assessment are divided into four main categories: PA spectroscopy analysis, PA frequency analysis, PA time-frequency spectral analysis methods, and PA two-dimensional spectral analysis.

17.3.1 Photoacoustic Spectroscopy Analysis

The bone detection method based on PA spectroscopy analysis is mainly based on the principle that the light absorption spectrum of different bio-macromolecules in bone tissue is different. It analyzes the multi-wavelength PA signal to realize the characterization of the type and content information of characteristic bio-macromolecules in complex bone tissue. According to Eqs. (17.5)–(17.7), based on the light absorption properties of various bio-macromolecules in bone tissue, quantitative detection of the relative content of various biological macromolecules in bone tissue can be achieved by analyzing the corresponding PA signal amplitude or energy at the characteristic absorption wavelength of bone tissue. However, the optical absorption spectrum of various chemical components is mixing within the near-infrared I and II windows. In this condition, the concentration of different components in bone tissue can be calculated by unmixing the PA spectroscopy based on Lambert Beer's law.

Steinberg et al. [46] developed a multispectral PA bone detection system by combining a 750–950 nm tunable laser with an acousto-optic modulator, measured the PA spectroscopy characteristics of cortical bone, and obtained quantitative information of bone chemical composition through decoupling. Based on the optical absorption spectra of the main chemical components in cancellous bone in the near-infrared regions I and II band range, Feng et al. [47–49] performed spectral unmixing of the multi-wavelength PA signals generated from cancellous bone by least-squares fitting using Lambert Beer's, as shown in Fig. 17.3. The analysis of the relative content of the main components of cancellous bone, such as minerals, hemoglobin, lipid, and collagen, has been realized, thus enabling the classification of normal bone from osteoporotic bone. Because the multi-wavelength PA signal carries the characteristic information of bio-macromolecules, it was found that the PA spectra of bone tissue can be obtained not only by calculating the amplitude or energy of PA signals but also by analyzing the statistical information of PA signals.

Collagen is an important biomarker of osteoporosis progression, which is hard to be detected by using DEXA, CT, or MRI techniques. Feng et al. studied the feasibility of multi-wavelength PA (MWPA) measurement of the collagen in bone, using the wavelength range of 1300–1800 nm, as shown in Fig. 17.4. First, the feasibility of this approach for detecting the collagen content of bone was demonstrated by means of numerical simulation. Then, ex vivo experiments were conducted on both animal and human bone specimens with different bone densities using the MWPA method. The relative collagen content was extracted and compared with the results of micro-computed tomography (micro-CT) and histology. The results showed that the "relative collagen content" parameter obtained using the MWPA method correlated well with the bone volume ratio obtained from micro-CT images and histological analysis results. This study highlights the potential of the proposed PA technique for determining the collagen content of bones as a biomarker for bone health assessment.

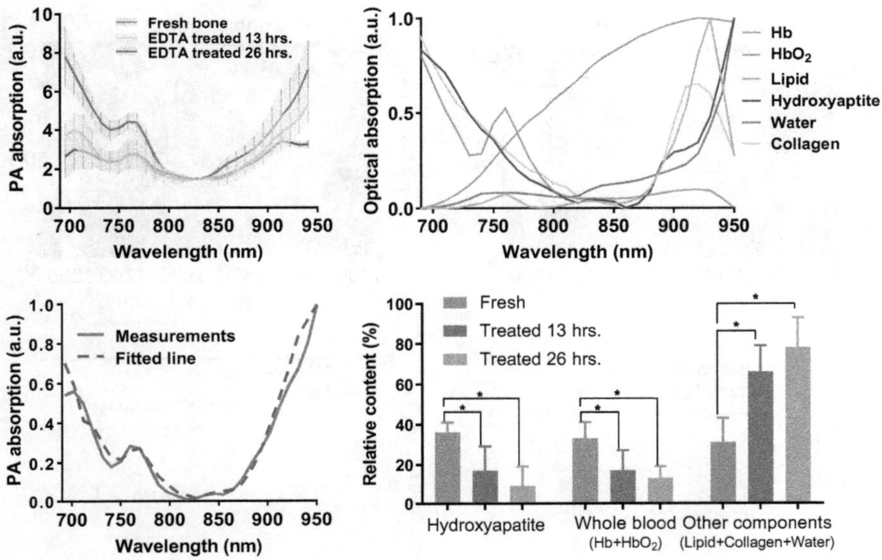

Fig. 17.3 Photoacoustic spectral unmixing for assessment of bone chemical composition [48]

Recently, Feng et al. [50] studied the PA spectroscopy analysis method based on Nakagami mathematical statistics, extracted the statistical information of PA signals excited by different light wavelengths with the help of statistical models, and obtained the PA spectroscopy with statistical characteristics, which can intelligently identify the differences in biochemical components of bone tissue, as shown in Fig. 17.5. Compared with the traditional PA spectroscopy method, this method is not sensitive to changes in laser energy, so it is more stable, having considerable potential for early and rapid detection of skeletal diseases.

17.3.2 Photoacoustic Spectrum Analysis

The acoustic spectrum of PA signals excited by a single wavelength of light is related to the physical characteristics of the chemical components that absorb light. Thus, the use of PA spectroscopy analysis can achieve the detection of information such as the content of chemical components, mechanical state, and microstructure of bone tissue.

Mandelis et al. [51–53] attempted to use PA technology for osteoporosis detection in the early 2010s and studied the feasibility of PA parameter imaging based on single frequency rectangular pulse laser in detecting changes in mineral and collagen content caused by osteoporosis. The proposed normalized backscatter parameter has a good correlation with bone volume fraction and collagen content. Steinberg et al.

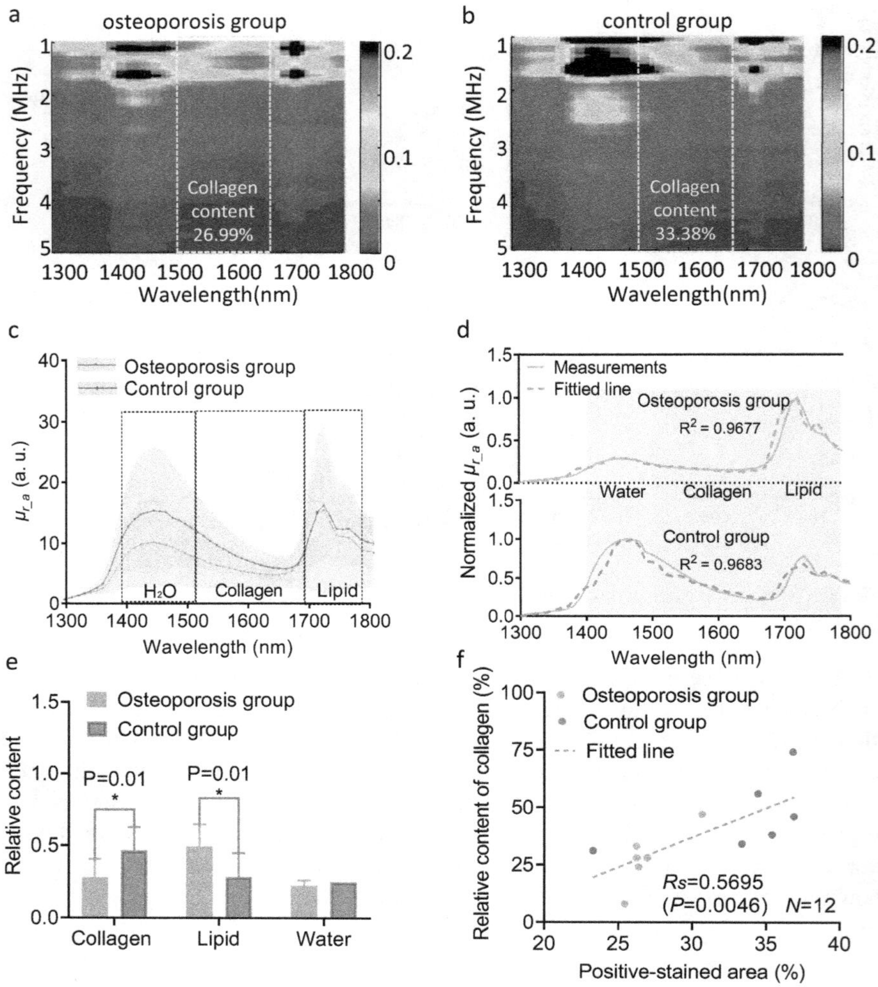

Fig. 17.4 Multiwavelength photoacoustic (MWPA) results for the osteoporosis and control rabbit model groups [47]

[46, 54] developed a PA and US dual-mode multispectral PA system, which provides information on bone chemical composition and mechanical information related to bone strength by analyzing the amplitude and phase of different frequency PA signals at multiple wavelengths. Xu et al. [55, 56] verified that the spectral energy distribution of broadband PA signals generated by soft tissues can also reflect the microstructure information of the region of interest. Feng et al. [57] further applied PA spectrum analysis to evaluate the microstructure of cancellous bone. Using rat tibias with different BMDs as models, the power spectral quantification parameter, slope, of PA signals was extracted to evaluate the BMD and average trabecular thickness of cancellous bone, as shown in Fig. 17.6. Taking advantages of no

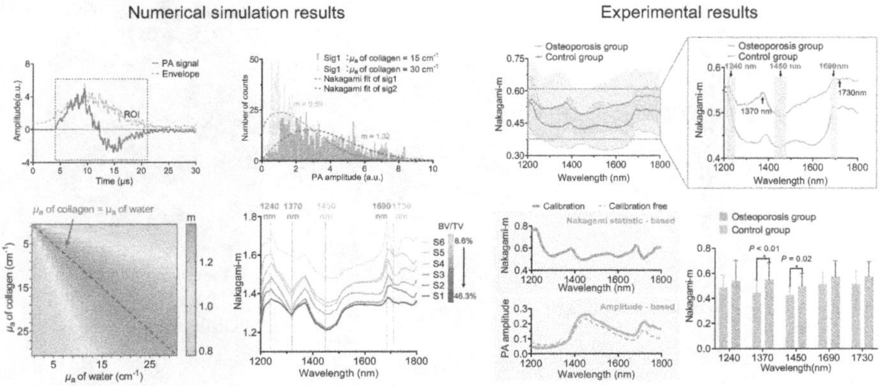

Fig. 17.5 Nakagami statistics-based photoacoustic spectroscopy used for label-free assessment of bone tissue [50]

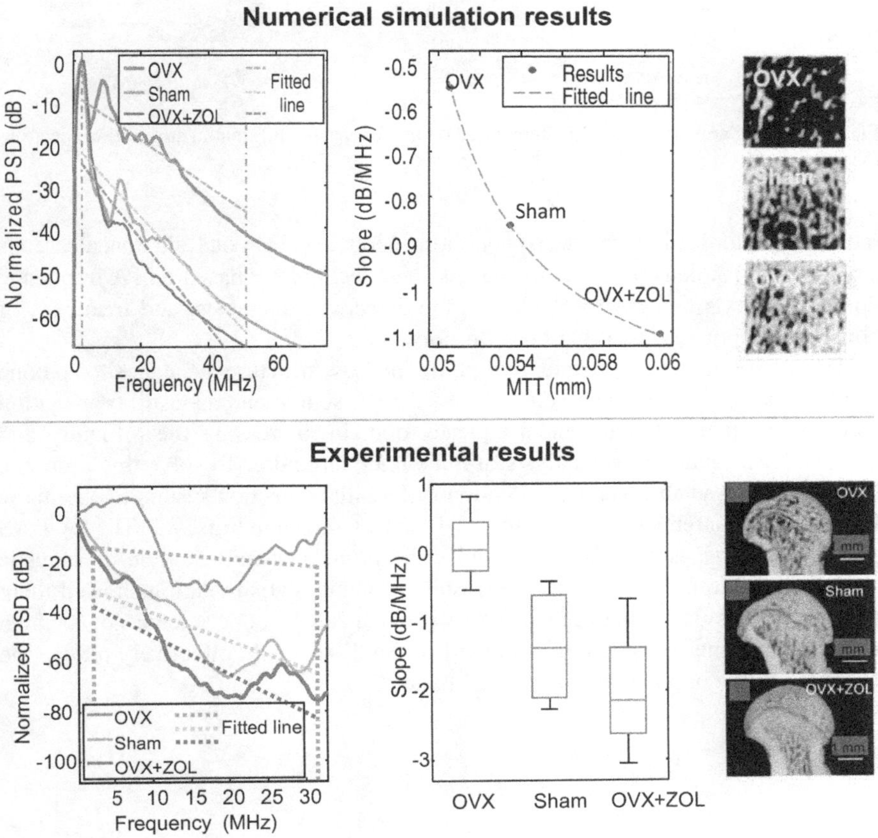

Fig. 17.6 Characterization of bone microstructure using photoacoustic spectrum analysis [57]

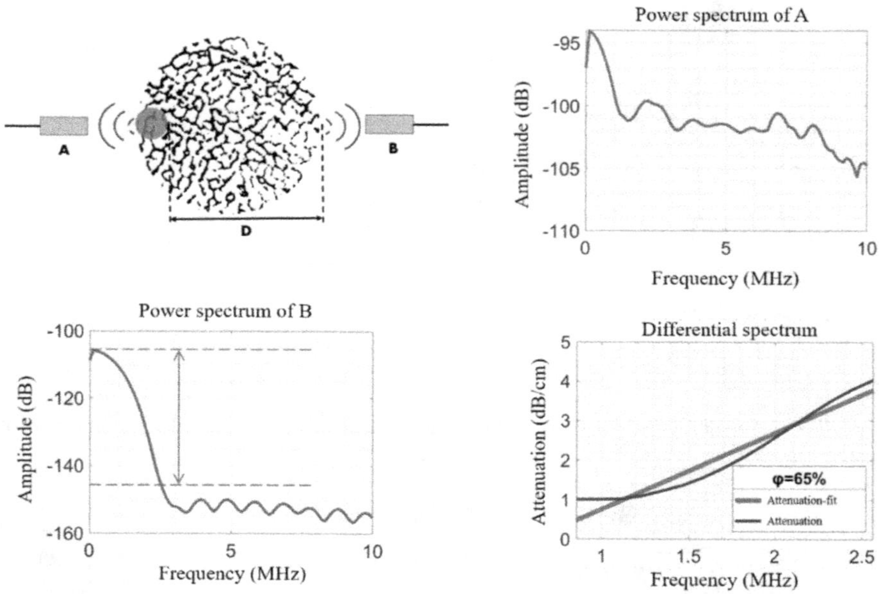

Fig. 17.7 Simulation of bone evaluation based on photoacoustic differential attenuation spectrum [58]

radiation, no ionization, and deep penetration depth in calcareous and noncalcareous structures of PA detection, the bone assessment technology based on PA frequency domain analysis has great potential in the detection, diagnosis, and treatment of clinical osteoporosis and other bone diseases.

The distributed solid–liquid two-phase porous structure of cancellous bone results in aliasing of the characteristics of the PA source and acoustic propagation medium in spatial domain and frequency domain, increasing the difficulty and complexity of quantitative analysis of bone microstructure. To solve this problem, Yu et al. proposed an eccentric excitation-differential detection system to obtain the PA differential attenuation spectrum (PA-DAS), as shown in Fig. 17.7. The PA-DAS parameter, *slope*, is extracted to quantify the propagation attenuation of acoustic signals of different frequencies. The results of numerical simulation showed there exits strong correlation between the porosity of cancellous bone and the slope, which indicated that the analysis method based on PA-DAS could effectively realize the quantitative evaluation and diagnosis of bone.

17.3.3 Photoacoustic Two-Dimensional Spectral Analysis

Cancellous bone is a solid–liquid two-phase porous medium, which is mainly composed of a solid reticular trabecular framework and fluid bone marrow filled inside. The complex chemical composition and porous structure lead to a nonstationary PA signal received by the sensor. It includes not only the spectrum information of the PA source itself but also the attenuation information of the PA signal propagation in porous cancellous bone and also includes comparative information on light absorption related to molecular metabolism in bone tissue. Therefore, through conjoint analysis of PA signals of bone tissue in two or more dimensions, such as time domain, frequency domain, laser wavelength, and ultrasonic number, the quantitative characterization parameter information highly related to bone tissue health can be obtained.

17.3.3.1 Photoacoustic Time-Frequency Spectral Analysis

The time-frequency spectrum contains rich information about the frequency spectrum distribution at each time and the change of each frequency with time, which reflects that the space distribution of the absorption source and the change information of the PA signal during propagation can be characterized. It can be used for the study of complex biological tissues with spatial distribution of optical absorption sources and acoustic impedance mismatch.

Xie et al. [59–61] proposed a PA time-frequency spectral analysis method for joint time-frequency domain analysis of bone PA signals so as to simultaneously obtain the spectrum of the PA signals generated by bone tissue as the PA source and the acoustic propagation characteristics of bone tissue, thus realizing the quantitative evaluation of multiple bone parameters. Xie et al. carried out a PA experiment on a rabbit model of osteoporosis, as shown in Fig. 17.8. Based on the quantitative relationship between the frequency spectrum of the PA source and the characteristic size of the PA source, the relationship between the acoustic propagation characteristics of the cancellous bone and the bone microstructure and BMD, and the relationship between the energy of the PA signal and the chemical component content, the time-spectrum analysis of MWPA signals was carried out, and multiple PA time-spectrum quantification parameters were extracted. Including different wavelength energy ratios, power-weighted mean frequency (PWMF) of the first 1 μs, and the PWMF curve fitting parameter *slope*, we can simultaneously quantitatively evaluate the content of multiple chemical components, the characteristic size of bone trabeculae and bone marrow clusters, and BMD of cancellous bone. The multiwavelength PA time-frequency spectral analysis method has the potential to provide a comprehensive evaluation of bone tissue physicochemical information.

In order to comprehensively characterize the microstructure and content of various molecular biomarkers related to bone tissue health, as shown in Fig. 17.9, Feng et al. [62] used PA physicochemical spectrogram (PCS) analysis methods

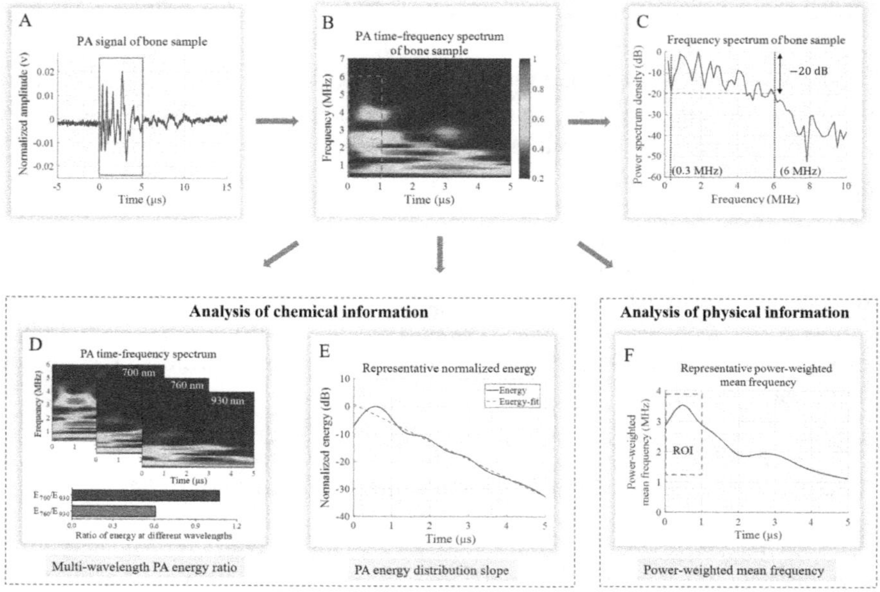

Fig. 17.8 Photoacoustic time-frequency spectral analysis method for bone evaluation [61]

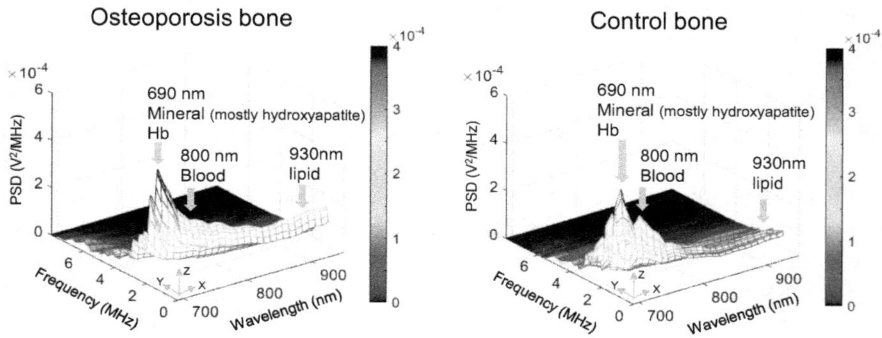

Fig. 17.9 Photoacoustic physicochemical spectrum acquired from osteoporosis bone and control bone [62]

to obtain spectral and acoustic features from PA signals. It can simultaneously characterize the microstructure and the content of different biomarkers making up the bone matrix, lipid clusters, collagen, and bone marrow clusters in bone tissue, without the need for multiple clinical imaging examinations. By combining the two-dimensional quantification results with machine-learning classification methods, the accuracy of bone tissue health assessment based on PA spectroscopy is further improved. The results showed that the quantified PA parameters from the PCS were highly correlated with DEXA, micro-CT, and MRI results, which can provide rich information on BMD, BMA, and lipid content, respectively. In addition to this, PA technique can provide better classification accuracy based on the assessment of the microstructure and chemical composition of bone, together with an appropriate categorization algorithm.

The existing research shows that the PA-guided wave generated by laser irradiation has the ability, similar to the piezoelectric ultrasonic guided wave, to evaluate the geometric structure and elastic mechanic properties and also has the sensitivity to the optical properties. Therefore, the PA guided wave method helps characterize the characteristics of cortical bone, such as structure, elasticity, and chemical composition. Chen et al. [63] studied the effect of cortical bone light transmittance on the characteristics of photoacoustic-guided waves (PAGW), and the results showed that the signal power spectrum parameters of PAGW have a good correlation with light transmittance. By analyzing the acoustic frequency-light wave number spectrum of PA signals, the propagation of PA signal modes in bone tissue can be distinguished, which helps to develop a PAGW-based evaluation method for osteoporosis in cortical bone.

17.4 Photoacoustic Bone Imaging

The spectral analysis method mainly helps extract the physical and chemical properties of bone tissue for accurate diagnosis. On the other hand, there is a strong demand for intuitive imaging of bone structure and characteristic distribution in clinical settings. Currently, PA bone imaging research can be broadly classified into three categories: The first is PA bone tissue functional imaging, such as imaging based on vascular repair in bone tissue after defect trauma, including stem cell repair, involving finger joints, bone joints, or defects in bone itself. The second is to directly image the microstructure of bone tissue, such as cortical and cancellous bone, using PA signals to obtain the structural distribution map of the bone. The third is to image the interior of the brain through the skull, which is called transcranial brain imaging.

17.4.1 Photoacoustic Bone Tissue Functional Imaging

PAI has been widely studied in soft tissue imaging due to its advantages in molecular metabolic function imaging. In recent years, its application has progressively extended to bone tissue function image, such as imaging of periosteous vessels and blood oxygen saturation related to bone tissue health, including blood flow distribution related to osteoarthritis, angiogenesis, and blood oxygen saturation monitoring during bone wound healing. Es et al. [64] utilized PA computed tomography (PACT) for imaging healthy human fingers, achieving high-resolution imaging of blood vessels in two interphalangeal joints and imaging of inflamed synovium based on this imaging method. Nishiyama et al. [65] developed a PAI system using a circular US transducer for small finger vascular imaging (i.e., diameter 0.1–0.5 mm), designed the transducer based on simulation, and imaged healthy human index finger vascular systems. Liu et al. [66] applied PAI to the experimental model of post-traumatic osteoarthritis and realized PAI of angiogenesis and hypoxia in synovial tissue. In addition, PAI has been employed for diagnose *Staphylococcus aureus*-induced acute osteomyelitis in mice [67]. Li et al. [68] used optical resolution photoacoustic microscopy (OR-PAM) to provide high-resolution, unlabeled, and long-term in vivo observation of bone vascularization during bone defect healing. Chen et al. [69] used OR-PAM for unlabeled, in vivo, and long-term observation of bone vascular dynamics in a mouse tibia defect model, achieving visualization of the vascular network during bone regeneration in the mouse tibia defect model. Menger et al. [70] applied PAI to analyze vascular formation and tissue oxygen saturation during fracture healing and nonunion formation. Zafar et al. [71] created a bone defect in the rat femur and evaluated the ability of differentiated mesenchymal stem cells to repair defects using high-frequency PA and US bimodal imaging and micro-CT. Liu et al. [72], in order to improve the imaging contrast and resolution, introduced exogenous contrast agents and used the PA fluorescence positron emission scanning three mode imaging system to achieve the diagnosis and localization of cerebral plaques. Zhang and Liu et al. [73, 74] proposed a dual-scale and multi-wavelength PA molecular imaging method for multifunctional detection of glioblastoma, introduced exogenous contrast agents, and quantified the multifeatures of neovascularization at macro- and microscales, realizing the characterization of glioblastoma microvessels, the blood–brain barrier, and immune activity.

In addition to the study of blood vessels around bone tissue, PAI can also focus on the imaging of bone tissue's intrinsic function, which is crucial in guiding orthopedic surgery and assessing bone tissue health. Shubert et al. [75] demonstrated that 3D PAI can elucidate the differences between cortical and cancellous bone. They further explored the use of 3D PAI on human vertebral samples to noninvasively distinguish cortical and cancellous bone, thereby aiding in surgical procedures. This has broad benefits for guiding spinal fusion surgery. Yang et al. [51] utilized US and PA techniques to perform grating scanning imaging on the surface of trabecular bone samples, which can detect changes in bone demineralization and decalcification. In PA bone imaging, the PA signal on the bone surface is intense, and its reflected signal often overlaps with the PA signal in other parts. In order to remove the problem of artifact mixing caused by strong reflection on the bone surface, Biswal et al. [76] treated the PA signal from the bone surface as a virtual US source in finger PA tomography and reconstructed the signal separately using a method similar to US reflection imaging to identify bone epidermal signals and the artifacts associated with them. Park et al. [77] applied near-infrared light excitation and acoustic detection methods to successfully present the spine and ribs of mice in high-resolution PA images and track the rib fracture of mice through PAI. Yang et al. [78] used a fully optical noncontact phase domain PA elastic microscopy to distinguish between bone tissue and soft tissue by measuring their elastic parameters.

17.4.2 Photoacoustic Bone Microstructure Imaging

The microstructure characteristics of bone are crucial for clinical bone health assessment. However, due to its complex bone tissue structure, imaging it poses a challenge in PAI. The spectral parameter imaging of PA signals can provide valuable physiological information beneficial to related medical diagnosis at the tissue microstructure level. Cheng et al. [79] studied slope PAI in detecting early dental lesions, demonstrating that PA parametric imaging has better sensitivity in detecting dental lesions compared to US imaging. Feng et al. [80] first explored the feasibility of PA three-dimensional bone microstructure imaging and obtained qualitative experimental results consistent with micro-CT images, as shown in Fig. 17.10. Nevertheless, due to the complexity of bone microstructure, quantitative

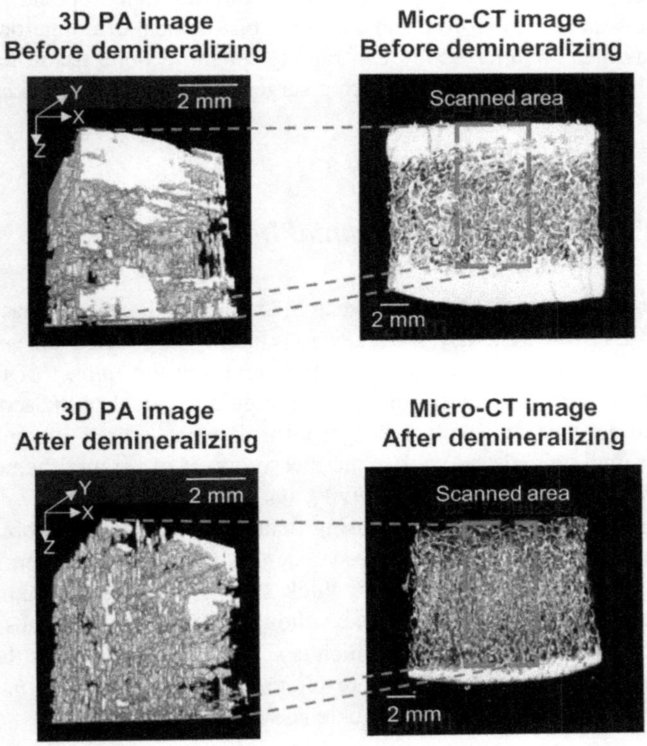

Fig. 17.10 Photoacoustic 3D imaging for cancellous bone tissue [80]

imaging experiments are challenging. Currently, more work mainly focuses on using deep learning to address the scattering issue of pore structures. Chen et al. [81] introduced the attention mechanism into the U-shaped convolutional neural network (CNN) structure and reconstructed a clearer bone microstructured PA image based on micro-CT data. Their research shows that the reconstruction imaging results of cancellous bone samples have a high similarity index with the bone structure model.

Bone tumor surgery typically relies on preoperative CT or MRI results to determine the boundary of the removed bone tissue. However, these techniques cannot accurately diagnose or confirm the tumor edge during surgery, often leading to the excision of more tissue than the actual tumor area. A study by Cao et al. [14] conducted real-time three-dimensional contour scanning of tissues using ultraviolet PA microscopy, which can be used for intraoperative evaluation of noncalcified and decalcified thick bone specimens without the need for tissue sectioning. According to the characteristics of the Haematoxylin and eosin (H&E) gold standard histological staining image of a traditional optical microscopy, they achieved virtual staining of the PA microscopic image of unmarked bone specimens with the help of deep learning. The tissue imaging results of the same bone specimen in the same area demonstrated a high correlation between the virtual staining images of the PA microscopy and the H&E images of the traditional optical microscope. This indicates that the label-free and slide-free histological examination conducted by the ultraviolet PA microscopy can rapidly diagnose bone tissue lesions, help determine the edge of bone tumors during surgery, and make cancer features easier to recognize.

17.4.3 Photoacoustic Transcranial Imaging

Due to the requirements of brain research, transcranial imaging has become a hot spot of PAI research. Early work was primarily conducted on animals with extremely thin skulls, such as mice and rats, ignoring the impact of the skull or simplifying the skull into a uniform flat plate in fluid to calculate the acoustic insertion loss and develop deconvolution algorithms to achieve three-dimensional brain imaging of small animals, as well as the measurement of cerebral hemodynamics, hemoglobin concentration, neural activity, and other parameters [73, 82–96]. In recent years, there has been an increasing demand for noninvasive brain research on larger mammals. Consequently, researchers worldwide have been conducting a series of research on the impact of thick skulls. Usually, the human skull is composed of three layers of dense bone solid–liquid two-phase porous cancellous bone, with a thickness of 3–11 mm, which is approximately ten times the thickness of the mouse skull (0.3–0.8 mm). As a result, the attenuation of PA signal scattering is extremely strong, and the image speckle is extremely severe.

Chen et al. [97] integrated a scanning acoustic microscopy with traditional PAM, and based on the similar propagation characteristics of PA signals and US echoes in the bone layer, they used US echoes as filters to remove multiple reflection artifacts in the PA signal. This effectively addressed the problem of artifacts in transcranial PAI, obtaining a PA image without artifacts. Hosseini et al. [98] studied the effects of phase aberrations, mode-converted transverse waves, and multiple scattering in skull formation and used refractive correction image reconstruction methods to form transcranial PA images, separating the effects of these effects.

Ni et al. [99] first validated the basic feasibility of visualizing transcranial structures and cerebrovascular systems through healthy adult temporal bones based on the developed spherical array PA tomography system and further validated it using MRI. Ben et al. [100] proposed the PA memory effect to address the issue of acoustic distortion caused by the skull, considering that nearby US experiences the same distortion. Building upon this, a PA transmission model was established and the light absorption distribution was reconstructed based on the PA signal, optimizing the quality of transcranial PAI.

Na et al. [101] proposed a transcranial PACT distortion removal reconstruction method based on layered universal back projection (L-UBP). Based on the traditional UBP, the acoustic refraction and wave conversion at the layer interface were considered, and the skull attenuation and transmission coefficient were compensated. The method's efficacy was initially validated through computer simulation, and a 3D imaging system was designed and constructed. The effectiveness of the proposed method was verified in the study of adult skulls in vitro. In order to compensate for the aberration and attenuation in the measured data caused by the propagation of the PA field through the skull, Poudel et al. [102] used the inversion method based on the wave equation to model the heterogeneous elastic properties of the medium. Based on the elastic wave equation, a three-dimensional transcranial PACT image reconstruction method based on optimization was developed, and a method of joint reconstruction of initial pressure distribution and low dimensional representation of skull elastic parameters was studied [103].

Transcranial PAI often utilizes array transducers within a limited angle to collect PA speckle signals and carry out computational reconstruction, posing a typical ill-posed inverse diffusion problem. Gao et al. [104] adopted a data-driven strategy and developed a U-Net based convolutional neural network for the multiple Mie scattering and significant attenuation of PA signals from multilayer porous structure of thick skulls. This approach aimed to deeply mine and extract hidden information in weak PA speckle signals, as shown in Fig. 17.11. PAI exhibits high sensitivity to hemoglobin, but the transcranial PA signals are overlaid with dual-domain information—frequency and spatial domains. In the frequency domain, the PA signals generated from the distributed vascular size are broadband, meaning the wavelength of PA signals is distributed. In the spatial domain, the transcranial signal field, scattered and superimposed by the size-distributed holes, exhibits a complex spatial distribution.

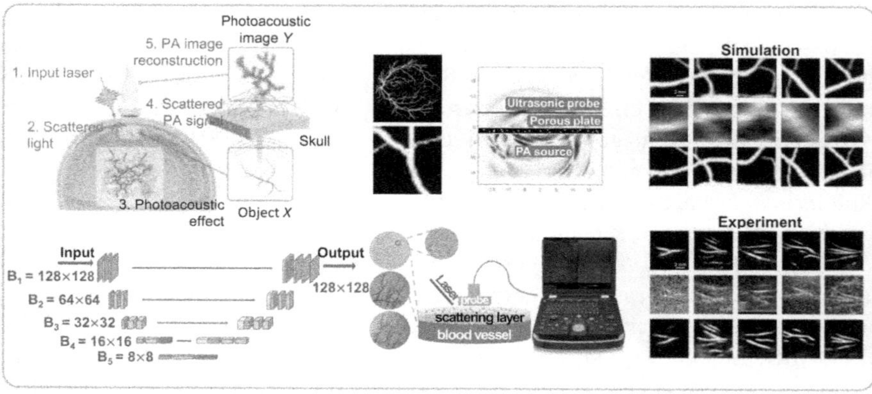

Fig. 17.11 Deep learning-based photoacoustic imaging of vascular network through thick porous media [104]

To address these characteristics, the shallow and deep layers of U-Net could acquire high- and low-frequency information from the image, respectively, and the information of each level is well preserved through skip-connections. Additionally, they incorporated the batch normalization (BN) method after each down-sampling convolutional layer to counteract the low sensitivity of the U-Net structure to ultrahigh frequency information, making it suitable for the distributed size of cerebral vessels. They collected a large number of PA images of chicken embryo vasculature for training the neural network and employed a variety of evaluation metrics, including error and similarity measures, to compare their method with other advanced reconstruction algorithms and neural network models. Experimental results revealed that this neural network can effectively fit the mapping relationship between speckle images and imaging targets, extract effective information from speckle images with low signal-to-noise ratio, severe speckle artifacts, and missing details, and reconstruct PA images of intracranial blood vessels with high accuracy. And it provides an effective tool for the image quality improvement of transcranial PA brain imaging.

17.5 Preclinical and Clinical Application Research

17.5.1 Preclinical Animal Experiments

Wood et al. [105] from the University of Texas, USA, used a multiwavelength PAI method to monitor changes in blood oxygen content in the bone marrow of leukemia rats in vivo. The study showed that PAI can monitor changes in blood oxygen saturation in the bone marrow cavity of mice, indicating that PAI has great potential in noninvasive and unlabeled monitoring of diseases such as bone marrow failure.

17.5.2 Preclinical Human Experiments

The above experimental studies based on in-vitro samples or in vivo animal have preliminarily verified the feasibility of PA evaluation of bone microstructure and metabolic information. However, there are still significant differences between in-vitro samples and in vivo human detection. The biggest difference is that the scattering and attenuation of light and US differ greatly, specifically manifested in the bone size of in vitro or small animals, and there is no (or very thin) soft tissue encapsulation. The attenuation of light and US can be simplified or ignored. However, the human body has a large bone size and is wrapped in thick soft tissue, which leads to severe scattering and attenuation of light and US, making the physical problems of the generation and propagation of PA signals more complex. If the traditional PAI method used for in vivo PA detection directly, it will result in significant errors. Therefore, it is necessary to develop inversion and quantification methods for noninvasive evaluation of bone microstructure and metabolic information in vivo.

To solve this problem, Steinberg et al indicated that the energy ratio of dual wavelength PA signals exhibits a good linear relationship with bone marrow blood concentration. This method achieves noninvasive evaluation of human tibial US velocity and intramedullary lipid ratio. This study reflects the great potential of PA technology in the early diagnosis of bone tissue and bone marrow metabolic disorder.

Based on the above research status, Feng et al. [106] preliminarily developed a PA and US dual-mode penetrating bone evaluation prototype device, as shown in Fig. 17.12. The characteristics of this prototype are the following: (1) Dual mode is not a simple physical integration, but a coaxial fusion form of light and US propagation paths, ensuring that the ultrasonic mode can provide important compensation and guidance information such as ultrasonic velocity and attenuation for the PA mode. (2) Using multi-wavelength PA spectroscopy (power spectrum and light absorption spectrum) to detect the dual information of microstructure and metabolic function in bone tissue in vivo. (3) Obtaining the original information of bone tissue to the maximum extent possible through penetrating PA detection. Based on this platform, the team conducted a clinical study on PA detection for the evaluation of osteoporosis. Clinical studies have shown that PA detection can achieve dual evaluation of chemical molecular changes related to bone microstructure and metabolic function in a noninvasive and nonradiative manner, providing a new approach for the diagnosis of degenerative and metabolic bone diseases. The above research is based on the combination of PA and US technology to evaluate the characteristics of bone microstructure and chemical composition metabolism information (minerals, blood, lipid content, and blood oxygen saturation), which

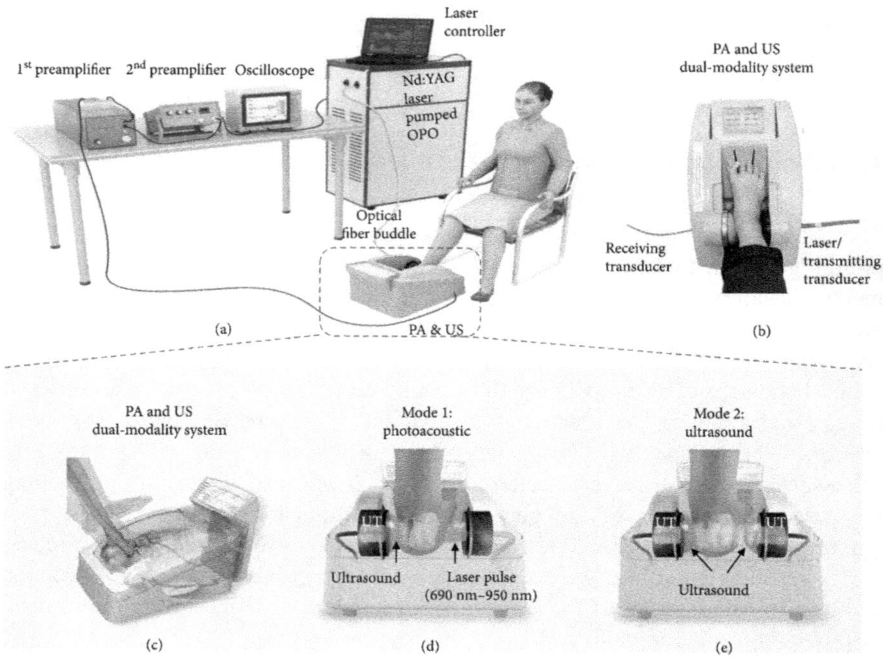

Fig. 17.12 Functional photoacoustic and ultrasonic assessment of cancellous bone [17]

preliminarily meets the dual needs of in vivo evaluation of bone microstructure and metabolic function.

In this study, the author proposed the theory and method of adaptive PA spectroscopy inversion, which solved the problem of nonlinear spectral coloring in multilayer complex media. Spectral coloring is due to the different light attenuation corresponding to the transmission of light at different wavelengths in biological tissues, which causes the nonlinear problem of light energy deposition (also known as spectral deformation), which directly affects the accuracy of PA spectroscopy unmixing. Traditional solutions often use empirical values to compensate for light attenuation, but the individual differences in light attenuation in bone tissue are very significant (the thickness and structure of different human bone tissues vary greatly), which poses a huge challenge for traditional methods in solving spectral coloring problems. In view of the influence of the nonlinear difference in the penetrability of light at different wavelengths on PA spectroscopy, the author proposed an adaptive PA spectroscopy extraction and correction method. Without knowing the prior value of light attenuation, the author adaptively extracted the characteristic information of light attenuation of bone tissue from the change of PA signal attenuation with time and further deduced the analytical formula of light absorption spectrum of bone tissue based on Mie scattering theory. The results show that this method solved the impact of spectral coloring on the PA spectroscopy and can efficiently and

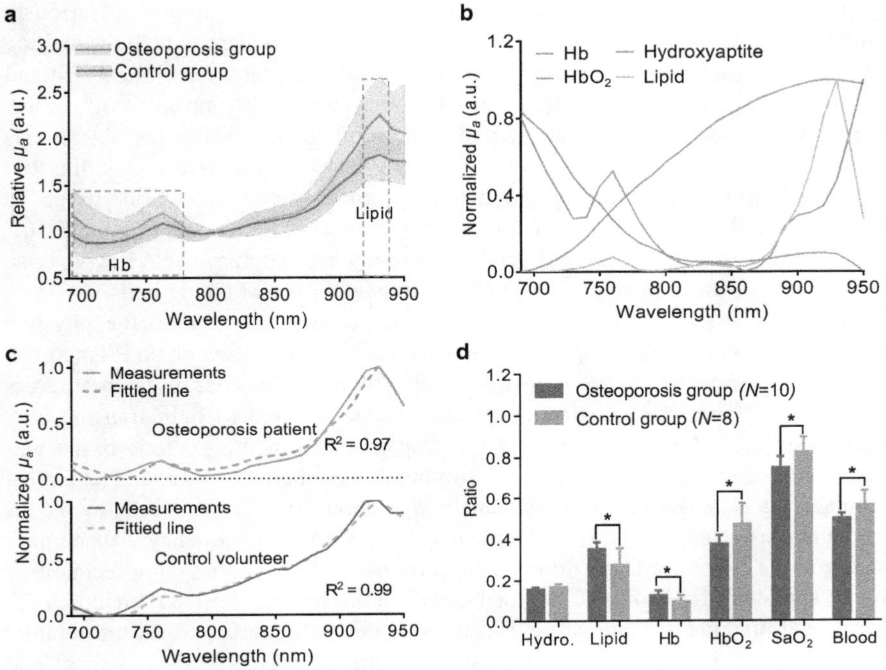

Fig. 17.13 Multiwavelength PA measurement of the optical absorption spectrum of the trabecular bone in the human calcaneus in vivo [17]

accurately reflect the optical absorption spectrum information directly related to the molecular metabolism of bone tissue, as shown in Fig. 17.13. Due to the fact that this method does not require the *a priori* knowledge or empirical values of light attenuation as a reference and has low algorithm complexity and good universality, it improves the accuracy and robustness of PA spectroscopy inversion and unmixing.

17.6 Summary and Outlook

Biomedical PAI can simultaneously provide histomorphology, metabolic function, and other information without radiation and ionization. It has the advantage of high penetration depth in both calcareous and noncalcareous structures. Bone assessment technology based on PA technology has great potential in the detection, diagnosis, and treatment monitoring applications of clinical osteoporosis and other metabolic and degenerative bone diseases. Besides, it has reusable academic value and application prospects in the field of noninvasive clinical precision diagnosis and treatment. However, in the quantitative detection and imaging applications of bone diseases, the complex structure of bone makes light and US have both strong scattering and attenuation characteristics when propagating in it, seriously affecting the accuracy of bone microstructure and metabolism detection, making its clinical application face huge bottlenecks. In recent years, the clinical demand for noninvasive precision medicine has become increasingly prominent, and PA bone imaging technology is facing new opportunities and challenges. PA signals have complex propagation modes in multilayer porous bone tissue, including but not limited to strong scattering and attenuation of light and US in bone tissue, as well as the influence of soft tissue on the propagation of PA signals in bone tissue. These complex factors directly affect the sensitivity and accuracy of PA in vivo bone assessment, posing a huge challenge to traditional PAI detection methods.

In recent years, many scholars have begun to pay attention to the physical problems of light and US propagation in bone tissue and the research on PA methods of bone detection ex vivo and in vivo. This chapter introduces our research progress in this area. First, the research on physical issues related to light transmission, light absorption, PA excitation, and PA propagation in complex bone tissue was elaborated. Second, the research of solving bone microstructure and chemical composition by various PA spectrum analysis methods, such as PA spectroscopy, PA frequency spectrum, PA physicochemical spectrum, and PA frequency wavenumber spectrum, is introduced. In addition, the progress of PA bone imaging technology in the application research of various bone diseases was emphasized, including PA functional imaging of bone tissue, PAI of bone microstructure, and PA transcranial brain imaging. Finally, the current research status of prototype development and clinical application of PA imaging and detection in clinical applications was reviewed. We believe that these studies will promote the in-depth development of bone PA detection and diagnosis technology and help solve some of the core technology bottlenecks in noninvasive bone detection using PA techniques in vivo. However, due to the complex dielectric properties of bone tissue, the current theories and mechanisms of PA bone detection and imaging are still not perfect. The resolution of related issues first relies on in-depth research on the physical issues of PA bone detection and in-depth exploration of PA detection methods for in vivo bone evaluation. Furthermore, a combination of US or MRI detection techniques can be considered to improve the sensitivity and accuracy of PA bone imaging and detection.

References

1. *Editorial Committee of Chinese Health Promotion Foundation Osteoporosis Prevention and Treatment White Paper. Osteoporosis in China: A White Paper.* Vol. 3. 2009: Chinese Journal of Health Management.
2. Kanis, J.A., *Diagnosis and clinical aspects of osteoporosis.* Pocket reference to osteoporosis, 2019: p. 11–20.
3. Albanese, C.V., F. De Terlizzi, and R. Passariello, *Quantitative ultrasound of the phalanges and DXA of the lumbar spine and proximal femur in evaluating the risk of osteoporotic vertebral fracture in postmenopausal women.* La radiologia medica, 2010. **116**(1): p. 92–101.

4. Takahashi, H.E., D.B. Burr, and N. Yamamoto, *Osteoporotic Fracture and Systemic Skeletal Disorders: Mechanism, Assessment, and Treatment*. 2021: Springer Nature.
5. Li, N., et al., *Comparison of QCT and DXA: osteoporosis detection rates in postmenopausal women.* International journal of endocrinology, 2013. **2013**: p. 1–5.
6. Li, X., L. Na, and C. Xiaoguang, *Update on the clinical application of quantitative computed tomography (QCT) in osteoporosis.* Current Radiology Reports, 2014. **2**: p. 1–5.
7. Wear, K.A., *Mechanisms of interaction of ultrasound with cancellous bone: A review.* IEEE transactions on ultrasonics, ferroelectrics, and frequency control, 2019. **67**(3): p. 454–482.
8. Griffith, J.F., et al., *Vertebral marrow fat content and diffusion and perfusion indexes in women with varying bone density: MR evaluation.* Radiology, 2006. **241**(3): p. 831–838.
9. Draper, E.R., et al., *Novel assessment of bone using time-resolved transcutaneous Raman spectroscopy.* Journal of Bone and Mineral Research, 2005. **20**(11): p. 1968–1972.
10. Morris, M.D. and G.S. Mandair, *Raman assessment of bone quality.* Clinical Orthopaedics and Related Research®, 2011. **469**: p. 2160–2169.
11. Pifferi, A., et al., *Optical biopsy of bone tissue: a step toward the diagnosis of bone pathologies.* Journal of biomedical optics, 2004. **9**(3): p. 474–480.
12. Wang, L.V. and S. Hu, *Photoacoustic tomography: in vivo imaging from organelles to organs.* Science, 2012. **335**(6075): p. 1458–1462.
13. Li, Y., et al., *Snapshot photoacoustic topography through an ergodic relay for high-throughput imaging of optical absorption.* Nature Photonics, 2020. **14**(3): p. 164–170.
14. Cao, R., et al., *Label-free intraoperative histology of bone tissue via deep-learning-assisted ultraviolet photoacoustic microscopy.* Nature Biomedical Engineering, 2023. **7**(2): p. 124–134.
15. Kim, M., et al., *Correction of wavelength-dependent laser fluence in swept-beam spectroscopic photoacoustic imaging with a hand-held probe.* Photoacoustics, 2020. **19**: p. 100192.
16. Wang, L.V. and H.-i. Wu, *Biomedical optics: principles and imaging*. 2012: John Wiley & Sons.
17. Feng, T., et al., *The feasibility study of the transmission mode photoacoustic measurement of human calcaneus bone in vivo.* Photoacoustics, 2021. **23**: p. 100273.
18. Qian, C., et al., *LED-Based Photoacoustic Imaging*. Diagnosis and Treatment Monitoring of Port-Wine Stain Using LED-Based Photoacoustics: Theoretical Aspects and First In-Human Clinical Pilot Study, ed. M.K.A. Singh. Vol. 7. 2020: Springer Singapore.
19. Chen, Y., *Research and Application of Quantitative Pulsed Photoacoustic Spectrum and Photoacoustic Imaging [D]*. 2021, Tongji University.
20. Cheng, Q., et al., *Diagnosis and treatment monitoring of port-wine stain using led-based photoacoustics: Theoretical aspects and first in-human clinical pilot study*, in *LED-Based Photoacoustic Imaging: From Bench to Bedside*. 2020, Springer Nature. p. 351–377.
21. Miller, J.G., et al., *Cancellous bone as a poroelastic medium: Extracting underlying material properties from improved ultrasonic measurements of frequency dependent attenuation and phase velocity.* The Journal of the Acoustical Society of America, 2009. **125**(4_Supplement): p. 2641–2641.
22. Biot, M.A., *Theory of propagation of elastic waves in a fluid-saturated porous solid. I. Low-Frequency Range.* The Journal of the acoustical Society of america, 1956. **28**(2): p. 168–178.
23. Biot, M.A., *Theory of propagation of elastic waves in a fluid-saturated porous solid. II. Higher frequency range.* The Journal of the acoustical Society of America, 1956. **28**(2): p. 179–191.
24. Biot, M.A. and D.G. Willis, *The elastic coefficients of the theory of consolidation.* 1957.
25. Biot, M.A., *Theory of deformation of a porous viscoelastic anisotropic solid.* Journal of Applied physics, 1956. **27**(5): p. 459–467.
26. Biot, M.A., *Theory of elasticity and consolidation for a porous anisotropic solid.* Journal of applied physics, 1955. **26**(2): p. 182–185.
27. Biot, M.A., *General theory of three-dimensional consolidation.* Journal of applied physics, 1941. **12**(2): p. 155–164.

28. Rice, J.R. and M.P. Cleary, *Some basic stress diffusion solutions for fluid-saturated elastic porous media with compressible constituents.* Reviews of Geophysics, 1976. **14**(2): p. 227–241.
29. Cleary, M.P., *Fundamental solutions for a fluid-saturated porous solid.* International Journal of Solids and Structures, 1977. **13**(9): p. 785–806.
30. Cleary, M.P., *Elastic and dynamic response regimes of fluid-impregnated solids with diverse microstructures.* International Journal of Solids and Structures, 1978. **14**(10): p. 795–819.
31. Rudnicki, J., *Plane strain dislocations in linear elastic diffusive solids.* 1987.
32. Fry, F.J. and J.E. Barger, *Acoustical properties of the human skull.* The Journal of the Acoustical Society of America, 1978. **63**(5): p. 1576–1590.
33. Ashman, R.B., J.D. Corin, and C.H. Turner, *Elastic properties of cancellous bone: measurement by an ultrasonic technique.* Journal of Biomechanics, 1987. **20**(10): p. 979–986.
34. Fellah, Z.E.A., et al., *Ultrasonic wave propagation in human cancellous bone: Application of Biot theory.* The Journal of the acoustical Society of America, 2004. **116**(1): p. 61–73.
35. Fellah, Z.E., et al., *Application of the Biot model to ultrasound in bone: Direct problem.* IEEE transactions on ultrasonics, ferroelectrics, and frequency control, 2008. **55**(7): p. 1508–1515.
36. Fellah, Z.E., et al., *Application of the Biot model to ultrasound in bone: inverse problem.* ieee transactions on ultrasonics, ferroelectrics, and frequency control, 2008. **55**(7): p. 1516–1523.
37. Sadouki, M., et al. *Ultrasonic propagation of reflected waves in cancellous bone: Application of Biot theory.* in *2015 6th European Symposium on Ultrasonic Characterization of Bone.* 2015. Corfu, Greece: IEEE.
38. Pakula, M., et al., *Application of Biot's theory to ultrasonic characterization of human cancellous bones: Determination of structural, material, and mechanical properties.* The Journal of the acoustical Society of America, 2008. **123**(4): p. 2415–2423.
39. Wyllie, M., G. Gardner, and A. Gregory, *Studies of elastic wave attenuation in porous media.* Geophysics, 1962. **27**(5): p. 569–589.
40. Hughes, E.R., et al., *Ultrasonic propagation in cancellous bone: a new stratified model.* Ultrasound in medicine & biology, 1999. **25**(5): p. 811–821.
41. Hughes, E.R., et al., *Estimation of critical and viscous frequencies for Biot theory in cancellous bone.* Ultrasonics, 2003. **41**(5): p. 365–368.
42. Bennamane, A. and T. Boutkedjirt, *Theoretical and experimental study of the ultrasonic attenuation in bovine cancellous bone.* Applied Acoustics, 2017. **115**: p. 50–60.
43. Cox, B.T., et al., *k-space propagation models for acoustically heterogeneous media: Application to biomedical photoacoustics.* The Journal of the Acoustical Society of America, 2007. **121**(6): p. 3453–3464.
44. Xie, W., *Research of photoacoustic quantitative assessment of bone quality [D].* 2023, Tongji University.
45. Feng, T., *The Research of Photoacoustic Technology in Bone Qualitative and Quantitative Assessment [D].* 2016, Nanjing University.
46. Steinberg, I., A. Eyal, and I. Gannot. *Multispectral photoacoustic method for the early detection and diagnosis of osteoporosis.* in *Photonic Therapeutics and Diagnostics Ix.* 2013. San Francisco, CA: SPIE.
47. Feng, T., et al., *Detection of collagen by multi-wavelength photoacoustic analysis as a biomarker for bone health assessment.* Photoacoustics, 2021. **24**: p. 100296.
48. Feng, T., et al., *Bone chemical composition assessment with multi-wavelength photoacoustic analysis.* Applied Sciences, 2020. **10**(22): p. 8214.
49. Feng, T., et al., *Bone chemical composition analysis using photoacoustic technique.* Frontiers in Physics, 2020. **8**: p. 601180.
50. Feng, T., et al., *Nakagami statistics-based photoacoustic spectroscopy used for label-free assessment of bone tissue.* Optics Letters, 2023. **48**(3): p. 656–659.
51. Yang, L., et al., *Photoacoustic and ultrasound imaging of cancellous bone tissue.* Journal of Biomedical Optics, 2015. **20**(7): p. 076016.
52. Kaiplavil, S., et al., *Photothermal tomography for the functional and structural evaluation, and early mineral loss monitoring in bones.* Biomedical optics express, 2014. **5**(8): p. 2488–2502.

53. Lashkari, B., L. Yang, and A. Mandelis, *The application of backscattered ultrasound and photoacoustic signals for assessment of bone collagen and mineral contents.* Quantitative imaging in medicine and surgery, 2015. **5**(1): p. 46.
54. Steinberg, I., et al., *First-in-human study of bone pathologies using low-cost and compact dual-wavelength photoacoustic system.* IEEE Journal of Selected Topics in Quantum Electronics, 2018. **25**(1): p. 1–8.
55. Xu, G., et al., *The functional pitch of an organ: quantification of tissue texture with photoacoustic spectrum analysis.* Radiology, 2014. **271**(1): p. 248–254.
56. Xu, G., et al., *Photoacoustic spectrum analysis for microstructure characterization in biological tissue: A feasibility study.* Applied physics letters, 2012. **101**(22).
57. Feng, T., et al., *Characterization of bone microstructure using photoacoustic spectrum analysis.* Optics express, 2015. **23**(19): p. 25217–25224.
58. Yu, D., et al., *Simulation of bone evaluation based on photoacoustic differential attenuation spectrum*. Technical Acoustics, 2023. 42(4): p.484–488.
59. Xie, W., et al., *Wavelet transform-based photoacoustic time-frequency spectral analysis for bone assessment.* Photoacoustics, 2021. **22**: p. 100259.
60. Xie, W., et al. *Bone microstructure evaluation by photoacoustic time-frequency spectral analysis*. in *2020 IEEE International Ultrasonics Symposium (IUS)*. 2020. Las Vegas, NV: IEEE.
61. Xie, W., et al., *Photoacoustic characterization of bone physico-chemical information.* Biomedical Optics Express, 2022. **13**(5): p. 2668–2681.
62. Feng, T., et al., *Characterization of multi-biomarkers for bone health assessment based on photoacoustic physicochemical analysis method.* Photoacoustics, 2022. **25**: p. 100320.
63. Chen, H., et al., *Influence of optical transmissivity on signal characteristics of photoacoustic guided waves in long cortical bone.* Ultrasonics, 2022. **126**: p. 106816.
64. van Es, P., et al., *Initial results of finger imaging using photoacoustic computed tomography.* Journal of biomedical optics, 2014. **19**(6): p. 060501–060501.
65. Nishiyama, M., et al., *Ring-array photoacoustic tomography for imaging human finger vasculature.* Journal of biomedical optics, 2019. **24**(9): p. 096005–096005.
66. Liu, Z., et al., *Photoacoustic imaging of synovial tissue hypoxia in experimental post-traumatic osteoarthritis*. Progress in Biophysics and Molecular Biology, 2019. **148**: p. 12–20.
67. Lu, X., et al., *High-resolution bimodal imaging and potent antibiotic/photodynamic synergistic therapy for osteomyelitis with a bacterial inflammation-specific versatile agent.* Acta biomaterialia, 2019. **99**: p. 363–372.
68. Li, B., Q. Chen, and L. Xi, *An artificial bone window for long-term photoacoustic monitoring of bone recovery.* Journal of Biophotonics, 2022. **15**(12): p. e202200196.
69. Chen, Q., et al., *High resolution intravital photoacoustic microscopy reveals VEGF-induced bone regeneration in mouse tibia.* Bone, 2023. **167**: p. 116631.
70. Menger, M.M., et al., *Photoacoustic imaging for the study of oxygen saturation and total hemoglobin in bone healing and non-union formation.* Photoacoustics, 2022. **28**: p. 100409.
71. Zafar, H., et al. *To assess the reparative ability of differentiated mesenchymal stem cells in a rat critical size bone repair defect model using high frequency co-registered photoacoustic/ultrasound imaging and micro computed tomography*. in *Photons Plus Ultrasound: Imaging and Sensing 2016*. 2016. San Francisco, CA: SPIE.
72. Liu, Y., et al., *Highly specific noninvasive photoacoustic and positron emission tomography of brain plaque with functionalized croconium dye labeled by a radiotracer.* Chemical Science, 2017. **8**(4): p. 2710–2716.
73. Liu, Y., et al., *Aggregation-Induced Absorption Enhancement for Deep Near-Infrared II Photoacoustic Imaging of Brain Gliomas In Vivo.* Advanced Science, 2019. **6**(8): p. 1801615.
74. Zhang, J., et al., *In vivo characterization and analysis of glioblastoma at different stages using multiscale photoacoustic molecular imaging.* Photoacoustics, 2023. **30**: p. 100462.
75. Shubert, J. and M.A. Lediju Bell, *Photoacoustic imaging of a human vertebra: implications for guiding spinal fusion surgeries.* Physics in Medicine & Biology, 2018. **63**(14): p. 144001.

76. Biswas, S.K., et al., *A method for delineation of bone surfaces in photoacoustic computed tomography of the finger.* Ultrasonic imaging, 2016. **38**(1): p. 63–76.
77. Park, E.-Y., et al., *Non-ionizing label-free photoacoustic imaging of bones.* IEEE Access, 2020. **8**: p. 160915–160920.
78. Yang, F., Z. Chen, and D. Xing, *All-optical noncontact phase-domain photoacoustic elastography.* Optics Letters, 2021. **46**(19): p. 5063–5066.
79. Cheng, R., et al., *Noninvasive assessment of early dental lesion using a dual-contrast photoacoustic tomography.* Scientific reports, 2016. **6**(1): p. 21798.
80. Feng, T., et al., *Feasibility study for bone health assessment based on photoacoustic imaging method.* Chinese Optics Letters, 2020. **18**(12): p. 121704.
81. Chen, P., et al., *Improved photoacoustic imaging of numerical bone model based on attention block U-net deep learning network.* Applied Sciences, 2020. **10**(22): p. 8089.
82. Estrada, H.c., et al., *Virtual craniotomy for high-resolution optoacoustic brain microscopy.* Scientific reports, 2018. **8**(1): p. 1459.
83. Estrada, H., et al., *Real-time optoacoustic brain microscopy with hybrid optical and acoustic resolution.* Laser Physics Letters, 2014. **11**(4): p. 045601.
84. Gottschalk, S., et al., *Noninvasive real-time visualization of multiple cerebral hemodynamic parameters in whole mouse brains using five-dimensional optoacoustic tomography.* Journal of Cerebral Blood Flow & Metabolism, 2015. **35**(4): p. 531–535.
85. Yao, J. and L.V. Wang, *Photoacoustic brain imaging: from microscopic to macroscopic scales.* Neurophotonics, 2014. **1**(1): p. 011003–011003.
86. Li, W., et al., *In vivo photoacoustic imaging of brain injury and rehabilitation by high-efficient near-infrared dye labeled mesenchymal stem cells with enhanced brain barrier permeability.* Advanced Science, 2018. **5**(2): p. 1700277.
87. Nie, L., et al., *Photoacoustic tomography through a whole adult human skull with a photon recycler.* Journal of biomedical optics, 2012. **17**(11): p. 110506–110506.
88. Subochev, P., et al., *Toward whole-brain in vivo optoacoustic angiography of rodents: modeling and experimental observations.* Biomedical optics express, 2020. **11**(3): p. 1477–1488.
89. Gottschalk, S., et al., *Rapid volumetric optoacoustic imaging of neural dynamics across the mouse brain.* Nature biomedical engineering, 2019. **3**(5): p. 392–401.
90. Tang, J., et al., *Wearable 3-D photoacoustic tomography for functional brain imaging in behaving rats.* Scientific reports, 2016. **6**(1): p. 25470.
91. Wang, X., et al., *Noninvasive photoacoustic angiography of animal brains in vivo with near-infrared light and an optical contrast agent.* Optics letters, 2004. **29**(7): p. 730–732.
92. Wang, X., et al., *Noninvasive imaging of hemoglobin concentration and oxygenation in the rat brain using high-resolution photoacoustic tomography.* Journal of biomedical optics, 2006. **11**(2): p. 024015-024015-9.
93. Wang, X., et al., *Noninvasive laser-induced photoacoustic tomography for structural and functional in vivo imaging of the brain.* Nature biotechnology, 2003. **21**(7): p. 803–806.
94. Wang, X., et al., *Three-dimensional laser-induced photoacoustic tomography of mouse brain with the skin and skull intact.* Optics letters, 2003. **28**(19): p. 1739–1741.
95. Yang, X., et al., *Skull optical clearing solution for enhancing ultrasonic and photoacoustic imaging.* IEEE transactions on medical imaging, 2016. **35**(8): p. 1903–1906.
96. Zhang, P., et al., *High-resolution deep functional imaging of the whole mouse brain by photoacoustic computed tomography in vivo.* Journal of biophotonics, 2018. **11**(1): p. e201700024.
97. Chen, W., C. Tao, and X. Liu, *Artifact-free imaging through a bone-like layer by using an ultrasonic-guided photoacoustic microscopy.* Optics Letters, 2019. **44**(5): p. 1273–1276.
98. Hosseini, F., et al. *Numerical Investigation of Multiple Scattering and Mode-Converted Shear Waves Caused by Temporal Bone in Transcranial Photoacoustic Imaging.* in *2022 IEEE International Ultrasonics Symposium (IUS)*. 2022. Venice, Italy: IEEE.
99. Ni, R., et al., *Coregistered transcranial optoacoustic and magnetic resonance angiography of the human brain.* Optics letters, 2023. **48**(3): p. 648–651.

100. Deán-Ben, X.L. and D. Razansky. *Transcranial imaging with the optoacoustic memory effect.* in *Photons Plus Ultrasound: Imaging and Sensing 2022*. 2022. San Francisco, CA: SPIE.
101. Na, S., et al., *Transcranial photoacoustic computed tomography based on a layered back-projection method.* Photoacoustics, 2020. **20**: p. 100213.
102. Poudel, J., et al., *Iterative image reconstruction in transcranial photoacoustic tomography based on the elastic wave equation.* Physics in Medicine & Biology, 2020. **65**(5): p. 055009.
103. Poudel, J. and M.A. Anastasio, *Joint reconstruction of initial pressure distribution and spatial distribution of acoustic properties of elastic media with application to transcranial photoacoustic tomography.* Inverse Problems, 2020. **36**(12): p. 124007.
104. Gao, Y., et al., *Deep learning-based photoacoustic imaging of vascular network through thick porous media.* IEEE Transactions on Medical Imaging, 2022. **41**(8): p. 2191–2204.
105. Wood, C., et al., *Photoacoustic-based oxygen saturation assessment of murine femoral bone marrow in a preclinical model of leukemia.* Photoacoustics, 2019. **14**: p. 31–36.
106. Feng, T., et al., *Functional photoacoustic and ultrasonic assessment of osteoporosis: a clinical feasibility study.* BME frontiers, 2020. **2020**: p. 1081540.

Chapter 18
Miniaturized Brain Imaging Apparatus Employing Light, Sound, and Magnetic Fields

Shuai Na, Jinyan Zhang, and Bangxu Fan

Abstract Brain imaging technologies have expanded our comprehension of cerebral function and dysfunction, providing insights into brain structure, electrophysiology, blood flow, and metabolic processes. Recent strides in miniaturizing brain imaging apparatus have sparked a paradigm shift in neuroscience research. These innovations offer real-time brain visualization that is cost-effective and portable. This chapter will review miniaturized brain imaging apparatus across various modalities and discuss the limitations and potential avenues for future advances.

Keywords Miniaturized brain imaging apparatus · Multiphoton fluorescence microscopy · Near-infrared spectroscopy · Functional ultrasound imaging · photoacoustic imaging · Magnetic resonance imaging · Optically pumped magnetometers

18.1 Introduction

The substantial size and limited mobility of many conventional brain imaging systems impede their routine use in crowded hospital wards and surgical suites. In scenarios necessitating functional studies of subjects or animals in motion, there is an immediate need for devices that offer improved portability, enhanced spatiotemporal resolution, and a broad field of view (FOV). This need has catalyzed the advancement of miniaturized brain imaging technologies, which promise to broaden the horizon for biomedical applications.

Functional brain imaging can be categorized into invasive and noninvasive techniques [1]. Invasively implanted electrode arrays offer cellular-level resolution in

S. Na (✉)
National Biomedical Imaging Center, College of Future Technology & Academy for Advanced Interdisciplinary Studies, Peking University, Beijing, China
e-mail: shuai@pku.edu.cn

J. Zhang · B. Fan
College of Future Technology, Peking University, Beijing, China

W. Xia (ed.), *Biomedical Photoacoustics*,
https://doi.org/10.1007/978-3-031-61411-8_18

studying neuronal electrophysiology [2]. In contrast, noninvasive techniques, such as electroencephalography (EEG), monitor the collective postsynaptic potentials of neuronal groups with lower spatial resolution. Advances in portable, user-friendly EEG devices now include wireless capabilities, enabling continuous data acquisition and remote monitoring. This technological leap enhances the feasibility of long-term and dynamic studies [3]. Furthermore, developments in signal processing have facilitated neural signal decoding, paving the way for brain-computer interface (BCI) applications [4, 5]. Given its established nature, EEG technology will not be the focus of subsequent sections.

Fluorescence imaging, known for its high sensitivity and specificity, is a principal method for visualizing cellular or molecular dynamics, often aided by immunofluorescence staining [6]. Single-photon microscopy, used alongside neuronal probes, can effectively track neuronal activity at cellular resolution to a depth of approximately 100 μm [7]. In contrast, multiphoton fluorescence microscopy penetrates deeper into tissue—several hundreds of micrometers—but at the expense of a reduced FOV and temporal resolution [8]. Near-infrared spectroscopy (NIRS) provides molecular specifics to a depth of about 10 mm but has lower resolution due to scattering [9, 10].

Functional ultrasound (fUS) imaging has emerged as a promising alternative, thanks to its ability to capture the entire FOV at a framerate surpassing one kilohertz, ensuring a high signal-to-noise ratio (SNR) for blood flow detection [11–13]. Photoacoustic (PA) imaging melds the best of light and sound, offering the ability to image beyond the optical mean free path (~1 mm) and thus sitting at the nexus of resolution and molecular specificity [14]. Functional magnetic resonance imaging (fMRI) remains a staple in both clinical and preclinical research [15]. However, its high magnetic field and bulky setup limit its flexibility. In comparison, magnetoencephalography (MEG) noninvasively measures the magnetic fields produced by neural currents [16], and advancements in optically pumped magnetometers (OPMs) have brought significant improvements in miniaturization over traditional superconducting quantum interference devices (SQUIDs) [17, 18]. OPMs, which do not require liquid helium, offer greater sensitivity and a wider dynamic range, facilitating cognitive research including perception, attention, memory, and decision-making [19, 20]. In the subsequent sections, we will explore the cutting-edge developments in the aforementioned imaging technologies, evaluate their limitations, and discuss potential roadmaps for further advancement.

18.2 Miniaturized Brain Imaging Apparatus

18.2.1 Miniaturized Multiphoton Fluorescence Microscopy

Conventional fluorescence microscopy excites fluorophores with a single photon, causing the emission of a photon with lower energy [7]. Multiphoton fluores-

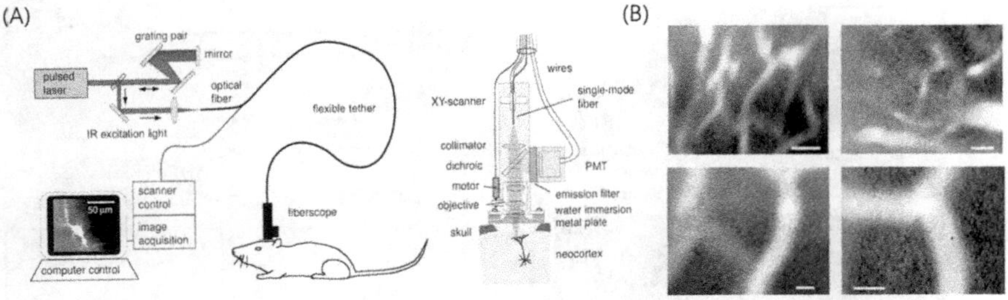

Fig. 18.1 (**a**) First miniaturized two-photon fluorescence microscope by Max Planck Institute for Medical Research and Bell Laboratories; (**b**) capillary images and their close-up views

cence microscopy utilizes a fluorophore's simultaneous absorption of two or more photons, typically infrared photons. This nonlinear excitation enables deeper penetration, reduced scattering and photodamage, and a decrease in out-of-focus background. The first miniaturized two-photon microscope was co-developed by the Max Planck Institute for Medical Research and Bell Laboratories in 2001, measuring 7.5 cm in length and 25 g in weight, as depicted in Fig. 18.1a [21]. The device was used to image the mouse cortex using a scanning optical fiber tip, and the results are displayed in Fig. 18.1b.

In a later study, Swinburne University of Technology employed a double-clad photonic crystal fiber to combine photon excitation and collection, as shown in Fig. 18.2a [22]. Double-clad fibers proved advantageous not only for multiphoton excitation but also for second harmonic generation (SHG) imaging. In SHG imaging, collagen-rich tissues produce a photon with half the wavelength of the incident light when two photons of an identical frequency interact. Johns Hopkins University showcased SHG endomicroscopy with an axial resolution of 0.76 μm and a lateral resolution of 4.36 μm, as presented in Fig. 18.2b [23]. Further advancements came from Cornell University, which introduced an assembly combining a raster scanner with a multielement gradient-index lens for enhanced two-photon and SHG imaging, detailed in Fig. 18.2c [24]. This configuration provided a frame rate of 4 Hz, a FOV of 110×110 um^2, an axial resolution of 0.8 um, and a lateral resolution of 10 um. Moreover, researchers have also adopted microelectromechanical systems (MEMSs) for laser scanning, providing real-time imaging and device miniaturization, as displayed in Fig. 18.2d [25, 26].

In 2017, researchers from Peking University reported a wearable two-photon microscope with improved imaging speed, FOV, and resolution. It remarkably weighs 2.15 g and successfully revealed neuronal activity in free-moving mice in real time [27]. They implemented a hollow-core optical fiber to reduce dispersion, designed a compact compound objective lens to increase excitation efficiency, and used a MEMS mirror to expedite the scanning process, as illustrated in Fig. 18.3a. The device exhibits an axial resolution of 3.35 um, a lateral resolution of 0.64 um, a raster scanning speed of 40 Hz, and a single-line scanning speed of 10,000 Hz, as depicted in Fig. 18.3b. The device is mounted onto a lightweight metal frame (Fig.

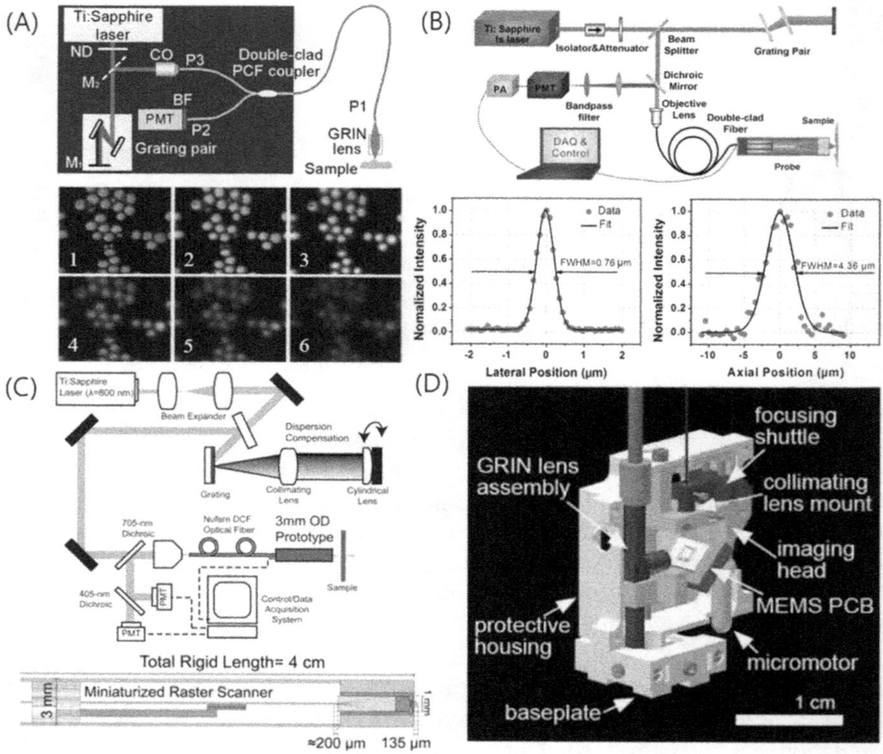

Fig. 18.2 Two-photon and SHG imaging apparatus employing double-clad photonic fibers. (**a**) A two-photon microscope utilizing double-clad photonic crystal fibers designed by Swinburne University of Technology; (**b**) an SHG endomicroscope developed by Johns Hopkins University; (**c**) a two-photon endoscope designed by Cornell University; (**d**) a schematic diagram of a two-photon imaging system incorporating MEMS for laser scanning

18.3c). An axial scanner was also incorporated to broaden the imaging volume to $420 \times 420 \times 180\ \mu m^3$, as shown in Fig. 18.3d [28].

Compared with two-photon microscopy, three-photon fluorescence microscopy, with its stronger nonlinear response, offers superior imaging depth, reduces photodamage and phototoxicity, and mitigates the effects of photon scattering [29]. Progress in this field has spurred elaborate research into imaging probes and advanced laser optics [30–33]. A study from Heidelberg University showcased an adaptive three-photon excitation technique synchronized with an electrocardiogram (ECG) to compensate for motion artifacts, as depicted in Fig. 18.4a, b [34]. Cornell University further advanced three-photon microscopy by employing a 1700 nm wavelength laser, allowing for deeper subcortical imaging. They enhanced the resolution through pulse compression to one-sixth the original pulse width—

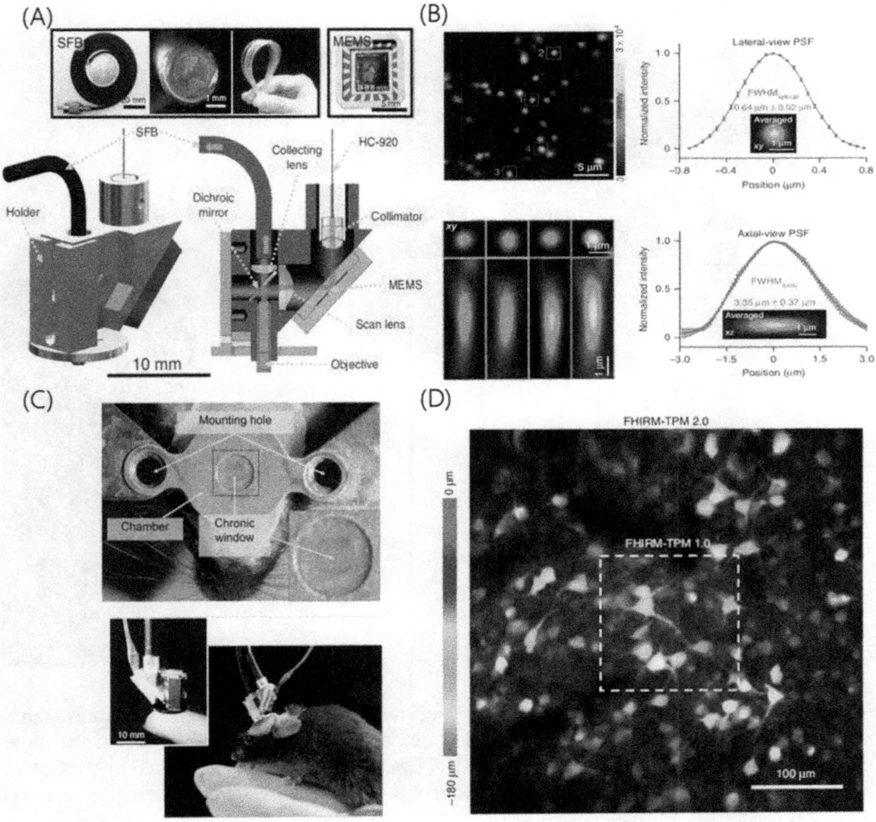

Fig. 18.3 Miniaturized two-photon microscope by Peking University. (**a**) System schematic diagram and optical path; (**b**) axial and lateral resolution; (**c**) installation; (**d**) fired neurons being imaged

yielding a horizontal resolution of 0.9 μm and an axial resolution of 4.4 μm, as presented in Fig. 18.4c, d [35].

Innovation continued in 2021 when the Max Planck Institute for Neurobiology of Behavior introduced a pioneering three-photon head-mounted microscope capable of scanning the entire mouse cerebral cortex with adjustable focal distance, hence facilitating volumetric imaging as shown in Fig. 18.4e [36]. However, this design traded off the resolution within a single plane for broader volumetric coverage and could not image three-dimensional volumes in a fast manner. Addressing this limitation, researchers from Peking University later engineered a system employing a simplified infinite miniature objective (SIMO) and a miniature Abbe condenser—optimizing fluorescence collection, featured in Fig. 18.4f [29]. This system was shown to image the neuronal somata from the posterior parietal cortex to the stratum pyramidale at depths down to 970 μm, as evident in Fig. 18.4g.

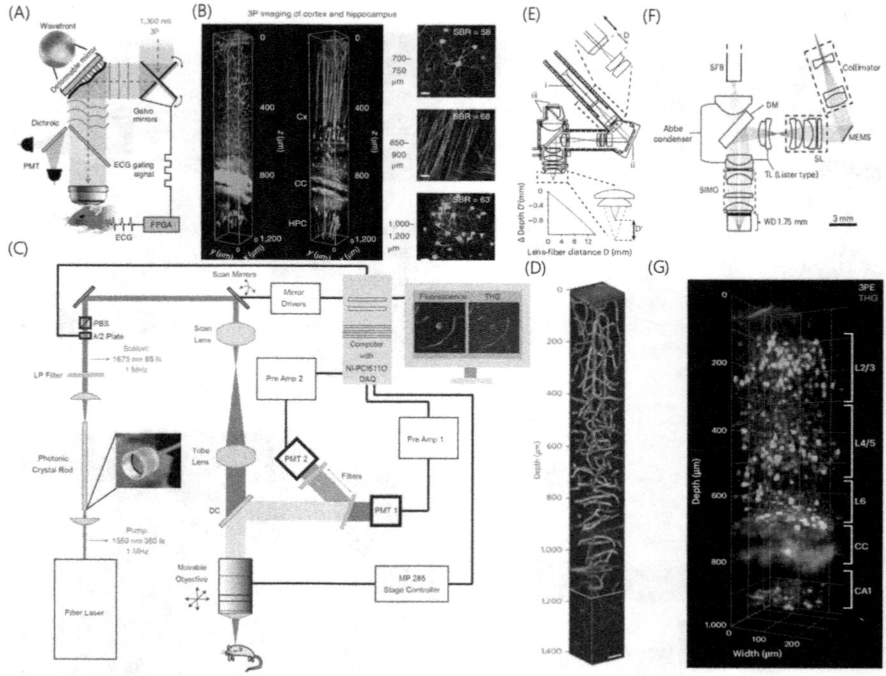

Fig. 18.4 Miniaturized three-photon fluorescence microscopy. (**a**) ECG-gated adaptive three-photon microscope; (**b**) images of the visual cortex and hippocampus in mouse; (**c**) the three-photon fluorescence microscope from Cornell University; (**d**) images demonstrated by the Cornell device; (**e**) microscope structure; (**f**) optical layout of the system of Peking University; (**g**) an image of GCaMP6s-labeled neurons

Balancing the trade-off between resolution and FOV remains a substantial challenge in multiphoton fluorescence microscopy. Wang et al. developed an objective-lens-free fiber-optic two-photon endomicroscope that expands the FOV, as shown in Fig. 18.5a [37]. The system provides a FOV of 750 × 750 um^2 with a lateral resolution of 3.03 μm. Additionally, the system's ability to fine-tune the excitation light enables precise activation of fluorophores, resulting in stronger fluorescence signals and improved resolution. Hoang et al. harnessed a broadband supercontinuum spectrum that spans various absorption peaks and employed a particle swarm algorithm to selectively refine the fluorophore's excitation wavelength, as illustrated in Fig. 18.5b [38].

In summary, the multiphoton fluorescence microscope represents a sensitive and noninvasive imaging technology offering high-resolution insights. Its ability to visualize cellular and subcellular structures, along with neural activities, positions it as a highly prospective instrument for cell biology and neuroscience.

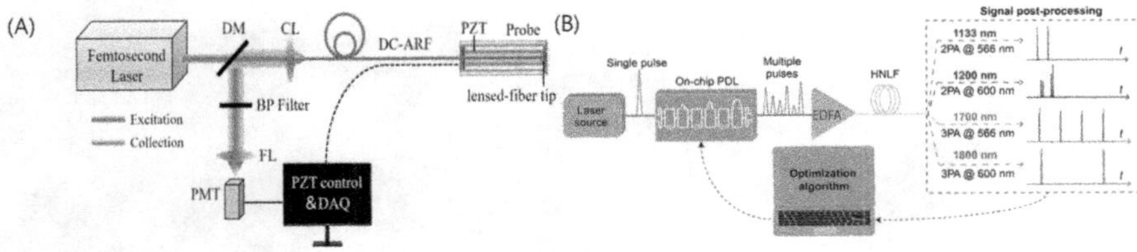

Fig. 18.5 Two-photon fluorescence microscope with improved performance. (**a**) Objective-lens-free fiber-optic two-photon endomicroscopy; (**b**) supercontinuum spectroscopy

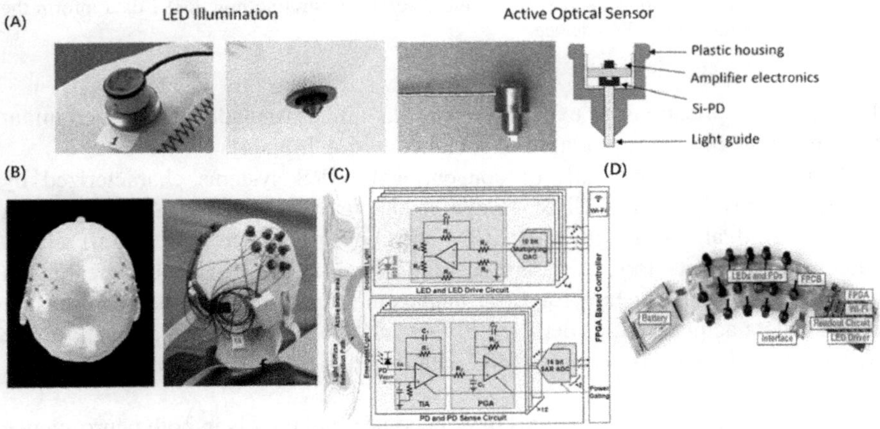

Fig. 18.6 Miniaturized NIR imaging systems. (**a**) LED sources and optical sensors; (**b**) NIRS light emitters and sensors on the head; (**c**) block diagram of the adaptive gating system; (**d**) the electronics of the wearable NIRS device

18.2.2 Miniaturized NIRS Brain Imaging Apparatus

Sophie K et al. designed a wearable NIRS device that utilizes near-infrared light's depth penetration and differential absorption by oxy- and deoxyhemoglobin to measure tissue oxygen saturation. This device features dual-wavelength LEDs at 760 and 850 nm, as depicted in Fig. 18.6a [39]. Its design prioritizes wearer comfort by replacing traditional fiber optics with flexible braided shielded cables, enhancing the passive coupling between the transmitter and receiver, as demonstrated on a subject in Fig. 18.6b. Additionally, Chen introduced a spatially adaptive sampling (SAS) technique, detailed in Fig. 18.6c, d, which dynamically modulates acquisition parameters to track real-time cerebral activity [40].

Gevins et al. validated the efficacy of wearable NIRS systems and demonstrated their utility beyond conventional laboratory environments [41]. Muehlemann et al. harnessed a bespoke wireless micro-NIRS device to assess the cerebral impact from sudden decompression on pilots in-flight [42]. In addition, Holper et al. implemented wearable NIRS technology in tracking the rehabilitative progress [41,

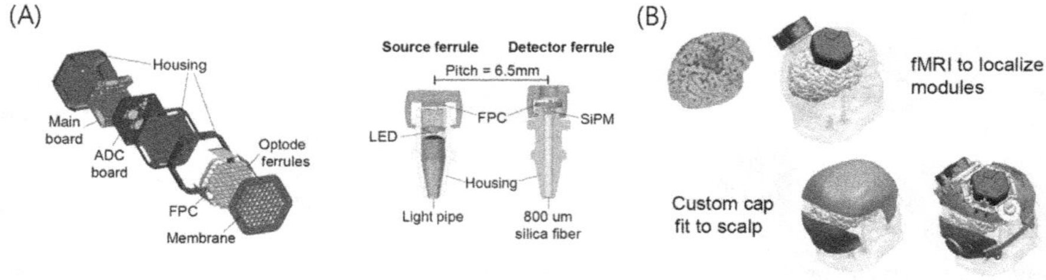

Fig. 18.7 Advanced NIRS imaging systems. (**a**) The assembly showcasing the integrated control and data acquisition components; (**b**) schematic diagram illustrating how fMRI data inform the precise positioning of the NIRS device

42]. Further applications, as explored in various studies, include NIRS-based infant brain imaging and the exploration of social cognition [43, 44].

Addressing the constraints of conventional NIRS systems characterized by low spatial resolution and SNR, Anaya et al. introduced an NIRS configuration utilizing 41 dual-wavelength NIR LEDs and 39 closely spaced (6.5 mm) silicon photomultiplier detectors, as described in their work [45]. This system features a main control board per unit for LED management and an ADC board to ensure accurate data acquisition, as depicted in Fig. 18.7a. Additionally, placement of the NIRS helmet is optimized using fMRI data to guide positioning, as shown in Fig. 18.7b.

Overall, wearable NIRS systems have become crucial tools in both neuroscience research and clinical settings [46]. Despite their growing significance, contemporary NIRS methodologies continue to grapple with the challenge of signal contamination from extracranial sources, largely attributed to the restricted optical power and the significant scattering characteristics of the skin [47].

18.2.3 Miniaturized Ultrasound Brain Imaging Apparatus

Ultrasound imaging utilizes high-frequency mechanical waves emitted and subsequently detected by an ultrasonic transducer to image the brain structures and functions [48, 49]. It has been used for evaluating intracranial hemorrhages [50], detecting brain abnormalities in infants [51], and guiding neurosurgical procedures [52, 53]. Alan et al. engineered a compact ultrasound system with a miniaturized transducer array (15 MHz, 64 elements, weighing less than 10 g) for rat brain imaging, as depicted in Fig. 18.8a [54]. This system achieves a FOV of $9.6 \times 9\ \text{mm}^2$, encompassing two-thirds of the coronal plane of the brain, with an imaging depth of 500 μm (Fig. 18.8b, c). Figure 18.8d captures the cerebral blood volume (CBV) fluctuations in reaction to whisker and visual stimuli. The employed algorithms for peak detection and categorization, in conjunction with decoding techniques, differentiated behavioral responses initiated by varied stimuli. Norman et al.

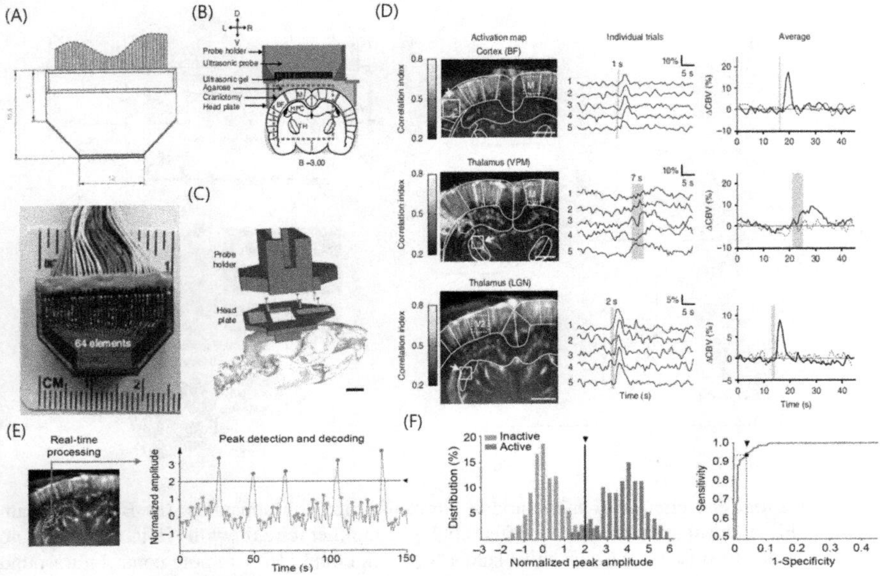

Fig. 18.8 Miniaturized ultrasound imaging apparatus. (**a**) Ultrasonic probe; (**b**) FOV; (**c**) probe installation; (**d**) CBV signal changes in response to whisker and visual stimuli. Top to bottom: whisker, beard, and visual stimuli, respectively; (**e**) decoding brain function in real time; (**f**) the ROC analysis

utilized a miniaturized linear array probe (15 MHz) to quantify CBV alterations and executed single-shot decoding of primate behaviors [55], achieving a spatial resolution of 100 μm, with slice thicknesses of around 400 μm and a maximum penetration depth of 16 mm.

Miniaturizing ultrasound imaging apparatus also involves the compact design of data acquisition and streaming electronics [56–58]. Peyton et al. proposed a quadrature synthetic aperture that utilizes a single channel to process signals sequentially, thereby minimizing system size, power consumption, and cost [59]. As shown in Fig. 18.9a, the detected signals are multiplexed and beamformed in the FPGA. The analog front-end (AFE) consists of a preamplifier with variable gains, programmable gains, and a low-pass filter, allowing for exponential gain compensation over time to counteract the attenuation of deep tissue and enhance the dynamic range. In Fig. 18.9b, the differential preamplifier is capable of adjusting the gain by changing the gate voltage while stabilizing the voltage output. Time gain compensation (TGC), a conventional compensation approach, uses discrete resistors and switches to directly modify the gain of the feedback amplifier [59–61]. The drawback lies in the inability to adjust the gain on the go, necessitating compensation for quantization errors. The continuous-time gain compensation circuit primarily adjusts the current ratio but with a relatively narrow linear range [62–65]. The common solution is to use a capacitor ladder as the

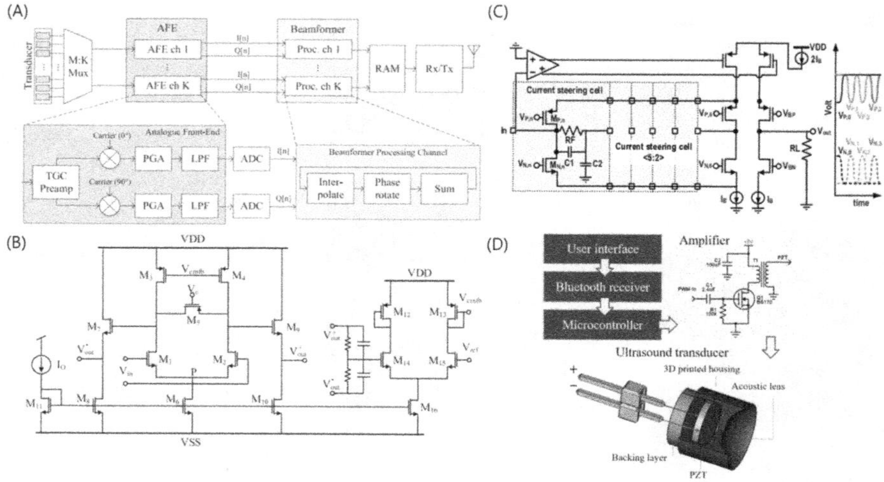

Fig. 18.9 Front-end circuits of miniaturized ultrasound imaging apparatus. (**a**) Block diagram of the AFE and subsystems; (**b**) a differential preamplifier circuit with variable gains and common-mode feedback; (**c**) a capacitive ladder feedback circuit; (**d**) a remote-control ultrasound stimulation system

feedback network and a current steering circuit in the anti-impedance amplifier to achieve continuous gain control, as shown in Fig. 18.9c. Figure 18.9d depicts a wireless, wearable ultrasound stimulation device operational in freely moving animals [56]. This prototype shares its wireless paradigm with ultrasound imaging systems, though wireless data transmission rates are a critical bottleneck for real-time communication.

Miniaturized ultrasound devices currently grapple with obtaining high-resolution 3D data in real time. Advances are anticipated through the development of miniaturized matrix arrays paired with parallel data acquisition. Skull remains an obstacle for ultrasound brain imaging, prompting the development of physics-based reconstruction algorithms to correct for the induced attenuation and aberration [66, 67]. The emergence of two-dimensional matrix arrays and sparse matrix technologies has supplanted mechanical scanning in traditional linear arrays, optimizing 3D data capture [68, 69]. Allevato et al. introduced an 8×8 phased-array ultrasound system with integrated waveguides and achieved half-wavelength inter-element spacing, enabling precise real-time 3D localization of targets, as illustrated in Fig. 18.10a [70]. To balance penetration depth with image resolution, multifrequency techniques are being employed. For instance, Gessner' s use of 2.5 and 30 MHz confocal transducers resulted in a 12.3 dB SNR boost without sacrificing the resolution of high-frequency imaging, as shown in Fig. 18.10b [71, 72]. Since the SNR is proportional to the piezoelectric element size, plane wave excitation may be utilized to counteract SNR reductions inherent in smaller transducer designs [73].

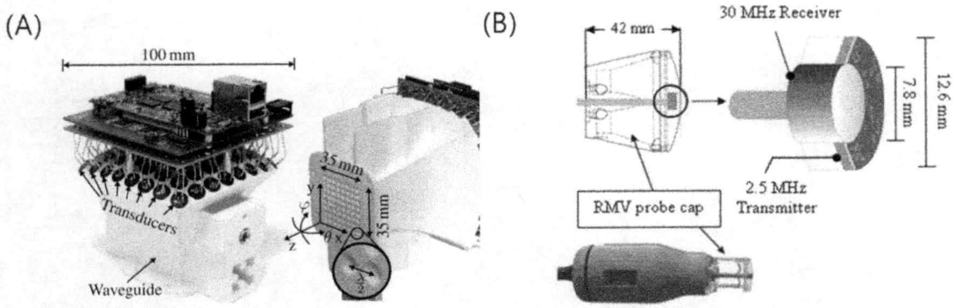

Fig. 18.10 Advancement in ultrasonic transducers. (**a**) A uniformly designed 8 × 8 phased-array showcasing the compact electronics; (**b**) schematic diagram of a dual-frequency ultrasonic transducer

To enhance the image quality, a standard practice is to conduct multiple scans and do averaging to improve the SNR. Advances such as parallel data acquisition and the use of GPUs have effectively overcome the decrease in frame rate typically associated with such scanning techniques. In addition, the development of data processing algorithms has streamlined the computations needed for image reconstruction. For instance, Rabut et al. introduced a Hadamard coding strategy that boosted the SNR by approximately 9 dB beyond what traditional plane wave imaging affords, as illustrated in Fig. 18.11a [74]. Jeon et al. harnessed deep learning algorithms to correct for the speed of sound distortions, reduce streaking artifacts, and enhance temporal resolution, with the results showcased in Fig. 18.11b [75]. Furthermore, Khan et al. developed an adaptive beamforming technique, embedded with a deep neural network, to enhance both the resolution and the SNR [76]. Deep learning techniques have been applied to extract full-bandwidth signals from bandwidth-constrained raw data and to eliminate reconstruction artifacts [77]. Full-wave inversion (FWI) is at the cutting edge of physics-based reconstruction techniques, modeling sound propagation through tissues and inversely reconstructing various tissue parameters such as sound speed, density, and attenuation, as illustrated in Fig. 18.11c [66]. However, the widespread integration of deep learning into compact imaging devices hits a snag, primarily due to a dearth of comprehensive raw training datasets and the restricted computational capacity that comes with the reduced scale of such devices. A research team at Zhejiang University has engineered a metamaterial that can reroute ultrasonic waves. This innovation minimizes the reflection of sound waves off the skull, facilitating high transmission. Norman et al. employed a 2-mm-thick layer of specialized acoustic material as an "acoustic window" for transcranial imaging. This application underscored the potential of ultrasound imaging capabilities, as highlighted in Fig. 18.11d [78].

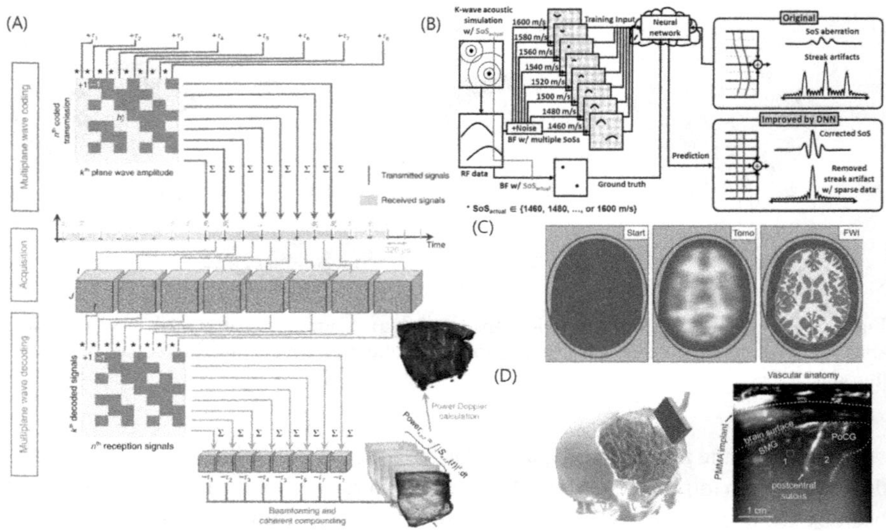

Fig. 18.11 Enhanced ultrasound imaging and data processing techniques. (**a**) Hadamard coding technique; (**b**) deep learning for acoustic aberration correction; (**c**) FWI flowchart; (**d**) use of acoustic-friendly coupling materials for transcranial imaging

18.2.4 Miniaturized PA Imaging Apparatus

Photoacoustic imaging is a noninvasive imaging modality, which generates ultrasonic waves by irradiating the tissue with pulsed laser and reconstructs the optical absorption distributions. The major challenge of miniaturizing PA imaging apparatus is to widely and densely distribute the transducer elements in a limited space. In 2016, Tang et al. designed a 2.5D wearable photoacoustic brain imaging device for animals, as shown in Fig. 18.12a [79]. The device consists of three 9.6 MHz 64-element elevationally focused transducer arrays to concurrently capture images at three discrete transverse planes, as shown in Fig. 18.12b, c. Figure 18.12d shows an annular ultrasonic transducer array design with an outer diameter of 2.5 mm and an inner diameter of 0.5 mm, which also holds potential for brain imaging [80]. Conventional mechanical scanning is increasingly being replaced by the use of compact ultrasonic arrays [81–83]. The use of a piezoelectric micromachined ultrasonic transducer (PMUT) marks a step forward; this configuration consists of six segments, with each one containing 17 individual cells, as depicted in Fig. 18.12e. Alongside this, there is a shift toward 3D hemispherical arrays, which provide a 2π solid angle coverage and significantly enhanced sensitivity and spatial resolution [84]. In Fig. 18.12f, Li et al. introduced a high-throughput PA imaging apparatus employing a single-element ultrasonic transducer and an ergodic relay [85]. The ergodic relay encodes the incident PA waves on its boundary surface, with the encoded signals collected by a single-element transducer. Reconstructing

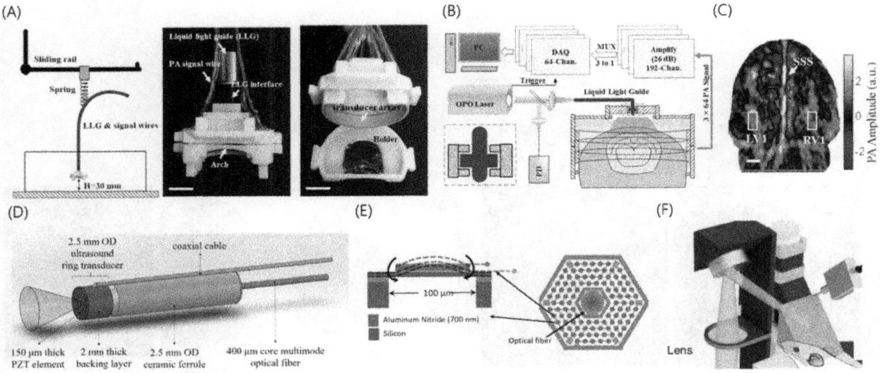

Fig. 18.12 Miniaturized photoacoustic imaging apparatus. (**a**) A 2.5D wearable PA imaging device; (**b**) system architecture; (**c**) an image of the mouse brain cortex; (**d**) miniaturized annular ultrasonic transducer; (**e**) a ring-shaped PMUT array; (**f**) single-pulse 2D PA imaging system based on an ergodic relay

a 2D image involves deconvolving the energy contributions at each pixel on the relay's boundary surface, using pre-calibrated temporal responses. This approach to single-channel detection paves the way for compact, wearable PA imaging systems, significantly enhancing the potential for real-time monitoring of rapid functional changes.

Human brain photoacoustic imaging signifies vast potential for clinical applications. In 2022, Na et al. introduced a parallelized ultrasonic transducer array, capable of generating tomographic images of the brain with a broad FOV and high spatial-temporal resolution [86]. The system comprises four main components: a dual-wavelength laser system emitting at 694 and 1064 nm for PA excitation; a 1024-element transducer array distributed evenly around a hemisphere; preamplifiers and data acquisition (DAQ) systems for signal amplification and acquisition; and a scanning mechanism offering azimuthal sampling, as illustrated in Fig. 18.13a, b. The functional photoacoustic computed tomography (fPACT) was validated against blood-oxygen-level-dependent (BOLD) fMRI. Figures 18.13c, d exhibit images of the cerebral blood vessels, along with measurements of their diameters, demonstrating a high correlation between the two modalities. Figure 18.13e shows the functional responses of fPACT and fMRI, which demonstrates a high spatial correlation. Furthermore, PACT can be combined with ultrasound computed tomography (USCT) to map out sound speed and attenuation distributions, which are instrumental in correcting acoustic distortion [87].

Advancing image reconstruction algorithms is critical for spatiotemporal resolution improvement. Wang et al. leveraged OpenCL to develop an efficient data compression algorithm [88]. When integrated with GPU acceleration, this algorithm compresses 33 MB of ultrasound data to 6.6 MB within 0.29 s, achieving an 80% compression rate that meets the demands of real-time imaging. Jeon et al. introduced an advanced imaging algorithm termed delay multiply and sum with a modified

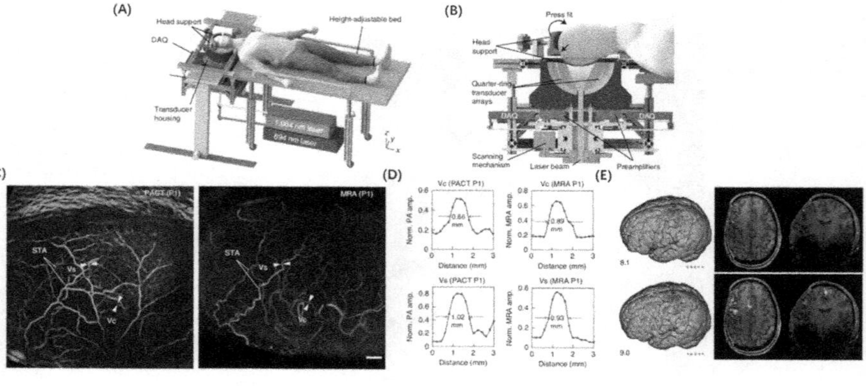

Fig. 18.13 Advanced photoacoustic imaging techniques applied to the human brain. (**a**) Perspective view of the photoacoustic imaging system; (**b**) sectional view of the imager; (**c**) vasculature imaging using fPACT (left) and MRA (right); (**d**) the measured diameters of two blood vessels; (**e**) brain activation maps obtained by fPACT (top) and MRI (bottom)

coherence factor (DMAS-CF) [67]. It demonstrates a noteworthy improvement in SNR (exceeding 40 dB compared to that obtained with DAS) and superior full width at half maximum (FWHM), as illustrated in Fig. 18.14a, b. Furthermore, techniques, such as convolutional neural networks (CNNs) and generative adversarial networks (GANs), have proven their efficacy in reconstructing high-fidelity images from partial or noisy datasets [89, 90], effectively compensating for the diminished SNR that may result from limited-view detection.

18.2.5 Miniaturized Magnetic-Based Brain Imaging Apparatus

MRI harnesses a strong gradient magnetic field and radiofrequency (RF) pulses to manipulate the nuclear spins in biological tissues. As these atomic nuclei return to their equilibrium state, they emit RF signals that correlate with different relaxation times. BOLD fMRI tracks changes in brain activity by detecting signal fluctuations resulting from alterations in deoxyhemoglobin concentration [91, 92]. Despite the remarkable capabilities of MRI, the development of portable MRI systems has been challenging due to the need for strong magnetic fields and components that are magnetically compatible. However, advances have been made in creating more compact and portable units. For example, Peng et al. successfully developed a portable MRI machine with dimensions of 19 × 16 cm, capable of monitoring changes in the magnetic susceptibility of human blood cells—a potential indicator for disease diagnosis, as shown in Fig. 18.15a [93]. Additionally, Kim et al. engineered an MRI-compatible ultrasound system, incorporating a miniaturized MRI coil to observe functional responses to acoustic stimulation in a mouse brain,

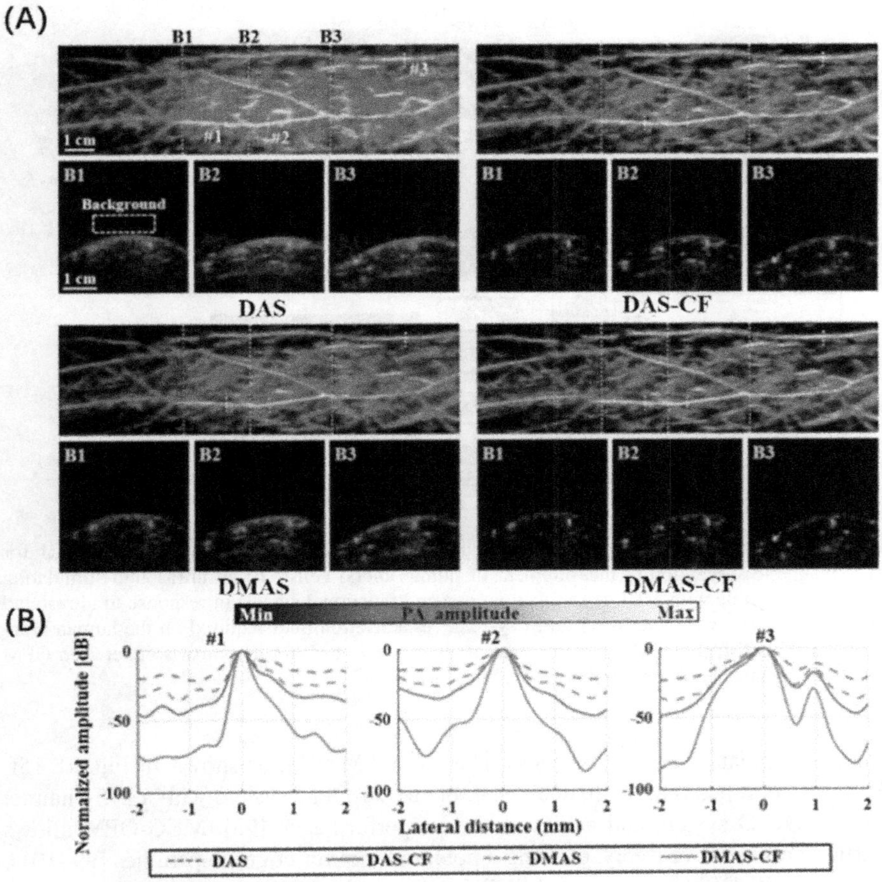

Fig. 18.14 Results of different PA reconstruction algorithms. (**a**) Photoacoustic images of a human forearm, with inserted panels B1, B2, and B3 highlighting views along the blue dashed line; (**b**) representation of the FWHM measurements

as shown in Fig. 18.15b, c [94]. These innovations signify critical steps toward more accessible and versatile MRI technologies.

For MEG, different from the conventional SQUID technology, OPM utilizes the interaction of light and atoms to detect ultra-low magnetic fields. Its small cross section (1–2 cm) and high sensitivity (10–20$fT/\sqrt{Hz}$) make it ideal for miniaturization [95]. Roberts et al. used a feedback loop to control a set of double-plane zeroing coils to compensate for the ambient static magnetic field, as shown in Fig. 18.15d [96]. This setup enables participants to move freely during data collection while maintaining the minimum magnetic background. Figure 18.15e displays the brain activation map acquired using the MEG-OPM. The high correlation between EEG and MEG-OPM signals, acquired across different brain

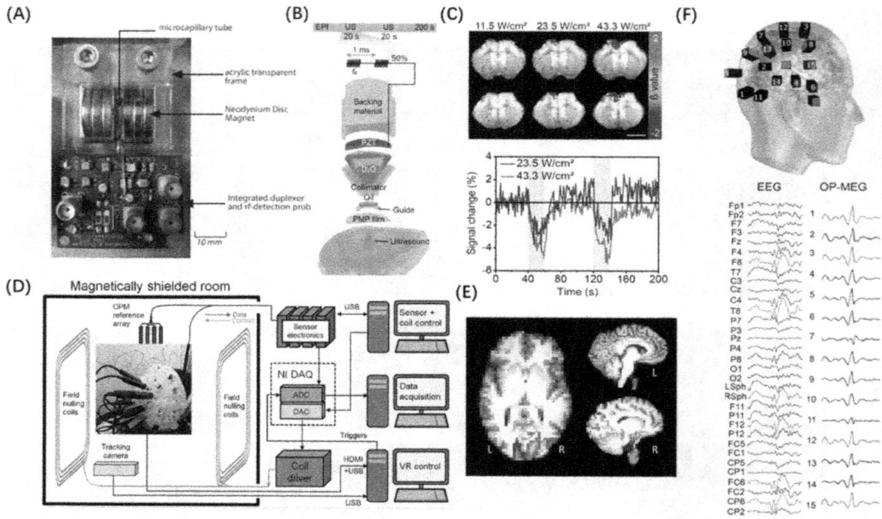

Fig. 18.15 Miniaturized magnetic-based brain imaging apparatus. (**a**) Miniaturized MRI for bulk magnetic susceptibility measurement of human blood cells; (**b**) an ultrasound stimulation system compatible with MRI; (**c**) activation map and functional signals in response to ultrasound stimulation; (**d**) a wearable MEG-OPM system; (**e**) activation map acquired on the human brain using MEG-OPM; (**f**) OPM sensors placed on the human head (top); comparison between OPM and EEG signals in different brain regions (bottom)

regions, validated the effectiveness of MEG-OPM [97], as shown in Fig. 18.15f. Hill et al. compared a 49-channel wearable MEG-OPM system with a 275-channel MEG-SQUID system and achieved similar performance [98]. MEG-OPM allows participants to move freely, offering a potential tool for cognition studies [99–101]. However, for OPM, an improved SNR is still required for deep brain imaging [19]. Currently, addressing the problems of OPM heating and magnetic field deviations also remains a challenge.

18.3 Conclusion

This chapter reviews the miniaturized medical imaging devices tailored for brain imaging—ranging from multiphoton fluorescence microscopy and NIRS to US imaging, PA imaging, fMRI, and MEG—laying bare their respective strengths, limitations, and technical hurdles. The quest for miniaturization in medical imaging is confronted by two challenges: firstly, improving image resolution, FOV, and SNR without expanding the physical size of devices, and secondly, the exigency for real-time data processing capabilities and high imaging throughput. It is envisaged that forthcoming breakthroughs will make significant impacts across neuroscience, BCI, and other related fields.

The shift toward multimodal imaging technology represents a significant trend in medical imaging, for instance, integrating fMRI with EEG [102, 103], EEG with functional transcranial Doppler [104], functional near-infrared spectroscopy (fNIRS) with EEG [105–108], and fNIRS with MEG [109], along with tri-modal MEG-EEG-fNIRS configurations [110]—to address the limitations inherent in single-modality imaging. These multimodal approaches are finding extensive applications in neuroscientific endeavors, including neural decoding [109], sleep research [102], and epilepsy studies [103].

Despite these advancements, current methods of modality fusion often rest on simplistic linear combinations, which fail to exploit the full potential of integrated data. Critical breakthroughs are needed in system design and integration to successfully navigate the technical complexities inherent in this field. To fully harness the benefits of multimodal technologies within compact imaging devices, there is a need to establish robust information linkages among different imaging modalities. Moreover, compensatory strategies must be devised to address the reductions in spatiotemporal resolution, sensitivity, and FOV that commonly accompany device miniaturization. These issues require in-depth research and innovative thinking to unlock the full potential of miniaturized medical brain imaging.

References

1. M. Cantone, G. Lanza, F. Ranieri, G. M. Opie, and C. Terranova, "Non-invasive brain stimulation in the study and modulation of metaplasticity in neurological disorders," vol. 12, ed: Frontiers Media SA, 2021, p. 721906.
2. D. Tsai, D. Sawyer, A. Bradd, R. Yuste, and K. L. Shepard, "A very large-scale microelectrode array for cellular-resolution electrophysiology," *Nature Communications,* vol. 8, no. 1, p. 1802, 2017.
3. E. Ahmed, I. Yaqoob, A. Gani, M. Imran, and M. Guizani, "Internet-of-things-based smart environments: state of the art, taxonomy, and open research challenges," *IEEE Wireless Communications,* vol. 23, no. 5, pp. 10–16, 2016.
4. B. He, H. Yuan, J. Meng, and S. Gao, "Brain–computer interfaces," *Neural engineering,* pp. 131–183, 2020.
5. B. He, B. Baxter, B. J. Edelman, C. C. Cline, and W. Y. Wenjing, "Noninvasive brain-computer interfaces based on sensorimotor rhythms," *Proceedings of the IEEE,* vol. 103, no. 6, pp. 907–925, 2015.
6. K. Wang, X. He, X. Yang, and H. Shi, "Functionalized silica nanoparticles: a platform for fluorescence imaging at the cell and small animal levels," *Accounts of chemical research,* vol. 46, no. 7, pp. 1367–1376, 2013.
7. F. Helmchen and W. Denk, "Deep tissue two-photon microscopy," *Nature methods,* vol. 2, no. 12, pp. 932–940, 2005.
8. M. J. Levene, D. A. Dombeck, K. A. Kasischke, R. P. Molloy, and W. W. Webb, "In vivo multiphoton microscopy of deep brain tissue," *Journal of neurophysiology,* vol. 91, no. 4, pp. 1908–1912, 2004.

9. O. V. Olesen, R. R. Paulsen, L. Hojgaard, B. Roed, and R. Larsen, "Motion tracking for medical imaging: a nonvisible structured light tracking approach," *IEEE transactions on medical imaging,* vol. 31, no. 1, pp. 79–87, 2011.
10. T. Muehlemann, D. Haensse, and M. Wolf, "Wireless miniaturized in-vivo near infrared imaging," *Optics express,* vol. 16, no. 14, pp. 10323–10330, 2008.
11. M. Tanter and M. Fink, "Ultrafast imaging in biomedical ultrasound," *IEEE transactions on ultrasonics, ferroelectrics, and frequency control,* vol. 61, no. 1, pp. 102–119, 2014.
12. J. Jensen, M. B. Stuart, and J. A. Jensen, "Optimized plane wave imaging for fast and high-quality ultrasound imaging," *IEEE transactions on ultrasonics, ferroelectrics, and frequency control,* vol. 63, no. 11, pp. 1922–1934, 2016.
13. C. Errico *et al.*, "Ultrafast ultrasound localization microscopy for deep super-resolution vascular imaging," *Nature,* vol. 527, no. 7579, pp. 499–502, 2015.
14. L. V. Wang, "Multiscale photoacoustic microscopy and computed tomography," *Nature photonics,* vol. 3, no. 9, pp. 503–509, 2009.
15. P. M. Matthews and P. Jezzard, "Functional magnetic resonance imaging," *Journal of Neurology, Neurosurgery & Psychiatry,* vol. 75, no. 1, pp. 6–12, 2004.
16. M. Hämäläinen and R. Hari, "Magnetoencephalographic (MEG) characterization of dynamic brain activation," *Brain mapping: the methods, Ed,* vol. 2, pp. 227–254, 2002.
17. D. Robbes, "Highly sensitive magnetometers—a review," *Sensors and Actuators A: Physical,* vol. 129, no. 1–2, pp. 86–93, 2006.
18. T. M. Tierney *et al.*, "Optically pumped magnetometers: From quantum origins to multi-channel magnetoencephalography," *NeuroImage,* vol. 199, pp. 598–608, 2019.
19. M. J. Brookes *et al.*, "Magnetoencephalography with optically pumped magnetometers (OPM-MEG): the next generation of functional neuroimaging," *Trends in Neurosciences,* 2022.
20. I. Savukov, Y. Kim, V. Shah, and M. Boshier, "High-sensitivity operation of single-beam optically pumped magnetometer in a kHz frequency range," *Measurement Science and Technology,* vol. 28, no. 3, p. 035104, 2017.
21. F. Helmchen, M. S. Fee, D. W. Tank, and W. Denk, "A miniature head-mounted two-photon microscope: high-resolution brain imaging in freely moving animals," *Neuron,* vol. 31, no. 6, pp. 903–912, 2001.
22. L. Fu and M. Gu, "Double-clad photonic crystal fiber coupler for compact nonlinear optical microscopy imaging," *Optics letters,* vol. 31, no. 10, pp. 1471–1473, 2006.
23. Y. Zhang *et al.*, "A compact fiber-optic SHG scanning endomicroscope and its application to visualize cervical remodeling during pregnancy," *Proceedings of the National Academy of Sciences,* vol. 109, no. 32, pp. 12878–12883, 2012.
24. D. R. Rivera *et al.*, "Compact and flexible raster scanning multiphoton endoscope capable of imaging unstained tissue," *Proceedings of the National Academy of Sciences,* vol. 108, no. 43, pp. 17598–17603, 2011.
25. W. Piyawattanametha *et al.*, "In vivo brain imaging using a portable 2.9 g two-photon microscope based on a microelectromechanical systems scanning mirror," *Optics letters,* vol. 34, no. 15, pp. 2309–2311, 2009.
26. W. Jung *et al.*, "Miniaturized probe based on a microelectromechanical system mirror for multiphoton microscopy," *Optics letters,* vol. 33, no. 12, pp. 1324–1326, 2008.
27. W. Zong *et al.*, "Fast high-resolution miniature two-photon microscopy for brain imaging in freely behaving mice," *Nature methods,* vol. 14, no. 7, pp. 713–719, 2017.
28. W. Zong *et al.*, "Miniature two-photon microscopy for enlarged field-of-view, multi-plane and long-term brain imaging," *Nature methods,* vol. 18, no. 1, pp. 46–49, 2021.
29. C. Zhao *et al.*, "Miniature three-photon microscopy maximized for scattered fluorescence collection," *Nature Methods,* vol. 20, no. 4, pp. 617–622, 2023.
30. T. Wang *et al.*, "Quantitative analysis of 1300-nm three-photon calcium imaging in the mouse brain," *Elife,* vol. 9, p. e53205, 2020.
31. J. P. Zinter and M. J. Levene, "Maximizing fluorescence collection efficiency in multiphoton microscopy," *Optics express,* vol. 19, no. 16, pp. 15348–15362, 2011.

32. D. G. Ouzounov *et al.*, "In vivo three-photon imaging of activity of GCaMP6-labeled neurons deep in intact mouse brain," *Nature methods,* vol. 14, no. 4, pp. 388–390, 2017.
33. M. Yildirim, H. Sugihara, P. T. So, and M. Sur, "Functional imaging of visual cortical layers and subplate in awake mice with optimized three-photon microscopy," *Nature communications,* vol. 10, no. 1, p. 177, 2019.
34. L. Streich *et al.*, "High-resolution structural and functional deep brain imaging using adaptive optics three-photon microscopy," *Nature methods,* vol. 18, no. 10, pp. 1253–1258, 2021.
35. N. G. Horton *et al.*, "In vivo three-photon microscopy of subcortical structures within an intact mouse brain," *Nature photonics,* vol. 7, no. 3, pp. 205–209, 2013.
36. A. Klioutchnikov, D. J. Wallace, J. Sawinski, K.-M. Voit, Y. Groemping, and J. N. Kerr, "A three-photon head-mounted microscope for imaging all layers of visual cortex in freely moving mice," *Nature Methods,* vol. 20, no. 4, pp. 610–616, 2023.
37. C. Wang *et al.*, "Lensed Fiber-Optic Two-Photon Endomicroscopy for Field-of-View Enhancement," in *Photonics*, 2023, vol. 10, no. 3: MDPI, p. 342.
38. V. T. Hoang, Y. Boussafa, L. Sader, S. Février, V. Couderc, and B. Wetzel, "Optimizing supercontinuum spectro-temporal properties by leveraging machine learning towards multi-photon microscopy," *Frontiers in Photonics,* vol. 3, p. 940902, 2022.
39. S. K. Piper *et al.*, "A wearable multi-channel fNIRS system for brain imaging in freely moving subjects," *Neuroimage,* vol. 85, pp. 64–71, 2014.
40. C. Chen *et al.*, "An energy-efficient wearable functional near-infrared spectroscopy system employing dual-level adaptive sampling technique," *IEEE Transactions on Biomedical Circuits and Systems,* vol. 16, no. 1, pp. 119–128, 2022.
41. L. Holper, T. Muehlemann, F. Scholkmann, K. Eng, D. Kiper, and M. Wolf, "Testing the potential of a virtual reality neurorehabilitation system during performance of observation, imagery and imitation of motor actions recorded by wireless functional near-infrared spectroscopy (fNIRS)," *Journal of neuroengineering and rehabilitation,* vol. 7, no. 1, pp. 1–13, 2010.
42. M. Mihara, I. Miyai, M. Hatakenaka, K. Kubota, and S. Sakoda, "Sustained prefrontal activation during ataxic gait: a compensatory mechanism for ataxic stroke?" *Neuroimage,* vol. 37, no. 4, pp. 1338–1345, 2007.
43. Y. Hoshi and S.-J. Chen, "Regional cerebral blood flow changes associated with emotions in children," *Pediatric neurology,* vol. 27, no. 4, pp. 275–281, 2002.
44. J. Mehnert *et al.*, "Developmental changes in brain activation and functional connectivity during response inhibition in the early childhood brain," *Brain and Development,* vol. 35, no. 10, pp. 894–904, 2013.
45. D. Anaya *et al.*, "Scalable, modular continuous wave functional near-infrared spectroscopy system (Spotlight)," *Journal of Biomedical Optics,* vol. 28, no. 6, pp. 065003–065003, 2023.
46. P. M. Koster, "Near Infrared Light Penetration in Human Tissue: An Analysis of Tissue Structure and Heterogeneities," Marquette University, 2022.
47. V. Quaresima and M. Ferrari, "A mini-review on functional near-infrared spectroscopy (fNIRS): where do we stand, and where should we go?," in *Photonics*, 2019, vol. 6, no. 3: MDPI, p. 87.
48. D. Jasaitiene, S. Valiukeviciene, G. Linkeviciute, R. Raisutis, E. Jasiuniene, and R. Kazys, "Principles of high-frequency ultrasonography for investigation of skin pathology," *Journal of the European Academy of Dermatology and Venereology,* vol. 25, no. 4, pp. 375–382, 2011.
49. T. L. Szabo, *Diagnostic ultrasound imaging: inside out*. Academic press, 2004.
50. E. Sauerbrei, M. Digney, P. B. Harrison, and P. L. Cooperberg, "Ultrasonic evaluation of neonatal intracranial hemorrhage and its complications," *Radiology,* vol. 139, no. 3, pp. 677–685, 1981.
51. L. M. Leijser, L. S. de Vries, and F. M. Cowan, "Using cerebral ultrasound effectively in the newborn infant," *Early human development,* vol. 82, no. 12, pp. 827–835, 2006.
52. A. Melada, Z. Heinrich, D. Chudy, and K. Rotim, "The difference between ultrasound-guided and stereotactic-guided neurosurgical procedures," *min-Minimally Invasive Neurosurgery,* vol. 43, no. 03, pp. 149–152, 2000.

53. M. M. J. Letteboer, P. W. Willems, M. A. Viergever, and W. J. Niessen, "Brain shift estimation in image-guided neurosurgery using 3-D ultrasound," *IEEE Transactions on Biomedical Engineering,* vol. 52, no. 2, pp. 268–276, 2005.
54. A. Urban, C. Dussaux, G. Martel, C. Brunner, E. Mace, and G. Montaldo, "Real-time imaging of brain activity in freely moving rats using functional ultrasound," *Nature methods,* vol. 12, no. 9, pp. 873–878, 2015.
55. S. L. Norman *et al.*, "Single-trial decoding of movement intentions using functional ultrasound neuroimaging," *Neuron,* vol. 109, no. 9, pp. 1554–1566. e4, 2021.
56. E. Kim *et al.*, "Wearable transcranial ultrasound system for remote stimulation of freely moving animal," *IEEE Transactions on Biomedical Engineering,* vol. 68, no. 7, pp. 2195–2202, 2020.
57. J. Lim, "Circuits on miniaturized ultrasound imaging system-on-a-chip: a review," *Biomedical Engineering Letters,* vol. 12, no. 3, pp. 219–228, 2022.
58. D. Chen *et al.*, "A survey on analog-to-digital converter integrated circuits for miniaturized high resolution ultrasonic imaging system," *Micromachines,* vol. 13, no. 1, p. 114, 2022.
59. G. Peyton, B. Farzaneh, H. Soleimani, M. G. Boutelle, and E. M. Drakakis, "Quadrature synthetic aperture beamforming front-end for miniaturized ultrasound imaging," *IEEE Transactions on Biomedical Circuits and Systems,* vol. 12, no. 4, pp. 871–883, 2018.
60. J. Lee *et al.*, "A 36-channel auto-calibrated front-end ASIC for a pMUT-based miniaturized 3-D ultrasound system," *IEEE Journal of Solid-State Circuits,* vol. 56, no. 6, pp. 1910–1923, 2021.
61. Y. Wang, M. Koen, and D. Ma, "Low-noise CMOS TGC amplifier with adaptive gain control for ultrasound imaging receivers," *IEEE Transactions on Circuits and Systems II: Express Briefs,* vol. 58, no. 1, pp. 26–30, 2010.
62. J.-Y. Um, "A compact variable gain amplifier with continuous time-gain compensation using systematic predistorted gain control," *IEEE Transactions on Circuits and Systems II: Express Briefs,* vol. 69, no. 2, pp. 274–278, 2021.
63. Q.-H. Duong, Q. Le, C.-W. Kim, and S.-G. Lee, "A 95-dB linear low-power variable gain amplifier," *IEEE Transactions on Circuits and Systems I: Regular Papers,* vol. 53, no. 8, pp. 1648–1657, 2006.
64. H.-H. Nguyen, H.-N. Nguyen, J.-S. Lee, and S.-G. Lee, "A binary-weighted switching and reconfiguration-based programmable gain amplifier," *IEEE Transactions on Circuits and Systems II: Express Briefs,* vol. 56, no. 9, pp. 699–703, 2009.
65. E. Kang *et al.*, "A variable-gain low-noise transimpedance amplifier for miniature ultrasound probes," *IEEE Journal of Solid-State Circuits,* vol. 55, no. 12, pp. 3157–3168, 2020.
66. L. Guasch, O. Calderón Agudo, M.-X. Tang, P. Nachev, and M. Warner, "Full-waveform inversion imaging of the human brain," *NPJ digital medicine,* vol. 3, no. 1, p. 28, 2020.
67. S. Jeon, E.-Y. Park, W. Choi, R. Managuli, K. Jong Lee, and C. Kim, "Real-time delay-multiply-and-sum beamforming with coherence factor for in vivo clinical photoacoustic imaging of humans," *Photoacoustics,* vol. 15, p. 100136, 2019.
68. J. Sauvage *et al.*, "4D functional imaging of the rat brain using a large aperture row-column array," *IEEE transactions on medical imaging,* vol. 39, no. 6, pp. 1884–1893, 2019.
69. M. Gesnik *et al.*, "3D functional ultrasound imaging of the cerebral visual system in rodents," *Neuroimage,* vol. 149, pp. 267–274, 2017.
70. G. Allevato *et al.*, "Real-time 3-D imaging using an air-coupled ultrasonic phased-array," *IEEE transactions on ultrasonics, ferroelectrics, and frequency control,* vol. 68, no. 3, pp. 796–806, 2020.
71. R. Gessner, M. Lukacs, M. Lee, E. Cherin, F. S. Foster, and P. A. Dayton, "High-resolution, high-contrast ultrasound imaging using a prototype dual-frequency transducer: in vitro and in vivo studies," *IEEE transactions on ultrasonics, ferroelectrics, and frequency control,* vol. 57, no. 8, pp. 1772–1781, 2010.

72. H.-L. Liu *et al.*, "Design and experimental evaluation of a 256-channel dual-frequency ultrasound phased-array system for transcranial blood–brain barrier opening and brain drug delivery," *IEEE Transactions on Biomedical Engineering,* vol. 61, no. 4, pp. 1350–1360, 2014.
73. C. Bimbard *et al.*, "Multi-scale mapping along the auditory hierarchy using high-resolution functional UltraSound in the awake ferret," *Elife,* vol. 7, p. e35028, 2018.
74. C. Rabut *et al.*, "4D functional ultrasound imaging of whole-brain activity in rodents," *Nature methods,* vol. 16, no. 10, pp. 994–997, 2019.
75. S. Jeon, W. Choi, B. Park, and C. Kim, "A deep learning-based model that reduces speed of sound aberrations for improved in vivo photoacoustic imaging," *IEEE Transactions on Image Processing,* vol. 30, pp. 8773–8784, 2021.
76. S. Khan, J. Huh, and J. C. Ye, "Adaptive and compressive beamforming using deep learning for medical ultrasound," *IEEE transactions on ultrasonics, ferroelectrics, and frequency control,* vol. 67, no. 8, pp. 1558–1572, 2020.
77. S. Gutta, V. S. Kadimesetty, S. K. Kalva, M. Pramanik, S. Ganapathy, and P. K. Yalavarthy, "Deep neural network-based bandwidth enhancement of photoacoustic data," *Journal of biomedical optics,* vol. 22, no. 11, pp. 116001–116001, 2017.
78. C. Rabut *et al.*, "A window to the brain: ultrasound imaging of human neural activity through a permanent acoustic window," *bioRxiv,* p. 2023.06. 14.544094, 2023.
79. J. Tang, J. E. Coleman, X. Dai, and H. Jiang, "Wearable 3-D photoacoustic tomography for functional brain imaging in behaving rats," *Scientific reports,* vol. 6, no. 1, p. 25470, 2016.
80. C. V. Bourantas *et al.*, "Hybrid intravascular imaging: recent advances, technical considerations, and current applications in the study of plaque pathophysiology," *European Heart Journal,* vol. 38, no. 6, pp. 400–412, 2017, doi: https://doi.org/10.1093/eurheartj/ehw097.
81. J. Gamelin *et al.*, "A real-time photoacoustic tomography system for small animals," *Optics express,* vol. 17, no. 13, pp. 10489–10498, 2009.
82. A. Dangi *et al.*, "Ring PMUT array based miniaturized photoacoustic endoscopy device," in *Photons Plus Ultrasound: Imaging and Sensing 2019*, 2019, vol. 10878: SPIE, pp. 46–53.
83. X. Wang, Y. Pang, G. Ku, X. Xie, G. Stoica, and L. V. Wang, "Noninvasive laser-induced photoacoustic tomography for structural and functional in vivo imaging of the brain," *Nature biotechnology,* vol. 21, no. 7, pp. 803–806, 2003.
84. S. Na and L. V. Wang, "Photoacoustic computed tomography for functional human brain imaging," *Biomedical Optics Express,* vol. 12, no. 7, pp. 4056–4083, 2021.
85. Y. Li *et al.*, "Snapshot photoacoustic topography through an ergodic relay for high-throughput imaging of optical absorption," *Nature Photonics,* vol. 14, no. 3, pp. 164–170, 2020.
86. S. Na *et al.*, "Massively parallel functional photoacoustic computed tomography of the human brain," *Nature Biomedical Engineering,* vol. 6, no. 5, pp. 584–592, 2022.
87. E. Merčep, J. L. Herraiz, X. L. Deán-Ben, and D. Razansky, "Transmission–reflection optoacoustic ultrasound (TROPUS) computed tomography of small animals," *Light: Science & Applications,* vol. 8, no. 1, p. 18, 2019.
88. B. Wang, P. Govindan, T. Gonnot, and J. Saniie, "Acceleration of ultrasonic data compression using OpenCL on GPU," in *2015 IEEE International Conference on Electro/Information Technology (EIT)*, 2015: IEEE, pp. 305–309.
89. M. T. McCann, K. H. Jin, and M. Unser, "Convolutional neural networks for inverse problems in imaging: A review," *IEEE Signal Processing Magazine,* vol. 34, no. 6, pp. 85–95, 2017.
90. M.-Y. Liu, X. Huang, J. Yu, T.-C. Wang, and A. Mallya, "Generative adversarial networks for image and video synthesis: Algorithms and applications," *Proceedings of the IEEE,* vol. 109, no. 5, pp. 839–862, 2021.
91. G. Katti, S. A. Ara, and A. Shireen, "Magnetic resonance imaging (MRI)–A review," *International journal of dental clinics,* vol. 3, no. 1, pp. 65–70, 2011.
92. P. Armstrong and S. F. Keevil, "Magnetic resonance imaging–1: Basic principles of image production," *BMJ: British Medical Journal,* vol. 303, no. 6793, p. 35, 1991.

93. W. K. Peng, L. Chen, and J. Han, "Development of miniaturized, portable magnetic resonance relaxometry system for point-of-care medical diagnosis," *Review of Scientific Instruments,* vol. 83, no. 9, 2012.
94. S. Kim *et al.*, "Miniaturized MR-compatible ultrasound system for real-time monitoring of acoustic effects in mice using high-resolution MRI," *NeuroImage,* p. 120201, 2023.
95. J. Osborne, J. Orton, O. Alem, and V. Shah, "Fully integrated standalone zero field optically pumped magnetometer for biomagnetism," in *Steep dispersion engineering and opto-atomic precision metrology XI*, 2018, vol. 10548: SPIE, pp. 89–95.
96. G. Roberts *et al.*, "Towards OPM-MEG in a virtual reality environment," *NeuroImage,* vol. 199, pp. 408–417, 2019.
97. U. Vivekananda *et al.*, "Optically pumped magnetoencephalography in epilepsy," *Annals of clinical and translational neurology,* vol. 7, no. 3, pp. 397–401, 2020.
98. R. M. Hill *et al.*, "Multi-channel whole-head OPM-MEG: Helmet design and a comparison with a conventional system," *NeuroImage,* vol. 219, p. 116995, 2020.
99. E. Boto *et al.*, "Moving magnetoencephalography towards real-world applications with a wearable system," *Nature,* vol. 555, no. 7698, pp. 657–661, 2018.
100. R. A. Seymour *et al.*, "Using OPMs to measure neural activity in standing, mobile participants," *NeuroImage,* vol. 244, p. 118604, 2021.
101. N. Holmes *et al.*, "A bi-planar coil system for nulling background magnetic fields in scalp mounted magnetoencephalography," *Neuroimage,* vol. 181, pp. 760–774, 2018.
102. N. E. Fultz *et al.*, "Coupled electrophysiological, hemodynamic, and cerebrospinal fluid oscillations in human sleep," *Science,* vol. 366, no. 6465, pp. 628–631, 2019.
103. F. Moeller, U. Stephani, and M. Siniatchkin, "Simultaneous EEG and f MRI recordings (EEG-fMRI) in children with epilepsy," *Epilepsia,* vol. 54, no. 6, pp. 971–982, 2013.
104. A. Khalaf, M. Sybeldon, E. Sejdic, and M. Akcakaya, "A brain-computer interface based on functional transcranial doppler ultrasound using wavelet transform and support vector machines," *Journal of neuroscience methods,* vol. 293, pp. 174–182, 2018.
105. J. Uchitel, E. E. Vidal-Rosas, R. J. Cooper, and H. Zhao, "Wearable, integrated eeg–fnirs technologies: A review," *Sensors,* vol. 21, no. 18, p. 6106, 2021.
106. Z. Liu, J. Shore, M. Wang, F. Yuan, A. Buss, and X. Zhao, "A systematic review on hybrid EEG/fNIRS in brain-computer interface," *Biomedical Signal Processing and Control,* vol. 68, p. 102595, 2021.
107. A. Kassab *et al.*, "Multichannel wearable f NIRS-EEG system for long-term clinical monitoring," Wiley Online Library, 1065–9471, 2018.
108. S. Ahn and S. C. Jun, "Multi-modal integration of EEG-fNIRS for brain-computer interfaces–current limitations and future directions," *Frontiers in human neuroscience,* vol. 11, p. 503, 2017.
109. Y. Seki, T. Miyashita, A. Kandori, A. Maki, and H. Koizumi, "Simultaneous measurement of neuronal activity and cortical hemodynamics by unshielded magnetoencephalography and near-infrared spectroscopy," *Journal of Biomedical Optics,* vol. 17, no. 10, pp. 107001–107001, 2012.
110. X. Ru *et al.*, "Multimodal neuroimaging with optically pumped magnetometers: A simultaneous MEG-EEG-fNIRS acquisition system," *Neuroimage,* vol. 259, p. 119420, 2022.

Chapter 19
Ultraviolet Photoacoustic Microscopy for Histopathology

Bingxin Huang, Ivy H. M. Wong, Yan Zhang, and Terence T. W. Wong

Abstract This chapter provides an overview of the use of ultraviolet photoacoustic microscopy (UV-PAM) in histopathology. After introducing the basics of histopathology, we review the principles of PAM and its application in imaging biological tissues. Then, we discuss the different designs of UV-PAM systems and their respective advantages and limitations in histopathology applications. Finally, we explore the recent advances in deep learning-assisted UV-PAM for virtual staining, which has the potential to revolutionize histopathology by enabling the analysis of tissue samples without the need for traditional staining methods. Overall, this chapter offers a comprehensive overview of the use of UV-PAM in histopathology and highlights its potential to advance the field through novel imaging techniques.

Keywords Photoacoustic microscopy · Multispectral imaging · Histological imaging · Deep learning · Virtual staining

19.1 Introduction

Histopathology plays a critical role in studying tissue microarchitectures, identifying abnormalities, subtyping and grading tumors, planning treatment strategies,

---

B. Huang · I. H. M. Wong · T. T. W. Wong (✉)
Translational and Advanced Bioimaging Laboratory, Department of Chemical and Biological Engineering, Hong Kong University of Science and Technology, Clear Water Bay, Hong Kong, China
e-mail: bhuangao@connect.ust.hk; hmwongal@connect.ust.hk; yzhangid@connect.ust.hk; yz3232@cornell.edu; ttwwong@ust.hk

Y. Zhang
Translational and Advanced Bioimaging Laboratory, Department of Chemical and Biological Engineering, Hong Kong University of Science and Technology, Clear Water Bay, Hong Kong, China

School of Applied and Engineering Physics, Cornell University, Ithaca, NY, USA
e-mail: yzhangid@connect.ust.hk; yz3232@cornell.edu

W. Xia (ed.), *Biomedical Photoacoustics*,
https://doi.org/10.1007/978-3-031-61411-8_19

and predicting patient outcomes. It has remained the gold standard for surgical margin assessment (SMA). However, routine pathological examination based on formalin-fixed and paraffin-embedded (FFPE) tissues involves a series of lengthy and laborious procedures, which is unable to guide surgeons intraoperatively. Although frozen sectioning can serve as a rapid alternative to FFPE histology, it still requires a turnaround time of 20–30 min. Besides, the frozen section is subjected to inadequate sampling of resection margins as only a small portion of the resected tissue is chosen and frozen. Furthermore, the diagnostic accuracy based on the analysis of frozen sections is lower than that of FFPE histology since freezing artifacts are inevitable when dealing with edematous and lipid-rich tissues [1]. To this end, developing a rapid and accurate tissue evaluation tool that can assist real-time image-guided surgery is a great medical pursuit.

Lots of efforts have been spent to improve routine pathological practice. Among them, optical imaging has demonstrated significant success due to its rapid and noninvasive nature. Advanced microscopy techniques with optical sectioning capability can image a thin layer of freshly excised tissue without the need for physically sectioning the specimen, greatly simplifying the procedures associated with slide preparation in conventional FFPE histology. These techniques enable high-resolution visualization of anatomical features that are comparable to the clinical standard, and the generated digital images can be stored locally and evaluated by pathologists to support surgeons intraoperatively. Imaging modalities based on exogenous fluorophores, including fluorescence confocal microscopy [2–5], structured illumination microscopy [6–8], and light sheet microscopy [9, 10], can reveal cellular features with enhanced image contrast and molecular specificity. The diversity of commercially available fluorescent probes allows various biomolecules to be identified simultaneously. These imaging platforms can provide highly specific content for diagnosis and sufficient sampling of resection margins within a clinically relevant time. However, they pose a threat to intraoperative procedures as some new fluorescent contrast agents might introduce toxicities to patients, and this effect is still unknown. Besides, the staining process may interfere with the subsequent molecular assays such as fluorescence in situ hybridization and DNA/RNA sequencing [11]. In comparison, imaging modalities based on intrinsic contrast mechanisms arising from reflection, absorption, intrinsic fluorescence, or molecular vibration are highly desired in modern clinical settings. Optical coherence tomography (OCT) [12–15], reflectance confocal microscopy (RCM) [16, 17], photoacoustic microscopy (PAM) [18], and nonlinear microscopy (NLM), including stimulated Raman scattering [19, 20] and multiphoton microscopy [21, 22], are falling into this category. These label-free imaging platforms demonstrate great compatibility with the current clinical practice in operating rooms. Some of these methods have been successfully translated and used in clinical practice for years. For instance, full-field OCT has gained wide applications in ophthalmology due to its high sensitivity, deep penetration, and noninvasiveness. RCM has demonstrated superiority in dermatology, which enables in vivo visualization of human skin layers from the epidermis down to the papillary dermis in real time. PAM can probe a variety of specific molecules with distinct optical absorption properties, such as hemoglobin and melanin, which has shown promising results in preclinical research such as tumor imaging and cardiovascular studies. NLM has been extensively studied in the past few decades. The multimodal and multispectral features of NLM can achieve label-free in vivo visualization of a variety of metabolic processes in an unperturbed and nondestructive way, demonstrating wide applications in tissue engineering and oncological research such as tumor infiltration and growth.

PAM is an advanced imaging technique that combines the strengths of both optical and acoustic imaging modalities. PAM offers unique advantages in terms of label-free, high-resolution imaging with deep tissue penetration and functional insights. In this chapter, we will focus on the application of ultraviolet PAM (UV-PAM) in intraoperative tissue assessment, which aims to identify remaining tumors at resection margins and provide on-site feedback to surgeons and pathologists for diagnostic decision-making.

19.1.1 Introduction to Photoacoustic Microscopy (PAM)

Photoacoustic tomography (PAT) is a rapidly growing imaging technique, which involves the use of absorption of pulse light energy to generate ultrasonic signals from the excited biomolecules. Typically, as shown in Fig. 19.1, when a pulsed or modulated light beam is irradiated on the tissue, endogenous or exogenous absorbers in the tissue absorb light and convert the photon energy into heat, causing a rapid temperature rise. Due to thermoelastic expansion, this temperature change leads to an initial pressure rise and the generation of acoustic waves, which are commonly named photoacoustic (PA) waves [23, 24]. A single-element ultrasonic transducer or a transducer array is applied to detect the PA waves. Finally, the detected PA signals are used to reconstruct an image that represents the absorbed optical energy distribution on the biological tissue. The PA signal (P) can be expressed as [25]

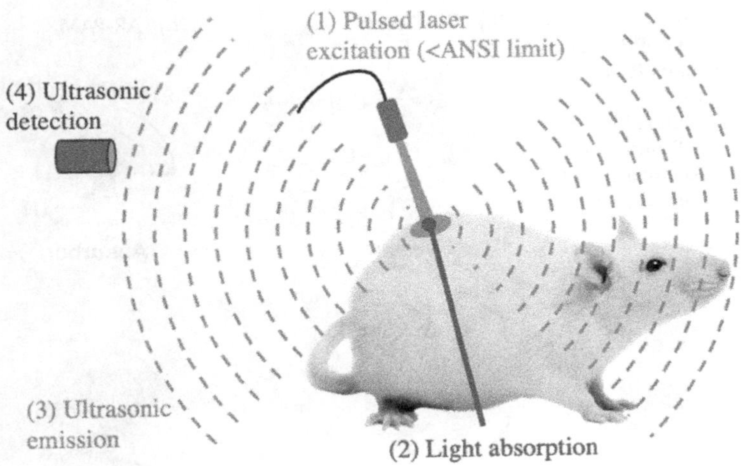

Fig. 19.1 Working principle of PAT

$$P = \Gamma \eta_{\text{th}} \mu_a F \tag{19.1}$$

where Γ is the Grueneisen parameter, which is proportional to the local temperature, η_{th} is the heat conversion efficiency, which represents the portion of absorbed light energy converted into heat, μ_a is the optical absorption coefficient (cm^{-1}) of the excited biomolecules, and F is the optical fluence (J/cm^2).

Over the past few decades, PAT has emerged as a promising imaging modality in various fields, including cardiology and oncology, due to its ability to combine the benefits of wavelength-specific optical absorption contrast and deep penetration depth of ultrasound, making it useful from laboratory research to clinical applications [26].

PAT can be categorized into two implementations based on the arrangements of the light focusing and ultrasonic transducer: photoacoustic computed tomography (PACT) and PAM. In a PACT configuration, high-power pulsed lasers are used to form a broad beam for illuminating the biological tissue. Multielement ultrasonic transducer array is typically employed for parallel detection. In general, the data acquisition for each PACT image is finished with each laser shot. PAM systems typically utilize confocal geometry (i.e., coaxially focus the excitation light beam and acoustic detection), resulting in high spatial resolution and signal-to-noise ratio. The lateral resolution of PAM systems is determined by either the focusing of the excitation light beam or the acoustic detection method, which are referred to as optical-resolution PAM (OR-PAM) and acoustic-resolution PAM (AR-PAM), respectively.

In OR-PAM (Fig. 19.2a), the laser beam is focused on a diffraction-limited spot on the sample by a focusing lens to induce PA signals within the optical focal spot. Thus, the lateral resolution of the OR-PAM system is optically determined, which can be expressed as [27]

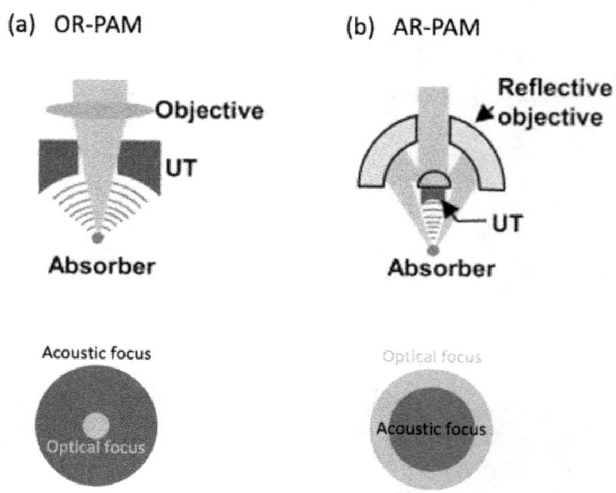

Fig. 19.2 Two implementations of PAM. (**a**) Optical-resolution PAM (OR-PAM). (**b**) Acoustic-resolution PAM (AR-PAM). UT: ultrasonic transducer. (Reproduced with permission from Reference [28])

$$R_{L,OR} = 0.51\frac{\lambda}{\text{NA}} \tag{19.2}$$

where λ is the wavelength of the excitation beam and NA is the numerical aperture (NA) of the focusing lens.

In AR-PAM (Fig. 19.2b), the laser beam is roughly focused on the sample. The acoustic focal spot of the ultrasound transducer is smaller than the optical focal spot. Therefore, the lateral resolution of the AR-PAM system is acoustically determined, which can be expressed as [27]

$$R_{L,AR} = 0.71\frac{\nu_s}{\text{NA}_{UT}\cdot f_0} \tag{19.3}$$

where NA_{UT} is the NA of the ultrasonic transducer, ν_s is the speed of sound in tissues, and f_0 represents the center frequency of PA signals, which is usually determined by the laser pulse width, targeted tissue depth, and ultrasonic transducer's frequency response.

Both OR-PAM and AR-PAM employ the same equation for the axial resolution [27]:

$$R_{A,OR/AR} = 0.88\frac{\nu_s}{\Delta f} \tag{19.4}$$

where Δf is the bandwidth of PA signals, which is usually determined by the detection bandwidth of the ultrasonic transducer and is often proportional to f_0. To obtain an entire tissue image, both OR-PAM and AR-PAM utilize a point-by-point scanning approach.

19.1.2 Multiwavelength Illumination for PAM

PAM shows superiority to conventional optical imaging modalities because of its rich optical absorption contrast, which has better specificity than optical scattering contrast. PAM can selectively highlight specific targets by utilizing appropriate wavelength excitation, as different biomolecules have distinct optical absorption spectra. The optical absorption spectra of major biomolecules are shown in Fig. 19.3 [28]. With the benefit of rich endogenous contrast agents, PAM has found wide-range applications in both clinical and basic research studies, providing label-free images for structural, histological, functional, and metabolic analysis.

Hemoglobin, which plays a crucial role in delivering oxygen through the bloodstream, serves as an excellent endogenous contrast agent. Visualizing hemoglobin with PAM can assist in the detection of early tumor development by facilitating the observation of the growth of new vasculature. Various studies have verified the potential of PAM for early cancer detection by monitoring the vasculature variations, including ear [29, 30], skin [31], and brain vasculature [32]. According

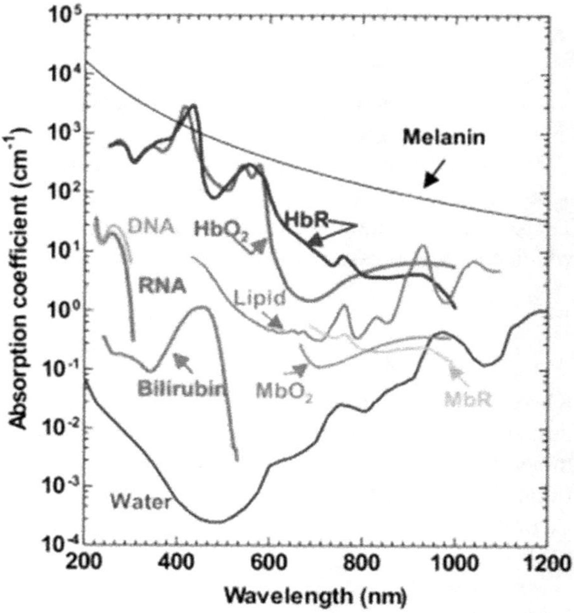

Fig. 19.3 Optical absorption spectra of major biomolecules in biological tissues. (Reproduced with permission from Reference [28])

to the optical absorption spectra, the visible light wavelengths (e.g., from 450 to 650 nm) are commonly used for PAM to image vasculature. Figure 19.4a, b shows OR-PAM images of the mouse ear and brain, respectively. In addition to anatomical information, by measuring both oxyhemoglobin and deoxyhemoglobin with two or more different wavelengths of light sources, the essential diagnostic parameters (e.g., blood oxygen saturation (Fig. 19.5a) and blood flow speed (Fig. 19.5b)) can be estimated for functional imaging [33, 34].

Melanin is the dominant light absorber found in eyes, hair, and skin. The loss of melanin in the retinal pigment epithelium is related to ocular senescence and age-related macular degeneration [35]. PAM, with 532 nm visible excitation light, has been investigated for quantifying the retinal pigment epithelium melanin concentration, holding great promise for the clinical diagnosis of eye diseases [36, 37].

Lipid, as one of the major absorbing molecules in the near-infrared region (NIR, 700–1400 nm), is closely related to energy storage. Abnormal distribution of lipids is likely a risk factor for coronary and peripheral artery diseases or atherosclerosis. As the NIR excitation light has a deeper penetration depth, PAM can image lipids with high resolution in relatively deep tissue [38]. Besides, PAM with NIR excitation is also a promising tool for glucose monitoring. Monitoring and controlling blood glucose levels is critical for diabetes diagnosis and treatment. PAM, as a noninvasive imaging technique, has been widely developed to measure glucose concentration both in in-vitro phantom tests and in vivo experiments, showing good correlations with the benchmark [39–41].

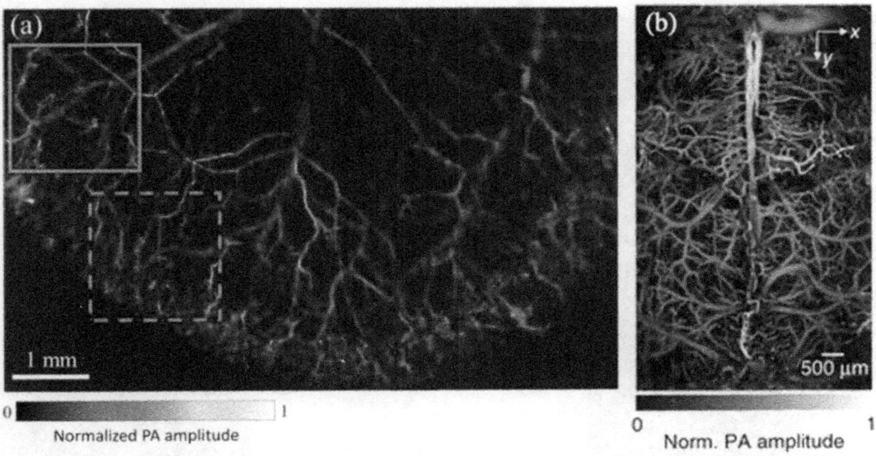

Fig. 19.4 OR-PAM images of blood vessels in the (**a**) ear and (**b**) brain. (Reproduced with permission from Reference [30, 32])

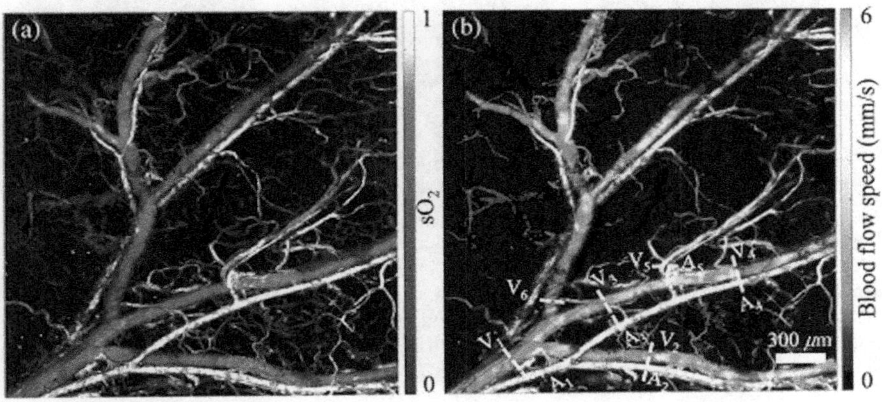

Fig. 19.5 Representative multiparameter PAM images of the blood vessels in the mouse ear. (**a**) OR-PAM of oxygen saturation (sO_2). (**b**) OR-PAM of blood flow speed. (Reproduced with permission from Reference [34])

DNA and RNA are genetic molecules that are rich in every cell nucleus and ubiquitous in every biological tissue. Imaging cell nuclei can provide histological images, which is useful for SMA by distinguishing between normal and tumor cells based on cell density and nuclear morphology. SMA is a critical clinical application that requires a diagnostic tool to confirm the complete excision of cancerous tissue accurately and swiftly. Figure 19.3 shows that DNA/RNA have high optical absorption under the UV wavelength range (210–280 nm); thus, when biological tissue is excited by UV pulsed laser light, PAM can highlight the cell nucleus due to the optical absorption of DNA/RNA, providing positive contrast

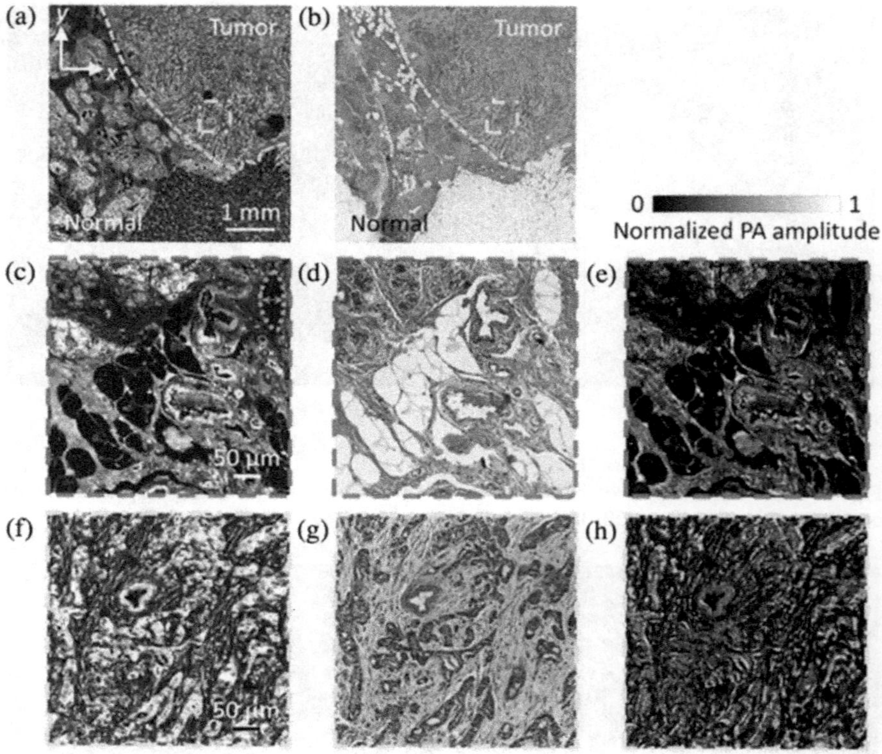

Fig. 19.6 Comparison between UV-PAM and H&E-stained images of a breast tissue slide. (**a**) UV-PAM image of unstained paraffin-embedded breast tissue. (**b**) Corresponding H&E image of the sample in (**a**). (**c**) and (**d**) Closed-up UV-PAM and H&E-stained images of the normal regions marked in (**a**) and (**b**). (**e**) Overlaid image of (**c**) and (**d**). (**f**) and (**g**) Closed-up UV-PAM and H&E-stained images of the normal regions marked in (**a**) and (**b**). (**h**) Overlaid image of (**f**) and (**g**). (Reproduced with permission from Reference [18])

of cellular imaging without tissue processing and labeling [42]. According to Eq. (19.2), UV-PAM can achieve subcellular imaging with a standard objective lens (NA: 0.1–0.3), providing an estimated lateral resolution of <1 μm. Hence, UV-PAM exhibits great promise for accurate and rapid histological analysis of biological tissues. For example, Wong et al. [18] developed a high lateral resolution (330 nm) UV-PAM system to obtain histology-like images of unprocessed human breast cancer tissues. Figure 19.6 demonstrates the ability of the UV-PAM system to provide similar histological features as the gold standard hematoxylin and eosin (H&E)-stained images for margin assessment.

19.2 Different Designs of UV-PAM Systems

As mentioned in the previous section, UV-PAM is efficacious to provide similar nuclear information compared to that provided by the traditional H&E-stained images. Thus, in recent years, different designs of UV-PAM systems have been proposed and developed to enhance imaging performance and achieve specific imaging goals. In this section, we will discuss various design strategies that have been implemented in UV-PAM systems and their applications in histopathology. We will also explore the advantages and limitations of different UV-PAM designs and provide insights into the future development of UV-PAM technology.

19.2.1 Transmission-Mode UV-PAM

By taking advantage of the intrinsic optical absorption contrast of biological tissues, Wong et al. [18] developed a UV-PAM system to achieve label-free histology-like imaging of breast tissue surface, visualizing diagnostic features for enabling the identification of margin between normal and cancer tissues. Figure 19.7a shows the transmission-mode UV-PAM system for SMA. The UV pulsed laser light is tightly focused on the sample by a high NA (= 0.48) aspheric lens to achieve a high

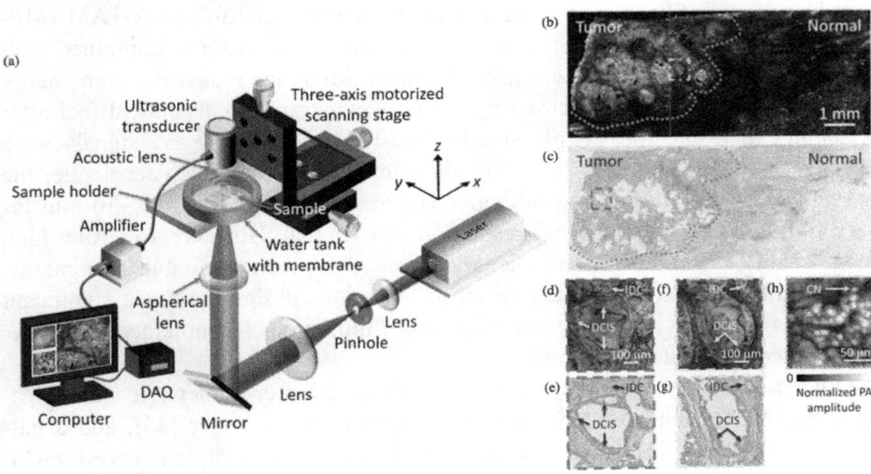

Fig. 19.7 Set up of UV-PAM and imaging of a breast tumor. (**a**) Schematic of the UV-PAM system for SMA. (**b**) UV-PAM image of fixed, unprocessed breast tissue. (**c**) Corresponding H&E-stained image of the sample in (**b**). (**d, e**) Closed-up UV-PAM and H&E-stained images of the normal regions marked in (**b**) and (**c**). (**f, g**) Closed-up UV-PAM and H&E-stained images of the normal regions marked in (**b**) and (**c**). (**h**) Closed-up UV-PAM images marked in (**b**) to show cell nuclei. IDC, invasive ductal carcinoma; DCIS, ductal carcinoma in situ; CN, cell nuclei. (Reproduced with permission from Reference [18])

resolution of ~330 nm. One-dimensional (1D) depth-resolved PA signals are first detected by a focused ultrasonic transducer, and then, a motorized scanning stage is used to obtain cross-sectional or volumetric images by linear or raster scanning. After data acquisition, a maximum amplitude projection (MAP) image from the acquired 3D data cube is displayed, revealing individual cell nuclear information for further histological analysis. A formalin-fixed but unprocessed human breast tumor was imaged by this UV-PAM system (Fig. 19.7b). Due to the intrinsic optical absorption properties, cell nuclei are typically the brightest structures in UV-PAM images (Fig. 19.7h), which can be imaged individually and clearly by the UV-PAM system. In addition, histologic features, such as invasive ductal carcinoma and ductal carcinoma in situ, can be discerned in the UV-PAM images (Fig. 19.7d, f), showing a high consistency (a correlation coefficient of 0.74) with the corresponding H&E-stained images (Fig. 19.7e, g). Therefore, the interface between normal and cancer regions can be outlined in both UV-PAM and H&E-stained images (dashed line in Fig. 19.7b, c). This work is the first and critical step in the development of UV-PAM in SMA.

The potential of UV-PAM for accurate SMA was demonstrated in the above work. Ongoing technical advancements can enable UV-PAM to realize its full potential in histopathology. For instance, the UV-PAM system acquires PA signals on a point-by-point basis by raster scanning, which limits the imaging time to ~100 min with a field of view (FOV) of 25 mm^2. Thus, improvement in imaging speed has far-reaching consequences for UV-PAM in practical and clinical applications.

To improve the imaging speed, a microlens array and a 1D array ultrasonic transducer were utilized to develop a high-throughput multifocal UV-PAM (MF-UV-PAM) [43]. Compared with conventional single optical focus combined with a single-element ultrasonic transducer, the main difference is that the high energy pulsed UV laser light was focused by a microlens array into 40 optical foci after expanding to excite the sample simultaneously, and then, the PA signals were detected by a high-frequency 1D array ultrasonic transducer, which accelerates the imaging speed by 40 times accordingly. The total imaging time was ~16 min for a ~100 mm^2 FOV with a spatial resolution of ~1.6 μm. However, to ensure high image quality, the number of microlenses is limited due to the available laser energy, which limits the maximum imaging speed. Also, this design may face significant challenges for in vivo imaging due to practical limitations in implementing a UV-PAM system in a reflection mode.

Other advances in scanning protocol, such as a water-immersible microelectromechanical system (MEMS) [44], a hexagon-mirror scanner [45], and a galvanometer mirror [46], have also been exploited to achieve high-speed PAM. Among them, Li et al. [46] developed a galvanometer mirror-based UV-PAM (GM-UV-PAM) system for high-speed histology-like imaging (Fig. 19.8). The galvanometer mirror was used to reflect and rapidly scan the laser beam by rotating rapidly within a small angle (~$\pm 9°$). The scanning beam was subsequently focused on the samples to induce PA signals by an objective lens. Figure 19.8b indicates the scanning trajectory of the GM-UV-PAM system. By utilizing a sine function to drive the 1D galvanometer mirror, the incident laser pulses can be reflected and

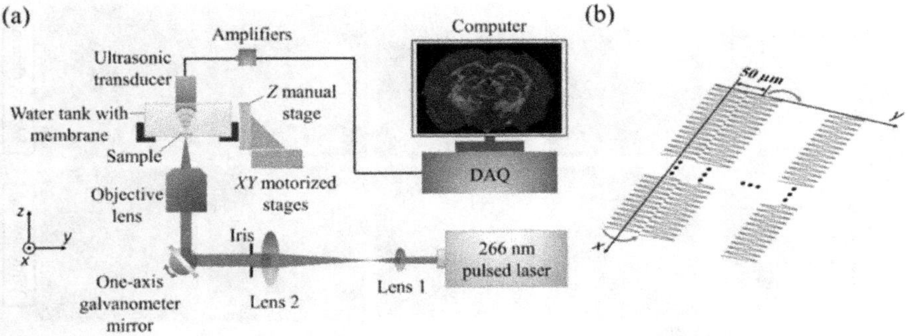

Fig. 19.8 (**a**) Schematic of the galvanometer mirror-based UV-PAM (GM-UV-PAM) system. (**b**) Scanning trajectory of the GM-UV-PAM system. DAQ data acquisition card. (Reproduced with permission from Reference [46] © The Optical Society)

repeatedly scanned along the *Y*-axis, ensuring a similar high detection sensitivity with a line-scanning interval of approximately 50 μm (60 scanning points). The *X*-axis motorized stage moves forward with a step size of 0.63 μm (determined by the minimum step size of the motorized stage) when the galvanometer mirror finishes a line scanning, and this synchronization and scanning repeat until it covers the whole sample along the *X*-axis. After that, the *Y*-axis motorized stage moves along the *Y*-axis with a step size of 50 μm. The process is repeated to scan the whole sample completely. GM-UV-PAM system can achieve histology-like imaging of a ~25 mm^2 FOV with a relatively high resolution of 1 μm within 15 min.

In addition to accelerating the imaging speed of UV-PAM, improving the image contrast and image depth of UV-PAM is also essential to enhance its capability for revealing fine details in organs. The idea of UV-PAM with tissue clearing was proposed to achieve high-contrast histological imaging [47]. This work applied an innovative optical clearing technique, CUBIC (clear, unobstructed brain/body imaging cocktails and computational analysis), to UV-PAM imaging by washing out the lipids and pigments, which have a high absorption of UV light. Figure 19.9 exhibits a comparison of a mouse brain sample imaged by UV-PAM without and with the tissue-clearing method. In UV-PAM images before clearing, the high background signals from lipids and pigments result in low image contrast, which makes it difficult to distinguish cell nuclei from the background. In contrast, most individual cell nuclei can be resolved after clearing (Fig. 19.9g–j) with about three times the improvement in contrast-to-noise ratio compared with that before clearing (Fig. 19.9c–f). Besides, the imaging depth of the UV-PAM was also increased. Six different depths (Fig. 19.9k, n) were extracted from the depth-resolved PA signals of the non-cleared region (Fig. 19.9f) and cleared region (Fig. 19.9j) to provide the depth-encoded image without clearing (Fig. 19.9l) and with clearing (Fig. 19.9m), showing that multilayers of cell nuclei can be observed sufficiently only in the cleared sample. Therefore, clearing-enhanced UV-PAM can provide fine details for organ imaging by improving the image contrast and imaging depth.

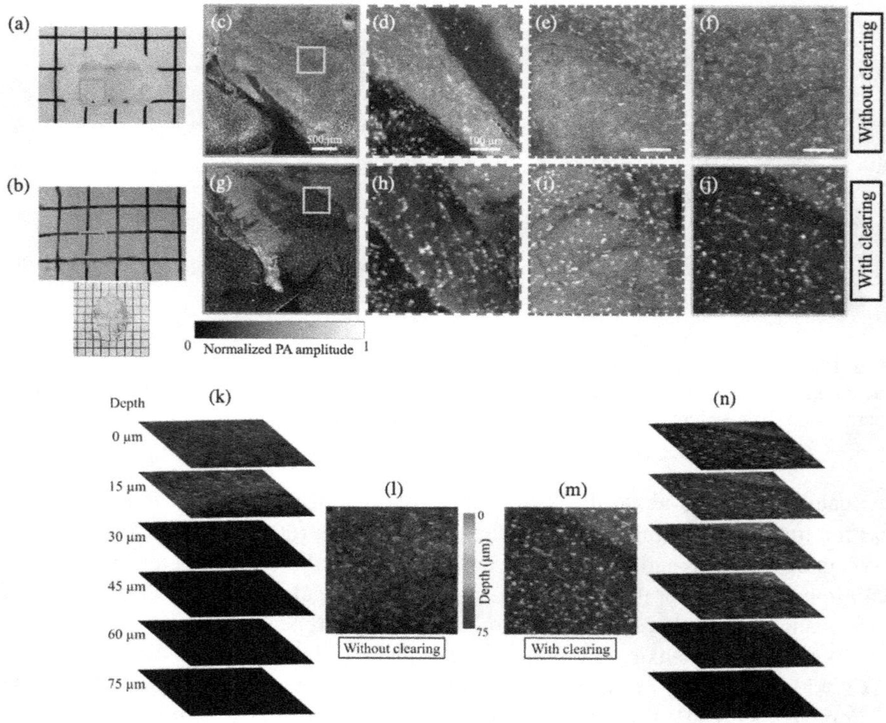

Fig. 19.9 Comparison of UV-PAM images of a mouse brain slice without and with clearing. (**a**) A slice sectioned from a formalin-fixed mouse brain without clearing. (**b**) A similar position slice sectioned from an intact and cleared mouse brain. (**c**) UV-PAM image of the solid box region marked in (**a**). (**d–f**) Closed-up UV-PAM images of the corresponding no-clearing brain slice regions marked in (**c**). (**g**) UV-PAM image of the solid box region marked in (**b**). (**h–j**) Closed-up UV-PAM images of the corresponding cleared brain slice regions marked in (**g**). (**k**) and (**l**) Image stacks and depth-encoded images of the mouse brain slice without clearing, respectively. (**m**) and (**n**) Depth-encoded image and image stacks of the mouse brain slice with clearing, respectively. (Reproduced with permission from Reference [47])

Conventional single-view OR-PAM, especially UV-OR-PAM, suffers from severe resolution anisotropy because the lateral and axial resolutions are determined optically and acoustically, respectively. The shorter laser wavelength (e.g., UV) with a tightly focused optical focal spot achieves a sub-micrometer-level lateral resolution, while the axial resolution determined by the bandwidth of the ultrasonic transducer is usually limited to tens of micrometers. To address this issue, a dual-view OR-PAM (DV-OR-PAM) with UV laser illumination was developed to enhance the axial resolution and thus improve resolution isotropy, as illustrated in Fig. 19.10 [48]. Two UV beams were focused on the sample orthogonally to generate the PA signals. The optical axes of the dual-view beams and the detection axis of the ultrasonic transducer are coplanar, positioned on the *X–Z*plane. The

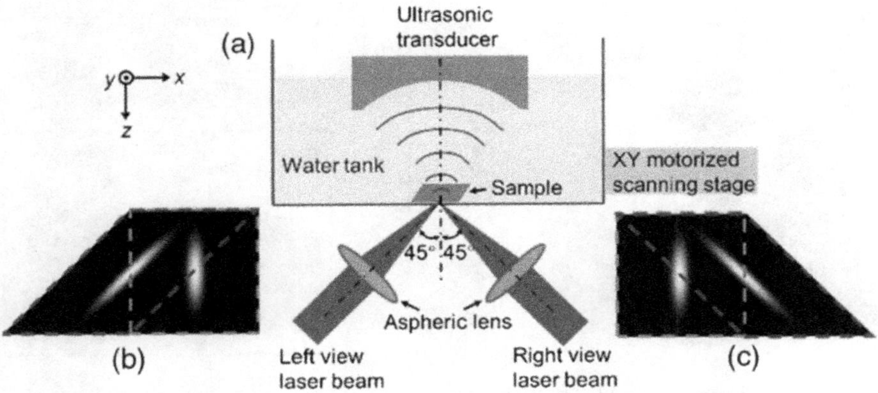

Fig. 19.10 (**a**) Schematic of the dual-view UV-PAM (DV-UV-PAM) system. (**b**) and (**c**) The left and right views display shear-transformed images enclosed in parallelogram dashed boxes, while the corresponding original images for comparison are enclosed in square dashed boxes. (Reproduced with permission from Reference [48] © The Optical Society)

sample was scanned twice with the same configuration but with the left-view laser beam and the right-view laser beam, respectively, to obtain two orthogonal-view 3D images. To register the left- and right-view images with the sample's coordinate system, an affine transformation was applied, which involves both shear (Fig. 19.10b, c) and translation transformations. Then, the co-registered two-view images were used for dual-view Richardson–Lucy deconvolution with several iterations to obtain a single dual-view image. The axial resolution improvement and resolution isotropy have been validated in this work, suggesting that the DV-OR-PAM with UV laser illumination provides a promising tool for quantitative 3D cell nuclear information analysis.

19.2.2 *Reflection-Mode UV-PAM*

Transmission-mode configuration is common in PAM systems due to their simple system design and high sensitivity. However, transmission-mode UV-PAM is limited to imaging relatively thin or transparent samples. Therefore, a reflection-mode UV-PAM system is more practical for various biomedical and clinical applications. Many attempts have tried to achieve the reflection-mode configuration by incorporating the opaque ultrasonic transducer along the illumination path without obstructing the light beam, e.g., using optical-acoustic beam combiners or ring-shaped ultrasonic transducers. Meanwhile, a promising all-optical detection method has also been configured into reflection-mode UV-PAM systems. In the following sections, we will discuss more details on these UV-PAM systems and their practical applications.

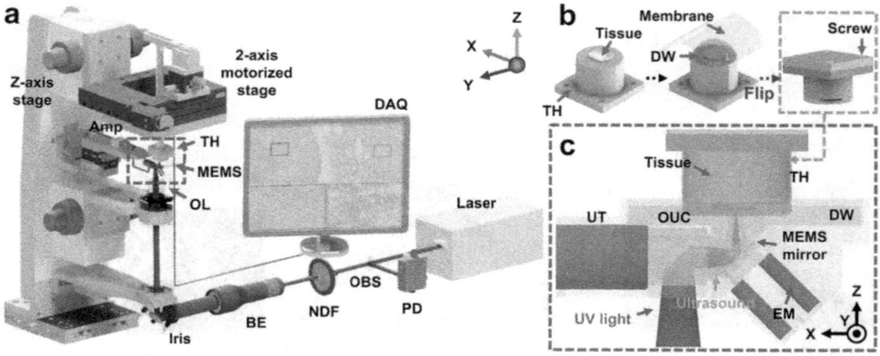

Fig. 19.11 Schematic of the MEMS-based reflection-mode UV-PAM (MEMS-UV-PAM) system. PD photodiode, OBS optical beam splitter, NDF neutral density filter, BE beam expander, OL objective lens, TH tissue holder, Amp amplifier, DAQ data acquisition board, UT ultrasound transducer, OUC opto-ultrasound beam combiner, EM electromagnets, DW deionized water. (Reproduced with permission from Reference [49])

19.2.2.1 Opto-Ultrasound Beam Combiner

Opto-ultrasound beam combiners (OUC) are the most used in reflection-mode PAM configuration. Recently, Baik et al. [49] applied MEMS into a reflection-mode UV-PAM system (MEMS-UV-PAM), achieving high-speed, nondestructive, and label-free imaging of human tissues with a lateral resolution of 1.2 μm. As shown in Fig. 19.11, a 1-axis waterproofing MEMS scanner combined with a 2-axis linear motorized stage is used for high-speed imaging (~4 min for scanning a 10 mm × 10 mm area). The UV light beam illuminated the samples after beam expansion, filtering, and reflection by the dielectric material-coated hypotenuse of the OUC and the MEMS scanner. In the backward path, the generated PA signals go through the OUC and are detected and converted into electrical signals by the ultrasonic transducer for mapping PA images. This reflection-mode UV-PAM system makes it possible for imaging whole thick tissues without any complex process with relatively high sensitivity and resolution. The ~4-mm-thick colon tissue of a patient was imaged using the MEMS-UV-PAM system and pseudo-colored to demonstrate that the system can support intraoperative histopathology (Fig. 19.12). Pseudo-colored UV-PAM images of colon and liver tissues from six cancer patients show high consistency with the H&E-stained images in all noncancerous, boundary, and cancerous regions.

19.2.2.2 Ring-Shaped Ultrasonic Transducer

A ring-shaped ultrasonic transducer with a hollow structure in the center can be utilized in a reflection-mode UV-PAM system, making the whole system simple and flexible. The objective and ring-shaped ultrasonic transducer can be

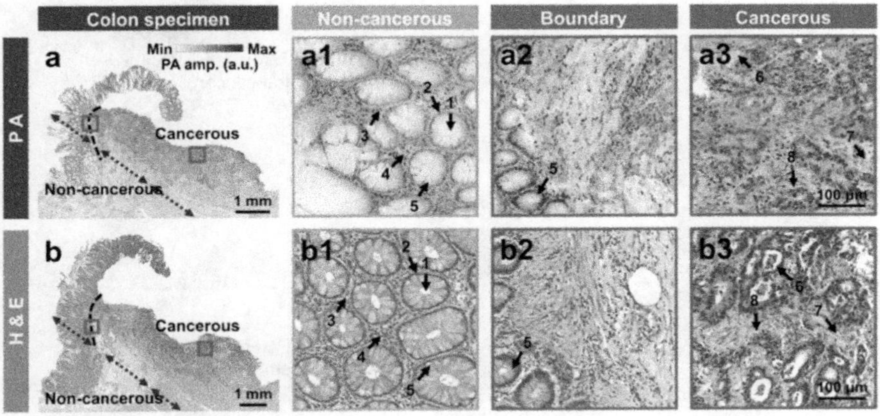

Fig. 19.12 Whole colon tissue imaging by the MEMS-UV-PAM system. (**a**) Label-free UV-PAM image of a colon specimen using the MEMS-UV-PAM system. (**b**) Corresponding H&E-stained image using the traditional bright-field microscope after UV-PAM imaging. (a1) and (b1) Closed-up histoarchitectural features of noncancerous region shown in (**a**) and (**b**). (a2) and (b2) Closed-up histoarchitectural features of boundary region shown in (**a**) and (**b**). (a3) and (b3) Closed-up histoarchitectural features of cancerous region shown in (**a**) and (**b**). 1, lumen; 2, epithelial cell nucleus; 3, stromal cell nucleus; 4, lamina propria; 5, mucin vacuole; 6, neoplastic gland; 7, desmoplastic reaction; 8, poorly differentiated cluster. (Reproduced with permission from Reference [49])

coaxially aligned on the same side, and the ring-shaped ultrasonic transducer detects backward propagating PA waves without affecting the illumination path because the laser beam can be focused on the sample through the hollow center of the ring-shaped ultrasonic transducer. In this configuration, a high NA ultrasonic transducer and objective can be employed to achieve high-resolution imaging. Li et al. [50] developed a reflection-mode and open-top UV-PAM system using a ring-shaped ultrasonic transducer to enable high-speed and high-resolution histological imaging intraoperatively for SMA, providing both original grayscale UV-PAM images and virtually stained images assisted by a deep learning algorithm. In addition, to apply histopathological interpretation from organelle to organ levels, Wong et al. [51] developed a microtomy-assisted PAM (mPAM) system with UV excitation to obtain label-free, distortion-free, and registration-free histopathological images of the whole organ. As illustrated in Fig. 19.13, the exposed tissue surface is imaged through raster scanning of the sample holder, resulting in a MAP image. Using a microtome, the imaged surface is then sectioned, and a new surface is automatically imaged. This cycle of imaging and sectioning is repeated until the sectioned layers reach the predetermined depth to obtain a 3D image shown in Fig. 19.14a. Some individual coronal sections, with their corresponding positions in the brain, are also shown in Fig. 19.14, providing a promising volumetric imaging tool for the histological analysis of a whole organ.

These high-resolution UV-PAM systems all suffer from a limited depth of field (DOF) due to the need for a tight optical focus, causing the degradation of lateral

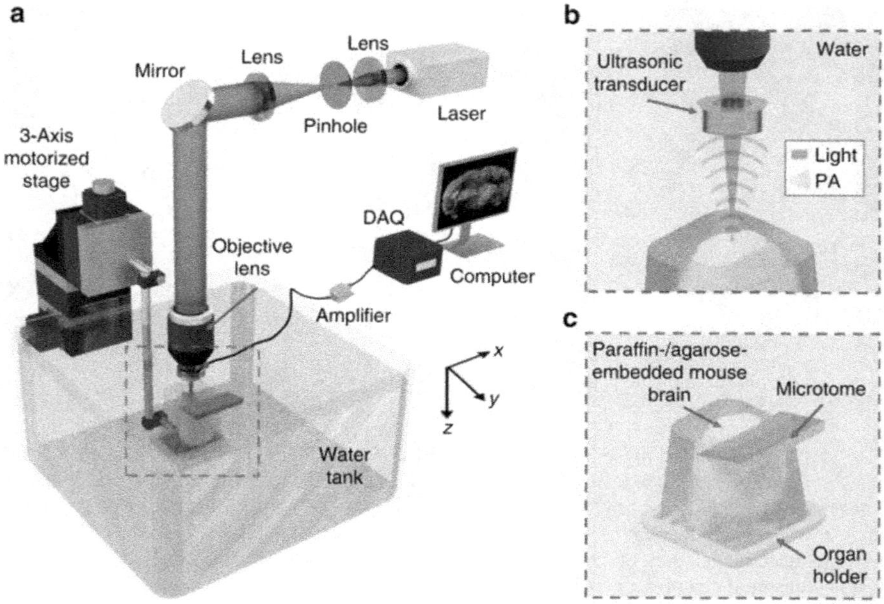

Fig. 19.13 Schematic of the mPAM system with UV excitation for whole-organ imaging and sectioning. (Reproduced with permission from Reference [51])

resolution at a distance away from the optical focal plane. In this case, for most UV-PAM systems, axial scanning at different planes is needed for high-resolution imaging of 3D volumetric imaging or samples with irregular surfaces, which is time-consuming. To address this issue, a recent and promising needle-shaped beam PAM (NB-PAM) via customized diffractive optical elements (DOEs) has been developed to extend DOF [52]. Cao et al. designed DOEs to create multiple closely adjacent foci along the axial direction, thus forming the NB. The beam length is determined by the number of foci. The DOE phase is formulated as

$$P_{\mathrm{DOE}}(x, y) = \sum\nolimits_{m=1}^{M} \left\{ \left[-\frac{\pi n \left(x^2 + y^2\right)\left(\frac{1}{f_m} - \frac{1}{f}\right)}{\lambda} - \pi m A \right] Loc_m(x, y) \right\} \tag{19.5}$$

where (x, y) is the planar coordinate, f is the focal length of the objective, f_m is the chosen focal position, m is the focus index, M is the foci number, λ is the light wavelength, n is the refractive index of the surrounding medium, $Loc_m(x, y)$ is a binary matrix to map the DOE pixels, A is a coefficient, πmA is the phase regulator related to the beam diameter, and the item in the square bracket means to shift the focus from f to f_m. The fabricated DOE is placed on the front surface of the objective lens, as shown in Fig. 19.15e, extending the DOF to around a 28-fold Rayleigh

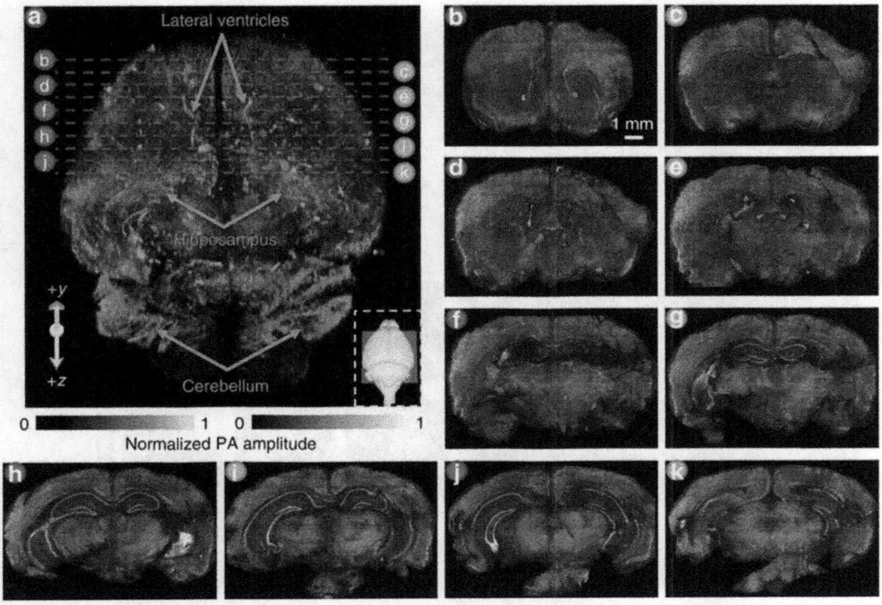

Fig. 19.14 3D label-free mPAM visualization of a mouse brain embedded in an agarose block and ten individual coronal views in the whole mouse brain. (Reproduced with permission from Reference [51])

length. A positive 1951 USAF resolution target, carbon particles, and a fresh mouse lung were imaged by the UV-NB-PAM system to demonstrate the perspective for slide-free intraoperative pathological imaging.

19.2.2.3 All-Optical Detection

Piezoelectric ultrasonic transducers remain the most widely used detectors in UV-PAM due to their stable performance and wide range of configurations. However, an all-optical detection method, UV photoacoustic remote sensing (UV-PARS) microscopy, has been proposed and gained increasing attention for label-free imaging of bulky tissues without ultrasonic transducers and the need for acoustic coupling [53, 54], which is based on a non-interferometric method to measure the initial pressures before wave propagation [55].

Unlike conventional UV-PAM systems, Fig. 19.16 shows the UV-PARS microscope, which consists of two optical beams. A nanosecond excitation beam and a continuous-wave probe beam are co-focused on the target. The excitation beam from a 0.5-ns pulsed 266-nm laser source provides the optical absorption contrast of cell nuclei in tissues. The absorbed optical energy in absorbers is converted to pressure through thermoelastic expansion. The absorbers experience elasto-optic

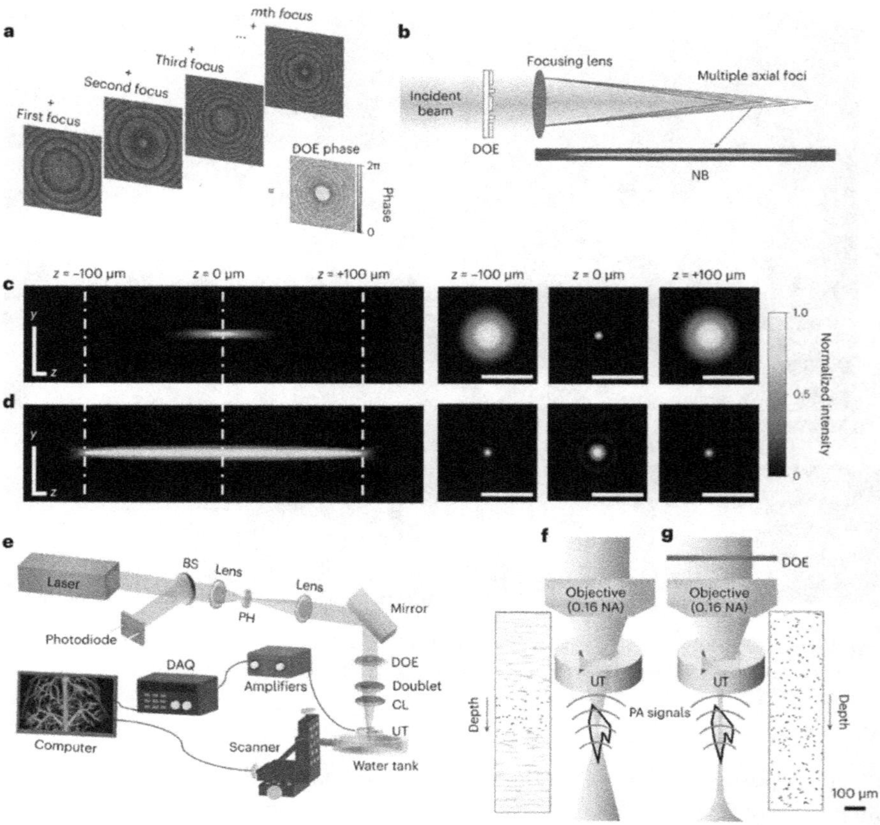

Fig. 19.15 (**a**) DOE phase pattern components. (**b**) Principle of a DOE to form multiple axial foci inducing the desired NB. (**c**) and (**d**) *y*–*z* profile (left) and *x*–*y* profiles (right) at different *z* positions of a Gaussian beam and a needle beam, respectively. Scale bar: 100 μm. (**e**) Schematic of the needle-shaped beam PAM (NB-PAM) system. (**f**) and (**g**) Simulated *y*–*z* projection images of uniformly distributed microspheres with a diameter of 7 μm. (Reproduced with permission from Reference [52])

modulations due to the pressure rise, leading to a change in the intensity of the back-reflected probe beam, which is proportional to the optical absorption at the excitation wavelength. Therefore, both the illumination and detection are based on an optical method and do not require any water coupling medium for the transmission of acoustic signals. Abbasi et al. [53] and Bell et al. [56] demonstrated comprehensive studies on the efficiency of UV-PARS microscopy for label-free noncontact histology-like imaging of human breast, gastrointestinal, and skin tissues from FFPE blocks, frozen sectioned human skin, and unprocessed murine kidney. Figure 19.17 shows a representative result of a tonsil tissue specimen. Moreover, Haven et al. [57] enhanced the lateral resolution of the UV-PARS microscopy to 0.39 μm by using a 0.5 NA reflective objective, and Cikaluk et al. [58] accelerated

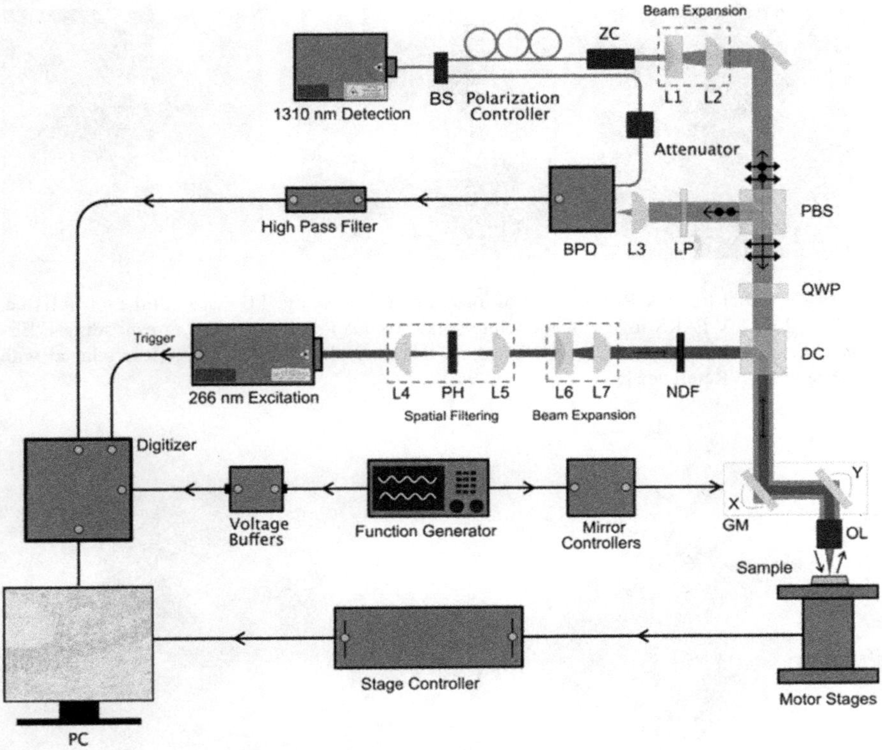

Fig. 19.16 Schematic of UV photoacoustic remote sensing (UV-PARS) microscopy system. BS beam splitter, ZC zoom collimator, LP long-pass filter, PBS polarizing beam splitter, QWP quarter-wave plate, DC dichroic mirror, BPD balanced photodiode, PH pinhole, NDF neutral density filter, GM galvanometer mirrors, OL objective lens. (Reproduced with permission from Reference [53])

the imaging speed of the UV-PARS microscopy by a voice-coil stage system, achieving finely resolved images for 2 × 2 mm^2 areas in 1.33 min and coarsely resolved images for 4 × 4 mm^2 areas in 2.5 min.

It is noteworthy that the absorbed optical energy by absorbers is relaxed in two ways (Fig. 19.18): emitting as photons (radiative relaxation) and converting to heat (non-radiative relaxation). An all-optical noncontact total-absorption PARS (TA-PARS) microscope was developed to detect multiple contrasts of radiative absorption, non-radiative absorption, and optical scattering contrast simultaneously [54, 59]. As shown in Fig. 19.19, besides capturing the optical scattering and non-radiative absorption contrast by a photodiode after the circulator, radiative fluorescence contrast is also captured by another photodiode. The technique provides complementary information, which shares convincing analog to structures in the conventional H&E-stained images (Fig. 19.20). This method will facilitate rapid label-free histological imaging of tissues, representing a crucial advancement toward intraoperative microscopy for diagnosis and SMA.

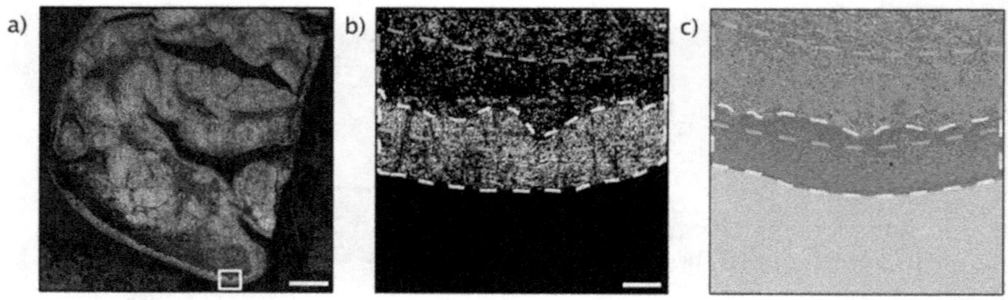

Fig. 19.17 Label-free UV-PARS image of tonsil tissue. (**a**) A large FOV scan of the tonsil tissue. (**b**) Closed-up UV-PARS image of the white solid box marked in (**a**). (**c**) Corresponding H&E-stained image of the adjacent section of tonsil tissue. Scale bars: 100 μm. (Reproduced with permission from Reference [53])

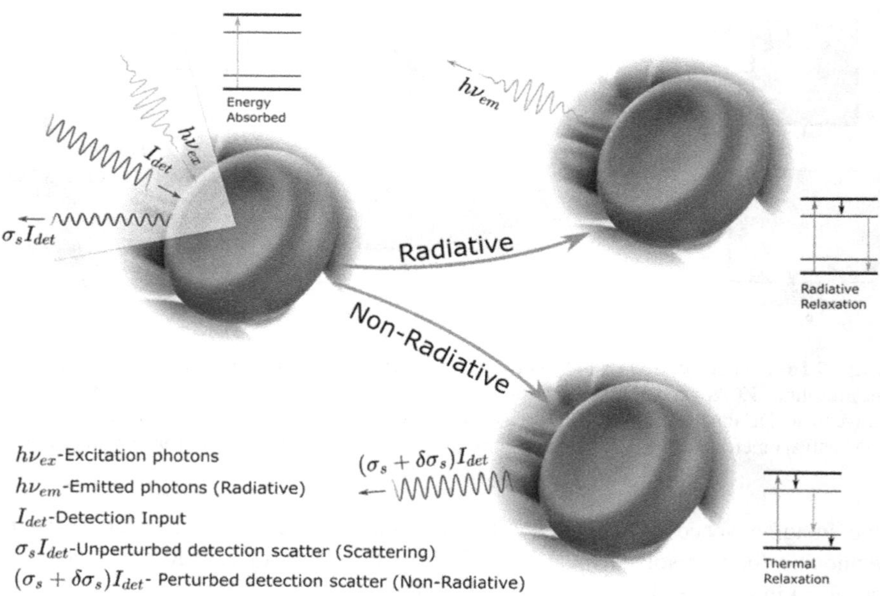

Fig. 19.18 Total-absorption photoacoustic remote sensing (TA-PARS) contrast mechanisms. (Reproduced with permission from Reference [54])

19.3 Development of UV-PAM Enabled by Deep Learning

19.3.1 Introduction to Deep Learning

Deep learning is a subset of machine learning that uses neural networks to model and solve complex problems. While traditional machine learning highly depends

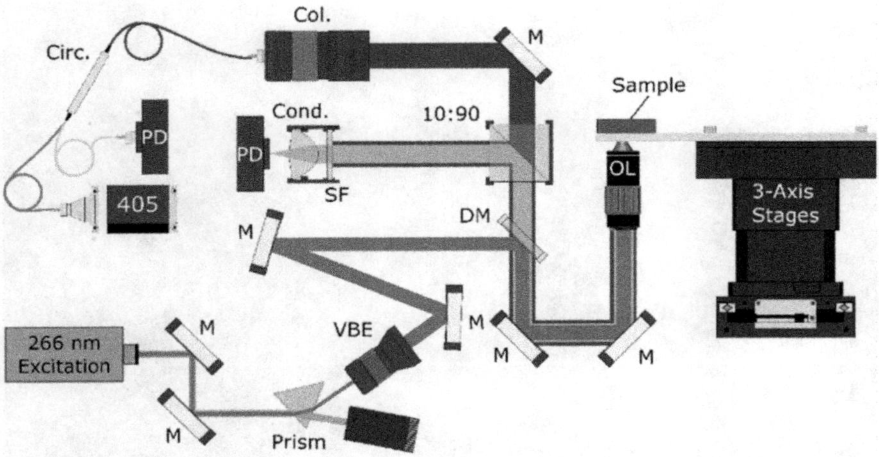

Fig. 19.19 Schematic of the TA-PARS system. DM dichroic mirror, VBE variable beam expander, Col. collimator, Circ. circulator, SF spectral filter, Cond. condenser lens, PD photodiode, 90:10 10:90 splitter, M mirror, OL objective lens. (Reproduced with permission from Reference [59])

on human intervention in designing the feature extraction method for classification or prediction tasks, deep learning is capable of learning useful representations from large amounts of data with minimal need for manual specification. Inspired by the strong processing ability of the human brain, the neural networks used in deep learning are organized into multiple layers of interconnected nodes, which are termed neurons. The simplest form of a neural network consists of an input layer, one or more hidden layers, and an output layer. In each layer, neurons are usually connected to the neurons in the adjacent layers. When a neuron in hidden layers receives one or more inputs, the weighted sum of inputs from the previous layer will be passed through a nonlinear function known as an activation function to generate the output. The output of the neuron is then passed as input to neurons in the subsequent layers. Connecting many of these simple units allows the neural network to learn increasingly complex representations of the input data as it moves through the layer.

The capability of deep learning in modeling complex problems with minimal human intervention has inspired lots of research interest in employing the data-driven deep learning method to foster the development of PAM for a wide range of applications [60–63]. With the power of deep learning, what are the opportunities and research gaps that deep learning can bring in for the development of UV-PAM in biomedical applications? In the upcoming section, we will delve into this topic and discuss the cutting-edge applications of deep learning in UV-PAM.

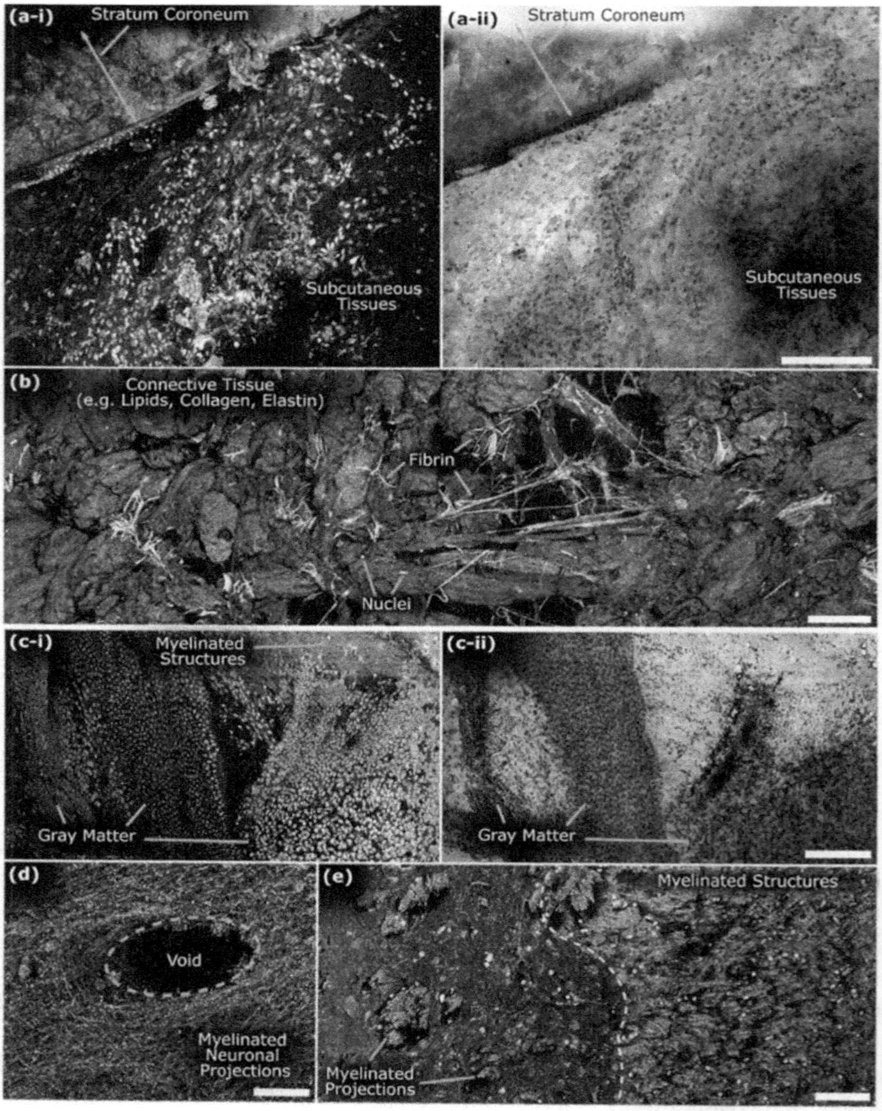

Fig. 19.20 Label-free TA-PARS images of unprocessed resected tissue specimens. (**a**) Label-free TA-PARS non-radiative absorption contrast (a-i) and radiative absorption contrast (a-ii) of human skin tissues. Scale bar: 200 μm. (**b**) Label-free TA-PARS non-radiative absorption contrast image of elongated fibrin strings in the connective tissues. Scale bar: 100 μm. (**c**) Label-free TA-PARS non-radiative absorption contrast (c-i) and radiative absorption contrast (c-ii) of Rattus brain tissues. Scale bar: 200 μm. (**d**) Label-free TA-PARS non-radiative absorption contrast image of myelinated structures in brain tissues. Scale bar: 50 μm. (**e**) Label-free TA-PARS non-radiative absorption contrast image of a boundary region in brain tissues. Scale bar: 100 μm. (Reproduced with permission from Reference [54])

19.3.2 *Virtual Staining for Histology*

As explained in the previous section, the utilization of UV light for excitation in PAM results in high and positive contrast of cell nuclei based on optical absorption, as unlabeled DNA/RNA strongly absorb light at 266 nm [42], generating a strong PA signal. This optical contrast provides label-free contrast analog to the conventional histology stained by H&E. Previous work has already proved the great promise of UV-PAM in providing histology-like images from unprocessed tissue for histopathology applications such as intraoperative cancer margin analysis [18, 49].

Since pathologists are trained to interpret H&E-stained images based on both structural and color information, it is important to enhance the interpretability of UV-PAM images, which were typically displayed in grayscale based on normalized PA amplitude (Fig. 19.6), to make it more applicable in clinical settings. Therefore, various techniques have been employed to process UV-PAM images into images that resemble histologically stained images, which can be classified into two main approaches: (1) linear color mapping and (2) deep learning-based approaches.

19.3.2.1 Color Mapping Approach

For the color mapping approach, Baik et al. [49] performed pseudo-coloring by applying a manually specified color map that resembles H&E staining based on the PA amplitude (Fig. 19.21). Ecclestone et al. attempted to highlight the nuclear and cytoplasm information provided by H&E by using multiple wavelengths for illumination. False coloring on UV-PARS images was generated by further image processing, including cell nuclear segmentation and weighted mixing function between images obtained from different illumination wavelengths [64].

19.3.2.2 Deep Learning-Based Virtual Staining Approach

In contrast with the color mapping approach, the deep learning-based approach does not require much manual design and prior knowledge about the absorption property of different biomolecules at specific illumination wavelengths in the biological tissues, which is usually hard to predict in practice due to the complex biological composition in tissue. Virtual staining using deep learning is essentially a data-driven color transformation approach to generate virtually stained histology images by learning the mapping function from a dataset of label-free images and the target chemically stained histology images [65]. With the nonlinearity introduced in deep learning training, virtual staining using deep learning has become a promising direction to exploit the label-free UV-PAM contrast not only limited to nuclei but also from various biomolecules for generating histology-like images without manual tuning.

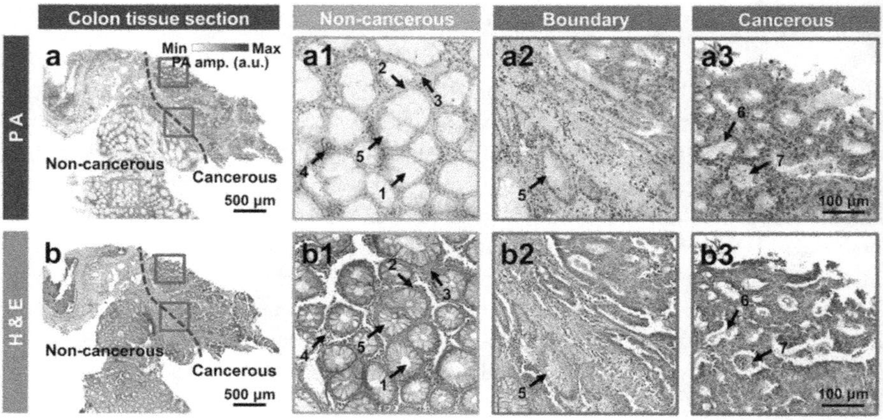

Fig. 19.21 Histology-like UV-PAM images with color mapping and the corresponding H&E-stained image of a thin FFPE human colon tissue section. (**a**) Label-free UV-PAM image with color mapping of an unstained and deparaffinized FFPE colon tissue section. (**b**) H&E-stained FFPE colon tissue section image acquired after the PA imaging experiment. (a1–a3,b1–b3) Zoomed-in UV-PAM images with color mapping and H&E-stained images corresponding to the solid regions in (**a**) and (**b**), respectively. (Reproduced with permission from Reference [49])

A generative adversarial network (GAN) is one of the commonly used deep learning frameworks for image-to-image translation [66]. It typically consists of two deep convolutional neural networks, one as a generator to create generated "fake" images and one as a discriminator to distinguish real and fake data. The generator and discriminator are trained in an adversarial manner, with the generator attempting to fool the discriminator into classifying its generated images as real data, whereas the discriminator attempts to classify the real data and fake data correctly. As this training process continues iteratively, eventually the generator should tend to generate images that resemble the real data that is indistinguishable from the discriminator ideally. Therefore, a loss function that measures the difference between the generated output and the real data and the ability of the discriminator to classify the data as real or fake correctly is used to guide the training process. By optimizing an objective function that incorporates different loss functions, the model learns to generate images that can better resemble the real images.

Existing virtual staining models can be categorized into supervised learning and unsupervised learning.

19.3.2.2.1 Supervised Learning

Conditional GAN, which is also known as Pix2pix, is a state-of-the-art example of supervised learning [67]. Boktor et al. [59] have demonstrated the use of Pix2pix for virtual H&E staining on the TA-PARS images of the FFPE tissue slice. As mentioned previously, the TA-PARS is an extended work of UV-PARS that exploits

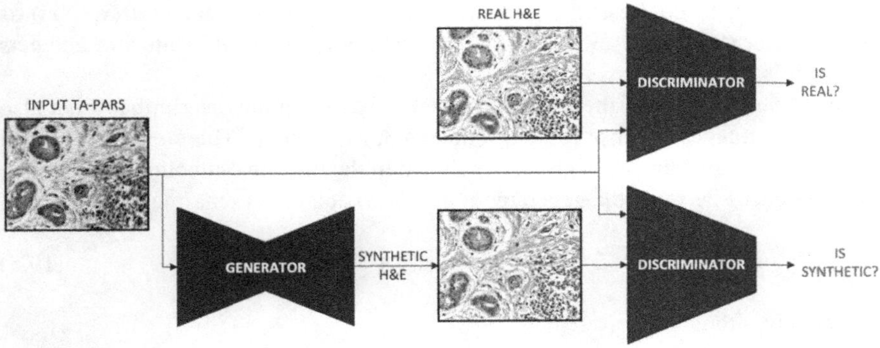

Fig. 19.22 An illustration of the Pix2pix architecture for generating virtually stained H&E images from the input TA-PARS images and real H&E-stained images. (Reproduced with permission from Reference [59])

the multiple contrasts of both non-radiative absorption, radiative absorption, and optical scattering contrast. Figure 19.22 gives an illustration of using the Pix2pix algorithm for virtual staining.

The architecture of Pix2pix consists of one generator G and one discriminator D. Taking the virtual staining of TA-PARS images as an example, the generator G is responsible for translating the image domain X (TA-PARS images) into image domain Y (H&E-stained images). The TA-PARS image x is input into the generator G, which generates the synthetic (fake) virtual H&E-stained image $\hat{y}$ where $\hat{y} = G(x)$. However, the discriminator D is responsible for distinguishing the TA-PARS image x from the fake virtual H&E image $\hat{y}$ or real H&E-stained image y. Mathematically, the objective of the training can be regarded as generator G tries to minimize the likelihood of the discriminator successfully distinguishing the fake image $\hat{y}$ while the discriminator tries to maximize the likelihood of successfully classifying a real image as real and a generated image as fake. This gives the general objective as

$$\min_{G}\max_{D}\mathcal{L}_{\text{cGAN}}\left(G, D\right) \tag{19.6}$$

$\mathcal{L}_{cGAN}$ is the loss function of cGAN which can be expressed as

$$\mathcal{L}_{\text{cGAN}}\left(G, D\right) = \mathbb{E}_{x,y}\left[\log D\left(x, y\right)\right] + \mathbb{E}_{x}\left[\log\left(1 - D\left(x, G(x)\right)\right)\right] \tag{19.7}$$

where $\mathbb{E}_{x,y}\left[\cdot\right]$ denotes the expectation operator which calculates the expected value, or in other words, the average value of a function over a probability distribution with respect to x and y. $\log D(x, y)$ describes the probability that the discriminator is correctly classifying the real image. Similarly, maximizing $\log(1 - D(x, G(x))$ encourages the discriminator to correctly classify the fake image as fake while penalizing when the discriminator incorrectly classifies the fake image as real. In contrast, the generator should minimize $\log(1 - D(x, G(x)))$ as the generator should get rewarded if it successfully fools the discriminator and gets penalized otherwise.

Moreover, the task of the generator is not only to fool the discriminator but also to generate images as close to the ground truth as possible. Therefore, a weighted L1 loss (mean absolute error) is also included in the objective function to minimize the differences between the generated image $G(x)$ and target real image y:

$$\mathcal{L}_{L1}(G) = \mathbb{E}_{x,y}\left[\|y - G(x)\|_1\right] \tag{19.8}$$

This gives the final objective as

$$\min_{G}\max_{D}\mathcal{L}_{\text{cGAN}}(G, D) + \lambda\mathcal{L}_{L1}(G) \tag{19.9}$$

where λ is the weighting assigned to describe the contribution of L1 loss in the objective function.

From its network design, we can see Pix2pix requires rigidly paired label-free images and the corresponding H&E-stained images with pixel-level image alignment for training. Therefore, a precise image registration between the UV-PAM images and the H&E-stained images obtained after UV-PAM image acquisition is needed to prepare the paired images for training. Sometimes, this image registration process between different image modalities could be complicated. Tissue could be slightly deformed when the FFPE tissue slides undergo the H&E staining process (Fig. 19.6d, g). It is even more challenging to obtain the well-aligned corresponding H&E images as the UV-PAM images of unprocessed thick specimens owing to the morphology changes induced during the FFPE sample preparation process. Therefore, the use of Pix2pix for virtual staining is currently limited to thin tissue slices where sufficient alignment level could be achieved by acquiring the label-free and H&E image pairs of the same tissue slice before and after staining.

19.3.2.2.2 Unsupervised Learning

To resolve the challenge of preparing well-aligned paired data for virtual staining thick specimens, an unsupervised cycle-consistent generative adversarial network (CycleGAN) [68] has been adopted to bypass the need for well-aligned UV-PAM and H&E-stained images for training. Kang et al. [69] have successfully implemented CycleGAN for virtual staining of the UV-PAM images of unprocessed thick mouse brain tissue (Fig. 19.23). Cao et al. [70] recently have also demonstrated a similar virtual histology idea using CycleGAN on human bone tissue sections on a glass slide.

The network architecture of a CycleGAN framework consists of two generators (G and F) and two discriminators (D_X and D_Y). While the goal of virtual staining is to translate the image domain X (UV-PAM images) into image domain Y (H&E-

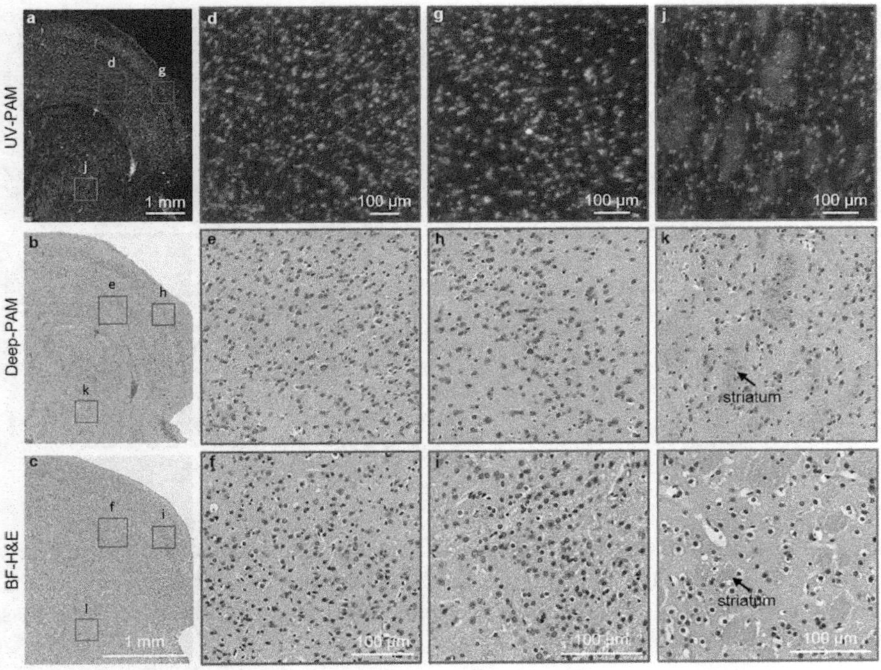

Fig. 19.23 UV-PAM, Deep-PAM (virtually stained H&E), and the adjacent H&E images of a 2-mm thick and fresh mouse brain. (**a, b**) UV-PAM and Deep-PAM images of the thick and fresh mouse brain's surface, respectively. (**c**) Bright-field H&E (BF-H&E) image of the adjacent thin section to the fresh mouse brain's surface for comparison. (**d, e, f**) Zoomed-in UV-PAM, Deep-PAM, and BF-H&E images of the regions in (**a**), (**b**), and (**c**), respectively. (**g, h, i**) Zoomed-in UV-PAM, Deep-PAM, and BF-H&E images of the regions in (**a**), (**b**), and (**c**), respectively. (**j, k, l**) Zoomed-in UV-PAM, Deep-PAM, and BF-H&E images of the regions in (**a**), (**b**), and (**c**), respectively. (Reproduced with permission from Reference [69])

stained images), a cycle-consistency loss is introduced by coupling an inverse mapping from Y (H&E-stained images) to X (UV-PAM images), which enables a constraint to enforce the translation from one domain to the other and back to the original domain, resulting in the original image. In such a case, the cycle-consistency loss enables CycleGAN to learn a mapping between two domains without explicit image alignment.

An example of network architecture implementing CycleGAN for virtual staining UV-PAM images is shown in Fig. 19.24. In particular, the generator G learns to transform the UV-PAM images (image domain X) to H&E images (image domain Y), while generator F learns to transform the H&E images (image domain Y) to UV-PAM images (image domain X). The discriminator D_X aims to classify real UV-PAM image x and fake UV-PAM image $\hat{x} = F(y)$, whereas the discriminator D_Y classifies real H&E-stained image y and fake virtual H&E image $\hat{y} = G(x)$. All the generators and discriminators are also trained in an adversarial manner. The

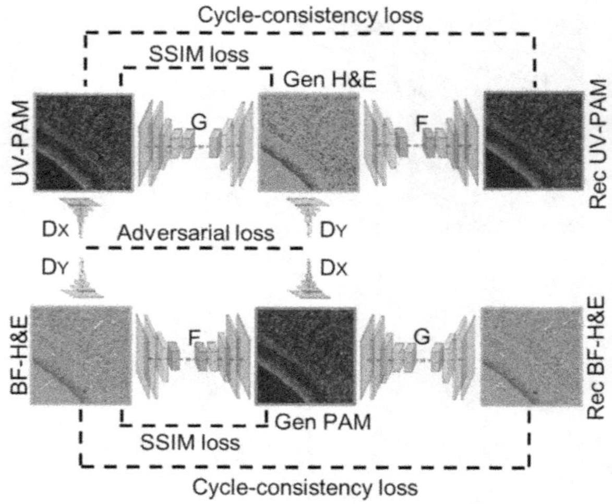

Fig. 19.24 An illustration of the CycleGAN architecture for generating virtually stained H&E images from the training with input UV-PAM images and real H&E-stained images. (Reproduced with permission from Reference [69])

training lasts until the discriminator D_Y no longer can distinguish the virtual H&E output from generator G with the real H&E-stained images (and vice versa for the generator F and D_X). With the goal of bidirectional mapping, the adversarial loss should then describe both transformations:

$$\mathcal{L}_{\text{GAN}}\left(G, D_Y, X, Y\right) = \mathbb{E}_{y\sim p_{\text{data}}(y)}\left[D_Y(y)^2\right] + \mathbb{E}_{x\sim p_{\text{data}}(x)}\left[\left(D_Y\left(G(x)\right) - 1\right)^2\right] \quad (19.10)$$

$$\mathcal{L}_{\text{GAN}}\left(F, D_X, Y, X\right) = \mathbb{E}_{x\sim p_{\text{data}}(x)}\left[D_X(x)^2\right] + \mathbb{E}_{y\sim p_{\text{data}}(y)}\left[\left(D_X\left(F(y)\right) - 1\right)^2\right] \quad (19.11)$$

The two equations above share the same central idea of GAN loss as introduced previously. The first term measures the ability of the discriminator to correctly classify real images. The second term measures the squared difference between the discriminator's output for fake images generated and a target value of 1. This again encourages the discriminator to classify the generated image as fake and penalize it when the discriminator incorrectly classifies the generated image as real. While the idea of adversarial training is the same, the above loss functions only defer by using the least square to describe the distance to the target value instead of using the logarithmic function. This least square variant is sometimes preferred in CycleGAN because it avoids the vanishing gradients that could occur when the generated images are very different from real images [71]. Having the least square loss can give more penalty to larger errors, which provides a large correction rather than a vanishing gradient that leads to minimal model update.

As illustrated in Fig. 19.24, after inputting a real UV-PAM image x to generator G, a fake virtual H&E image $\hat{y}$ will be generated. The discriminator D_Y will determine if this generated H&E $\hat{y}$ is real or fake. This generated H&E will be transformed back to a recovered UV-PAM image $F\left(\hat{y}\right) = F\left(G(x)\right)$ by generator F, which in the ideal case should be consistent with the input real UV-PAM image. In the reverse direction, the generator F is responsible for generating the fake UV-PAM image $\hat{x}$. The discriminator D_X will determine if this generated UV-PAM image $\hat{x}$ is real or fake. This generated UV-PAM will also be transformed back to a recovered H&E image $G\left(\hat{x}\right) = G\left(F(y)\right)$ by generator G, which should be consistent with the input real H&E image. A cycle-consistency loss calculated using mean absolute error can describe this forward and backward consistent relationship between the input real image and the recovered image for both UV-PAM and H&E images:

$$\mathcal{L}_{\text{cyc}}\left(G, F\right) = \mathbb{E}_{x \sim p_{\text{data}}(x)}\left[\left\|F\left(G(x)\right) - x\right\|_1\right] + \mathbb{E}_{y \sim p_{\text{data}}(y)}\left[\left\|G\left(F(y)\right) - y\right\|_1\right] \tag{19.12}$$

This gives the final objective of basic CycleGAN training as

$$\min_{G,F} \max_{D_X,D_Y} \mathcal{L}\left(G, F, D_X, D_Y\right) \tag{19.13}$$

where the total loss $\mathcal{L}$ is described as

$$\mathcal{L}\left(G, F, D_X, D_Y\right) = \mathcal{L}_{\text{GAN}}\left(G, D_Y, X, Y\right) + \mathcal{L}_{\text{GAN}}\left(F, D_X, Y, X\right) + \lambda \mathcal{L}_{\text{cyc}}\left(G, F\right) \tag{19.14}$$

To better constrain the mapping, Kang et al. also incorporated a weighted structural similarity loss to the total loss function to impose restrictions on the transformation to ensure the feature details originate from the input image instead of random structures. As the value of the structural similarity index measure (SSIM) ranges from zero to one, the structural similarity loss is constructed as

$$\begin{aligned}\mathcal{L}_{\text{ssim}}\left(G, F\right) = 1 - \mathbb{E}_{x \sim p_{\text{data}}(x)}\left[\text{SSIM}\Big(x, G(x)\Big)\right] \\ + \left(1 - \mathbb{E}_{y \sim p_{\text{data}}(y)}\left[\text{SSIM}\Big(y, F(y)\Big)\right]\right)\end{aligned} \tag{19.15}$$

In conclusion, deep learning-based virtual staining methods offer an alternative to traditional image processing techniques, enabling the color transformation of UV-PAM images to virtually stained H&E images with minimal manual tuning. While deep learning offers opportunity, one limitation of deep learning is that there is not much clinically recorded PAM data for training, which will take time for data collection. With the great promise of UV-PAM in providing label-free, slide-free, and multilayer histology-like images for histopathology analysis, the color transformation provided by virtual staining should further enhance the ease of interpretation by clinical professionals, facilitating the clinical translation of UV-PAM.

19.4 Conclusion

In conclusion, UV-PAM is a promising imaging modality for histopathological analysis of tumor tissue, offering high-resolution images with excellent label-free contrast and multilayer information without the need for physical sectioning, staining, or labeling. Over the years, different efforts have been made to significantly improve the clinical translational value in the aspect of imaging speed and ease of image interpretation. UV-PAM has been explored for intraoperative SMA application, particularly on unprocessed tissue in breast, liver, colon, and bone tumor tissues. Further research and development in UV-PAM are expected to expand its application and impact with more clinical studies.

References

1. Preeti A, Sameer G, Kulranjan S, Abhinav SA, Preeti R, Sunita Y, Mati GM (2016) Intra-operative frozen sections: Experience at a tertiary care centre. Asian Pacific J Cancer Prev 17:5057–5061. https://doi.org/10.22034/APJCP.2016.17.12.5057
2. Krishnamurthy S, Ban K, Shaw K, Mills G, Sheth R, Tam A, Gupta S, Sabir S (2019) Confocal Fluorescence Microscopy Platform Suitable for Rapid Evaluation of Small Fragments of Tissue in Surgical Pathology Practice. Arch Pathol Lab Med 143:305–313. https://doi.org/10.5858/arpa.2018-0352-OA
3. Dobbs J, Krishnamurthy S, Kyrish M, Benveniste AP, Yang W, Richards-Kortum R (2015) Confocal fluorescence microscopy for rapid evaluation of invasive tumor cellularity of inflammatory breast carcinoma core needle biopsies. Breast Cancer Res Treat 149:303–310. https://doi.org/10.1007/s10549-014-3182-5
4. Thiberville L, Salaun M, Lachkar S, Dominique S, Moreno-Swirc S, Vever-Bizet C, Bourg-Heckly G (2009) Human in vivo fluorescence microimaging of the alveolar ducts and sacs during bronchoscopy. Eur Respir J 33:974–985. https://doi.org/10.1183/09031936.00083708
5. Unnersjö-Jess D, Scott L, Sevilla SZ, Patrakka J, Blom H, Brismar H (2018) Confocal super-resolution imaging of the glomerular filtration barrier enabled by tissue expansion. Kidney Int 93:1008–1013. https://doi.org/10.1016/j.kint.2017.09.019
6. Schlichenmeyer TC, Wang M, Elfer KN, Brown JQ (2014) Video-rate structured illumination microscopy for high-throughput imaging of large tissue areas. Biomed Opt Express 5:366. https://doi.org/10.1364/BOE.5.000366
7. Mertz J (2011) Optical sectioning microscopy with planar or structured illumination. Nat Methods 8:811–819. https://doi.org/10.1038/nmeth.1709
8. Wang M, Tulman DB, Sholl AB, Kimbrell HZ, Mandava SH, Elfer KN, Luethy S, Maddox MM, Lai W, Lee BR, Brown JQ (2016) Gigapixel surface imaging of radical prostatectomy specimens for comprehensive detection of cancer-positive surgical margins using structured illumination microscopy. Sci Rep 6:27419. https://doi.org/10.1038/srep27419
9. Glaser AK, Reder NP, Chen Y, McCarty EF, Yin C, Wei L, Wang Y, True LD, Liu JTC (2017) Light-sheet microscopy for slide-free non-destructive pathology of large clinical specimens. Nat Biomed Eng 1:1–19. https://doi.org/10.1038/s41551-017-0084
10. Xie W, Glaser AK, Vakar-Lopez F, Wright JL, Reder NP, Liu JTC, True LD (2020) Diagnosing 12 prostate needle cores within an hour of biopsy via open-top light-sheet microscopy. J Biomed Opt 25:1–10. https://doi.org/10.1117/1.JBO.25.12.126502
11. Chen Y, Xie W, Glaser AK, Reder NP, Mao C, Dintzis SM, Vaughan JC, Liu JTC (2019) Rapid pathology of lumpectomy margins with open-top light-sheet (OTLS) microscopy. Biomed Opt Express 10:1257. https://doi.org/10.1364/BOE.10.001257

12. Assayag O, Grieve K, Devaux B, Harms F, Pallud J, Chretien F, Boccara C, Varlet P (2013) Imaging of non-tumorous and tumorous human brain tissues with full-field optical coherence tomography. NeuroImage Clin 2:549–557. https://doi.org/10.1016/J.NICL.2013.04.005
13. Bus MTJ, de Bruin DM, Faber DJ, Kamphuis GM, Zondervan PJ, Laguna-Pes MP, van Leeuwen TG, de Reijke TM, de la Rosette JJMCH (2016) Optical Coherence Tomography as a Tool for In Vivo Staging and Grading of Upper Urinary Tract Urothelial Carcinoma: A Study of Diagnostic Accuracy. J Urol 196:1749–1755. https://doi.org/10.1016/J.JURO.2016.04.117
14. Mukherjee S, Jain M, Narula N, Salamoon B, Shevchuk M, Aggarwal A, Altorki N, Stiles B, Boccara C (2013) Full-field optical coherence tomography for the analysis of fresh unstained human lobectomy specimens. J Pathol Inform 4:26. https://doi.org/10.4103/2153-3539.119004
15. Assayag O, Antoine M, Sigal-Zafrani B, Riben M, Harms F, Burcheri A, Grieve K, Dalimier E, De Poly BLC, Boccara C (2014) Large field, high resolution full-field optical coherence tomography: A pre-clinical study of human breast tissue and cancer assessment. Technol Cancer Res Treat 13:455–468. https://doi.org/10.7785/tcrt.2013.600254
16. Yin C, Wei L, Abeytunge S, Peterson G, Rajadhyaksha M, Liu JTC (2019) Label-free in vivo pathology of human epithelia with a high-speed handheld dual-axis confocal microscope. J Biomed Opt 24:1. https://doi.org/10.1117/1.JBO.24.3.030501
17. Pezzini C, Kaleci S, Chester J, Farnetani F, Longo C, Pellacani G (2020) Reflectance confocal microscopy diagnostic accuracy for malignant melanoma in different clinical settings: systematic review and meta-analysis. J Eur Acad Dermatology Venereol 34:2268–2279. https://doi.org/10.1111/jdv.16248
18. Wong TTW, Zhang R, Hai P, Zhang C, Pleitez MA, Aft RL, Novack D V., Wang L V. (2017) Fast label-free multilayered histology-like imaging of human breast cancer by photoacoustic microscopy. Sci Adv 3:1602168. https://doi.org/10.1126/sciadv.1602168
19. Lu F-K, Basu S, Igras V, Hoang MP, Ji M, Fu D, Holtom GR, Neel VA, Freudiger CW, Fisher DE, Xie XS (2015) Label-free DNA imaging in vivo with stimulated Raman scattering microscopy. Proc Natl Acad Sci 112:11624–11629. https://doi.org/10.1073/pnas.1515121112
20. Liao C-S, Wang P, Huang CY, Lin P, Eakins G, Bentley RT, Liang R, Cheng J-X (2018) In Vivo and in Situ Spectroscopic Imaging by a Handheld Stimulated Raman Scattering Microscope. ACS Photonics 5:947–954. https://doi.org/10.1021/acsphotonics.7b01214
21. You S, Tu H, Chaney EJ, Sun Y, Zhao Y, Bower AJ, Liu Y-Z, Marjanovic M, Sinha S, Pu Y, Boppart SA (2018) Intravital imaging by simultaneous label-free autofluorescence-multiharmonic microscopy. Nat Commun 9:2125. https://doi.org/10.1038/s41467-018-04470-8
22. Tao YK, Shen D, Sheikine Y, Ahsen OO, Wang HH, Schmolze DB, Johnson NB, Brooker JS, Cable AE, Connolly JL, Fujimoto JG (2014) Assessment of breast pathologies using nonlinear microscopy. Proc Natl Acad Sci 111:15304–15309. https://doi.org/10.1073/pnas.1416955111
23. Wang L V., Hu S (2012) Photoacoustic tomography: In vivo imaging from organelles to organs. Science 335:1458–1462. https://doi.org/10.1126/science.1216210
24. Wang L V., Yao J (2016) A practical guide to photoacoustic tomography in the life sciences. Nat Methods 13:627–638. https://doi.org/10.1038/nmeth.3925
25. Zhou Y, Yao J, Wang L V. (2016) Tutorial on photoacoustic tomography. J Biomed Opt 21:061007. https://doi.org/10.1117/1.jbo.21.6.061007
26. Wang L V., Gao L (2014) Photoacoustic Microscopy and Computed Tomography: From Bench to Bedside. Annu Rev Biomed Eng 16:155–185. https://doi.org/10.1146/annurev-bioeng-071813-104553
27. Yao J, Wang L V. (2013) Photoacoustic microscopy. Laser Photon Rev 7:758–778. https://doi.org/10.1002/lpor.201200060
28. Yao J, Wang L V. (2014) Sensitivity of photoacoustic microscopy. Photoacoustics 2:87–101. https://doi.org/10.1016/j.pacs.2014.04.002
29. Hu S, Wang L V. (2010) Photoacoustic imaging and characterization of the microvasculature. J Biomed Opt 15:011101. https://doi.org/10.1117/1.3281673

30. Li X, Tsang VTC, Kang L, Zhang Y, Wong TTW (2021) High-speed high-resolution laser diode-based photoacoustic microscopy for in vivo microvasculature imaging. Vis Comput Ind Biomed Art 4:1. https://doi.org/10.1186/s42492-020-00067-5
31. Li D, Humayun L, Vienneau E, Vu T, Yao J (2021) Seeing through the Skin: Photoacoustic Tomography of Skin Vasculature and Beyond. JID Innov 1:100039. https://doi.org/10.1016/j.xjidi.2021.100039
32. Yao J, Wang L, Yang JM, Maslov KI, Wong TTW, Li L, Huang CH, Zou J, Wang L V. (2015) High-speed label-free functional photoacoustic microscopy of mouse brain in action. Nat Methods 12:407–410. https://doi.org/10.1038/nmeth.3336
33. Li X, Yeung K, Tsang VTC, Huang B, Lo CTK, Wong TTW (2022) Low-cost high-resolution photoacoustic microscopy of blood oxygenation with two laser diodes. Biomed Opt Express 13:3893. https://doi.org/10.1364/BOE.458645
34. Liu C, Chen J, Zhang Y, Zhu J, Wang L (2021) Five-wavelength optical-resolution photoacoustic microscopy of blood and lymphatic vessels. Adv Photonics 3:016002. https://doi.org/10.1117/1.ap.3.1.016002
35. Sarna T, Burke JM, Korytowski W, Różanowska M, Skumatz CMB, Zaręba A, Zaręba M (2003) Loss of melanin from human RPE with aging: possible role of melanin photooxidation. Exp Eye Res 76:89–98. https://doi.org/10.1016/S0014-4835(02)00247-6
36. Zhang X, Zhang HF, Puliafito CA, Jiao S (2011) Simultaneous in vivo imaging of melanin and lipofuscin in the retina with photoacoustic ophthalmoscopy and autofluorescence imaging. J Biomed Opt 16:080504. https://doi.org/10.1117/1.3606569
37. Shu X, Li H, Dong B, Sun C, Zhang HF (2017) Quantifying melanin concentration in retinal pigment epithelium using broadband photoacoustic microscopy. Biomed Opt Express 8:2851. https://doi.org/10.1364/boe.8.002851
38. Matthews TP, Zhang C, Yao D-K, Maslov KI, Wang L V. (2014) Label-free photoacoustic microscopy of peripheral nerves. J Biomed Opt 19:1. https://doi.org/10.1117/1.JBO.19.1.016004
39. Kottmann J, Rey JM, Sigrist MW (2016) Mid-infrared photoacoustic detection of glucose in human skin: Towards non-invasive diagnostics. Sensors (Switzerland) 16:1–14. https://doi.org/10.3390/s16101663
40. Yang L, Chen C, Zhang Z, Wei X (2022) Glucose Determination by a Single 1535 nm Pulsed Photoacoustic Technique: A Multiple Calibration for the External Factors. J Healthc Eng 2022:9593843. https://doi.org/10.1155/2022/9593843
41. Zhang R, Gao F, Feng X, Liu S, Kishor R, Luo Y, Zheng Y (2017) Noninvasive photoacoustic measurement of glucose by data fusion. Analyst 142:2892–2896. https://doi.org/10.1039/c7an00743d
42. Yao D-K, Maslov K, Shung KK, Zhou Q, Wang L V. (2010) In vivo label-free photoacoustic microscopy of cell nuclei by excitation of DNA and RNA. Opt Lett 35:4139. https://doi.org/10.1364/OL.35.004139
43. Imai T, Shi J, Wong TTW, Li L, Zhu L, Wang L V. (2018) High-throughput ultraviolet photoacoustic microscopy with multifocal excitation. J Biomed Opt 23:1. https://doi.org/10.1117/1.jbo.23.3.036007
44. Yao J, Wang L, Yang J-M, Gao LS, Maslov KI, Wang L V., Huang C-H, Zou J (2012) Wide-field fast-scanning photoacoustic microscopy based on a water-immersible MEMS scanning mirror. J Biomed Opt 17:1. https://doi.org/10.1117/1.jbo.17.8.080505
45. Lan B, Liu W, Wang Y, Shi J, Li Y, Xu S, Sheng H, Zhou Q, Zou J, Hoffmann U, Yang W, Yao J (2018) High-speed widefield photoacoustic microscopy of small-animal hemodynamics. Biomed Opt Express 9:4689. https://doi.org/10.1364/boe.9.004689
46. Li X, Kang L, Zhang Y, Wong TTW (2020) High-speed label-free ultraviolet photoacoustic microscopy for histology-like imaging of unprocessed biological tissues. Opt Lett 45:5401. https://doi.org/10.1364/ol.401643

47. Li X, Kot JCK, Tsang VTC, Lo CTK, Huang B, Tian Y, Wong IHM, Cheung HHY, Kang L, Chang ACY, Wong TTW (2022) Ultraviolet photoacoustic microscopy with tissue clearing for high-contrast histological imaging. Photoacoustics 25:100313. https://doi.org/10.1016/j.pacs.2021.100313
48. Cai D, Wong TTW, Zhu L, Shi J, Chen S-L, Wang L V. (2018) Dual-view photoacoustic microscopy for quantitative cell nuclear imaging. Opt Lett 43:4875. https://doi.org/10.1364/OL.43.004875
49. Baik JW, Kim H, Son M, Choi J, Kim KG, Baek JH, Park YH, An J, Choi HY, Ryu SY, Kim JY, Byun K, Kim C (2021) Intraoperative Label-Free Photoacoustic Histopathology of Clinical Specimens. Laser Photonics Rev 15:1–11. https://doi.org/10.1002/lpor.202100124
50. Li X, Kang L, Lo CTK, Tsang VTC, Wong TTW (2022) High-Speed Ultraviolet Photoacoustic Microscopy for Histological Imaging with Virtual-Staining assisted by Deep Learning. J Vis Exp 2022:63649. https://doi.org/10.3791/63649
51. Wong TTW, Zhang R, Zhang C, Hsu HC, Maslov KI, Wang L, Shi J, Chen R, Shung KK, Zhou Q, Wang L V. (2017) Label-free automated three-dimensional imaging of whole organs by microtomy-assisted photoacoustic microscopy. Nat Commun 8:1386. https://doi.org/10.1038/s41467-017-01649-3
52. Cao R, Zhao J, Li L, Du L, Zhang Y, Luo Y, Jiang L, Davis S, Zhou Q, de la Zerda A, Wang L V. (2023) Optical-resolution photoacoustic microscopy with a needle-shaped beam. Nat Photonics 17:89–95. https://doi.org/10.1038/s41566-022-01112-w
53. Abbasi S, Le M, Sonier B, Dinakaran D, Bigras G, Bell K, Mackey JR, Haji Reza P (2019) All-optical Reflection-mode Microscopic Histology of Unstained Human Tissues. Sci Rep 9:13392. https://doi.org/10.1038/s41598-019-49849-9
54. Ecclestone BR, Bell K, Sparkes S, Dinakaran D, Mackey JR, Haji Reza P (2022) Label-free complete absorption microscopy using second generation photoacoustic remote sensing. Sci Rep 12:8464. https://doi.org/10.1038/s41598-022-11235-3
55. Hajireza P, Shi W, Bell K, Paproski RJ, Zemp RJ (2017) Non-interferometric photoacoustic remote sensing microscopy. Light Sci Appl 6:e16278–8. https://doi.org/10.1038/lsa.2016.278
56. Bell K, Abbasi S, Dinakaran D, Taher M, Bigras G, van Landeghem FKH, Mackey JR, Haji Reza P (2020) Reflection-mode virtual histology using photoacoustic remote sensing microscopy. Sci Rep 10:19121. https://doi.org/10.1038/s41598-020-76155-6
57. Haven NJM, Kedarisetti P, Restall BS, Zemp RJ (2020) Reflective objective-based ultraviolet photoacoustic remote sensing virtual histopathology. Opt Lett 45:535. https://doi.org/10.1364/OL.382415
58. Cikaluk BD, Restall BS, Haven NJM, Martell MT, McAlister EA, Zemp RJ (2023) Rapid ultraviolet photoacoustic remote sensing microscopy using voice-coil stage scanning. Opt Express 31:10136. https://doi.org/10.1364/OE.481313
59. Boktor M, Ecclestone BR, Pekar V, Dinakaran D, Mackey JR, Fieguth P, Haji Reza P (2022) Virtual histological staining of label-free total absorption photoacoustic remote sensing (TA-PARS). Sci Rep 12:10296. https://doi.org/10.1038/s41598-022-14042-y
60. Hariri A, Alipour K, Mantri Y, Schulze JP, Jokerst J V. (2020) Deep learning improves contrast in low-fluence photoacoustic imaging. Biomed Opt Express 11:3360. https://doi.org/10.1364/BOE.395683
61. He D, Zhou J, Shang X, Tang X, Luo J, Chen S-L (2023) De-Noising of Photoacoustic Microscopy Images by Attentive Generative Adversarial Network. IEEE Trans Med Imaging 42:1349–1362. https://doi.org/10.1109/TMI.2022.3227105
62. Zhou J, He D, Shang X, Guo Z, Chen S-L, Luo J (2021) Photoacoustic microscopy with sparse data by convolutional neural networks. Photoacoustics 22:100242. https://doi.org/10.1016/j.pacs.2021.100242
63. Cheng S, Zhou Y, Chen J, Li H, Wang L, Lai P (2022) High-resolution photoacoustic microscopy with deep penetration through learning. Photoacoustics 25:100314. https://doi.org/10.1016/j.pacs.2021.100314

64. Ecclestone BR, Bell K, Abbasi S, Dinakaran D, van Landeghem FKH, Mackey JR, Fieguth P, Haji Reza P (2020) Improving maximal safe brain tumor resection with photoacoustic remote sensing microscopy. Sci Rep 10:17211. https://doi.org/10.1038/s41598-020-74160-3
65. Chen Z, Yu W, Wong IHM, Wong TTW (2021) Deep-learning-assisted microscopy with ultraviolet surface excitation for rapid slide-free histological imaging. Biomed Opt Express 12:5920. https://doi.org/10.1364/BOE.433597
66. Labaca-Castro R (2023) Generative Adversarial Nets. In: Ghahramani Z, Welling M, Cortes C, Lawrence N, Weinberger KQ (eds) Machine Learning under Malware Attack. Springer Fachmedien Wiesbaden, Wiesbaden, pp 73–76
67. Isola P, Zhu J-Y, Zhou T, Efros AA (2017) Image-to-Image Translation with Conditional Adversarial Networks. In: 2017 IEEE Conference on Computer Vision and Pattern Recognition (CVPR). IEEE, pp 5967–5976
68. Zhu J-Y, Park T, Isola P, Efros AA (2017) Unpaired Image-to-Image Translation Using Cycle-Consistent Adversarial Networks. In: 2017 IEEE International Conference on Computer Vision (ICCV). IEEE, pp 2242–2251
69. Kang L, Li X, Zhang Y, Wong TTW (2022) Deep learning enables ultraviolet photoacoustic microscopy based histological imaging with near real-time virtual staining. Photoacoustics 25:100308. https://doi.org/10.1016/j.pacs.2021.100308
70. Cao R, Nelson SD, Davis S, Liang Y, Luo Y, Zhang Y, Crawford B, Wang L V (2022) Label-free intraoperative histology of bone tissue via deep-learning-assisted ultraviolet photoacoustic microscopy. Nat Biomed Eng 7:124–134. https://doi.org/10.1038/s41551-022-00940-z
71. Mao X, Li Q, Xie H, Lau RYK, Wang Z, Smolley SP (2019) On the Effectiveness of Least Squares Generative Adversarial Networks. IEEE Trans Pattern Anal Mach Intell 41:2947–2960. https://doi.org/10.1109/TPAMI.2018.2872043

Chapter 20
Recent Advances in Photoacoustic Imaging of Breast Cancer

Huijuan Zhang, Emily Zheng, and Jun Xia

Abstract Photoacoustic imaging (PAI) has emerged as a promising modality for enhancing breast cancer diagnosis and treatment monitoring. By leveraging the distinct optical absorption properties of hemoglobin, PAI combines functional and anatomical information, offering a noninvasive and high-resolution approach to visualizing breast tumors. Numerous clinical trials and studies have demonstrated the potential of PAI in detecting breast tumors and guiding therapeutic decisions based on physiological markers. This book chapter reviews recent advances in PA-based breast imaging. We classified the imaging system based on the patient's pose and summarized the key features and capabilities of PAI for breast cancer imaging. The integration of ultrasound and PAI provides comprehensive insights into tumor characteristics, enabling accurate assessment of morphology, vascularity, and functional parameters. The fusion of functional and anatomical insights makes PAI to be an innovative tool for early screening, diagnosis, and treatment monitoring of breast cancer. We believe that PAI has the potential to revolutionize breast cancer management, ultimately improving patient outcomes and transforming the landscape of breast cancer diagnostics and treatment.

Keywords Photoacoustic · Optoacoustic · Breast cancer · Tumor · Diagnosis · Treatment · Blood oxygenation · Ultrasound · Vasculature

20.1 Introduction

Breast cancer is the leading cause of cancer among women worldwide [1]. As evidence shows, 1 in 8 women will develop invasive breast cancer in their lifetime, and early detection can reduce mortality rates [2, 3]. Therefore, medical imaging technologies play an important role in identifying tumors at early stage

---

H. Zhang · E. Zheng · J. Xia (✉)
Department of Biomedical Engineering, University at Buffalo, The State University of New York, Buffalo, NY, USA
e-mail: junxia@buffalo.edu

W. Xia (ed.), *Biomedical Photoacoustics*,
https://doi.org/10.1007/978-3-031-61411-8_20

of the disease. X-ray mammography, ultrasound imaging, and magnetic resonance imaging (MRI) are the three standard radiological imaging technologies for breast cancer imaging [4, 5]. Mammography is the most widely used screening technique for breast cancer. However, it has a lower sensitivity in dense breast tissue (24–47% sensitivity in women with heterogeneously dense breasts compared with 71–78% sensitivity in women with fatty breasts [6, 7]). In addition, mammography also involves painful breast compression and ionizing radiation [8, 9]. Ultrasonography (US) is an adjunct technique to mammography, but it has low specificity (65–89%) [10]. The imaging results are highly dependent on the experience and skills of the operators. MRI has a high sensitivity than mammography, but it shows a relatively low specificity, because sometimes both benign and malignant lesions get enhanced signals [11]. In addition, breast MRI requires a metal-based contrast agent, gadolinium, which can cause an allergic reaction or damage the nervous system in the body [12]. Moreover, it has a high cost and longer imaging time and is not universally applicable to developing countries. Therefore, a new breast imaging modality is urgently needed to overcome the aforementioned limitations [13].

Photoacoustic imaging (PAI) has great potential to meet this requirement. PAI possesses several unique characteristics, such as the absence of ionizing radiation and the elimination of the need for exogenous contrast agents [14]. It exploits the absorption of near-infrared (NIR) light by oxy-(HbO_2) and deoxy-hemoglobin (Hb) molecules, thereby enabling the capture of both functional and vascular features of the breast tissue [15]. The capacity to reveal microvascular and macrovascular features is particularly valuable in the identification of malignant tissue, given the correlation between tumor growth and angiogenesis [16, 17]. Moreover, the functional information provided by PAI is equally significant since tumor regions frequently exhibit reduced blood oxygenation due to hypoxia [18]. By harnessing the natural contrast properties of hemoglobin molecules in breast tissue and utilizing NIR light, PAI overcomes the limitations of existing breast imaging modalities and can be applied effectively regardless of breast density [19]. The use of endogenous imaging contrast facilitates enhanced diagnostic accuracy and minimizes potential risks associated with contrast agents or ionizing radiation. As a result, PAI holds immense promise as a safe, efficient, and highly informative technique for breast imaging and could significantly advance the field of breast cancer diagnosis and monitoring.

This book chapter reviews recent advances in PA-based breast imaging. The first section classified the imaging system based on the patient's pose. The second section presents PA features of breast malignancy, and the last section reviews recent studies in PA-based treatment monitoring. As there are already a few review articles on PA breast imaging [15, 20], this book chapter tried to focus on studies reported in the last 5 years.

20.2 PAI Systems for Breast Cancer Screening

Based on the patient's imaging position, PAI breast systems can be classified into three categories: prone position, supine position, and upright position. Each position has its unique advantages and disadvantages, as discussed below.

20.2.1 Prone Position

The prone position is commonly seen in the design of PAI systems. Such designs often mimic the MRI examination, where the patient lies prone on a patient bed, and one side of the breast is pendant into an aperture imaging window for scanning. For good acoustic coupling, the breast will be either immersed in the water tank or held by a bowl-shaped mold. A demonstration of the prone imaging system is shown in Fig. 20.1a [21]. This imaging position allows for 3D scanning of the breast, thus providing a volumetric image of the whole breast. These systems typically used ring or hemispherical arrays, which helped to address the limited view problem in PAI [22]. As a trade-off, the light fluence intensity might be reduced when the laser beam is expanded to illuminate the whole breast. In the following sections, we review four prone position PAI systems.

The PAI-04 system used a hemispherical detector array (HDA) [23], which was initially developed by Kruger et al. [24]. The HDA has a central frequency of 4 MHz and 500 elements spirally distributed in the hemisphere. Two laser wavelengths (756 and 797 nm) were used for HbO_2 and Hb detection. To increase the sampling density and field of view, the HDA performed scanning in a spiral pattern. As the HAD was designed in receiver-only mode, another 256-element linear array, placed adjacent to HAD, was used for ultrasound image acquisition. Image co-registration was needed as PA and ultrasound were acquired using two different transducers. This system provided the imaging of tumor-related blood vessels and blood oxygenation levels. An overlaid view of the 3D PA and US

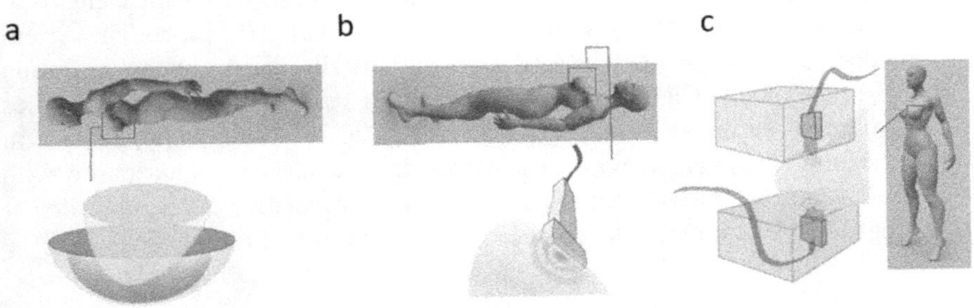

Fig. 20.1 Different breast imaging positions in PAI. (**a**) The prone imaging position. (**b**) The supine imaging position. (**c**) The upright imaging position

images is shown in Fig. 20.2a. Fine blood vessels were observed in the vicinity of the tumor. A B-mode ultrasound image can be found in Fig. 20.2b, where a hypoechoic region can be seen. The selected patient case has a tumor located near the surface (depth <10 mm). Therefore, it is unclear whether fine vessels from the tumor can be revealed at deeper depths.

Single-breath-hold photoacoustic computed tomography (SBH-PACT) system used a 512-element full-ring transducer array to acquire the tomographic image of the breast [25]. The array had a central frequency of 2.25 MHz, and the excitation laser had a single wavelength of 1064 nm. The patient lay in a prone position with the breast being slightly compressed. The breast was then scanned perpendicularly by the ring array through the open hole of the table. With the help of a 512-channel data acquisition system, the system could acquire a volumetric image in 15 s (within a single breath-hold). This system allows visualization of breast vasculature at 0.25-mm resolution and up to a depth of 4 cm. The PA elastography technique allows the system to identify missed tumors in vascular images. Exemplary PA features are shown in Fig. 20.2c. The left panel indicates regions with a high vessel density, while the right panel displays the PA elastography map, where a red oval marks the tumor region. However, the SBH-PACT system lacks US imaging capabilities, which makes it impossible to provide co-registered PA and US images.

The Twente Photoacoustic Mammoscope 2 system used 12 1-MHz 32-element-arc-shaped arrays in a bowl shape to acquire the volumetric PA breast image [5]. The patient lay in a prone position and suspended the breast inside a water tank for imaging. A dual-wavelength (755 and 1064 nm) laser illuminated the tissue with uniform distribution on both the bottom and side of the breast. This system can visualize the 3D contour shape of the breast from the nipple to the chest wall, as shown in Fig. 20.3d. The vascular image indicated a 22 mm imaging depth. While the system is capable of functional imaging, recent studies focused on investigating the vascular features related to malignancy [26]. However, this system lacked co-registered ultrasound imaging capabilities.

Laser Optoacoustic Ultrasonic Imaging System Assembly (LOUISA-3D) provided full-view photoacoustic imaging via an arc-shaped wideband transducer (50 kHz to 6 MHz) [27]. The patient lay prone for examination with the breast placed in a hemispherical mold. The arc-shaped optical fiber had binary laser wavelengths of 757 and 797 nm, rotationally illuminating the breast within the probe. The system provided functional images of total hemoglobin (HbT) and oxygen saturation (sO_2) in the blood. The visualized deoxygenated veins and oxygenated arteries of the breast are shown in the top image of Fig. 20.2e. Additionally, the system can perform ultrasound imaging by using another arc-shape transducer array with 192 elements and 7 MHz central frequency. A representative 2D sagittal slice of co-registered PA and US images is shown at the bottom of Fig. 20.2e.

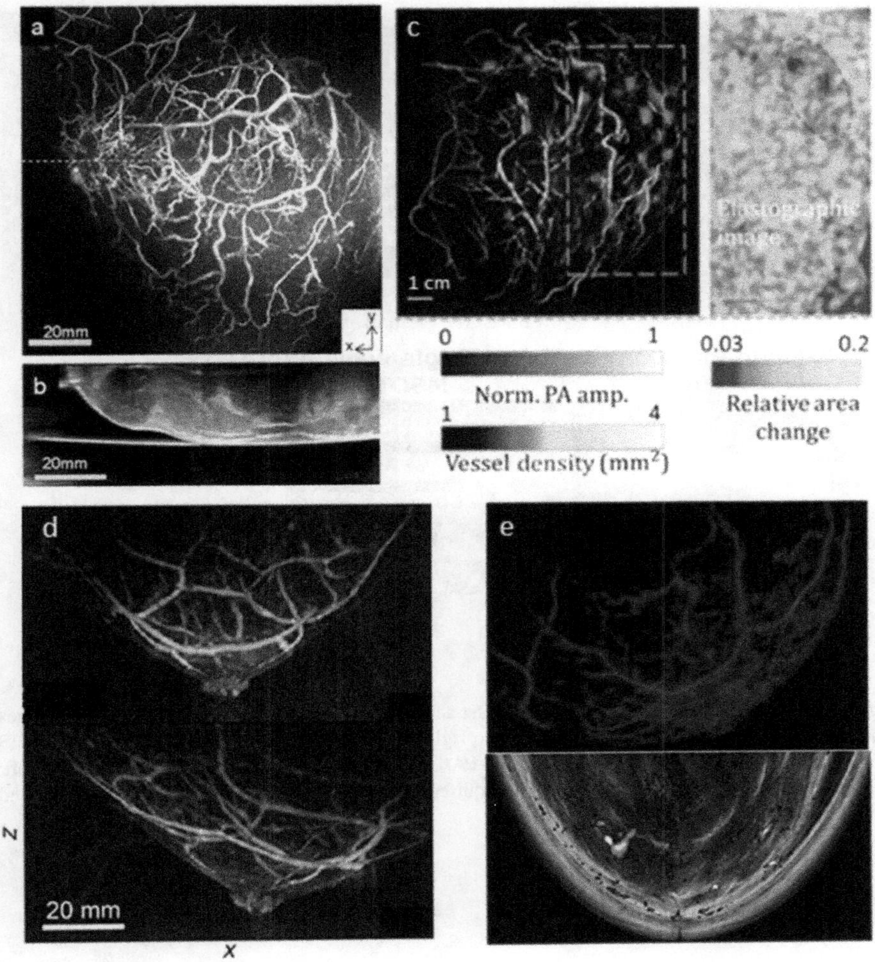

Fig. 20.2 Exemplary PA images of the breast acquired in the prone position. (**a**) A fusion image of the PA (gray) and US (red) acquired by the PAI-04 system. Several clustered fine blood vessels were observed in the vicinity of the tumor. (**b**) US B-mode image along the yellow dashed line of a. A hypoechoic region is visible in the cross section. (**c**) PAI images acquired by the SBH-PACT system. Left: vessel density map overlaid on the grayscale vessel image. Right: PA elastography map of the region outlined by the blue dash box in the vessel density map. (**d**) PA images acquired by the Twente photoacoustic mammoscope system. The top and bottom images represent two sides of the breast of a healthy volunteer. The 3D contour shape of the breast was depicted. (**e**) PA images acquired by the LOUISA-3D system. The top image is a MAP PA image demonstrating the breast vascular structure. The bottom image is a 2D sagittal slice of PA and ultrasound overlay. (Images were adapted with permission from [5, 23, 25, 27])

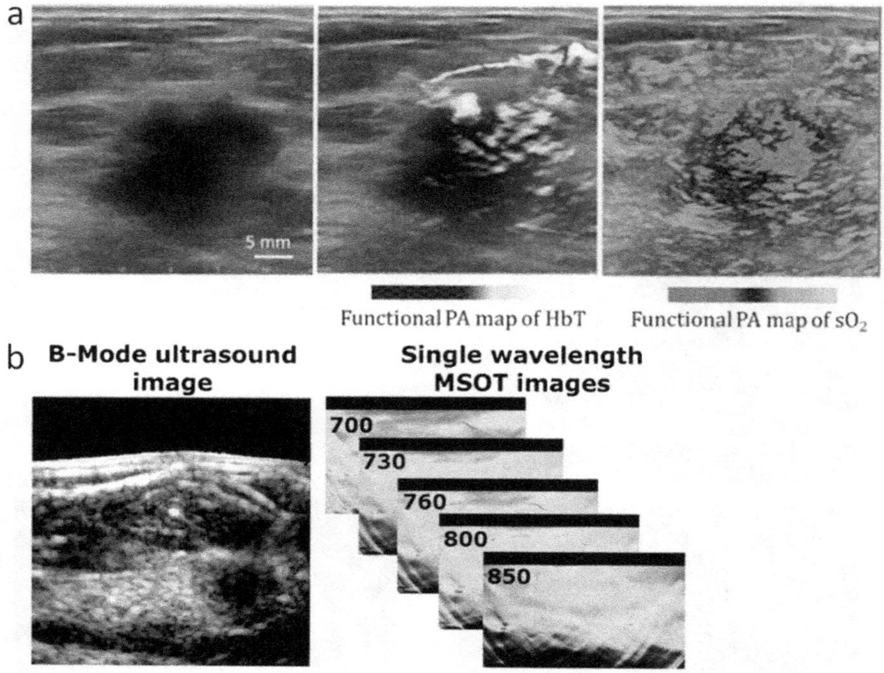

Fig. 20.3 Exemplary PA images of the breast acquired in the supine position. (**a**) PA/US images acquired by the Imagio system. From left to right: B-mode US image, HbT map fused with US image, SO_2 map fused with US image. (**b**) PA/US images acquired by the MSOT system. Left: B-mode US image. Right: MSOT images acquired at different wavelengths. (Images were adapted with permission from [3, 30])

20.2.2 Supine Position

The supine position is typically used in handheld PAI systems, mimicking the clinical breast ultrasound. A demonstration of the supine imaging system is shown in Fig. 20.1b. Due to the nature of handheld imaging, these systems cannot provide volumetric imaging. Two supine position imaging systems are discussed in the section below.

Imagio system from *Seno Medical* is a handheld PA/US device, which is also the first Food and Drug Administration (FDA)-approved PAI system for breast cancer imaging [28]. Imagio consists of a 128-element linear array (bandwidth from 0.1 to 12 MHz) with an integrated optical fiber bundle on both sides of the probe [29, 30]. The system used lasers with two wavelengths (757 and 1064 nm) to visualize HbO_2, Hb, and HbT distributions in human breasts. Due to its handheld design, co-registered functional PA and ultrasound images can be displayed in real time up to an imaging depth of 3 cm. Representative images are shown in Fig. 20.3a.

The grayscale ultrasound image shows the hypoechoic malignant mass. The HbT map in yellow indicates a high density of angiogenesis in the tumor. The oxygen saturation map further shows the distribution of Hb (red) and HbO_2 (green) around the suspicious region. As handheld ultrasound is already widely used for breast imaging, the Imagio system has a high potential for clinical translation. However, the 2D handheld system is operator-dependent and cannot provide 3D information about the breast.

Multispectral optoacoustic tomography (MSOT) is a handheld imaging system developed by Becker et al. [3]. MSOT utilized a 256-element curved array probe with 3 MHz central frequency. An integrated fiber bundle delivered the laser beam at five wavelengths of 700 nm, 730 nm, 760 nm, 800 nm, and 850 nm. The patients were scanned in a supine position. With this setup, the penetration depth could reach up to 3 cm. HbO_2, Hb, HbT, and oxygen saturation features were assessed at imaging depths of 5 mm, 10 mm, and 15 mm. Deeper depths were not quantified, likely due to light attenuation. The team demonstrated that MSOT could constantly provide stable and reproducible breast images at all wavelengths. The hybrid US/MSOT system provided co-registered PA-US images simultaneously. Representative results are shown in Fig. 20.3b, where the left panel displays the B-mode ultrasound image. The right panel shows multiwavelength MSOT images, which present the absorption profile of tissue chromophores. Similar to the Imagio system, the MSOT system is also operator-dependent and cannot provide 3D vascular information about the breast.

20.2.3 Upright Position

The upright systems image the patient in the standing or sitting pose, as shown in Fig. 20.1c. It mimics the imaging procedure of mammography but only requires mild compression, which helps to reduce the optical tissue thickness and increases the chance of capturing the tumor deep inside the breast. Moreover, the upright design saves clinical space and makes the system compact.

The dual-scan mammoscope (DSM) is the only PAI system that images breast in the standing pose [32]. During imaging, the patient stayed in a standing pose and the breast was compressed between two water tanks with mild compression. DSM employed two 128-element linear arrays with a central frequency of 2.25 MHz and a bifurcated fiber bundle for the double-side illumination (one for each transducer). A 1064 nm wavelength is used to visualize the total hemoglobin in the breast. The two linear arrays scanned the breast along the elevational direction, thus achieving volumetric visualization of the breast. Data from both transducers were combined and aligned for complete visualization of vascular structures in the compressed breast region. With the help of double-sided illumination, DSM achieved a depth of 7 cm, which represents the deepest PAI breast system to date. In addition, the DSM system allowed US data acquisition between PA pulses to enable naturally

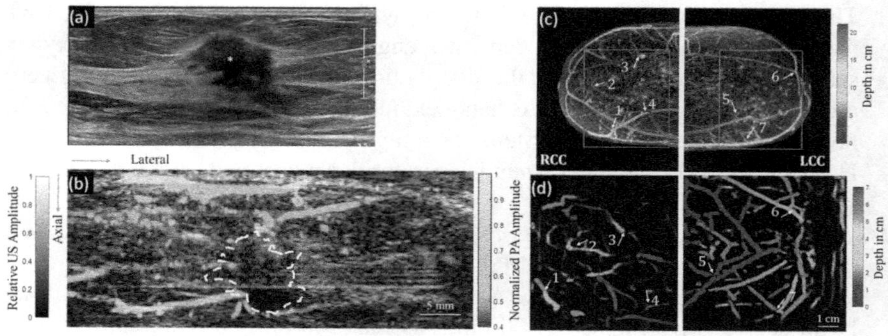

Fig. 20.4 Exemplary PA images of the breast acquired in the upright position. (**a**) Clinical US image of a subject. (**b**) Fused PA and US cross-sectional images acquired by the DSM system. The tumor is marked by dashed lines. (**c**) Clinical MRI image of another subject. RCC: right craniocaudal view. LCC: left craniocaudal view. (**d**) DSM images of the same subject in (**c**) displayed in RCC and LCC views. Breast vessels can be seen through 7 cm of tissue thickness. The white arrows show identical vessels in MRI and PAI. (Images were adapted with permission from [31, 32])

co-registered PA and US acquisition. Representative imaging results are depicted in Fig. 20.4. Recent developments in the system further allow a larger view of view and multimodal PA and US elastography imaging [33]. The system has limitations, including the use of a single wavelength (1064 nm), which precludes functional characterization, and limited view problems due to the use of linear transducer arrays. These can potentially be addressed by deep learning methods [34].

20.3 PAI for Breast Cancer Identification

Most systems mentioned in the section above have been tested in the clinic to study PA features of breast cancer. These studies provided strong evidence that PAI could be used for breast cancer screening and diagnosis. In Table 20.1, we compiled a list of PAI characteristics and highlighted their association with breast cancer.

Oxygenation saturation is a crucial functional characteristic in cancer diagnosis, as evidenced by several studies [18, 29, 30]. The rapid growth of tumors often leads to inadequate nourishment, resulting in hypoxia within the malignant tissue. A promising approach to assess hemoglobin saturation status involves using two wavelengths in the near-infrared (NIR) region. In this regard, Fig. 20.5a illustrates the detection of deoxygenated blood signals (red) at the border of the malignant mass, while oxygenated blood (green) predominantly surrounds the normal tissue [30]. This tumor feature has been successfully demonstrated in multiple studies [18, 30].

HbT concentration serves as another essential functional indicator for photoacoustic (PA) breast cancer imaging, as demonstrated by multiple groups [18,

Table 20.1 PA features of breast tumor

PAI features	Correlation with breast malignancy
Oxygen saturation	Tumor regions are prone to show deoxygenated blood signal.
HbT concentration	Higher level of total hemoglobin concentration is seen at tumor regions
Vascular density	Higher number of vessels in the malignant breast
Vascular branching points	Larger number of VBP
Centripetal vasculature/Peripheral vascularity	Vasculature around the tumor region grows centripetally around it and becomes narrower at the tumor edge
Vessel intensity/Vessel contrast	Higher intensity/Stronger contrast
PA elastography	Tumor region shows smaller relative area change from motion

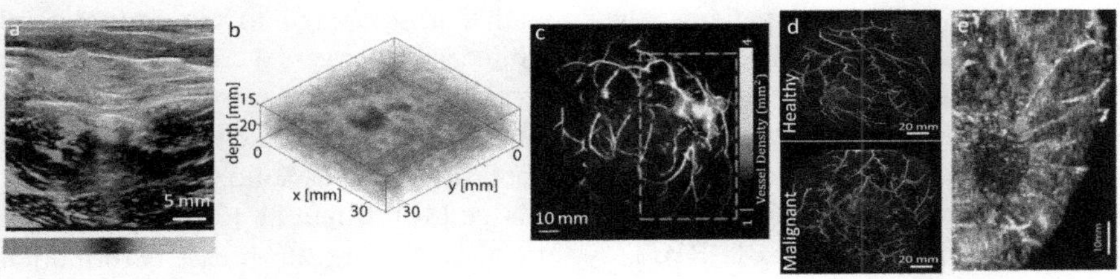

Fig. 20.5 PA features of breast malignancy. (**a**) PA blood oxygenation saturation map overlaid on the gray scale ultrasound. Red denotes deoxygenated blood, and green denotes normally oxygenated blood [30]. (**b**) The HbT image of infiltrating lobular carcinoma (ILC, grade 2) of 22 mm in a 65-year-old patient present [36]. (**c**) Vessel density map of the breast with invasive ductal carcinoma (IDC) in a 69-year-old female patient shows the tumor, marked in a green circle [25]. (**d**) VBPs (red dots) overlaid on the grayscale vascular map. Compared to the healthy breast (top), increased VBPs are seen in the breast with tumor (bottom) [37]. (**e**) An example of centripetal vasculature is shown in the PA/US co-registered image of a lesion of invasive ductal carcinoma [21]

29, 30]. The hypoxic condition in tumor stimulates the expression of vascular endothelial growth factor, leading to the growth of leaky capillaries [35]. As a result, malignant tissue exhibits a higher HbT concentration compared to normal tissue. A representative HbT map is shown in Fig. 20.5b, where the tumor lesion can be clearly identified.

Vascular density could be a feature associated with rapidly growing tumors. As the literature suggests, tumor angiogenesis is a biomarker for malignancy [16]. The vessel density map provides valuable information about the distribution and density of blood vessels in the breast tissue. Studies by Heijblom et al. (lower resolution) [36] and Lin et al. (higher resolution) [25] have both identified regions with increased vessel density that were confirmed to be tumors. Figure 20.5c illustrates a representative case where a green circle marks high vessel density at the tumor site [25].

Vascular branching points (VBPs) have been proposed as a possible biomarker for breast cancer by Yamaga et al. [37]. A representative image is shown in Fig. 20.5d. Compared to the contralateral healthy breast, a higher number of VBPs were observed in the tumor-bearing breast, reflecting sustained angiogenesis in malignant tumors.

Centripetal vasculature/peripheral vascularity is also a tumor-related feature. This unique characteristic signifies tumor-related blood vessels converging from the surrounding normal breast tissue toward the center of the tumor. As depicted in Fig. 20.5e, the PA mammography system [21, 38] observed the centripetal growth of blood vessels around the tumor region, with the PA intensity weakening at the tumor boundary. In addition, Diot et al. [18] reported the presence of high peripheral vasculature around the tumor, observed across various tumor types.

Vessel diameter/vessel contrast is a feature proposed by Nikhila et al. [31]. After analyzing data from 38 patients, they found that compared to the contralateral healthy breast, the tumor-bearing breast exhibited stronger vessel contrast and larger variations in background signals. These observations are likely caused by the tumor-related increase in blood volume, vessel diameter, and vascular activity [39, 40].

PA elastography can be used to identify breast tumors, as solid tumors typically exhibit higher stiffness than normal breast tissue [41]. Lin et al. [25] implemented PA elastography in the SBH-PACT system by analyzing the breast deformation during patient breathing. Tumor regions, being more rigid, showed relatively lower area changes compared to the surrounding normal breast tissue. As shown in Fig. 20.2c, the team was able to quantify regions with less deformation (higher stiffness). The tumor was marked by a red dashed oval.

20.4 PAI for Treatment Monitoring

Chemotherapy is commonly used for breast cancer treatment, but its effectiveness varies among patients. Monitoring the response to chemotherapy is crucial to treatment optimization and dosage adjustment. PAI has been investigated as an alternative approach for breast cancer treatment monitoring.

Vascular morphology has been proven to be a viable biomarker for monitoring chemotherapy-induced tumor downstaging [42]. As shown in Fig. 20.6a–c, PAT imaging conducted at time points before, during, and after the neoadjuvant chemotherapy (NAC) [43] reveals a decrease in vascular density, entropy, and anisotropy. In comparison, the same parameters are almost unchanged for the healthy side of the same patient. Other studies [15, 44] observed increased intratumoral blood flow in PA images after chemotherapy. Researchers believe this effect is caused by the increased blood flow following the therapy and can be interpreted as a sign of normalization of tumor vasculature. Based on that, the adaptation of PA

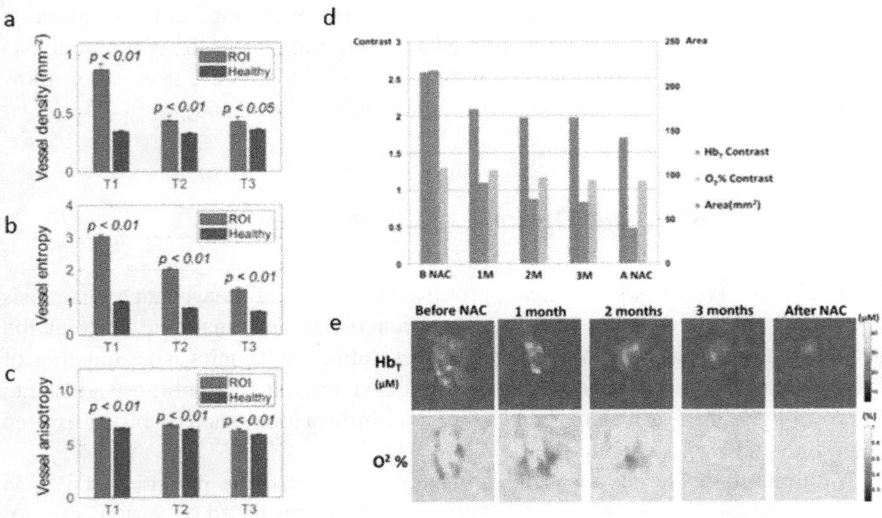

Fig. 20.6 PAI features used for breast treatment monitoring. (**a–c**) The regional vessel density, vessel entropy, and anisotropy distribution of both breasts of the same patient. T1–T3 referred to different time points. T1: before initiation of NAC; T2: during NAC; and T3: after completion of NAC. For each imaging, the malignant side is compared to the healthy side of the same patient, and the corresponding p-value is calculated. (**d**) Quantification of hemoglobin concentration, oxygen saturation status, and the size of the malignant lesion measured before NAC (B NAC), one month (1M), two months (2M), and three months (3M) after the first NAC treatment till complete NAC treatment and before surgery (A NAC). (**e**) PA images of a patient with high-grade ductal carcinoma in situ (DCIS) and invasive mammary carcinoma. Changes in hemoglobin concentration and oxygen saturation status can be seen over the treatment period. (Adapted with permission from [43, 45])

can be extended to not only chemotherapy but also molecular targeting therapies such as anti-HER2 antibody treatment.

Hemoglobin and oxygen saturation can be evaluated by multispectral PA modality for treatment monitoring. PAI was conducted before the treatment and right after the first treatment for three consecutive months [45]. Fig. 20.6d, e show the experiment results, which revealed a gradual decrease in the total hemoglobin and oxygen saturation levels. Besides tomographic imaging, microscopic PAI has also been proposed for treatment guidance. The study led by Hai et al. developed a single-cell PA microscopy that can image intratumoral metabolic heterogeneity from excised tissue [46]. They found that the tumor cell suspensions show higher cellular metabolic heterogeneity compared to the healthy cells extracted from the same patient, and this characteristic can potentially be adapted to monitor the resistance or sensitivity of specific cell populations to the therapy.

The size of the malignant lesion in response to treatment can also be visualized through the PAI. Several methods have been proposed to better reveal the structural information of tumor. For instance, multiwavelength imaging and lesion segmentation have been conducted to detect regions with higher hemoglobin concentration [45], and the entropy map has been utilized as a weighting factor to highlight regions with higher vascular irregularity due to angiogenesis [43]. These techniques allow the visualization of tumor shrinkage throughout NAC (Fig. 20.6d).

20.5 Conclusion and Outlook

In conclusion, PAI shows significant promise in the field of breast cancer diagnosis and treatment monitoring. Utilizing both functional and anatomic information provided by hemoglobin, PAI offers high-resolution and noninvasive imaging of breast tumors. The combination of US and PAI provides valuable insights into tumor morphology, vascularity, and functional information, such as blood oxygen saturation and HbT concentration.

Multiple clinical trials and studies have demonstrated the potential of PAI in detecting breast tumors and guiding therapeutic decisions based on various physiological parameters. However, the clinical translation of PA breast imaging still poses several challenges. Standardizing technology and operations is crucial to improving reliability and reproducibility. Radiologists and medical professionals will require specific training to interpolate new information and features obtained from PAI systems. To facilitate the clinical translation of PAI, large-scale prospective clinical trials are essential in identifying improvements and addressing challenges associated with the integration of PAI into existing diagnostic workflows.

Overall, PAI breast imaging holds great promise as a valuable tool for early screening, diagnosis, and treatment monitoring of breast cancer, with the potential to revolutionize breast cancer management and improve patient outcomes.

References

1. Arnold, Melina, et al., *Current and future burden of breast cancer: Global statistics for 2020 and 2040.* The Breast, 2022. **66**: p. 15–23.
2. Sung, H., et al., *Global cancer statistics 2020: GLOBOCAN estimates of incidence and mortality worldwide for 36 cancers in 185 countries.* CA: a cancer journal for clinicians, 2021. **71**(3): p. 209–249.
3. Becker, A., et al., *Multispectral optoacoustic tomography of the human breast: characterisation of healthy tissue and malignant lesions using a hybrid ultrasound-optoacoustic approach.* European radiology, 2018. **28**: p. 602–609.
4. Kuhl, C.K., et al., *Mammography, breast ultrasound, and magnetic resonance imaging for surveillance of women at high familial risk for breast cancer.* Journal of clinical oncology, 2005. **23**(33): p. 8469–8476.
5. Schoustra, S.M., et al., *Twente Photoacoustic Mammoscope 2: system overview and three-dimensional vascular network images in healthy breasts.* Journal of biomedical optics, 2019. **24**(12): p. 121909–121909.

6. von Euler-Chelpin, M., et al., *Sensitivity of screening mammography by density and texture: a cohort study from a population-based screening program in Denmark.* Breast Cancer Research, 2019. **21**: p. 1–7.
7. Devolli-Disha, E., et al., *Comparative accuracy of mammography and ultrasound in women with breast symptoms according to age and breast density.* Bosnian journal of basic medical sciences, 2009. **9**(2): p. 131.
8. Lee, C.H., et al., *Breast cancer screening with imaging: recommendations from the Society of Breast Imaging and the ACR on the use of mammography, breast MRI, breast ultrasound, and other technologies for the detection of clinically occult breast cancer.* Journal of the American college of radiology, 2010. **7**(1): p. 18–27.
9. Pinsky, R.W. and M.A. Helvie, *Mammographic breast density: effect on imaging and breast cancer risk.* Journal of the National Comprehensive Cancer Network, 2010. **8**(10): p. 1157–1165.
10. Cheng, H.D., et al., *Automated breast cancer detection and classification using ultrasound images: A survey.* Pattern Recognition, 2010. **43**(1): p. 299–317.
11. Shahid, H., et al., *An overview of breast MRI.* Applied Radiology, 2016. **45**(10): p. 7.
12. Grubstein, A., M. Yepes, and R. Kiszonas, *Magnetic resonance imaging of breast vascularity in medial versus lateral breast cancer.* Eur J Radiol, 2010. **75**(2): p. e9–11.
13. Vedantham, S. and A. Karellas. *Emerging Breast Imaging Technologies on the Horizon.* in *Seminars in Ultrasound, CT and MRI.* 2018. Elsevier.
14. Xia, J., J. Yao, and L.H.V. Wang, *Photoacoustic Tomography: Principles and Advances.* Progress In Electromagnetics Research, 2014. **147**: p. 1–22.
15. Manohar, S. and M. Dantuma, *Current and future trends in photoacoustic breast imaging.* Photoacoustics, 2019. **16**: p. 100134.
16. Bielenberg, D.R. and B.R. Zetter, *The contribution of angiogenesis to the process of metastasis.* The Cancer Journal, 2015. **21**(4): p. 267–273.
17. McDougall, S.R., A.R. Anderson, and M.A. Chaplain, *Mathematical modelling of dynamic adaptive tumour-induced angiogenesis: clinical implications and therapeutic targeting strategies.* Journal of theoretical biology, 2006. **241**(3): p. 564–589.
18. Diot, G., et al., *Multispectral Optoacoustic Tomography (Msot) of Human Breast Cancermsot Signatures of Human Breast Cancer.* Clinical Cancer Research, 2017. **23**(22): p. 6912–6922.
19. Kratkiewicz, K., et al., *Ultrasound and photoacoustic imaging of breast cancer: Clinical systems, challenges, and future outlook.* Journal of Clinical Medicine, 2022. **11**(5): p. 1165.
20. Nyayapathi, N. and J. Xia, *Photoacoustic imaging of breast cancer: a mini review of system design and image features.* Journal of Biomedical Optics, 2019. **24**(12): p. 121911.
21. Shiina, T., M. Toi, and T. Yagi, *Development and clinical translation of photoacoustic mammography.* Biomedical Engineering Letters, 2018. **8**: p. 157–165.
22. Huang, B., et al., *Improving limited-view photoacoustic tomography with an acoustic reflector.* Journal of biomedical optics, 2013. **18**(11): p. 110505.
23. Matsumoto, Y., et al., *Visualising peripheral arterioles and venules through high-resolution and large-area photoacoustic imaging.* Scientific reports, 2018. **8**(1): p. 14930.
24. Kruger, R.A., et al., *Dedicated 3D photoacoustic breast imaging.* Medical physics, 2013. **40**(11).
25. Lin, L., et al., *Single-breath-hold photoacoustic computed tomography of the breast.* Nature communications, 2018. **9**(1): p. 2352.
26. Schoustra, S.M., et al., *Imaging breast malignancies with the Twente Photoacoustic Mammoscope 2.* Plos one, 2023. **18**(3): p. e0281434.
27. Oraevsky, A., et al. *Full-view 3D imaging system for functional and anatomical screening of the breast.* in *Photons Plus Ultrasound: Imaging and Sensing 2018.* 2018. SPIE.
28. Stephens, K., *FDA Approves Seno Medical's Breast Cancer Diagnostic Technology.* AXIS Imaging News, 2021.
29. Neuschler, E.I., et al., *A pivotal study of optoacoustic imaging to diagnose benign and malignant breast masses: a new evaluation tool for radiologists.* Radiology, 2018. **287**(2): p. 398–412.

30. Oraevsky, A., et al., *Clinical optoacoustic imaging combined with ultrasound for coregistered functional and anatomical mapping of breast tumors.* Photoacoustics, 2018. **12**: p. 30–45.
31. Nyayapathi, N., et al., *Photoacoustic dual-scan mammoscope: results from 38 patients.* Biomedical Optics Express, 2021. **12**(4): p. 2054–2063.
32. Nyayapathi, N., et al., *Dual scan mammoscope (DSM)—a new portable photoacoustic breast imaging system with scanning in craniocaudal plane.* IEEE Transactions on Biomedical Engineering, 2019. **67**(5): p. 1321–1327.
33. Zheng, E., et al., *Second-generation dual scan mammoscope with photoacoustic, ultrasound, and elastographic imaging capabilities.* Frontiers in Oncology, 2021. **11**: p. 779071.
34. Zheng, W., et al., *Deep Learning Enhanced Volumetric Photoacoustic Imaging of Vasculature in Human.* Advanced Science, 2023. **n/a**(n/a): p. 2301277.
35. Hicklin, D.J. and L.M. Ellis, *Role of the vascular endothelial growth factor pathway in tumor growth and angiogenesis.* Journal of clinical oncology, 2005. **23**(5): p. 1011–1027.
36. Heijblom, M., et al., *The state of the art in breast imaging using the Twente Photoacoustic Mammoscope: results from 31 measurements on malignancies.* European radiology, 2016. **26**: p. 3874–3887.
37. Yamaga, I., et al., *Vascular branching point counts using photoacoustic imaging in the superficial layer of the breast: A potential biomarker for breast cancer.* Photoacoustics, 2018. **11**: p. 6–13.
38. Asao, Y., et al., *Photoacoustic mammography capable of simultaneously acquiring photoacoustic and ultrasound images.* Journal of biomedical optics, 2016. **21**(11): p. 116009–116009.
39. Mikalsen, L. T. G., Dhakal, H. P., Bruland, Ø. S., Naume, B., Borgen, E., Nesland, J. M., & Olsen, D. R., 2013. *The clinical impact of mean vessel size and solidity in breast carcinoma patients.* PLoS One, **8**(10): p. e75954.
40. Folkman, J., *Tumor angiogenesis.* Advances in cancer research, 1985. **43**: p. 175–203.
41. Varghese, T., *Quasi-Static Ultrasound Elastography.* Ultrasound Clin, 2009. **4**(3): p. 323–338.
42. Luengo-Gil, G., et al., *Effects of conventional neoadjuvant chemotherapy for breast cancer on tumor angiogenesis.* Breast Cancer Res Treat, 2015. **151**(3): p. 577–87.
43. Lin, L., et al., *Photoacoustic Computed Tomography of Breast Cancer in Response to Neoadjuvant Chemotherapy.* Adv Sci (Weinh), 2021. **8**(7): p. 2003396.
44. Toi, M., et al., *Visualization of tumor-related blood vessels in human breast by photoacoustic imaging system with a hemispherical detector array.* Sci Rep, 2017. **7**: p. 41970.
45. Li, X., C. Heldermon, and H. Jiang. *Monitoring neoadjuvant chemotherapy in breast cancer using quantitative photoacoustic tomography.* in *Biomedical Optics*. 2014. Optica Publishing Group.
46. Hai, P., et al., *High-throughput, label-free, single-cell photoacoustic microscopy of intratumoral metabolic heterogeneity.* Nat Biomed Eng, 2019. **3**(5): p. 381–391.

Chapter 21
Photoacoustic Imaging of Interventional Devices for Guiding Minimally Invasive Medical Procedures

Mengjie Shi, Simeon J. West, Tom Vercauteren, Sacha Noimark, Adrien E. Desjardins, and Wenfeng Xia

Abstract Despite rapid advancements in medical technology, accurate identification and assessment of abnormalities and precise visualisation of interventional devices are imperative for successful outcomes in minimally invasive medical procedures. In the past few decades, photoacoustic (PA) imaging has emerged as a promising intraoperative imaging modality for surgical guidance, offering rich optical absorption contrast derived from intrinsic tissue chromophores, extrinsic contrast agents, and interventional devices. This chapter reviews recent developments in PA imaging for guiding minimally invasive medical procedures, with a particular focus on its capability of visualising interventional devices including clinical metallic needles, catheters, and metal implants, and signal and image processing methods for enhanced visualisation.

Keywords Photoacoustic imaging · Minimally invasive medical procedures · Interventional devices

---

M. Shi · T. Vercauteren · W. Xia (✉)
School of Biomedical Engineering & Imaging Sciences, King's College London, London, UK
e-mail: mengjie.shi@kcl.ac.uk; tom.vercauteren@kcl.ac.uk; wenfeng.xia@kcl.ac.uk

S. J. West
Department of Anaesthesia, University College Hospital, London, UK
e-mail: simeon.west@nhs.net

S. Noimark · A. E. Desjardins
Department of Medical Physics and Biomedical Engineering, University College London, London, UK

Wellcome/EPSRC Centre for Interventional and Surgical Sciences, University College London, London, UK
e-mail: s.noimark@ucl.ac.uk; a.desjardins@ucl.ac.uk

W. Xia (ed.), *Biomedical Photoacoustics*,
https://doi.org/10.1007/978-3-031-61411-8_21

21.1 Introduction

Minimally invasive medical procedures such as tumour biopsies, peripheral nerve blocks, foetal surgery, and brachytherapy (BT) require image guidance for identifying procedure targets and visualising interventional medical devices. Several medical imaging techniques are commonly used in clinical practice including endoscopy [1], magnetic resonance imaging (MRI) [2], X-ray computed tomography (CT) [3], fluoroscopy [4], and ultrasound (US) [5]. However, each of these modalities has certain advantages and limitations. For instance, MRI and X-ray CT are usually used for preoperative assessments, necessitating registration, typically with fluoroscopy, during minimally invasive medical procedures. Nonetheless, fluoroscopy exposes both patients and operators to ionising radiation. The potential harmfulness of the radiation may accumulate with an increased dose and duration. Although US is one of the most frequently used intraoperative image modalities for minimally invasive procedures (specifically percutaneous needle techniques), it suffers from insufficient tissue contrast for target identification. Besides, the presence of acoustic artefacts can impede the visualisation of interventional devices.

Photoacoustic (PA) imaging is an emerging imaging modality, featuring spectroscopic contrast, sub-millimetre ultrasonic resolution, and centimetre-scale tissue penetration depth [6]. With conventional PA imaging, biological tissues are illuminated by pulsed or modulated light, causing local temperature rises due to light absorption by tissue chromophores. This process leads to the generation of US waves via thermal expansion, which can be collected at the tissue surface with a US probe to form images representing the optical absorption distributions. PA imaging can offer high tissue specificity as the contrast is derived from either the optical absorption of endogenous tissue chromophores or exogenous contrast agents. Moreover, PA imaging can be readily integrated with US imaging to enable a dual-modal functionality that combines structural information and molecular contrast by sharing a US detector. Notably, its potential to be complementary with US imaging for guiding several minimally invasive procedures has been comprehensively reviewed in [7], which aimed to provide an overview of various instrumentations on PA imaging to identify peripheral tissue targets, including sentinel lymph nodes (SLNs), blood vessels, and nerves. In addition to visualising tissue targets, PA imaging is also capable of visualising interventional devices. Minimally invasive medical procedures typically involve the insertion of elongated rigid or flexible metal devices through small incisions into the body. As shown in Fig. 21.1, for clinical applications such as regional anaesthesia and pain management [8], a metallic needle featuring a relatively thin, hollow tube with a sharp extremity was percutaneously inserted. Radiofrequency ablation (RFA) utilises a metallic needle/catheter with multiple prongs at its tip [9]. Prostate BT involves the implantation of radioactive BT seeds in the prostate gland [10]. The seeds are stainless steel or titanium capsules and have common sizes of 0.8 mm × 4.5 mm. Metallic stents, such as coronary stents, are expandable metal tubes typically used in percutaneous coronary intervention, a standard therapy for atherosclerosis diseases [11]. Since the optical absorption

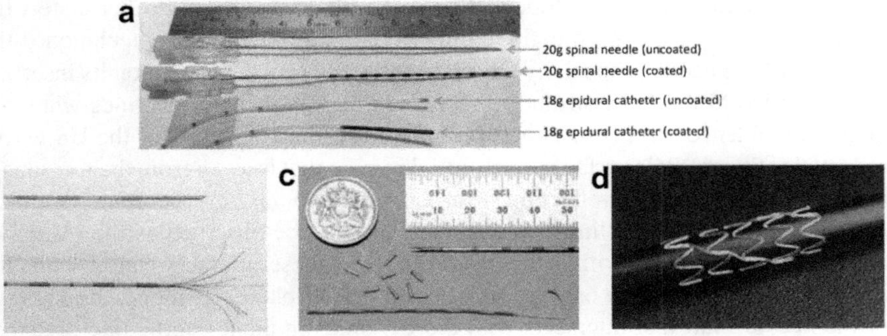

Fig. 21.1 Interventional devices utilised during minimally invasive medical procedures. (**a**) Spinal needles and epidural catheters (Reprinted (adapted) with permission from Ref. [13]. © 2019 Society of Photo-Optical Instrumentation Engineers (SPIE)). (**b**) Radiofrequency ablation needles/catheters (Reprinted (adapted) with permission from Ref. [9]. © 2006 Springer-Verlag Berlin Heidelberg). (**c**) Prostate brachytherapy seeds and needles (Reprinted (adapted) with permission from The Prostate Cancer Centre PBAG. © 2008 Wikipedia). (**d**) Coronary artery stents. (Reprinted (adapted) with permission from Ref. [14]. © 2021 MDPI)

coefficients of metals are orders of magnitude higher than most of the tissue chromophores in the visible and Near-infrared (NIR) wavelength ranges [12], PA imaging is presumably beneficial for visualising interventional devices inserted into tissues due to its superior contrast compared to US imaging. This chapter focuses on presenting a comprehensive review of relevant research concerning PA imaging of interventional devices for US-guided minimally invasive medical procedures. This encompasses interventional devices such as clinical metallic needles and ablation catheters, as well as both temporary and permanent metal implants such as BT seeds and coronary stents.

21.2 PA Visualisation of Interventional Devices

21.2.1 Clinical Metallic Needles

Metallic needles are involved in a wide range of minimally invasive medical procedures such as regional anaesthesia and pain management [8], tumour biopsies [15], and foetal interventions [16]. US imaging stands out as one of the most commonly used intraoperative modalities, offering real-time visual feedback on the needle placement and anatomical information of tissues. However, accurately identifying the needle including the tip and shaft with US imaging remains challenging due to the poor contrast of the needle [17]. The visibility of the needle tip can be resolved with US imaging due to the US reflection from the irregular surface of the bevel, but further identification can be non-trivial because of its diminutive dimension; additionally, side-lobe and beam-width artefacts can further distort the US appearance of the needle tip. Furthermore, for the in-plane technique, the visibility of the needle, especially the shaft, is moderately dependent on its insertion angle relative to the US transducer. The visibility significantly declines when the angle of incidence increases. This is because a substantial portion of the US waves reflected at the needle's surface travels in a direction that is away from the transducer aperture due to specular reflection. Moreover, ultrasonic "reverberation" artefacts such as multiple parallel lines beneath the shaft can be observed as the results of acoustic reverberations within the needle lumen. These artefacts can obscure the tissue interpretation in the background. For the out-of-plane technique, the segment of the needle shaft that intersects with the US imaging plane can be interpreted as the needle tip in US images during needle insertions. Misinterpretation of the needle tip location can lead to significant complications including stroke [18].

Various methods have been proposed to improve the US visibility of clinical metallic needles. One such method involves the use of echogenic needles, which feature specially crafted surfaces with increased roughness. This roughness enhances US scattering, thereby improving needle visibility even at steep insertion angles. However, these echogenic needles are often costly and still susceptible to reflection artefacts [19–21]. Recently, an ultrasonic tracking technique has been investigated for needle tip localisation with an external US probe. Two typical implementations include transmitting US waves from the needle tip, which are then detected by the external imaging probe [24], or reciprocally receiving US waves from the probe with a US detector equipped at the needle tip [22, 23, 25–27]. However, visualisation of the needle shaft remains challenging with this technique, which is crucial for precise needle manipulation, especially during steep insertions.

Emelianov's group was the first to investigate the feasibility of PA imaging for visualising metallic hypodermic needles [28, 29]. A Cortex US imaging system and an Nd:YAG laser were utilised for performing US and PA imaging. Porcine tissue samples were illuminated by the laser delivered through fibre bundles attached to both sides of the US probe. Longitudinal and transverse views of the needle insertions into porcine tissues ex vivo were imaged using the system. The needle was visible with both US and PA imaging when it was placed horizontally in the imaging plane. The PA image of the needle demonstrated a higher signal-to-noise ratio (SNR) compared to the corresponding US image where a reverberation artefact from the needle shaft was present. Furthermore, the angle of incidence had minimal impact on the PA visualisation of the needle while the needle became barely visible with US imaging as the insertion angle increased. This was also observed in the PA and US images depicting the transverse section of the needle.

The potential of PA imaging for guiding various interventional procedures involving needle placement was later studied by several research groups. The instrumentation usually comprised a dual-modal PA/US imaging system built by leveraging their inherent co-registration. In 2010, Kim et al. reported the feasibility of applying PA imaging for guiding needle biopsy of SLNs [30]. SLN biopsy is a standard minimally invasive surgical procedure used for breast cancer diagnosis. SLNs receive drainage from nearby tumour sites, which can be sampled to identify the spread of the cancer cells for cancer staging. In practice, a tracer such as an indocyanine green angiography (ICG) is injected near the tumour site, which is taken up by the surrounding SLNs. A needle is inserted percutaneously under imaging guidance to sample the ICG-dyed SLNs for examination, referred to as fine-needle aspiration. In this work, a handheld PA/US probe was developed based on the augmentation of a clinical US transducer with compact optical fibre bundles for PA light delivery. PA excitation light with a wavelength of 810 nm was used to target the optical absorption peak of ICG. The needle inserted into the left axillary region of a rat was tracked in vivo with the system. PA imaging was able to detect the needle position with an enhanced mean contrast of 8.5 compared to 1.2 with US imaging. In the meantime, the SLN was scanned both photoacoustically and ultrasonically before and after the ICG injection. ICG accumulation at depths of up to ~2.1 cm was visualised with PA imaging in vivo at a frame rate of ~1 fps, which was consistent with observations in fluorescence imaging. Therefore, with the improved visualisation of the needle and specificity of the contrast agents, PA imaging can be potentially helpful for guiding needle biopsies. However, with PA imaging in 2D, the needle visibility can be affected when there is an alignment deficiency between the needle and the imaging plane. In 2019, Wang et al. extended the needle visualisation from 2D to 3D for SLN biopsy [31]. 3D visualisation was achieved by rapidly scanning a linear US transducer to obtain the slices in 2D, followed by image stitching. However, the temporal resolution for visualising needles in 3D was insufficient for guiding real-time operations. The reported maximum imaging rate was around 4s per volume measuring 13 mm × 10 mm × 11 mm for in vivo imaging. In 2015, Wei et al. proposed an integrated PA/US imaging system for guiding interventional procedures in real-time [32]. PA images were reconstructed through the aggregation of the sub-region images, which were obtained by scanning a low-energy laser with a narrow beam. The performance of the system was validated by needle insertions into chicken breast tissue ex vivo. PA images of the needle achieved an SNR of at least 35 dB at a depth of up to 1.2 cm with a frame rate of up to 30 Hz. This system utilised PA excitation at a single wavelength of 1053 nm. Later, Jeng et al. proposed a real-time interleaved spectroscopic PA/US imaging system [33]. The system was then applied for guiding nanoparticle injection in chicken breast tissue ex vivo. As shown in Fig. 21.2a, neither the injected contrast agent nor the needle was poorly visualised with US imaging. In contrast, both the contrast agent and the needle were visible with PA imaging. The needle visualisation can be further enhanced by weighting the PA images acquired with multiple wavelengths with the absorption spectra of the needle (results are not shown here). PA imaging has also been demonstrated viable for guiding stem cell delivery into the spinal cord by Donnelly et al. in 2018 [34]. In the stem cell delivery procedure, precise needle placement is imperative for successfully administering the cells into the cord. Stem cells were labelled by exogenous contrast agents, i.e. plasmonic gold nanospheres (AuNSs) for distinct imaging and subsequent tracking. As shown in Fig. 21.2b, PA imaging offered enhanced distinguishability of the spinal needle as well as clear visualisation of the cells' distribution within the rodent

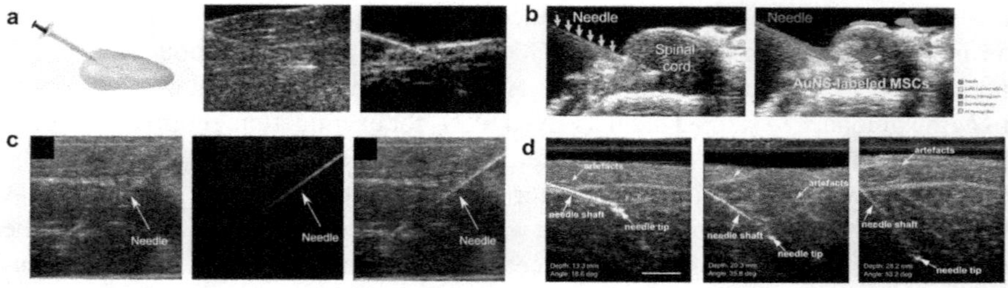

Fig. 21.2 Photoacoustic (PA) visualisation of clinical metallic needles for guiding minimally invasive medical procedures. (**a**) Ultrasound (US)/PA visualisation of needle injections into chicken breast tissue ex vivo (Reprinted (adapted) with permission from Ref. [33]. © 2021 Nature Publishing Group). (**b**) US/PA visualisation of needle injections of labelled stem cells into the rodent spinal cord (Reprinted (adapted) with permission from Ref. [34]. © 2018 American Chemical Society). (**c**) PA/US visualisation of a 14 G metallic needle inserted into chicken breast tissue ex vivo using a LED-based PA/US imaging system (left to right: US, PA, and PA and US overlay) (Reprinted (adapted) with permission from Ref. [41]. © 2018 MDPI). (**d**) Enhanced US/PA visualisation of a spinal needle inserted into chicken breast tissue ex vivo by combing nanocomposite coatings with extracorporeal and interstitial illuminations. (Reprinted (adapted) with permission from Ref. [42]. © 2022 MDPI)

cord. Therefore, it can be potentially helpful to improve the effectiveness of stem cell therapies in clinical practice.

In the above discussion, light illumination for PA excitation was delivered at the tissue surface. However, light attenuation including absorption and scattering within the tissue can impede the imaging performance of PA at deep depths, leading to a reduction in the visibilities of tissue targets and interventional devices. To further explore the PA's capability for tissue and tool imaging at deep depths, PA excitation light can be delivered interstitially by integrating an optical fibre inside the working channels of interventional devices. This was initially conceptualised by Piras et al. in 2013 where the PA needle was proposed as capable of imaging tissue samples and visualising the needle progression with interstitial illumination and extracorporeal US detection [35]. The imaging capability of the PA needle was demonstrated using a chicken breast phantom with an inclusion of a fish heart perfused with blood. PA signals with relatively high intensities were observed in the needle tip area and the vicinity of the inclusion when the needle was approaching. The needle trajectory was better visible in the accumulation of the corresponding PA images, with a 17% higher contrast compared to that of co-registered US images. However, the system had a single wavelength of 1064 nm, which limited its applicability to obtaining contrast from haemoglobins. In 2015, Xia et al. developed an interventional multispectral PA imaging platform with two wavelength ranges in the NIR (750–900 nm and 1150–1300 nm), which could be applicable in many minimally invasive medical procedures [36]. The PA excitation light was delivered to biological tissues with a bare optical fibre positioned within the needle cannula. The capability of the interventional multispectral PA imaging was demonstrated by imaging fat and blood samples buried in chicken breast tissues ex vivo [36] and nerves and tendons ex vivo [37]. The potential of the system for guiding nerve blocks was further investigated with needle insertions to the brachial plexus in a swine model in vivo [38]. In addition to the enhanced tip visibility, interventional multispectral PA imaging offers high tissue specificity, which can be used for tissue identification during minimally invasive medical procedures [37]. However, this falls beyond the scope of this review and will not be discussed in detail herein.

In recent years, there has been growing interest in light-emitting-diodes (LEDs) and laser diodes (LDs) as promising alternatives to solid-state lasers due to their lower cost, higher portability, and comparable excitation efficiency [39], which further benefit the clinical translation of PA imaging. Agano et al. (2016) imaged a 20 G Cattelan needle inserted into a tissue-mimicking phantom with an LED-based PA/US imaging system, AcousticX [40]. With a black coating at the exterior, the needle can be visualised at depths up to 30 mm. Xia et al. (2018) later demonstrated the feasibility of this system for guiding minimally invasive medical procedures [41]. The dual-modal handheld imaging probe was achieved with a clinical US probe sandwiched by two narrow LED bars. The potential of the system for guiding peripheral nerve blocks was demonstrated by in-plane insertions of a 14 G needle into ex vivo chicken breast with real-time image guidance. The results showed that the needle in US images was almost invisible and hardly differentiated from the hyperechoic background. In contrast, the needle could be visualised with PA imaging at depths of up to 2 cm, with an SNR of around 10 dB (<5 dB with US imaging) at an insertion angle of 35°, as shown in Fig. 21.2c. The capability of guiding the needle placement was further validated on a vessel-mimicking phantom. With the excitation wavelength of 850 nm, the targeted vessels and needle were both apparent in PA images, showcasing its potential usefulness for intraoperative real-time guidance of minimally invasive medical procedures. However, it is noticeable that the needle visibility in PA images degraded at deep depths due to light attenuation within the tissue. This can pose a challenge in determining the tip location and the shaft orientation during the placement. Shi et al. (2022) proposed a method for needle enhancement by combining extracorporeal and interstitial illumination with elastomeric nanocomposite coatings (shown in Fig. 21.2d) [42]. The coatings, encompassing high optical absorption coefficient and US wave generation efficiency, were applied onto the exterior walls of the shaft of a bare needle and the end face of an optical fibre that was placed inside the needle lumen. Later, the shaft coating was excited extracorporeally by the LED arrays (at 850 nm) attached to the scanning US probe, while the tip coating was illuminated by the fibre-coupled light from an LD (at 1550 nm). With the in-plane needle insertion technique using chicken breast tissues ex vivo, the system provided the consistently enhanced PA contrast of the needle shaft over insertion angles ranging from 20° to 55°. The needle tip could be unambiguously identified up to the maximum displaying depth of the system, i.e. 38 mm. The SNR in PA images was measured both at the tip and shaft, demonstrating an improvement by a factor of 1.7 compared to the unmodified needle.

21.2.2 Catheters

Catheters are basic and critical interventional instruments in minimally invasive medical procedures such as RFA [43] and endovascular surgeries [44]. Supervision of the catheter placement, which includes the identification of the catheters especially the tips, and the procedure targets (i.e. the ablation lesions in RFA) is of utmost importance during these procedures. Several imaging modalities have been employed to assist catheter-based minimally invasive medical procedures in an intraoperative way, e.g. fluoroscopy [45, 46] and US imaging [47, 48]. However, X-ray-based fluoroscopy is associated with exposure to harmful ionising radiation and lacks depth information with the planar projections. US imaging is widely adopted for guiding RFA surgeries in clinical practice due to its radiation-free nature and the ability to offer morphological information at deep depths in real-time. Nonetheless, US imaging has image quality deficiencies. For instance, catheters and soft tissues exhibit insufficient contrast with US imaging, which could lead to inaccurate catheter placement and incomplete tissue ablation. In addition, intravascular US imaging can be used for catheter tracking. However, it cannot provide a global reference for catheter navigation and thus necessitates the incorporation of fluoroscopy for a comprehensive assessment.

Francis et al. (2019) investigated PA imaging as an adjunction to US imaging for guiding percutaneous RFA [49, 50]. The RFA device, referred to as the RFA needle/catheter, used for delivering heat to the tumour area, consisted of a main trocar with a diameter of 2.1 mm and multiple electrodes (tines) with a diameter of 0.4 mm for each. For PA imaging, laser-coupled fibre bundles, operating at a wavelength of 720 nm, were affixed to a commercial US probe on both sides. The capability of PA imaging to visualise the needle electrodes was investigated with chicken breast tissue ex vivo. Black ink was coated on the surface of the tines to enhance the PA signals. The results showed that the RFA instrument including the trocar and tines was visualised using PA imaging at different insertion angles (a clinically relevant angle of 30 degrees was shown in Fig. 21.3a). PA imaging demonstrated improved visibility at an insertion angle of 30 degrees, with an increase in contrast from 5.3 to 7.5 for the tines and 3.2 to 3.7 for the trocar, compared to US imaging whose performance was severely degraded due to specular reflection.

Likewise, in the context of endovenous laser ablation (EVLA), an ablation catheter needs to be precisely tracked within diseased veins. US imaging is commonly used for intraoperatively guiding EVLA but may result in the loss of the visibility of the angled catheter as the US waves are reflected away from the transducers. This can lead to severe complications such as cauterisation and increase the risk of varicose veins recurrence. Yan et al. (2021) proposed a PA-guided EVLA system where a pulsed laser source at a wavelength of 532 nm was coupled into the ablation catheter for PA imaging [51, 52]. The catheter tip was highlighted with the external US detection of the PA signals that were generated due to the optical absorption of blood at the end. When characterising the system with a blood-vessel

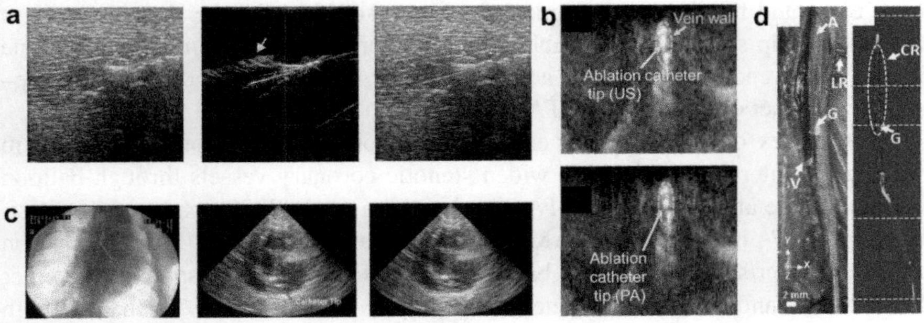

Fig. 21.3 Photoacoustic (PA) visualisation of catheters for guiding minimally invasive medical procedures. (**a**) Ultrasound (US), PA, and US/PA of a radiofrequency ablation catheter inserted into chicken breast tissue ex vivo (Reprinted (adapted) with permission from Ref. [50]. © 2019 Institute of Physics and Engineering in Medicine). (**b**) US and PA imaging of an ablation catheter during endovenous laser ablation procedures. Transverse view of an ablation catheter placed inside the jugular vein (Reprinted (adapted) with permission from Ref. [51]. © 2021 Elsevier GmbH). (**c**) Fluoroscopic, US, and PA (overlaid with US) images of the catheter tip within the right atrium (Reprinted (adapted) with permission from Ref. [53]. © 2019 IEEE). (**d**) Photograph of a balloon catheter inserted into the aorta at an anesthetised rate. Maximum intensity projection of 3D PA microscopy imaging of the catheter without air injection. (Reprinted (adapted) with permission from Ref. [55]. © 2020 MDPI. Blue arrows in (**a**) and (**c**) denote PA artefacts)

mimicking phantom using a porcine tissue ex vivo, the PA signals generated at the ablation catheter tip were detectable with a pulse energy of 50 μJ at an imaging depth of up to 3 cm. Further, the amplitudes of PA signals were consistent under different bending conditions of the catheters, demonstrating superiority over the angular-dependent visibility of the catheter tip with US imaging. The performance of catheter tracking with PA imaging was further evaluated in vivo using a canine model. As shown in Fig. 21.3b, the ablation catheter tip inside the jugular vein can be better visualised with PA imaging, achieving an SNR and contrast-to-noise ratio (CNR) of 21.6 dB and 18.2 dB compared to 9 dB and 2.1 dB with US imaging.

The enhanced visualisation of the catheters using PA imaging has the potential to aid in catheter tip tracking by reducing ambiguity. Michelle et al. (2020) proposed the application of PA imaging in conjunction with a robotic visual serving system for catheter navigation in cardiac interventions [53, 54]. PA imaging was achieved based on interstitial illumination where an optical fibre was utilised to deliver laser at a wavelength of 750 nm and inserted into a cardiac catheter. Mechanical scanning of the US probe was operated by a robot, which maintained visual contact with the catheter tip by continuously tracking the PA signals emitted from the catheter tip. The results in Fig. 21.3c revealed that compared to US and 2D fluoroscopy, PA imaging offered enhanced visualisation of the catheter tip in the presence of in vivo heart tissue, with a mean contrast of up to 48.4 dB. The tip localisation within the in vivo heart using PA imaging demonstrated a high level of accuracy, reaching mean square errors ranging from 0.25 to 2.28 mm in 1D compared to the ground truth provided by fluoroscopy, and 1.24 to 1.54 mm for 3D compared to ground truth estimated by electro-anatomical tracking. Additionally, the PA amplitude at the catheter tip showed a noticeable increase upon contact with myocardial tissue including the endocardium. This augmentation can serve as an indicator of tissue–catheter contact during cardiac RFA.

Other types of catheters, for example a balloon catheter, commonly used in vasculature interventions, e.g. to widen stenotic coronary vessels through balloon inflation, were also imaged with PA imaging. Kim et al. (2020) explored the feasibility of fast PA microscopy (f-PAM) for balloon catheter tracking [55]. The balloon catheter comprised two parts, a balloon made of optically transparent polyether block amide and a core constructed from polyvinyl chloride (PVC) marked with platinum. Although the catheter core exhibited a higher PA amplitude compared to whole blood at 532 nm, the visibility of the catheter core was overwhelmed by the optical absorption of blood vessels. To delineate the catheter in the blood vessels, the PA A-line signals were post-processed concerning the distance differences towards the probe. The PA signals of the catheter were investigated under different injection conditions with a rat model ex vivo and in vivo. The results indicated that acoustic attenuation caused by air bubbles or water can render complete invisibility or reduced visibility of the catheter segment during the inflation stage (shown in Fig. 21.3d). This can inform the functionality of the balloon. It is worth noticing that the visibility of the catheter core with PA imaging was also affected by its proximity to the vessel wall due to the high blood absorption. This could be improved by exploring specific wavelengths where the balloon catheter can be selectively enhanced.

PA imaging instruments can be miniaturised to image intravascularly, referred to as intravascular PA (IVPA) [56], e.g. focusing on visualising vessel walls for assessing atherosclerotic plaques [57]. However, this topic falls beyond the scope of the current discussion, which is centred on PA-enhanced catheter visualisation and will not be covered in detail.

21.2.3 Brachytherapy Seeds

BT is a type of localised radiation therapy involving (temporarily or permanently) implanting radioactive sources in or near the cancerous area [10]. In prostate BT, BT seeds made of stainless steel or titanium are filled with a radioisotope such as iodine-125 and palladium-103 and injected into the prostate tissue through a needle. The positions of the seeds can be intraoperatively evaluated by surgeons with the help of image guidance such as transrectal US (TRUS) [58]. However, it remains challenging to unambiguously identify seeds under US guidance [59]. For instance, the seeds' visibility can be perturbed by the off-plane placement, hyperechoic background, and acoustic artefacts from microcalcifications. Alternatively, other imaging modalities including fluoroscopy [60], CT [61], and MRI [62] showcased their respective advantages and limitations for BT application.

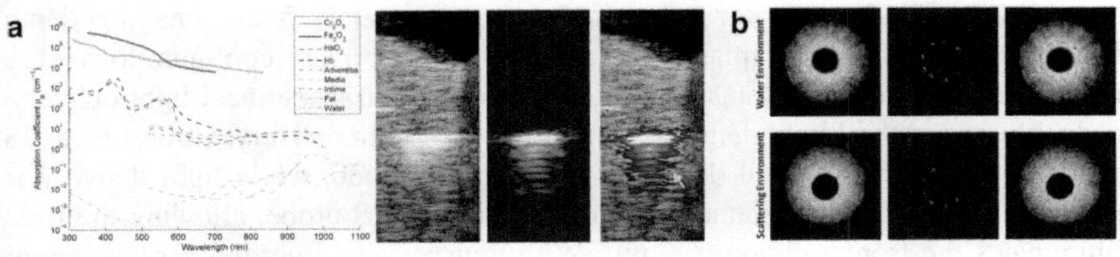

Fig. 21.4 Photoacoustic (PA) visualisation of brachytherapy seeds and coronary stents for guiding minimally invasive medical procedures. (**a**) Optical absorption spectra of various tissue chromophores and steels (left); Ultrasound (US), PA visualisation, and their combination of a non-reactive brachytherapy seed embedded in bovine prostate tissues ex vivo at the long-axis orientation. The blue arrow denotes comet-tail or acoustic reverberation artefacts (Reprinted (adapted) with permission from Ref. [63]. © 2011 Optical Society of America). (**b**) Intravascular US, PA visualisation, and their combination of a tissue-mimicking phantom with a stent embedded. (Reprinted (adapted) with permission from Ref. [77]. © 2009 OSA)

In 2011, Su et al. investigated the feasibility of BT seed visualisation with PA imaging [63]. The optical absorption spectra of stainless steels, one of the common constituents of BT seeds, were compared with tissue chromophores as shown in Fig. 21.4a. Significantly higher levels of absorption were reached by the steels including Cr_2O_3 and Fe_2O_3, inferring the potential utility of PA imaging for visualising seeds. Different seed orientations were imaged with PA and US imaging to evaluate their angular dependency. Comet-tail or acoustic reverberation artefacts caused by the highly reflective surfaces of the seeds appeared on both PA and US images (denoted by the blue arrow in Fig. 21.4a). For a small rotation angle of 15°, a large portion of the seed was visible with PA imaging while only the proximal tip area was visible with US imaging. After that, the middle portion of the seed became indiscernible with both PA and US imaging. Furthermore, the seed was embedded in bovine prostate tissue ex vivo and imaged using both PA and US imaging. The seeds can be visualised at depths of up to 13 mm with PA imaging, achieving an improvement of up to 27.9 dB in contrast compared to US imaging. Additionally, by analysing the PA spectrum derived from seeds signals ex vivo, a long wavelength of 1047 nm was potentially advantageous for seed visualisation because of the trade-off between the light attenuation within the prostate tissue and the optical absorption of the metals. Meanwhile, Tyler et al. (2011) optimised the wavelength of PA imaging in terms of seed detection using chicken breast tissue ex vivo [64, 65]. Seed contrast in PA and US was compared with an endogenous contrast (rabbit blood) at multiple wavelengths and imaging depths. A long wavelength of 1064 nm achieved unique seed identification with PA imaging, with an improved classification performance compared to US imaging (AUC was 0.80 for PA and 0.59 for US). Kuo et al. (2014) reported the PA images of multiple palladium seeds implanted in a dog prostate tissue ex vivo, showing the potential feasibility of applying PA for BT in clinical practice [66, 67].

As the excitation light encounters drastic reduction during propagation in tissue due to optical attenuation (i.e. optical absorption and scattering), to fulfil the practical requirement for achieving an approximately 5 cm imaging depth during whole-field seed implantation, different light delivery configurations were investigated including (i) transrectal light delivery, (ii) transperineal light delivery, and (iii) transurethral light delivery. With transrectal light delivery, a transrectal US probe is used for PA signal detection in backward mode, while light delivery is provided by optical fibres attached to the transrectal US probe, allowing the light illuminates the frontal region [68, 69]. With transperineal light delivery, an open-end optical fibre, serving as a conduit for light transmission, was integrated with a BT needle. This arrangement permits targeted illumination in proximity to the seeds during their implantation with the BT needle [70]. For transurethral light delivery, the light excitation is delivered through an optical fibre positioned inside the urinary catheter, which is inserted through the urethra to drain the bladder during the prostate seed BT. In contrast to the radial illumination featured by transperineal light delivery, transurethral light can be directed parallel to the fibre axis at the end face, perpendicular to the fibre axis, or in an omnidirectional manner [71, 72].

Yaseen et al. (2008) compared transrectal and transurethral light delivery methods in a hybrid laser optoacoustic and ultrasonic imaging system named LOUIS-P, dedicated to prostate imaging [73]. Light fluence distributions in prostate tissue were studied by Monte Carlo simulations [74]. The results revealed that the transrectal delivery method had limited light penetration depths, attributed to the optical attenuation at the rectal wall. Bell et al. (2013) developed a transperineal interstitial light delivery method for PA imaging of prostate BT seeds [70]. The performance was further validated using canine prostates in vivo [75]. The results showed that the seeds were better visualised with PA imaging compared to US imaging. A seed located farthest from the light source was found visible with an energy density at the fibre tip of 123 mJ/cm^2, which slightly exceeded the safe laser exposure (100 mJ/cm^2 at 1064 nm) according to the American National Standards Institute (ANSI). Bell et al. (2014) later proposed a transurethral arrangement with two types of light-emitting implementations (i.e. end-firing and side-firing) and demonstrated their feasibility using canine prostate tissues ex vivo [71]. End-firing fibre features radial illumination capable of exciting multiple seeds suited around the urethra. PA imaging was able to visualise BT seeds, with a maximum visible radius of up to 16 mm from the fibre tip. In contrast, the side-firing implementation enables selective illumination of the seeds through catheter rotation. Bell et al. (2015) developed a side-firing fibre prototype for PA imaging of prostates [76]. Its feasibility was studied by imaging BT seeds with plastisol phantoms and canine prostates ex vivo and in vivo. In B-mode curvilinear US images, the seeds close to the centre of the urethra with radial distances of 6–9 mm were easier to recognise compared to those located outside this range. However, all seeds were visualised with PA imaging. Moreover, the B-mode linear US images revealed the longitude view of the seeds. In the corresponding PA images, an out-of-plane seed became discernible following precise light adjustment via catheter rotation.

21.2.4 Coronary Stents

Coronary stents are small and expandable metal tubes and are used in treating coronary heart diseases by keeping narrowed or blocked arteries open. Image guidance is required during stent placement and postoperative assessment to accurately identify the stent positions. X-ray coronary fluoroscopy is most used for guiding this procedure, but it involves radiation exposure and lacks the depth perception necessary for determining stent apposition within the vessel lumen. MRI and multi-slice computed tomographic angiography have also been investigated in imaging the stents. However, the metallic structures can generate image artefacts, hindering the accurate interpretation of the stent locations. Moreover, optical coherence tomography can visualise stents at a high spatial resolution of 10–20 μm, but it is constrained by its limited imaging depth, approximately 2 mm. In contrast, intravascular US imaging facilitates a deep imaging depth enabling a morphological reconstruction of the vessel walls, but its contrast is inadequate to distinguish the stent struts due to the interference of acoustic clutters and speckle noise. The capability of intravascular PA imaging, in combination with intravascular US imaging, was first investigated for guiding the stenting procedure by Su et al. (2009) [77]. Experiments using a PVC-based tissue-mimicking phantom showed that the stents could be visualised in PA images (shown in Fig. 21.4b). This was evidenced by an accurate estimation of the diameter of the stent with PA imaging. Besides, the combination of intravascular US and intravascular PA imaging permits the visualisation of the stent with high specificity, along with the structural information of the surrounding tissue. This capability holds great promise for benefiting stenting procedures that demand precise registration of the stent within the implanted environment.

21.3 Enhanced PA Visualisation

21.3.1 Medical Device Modifications

Although PA imaging can substantially enhance the contrast of interventional devices as compared to US imaging, the image quality can be degraded by insufficient light energy, acoustic clutters, reconstruction artefacts, etc. For interstitial PA imaging of BT seeds, transrectal light delivery suffers from substantial light attenuation when passing through the rectal wall. In contrast, transperineal and transurethral light delivery methods are more effective as the light can be delivered closer to the seeds. However, the light emitted from the fibre tip undergoes attenuation as it travels through the tissue, resulting in a maximum tissue penetration depth of around 1–2 cm in vivo. This is limited for visualising the deep-embedded devices, e.g. a typical scanning depth for seed implantation in practice is about 5 cm.

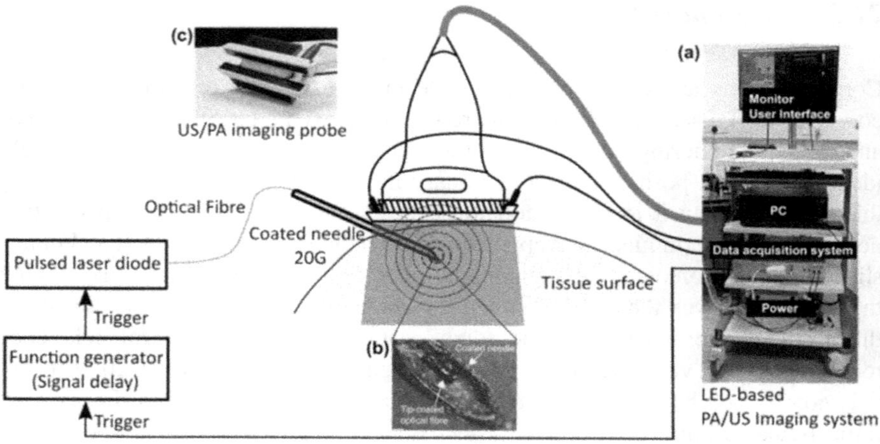

Fig. 21.5 Enhanced photoacoustic (PA) visualisation of clinical metallic needles by combining nanocomposite coatings with extracorporeal and interstitial illuminations. A light-emitting-diode-based dual-modal PA/Ultrasound imaging system for needle enhancement. (Reprinted (adapted) with permission from Ref. [42]. © 2022 MDPI)

This is also the case when imaging other interventional devices with extracorporeal surface illumination.

Optical absorption of interventional devices can be altered to improve PA signal generation, leading to higher image contrast. Coatings with high optical absorption have been investigated by applying them to the exteriors of the devices, including metallic spinal needles, non-metallic catheters [13], and BT seeds [78]. In a study by Pan et al. (2013) [78], a black dye was chosen as the coating material for the seeds, resulting in a maximum visibility depth of up to 50 mm with a laser spot illumination (max. local fluence: 20 mJ/cm^2) in a PVC phantom with chicken breast tissue ex vivo on the top. The maximum increase in SNR with PA imaging was measured as 18.5 dB compared to bare seeds. The use of an elastomeric nanocomposite coating consisting of multi-walled carbon nanotubes and polydimethylsiloxame (MWCNT/PDMS) demonstrated visibility enhancements for metallic spinal needles and epidural catheters [13]. Using an LED-based PA imaging system (max. local fluence: 0.11 mJ/cm^2) with extracorporeal illumination, the coated devices were visualised up to a depth of 35 mm in chicken breast tissue ex vivo. Furthermore, the tips of metallic needles and catheters can be highlighted by integrating tip-coated optical fibres within the devices. Shi et al. (2022) coated the tip of an optical fibre with candle soot nanoparticles and polydimethylsiloxame (CSNP/PDMS) nanocomposites and placed it inside a metallic spinal needle exteriorly coated with the same material [42]. As shown in Fig. 21.5, the coated needle was imaged by an LED-based PA/US imaging system. Strong PA signals were observed at the tip due to the PA excitation of the tip coating with the interstitial illumination from an LD, serving as a beacon for tracking the needle tip during the placement. The potential translation of this technique into clinical practice was further discussed regarding the biological safety of the CSNP/PDMS composites and their adhesions to the fibre tip and needle exteriors. Mitcham et al. (2013) introduced different approaches to modifying the surfaces of metallic implants and investigated their effects on PA signal generation with a wire phantom [79]. The modifications involved surface roughness by sanding or electropolishing, and the application of various physical vapour deposition (PVD) coatings. In general, either increasing the surface roughness or employing PVD coatings was effective in manipulating PA signal generation. The former could enhance the PA signals because of the increased light absorption, resulting from the larger surface area, more diffuse scattering, and strengthened transmittance. The latter, PVD coatings, provided the opportunity to either increase or suppress the implant visibility, with 4-dB improvement using MoS_2 and 26-dB reduction using TiN.

21.3.2 Device Identification, Denoising, and Artefact Removal

Device-related artefacts can be observed with comparable PA intensities to those of interventional devices and tissue targets, obstructing peripheral tissue targets. For example, reconstruction artefacts associated with metallic needles and RFA catheters were observed with the in-plane insertions into chicken breast tissue ex vivo (as denoted by blue arrows in Figs. 21.2 and 21.3a). It was noted that the artefacts were dominant at larger insertion angles. Image thresholding is easily implemented and effective for suppressing the artefacts, but an inappropriate threshold value could also eliminate the needle signals, leading to potential signal loss. Shi et al. (2022) proposed a deep learning (DL)-based framework to enhance the needle visibility with an LED-based PA/US imaging system [80]. Free from demanding manually annotated datasets for model training, a semi-synthetic dataset consisting of needle simulations and PA measurements of human vasculature in vivo was established. The networks were trained to identify the needles from backgrounds with the synthetic needles as the ground truth. The experimental results using ex vivo, phantom, and in vivo needle data showed that the networks were able to effectively separate the needles from the backgrounds (representative in vivo results are shown in Fig. 21.6a). The metrics, including SNR and the modified Hausdorff distance measuring the distance between the reconstructed needle and its ground truth, exhibited improvements of 5.8 times and 4.5 times, respectively, with in vivo data, compared to conventional reconstructions. Furthermore, the proposed method is versatile and could be tailored for various interventional devices, such as RFA needles and catheters, which possess distinctive geometries that can be realistically simulated.

Besides, PA signals generated under conditions of low optical fluence are readily contaminated by noise, resulting in poor SNRs. Bell et al. (2013) employed a short-lag spatial coherence (SLSC) beamformer to enhance the seed contrast with PA imaging [70]. The SLSC beamformer calculated the spatial coherence of the PA

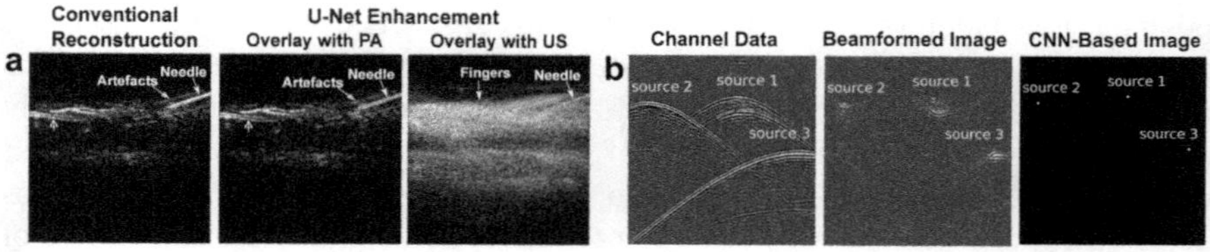

Fig. 21.6 Enhanced photoacoustic (PA) visualisation of clinical metallic needles and point sources with deep learning (DL). (**a**) PA visualisation of in-plane insertions in vivo with conventional reconstruction and DL-based needle enhancement (Reprinted (adapted) with permission from Ref. [80]. © 2022 Elsevier GmbH). (**b**) PA visualisation of point sources. From the left to right: PA channel data, PA reconstruction, and DL-based point source enhancement. (Reprinted (adapted) with permission from Ref. [87]. © 2018 IEEE)

signals received across the US aperture. Compared to background noise, PA signals from the seeds presented higher spatial correlations due to their regular and specific geometries. Thus, the seeds can be effectively enhanced with the SLSC beamformer. The performance of the SLSC beamformer was validated using a tissue phantom with non-radioactive seeds embedded. In contrast to a conventional delay-and-sum beamformer, the use of the SLSC beamformer resulted in enhancements in the SNR and CNR of the seeds by 24.4 dB and 2.1 dB, respectively. PA imaging systems based on affordable light sources such as LDs and LEDs suffer from poor image quality due to their low output light energies. Different techniques such as wavelet transform [81], signal decomposition [82], and DL [83] have been investigated. Shi et al. (2022) proposed a spatiotemporal signal value decomposition method dedicated to low-fluence PA imaging systems [84]. This method leveraged the spatial and temporal correlations of random noise and tissue targets across multiple pre-beamformed PA frames. With simulations and in vivo tests, the method effectively suppressed random noise in the background, and more importantly, it demonstrated great sensitivity to slow motions, with the SNR of in vivo PA images improved by 56% compared to frame averaging. An average denoising speed of about 50 μm per frame was achieved with GPU acceleration, indicating its potential for real-time applications.

Moreover, a common artefact in PA imaging of BT seeds using interstitial excitation is caused by acoustic reflections. In this case, strong PA signals emanating from the fibre/needle tip were reflected off the seeds, resulting in reflection artefacts below the seeds. Singh et al. (2016) proposed a method called PA-guided focused US (PAFUSion) for reflection artefact identification and removal [85, 86]. US was firstly focused on the fibre/needle tip to produce an image with only reflections at deeper depths. The image was then scaled and subtracted from the corresponding PA image to acquire an artefact-free image. With the validations using tissue-mimicking phantoms and porcine tissue ex vivo, PAFUSion could successfully identify the reflected signals and remove them without compromising the SNRs of the seed signals. Allman et al. (2018) proposed a DL-based method to identify and remove reflection artefacts using simulated PA channel data (shown in Fig. 21.6b) [87]. The reflection artefacts were simulated by moving real point sources to deeper areas, aiming to create wavefront shapes resembling those signals distributed at shallower depths. Independent convolutional neural networks (CNNs) were established. The input channel data were fed into deep fully convolutional networks based on VGG16 and a region proposal network. A Fast R-CNN was subsequently employed to classify sources and artefacts and produce bounding boxes that indicated their locations and uncertainty via confidence scores. The trained model achieved 96.67% source classification accuracy on a phantom experiment with BT seeds, with a mean location error of 0.38 ± 0.25 mm. Allman et al. (2019) further extended the DL-based framework for catheter tip detection with PA imaging [88]. Two CNNs were independently trained for linear and phrase array US transducers and tested with in vivo data obtained from a swine catheterisation procedure. Classification accuracies of 88.8% and 91.4% were achieved during the procedure with linear and phrase array US transducers, respectively. Therefore, the proposed framework was potentially helpful for enhancing tool visibility in PA-guided minimally invasive medical procedures. Researchers also explored DL-based frameworks for PA point source identification. Johnstonbaugh et al. (2020) trained an encoder–decoder CNN using simulated PA waveforms [89]. Various spatial distributions and optical scattering conditions were considered during simulations. A mean localisation error of 1.06 mm was achieved at depths between 40 and 60 mm using a point absorber phantom (a 0.3-mm diameter pencil lead in aqueous solutions of Intralipid). This work was further expanded by Yazdani et al. (2021) [90], where deep neural networks were trained for simultaneous point source localisation and denoising. However, model evaluation using real PA measurements was missed. The clinically relevant data can be acquired by imaging the cross-sections of micro-vessels, metallic needles, and BT seeds embedded in ex vivo or in vivo tissues.

21.4 Discussion and Conclusions

Optimising patient outcomes of minimally invasive medical procedures often involves accurate visualisation of interventional devices, including intraoperative tools and metal implants. US imaging has served as one of the most used modalities for guiding minimally invasive medical procedures in clinical practice. However, the visibility of interventional devices such as metallic needles in US images can be poor due to specular reflections and acoustic artefacts. PA imaging has emerged as a promising biomedical imaging modality in the past two decades. This hybrid imaging modality combines optical absorption and US detection, inheriting advantages from both optical and US imaging, thus providing distinctive optical contrast, high ultrasonic resolution, and large imaging depths. PA imaging is potentially helpful in visualising interventional devices inserted into tissues due to the relatively high optical absorption of metals compared to most of the tissue chromophores in the visible and NIR wavelength ranges. Previous research has investigated the PA visualisation of interventional devices commonly utilised in minimally invasive medical procedures. These include clinical metallic needles, ablation catheters, and metallic implants such as BT seeds and coronary stents.

Pre-clinical results acquired with ex vivo tissues and in vivo animal models show that the visualisation of the interventional devices inserted into the tissues can be enhanced with PA imaging. However, there are still important translational challenges to be addressed before the widespread adoption of PA imaging in clinical practice for guiding minimally invasive medical procedures. Firstly, the image quality of PA imaging can be improved. PA visualisation of interventional devices can be suboptimal due to image artefacts and insufficient light fluence. The typical image artefacts consist of acoustic clutters such as acoustic reverberations and aberrations and out-of-plane artefacts associated with the limited view of linear array US transducers. On the one hand, advanced beamforming algorithms such as the SLSC beamformer were promising for enhancing low-fluence-based PA images. With the rapid development of machine learning techniques, DL-based methods demonstrated their effectiveness in improving the PA image quality of interventional devices, especially at the post-processing stage [80, 87, 91]. Nonetheless, for in-plane imaging, interventional devices such as metallic needles can only be partially resolved due to the limited light penetration depth (less than 2 cm in vivo). To enable the visibility of the needle tip for deep insertions, PA excitation light can be delivered to the tip through an optical fibre placed inside the needle lumen. The needle tip can be highlighted with PA signals that are generated through optical absorption either by tissues in front of the needle or tip coatings. However, in practice, the PA signals generated at the tip vicinity can be affected by acoustic clutters and tissue motions, which can degrade the accuracy of tip registration. On the other hand, device modifications such as coating strategies were studied. Nanocomposite coatings can be used for enhancing needle visualisation in highly scattering biological tissues because of their strong optical absorption and high thermal expansion efficiency [13, 42]. However, their mechanical stability such as the adhesion to the devices and biosafety needs to be further studied with animal and human models for acquiring regulatory approvals [92, 93]. Secondly, most commercially available PA/US imaging systems integrate bulky light delivery modules such as solid-state lasers and diode-pumped lasers. Comprehensive risk assessment is needed and specialised equipment such as interlocks is required to be implemented in the operating room for intraoperative surgical guidance. Meanwhile, eye protection gadgets such as goggles are required for operators including surgeons and nurses as well as patients during the operation. LEDs can be promising alternatives to solid-state laser sources because of their relatively low cost and output energy, eliminating the need for laser hazard protection [39, 94]. Further consideration involves the training of the operators that focuses on interpreting PA images using standardised phantoms. Overall, the clinical implementation of PA imaging can bring additional cost and operation complexity, which cannot be overlooked compared to its potential technical benefits.

In summary, PA imaging holds great potential for guiding a variety of minimally invasive medical procedures by offering real-time visualisation of not only the tissue composition of the procedure targets but also the interventional devices. Recent studies demonstrated the superiority of PA imaging for visualising a range of devices including metallic needles, catheters, BT seeds, and coronary stents in US-guided interventional procedures. Moreover, PA imaging offers high tissue specificity including nerves, blood vessels, and tumours. Indeed, all these factors collectively contribute to the strong potential of PA imaging in effectively guiding minimally invasive medical procedures across a broad spectrum of clinical scenarios.

Acknowledgements This work was supported by the Wellcome Trust, United Kingdom (203148/Z/16/Z), the Engineering and Physical Sciences Research Council (EPSRC), United Kingdom (NS/A000049/1), EPSRC UKRI Innovation Fellowship (EP/S001506/1), and King's–China Scholarship Council PhD Scholarship Program (K-CSC) (202008060071). TV owns shares in Hypervision Surgical Ltd and Mauna Kea Technologies Ltd. AED owns shares in Echopoint Medical Ltd. For the purpose of open access, the authors have applied a CC BY public copyright license to any author-accepted manuscript version arising from this submission.

References

1. Baron TH (2007) Natural orifice transluminal endoscopic surgery. British Journal of Surgery 94:1–2. https://doi.org/10.1002/bjs.5681
2. Van Gellekom MPR, Moerland MA, Battermann JJ, Lagendijk JJW (2004) MRI-guided prostate brachytherapy with single needle method—a planning study. Radiotherapy and Oncology 71:327–332. https://doi.org/10.1016/j.radonc.2004.03.002
3. Cha CH, Lee FT, Gurney JM, Markhardt BK, Warner TF, Kelcz F, Mahvi DM (2000) CT Versus Sonography for Monitoring Radiofrequency Ablation in a Porcine Liver. American Journal of Roentgenology 175:705–711. https://doi.org/10.2214/ajr.175.3.1750705
4. Yatziv L, Chartouni M, Datta S, Sapiro G (2012) Toward Multiple Catheters Detection in Fluoroscopic Image Guided Interventions. IEEE Transactions on Information Technology in Biomedicine 16:770–781. https://doi.org/10.1109/TITB.2012.2189407
5. Holm HH, Juul N, Pedersen JF, Hansen H, Strøyer I (1983) Transperineal 125Iodine Seed Implantation in Prostatic Cancer Guided by Transrectal Ultrasonography. The Journal of Urology 130:283–286. https://doi.org/10.1016/S0022-5347(17)51108-8
6. Beard P (2011) Biomedical photoacoustic imaging. Interface Focus 1:602–631. https://doi.org/10.1098/rsfs.2011.0028
7. Zhao T, Desjardins AE, Ourselin S, Vercauteren T, Xia W (2019) Minimally invasive photoacoustic imaging: Current status and future perspectives. Photoacoustics 16:100146. https://doi.org/10.1016/j.pacs.2019.100146
8. Chin KJ, Perlas A, Chan VWS, Brull R (2008) Needle Visualization in Ultrasound-Guided Regional Anesthesia: Challenges and Solutions. Reg Anesth Pain Med 33:532–544. https://doi.org/10.1136/rapm-00115550-200811000-00005
9. Pereira PL, Boss A, Clasen S, Gouttefangeas C, Schmidt D, Claussen CD (2006) Radiofrequency Ablation: The Percutaneous Approach. In: Stroszczynski C (ed) Minimally Invasive Tumor Therapies. Springer, Berlin, Heidelberg, pp 39–52
10. Peschel RE, Colberg JW (2003) Surgery, brachytherapy, and external-beam radiotherapy for early prostate cancer. The Lancet Oncology 4:233–241. https://doi.org/10.1016/S1470-2045(03)01035-0
11. Serruys PW, Morice M-C, Kappetein AP, Colombo A, Holmes DR, Mack MJ, Ståhle E, Feldman TE, van den Brand M, Bass EJ, Van Dyck N, Leadley K, Dawkins KD, Mohr

FW (2009) Percutaneous Coronary Intervention versus Coronary-Artery Bypass Grafting for Severe Coronary Artery Disease. New England Journal of Medicine 360:961–972. https://doi.org/10.1056/NEJMoa0804626
12. Karlsson B, Ribbing CG (1982) Optical constants and spectral selectivity of stainless steel and its oxides. Journal of Applied Physics 53:6340–6346. https://doi.org/10.1063/1.331503
13. Xia W, Noimark S, Maneas E, Brown NM, Singh MKA, Ourselin S, West SJ, Desjardins AE (2019) Enhancing photoacoustic visualization of medical devices with elastomeric nanocomposite coatings. In: Photons Plus Ultrasound: Imaging and Sensing 2019. International Society for Optics and Photonics, p 108783G
14. Wang L, Jiao L, Pang S, Yan P, Wang X, Qiu T (2021) The Development of Design and Manufacture Techniques for Bioresorbable Coronary Artery Stents. Micromachines 12:990. https://doi.org/10.3390/mi12080990
15. Helbich TH, Matzek W, Fuchsjäger MH (2004) Stereotactic and ultrasound-guided breast biopsy. Eur Radiol 14:383–393. https://doi.org/10.1007/s00330-003-2141-z
16. Daffos F, Capella-Pavlovsky M, Forestier F (1985) Fetal blood sampling during pregnancy with use of a needle guided by ultrasound: a study of 606 consecutive cases. Am J Obstet Gynecol 153:655–660. https://doi.org/10.1016/s0002-9378(85)80254-4
17. Reusz G, Sarkany P, Gal J, Csomos A (2014) Needle-related ultrasound artifacts and their importance in anaesthetic practice. British Journal of Anaesthesia 112:794–802. https://doi.org/10.1093/bja/aet585
18. Rathmell JP, Benzon HT, Dreyfuss P, Huntoon M, Wallace M, Baker R, Riew KD, Rosenquist RW, Aprill C, Rost NS, Buvanendran A, Kreiner DS, Bogduk N, Fourney DR, Fraifeld E, Horn S, Stone J, Vorenkamp K, Lawler G, Summers J, Kloth D, O'Brien D, Tutton S (2015) Safeguards to prevent neurologic complications after epidural steroid injections: consensus opinions from a multidisciplinary working group and national organizations. Anesthesiology 122:974–984. https://doi.org/10.1097/ALN.0000000000000614
19. Hebard S, Hocking G (2011) Echogenic technology can improve needle visibility during ultrasound-guided regional anesthesia. Reg Anesth Pain Med 36:185–189. https://doi.org/10.1097/aap.0b013e31820d4349
20. Hocking G, Mitchell CH (2012) Optimizing the safety and practice of ultrasound-guided regional anesthesia: the role of echogenic technology. Current Opinion in Anaesthesiology 25:603–609. https://doi.org/10.1097/ACO.0b013e328356b835
21. Nakagawa K, Kamiya T, Arakawa K, Akiyama S, Sakai K (2015) Objective and subjective comparison of the visibility of three echogenic needles and a nonechogenic needle on older ultrasound devices. Acta Anaesthesiologica Taiwanica 53:1–6. https://doi.org/10.1016/j.aat.2014.11.004
22. Xia W, Mari JM, West SJ, Ginsberg Y, David AL, Ourselin S, Desjardins AE (2015) In-plane ultrasonic needle tracking using a fiber-optic hydrophone. Med Phys 42:5983–5991. https://doi.org/10.1118/1.4931418
23. Xia W, Ginsberg Y, West SJ, Nikitichev DI, Ourselin S, David AL, Desjardins AE (2016) Coded excitation ultrasonic needle tracking: An in vivo study. Med Phys 43:4065. https://doi.org/10.1118/1.4953205
24. Xia W, Noimark S, Ourselin S, West SJ, Finlay MC, David AL, Desjardins AE (2017) Ultrasonic Needle Tracking with a Fibre-Optic Ultrasound Transmitter for Guidance of Minimally Invasive Fetal Surgery. In: Descoteaux M, Maier-Hein L, Jannin P, Collins DL, Duchesne S (eds) Medical Image Computing and Computer-Assisted Intervention – MICCAI 2017. Springer International Publishing, Cham, pp 637–645
25. Mung J, Han S, Yen JT (2011) Design and in vitro evaluation of a real-time catheter localization system using time of flight measurements from seven 3.5 MHz single element ultrasound transducers towards abdominal aortic aneurysm procedures. Ultrasonics 51:768–775. https://doi.org/10.1016/j.ultras.2011.03.005
26. Mung J, Vignon F, Erkamp R, Stanton D, Jain A (2013) Ultrasonically marked instruments for ultrasound-guided interventions. In: 2013 IEEE International Ultrasonics Symposium (IUS). pp 2053–2056

27. Baker C, Xochicale M, Lin F-Y, Mathews S, Joubert F, Shakir DI, Miles R, Mosse CA, Zhao T, Liang W, Kunpalin Y, Dromey B, Mistry T, Sebire NJ, Zhang E, Ourselin S, Beard PC, David AL, Desjardins AE, Vercauteren T, Xia W (2022) Intraoperative Needle Tip Tracking with an Integrated Fibre-Optic Ultrasound Sensor. Sensors 22:9035. https://doi.org/10.3390/s22239035
28. Su JL, Karpiouk AB, Wang B, Emelianov SY (2010) Photoacoustic imaging of clinical metal needles in tissue. JBO 15:021309. https://doi.org/10.1117/1.3368686
29. Su JL, Karpiouk AB, Chen Y-S, Emelianov SY (2011) Photoacoustic imaging to guide needle injections. In: Photons Plus Ultrasound: Imaging and Sensing 2011. SPIE, pp 665–671
30. Kim C, Erpelding TN, Maslov K, Jankovic L, Akers WJ, Song L, Achilefu S, Margenthaler JA, Pashley MD, Wang LV (2010) Handheld array-based photoacoustic probe for guiding needle biopsy of sentinel lymph nodes. J Biomed Opt 15:046010. https://doi.org/10.1117/1.3469829
31. Wang H, Liu S, Wang T, Zhang C, Feng T, Tian C (2019) Three-dimensional interventional photoacoustic imaging for biopsy needle guidance with a linear array transducer. Journal of Biophotonics 12:e201900212. https://doi.org/10.1002/jbio.201900212
32. Wei C-W, Nguyen T-M, Xia J, Arnal B, Wong EY, Pelivanov IM, O'Donnell M (2015) Real-time integrated photoacoustic and ultrasound (PAUS) imaging system to guide interventional procedures: ex vivo study. IEEE Transactions on Ultrasonics, Ferroelectrics, and Frequency Control 62:319–328. https://doi.org/10.1109/TUFFC.2014.006728
33. Jeng G-S, Li M-L, Kim M, Yoon SJ, Pitre JJ, Li DS, Pelivanov I, O'Donnell M (2021) Real-time interleaved spectroscopic photoacoustic and ultrasound (PAUS) scanning with simultaneous fluence compensation and motion correction. Nat Commun 12:716. https://doi.org/10.1038/s41467-021-20947-5
34. Donnelly EM, Kubelick KP, Dumani DS, Emelianov SY (2018) Photoacoustic Image-Guided Delivery of Plasmonic-Nanoparticle-Labeled Mesenchymal Stem Cells to the Spinal Cord. Nano Lett 18:6625–6632. https://doi.org/10.1021/acs.nanolett.8b03305
35. Piras D, Grijsen C, Schutte P, Steenbergen W, Manohar S (2013) Photoacoustic needle: minimally invasive guidance to biopsy. JBO 18:070502. https://doi.org/10.1117/1.JBO.18.7.070502
36. Xia W, Nikitichev DI, Mari JM, West SJ, Pratt R, David AL, Ourselin S, Beard PC, Desjardins AE (2015) Performance characteristics of an interventional multispectral photoacoustic imaging system for guiding minimally invasive procedures. JBO 20:086005. https://doi.org/10.1117/1.JBO.20.8.086005
37. Mari JM, Xia W, West SJ, Desjardins AE (2015) Interventional multispectral photoacoustic imaging with a clinical ultrasound probe for discriminating nerves and tendons: an ex vivo pilot study. J Biomed Opt 20:110503. https://doi.org/10.1117/1.JBO.20.11.110503
38. Xia W, West SJ, Nikitichev DI, Ourselin S, Beard PC, Desjardins AE (2016) Interventional multispectral photoacoustic imaging with a clinical linear array ultrasound probe for guiding nerve blocks. In: Photons Plus Ultrasound: Imaging and Sensing 2016. SPIE, pp 87–92
39. Kuniyil Ajith Singh M, Xia W (2020) Portable and Affordable Light Source-Based Photoacoustic Tomography. Sensors 20:6173. https://doi.org/10.3390/s20216173
40. Agano T, Sato N, Nakatsuka H, Kitagawa K, Hanaoka T, Morisono K, Shigeta Y, Tanaka C (2016) Photoacoustic imaging of clinical metal needle by a LED light source integrated transducer. In: Photons Plus Ultrasound: Imaging and Sensing 2016. SPIE, pp 678–682
41. Xia W, Kuniyil Ajith Singh M, Maneas E, Sato N, Shigeta Y, Agano T, Ourselin S, J. West S, E. Desjardins A (2018) Handheld Real-Time LED-Based Photoacoustic and Ultrasound Imaging System for Accurate Visualization of Clinical Metal Needles and Superficial Vasculature to Guide Minimally Invasive Procedures. Sensors 18:1394. https://doi.org/10.3390/s18051394
42. Shi M, Bodian S, West SJ, Sathasivam S, Gordon RJ, Collier P, Vercauteren T, Desjardins AE, Noimark S, Xia W (2022) Enhanced Photoacoustic Visualisation of Clinical Needles by Combining Interstitial and Extracorporeal Illumination of Elastomeric Nanocomposite Coatings. Sensors 22:6417. https://doi.org/10.3390/s22176417

43. Hinshaw JL, Lubner MG, Ziemlewicz TJ, Lee FT, Brace CL (2014) Percutaneous Tumor Ablation Tools: Microwave, Radiofrequency, or Cryoablation—What Should You Use and Why? RadioGraphics 34:1344–1362. https://doi.org/10.1148/rg.345140054
44. Fillinger MF (1999) New Imaging Techniques in Endovascular Surgery. Surgical Clinics of North America 79:451–475. https://doi.org/10.1016/S0039-6109(05)70018-9
45. Mahesh M (2001) Fluoroscopy: Patient Radiation Exposure Issues. RadioGraphics 21:1033–1045. https://doi.org/10.1148/radiographics.21.4.g01jl271033
46. Lickfett L, Mahesh M, Vasamreddy C, Bradley D, Jayam V, Eldadah Z, Dickfeld T, Kearney D, Dalal D, Lüderitz B, Berger R, Calkins H (2004) Radiation Exposure During Catheter Ablation of Atrial Fibrillation. Circulation 110:3003–3010. https://doi.org/10.1161/01.CIR.0000146952.49223.11
47. Bartel T, Müller S, Biviano A, Hahn RT (2014) Why is intracardiac echocardiography helpful? Benefits, costs, and how to learn. Eur Heart J 35:69–76. https://doi.org/10.1093/eurheartj/eht411
48. Sperandeo M, Rotondo A, Guglielmi G, Catalano D, Feragalli B, Trovato GM (2014) Transthoracic ultrasound in the assessment of pleural and pulmonary diseases: use and limitations. Radiol Med 119:729–740. https://doi.org/10.1007/s11547-014-0385-0
49. Francis KJ, Rascevska E, Manohar S (2019) Photoacoustic Imaging Assisted Radiofrequency Ablation: Illumination Strategies and Prospects. In: TENCON 2019 – 2019 IEEE Region 10 Conference (TENCON). pp 118–122
50. Francis KJ, Manohar S (2019) Photoacoustic imaging in percutaneous radiofrequency ablation: device guidance and ablation visualization. Phys Med Biol 64:184001. https://doi.org/10.1088/1361-6560/ab36a1
51. Yan Y, John S, Shaik T, Patel B, Lam MT, Kabbani L, Mehrmohammadi M (2021) Photoacoustic-guided endovenous laser ablation: Characterization and in vivo canine study. Photoacoustics 24:100298. https://doi.org/10.1016/j.pacs.2021.100298
52. Yan Y, John S, Ghalehnovi M, Kabbani L, Kennedy NA, Mehrmohammadi M (2019) Photoacoustic Imaging for Image-guided Endovenous Laser Ablation Procedures. Sci Rep 9:2933. https://doi.org/10.1038/s41598-018-37588-2
53. Graham M, Assis F, Allman D, Wiacek A, Gonzalez E, Gubbi M, Dong J, Hou H, Beck S, Chrispin J, Bell MAL (2020) In Vivo Demonstration of Photoacoustic Image Guidance and Robotic Visual Servoing for Cardiac Catheter-Based Interventions. IEEE Transactions on Medical Imaging 39:1015–1029. https://doi.org/10.1109/TMI.2019.2939568
54. Graham M, Assis F, Allman D, Wiacek A, Gonzalez E, Gubbi M, Dong J, Hou H, Beck S, Chrispin J, Bell M (2020) Photoacoustic image guidance and robotic visual servoing to mitigate fluoroscopy during cardiac catheter interventions. In: Mahadevan-Jansen A (ed) Advanced Biomedical and Clinical Diagnostic and Surgical Guidance Systems XVIII. SPIE, San Francisco, United States, p 45
55. Kim J, Mai TT, Kim JY, Min J-J, Kim C, Lee C (2020) Feasibility Study of Precise Balloon Catheter Tracking and Visualization with Fast Photoacoustic Microscopy. Sensors 20:5585. https://doi.org/10.3390/s20195585
56. Jansen K, van Soest G, van der Steen AF (2014) Intravascular photoacoustic imaging: a new tool for vulnerable plaque identification. Ultrasound in medicine & biology 40:1037–1048
57. Jansen K, van der Steen AFW, van Beusekom HMM, Oosterhuis JW, van Soest G (2011) Intravascular photoacoustic imaging of human coronary atherosclerosis. Opt Lett, OL 36:597–599. https://doi.org/10.1364/OL.36.000597
58. Nag S, Ciezki JP, Cormack R, Doggett S, DeWyngaert K, Edmundson GK, Stock RG, Stone NN, Yu Y, Zelefsky MJ (2001) Intraoperative planning and evaluation of permanent prostate brachytherapy: Report of the American Brachytherapy Society11Financial disclosure: Dr. Steven Doggett is a consultant for Burdette Medical Systems. Dr. Nelson Stone and Dr. Richard Stock have financial interest in ProSeed Inc. The software developed in the PIPER system has been licensed to RTek Medical Systems by the University of Rochester. The Research Corporation of the University of Rochester holds an equity interest in RTek Medical Systems

on behalf of the University of Rochester and Dr. Yan Yu. International Journal of Radiation Oncology*Biology*Physics 51:1422–1430. https://doi.org/10.1016/S0360-3016(01)01616-9
59. Han BH, Wallner K, Merrick G, Butler W, Sutlief S, Sylvester J (2003) Prostate brachytherapy seed identification on post-implant TRUS images. Med Phys 30:898–900. https://doi.org/10.1118/1.1568976
60. Gong L, Cho PS, Han BH, Wallner KE, Sutlief SG, Pathak SD, Haynor DR, Kim Y (2002) Ultrasonography and fluoroscopic fusion for prostate brachytherapy dosimetry. Int J Radiat Oncol Biol Phys 54:1322–1330. https://doi.org/10.1016/s0360-3016(02)03754-9
61. Fuller DB, Jin H, Koziol JA, Feng AC (2005) CT-ultrasound fusion prostate brachytherapy: a dynamic dosimetry feedback and improvement method. A report of 54 consecutive cases. Brachytherapy 4:207–216. https://doi.org/10.1016/j.brachy.2005.07.005
62. Patriciu A, Petrisor D, Muntener M, Mazilu D, Schär M, Stoianovici D (2007) Automatic Brachytherapy Seed Placement Under MRI Guidance. IEEE Trans Biomed Eng 54:1499–1506. https://doi.org/10.1109/TBME.2007.900816
63. Su JL, Bouchard RR, Karpiouk AB, Hazle JD, Emelianov SY (2011) Photoacoustic imaging of prostate brachytherapy seeds. Biomed Opt Express 2:2243–2254. https://doi.org/10.1364/BOE.2.002243
64. Harrison T, Zemp RJ (2011) Coregistered photoacoustic-ultrasound imaging applied to brachytherapy. JBO 16:080502. https://doi.org/10.1117/1.3606566
65. Harrison T, Zemp RJ (2011) Photoacoustic imaging of brachytherapy seeds using a channel-domain ultrasound array system. In: Photons Plus Ultrasound: Imaging and Sensing 2011. SPIE, pp 94–100
66. Kuo NP, Kang HJ, Song DY, Kang JU, Boctor EM (2012) Real-time photoacoustic imaging of prostate brachytherapy seeds using a clinical ultrasound system. JBO 17:066005. https://doi.org/10.1117/1.JBO.17.6.066005
67. Kuo N, Kang HJ, DeJournett T, Spicer J, Boctor E (2011) Photoacoustic imaging of prostate brachytherapy seeds in ex vivo prostate. In: Medical Imaging 2011: Visualization, Image-Guided Procedures, and Modeling. SPIE, pp 68–74
68. Yaseen M, Brecht H, Ermilov S, Gharieb R, Conjusteau A, Oraevsky A (2008) Hybrid optoacoustic and ultrasonic Imaging system for detection of prostate malignancies - art. no. 68560T. Progress in Biomedical Optics and Imaging - Proceedings of SPIE 6856:. https://doi.org/10.1117/12.769881
69. Valluru KS, Chinni BK, Rao NA, Bhatt S, Dogra VS (2011) Development of a c-scan photoacoutsic imaging probe for prostate cancer detection. In: Medical Imaging 2011: Ultrasonic Imaging, Tomography, and Therapy. SPIE, pp 97–103
70. Bell MAL, Kuo N, Song DY, Boctor EM (2013) Short-lag spatial coherence beamforming of photoacoustic images for enhanced visualization of prostate brachytherapy seeds. Biomed Opt Express, BOE 4:1964–1977. https://doi.org/10.1364/BOE.4.001964
71. Bell MAL, Guo X, Song DY, Boctor EM (2014) Photoacoustic imaging of prostate brachytherapy seeds with transurethral light delivery. In: Photons Plus Ultrasound: Imaging and Sensing 2014. SPIE, pp 113–118
72. Xie W, Li L, Li Z, Li H (2012) Photoacoustic imaging of prostate cancer using cylinder diffuse radiation. In: Optics in Health Care and Biomedical Optics V. SPIE, pp 509–514
73. Yaseen MA, Ermilov SA, Brecht H-P, Su R, Conjusteau A, Fronheiser M, Bell BA, Motamedi M, Oraevsky AA (2010) Optoacoustic imaging of the prostate: development toward image-guided biopsy. J Biomed Opt 15:021310. https://doi.org/10.1117/1.3333548
74. Wang L, Jacques SL, Zheng L (1995) MCML—Monte Carlo modeling of light transport in multi-layered tissues. Computer Methods and Programs in Biomedicine 47:131–146. https://doi.org/10.1016/0169-2607(95)01640-F
75. Lediju Bell MA, Kuo NP, Song DY, Kang JU, Boctor EM (2014) In vivo visualization of prostate brachytherapy seeds with photoacoustic imaging. J Biomed Opt 19:126011. https://doi.org/10.1117/1.JBO.19.12.126011
76. Bell MAL, Guo X, Song DY, Boctor EM (2015) Transurethral light delivery for prostate photoacoustic imaging. JBO 20:036002. https://doi.org/10.1117/1.JBO.20.3.036002

77. Su JL-S, Wang B, Emelianov SY (2009) Photoacoustic imaging of coronary artery stents. Opt Express, OE 17:19894–19901. https://doi.org/10.1364/OE.17.019894
78. Pan L, Baghani A, Rohling R, Abolmaesumi P, Salcudean S, Tang S (2013) Improving photoacoustic imaging contrast of brachytherapy seeds. In: Photons Plus Ultrasound: Imaging and Sensing 2013. SPIE, pp 750–755
79. Mitcham TM, Homan KA, Frey W, Chen Y-S, Emelianov SY, Hazle JD, Bouchard R (2013) Modulation of photoacoustic signal generation from metallic surfaces. JBO 18:056008. https://doi.org/10.1117/1.JBO.18.5.056008
80. Shi M, Zhao T, West SJ, Desjardins AE, Vercauteren T, Xia W (2022) Improving needle visibility in LED-based photoacoustic imaging using deep learning with semi-synthetic datasets. Photoacoustics 26:100351. https://doi.org/10.1016/j.pacs.2022.100351
81. Holan SH, Viator JA (2008) Automated wavelet denoising of photoacoustic signals for circulating melanoma cell detection and burn image reconstruction. Phys Med Biol 53:N227. https://doi.org/10.1088/0031-9155/53/12/N01
82. Hill ER, Xia W, Clarkson MJ, Desjardins AE (2017) Identification and removal of laser-induced noise in photoacoustic imaging using singular value decomposition. Biomed Opt Express, BOE 8:68–77. https://doi.org/10.1364/BOE.8.000068
83. Hariri A, Hariri A, Alipour K, Alipour K, Mantri Y, Schulze JP, Schulze JP, Jokerst JV, Jokerst JV, Jokerst JV (2020) Deep learning improves contrast in low-fluence photoacoustic imaging. Biomed Opt Express, BOE 11:3360–3373. https://doi.org/10.1364/BOE.395683
84. Shi M, Vercauteren T, Xia W (2022) Spatiotemporal singular value decomposition for denoising in photoacoustic imaging with a low-energy excitation light source. Biomed Opt Express, BOE 13:6416–6430. https://doi.org/10.1364/BOE.471198
85. Singh MKA, Parameshwarappa V, Hendriksen E, Steenbergen W, Manohar S (2017) Identification and removal of reflection artifacts in minimally invasive photoacoustic imaging for accurate visualization of brachytherapy seeds. In: Photons Plus Ultrasound: Imaging and Sensing 2017. SPIE, pp 49–55
86. Singh MKA, Parameshwarappa V, Hendriksen E, Steenbergen W, Manohar S (2016) Photoacoustic-guided focused ultrasound for accurate visualization of brachytherapy seeds with the photoacoustic needle. J Biomed Opt 21:120501. https://doi.org/10.1117/1.JBO.21.12.120501
87. Allman D, Reiter A, Bell MAL (2018) Photoacoustic Source Detection and Reflection Artifact Removal Enabled by Deep Learning. IEEE Trans Med Imaging 37:1464–1477. https://doi.org/10.1109/TMI.2018.2829662
88. Allman D, Assis F, Chrispin J, Bell MAL (2019) A deep learning-based approach to identify in vivo catheter tips during photoacoustic-guided cardiac interventions. In: Photons Plus Ultrasound: Imaging and Sensing 2019. SPIE, pp 454–460
89. Johnstonbaugh K, Agrawal S, Durairaj DA, Fadden C, Dangi A, Karri SPK, Kothapalli S-R (2020) A Deep Learning Approach to Photoacoustic Wavefront Localization in Deep-Tissue Medium. IEEE Transactions on Ultrasonics, Ferroelectrics, and Frequency Control 67:2649–2659. https://doi.org/10.1109/TUFFC.2020.2964698
90. Yazdani A, Agrawal S, Johnstonbaugh K, Kothapalli S-R, Monga V (2021) Simultaneous Denoising and Localization Network for Photoacoustic Target Localization. IEEE Transactions on Medical Imaging 40:2367–2379. https://doi.org/10.1109/TMI.2021.3077187
91. Shi M, Vercauteren T, Xia W (2023) Learning-based sound-speed correction for dual-modal photoacoustic/ultrasound imaging. In: 2023 IEEE International Ultrasonics Symposium (IUS). IEEE, Montreal, QC, Canada, pp 1–4
92. Pandey H, Saini S, Singh SP, Gautam NK, Singh S (2020) Candle soot derived carbon nanoparticles: An assessment of cellular and progressive toxicity using Drosophila melanogaster model. Comparative Biochemistry and Physiology Part C: Toxicology & Pharmacology 228:108646. https://doi.org/10.1016/j.cbpc.2019.108646

93. Esteves da Silva JCG, Gonçalves HMR (2011) Analytical and bioanalytical applications of carbon dots. TrAC Trends in Analytical Chemistry 30:1327–1336. https://doi.org/10.1016/j.trac.2011.04.009
94. Kuniyil Ajith Singh, M., & Xia, W. (2021). Biomedical photoacoustic imaging and sensing using affordable resources. Sensors, 21(7), 2572.

Chapter 22
The Evolution of LED-based Photoacoustic Imaging: From Labs to Clinics

Mithun Kuniyil Ajith Singh, Naoto Sato, Fumiyuki Ichihashi, Wenfeng Xia, and Yoshiyuki Sankai

Abstract Photoacoustic (PA) imaging, with its unique capability to provide simultaneous structural, functional, and molecular information, has emerged as one of the most rapidly advancing biomedical imaging modalities in recent years. As a hybrid technique, it offers superior penetration depth compared to purely optical imaging methods and grants optical contrast for molecular constituents within living tissues. Traditionally, photoacoustic imaging systems have relied on bulky and expensive Class IV lasers, presenting a significant hurdle to the clinical adoption of this promising modality. The integration of light-emitting diodes (LEDs), characterized by their portability and affordability, presents a transformative opportunity to expedite the clinical translation of photoacoustic imaging. In this chapter, we embark on a comprehensive exploration of the historical development of LEDs as an illumination source in biomedical photoacoustic imaging. We delve into key milestones, tracing the evolution from point-source measurements to the realization of high-power LED arrays. These innovations underscore the pivotal role of LEDs in shaping the future of photoacoustic imaging. Furthermore, this chapter encompasses a meticulous analysis of numerous studies, spanning phantom experiments, ex vivo investigations, in vivo animal studies, human in vivo applications, and clinical pilot studies. These examples collectively illuminate the exceptional preclinical and clinical potential harnessed by LED-based photoacoustic imaging, reaffirming its status as a groundbreaking modality with vast promise.

M. K. A. Singh (✉)
Research and Business Development Division, CYBERDYNE INC, Rotterdam, the Netherlands
e-mail: mithun_ajith@cyberdyne.jp

N. Sato · F. Ichihashi · Y. Sankai
Research and Development Division, CYBERDYNE INC, Tsukuba, Japan

W. Xia
School of Biomedical Engineering & Imaging Sciences, King's College London, London, UK

W. Xia (ed.), *Biomedical Photoacoustics*,
https://doi.org/10.1007/978-3-031-61411-8_22

Keywords LED-photoacoustic imaging · Light-Emitting Diodes (LEDs) · Photoacoustic tomography · Diagnostic imaging · Theranostics · Multimodal imaging · Optical imaging · Clinical translation · Preclinical studies

22.1 Introduction

In the realm of biomedical imaging, photoacoustic (PA) imaging stands as a beacon of promise, offering a multifaceted glimpse into tissue structure, function, and molecular constituents [1–3]. This emerging modality holds the capacity to transform our understanding of biological tissues while providing scalable resolution and imaging depth [1]. One of the central challenges in light-based biomedical imaging stems from the high optical scattering inherent to biological tissues. Beyond a depth of 1 mm, conventional ballistic optical microscopy techniques falter, rendering them ineffective. It is at this juncture that PA imaging shines. Its reliance on acoustic detection distinguishes it from purely optical methods, as sound scattering in tissues is orders of magnitude lower than that of light. The fundamental principle of PA imaging involves the emission of short-pulsed light onto tissue. Within the tissue, endogenous chromophores absorb this light, giving rise to a localized temperature increase [1–3]. This transient temperature elevation triggers thermoelastic expansion, leading to the generation of light-induced ultrasound (US) waves. These waves are subsequently detected by US sensors placed on the skin's surface, culminating in the creation of an optical absorption map with acoustic precision [4].

In a clinical context, PA imaging's greatest strength emerges from its seamless integration with US imaging, endowing it with the capability to deliver a comprehensive snapshot of anatomical, functional, molecular, and metabolic information [5]. This comprehensive insight leverages the unique optical absorption contrast provided by the vasculature, hemodynamics, oxygen metabolism, biomarkers, and gene expression, unleashing a multitude of preclinical and clinical applications across various medical domains, including vascular biology [6–9], oncology [10, 11], neurology [12, 13], ophthalmology [14, 15], dermatology [16, 17], gastroenterology [18–22], osteology [23–26], and cardiology [27, 28]. The advancement of PA imaging, however, has long been tethered to the limitations of traditional laser sources, primarily reliant on bulky and cost-prohibitive Class IV lasers [5]. These high-powered lasers, though effective, have been an impediment to the clinical adoption of this transformative imaging modality. A groundbreaking solution to this predicament comes in the form of light-emitting diodes (LEDs), distinguished by their portability, affordability, and clinical adaptability [8, 9, 29–41]. This transition to LED illumination sources heralds a significant stride toward making PA imaging accessible and practicable in a clinical environment. Laser diodes (LD) have also emerged as a contender for PA imaging illumination, offering advantages such as higher pulse repetition rates (PRRs), high power output, and optical energy [42–48]. However, LDs are predominantly available at wavelengths greater than 750 nm and

Table 22.1 Comparison of lasers, laser diodes and LEDs

	Energy(mJ)	PRR (Hz)	Pulse width(ns)	Cost	Advantages	Disadvantages
Laser	5~120	~10	<10	\$70–200 K	Powerful, ~5 cm penetration depth, tunable wavelength	Bulky size, eye protection and laser-safe rooms are needed
LD	0.5–2.5	~1 k–6 k	30–200	~\$10–25 K	Integration in a handheld probe is feasible, high PRR	Limited penetration depth, eye protection and laser-safe rooms needed, wavelength tuning is not possible
LED	0.2	~200-16 k	30–100	\$10–15 K	Integration in a handheld probe is feasible, high PRR, laser-safe rooms and eye-safety goggles are not needed	Limited penetration depth, wavelength tuning is not possible

Reproduced with permission from Ref. [39]

require laser-safe facilities and goggles. On the other hand, LEDs present a wider wavelength range (400–1000 nm) and provide lower optical energy but at higher repetition rates, allowing for more frame averaging without sacrificing temporal resolution. Further, the versatility in pulse widths makes LED and LD sources adaptable to the spatial resolution and imaging depth requirements. While LEDs have pulse widths in the range of 30–100 ns, LDs can be tuned from 30 ns to 200 ns. In terms of optical output, LED arrays, for instance, may reach up to 0.2 mJ per pulse (the highest reported optical output for 850 nm LED arrays), while LDs offer higher pulse energy but share similarities with Class IV lasers in terms of optical coherence and safety considerations. In light of these considerations, LED-based PA imaging is emerging as a portable, cost-effective, and safety-compliant alternative, poised to provide real-time functional and structural characterization of tissues across a spectrum of superficial and sub-surface imaging applications, especially in a clinical setting. Table 22.1 presents a comprehensive overview of the performance characteristics of commonly used lasers, LDs, and LEDs, underscoring the potential of LED-based PA imaging to expedite clinical translation.

As we delve into this chapter, we embark on a journey to understand the evolution and application of LED-based PA imaging, a path that converges LED light and sound to redefine the landscape of biomedical imaging.

22.2 Historical Developments in LED-Based Photoacoustic Imaging

PA imaging has already demonstrated its remarkable potential in diverse preclinical and clinical applications, establishing a solid foundation in research settings. As we stand at the threshold of a compelling shift from laboratory to clinical practice, the spotlight turns to the exploration of LD- and LED-based systems. These technologies are gaining substantial attention due to their inherent portability, affordability, and suitability for clinical deployment. Pulsed LDs have become a subject of intense investigation by various research groups, particularly for point-of-care applications [42–44]. However, it is worth noting that commercially available pulsed LDs are primarily confined to the near-infrared (NIR) wavelength range, and integrating multiple wavelengths in a handheld device can be a cumbersome endeavor [47, 48]. In contrast, LEDs offer a distinct advantage by spanning a broad wavelength range, ranging from 400 nm to 1000 nm while maintaining reasonable optical energy output. Achieving this broad coverage involves the development of arrays comprising multiple LED elements that can be strategically overdriven [36, 37]. Within this extensive wavelength spectrum, PA imaging reveals its strength, providing high contrast for crucial constituents such as melanin, hemoglobin, and, to some extent, fat. This makes LEDs an exceptional choice as an illumination source for multispectral PA imaging, enabling in-depth tissue imaging up to a depth of 1–1.5 cm. The convergence of LED technology and PA imaging holds the promise of widening the horizons of clinical diagnostics and translating the benefits of this cutting-edge imaging modality into routine medical practice.

LEDs operate on the fundamental principle of a semiconductor device known as a p–n junction diode. This two-terminal semiconductor device permits the flow of electric current in a singular direction while effectively blocking reverse current. When a voltage is applied, a process called forward biasing, the diode becomes conducive to the flow of electric current. Notably, even a small voltage increment can result in a substantial change in current flow. At the heart of this process is the movement of charge carriers, specifically holes from the P-type material and electrons from the N-type material, across the junction. This migration fosters robust electron-hole radiative recombination, ultimately leading to the emission of a multitude of photons [49]. LEDs are traditionally designed for continuous wave (CW) operation. However, it is worth highlighting that they can also be driven with pulsed current. In pulsed mode, the output energy of LEDs hinges on the peak current, often surpassing the CW-rated current, especially when the duty cycle is maintained at a low threshold (typically below 0.1%). This approach is adopted to prevent thermal damage to the LED. When overdriven in a pulsed mode, LEDs are capable of producing substantially higher optical output than in their conventional continuous operation, a variation often referred to as high-power LEDs. It is crucial to exercise caution in high-power operations, as excessive drive currents can lead to accelerated aging due to heat generation and, in extreme cases, immediate LED failure. Additionally, higher drive currents can diminish the quantum efficiency of the device [50]. In light of these considerations, the development of efficient and safe electronic drivers, along with effective heat dissipation systems, is paramount for harnessing the full potential of LEDs in high-power applications.

Approximately a decade ago, a groundbreaking proposal by Hansen introduced a novel concept—the utilization of LEDs emitting at a wavelength of 627 nm—as an affordable and compact excitation source for biomedical PA imaging [51]. This visionary work marked the first-ever demonstration of the feasibility of LEDs as primary light sources for PA imaging. The core principle underlying the creation of pulsed high-power LEDs hinges on the concept of overdriving low-power LEDs to generate optical output significantly surpassing their standard continuous-wave (CW) specifications [52–55]. In their pioneering work, Hansen and colleagues employed the Luxeon LXHL_PD09 LED, which, when subjected to a 1A DC-current, delivered approximately 250 mW of light output. To achieve pulsed illumination, they adopted a modified version of the MOSFET-based circuit initially presented by Alton and Raji [56]. By supplying the LED with 60 ns current pulses featuring a peak value of 40A, they achieved a pulse energy output of 400 nJ per pulse, along with a pulse width of 60 ns. This precision illumination allowed the generation of radiant energy essential for eliciting a PA response. Their initial proof-of-concept study averaged 50,000 A-line signals to detect the PA response, even employing a nonrealistic phantom. Nevertheless, it is worth noting that, owing to the pulse repetition rate (PRR) of the LED (200 Hz), this system necessitated 250 seconds to acquire an image, thereby falling short of real-time tissue imaging. Over the past decade, substantial advancements in solid-state device technology and collaborative efforts from various research groups have markedly enhanced the performance of LEDs [38]. These improvements encompass augmented pulse energy, increased pulse repetition rate (PRR), and other key parameters. The collective progress in LED technology has subsequently paved the way for the step-by-step development and commercialization of LED-based PAI systems that are now on par with laser-based counterparts.

In 2013, a pivotal development by Allen and colleagues marked a significant milestone in the integration of high-power LEDs as an illumination source for biomedical PA imaging [57]. The high-power LEDs in question, exemplified by the CBT-120 model from Luminus, were designed to emit light in the range of 400 nm to 650 nm wavelengths. Notably, this work led to an impressive enhancement in LED-PA imaging performance. Through a precise method of overdriving the LED elements, pushing them to operate at ten times their rated current, the pulse energy of the LEDs was elevated to an impressive 22 μJ per pulse, coupled with a pulse width of 500 ns. This achievement was underpinned by the same MOSFET-based driver that Hansen et al. used in their initial work on LED-PA imaging [51]. The augmentation in light energy output had a profound impact. Despite maintaining the same PRR of 200 Hz, the need for frame averaging was dramatically reduced by a factor of 1000. This reduction in frame averaging represented a noteworthy advancement, underscoring the potential of high-power LEDs to expedite imaging processes and improve the efficiency of biomedical PA imaging.

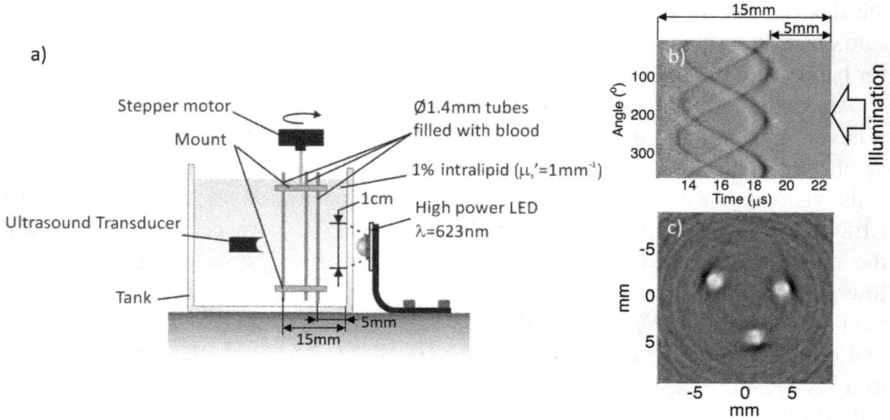

Fig. 22.1 (**a**) Photoacoustic imaging configuration. (**b**) Time-resolved photoacoustic signals from three 1.4 mm tubes containing human blood (35% hematocrit) submerged in a medium with 1% intralipid (μs' = 1 mm^{-1}). (**c**) Reconstructed photoacoustic image. Laser energy = 9 μJ, image frame averaging = 5000. (Reproduced with permission from Ref. [35])

In 2016, Allen and colleagues made significant enhancements to their system's performance [35]. They incorporated high-power LEDs (specifically, SST-90 from Luminus) operating within the wavelength of 400 nm to 650 nm. To achieve this, they utilized a commercial electronic driver (PCO-7120, Directed Energy, Inc.) that delivered 9 μJ per pulse for 620 nm, featured a pulse duration of 200 ns, a peak current of 50 A, and a PRR of 500 Hz. They also utilized 460 nm, 530 nm, and 590 nm LED elements and utilized coded excitation to demonstrate rapid multispectral PA imaging. Remarkably, they achieved these parameters by overdriving the LEDs by 20 times their nominal current. It is worth noting that they took great care to ensure the safety of the device. They maintained a duty cycle of 0.01%, which remained well below the previously reported safety limit of 1%. In their experimental work, they initiated imaging with a realistic tissue-mimicking phantom. This involved averaging 5000 image frames while adopting a wide-field illumination strategy. Their remarkable achievement in imaging depth was notable, reaching 15 mm within a medium containing 1% intralipid (Fig. 22.1).

The groundbreaking endeavors highlighted earlier served as the cornerstone for LED-based PA imaging, notably showcasing the viability of employing visible light for PA imaging—a feat unattainable with traditional lasers or laser diodes (LDs). A pivotal application within PA imaging is oxygen saturation imaging, a domain of utmost significance. To facilitate this, it is imperative to possess illumination wavelengths specifically tailored to this purpose. Given the distinct absorption peaks for deoxyhemoglobin (at 690 nm) and oxyhemoglobin (at 850 nm), the development of LEDs within these wavelength ranges emerged as a critical pursuit. Consequently, many researchers have actively invested their efforts in this direction, and their contributions are elaborated upon in the subsequent sections.

Table 22.2 Key milestones in the advancement of LED-based photoacoustic imaging technology

Year	Authors	Pulse width(ns)	Peak current(A)	Pulse energy(mJ)	Repetition rate (Hz)	Wavelength (nm)
2011	Hansen et al. [51]	60	40	0.0004	200	627
2013	Allen et al. [57]	500	200	0.0022	200	623
2016	Allen et al. [35]	200	50	0.0009	500	620
2016–2017	Agano et al. [36, 37, 46]	70	15–20	0.15–0.2	4000	850
2018	Zhu et al. [8]	70	20	0.2	4000–16,000	850

Adapted with permission from Ref. [39]

In 2016 and 2017, Agano and colleagues introduced a pioneering concept by utilizing high-power LED arrays operating at 850 nm [36, 37]. These LEDs delivered optical energy of 200 μJ per pulse with a pulse duration ranging from approximately 70–100 ns. Their achievement was made possible through a meticulous optimization process aimed at enhancing signal-to-noise ratio (SNR). This optimization involved amplifying the optical output and developing a highly noise-efficient front-end electronics system boasting multiple stages of amplification, totaling approximately 106 dB. Additionally, they ensured that the characteristics of the US transducer were harmoniously matched with the LED light pulse width [46]. To elevate the optical output to such impressive levels, Agano and the team engineered LED arrays, applying hybrid techniques to optimize their performance. A single LED element yielded an output energy of 0.024 μJ per pulse with a pulse duration of 70 ns at a 1A DC current. Through the ingenious creation of LED elements featuring a double-stack structure, these elements were methodically arranged in an array configuration. By boldly subjecting them to 14 times the rated current, they successfully attained an optical pulse energy of 200 μJ at a wavelength of 850 nm. Based on this, the first commercially available LED-based PA imaging (PAI) system, known as AcousticX, was developed, which boasted an impressive PRR of 4 KHz. In 2018, Zhu and colleagues further improved the PRR of LEDs, elevating it to 16 KHz [8]. They employed the same system to showcase its dynamic capabilities in structural and functional imaging. Key milestones in the advancement of LED-based PA imaging technology are listed in Table 22.2.

The AcousticX system underwent comprehensive characterization across multiple studies, evaluating its imaging depth, spatial resolution, frame rate, and accuracy in oxygen saturation imaging. In one set of investigations employing a 9 MHz US probe in conjunction with 850 nm LED arrays, Xia and colleagues reported mean axial and lateral resolutions of 220 μm and 460 μm, respectively, which proved consistent for both US and PA imaging [34]. They further detailed an imaging depth of 2.8 cm at an interleaved US and PA frame rate of 30 Hz in an ex vivo tissue study. Notably, they enhanced the imaging depth to 3.8 cm through frame averaging, resulting in a final frame rate of 1.5 Hz. In another study led by Hariri, a combination of a 7 MHz US probe and 850 nm LED arrays yielded mean axial and lateral resolutions of 268 μm and 570 μm, respectively [29]. Their research demonstrated an imaging depth of 3.2 cm at a frame rate of 15 Hz in a phantom study. Furthermore, Hariri's team conducted a molecular imaging study to evaluate the system's sensitivity in detecting commonly used molecular contrast agents. They reported limits of detection for ICG, MB, and DiR as 9 μM, 0.75 mM, and 68 μM, respectively. Oxygen saturation imaging stands as a pivotal application within PA imaging. Employing a two-wavelength (750/850 nm) approach, Kalloor Joseph and colleagues assessed the AcousticX system's capabilities in oxygen saturation imaging, conducting experiments on phantoms and in vivo small animals [58]. Their investigations, supported by 28 in vitro measurements using human blood, revealed a standard error of 8.4% between actual oxygen saturation values and those derived from the system's oxygen saturation images. Furthermore, they demonstrated the repeatability and reproducibility of oxygen saturation imaging by conducting an in vivo mouse oxygen breathing challenge experiment within the same study.

Since the start of 2017, there has been a remarkable surge in this field, with numerous studies showcasing the capabilities of LED-based PA imaging. In the following section, we will delve into a comprehensive review of the diverse preclinical and clinical applications that have been explored, all made possible using AcousticX—a commercially available LED-based PA and US imaging system.

22.3 Preclinical and Clinical Applications of LED-Based PAI

In this section, we undertake an extensive review of numerous preclinical and clinical studies that have harnessed the potential of LED-based PA imaging. By examining these studies, we aim to offer valuable insights and perspectives that can guide future endeavors in the realm of clinical translation.

22.3.1 Phantom and Ex Vivo Studies

22.3.1.1 Imaging of Intraocular Tumors

PA imaging has shown promise in the diagnosis and characterization of cancer, particularly in the context of head and neck tumor imaging [59–64]. Intraocular tumors, though relatively rare, pose a significant threat to life [64, 65]. Xu et al. showcased the capability of laser-based PA imaging in delineating intraocular tumors by unraveling molecular composition and structural heterogeneity [66]. Expanding on this, Zhu et al. ventured into exploring the applicability of LED-based PAI for imaging intraocular tumors within an intact human eyeball [8]. Figure 22.2a, b provides a glimpse into 2D US and PA images of a choroidal melanoma tumor within the eyeball. The US image is represented in gray scale, while the PA image is displayed in pseudo-color. Figure 22.2c presents a PA/US overlay, offering

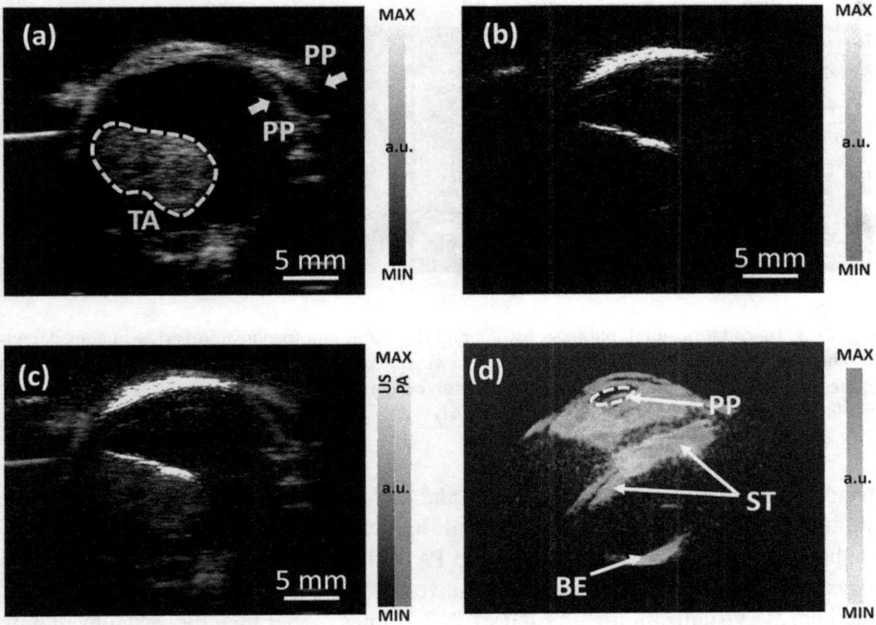

Fig. 22.2 (**a**) 2D B-scan ultrasound image and (**b**) photoacoustic image of an ex vivo ocular globe with a clinically confirmed tumor. (**c**) Combined photoacoustic and ultrasound image. (**d**) Perspective view of a 3D photoacoustic image of the ocular globe featuring a choroidal melanoma tumor. PP pupil, TA tumor area, ST surface of the tumor, BE back of the eye. (Reproduced with permission from Ref. [8])

a synergistic depiction of information. Notably, PA results underscore the capability of LED-based PA imaging to penetrate with sufficient depth, covering the entire tumor volume. The 3D PA image of the eyeball in Fig. 22.2d captures the intricate details. In this study, single-wavelength LED arrays (850 nm) were employed to visualize vasculature, with the prospect of utilizing LED arrays featuring multiple wavelengths to observe individual molecular components in future applications. These findings underscore the vast potential of LED-based PAI in the expansive field of cancer diagnosis and staging.

22.3.1.2 Guidance of Minimally Invasive Procedures with Peripheral Tissue Targets

Accurate guidance is paramount in minimally invasive vascular access procedures, where conventional US imaging is often utilized [67]. However, visualizing medical needles and tissue targets can be challenging. PA imaging, offering high contrast for metallic needles and vasculature, is emerging as a promising solution for guiding such procedures. The adoption of PA imaging for point-of-care applications has

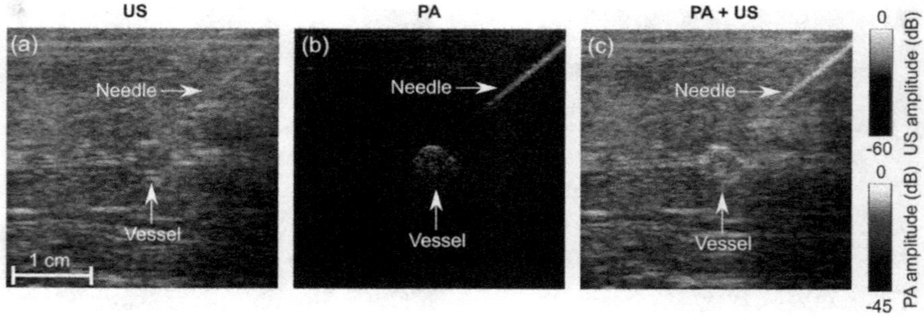

Fig. 22.3 (**a–c**) Ultrasound, photoacoustic, and ultrasound–photoacoustic overlay images depicting the insertion of a medical needle towards a vessel-mimicking phantom embedded in chicken tissue, respectively. (Note: The uppermost 5 mm, containing the ultrasound gel, is omitted in the display. Adapted with permission from Ref. [34])

been hindered by the need for bulky and expensive solid-state lasers for tissue illumination [68, 69]. In 2018, Xia et al. harnessed the capabilities of AcousticX to showcase the potential of LED-based PA and US imaging in guiding minimally invasive procedures [34]. Their results unveiled that LED-based PA imaging facilitated needle visualization with SNRs 1.2–2.2 times higher than those achieved with US imaging across needle insertion angles ranging from 26° to 51°. The phantom study illustrated an impressive imaging depth close to 4 cm, as depicted in Fig. 22.3a–c. Additionally, the study demonstrated real-time visualization of superficial vasculature in the finger and wrist of a human volunteer alongside conventional pulse-echo US. These promising outcomes suggest that the combination of LED-based PA and US imaging holds substantial potential for guiding minimally invasive procedures targeting peripheral tissues.

22.3.1.3 Imaging of Human Placental Vasculature

Precision in guidance holds significance not just in vascular access procedures but extends to minimally invasive fetal surgeries, such as those addressing twin-to-twin transfusion syndrome (TTTS) [70]. TTTS arises from imbalanced blood flow between twin vascular connections (anastomoses) in the placenta and, if untreated, poses a high risk of morbidity and mortality [71]. Current clinical practices for TTTS procedures involve in vivo imaging of the placenta using white-light fetal endoscopy and external B-mode US imaging. However, these methods lack sufficient contrast to visualize small anastomotic vessels beneath the surface of the chorionic placenta [72]. In 2018, Maneas et al. introduced LED-based PA imaging for wide-field imaging of the human placental vasculature ex vivo, leveraging the high optical absorption contrast offered by oxyhemoglobin and deoxyhemoglobin [73]. Previous studies indicated the detectability of PA signals from placental vessels using external clinical US probes [74–76]. Maneas et al. utilized the LED-based PA system (AcousticX) to image chorionic (fetal) superficial and subsurface vasculature in both normal and TTTS-treated human placentas. The clinical US probe-integrated LED-based PA system enabled rapid interleaved 2D PA and US imaging (see Fig. 22.4). This work not only confirms the feasibility of volumetric LED-based PA imaging for generating detailed 3D maps of surface and subsurface vasculature up to a depth of approximately 7 mm in the human placenta but also envisions the potential of interventional PA imaging for visualizing the chorionic surface and superficial placental blood vessels with high spatial resolution, thereby offering value in minimally invasive fetal treatments.

22.3.2 *In Vivo Small Animal Imaging Studies*

22.3.2.1 Imaging of Tumor Vasculature Using Contrast Enhancement

The utilization of contrast enhancement in vasculature imaging is a common practice. Xavierselvan et al. investigated the capabilities of the LED-PA system for imaging tumor vasculature with contrast enhancement [77]. In their study, subcutaneous head and neck tumor (FaDu) xenografts in nude mice were imaged when the tumor size reached approximately 100 mm^3, employing the AcousticX system. The images revealed heterogeneous vascular density within the tumor, showcasing the system's ability to extract vascular information from tissues at depths exceeding 1 cm. Their findings position LED-based PA imaging as a more cost-effective option to leverage the technology to noninvasively monitor nanoparticle or drug uptake in preclinical experiments. Figure 22.5 illustrates the LED-based PA imaging system's capacity to image a naphthalocyanine (NC) dye with strong absorption in the NIR region (absorption peak at 860 nm). The imaging of pre- and postinjection of 100 μL NC dye intratumorally into the FaDu tumor interstitium showcased the dye's distribution throughout the tumor, with a particularly strong PA signal observed from the top of the tumor—possibly attributed to these areas receiving higher light energy.

22.3.2.2 Noninvasive Monitoring of Angiogenesis

The incorporation of engineered constructs into the host blood supply necessitates vascularization for successful implant integration. Monitoring the neovascularization of implanted constructs is crucial for translational applications. Laser speckle contrast analysis (LASCA), commonly used in regenerative medicine, offers high spatial resolution but is limited in imaging depth and sensitivity to perfused blood vessels [78–80]. In 2019, Zhu et al. addressed this challenge using LED-based PA and US imaging [9]. They employed an LED-based PA-US dual-mode system to observe angiogenesis in fibrin-based scaffolds implanted subcutaneously in mice over 7 days. Scaffolds, with or without basic fibroblast growth factor (bFGF), were

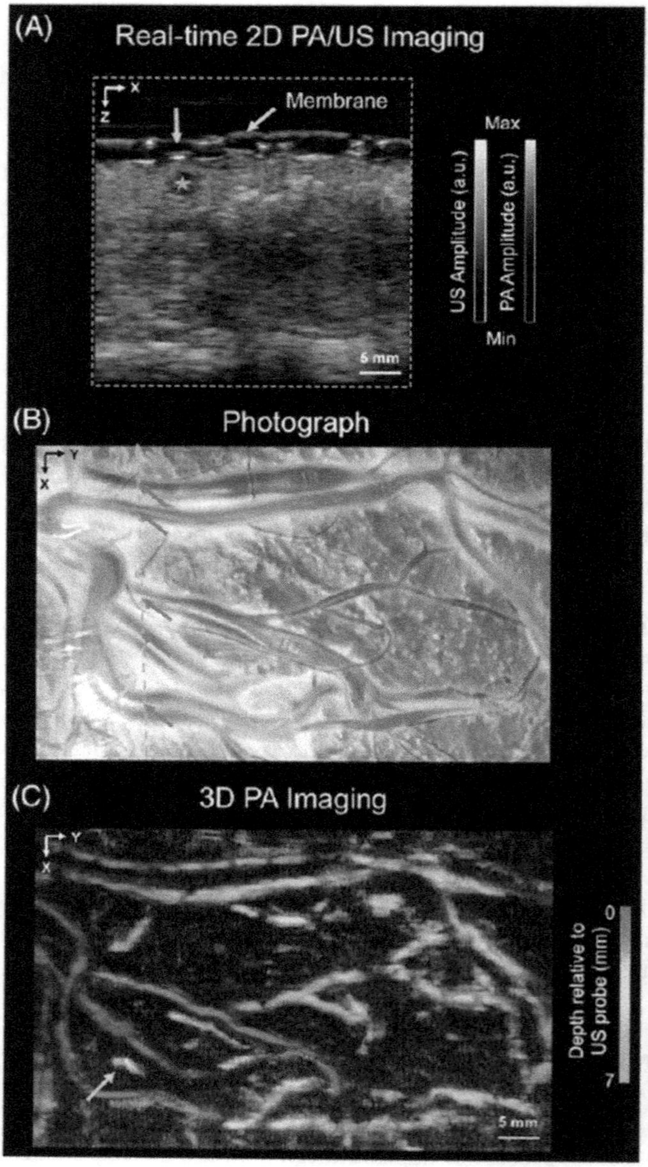

Fig. 22.4 Wide-field photoacoustic–ultrasound images capturing the chorionic placental vasculature in an untreated segment of a twin-to-twin transfusion syndrome (TTTS) photocoagulated placenta. (**a**) 2D photoacoustic and ultrasound overlay image obtained from a region corresponding to the green line in (**b**), a photograph of the imaged placenta. (**c**) 3D photoacoustic image presented as a maximum intensity projection (MIP) of the reconstructed image volume. (Reproduced with permission from Ref. [73])

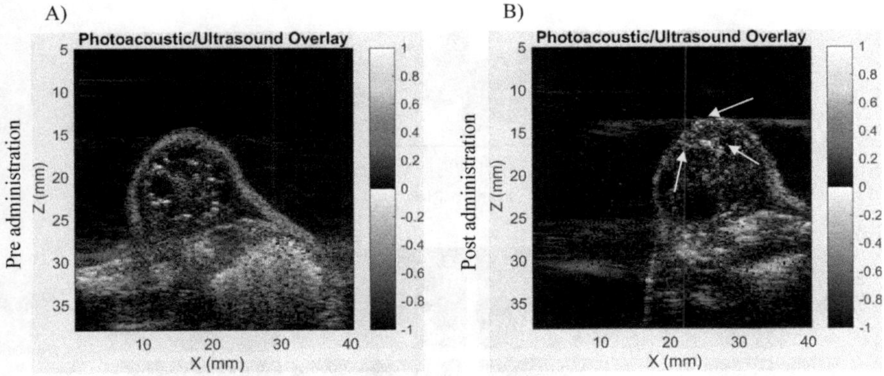

Fig. 22.5 Improving contrast in LED-based photoacoustic imaging through the use of contrast agents is explored. A composite display of photoacoustic and ultrasound images of subcutaneous FaDu tumor in mice is presented before (**a**) and after (**b**) the administration of the NC dye. The photoacoustic images were captured using an 850 nm LED light source, with areas exhibiting heightened contrast highlighted by white arrows. (Reproduced with permission from Ref. [77])

imaged on days 0 (postimplantation), 1, 3, and 7 using both LASCA and PA-US systems (Fig. 22.6). Quantified perfusion from LASCA and PA imaging was compared with histologically determined blood vessel density on day 7. Vessel density aligned with changes in perfusion measured by both LASCA and PA. PA imaging revealed distinctions in neovascularization between the upper and lower regions of the scaffold. This study highlights that LED-PA imaging can serve as a noninvasive and highly sensitive method for monitoring deep-seated vascularization in regenerative applications.

22.3.2.3 Oxygen Saturation Imaging in the Diagnosis of Rheumatoid Arthritis

Multiwavelength LED-based PA imaging proves to be a valuable tool for functionally characterizing tissue in various clinical applications, such as determining blood oxygen saturation. In conditions like rheumatoid arthritis (RA), hypoxia in the joints serves as a biomarker. The accurate estimation of blood oxygen concentration makes multiwavelength PA imaging a promising tool for the early detection of RA. Joseph et al. conducted an in vivo animal study to assess the system's capability to measure blood oxygen saturation [58]. Initially, they validated ex vivo PA oxygen saturation imaging using human blood against oximeter readings and further confirmed it through in vivo animal studies. The PA oxygen saturation estimation aligned with oximeter readings, as verified in the in vivo studies. The system achieved an imaging depth of approximately 5 mm for blood oxygen saturation, deemed sufficient for RA imaging of the finger joint. Figure 22.7 illustrates the results of blood oxygen saturation imaging. Figure 22.7a–c display an oxygen saturation map of two tubes—

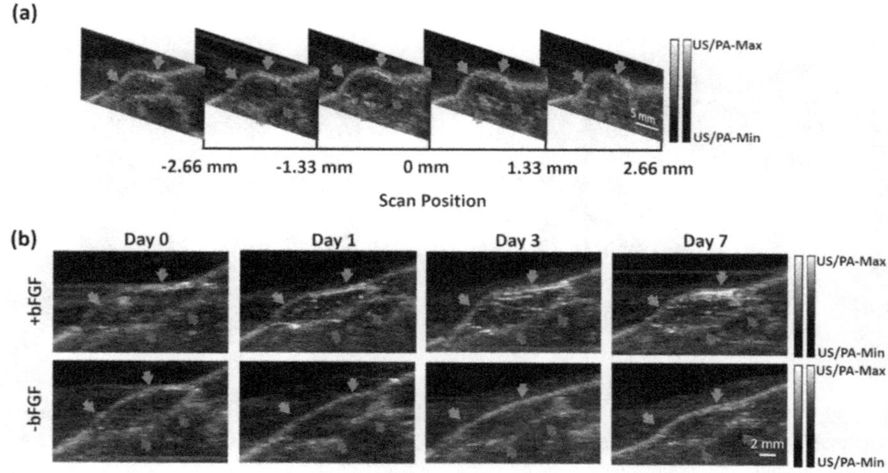

Fig. 22.6 In the longitudinal LED-based photoacoustic-ultrasound imaging of two subcutaneous implants, green and red arrows point to the upper and lower edges of the scaffold, identified through B-mode ultrasound. (**a**) Depicts a series of two-dimensional ultrasound-photoacoustic images from a + bFGF scaffold on day 7 at various scan positions, with images only within the range of −2.5 to 2.5 mm utilized for the maximum intensity projection (MIP). (**b**) Shows a sequence of longitudinal MIP ultrasound–photoacoustic images of +bFGF and − bFGF scaffolds from the same mouse. On day 7, the red representation of PA intensity displays the most significant difference. (Reproduced with permission from Ref. [9])

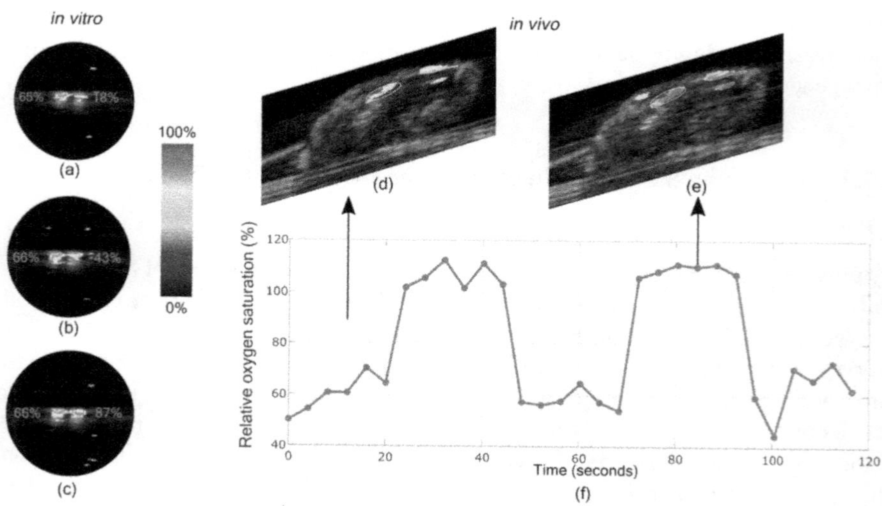

Fig. 22.7 In vitro and in vivo oxygen saturation imaging: (**a–c**) Oxygen saturation map from three different oxygen concentrations in the tube on the right while keeping it unchanged in the left tube during an in vitro study, (**d–e**) Photoacoustic and ultrasound imaging of a mouse's knee, (**f**) Average oxygen saturation in the region of interest (ROI) plotted over time during in vivo oxygen saturation imaging of the mouse's thigh muscle. (Reprinted with permission from Ref. [58])

one with normal blood (oxygen concentration around 65%) and another with blood having concentrations of 18%, 43%, and 87%, respectively. This corresponds to the estimated oxygen saturation generated by the imaging system. Additionally, Fig. 22.7d–f showcase in vivo imaging of the mouse thigh muscle. When the mouse breathed normal air, an average oxygen concentration of 61.1% was observed, and with 100% oxygen, it was 105.3% using the system. This indicates that the system can provide a reasonably accurate estimate of oxygen saturation with in vivo imaging. The results demonstrate that early-stage changes in rheumatoid arthritis, such as synovial angiogenesis and hypoxia, can be imaged using the LED-based PA imaging system. This suggests that multispectral LED-based PAI holds great potential for the early detection and staging of RA in animal studies.

22.3.2.4 Noninvasive Imaging of Pressure Ulcers

Hariri et al. utilized PA imaging as a noninvasive method to detect early tissue damage that may not be visible to the naked eye [81]. They employed a mouse model of pressure ulcers by implanting subdermal magnets in the dorsal flank and periodically applying an external magnet to the healed implant site. The application of magnet-induced pressure occurred in cycles with the extent of ulceration determined by the number of cycles. LED-based PA imaging was used to assess these ulcers. Figure 22.8 displays baseline (top) and stage I ulcers (bottom) through US, PA, and US/PA overlay images obtained using LED excitation. The insets in Fig. 22.8b, e feature photographs of mice without and with pressure ulcers, respectively. The LED-based PA imaging system detected early-stage (stage I) pressure ulcers, revealing a 2.5-fold increase in the PA signal. Significantly, the technique demonstrated the ability to detect dysregulated skin even before the eruption of stage I ulcers. Moreover, they observed notable changes in PA intensity during the healing process, indicating that this approach could monitor therapy. Although not shown here, these findings were validated with histology. This suggests that the PA-based approach could hold clinical value for monitoring skin diseases, including pressure ulcers.

22.3.2.5 Molecular Imaging: Detection and Monitoring of Reactive Oxygen and Nitrogen Species

Reactive oxygen and nitrogen substances (RONS) play crucial roles in regulating various functions in living systems. Endogenous RONS contribute to processes such as signal transduction, smooth muscle relaxation, and blood pressure regulation [61, 62]. Disorders in RONS levels can lead to diseases like cancer, making RONS detection valuable for diagnosing and treating infections and various illnesses [63–65].

In a 2019 study, Hariri et al. explored the molecular imaging of RONS using near-infrared absorbing small molecules (CyBA) and an LED-based PA imaging

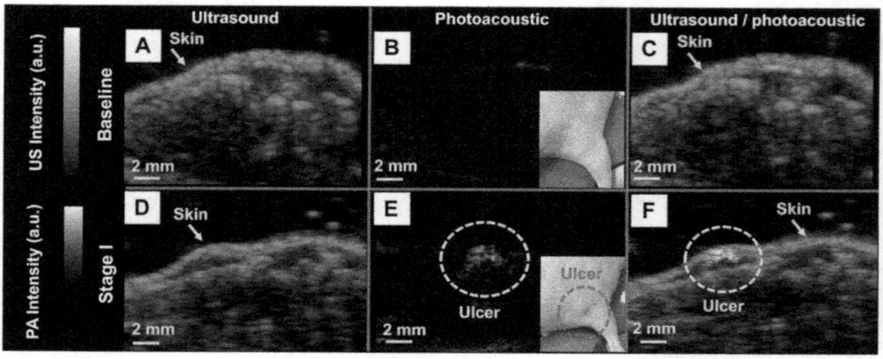

Fig. 22.8 LED-based photoacoustic evaluation of stage I pressure ulcers. (**a**) B-mode ultrasound image at baseline conditions without applied pressure, (**b**) B-mode photoacoustic image at baseline, showing minor photoacoustic signal from the epidermis. Inset: Photographic view of the mouse without an ulcer, (**c**) B-mode photoacoustic/ultrasound overlay at baseline conditions, (**d**) B-mode ultrasound image at stage I, (**e**) B-mode photoacoustic image at stage I, demonstrating a 2.5-fold increase in photoacoustic intensity compared to baseline. Inset: Photographic view of the mouse with a stage I ulcer, (**f**) B-mode photoacoustic/ultrasound overlay at stage I, Image depth: 1 cm, Scale bars: 2 mm. (Reprinted with permission from Ref. [81])

system [31]. They assessed CyBA's ability to measure inflammation in mice. Figure 22.9a–d depict US/PA images at different time points after the injection of CyBA and zymosan. These images show an increasing trend in PA signals from the diffused probe (CyBA) at the region of interest (ROI) marked by a yellow circle. Figure 22.9e presents the quantitative analysis of CyBA's in vivo PA imaging assessment, revealing a gradual increase in PA intensity, with a ~3.2-fold increase quantified 90 min after CyBA injection. The flat PA intensity trend of separated zymosan and the dye is also evident.

While LED output energy is considerably lower than that of conventional Nd:YAG lasers, it proves advantageous in this context. The higher power generated by Nd:YAG lasers can bleach dyes, unlike the light from LEDs. This work not only highlights the sensitivity of PA in detecting RONS but also underscores the practicality of LED-based PA imaging in clinical settings for molecular imaging applications involving dyes. The study demonstrates that the advantages of LEDs as an illumination source can expedite the clinical translation of PA imaging. The authors foresee the use of LED-based PA imaging and the reported probes for clinical monitoring of RONS, particularly in keloid diagnosis and drug toxicity studies [82].

22.3.2.6 Imaging of Molecular-Labeled Cells

In a few previous studies, PA imaging was used for stem cell imaging [83, 84]. In a recent study [29], Hariri et al. explored the in vivo molecular imaging performance

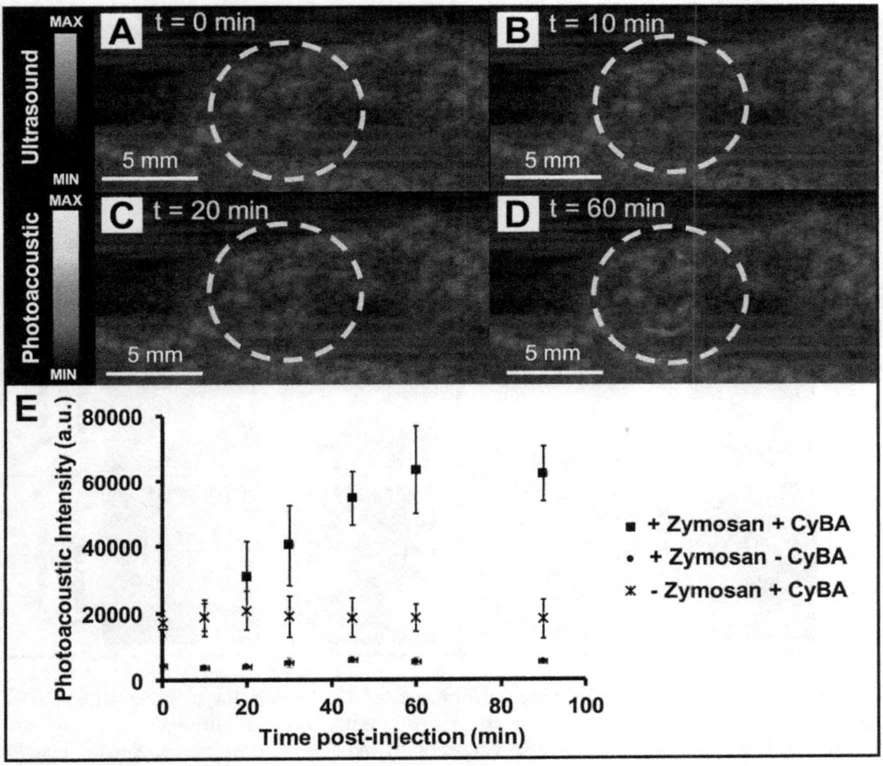

Fig. 22.9 In vivo photoacoustic evaluation of CyBA. (**a–d**) Ultrasound–photoacoustic images at different time points: (**a**) baseline, (**b**) 10 minutes, (**c**) 20 minutes, and (**d**) 60 minutes after CyBA injection, (**e**) Quantitative analysis of photoacoustic intensity as a function of time postinjection of CyBA. (Reprinted with permission from Ref. [31])

of AcousticX using labeled cells. DiR, an effective contrast agent for cellular imaging, was employed, and human mesenchymal stem cells (HMSCs; Lonza, PT-2501, NJ, USA) were used in this investigation. Figure 22.10 illustrates PA images before and after the injection of DiR, DiR @ HMSC, and HMSC, respectively. The needle generates a robust PA signal, and simultaneous US acquisition provides detailed structural information alongside functional details from DiR-labeled cells. Strong PA signals are visible in the presence of DiR (Fig. 22.10d). Additionally, an increase in signal is observed when HMSC is labeled using DiR (Fig. 22.10h). Unlabeled HMSCs were injected as a control, but no increase in PA signal was observed. This study underscores the potential of LED-based PA imaging in cellular imaging.

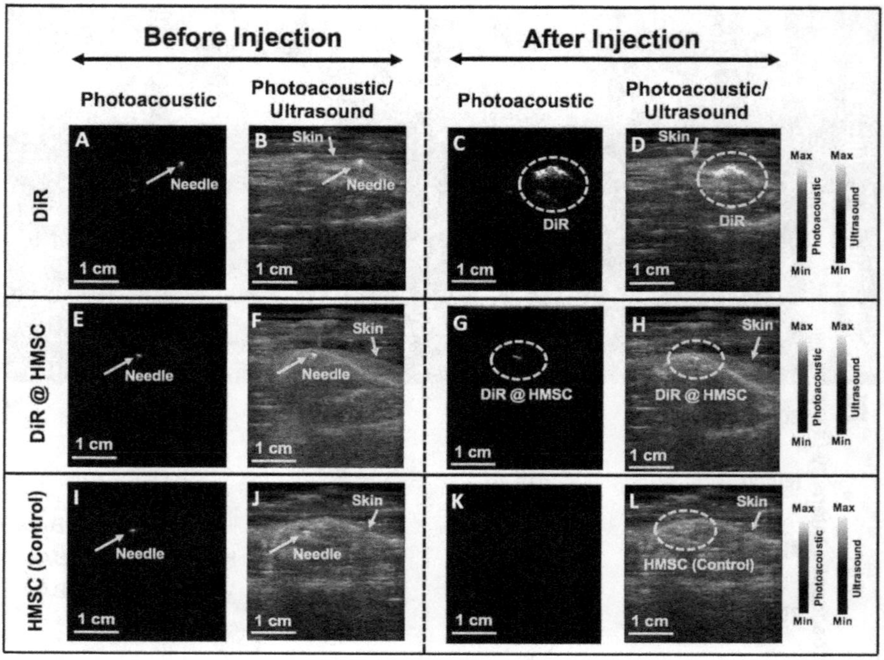

Fig. 22.10 In vivo molecular imaging performance of LED-based PA imaging with labeled cells. Images were categorized into three groups: (**a**–**d**) Injecting DiR—photoacoustic and overlaid photoacoustic-ultrasound images before (**a**, **b**) and after (**c**, **d**) injection. Needles marked in the images are subcutaneously injected into the spinal cord area before DiR injection. The needle exhibits a strong photoacoustic signal. (**e**–**h**) DiR @ HMSC—photoacoustic and overlaid photoacoustic-ultrasound images before (**e**, **f**) and after (**g**, **h**) injection. (**i**–**l**) HMSC—photoacoustic and overlaid photoacoustic-ultrasound images before (**i**, **j**) and after (**k**, **l**) injection. (Reproduced with permission from Ref. [29])

22.3.3 *In Vivo Human Volunteer Studies*

22.3.3.1 Capturing Images of Both Veins and Lymphatic Vessels Simultaneously in Real Time

Kuniyil Ajith Singh et al. showcased the capability of differentiating veins and lymphatic vessels in human volunteers using dual-wavelength LED arrays (820/940 nm) in LED-based PA imaging [85]. By injecting indocyanine green (ICG) as a contrast agent, they applied a straightforward image division algorithm, based on the relatively low optical absorption of ICG at 940 nm. This approach effectively distinguished veins and lymphatic vessels in real time, as depicted in Fig. 22.11. The utilization of this multispectral PA/US technique holds significant promise in guiding procedures such as lymphaticovenous anastomosis, where the differentiation between venous blood and lymphatic vessels is crucial.

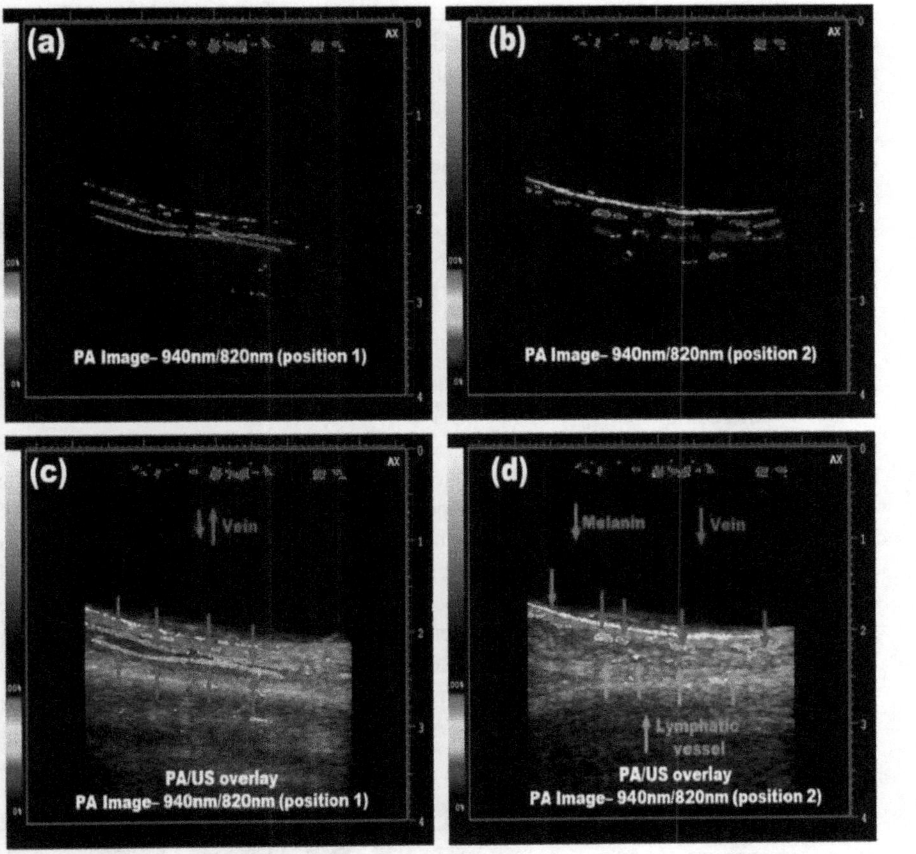

Fig. 22.11 Achieving real-time visualization of both lymphatic vessels and veins simultaneously using LED-based photoacoustic imaging. (**a**) Photoacoustic image at 940 nm/820 nm generated when the probe was aligned to a vein (position 1), (**b**) photoacoustic image at 940 nm/820 nm generated when the probe was aligned to a lymphatic vessel (position 2), (**c**) overlay of 940 nm/820 nm photoacoustic image on ultrasound image when the probe was in position 1, and (**d**) overlay of 940 nm/820 nm photoacoustic image on ultrasound image when the probe was aligned to position 2. (Reproduced with permission from Ref. [85])

22.3.3.2 Full-View Tomography of Finger Joints

It is widely recognized that 2D PA and US imaging using linear arrays can be hindered by limited-view artifacts. Additionally, the directional constraints of US transducers pose a risk of information loss. An effective solution to this limitation involves scanning around the object of interest to generate full-view 3D tomographic images, offering higher spatial resolution, and eliminating limited-view issues. However, many existing 3D PA/US tomography systems, whether commercially available or lab-made, rely on expensive and bulky pulsed lasers, making them unsuitable for point-of-care settings. In a recent breakthrough, Joseph et al. demonstrated the feasibility of producing full-view LED-based 3D PA and US tomographic images of human finger joints (Fig. 22.12) [86]. This proof-of-concept study affirms that cost-effective and portable LEDs, combined with commercially available linear array US probes, can generate high-quality 3D tomographic images with rich information suitable for various point-of-care clinical applications. In another promising development, Agrawal et al. designed a multispectral molecular LED-based 3D PA tomography system, showcasing its potential in unmixing and separating three optical absorbers (Melanin, methylene blue, and ICG) embedded inside a tissue-mimicking phantom [87].

22.3.3.3 Imaging of Peripheral Vascular Imaging and Function in 2D and 3D

Hypoxia serves as a crucial biomarker reflecting the onset and progression of various diseases, including cancer [88]. Multispectral PA imaging has proven effective in noninvasively detecting relative hemoglobin oxygen saturation and hypoxia in vivo, leveraging the spectral differences between oxygenated hemoglobin and deoxyhemoglobin [89–91]. In a study by Zhu et al., the authors demonstrated the potential of LED-PA in imaging peripheral vasculature and its oxygen saturation in real time [8]. Figure 22.13a displays LED-based PA and US B-scan images of a human finger along the sagittal section, highlighting arterial pulsation (indicated by the arrow) across four movie frames. The research delves into the feasibility of using

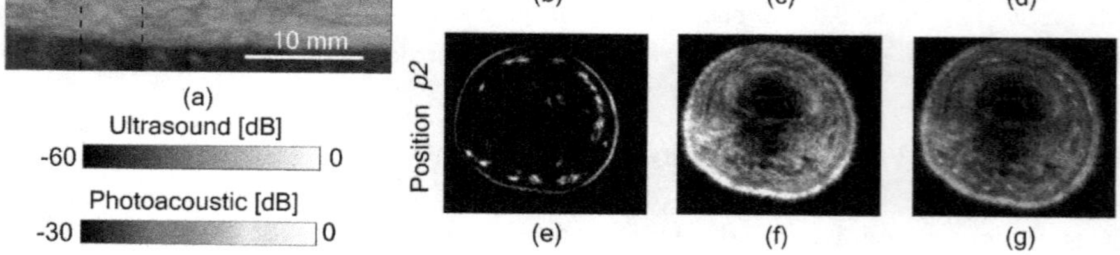

Fig. 22.12 In vivo tomographic imaging of finger joints employing multi-angle spatial compounding. (**a**) Overlay of LED-based photoacoustic and ultrasound maximum intensity projection (MIP) image presenting the finger joint from a linear scan, (**b–d**) tomographic images of the finger joint (p1) captured through photoacoustic, ultrasound, and the combination of both techniques, (**e–g**) tomographic images 5 mm in front of the joint (p2) obtained through photoacoustic, ultrasound, and the combined approach. (Reproduced with permission from Ref. [86])

LED-based PA imaging to measure blood oxygen content in human finger blood vessels, employing a pair of two-wavelength LED bars (850 nm and 690 nm), as depicted in Fig. 22.13b. Functional PA imaging of blood oxygen saturation in blood vessels was conducted on the index fingers of volunteers. Figure 22.13c provides an example of a PA 2D image resolving blood vessels in an axial view of a finger, overlaid on a gray scale US image. To quantify the functional information, the oxygenation saturation levels of pixels in the region of interest (ROI) are averaged, as indicated by the yellow dotted circle. The quantitative PA measurement of blood oxygenation saturation in the finger is then correlated with the reading of a pulse oximeter, considered the gold standard. Figure 22.13d illustrates a strong correlation (R-squared ~0.98) between the blood oxygenation measured by LED-based PA imaging and the pulse oximeter reading.

Visualizing vasculature in 3D is of paramount importance in different peripheral vascular imaging applications. By applying a skin-PA signal suppression algorithm and new colormaps, Kuniyil Ajith Singh et al. recently reported the feasibility of LED-based PA imaging in 3D imaging of human vasculature in quasi real time [92]. Figure 22.14 shows the 3D PA images of human foot dorsum and fingers acquired using the AcousticX system.

22.3.4 Clinical Pilot and Feasibility Studies

22.3.4.1 Diagnosis and Treatment Monitoring of Port Wine Stain

Port wine stain (PWS) is a vascular malformation that carries a substantial social and emotional impact. In a clinical pilot study led by Cheng et al., the application of LED-based PA imaging as a point-of-care tool for clinical evaluation and treatment monitoring of PWS disease was demonstrated [93]. This technology holds the potential to make a profound impact in this field, where vascular contrast with depth information (approximately 1 cm) is crucial for accurate diagnosis. PWS is characterized as a benign capillary vascular malformation, typically challenging to cure. While it commonly appears on the face, it can affect other areas of the body as well. The affected skin surface may exhibit slight thickening and develop an irregular, pebbled surface in adulthood, causing substantial mental stress for patients due to its cosmetic appearance. In the study by Cheng et al., 22 patients were enrolled and categorized into two groups based on their age (group 1: 3–6 years, group 2: above 6 years). Representative PA/US images displaying the PWS region and a normal region with clear differences in vascular contrast are depicted in Fig. 22.15, accompanied by masked photographs of the patients. The significant distinctions between the two age groups align well with existing knowledge of PWS disease. This study demonstrates, for the first time, that LED-based PA imaging can serve as an effective point-of-care tool for the clinical evaluation and photodynamic therapy (PDT) treatment monitoring of PWS disease.

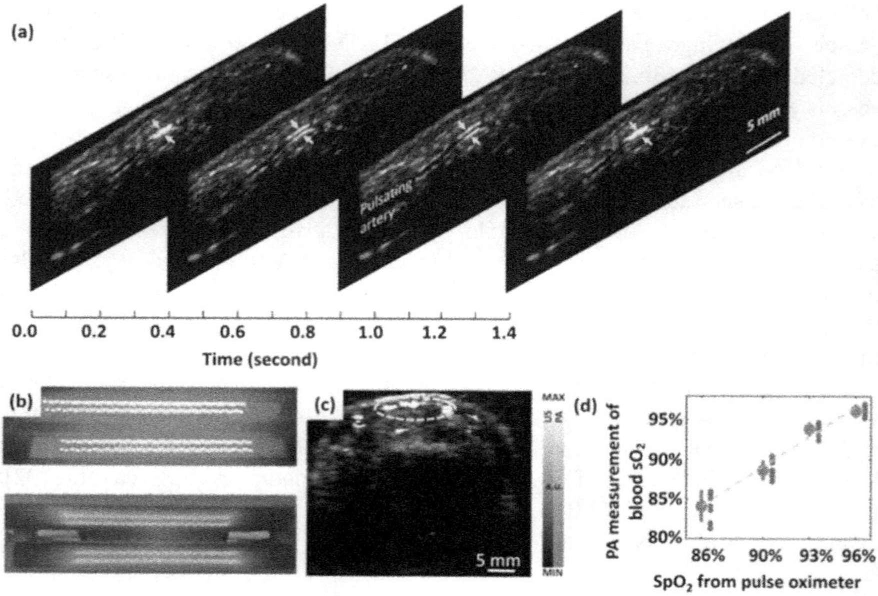

Fig. 22.13 (**a**) Presentation of four frames extracted from a photoacoustic imaging video of the longitudinal section of the index finger, highlighting arterial pulsation marked by arrows. (**b**) Image of a pair of dual-wavelength LED bars emitting 690-nm and 850-nm light alternatively. (**c**) Combined photoacoustic and ultrasound image revealing the cross-section of vasculature in a human finger. (**d**) Correlation plot demonstrating the relationship between blood oxygen saturation (SO_2) measured by LED-based PA imaging in the finger and the corresponding readings of oxygen saturation (SpO_2) from a pulse oximeter. (Reproduced with permission from Ref. [8])

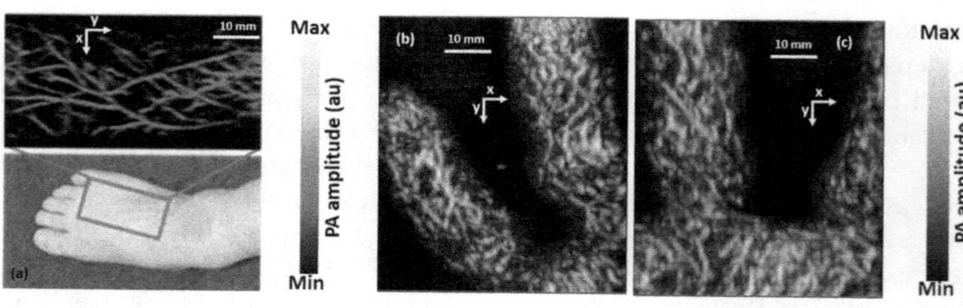

Fig. 22.14 (**a**) Depiction of the imaging location on the dorsum of the human foot, accompanied by the corresponding 3D LED-based photoacoustic image presented using the enhanced hot colormap. (**b**, **c**) showcase 3D maximum intensity projection (MIP) LED-based PA images of human fingers captured at different locations, presented using the filtered gray colormap in the AcousticX system. Notably, fine vascular features are discernible in all images, attaining a spatial resolution of approximately 200 μm. (Reproduced with permission from Ref. [92])

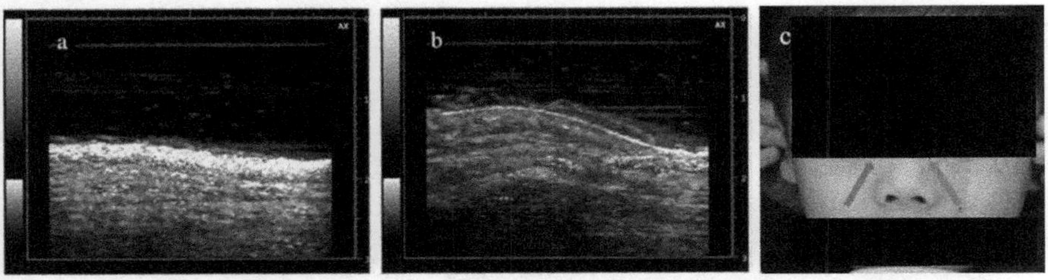

Fig. 22.15 (**a**) Depiction of a typical photoacoustic/ultrasound overlay image focused on the port wine stain (PWS) region (ROI). (**b**) Illustration of a typical photoacoustic/ultrasound overlay image focused on the control region (HR). (**c**) Photograph featuring a PWS patient with marks indicating the areas where imaging assistance was applied. (Reproduced with permission from Ref. [94])

22.3.4.2 Visualization of Inflammatory Arthritis

Due to its ability to detect hypervascularization and hypoxia, pivotal early markers of rheumatoid arthritis, PA imaging has undergone extensive exploration in various preclinical and early clinical pilot studies [95–101]. In the context of inflammatory arthritis, typically affecting relatively small joints first, the combination of PA imaging and conventional US imaging holds significant promise. Early findings from both animal models and human subjects strongly suggest that PA imaging is on the verge of translation into clinics as an affordable point-of-care tool for the early detection of inflammatory diseases [102–104]. Janggun et al. investigated the feasibility of employing an LED-based PA system to detect inflammation in soft tissues surrounding joints and the capability to differentiate arthritic joints from normal ones by assessing enhanced blood signals in synovial tissue [33].

Figure 22.16 presents representative US Doppler and PA images of three groups compared in the study: clinically active arthritis joints, subclinical active arthritis joints, and normal healthy joints. In Fig. 22.16a, a case of clinically active arthritis reveals hyperemia at the same location in both US Doppler and PA images. Figure 22.16b showcases a case of subclinical active arthritis, where hyperemia is visible only in the PA image of the patient's metacarpophalangeal (MCP) joint, not in the US Doppler image. Normal subjects, who underwent US Doppler scans, were also subjected to scanning the same joint locations using LED-based PA imaging. Compared with the results of arthritic joints, no significant hyperemia was observed in the synovium area of normal subjects, as confirmed by US Doppler imaging (Fig. 22.16c). Finally, Fig. 22.16d, e presents statistical results indicating a significant difference between different groups, confirmed through a two-tailed p-test. This clinical pilot study provides direct confirmation that LED-based PA imaging offers a higher sensitivity to angiogenic microvasculature than US Doppler imaging, the current gold standard.

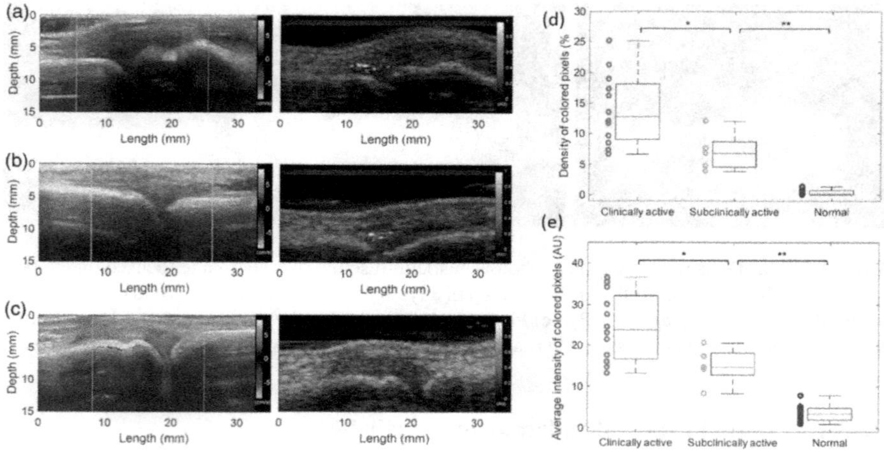

Fig. 22.16 (**a–c**) Sets of images displaying ultrasound Doppler (left) and photoacoustic (right) images obtained from an LED-based system for three groups: clinically active arthritis joints, subclinical active arthritis joints, and normal healthy joints, (**d**, **e**) Quantified results illustrating the density and average intensity of colored pixels in photoacoustic images for the three groups. Statistical significance is denoted by asterisks ($*p < 0.05$, $**p < 0.005$). (Reproduced with permission from Ref. [33])

22.3.4.3 Monitoring Angiogenesis in Wound Healing

Chronic wounds pose a significant healthcare challenge, costing billions of dollars annually and often proving difficult to heal, causing substantial discomfort. Despite the array of therapeutic options available to wound specialists, there is a lack of tools for three-dimensional mapping of wound bed physiology to guide therapy effectively. Current visual cues are limited to surface assessment, and clinicians rely on experience to predict therapy response. PA imaging is emerging as a promising noninvasive hybrid imaging modality that can address these limitations. PA imaging relies on the contrast provided by hemoglobin in the blood, enabling the mapping of local angiogenesis, tissue perfusion, and oxygen saturation—critical parameters for wound healing. In a pilot study from Mantri et al., the authors studied the use of combined LED-PA and US imaging to monitor angiogenesis and stratify patients based on their response to therapy [105]. Over a three-week period, they imaged 19 patients with 22 wounds weekly. The results suggest that PA imaging can directly visualize angiogenesis, providing valuable insights for patient stratification in terms of therapy response. Figure 22.17 shows an example in which LED-based PA imaging was used to monitor wound healing and angiogenesis in an 82-year-old female with a chronic left posterolateral ankle ulcer.

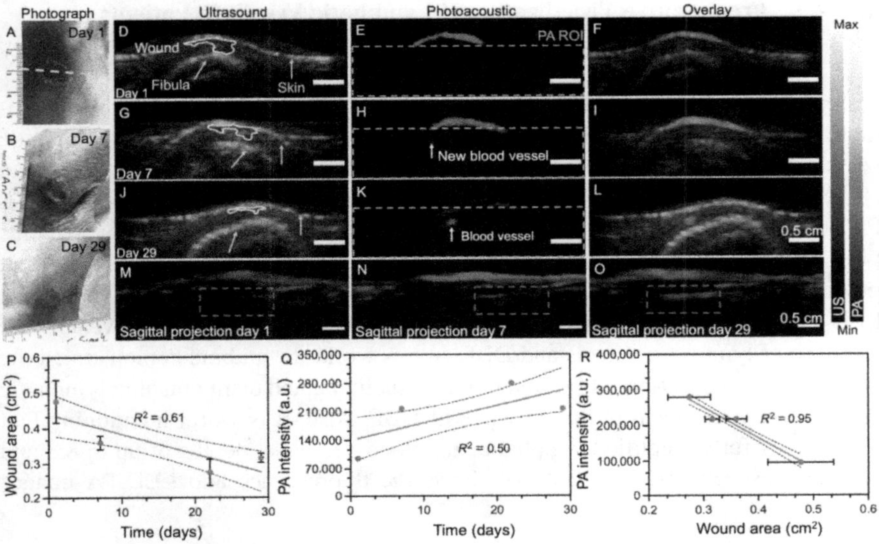

Fig. 22.17 PA imaging tracks angiogenesis in healing ankle wound over 29 days. Panels (**a–c**) showcase wound photographs on days 1, 7, and 29. Yellow ROI on ultrasound outlines the wound, with green and blue arrows marking the skin surface and fibula. Panels (**d–f**), (**g–i**), and (**j–l**) display ultrasound, photoacoustic, and overlay images on corresponding days. White arrows in (**h**, **k**) indicate angiogenesis. Sagittal maximum intensity projections (**m–o**) reveal new blood vessels invading the wound bed. Scale bars: 0.5 cm. (**p**) Negative correlation between wound area and time ($R^2 = 0.61$) suggests wound closure. (**q**) A significant positive correlation between photoacoustic intensity and time ($R^2 = 0.50$) indicates angiogenesis. (**r**) Linear increase in photoacoustic intensity with wound healing ($R^2 = 0.95$), correlating angiogenesis with closure. Scale bars represent 0.5 cm. Error bars show standard deviation in three representative frames from the wound center. Error bars for photoacoustic intensity in Q and R are too small to be shown. (Reproduced with permission from Ref. [105])

22.3.4.4 Monitoring Perfusion in Dialysis Patients

In chronic kidney disease patients undergoing hemodialysis (HD), intra-dialytic hypotension poses challenges, potentially affecting foot perfusion and wound healing. Mantri et al. reported a study in which 20 HD and 11 healthy subjects were imaged using LED-based PA imaging [106]. HD patients were scanned pre- and post-dialysis, while healthy subjects were scanned at rest and postexercise with elevated blood pressure (BP). Both groups demonstrated a correlation between PA intensity and systolic BP—healthy ($r = 0.70, p < 0.0001$) and HD subjects ($r = 0.43$, $p < 0.01$). Importantly, the HD cohort exhibited a diminished PA response to BP changes compared to healthy controls ($p < 0.0001$), emphasizing the potential of LED-PAI in monitoring hemodynamic shifts due to BP variation. Figure 22.18 shows an example case in which changes in tissue perfusion during hemodialysis were monitored using LED-PAI.

22.3.4.5 Preoperative Visualization of Lymphatic Vessels in Patients with Secondary Limb Lymphedema

Lymphedema, characterized by the accumulation of protein-rich fluid in the interstitium, involves dermal backflow (DBF). Accurate preoperative imaging of lymphatic vessels is essential for planning lymphovenous bypass surgery. In this first European clinical pilot study using AcousticX, Van Heumen et al. explored the use of LED-PA imaging to visualize lymphatic vessels and veins in lymphedema patients [40]. In this study, 15 patients with secondary limb lymphedema underwent indocyanine-green mediated near-infrared fluorescence lymphography (NIRF-L). LED-PA images were then acquired at locations where NIRF-L revealed the presence of lymphatic vessels and DBF. In this work, the authors demonstrated the effectiveness of LED-based PA imaging in visualizing and distinguishing lymphatic vessels and veins, even in the presence of DBF. Also, it is worth mentioning that they successfully identified lymphatic and blood vessels at depths of up to 8.3 and 8.6 mm, respectively. Figure 22.19 shows the fluorescence and LED-PA images

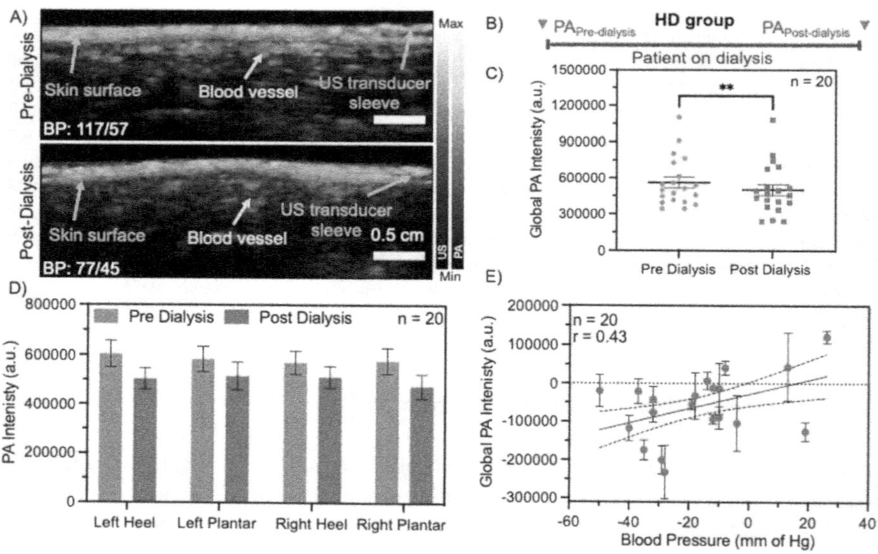

Fig. 22.18 Changes in peripheral tissue perfusion during hemodialysis monitored using LED-based photoacoustic imaging. (**a**) Photoacoustic and ultrasound overlay of the left heel in subject. Blood perfusion into the heel is considerably lower after dialysis when BP is low. The scale bar represents 0.5 cm. (**b**) HD patients were scanned just before and after a routine dialysis session. (**c**) A paired t-test showed a significant decrease in global photoacoustic signal and hence perfusion after dialysis (**, $p < 0.01$). (**d**) Photoacoustic intensity at individual imaging areas was lower (not significantly) after dialysis. **e.** Hemodynamic response of the HD group to changes in BP during dialysis showed a positive correlation ($r = 0.43$, $p < 0.01$). Error bars in panels (**c, d**) represent standard deviation among 20 patients and in panel E at least 90 frames. (Reproduced with permission from Ref. [106])

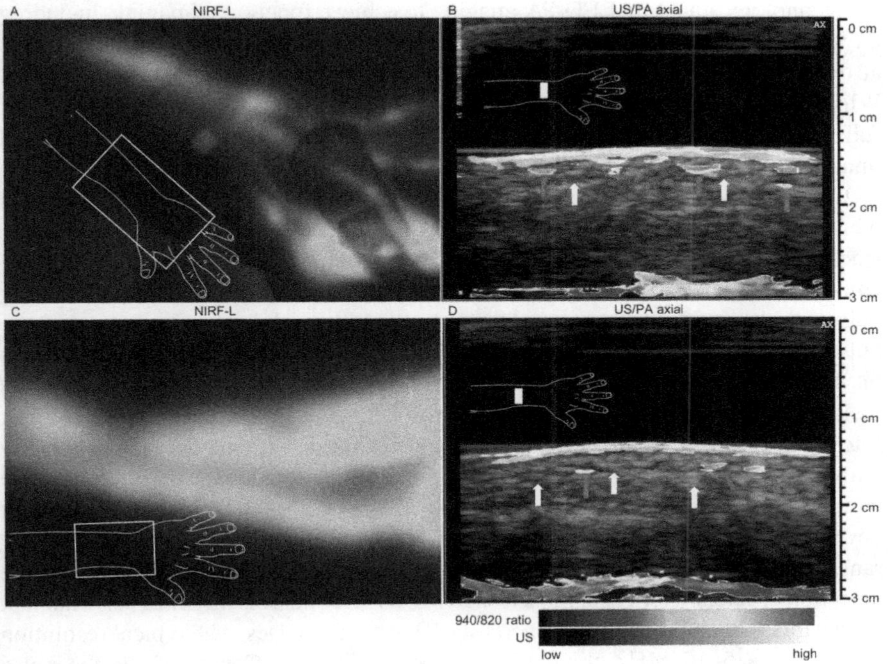

Fig. 22.19 Near-infrared fluorescence lymphography (NIRF-L) images (**a** and **c**) with linear patterns in a healthy arm (**a**) and lymphedema arm (**c**). (**b** and **c**) represent photoacoustic images of 940/820 nm ratio superimposed on gray scale ultrasound images of the corresponding NIRF-L regions. Red arrows indicate blood vessels, and white arrows indicate lymphatic vessels. More lymphatic vessels were observed with photoacoustic imaging compared to NIRF-L. Reflection artifacts originating from the skin surface can be observed at a depth of ~3 cm. (Reproduced with permission from Ref. [40])

from a healthy and lymphedema arm in which it is clear that LED-PA imaging offers better contrast for lymphatic vessels located close to or even beneath a blood vessel, a suitable bypass location.

22.4 Discussion and Outlook

In this chapter, we embark on a journey through the historical evolution of LED-based PA imaging, tracing its development from initial single-point measurements to the latest clinical pilot studies utilizing high-power LED arrays. Over the past decade, remarkable progress has been witnessed in this field, marked by advancements in pulse energy (from nJ to hundreds of μJ) and pulse repetition rate (from 200 Hz to 16,000 Hz), coupled with innovations in low-noise data acquisition electronics. These breakthroughs have led to the commercialization of the technology, enabling LED-PA imaging to achieve functional imaging, including oxygenation and blood flow imaging, in both superficial and sub-surface tissues at frame rates that were previously unattainable for laser-based systems (500 Hz). While not suitable for applications demanding greater imaging depth, such as full breast imaging, LED-PAI exhibits significant potential in various superficial imaging applications, particularly in rheumatology and dermatology.

Compared to solid-state lasers, LEDs exhibit energy levels that are two orders of magnitude lower. However, LEDs boast significantly higher PRR than lasers, leading to enhanced imaging quality through frame averaging. Beyond energy emission, a crucial distinction lies in the temporal pulse width, with LEDs featuring pulse widths in the tens of nanoseconds, whereas solid-state lasers may have pulse widths below ten nanoseconds. This temporal factor imposes constraints on spatial resolution, exemplified by the 70-ns pulse width of an 850-nm LED yielding a spatial resolution of 105 μm (calculated as 70 ns × 1500 μm/μs). Interestingly, this limitation proves advantageous for detection efficiency, especially with bandlimited US probes. PA signals from a solid-state laser with a 3.5-ns pulse width encompass frequency components up to almost 300 MHz, rendering anything beyond approximately 12 MHz undetectable with a conventional mid-frequency range US probe. Using a 5 MHz US probe with 80% detection bandwidth, the detection efficiency for PA signals from a 3.5-ns laser pulse is 40 times less than that achievable with an LED array generating 100 ns light pulses. The typical resolution offered by a 5–10 MHz clinical US probe is 200 μm. Consequently, the pulse width of LEDs can potentially be extended to 100 ns without compromising spatial resolution, with the 70-ns setting being employed in all the studies reported in this chapter.

The second part of this chapter provides an overview of reported preclinical and clinical applications utilizing LED-based PA imaging, categorized into phantom, ex vivo, small animal, and human in vivo studies (refer to Table 22.3). The results underscore the robust potential of multispectral LED-based PA and US imaging across various applications, including guiding minimally invasive procedures, blood oxygen saturation imaging, diagnosing and staging inflammatory arthritis, peripheral vascular assessment, angiogenesis monitoring, and guiding surgical procedures like lymphaticovenous anastomosis. These studies consistently achieved an imaging depth of 0.5–1 cm at high interleaved US and PA frame rates, catering to multiple clinical applications. However, addressing the need for improved imaging depth, particularly in breast imaging and cardiovascular medicine, is crucial. Advancements in reconstruction and image processing techniques [107–112] aim to tackle this challenge and expedite the clinical translation of LED-based PA imaging. Additionally, the utilization of clinically approved contrast agents, such as ICG, holds promise for enhancing imaging depth.

The described LED-based PA imaging system (AcousticX) ensures both skin and eye safety. As LED emissions are incoherent, the safety standards outlined by the International Electrotechnical Commission (IEC) 62471 are followed instead of the ANSI safety limits applicable to collimated laser beams. According to IEC 62471, the exposure limit for skin is based on preventing thermal injury caused

Table 22.3 Preclinical and clinical applications of LED-based PAI

Target	Application		Depth (mm)	Contrast agent	Wavelength (nm)
Medical needles, Vasculature	Guidance of minimally invasive procedures with peripheral tissue targets [34]	Phantom and ex vivo studies	38	N/A	850
Vasculature	Imaging of human placental vasculature [73]		7	N/A	850
Tumor	Imaging of intraocular tumors [8]		10	N/A	850
Vasculature	Noninvasive monitoring of angiogenesis [9]	Animal in vivo	10	N/A	850
Ulcer	Noninvasive imaging of pressure ulcers [81]		10	N/A	690
Oxygen saturation	Oxygen saturation imaging in rheumatoid arthritis [58]		5	N/A	750/850
Molecular	Detection and monitoring of reactive oxygen and nitrogen species [31]		10	CyBA	850
Tumor/contrast agents	Imaging of tumor using contrast enhancement [77]		10	NC	850
Cells/contrast agents	Imaging of molecular-labeled cells [29]		10	DiR	850
Vasculature	Imaging of peripheral microvasculature and function [8]	Healthy human in vivo	10	N/A	690/850
Vasculature	Simultaneous imaging of veins and lymphatic vessels [85]		10	ICG	940/820
Finger joints	Full view tomography of finger joints [86]		5	N/A	850
Finger joints	Imaging of inflammatory arthritis [33]	Patient in vivo	5	N/A	850
Skin	Imaging of port wine stain [94]	Patient in vivo	10	N/A	850
Lymphatic vessels	Lymphovenous anastomosis [40]	Patient in vivo	8	N/A	940/820
Vasculature	Wound healing monitoring [105]	Patient in vivo	5	N/A	850
Vasculature	Imaging perfusion in dialysis patients [106]	Patient in vivo	5	N/A	850

Adapted with permission from Ref. [39]

by temperature rise in tissue. Under the assumption of continuous illumination on the same skin area for 5 seconds using two 850-nm LED bars operating at a 4 KHz pulse repetition rate, the estimated exposure is 4.57×10^3 W·m^{-2}, which falls below the thermal hazard limit for skin of 5.98×10^3 W·m^{-2}. For eye safety, two considerations come into play: the retinal thermal hazard exposure limit (related to weak visual stimulus) and the infrared radiation eye safety limit. Assuming continuous illumination at the front of the eye for 5 seconds using two 850-nm LED bars with a 4 KHz pulse repetition rate, the estimated exposures are 2.92×10^3 W·m^{-2}·sr^{-1} for retinal thermal exposure and $4.57 \times$ 10^3 W·m^{-2} for infrared radiation exposure. These values are lower than the safety limits for the eye, which are 1.34×10^5 W·m^{-2}·sr^{-1} and 5.38×10^3 W·m^{-2}, respectively.

As of now, LED-based PA imaging has achieved a maximum imaging depth of 1 cm in an in vivo setting. While this is promising, given the low pulse energy, enhancing the optical output of LEDs is crucial for broader clinical applications. Signal averaging improves signal-to-noise ratio (SNR) to some extent without compromising the frame rate. However, if the acoustic signals from weak illumination are significantly below the noise-equivalent pressure level of the US probe, averaging may not effectively enhance SNR. Integrating PA imaging into an existing clinical US system will likely gain clinical acceptance more easily. Considering the acoustic bandwidth limitation of conventional pulse-echo US probes, the relatively low pulse width of LEDs (70 ns in the applications discussed in this chapter) is not a hindrance. Two main avenues for improving imaging depth are anticipated: (1) leveraging nano-stack technology to extract more light from LED elements, enhancing optical output by increasing the regions generating light, and (2) enhancing driving electronics to further increase the PRR, allowing for more frame averaging and improved SNR without sacrificing temporal resolution. While LED-based PA imaging may compete with laser diode (LD) systems during clinical translation, it is not expected to significantly overlap with Nd:YAG OPO-based PAI systems, which are designed for deep-tissue applications. However, LED-PA imaging is poised to be a suitable choice for point-of-care applications, including guiding peripheral vascular access procedures, rheumatoid arthritis screening, lymphatic imaging, port wine stain (PWS) diagnosis, and treatment monitoring.

While the use of LEDs as an illumination source in PA imaging offers many advantages, it presents certain limitations too. Firstly, LEDs lack spectral tunability, limiting their application in scenarios requiring spectral specificity for multiple chromophores. Additionally, the relatively large pulse widths of LEDs (30–100 ns) compared to lasers can impact stress confinement and influence acoustic wave generation efficiency. Moreover, LEDs exhibit lower optical output power, potentially restricting the penetration depth at higher frame rates. Nevertheless, LED-based PA imaging systems have notable benefits, such as a significant cost reduction, a smaller footprint, no need for laser calibration and monitoring, and the absence of requirements for optical goggles or light-tight shields. Consequently, these systems are well suited for point-of-care noninvasive applications that complement US imaging. Moreover, their characteristics make them ideal for personalized or wearable PA equipment. Looking ahead, there is anticipation that LED-based PA imaging technology could find broad utility in various therapeutic drug monitoring applications.

Looking ahead, the future of PA imaging is expected to witness increased utilization of high-power LED technology, primarily influenced by advancements in the lighting industry. Additionally, significant progress in machine/deep learning and signal processing algorithms is anticipated to further enhance the integration of LEDs in PA imaging applications. Novel LED arrays, tailored to different shapes, could be developed for diverse applications. For instance, a ring-shaped LED might be designed for breast 3D imaging, while light coupling into optical fibers could be explored for minimally invasive and endoscopic procedures. It is noteworthy that if sufficient focusing can be achieved, LEDs hold the potential for acoustic resolution photoacoustic microscopy (PAM). Reports in this direction include Dai et al.'s work about a PAM system based on a miniature LED operating at 405 nm [113, 114]. This system demonstrated in vivo mapping of vasculature networks in biological tissue using a high-power LED (1.2 W) with a pulse width of 200 ns and a repetition rate extended to 40 KHz. While this represents an encouraging proof-of-concept study for LED-based PAM, the temporal resolution achieved is not yet optimal for clinical studies. On a different note, the application of LEDs in optical resolution PAM may pose challenges, particularly in achieving the necessary micron-scale diffraction-limited spot sizes. Exploring the development of new low-frequency ultrasound probes (2–3 MHz) with ultra-high bandwidth and sensitivity could be an intriguing avenue to achieve higher imaging depth without compromising spatial resolution.

22.5 Conclusion

In recent years, LED-based PA imaging is emerging as a noteworthy advancement in functional and molecular imaging. Its practical features, such as a broad optical wavelength range (400–1000 nm), adaptable pulse-width settings, compact design, cost-effectiveness, and energy efficiency, position LED-based PA imaging as a valuable tool for preclinical and clinical applications. The versatility of LEDs, especially for point-of-care diagnostic imaging, makes them a compelling addition to conventional US imaging in clinical scanners. This integration is poised to enhance existing technologies and expedite the adoption of PA imaging in clinical settings. Looking ahead, LED-based PA imaging shows promise in contributing to medical diagnostics, with the potential to bring about significant advancements in patient care.

References

1. Wang, L., Yao, J. A practical guide to photoacoustic tomography in the life sciences. Nat Methods 13, 627–638 (2016).

2. Ntziachristos, V. Going deeper than microscopy: the optical imaging frontier in biology. Nat Methods 7, 603–614 (2010)
3. Wang, L.V. and S. Hu, Photoacoustic tomography: in vivo imaging from organelles to organs. Science, 2012. 335(6075): p. 1458–1462.
4. Beard, P., Biomedical photoacoustic imaging. Interface focus, 2011. 1(4): p. 602–631
5. Singh, M.K.A.; Steenbergen, W.; Manohar, S. Handheld Probe-Based Dual Mode Ultrasound/Photoacoustics for Biomedical Imaging. In Frontiers in Biophotonics for Translational Medicine; Progress in Optical Science and Photonics; Springer: Singapore, 2015; Volume 3, pp. 209–247
6. Oladipupo, S., et al., VEGF is essential for hypoxia-inducible factor-mediated neovascularization but dispensable for endothelial sprouting. Proceedings of the National Academy of Sciences, 2011. 108(32): p. 13264–13269.
7. Oladipupo, S.S., et al., Conditional HIF-1 induction produces multistage neovascularization with stage-specific sensitivity to VEGFR inhibitors and myeloid cell independence. Blood, The Journal of the American Society of Hematology, 2011. 117(15): p. 4142–4153.
8. Zhu, Y., et al., Light emitting diodes based photoacoustic imaging and potential clinical applications. Scientific reports, 2018. 8(1): p. 1–12.
9. Zhu, Y., et al., LED-Based Photoacoustic Imaging for Monitoring Angiogenesis in Fibrin Scaffolds. Tissue Engineering Part C: Methods, 2019. 25(9): p. 523–531.
10. Erpelding, T.N., et al., Sentinel lymph nodes in the rat: noninvasive photoacoustic and US imaging with a clinical US system. Radiology, 2010. 256(1): p. 102–110.
11. Kim, C., et al., In vivo molecular photoacoustic tomography of melanomas targeted by bioconjugated gold nanocages. ACS nano, 2010. 4(8): p. 4559–4564.
12. Hu, S. and L. Wang, Neurovascular photoacoustic tomography Front. 2010, Neuroenerg.
13. Hu, S., et al., Intravital imaging of amyloid plaques in a transgenic mouse model using optical-resolution photoacoustic microscopy. Optics letters, 2009. 34(24): p. 3899–3901.
14. Hu, S., et al., Label-free photoacoustic ophthalmic angiography. Optics letters, 2010. 35(1): p. 1–3.
15. Jiao, S., et al., Photoacoustic ophthalmoscopy for in vivo retinal imaging. Optics express, 2010. 18(4): p. 3967–3972.
16. Zhang, E.Z., et al., Multimodal photoacoustic and optical coherence tomography scanner using an all optical detection scheme for 3D morphological skin imaging. Biomedical optics express, 2011. 2(8): p. 2202–2215.
17. Favazza, C.P., et al., In vivo photoacoustic microscopy of human cutaneous microvasculature and a nevus. Journal of biomedical optics, 2011. 16(1): p. 016015
18. Yang, J.-M., et al. Toward dual-wavelength functional photoacoustic endoscopy: laser and peripheral optical systems development. in Photons Plus Ultrasound: Imaging and Sensing 2012. 2012. International Society for Optics and Photonics.
19. Yang, J.-M., et al., Photoacoustic endoscopy. Optics letters, 2009. 34(10): p. 1591–1593.
20. Zhu, Y., et al., Identifying intestinal fibrosis and inflammation by spectroscopic photoacoustic imaging: an animal study in vivo. Biomedical optics express, 2018. 9(4): p. 1590–1600.
21. Zhu, Y., et al., Strain-photoacoustic imaging as a potential tool for characterizing intestinal fibrosis. Gastroenterology, 2019. 157(5): p. 1196–1198.
22. Zhu, Y., et al. Characterization of intestinal fibrosis and inflammation with transcutaneous spectroscopic PA imaging (Conference Presentation). in Photons Plus Ultrasound: Imaging and Sensing 2018. 2018. International Society for Optics and Photonics.
23. Feng, T., et al., Characterization of bone microstructure using photoacoustic spectrum analysis. Optics express, 2015. 23(19): p. 25217–25224.
24. Kaiplavil, S., et al., Photothermal tomography for the functional and structural evaluation, and early mineral loss monitoring in bones. Biomedical optics express, 2014. 5(8): p. 2488–2502.
25. Feng, T., et al., Bone assessment via thermal photo-acoustic measurements. Optics letters, 2015. 40(8): p. 1721–1724.
26. Wang, X., et al. Photoacoustic measurement of bone health: a study for clinical feasibility. in 2016 IEEE International Ultrasonics Symposium (IUS). 2016. IEEE.

27. Jansen, K., et al., Intravascular photoacoustic imaging of human coronary atherosclerosis. Optics letters, 2011. 36(5): p. 597–599.
28. Wang, B., et al., Plasmonic intravascular photoacoustic imaging for detection of macrophages in atherosclerotic plaques. Nano Letters, 2009. 9(6): p. 2212–2217
29. Hariri, A., et al., The characterization of an economic and portable LED-based photoacoustic imaging system to facilitate molecular imaging. Photoacoustics, 2018. 9: p. 10–20.
30. Hariri, A., et al. Molecular imaging of oxidative stress sensing using LED-based photoacoustic imaging (Conference Presentation). in Photons Plus Ultrasound: Imaging and Sensing 2019. 2019. International Society for Optics and Photonics.
31. Hariri, A., et al., Molecular imaging of oxidative stress using an LED-based photoacoustic imaging system. Scientific reports, 2019. 9(1): p. 1–10.
32. Xia, W., et al. Imaging of human peripheral blood vessels during cuff occlusion with a compact LED-based photoacoustic and ultrasound system. in Photons Plus Ultrasound: Imaging and Sensing 2019. 2019. International Society for Optics and Photonics.
33. Jo, J., et al., Detecting joint inflammation by an LED-based photoacoustic imaging system: a feasibility study. Journal of biomedical optics, 2018. 23(11): p. 110501.
34. Xia, W., et al., Handheld real-time LED-based photoacoustic and ultrasound imaging system for accurate visualization of clinical metal needles and superficial vasculature to guide minimally invasive procedures. Sensors, 2018. 18(5): p. 1394.
35. Allen, T.J. and P.C. Beard, High power visible light emitting diodes as pulsed excitation sources for biomedical photoacoustics. Biomedical optics express, 2016. 7(4): p. 1260–1270.
36. Agano, T., et al. Attempts to increase penetration of photoacoustic system using LED array light source. in Photons Plus Ultrasound: Imaging and Sensing 2015. 2015. International Society for Optics and Photonics.
37. Agano, T., et al. Comparative experiments of photoacoustic system using laser light source and LED array light source. in Photons Plus Ultrasound: Imaging and Sensing 2015. 2015. International Society for Optics and Photonics.
38. Singh, M.K.A. LED-Based Photoacoustic Imaging; Springer Nature Singapore Pte Ltd.: Singapore, 2020.
39. Zhu Y, Feng T, Cheng Q, Wang X, Du S, Sato N, Yuan J, Kuniyil Ajith Singh M. Towards Clinical Translation of LED-Based Photoacoustic Imaging: A Review. Sensors. 2020; 20(9):2484.
40. Van Heumen S, Riksen J, Singh MKA, Van Soest G, Vasilic D. LED-based photoacoustic imaging for preoperative visualization of lymphatic vessels in patients with secondary limb lymphedema. Photoacoustics. 2022 Dec 28;29:100446.
41. Liu X, Kalva SK, Lafci B, Nozdriukhin D, Deán-Ben XL, Razansky D. Full-view LED-based optoacoustic tomography. Photoacoustics. 2023 Jun 8;31:100521.
42. Allen, T.J. and P.C. Beard, Pulsed near-infrared laser diode excitation system for biomedical photoacoustic imaging. Optics letters, 2006. 31(23): p. 3462–3464.
43. Kolkman, R.G., W. Steenbergen, and T.G. van Leeuwen, In vivo photoacoustic imaging of blood vessels with a pulsed laser diode. Lasers in medical science, 2006. 21(3): p. 134–139.
44. Daoudi, K., et al., Handheld probe integrating laser diode and ultrasound transducer array for ultrasound/photoacoustic dual modality imaging. Optics express, 2014. 22(21): p. 26365–26374.
45. Kalva, S.K., P.K. Upputuri, and M. Pramanik, High-speed, low-cost, pulsed-laser-diode-based second-generation desktop photoacoustic tomography system. Optics Letters, 2019. 44(1): p. 81–84.
46. Agano, T., et al., Effect of light pulse width on frequency characteristics of photoacoustic signal–an experimental study using a pulse-width tunable LED-based photoacoustic imaging system. International Journal of Engineering and Technology, 2018. 7(4): p. 4300–4303.
47. Upputuri, P.K. and M. Pramanik, Pulsed laser diode based optoacoustic imaging of biological tissues. Biomedical Physics & Engineering Express, 2015. 1(4): p. 045010.
48. Wang, T., et al., A low-cost photoacoustic microscopy system with a laser diode excitation. Biomedical optics express, 2014. 5(9): p. 3053–3058

49. Allen, T.J., High-Power Light Emitting Diodes; An Alternative Excitation Source for Photoacoustic Tomography, in LED-Based Photoacoustic Imaging. 2020, Springer. p. 23–43.
50. Willert, C., et al., Pulsed operation of high-power light emitting diodes for imaging flow velocimetry. Measurement Science and Technology, 2010. 21(7): p. 075402.
51. Hansen, R.S. Using high-power light emitting diodes for photoacoustic imaging. in Medical Imaging 2011: Ultrasonic Imaging, Tomography, and Therapy. 2011. International Society for Optics and Photonics.
52. Ikuta, T. and R. Shimizu, The dynamic response of magnetic domain walls to applied fields in yttrium orthoferrite, observed by a stroboscopic technique. Journal of Physics D: Applied Physics, 1974. 7(5): p. 726.
53. McFarlane, W., An inexpensive nanosecond light pulser for use in photomultiplier system testing. Review of Scientific Instruments, 1974. 45(2): p. 286–289.
54. Araki, T. and H. Misawa, Light emitting diode-based nanosecond ultraviolet light source for fluorescence lifetime measurements. Review of scientific instruments, 1995. 66(12): p. 5469–5472.
55. Araki, T., Y. Fujisawa, and M. Hashimoto, An ultraviolet nanosecond light pulse generator using a light emitting diode for test of photodetectors. Review of scientific instruments, 1997. 68(3): p. 1365–1368.
56. Chaney, A. and R. Sundararajan, Simple MOSFET-based high-voltage nanosecond pulse circuit. IEEE Transactions on Plasma science, 2004. 32(5): p. 1919–1924.
57. Allen, T.J. and P.C. Beard. Light emitting diodes as an excitation source for biomedical photoacoustics. in Photons Plus Ultrasound: Imaging and Sensing 2013. 2013. International Society for Optics and Photonics.
58. Kalloor Joseph, Francis et al. "LED-based photoacoustic imaging for early detection of joint inflammation in rodents: towards achieving 3Rs in rheumatoid arthritis research." BiOS (2020).
59. Mallidi, S., et al., Multiwavelength photoacoustic imaging and plasmon resonance coupling of gold nanoparticles for selective detection of cancer. Nano letters, 2009. 9(8): p. 2825–2831.
60. Krishna, V., et al., Polyhydroxy fullerenes for non-invasive cancer imaging and therapy. Small, 2010. 6(20): p. 2236–2241.
61. Kircher, M.F., et al., A brain tumor molecular imaging strategy using a new triple-modality MRI-photoacoustic-Raman nanoparticle. Nature medicine, 2012. 18(5): p. 829.
62. Li, M.-L., et al., Simultaneous molecular and hypoxia imaging of brain tumors in vivo using spectroscopic photoacoustic tomography. Proceedings of the IEEE, 2008. 96(3): p. 481–489.
63. Levi, J., et al., Molecular photoacoustic imaging of follicular thyroid carcinoma. Clinical Cancer Research, 2013. 19(6): p. 1494–1502.
64. Oyama, T., et al., Endoscopic submucosal dissection of early esophageal cancer. Clinical Gastroenterology and Hepatology, 2005. 3(7): p. S67-S70.
65. Polednak, A.P. and J.T. Flannery, Brain, other central nervous system, and eye cancer. Cancer, 1995. 75(S1): p. 330–337.
66. Xu, G., et al., Photoacoustic imaging features of intraocular tumors: Retinoblastoma and uveal melanoma. PloS one, 2017. 12(2).
67. Narouze, S.N., Atlas of ultrasound-guided procedures in interventional pain management. 2018: Springer.
68. Kim, C., et al., Handheld array-based photoacoustic probe for guiding needle biopsy of sentinel lymph nodes. Journal of biomedical optics, 2010. 15(4): p. 046010.
69. Su, J.L., et al., Photoacoustic imaging of clinical metal needles in tissue. Journal of biomedical optics, 2010. 15(2): p. 021309.
70. Pratt, R., et al., Computer-assisted surgical planning and intraoperative guidance in fetal surgery: a systematic review. Prenatal diagnosis, 2015. 35(12): p. 1159–1166.
71. Lewi, L., J. Deprest, and K. Hecher, The vascular anastomoses in monochorionic twin pregnancies and their clinical consequences. American journal of obstetrics and gynecology, 2013. 208(1): p. 19–30.

72. Guiot, C., et al., Is three-dimensional power Doppler ultrasound useful in the assessment of placental perfusion in normal and growth-restricted pregnancies? Ultrasound in obstetrics & gynecology, 2008. 31(2): p. 171–176.
73. Maneas, E., et al., Photoacoustic imaging of the human placental vasculature. Journal of biophotonics, 2019: p. e201900167.
74. Ntziachristos, V. and D. Razansky, Molecular imaging by means of multispectral optoacoustic tomography (MSOT). Chemical reviews, 2010. 110(5): p. 2783–2794.
75. Xia, W., et al. Interventional photoacoustic imaging of the human placenta with ultrasonic tracking for minimally invasive fetal surgeries. in International Conference on Medical Image Computing and Computer-Assisted Intervention. 2015. Springer.
76. Xia, W., et al., Performance characteristics of an interventional multispectral photoacoustic imaging system for guiding minimally invasive procedures. Journal of biomedical optics, 2015. 20(8): p. 086005.
77. Xavierselvan, M. and S. Mallidi, LED-Based Functional Photoacoustics—Portable and Affordable Solution for Preclinical Cancer Imaging, in LED-Based Photoacoustic Imaging: From Bench to Bedside, M. Kuniyil Ajith Singh, Editor. 2020, Springer Singapore: Singapore. p. 303–319.
78. Draijer, M., et al., Review of laser speckle contrast techniques for visualizing tissue perfusion. Lasers in medical science, 2009. 24(4): p. 639.
79. Briers, J.D., Laser speckle contrast imaging for measuring blood flow. 2007.
80. Dunn, A.K., Laser speckle contrast imaging of cerebral blood flow. Annals of biomedical engineering, 2012. 40(2): p. 367–377.
81. Hariri, A., et al., Noninvasive staging of pressure ulcers using photoacoustic imaging. Wound Repair and Regeneration, 2019. 27(5): p. 488–496.
82. Shuhendler, A.J., et al., Real-time imaging of oxidative and nitrosative stress in the liver of live animals for drug-toxicity testing. Nature biotechnology, 2014. 32(4): p. 373.
83. Jokerst, J.V., et al., Photoacoustic imaging of mesenchymal stem cells in living mice via silica-coated gold nanorods. ACS nano, 2012. 6(7): p. 5920–5930.
84. Nam, S.Y., et al., In vivo ultrasound and photoacoustic monitoring of mesenchymal stem cells labeled with gold nanotracers. PLoS One, 2012. 7(5).
85. Singh, M.K.A., et al. Real-time in vivo imaging of human lymphatic system using an LED-based photoacoustic/ultrasound imaging system. in Photons Plus Ultrasound: Imaging and Sensing 2018. 2018. International Society for Optics and Photonics
86. Joseph, F.K., et al., Tomographic imaging with an ultrasound andLED-based photoacoustic system. Biomedical Optics Express
87. Agrawal, S., et al., Light-Emitting-Diode-Based Multispectral Photoacoustic Computed Tomography System. Sensors, 2019. 19(22): p. 4861.
88. Bristow, R.G. and R.P. Hill, Hypoxia and metabolism: Hypoxia, DNA repair and genetic instability. Nature Reviews Cancer, 2008. 8(3).
89. Wang, X., et al., Noninvasive imaging of hemoglobin concentration and oxygenation in the rat brain using high-resolution photoacoustic tomography. Journal of biomedical optics, 2006. 11(2): p. 024015.
90. Yang, S., et al., Functional imaging of cerebrovascular activities in small animals using high-resolution photoacoustic tomography. Medical physics, 2007. 34(8): p. 3294–3301.
91. Kim, C., C. Favazza, and L.V. Wang, In vivo photoacoustic tomography of chemicals: high-resolution functional and molecular optical imaging at new depths. Chemical reviews, 2010. 110(5): p. 2756–2782.
92. Kuniyil Ajith Singh, M et al "Improving 3D imaging capability of LED-based photoacoustic imaging for high resolution human vascular imaging applications," Proc. SPIE 12379, Photons Plus Ultrasound: Imaging and Sensing 2023, 123790R
93. Zhang, H., et al. A novel light emitting diodes based photoacoustic evaluation method for port wine stain and its clinical trial (Conference Presentation). in Photons Plus Ultrasound: Imaging and Sensing 2019. 2019. International Society for Optics and Photonics.

94. Cheng, Q., et al., Diagnosis and Treatment Monitoring of Port-Wine Stain Using LED-Based Photoacoustics: Theoretical Aspects and First In-Human Clinical Pilot Study. LED-Based Photoacoustic Imaging: From Bench to Bedside, 2020: p. 351–377.
95. Wang, X., et al., Imaging of joints with laser-based photoacoustic tomography: An animal study. Medical physics, 2006. 33(8): p. 2691–2697.
96. Zhang, H.F., et al., Imaging of hemoglobin oxygen saturation variations in single vessels in vivo using photoacoustic microscopy. Applied physics letters, 2007. 90(5): p. 053901.
97. Sun, Y., E. Sobel, and H. Jiang, Quantitative three-dimensional photoacoustic tomography of the finger joints: an in vivo study. J Biomed Opt, 2009. 14(6): p. 064002.
98. Xu, G., et al., Photoacoustic and ultrasound dual-modality imaging of human peripheral joints. J Biomed Opt, 2013. 18(1): p. 10502.
99. van Es, P., et al., Initial results of finger imaging using photoacoustic computed tomography. J Biomed Opt, 2014. 19(6): p. 060501–060501.
100. Sun, Y., E.S. Sobel, and H. Jiang, First assessment of three-dimensional quantitative photoacoustic tomography for in vivo detection of osteoarthritis in the finger joints. Medical physics, 2011. 38(7): p. 4009–4017.
101. Rajian, J.R., G. Girish, and X. Wang, Photoacoustic tomography to identify inflammatory arthritis. J Biomed Opt, 2012. 17(9): p. 96013–1.
102. Rajian, J.R., et al., Characterization and treatment monitoring of inflammatory arthritis by photoacoustic imaging: a study on adjuvant-induced arthritis rat model. Biomed Opt Express, 2013. 4(6): p. 900–8.
103. Beziere, N., et al., Optoacoustic imaging and staging of inflammation in a murine model of arthritis. Arthritis Rheumatol, 2014. 66(8): p. 2071–8.
104. Jo, J., et al., A Functional Study of Human Inflammatory Arthritis Using Photoacoustic Imaging. Scientific reports, 2017. 7(1): p. 15026.
105. Mantri, Y, Tsujimoto, J, Donovan, B, et al. Photoacoustic monitoring of angiogenesis predicts response to therapy in healing wounds. Wound Rep Reg. 2022; 1–10.
106. Y. Mantri, T.R. Dorobek, J. Tsujimoto et al; "Monitoring peripheral hemodynamic response to changes in blood pressure via photoacoustic imaging," Photoacoustics, Volume 26, 100345 (2022)
107. Anas, E.M.A., et al., Enabling fast and high quality LED photoacoustic imaging: a recurrent neural networks based approach. Biomedical Optics Express, 2018. 9(8): p. 3852–3866.
108. Mozaffarzadeh, M., et al., The double-stage delay-multiply-and-sum image reconstruction method improves imaging quality in a led-based photoacoustic array scanner. Photoacoustics, 2018. 12: p. 22–29.
109. Singh, M.K.A., et al. Deep learning-enhanced LED-based photoacoustic imaging. in Photons Plus Ultrasound: Imaging and Sensing 2020. 2020. International Society for Optics and Photonics.
110. Paul S, Mulani S, Singh MKA, Singh MS. Improvement of LED-based photoacoustic imaging using lag-coherence factor (LCF) beamforming. Med Phys. 2023; 50: 7525–7538.
111. Shi M, Zhao T, West SJ, Desjardins AE, Vercauteren T, Xia W. Improving needle visibility in LED-based photoacoustic imaging using deep learning with semi-synthetic datasets. Photoacoustics. 2022 Apr 7;26:100351.
112. Shi M, Vercauteren T, Xia W. Spatiotemporal singular value decomposition for denoising in photoacoustic imaging with a low-energy excitation light source. Biomedical Optics Express. 2022 Dec 1;13(12):6416–30.
113. Dai, X., H. Yang, and H. Jiang, In vivo photoacoustic imaging of vasculature with a low-cost miniature light emitting diode excitation. Optics letters, 2017. 42(7): p. 1456–1459
114. Dai, X., H. Yang, and H. Jiang. Low-cost high-power light emitting diodes for photoacoustic imaging. in Photons Plus Ultrasound: Imaging and Sensing 2017. 2017. International Society for Optics and Photonics

Index

B
Bacteria, 198, 205, 406, 408, 411
Bacteriophage, 406–409, 411
Biomedical imaging, 3, 4, 52, 54, 75–77, 110, 115, 182, 334, 379, 435, 563, 574, 575
Blood oxygenation, 30, 41–43, 93, 146, 366, 534, 535, 541, 593
Blood oxygen saturation (SO_2), 38, 41, 134, 145, 146, 383, 387–388, 461, 466, 467, 504, 536, 538, 544, 585, 593, 595, 600
Bone characterization, 433–470
Bone metabolism, 434, 435
Bone microstructure, 435, 457–459, 463–464, 467, 468, 470
Bragg peak localization, 326
Breast cancer, 45–47, 86, 95, 98, 99, 170, 256, 334, 357, 361, 382, 385–388, 390, 392, 394, 395, 428–429, 533–544, 550

C
Cancer detection, 86, 110, 111, 170, 414, 421, 503
Clinical translation, 90, 97, 126, 156, 214, 219, 223, 245, 398, 527, 528, 539, 544, 575, 580, 588, 600, 602
Contrast agents, 8, 77, 111, 184, 232, 255, 366, 379, 462, 500, 534, 548, 580
Convolutional neural networks (CNNs), 157–166, 325, 427, 464, 465, 490, 522, 563
Curved array transducer, 81, 85, 86, 100

D
Deep learning (DL), 81, 85, 100, 155–174, 247, 382, 383, 427, 464, 487, 488, 513, 518–528, 540, 561, 562, 603
Diagnosis, 38, 52, 75, 87, 96, 110, 111, 114, 126, 127, 156, 256, 334, 366, 378, 414, 421–423, 425, 429, 434, 435, 458, 461–463, 467, 469, 470, 490, 500, 504, 517, 534, 540, 544, 550, 580, 581, 585–588, 593–595, 602
Diagnostic, 4, 5, 48, 94, 97, 114–116, 127, 186, 214, 256, 355, 357, 366, 368, 380, 406, 414, 421, 429, 500, 501, 504, 505, 507, 534, 544, 576, 603
Diagnostic imaging, 4, 603
Dual-modal, 243, 244, 327, 357, 358, 364, 366, 548, 550, 553, 560

E
Endoscopy, 22, 81, 109–127, 242, 243, 245, 246, 300, 308–310, 381, 548, 582

F
Fabry-Pérot interferometers, 301–302, 306
Fiber Bragg grating (FBG), 26, 28, 307
Fibre-optic, 243–246, 273, 275–287, 309, 340
Fluorescence, 7, 42, 76, 190, 192, 197, 198, 201, 218, 236, 237, 243, 244, 246, 255, 354, 359, 362, 379–382, 385, 396, 397, 406, 462, 477–483, 492, 500, 517, 551, 598, 599
Functional ultrasound (fUS) imaging, 478

W. Xia (ed.), *Biomedical Photoacoustics*,
https://doi.org/10.1007/978-3-031-61411-8

H
Histological imaging, 509, 513, 517
Hybrid system, 358, 359, 361–363

I
Image reconstruction, 22, 28, 83, 85, 88, 113, 143, 156–167, 169, 173, 174, 213, 324, 325, 328, 334, 340–341, 345, 465, 487, 489
Interventional devices, 547–565
Inverse problems, 134, 137, 138, 141, 144–147
In vivo dosimetry, 323, 326–327

L
Laser-induced ultrasound CT, 333, 349
LED-photoacoustic imaging, 573–603
Light-emitting diodes (LEDs), 82, 84, 85, 89, 90, 100, 284, 382, 384–386, 483, 484, 553, 560, 562, 564, 573–603
Linear array transducer, 81, 88–90, 381, 382, 386

M
Magnetic resonance imaging (MRI), 4, 43, 75, 76, 84, 96, 98, 99, 173, 182, 214, 223, 255, 258, 286, 328, 354, 363–367, 379, 380, 388, 423, 425, 434, 454, 460, 464, 465, 470, 471, 490–492, 534, 540, 548, 556
Micro-ring resonator (MRR), 27–28, 50, 301, 305–307, 311, 362
Miniaturized brain imaging apparatus, 477–493
Miniaturized probes, 110
Minimally invasive medical procedures, 547–565
Minimally invasive surgery (MIS), 254–255, 257, 260, 266, 277, 286
Molecular imaging, 37, 54, 97, 111, 182, 184, 334, 354, 364–366, 462, 580, 587–588, 590, 603
Multimodal imaging, 127, 258, 275, 287, 361, 365, 366, 493
Multimodality imaging, 268, 285, 381
Multiphoton fluorescence microscopy, 478–483, 492
Multispectral imaging, 126, 247
Murine models, 387–390

N
Nanocomposites, 283, 284, 552, 553, 560, 564
Near-infrared spectroscopy (NIRS), 41, 76, 478, 483, 484, 492, 493

O
Optical, 4, 76, 110, 134, 156, 182, 214, 231, 258, 300, 323, 334, 354, 379, 406, 414, 434, 478, 500, 536, 548, 574
Optical imaging, 4, 6, 7, 51, 76, 136, 156, 182, 205, 215, 232, 354, 359–363, 379, 380, 434, 500, 503
Optically pumped magnetometers (OPMs), 478, 491, 492
Optical ultrasound (OpUS) generation, 258–260, 263, 265, 270–274, 276, 278–281, 284–286
Optical ultrasound sensor, 299–313
Optoacoustic, 32, 86, 110, 365, 558
Optoacoustic imaging (OAI), 4, 29, 86, 353–368, 379

P
PA agents, 184, 186, 188–191
PA bone imaging, 461–466, 470
PA endomicroscopy, 243–245, 247, 309
Phantom, 27, 31, 84, 89, 91, 92, 94, 142, 146, 164, 165, 171, 172, 197, 202, 204, 213–223, 242, 247, 284–286, 327, 335–337, 339, 341–349, 504, 552, 553, 555, 557–564, 573, 577, 578, 580–583, 592, 600, 601
Photoacoustic (PA), 4, 76, 110, 133, 156, 182, 213, 231, 258, 299, 325, 334, 354, 379, 406, 414, 435, 478, 500, 534, 548, 574
Photoacoustic computed tomography (PACT), 5, 14, 21, 22, 75–100, 163, 164, 183, 192, 193, 202–205, 299, 425, 461, 465, 489, 502
Photoacoustic guidestar, 232, 236–241
Photoacoustic imaging (PAI), 4, 76, 110, 144, 156, 182, 213, 231, 275, 299, 334, 379, 435, 488, 534, 578
Photoacoustic microscopy (PAM), 5, 76, 183, 243, 299, 462, 500, 603
Photoacoustic spectrum analysis (PASA), 413–430, 455–458
Photoacoustic tomography (PAT), 76, 81, 133–148, 158, 163, 164, 200, 308, 312, 334–338, 342, 343, 345–349, 386, 390, 501, 502, 542

Planar matrix transducer, 91, 92
Positron emission tomography (PET), 4, 51, 75, 76, 182, 355, 366, 367, 379, 394, 414, 423
Preclinical, 20, 54, 85, 95, 125, 183, 215, 219, 364, 366, 377–398, 467–469, 573, 580, 583, 595, 601
Preclinical and clinical applications, 4, 45–52, 125, 363, 466–469, 574, 576, 580–601, 603
Preclinical research, 355, 357, 368, 378, 380, 381, 387, 388, 395, 478, 500
Preclinical studies, 54, 85, 219, 385
Protoacoustic imaging, 320, 322, 324, 325, 327, 328

Q

Quantitative imaging, 133–148

R

Radiation oncology, 319–328

S

Single element transducer (SET), 81–85, 100, 359, 381–383, 488
Single-photon-emission computed tomography (SPECT), 4, 75, 366, 367, 379, 423
Sound speed map, 334, 335, 340–341, 344–348
Spherical matrix transducer, 92
Standardization, 169, 170, 223, 398

T

Targeted contrast agents, 111, 112, 381, 384, 397
Technical validation, 213–223
Theranostics, 366
Therapy monitoring, 388–392
Tomographic reconstruction, 134, 355
Treatment, 38, 52, 75, 90, 110, 126, 127, 199, 320, 324–327, 355, 358, 379, 380, 383, 384, 388–393, 396, 398, 414, 423, 434, 435, 458, 469, 499, 504, 534, 542–544, 583, 593–595, 602
Tumor, 9, 90, 111, 170, 184, 320, 357, 378, 414, 435, 499, 533, 580
Tumor grading, 425–428, 499
Tumor identification, 98, 421, 427, 533, 536, 540–542
Tumor microenvironment (TME), 365, 379, 380, 384, 387, 394, 396, 431
Tumor vasculature, 379, 382, 383, 387, 388, 390, 397, 542, 583

U

Ultrasound (US), 4, 75, 110, 135, 156, 182, 219, 231, 255, 299, 320, 334, 354, 379, 414, 434, 478, 502, 534, 548, 574
Ultrasound imaging, 4, 76, 111, 156, 182, 255, 299, 334, 382, 484, 534, 560, 586
Ultrathin endoscope, 232

V

Vasculature, 39, 43, 47, 51, 85, 87, 89, 90, 93, 95, 125, 170, 191, 243, 245, 246, 255, 309–311, 313, 379, 382–384, 387–390, 394, 397, 466, 490, 503, 504, 536, 541, 542, 556, 561, 574, 581–584, 592–594, 601, 603
Virtual staining, 464, 521–528

W

Wavefront shaping, 126, 231–247, 309

X

X-ray computed tomography (CT), 4, 75, 255, 366, 548

GPSR Compliance
The European Union's (EU) General Product Safety Regulation (GPSR) is a set of rules that requires consumer products to be safe and our obligations to ensure this.

If you have any concerns about our products, you can contact us on

ProductSafety@springernature.com

In case Publisher is established outside the EU, the EU authorized representative is:

Springer Nature Customer Service Center GmbH
Europaplatz 3
69115 Heidelberg, Germany

www.ingramcontent.com/pod-product-compliance
Lightning Source LLC
Chambersburg PA
CBHW071307180926
51613CB00001B/145
* 9 7 8 3 0 3 1 6 1 4 1 0 1 *